Nuclear Structure

NUCLEAR STRUCTURE

NUCLEAR STRUCTURE

VOLUME II *Nuclear Deformations*

AAGE BOHR
BEN R. MOTTELSON

The Niels Bohr Institute and *NORDITA*, *Copenhagen*

1975

W. A. BENJAMIN, INC., Advanced Book Program
Reading, Massachusetts
London · Amsterdam · Don Mills, Ontario · Sydney · Tokyo

NUCLEAR STRUCTURE (In 3 volumes)

VOLUME **I**: *Single-Particle Motion*, 1969

VOLUME **II**: *Nuclear Deformations*, 1975

Library of Congress Cataloging in Publication Data

Bohr, Aage.
Nuclear structure.

Vol. 2-published by Advanced Book Program, W. A. Benjamin.
Bibliography: p.
Includes indexes.
CONTENTS—v. 1. Single-particle motion.—v. 2. Nuclear deformations.
1. Nuclear physics. I. Mottelson, Ben R., joint author. II. Title.
QC776.B55 539.7′4 68-57860
ISBN 0–8053–1016–9 (v. 2)

Manufactured in the United States of America

ABCDEFGHIJ-HA-798765

CONTENTS, VOLUME II

CONTENTS, VOLUME I

PREFACE

While the first volume of the present treatise was concerned primarily with nuclear properties associated with single-particle motion, the second volume deals with aspects of nuclear dynamics associated with the various types of collective deformations that may occur in the nucleus. The collective as well as the single-particle degrees of freedom constitute elementary modes of excitation, and a central theme in the developing understanding of nuclear structure has been the effort to achieve a proper balance in the use of concepts referring to these two contrasting facets of nuclear dynamics.

The earliest discussion of nuclei as built out of neutrons and protons was based on independent-particle motion in a collective central field, in analogy to the picture that had been successfully employed in the description of atomic structure.[1] The development took a new direction as a result of the discovery of the dense spectrum of narrow resonances in neutron-induced reactions, which drew attention to the strong coupling between the motion of the individual nucleons in the nucleus (see Vol. I, p. 156), and led to a description of nuclear dynamics in terms of collective degrees of freedom similar to the vibrational modes of a liquid drop (Bohr and Kalckar, 1937). A few years later, the discovery of the fission process provided a striking example of such collective motion.

A new major turn in the development resulted from the analysis of the accumulated evidence on nuclear binding energies and moments, which demonstrated convincingly the existence of nuclear shell structure (Haxel, Jensen, and Suess, 1949; Mayer, 1949; see also Vol. I, pp. 189 ff.). One was then faced with the problem of reconciling the simultaneous occurrence of single-particle and collective degrees of freedom and of exploring the

[1]A vivid impression of this early phase in the exploration of nuclear structure is conveyed by the discussions of the Seventh Solvay Meeting (Gamow, 1934; Heisenberg, 1934).

variety of phenomena that arise from their interplay (Rainwater, 1950; Bohr, 1952; Hill and Wheeler, 1953; Bohr and Mottelson, 1953).

In the evolution of an appropriate framework for the description of nuclear structure, the concept of a deformation in the nuclear density and potential has been the unifying element. The deformations represent the collective degrees of freedom and, at the same time, affect the motion of the individual nucleons, thereby providing the organizing agent responsible for the collective motion itself. The roles of the many different types of deformations that may occur in the nuclear systems constitute the topic of the present volume.

In the analysis of collective motion in the nucleus, the study of rotations has played a special role, due to the unique simplicity of these degrees of freedom. (Indeed, the response to rotational motion has been a key to the development of dynamical concepts ranging from celestial mechanics to the spectra of elementary particles.) The special position of the rotations and the extensive empirical evidence on rotational spectra motivate the treatment of this mode as the first topic of the present volume (Chapter 4).

The occurrence of rotational spectra is a feature of nuclei with equilibrium shapes that deviate from spherical symmetry (collective mode associated with spontaneously broken symmetry). The simple quantitative relationships that govern these spectra make possible a detailed study of single-particle motion in the nonspherical nuclei. This analysis, discussed in Chapter 5, constitutes a major extension of the evidence on single-particle motion obtained from the study of spherical nuclei, and at the same time provides a basis for exploring the coupling of rotational and single-particle motion.

Chapter 6 contains a discussion of the great variety of collective vibrational modes and of the many couplings between the different elementary modes of excitation. The writing of this chapter has gone through numerous drafts reflecting the increasing awareness of the vast scope of the field and of the possibilities for obtaining a highly unified theoretical framework in terms of the particle-vibration coupling.

In the original plan, it was envisaged that the analysis of the collective modes in terms of the motion of the individual nucleons would be a subject for Volume III, to be introduced through the study of interaction effects in configurations involving a few nucleons. However, the evolution of Volume II has led to the incorporation of the microscopic theory of collective motion following directly from the analysis of the particle-vibration coupling. With this change of plan, the first two volumes have emerged as self-contained and at the same time present a more comprehensive view of nuclear dynamics than was originally envisaged.[2]

[2]The reader encountering references to Volume III, which occur in Volume I, should consult the Subject Index of Volume II for a discussion of the relevant topics.

As in Volume I, the material in Volume II is divided into text (文), illustrative examples (圖), and appendices (附). The examples, especially in Chapter 6, play a somewhat expanded role, since it was deemed desirable to exploit this dimension in order to develop a number of topics without interrupting the coherent presentation of the text.

In the preparation of the present volume, we have had the benefit of criticisms, discussions, and direct help from a large circle of colleagues. We as well as the readers owe a special debt to Peter Axel, without whose patient and imaginative suggestions the presentation would have been even more impenetrable, and to Ikuko Hamamoto whose probing testing of the entire material has been both an inspiration and a major contribution to clarity and consistency. We should also like to express our appreciation to Bertel Lohmann Andersen, Sven Bjørnholm, Ricardo Broglia, Sven Gösta Nilsson, David Pines, John Rasmussen, Vilen Strutinsky, and Wladek Swiatecki for valuable discussions and suggestions.

It has been our immense good fortune, throughout the more than fifteen challenging years of struggle with the present treatise, to have the continued and never-failing support of the outstanding team comprising Lise Madsen, Henry Olsen, and Sophie Hellmann. A special tribute is due to Sophie Hellmann who, with undiminished strength and enthusiasm, has fulfilled her unique role even though the campaign was to extend beyond her eightieth year; to our gratitude and admiration, we should like to add our deep appreciation for the pleasure and inspiration it has been to work together with her.

AAGE BOHR

BEN R. MOTTELSON

Copenhagen
June 1975

CHAPTER 4

Rotational Spectra

4-1 OCCURRENCE OF COLLECTIVE ROTATIONAL MOTION IN QUANTAL SYSTEMS[1]

A common feature of systems that have rotational spectra is the existence of a "deformation", by which is implied a feature of anisotropy that makes it possible to specify an orientation of the system as a whole. In a molecule, as in a solid body, the deformation reflects the highly anisotropic mass distribution, as viewed from the intrinsic coordinate frame defined by the equilibrium positions of the nuclei. In the nucleus, the rotational degrees of freedom are associated with the deformations in the nuclear equilibrium shape that result from the shell structure. (Evidence for these deformations will be discussed on pp. 133ff. ($E2$ moments) and in Chapter 5 (deformed single-particle potential).) Rotational-like sequences are also observed in the hadron spectra and are referred to as Regge trajectories (see, for example, Fig. 1-13, Vol. I, p. 65), but the nature of the deformations involved has not yet been identified.

Collective motion having a structure similar to the rotations in space may occur in other dimensions, including isospace and particle-number space, if the system possesses a deformation that defines an orientation in these spaces. The rotational bands then involve sequences of states differing in the associated angular-momemtum-like quantum numbers, such as isospin and nucleon number. (Such sequences occur in superfluid systems (see pp. 393ff.) and may also occur as excitations of the nucleon (see pp. 20ff.).)

The deformation may be invariant with respect to a subgroup of rotations of the coordinate system, as for example in the case of axially

---

[1]Spectra associated with quantized rotational motion were first recognized in the absorption of infrared light by molecules (Bjerrum, 1912). The possible occurrence of rotational motion in nuclei became an issue in connection with the early attempts to interpret the evidence on nuclear excitation spectra (see, for example, Teller and Wheeler, 1938). The available data, as obtained, for example, from the fine structure of α decay, appeared to provide evidence against the occurrence of low-lying rotational excitations, but the discussion was hampered by the expectation that rotational motion would either be a property of all nuclei or be generally excluded, as in atoms, and by the assumption that the moment of inertia would have the classical value, as for rigid rotation. The establishment of the rotational mode in nuclei followed the recognition that such a mode was a necessary consequence of the existence of strongly deformed equilibrium shapes (Bohr, 1951); the occurrence of such deformations had been inferred at an early stage (Casimir, 1936) from the determination of nuclear quadrupole moments on the basis of atomic hyperfine structure. The analysis of $E2$ transitions gave additional evidence for collective effects associated with deformations in the nuclear shape (Goldhaber and Sunyar, 1951; Bohr and Mottelson, 1953a). The identification of rotational excitations came from the observation of level sequences proportional to $I(I+1)$ (Bohr and Mottelson, 1953b; Asaro and Perlman, 1953) and was confirmed by the evidence of the rotational intensity relations (Alaga *et al.*, 1955). The development of the Coulomb excitation process provided a powerful tool for the systematic study of the rotational spectra (Huus and Zupančič, 1953; see also the review by Alder *et al.*, 1956).

symmetric deformations. In such a situation, the deformation only partially defines the orientation of the intrinsic coordinate system, and the rotational degrees of freedom are correspondingly restricted. A first step in the analysis of the rotational spectra is therefore an analysis of the symmetry of the deformation and the resulting rotational degrees of freedom. This topic will be discussed in Sec. 4-2 for the case of axially symmetric systems, which are of special significance for nuclear spectra; systems without axial symmetry will be considered in Sec. 4-5. (The consequences of symmetry in the deformation represent a generalization of the well-known restriction of molecular rotational states imposed by identity of nuclei; see pp. 11 and 180.)

The occurrence of rotational degrees of freedom may thus be said to originate in a breaking of rotational invariance. In a similar manner, the translational degrees of freedom are based upon the existence of a localized structure. However, while the different states of translational motion of a given object are related by Lorentz invariance, there is no similar invariance applying to rotating coordinate frames. The Coriolis and centrifugal forces that act in such coordinate frames perturb the structure of a rotating object.

In a quantal system, the frequency of even the lowest rotational excitations may be so large that the Coriolis and centrifugal forces affect the structure in a major way. The condition that these perturbations be small (adiabatic condition) is intimately connected with the condition that the zero-point fluctuations in the deformation parameters be small compared with the equilibrium values of these parameters, and the adiabatic condition provides an alternative way of expressing the criterion for the occurrence of rotational spectra (Born and Oppenheimer, 1927; Casimir, 1931).

A simple illustration of this equivalence is provided by a system consisting of two particles bound together by a potential possessing a minimum for separation R (the equilibrium separation). The motion of the system can be described in terms of rotation and radial vibrations. The frequency of the rotational motion, for the lowest states, is

$$\omega_{\mathrm{rot}} \sim \frac{\hbar}{M_0 R^2} \tag{4-1}$$

where M_0 is the reduced mass. The frequency of the vibrational motion depends on the amplitude ΔR of zero-point vibration

$$\omega_{\mathrm{vib}} \sim \frac{\hbar}{M_0 (\Delta R)^2} \tag{4-2}$$

The condition that the fluctuations in shape be small compared with the average deformation, $\Delta R \ll R$, is therefore equivalent to the adiabatic condition $\omega_{\text{rot}} \ll \omega_{\text{vib}}$. This simple system illustrates the manner in which the rotational modes emerge as a low-frequency branch in the vibrational spectrum, in a situation where the vibrational potential energy possesses a minimum for an anisotropic shape.

The relationship between members of a rotational band manifests itself in the regularities of the energy spectra and in the intensity rules governing the transitions to different members of a band. For sufficiently small values of the rotational angular momentum, the analysis can be based on an expansion of energies and transition amplitudes in powers of the rotational frequency or angular momentum. These expressions acquire a special simplicity for systems with axially symmetric shape and in this form are found to provide a basis for the interpretation of an extensive body of data on nuclear spectra (Sec. 4-3).

The dependence of matrix elements on the rotational angular momentum reflects the response of the intrinsic motion to the Coriolis and centrifugal forces and can be analyzed in terms of the coupling between rotational bands based on different intrinsic structures, as discussed in Sec. 4-4 (see also pp. 111ff. and pp. 130ff.). For large values of the angular momentum, the rotational perturbations may strongly modify the intrinsic structure of the system. The structure of nuclear matter under these extreme conditions is a matter of considerable current interest (see pp. 41 ff.).

The discussion of rotational bands in the present chapter is based on the geometry of the deformed intrinsic structure. The states in a rotational band can also be characterized in terms of representations of symmetry groups; the group structure then expresses the symmetry of the rotating object. The bands described by representations of compact groups terminate after a finite number of states; thus, the U_3 symmetry group applying to particle motion in a harmonic oscillator potential has been exploited to illuminate features of nuclear rotational spectra associated with the finiteness of the number of nucleons that contribute to the anisotropy (Elliott, 1958; see also the discussion on pp. 93ff.). Bands continuing to indefinitely large values of the angular momentum can be associated with representations of noncompact symmetry groups (see the discussion in Chapter 6, p. 411).

4-2 SYMMETRIES OF DEFORMATION. ROTATIONAL DEGREES OF FREEDOM

A separation of the motion into intrinsic and rotational components corresponds to a Hamiltonian of the form

$$H = H_{\text{intr}}(q,p) + H_{\text{rot},\alpha}(P_\omega) \tag{4-3}$$

The intrinsic motion is described by the coordinates q and conjugate momenta p, which are measured relative to the body-fixed coordinate frame and are therefore scalars with respect to rotations of the external coordinate system. The orientation of the body-fixed frame, defined by the deformation of the system (see p. 1), is specified by angular variables denoted by ω. The rotational Hamiltonian does not depend on the orientation ω (if no external forces act on the system) and is a function of the conjugate angular momenta P_ω. The labeling of the rotational Hamiltonian in Eq. (4-3) indicates that the rotational motion may depend on the quantum numbers α specifying the intrinsic state.

The eigenstates of the Hamiltonian (4-3) are of the product form

$$\Psi_{\alpha,I} = \Phi_\alpha(q)\varphi_{\alpha,I}(\omega) \tag{4-4}$$

For each intrinsic state α, the spectrum involves a sequence of rotational levels, specified by a set of angular momentum quantum numbers denoted by I in Eq. (4-4).

The discussion of the consequence of symmetry in the present section is independent of the explicit relationship between the set of variables q, p, ω, P_ω, in which the Hamiltonian approximately separates, and the variables describing the position, momenta, and spins of the individual particles. This relationship is connected with the microscopic analysis of the collective rotational mode and is implicit in the treatment of this mode in terms of the intrinsic excitations, as discussed in Sec. 6-5h (see also p. 210).

4-2a Degrees of Freedom Associated with Spatial Rotations

Rotational motion in two dimensions (rotation about a fixed axis) has a very simple structure. The orientation is characterized by the azimuthal angle ϕ, and the state of motion by the eigenvalue M of the conjugate angular momentum. The associated rotational wave function is

$$\varphi_M(\phi) = (2\pi)^{-1/2}\exp\{iM\phi\} \tag{4-5}$$

The orientation of a body in three-dimensional space involves three angular variables, such as the Euler angles, $\omega = \phi, \theta, \psi$ (see Fig. 1A-1, Vol. I, p. 76), and three quantum numbers are needed in order to specify the state of motion. The total angular momentum I and its component $M = I_z$ on a space-fixed axis provide two of these quantum numbers; the third may be obtained by considering the components of $\mathbf{I}$ with respect to an intrinsic (or body-fixed) coordinate system with orientation ω (see Sec. 1A-6a). The intrinsic components $I_{1,2,3}$ commute with the external components $I_{x,y,z}$, because $I_{1,2,3}$ are independent of the orientation of the external system (are scalars). The commutation rules of the intrinsic components among themselves are similar to those of $I_{x,y,z}$, but involve an opposite sign (see Eq. (1A-91)). As a commuting set of angular momentum variables, we may thus choose $\mathbf{I}^2$, I_z, and I_3. The eigenvalues of I_3 are denoted by K (see Fig. 4-1), and have the same range of values as does M,

$$K = I, I-1, \ldots, -I \tag{4-6}$$

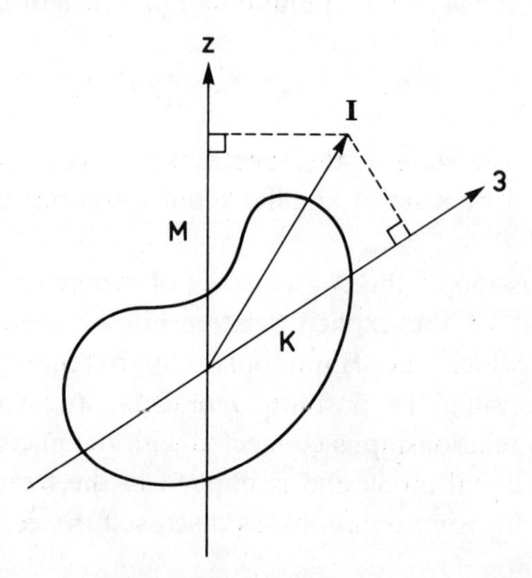

Figure 4-1 Angular momentum quantum numbers describing rotational motion in three dimensions. The z axis belongs to a coordinate system fixed in the laboratory, while the 3 axis is part of a body-fixed coordinate system (compare the $\mathcal{K}$ and $\mathcal{K}'$ systems defined in Fig. 1A-1, Vol. I, p. 76).

For specified values of the three quantum numbers I, K, and M, the rotational wave function is given by (see Eq. (1A-97))

$$\varphi_{IKM}(\omega) = \left(\frac{2I+1}{8\pi^2} \right)^{1/2} \mathscr{D}^I_{MK}(\omega) \tag{4-7}$$

where the functions $\mathscr{D}^I_{MK}$ are the rotation matrices. The result (4-7) can be obtained by a transformation from the fixed coordinate system to a rotated system coinciding with the intrinsic frame (see Vol. I, p. 89). For $K=0$, the

$\mathscr{D}$ functions reduce to spherical harmonics (see Eq. (1A-42))

$$\varphi_{I,K=0,M}(\omega)=(2\pi)^{-1/2}Y_{IM}(\theta,\phi) \tag{4-8}$$

The wave function (4-8) is independent of ψ, but normalized, as in Eq. (4-7), with respect to integration over all three Euler angles.

For $K=0$, the rotational wave function is the same as for the angular motion of a point particle with no spin. For finite K, the rotational motion corresponds to the angular motion of a particle with helicity $h=K$ (see Eq. (3A-5)).

While $\mathbf{I}^2$ and I_z are constants of the motion for any rotationally invariant Hamiltonian, the commutator of I_3 with the Hamiltonian depends on intrinsic properties of the system. In general, therefore, the stationary states involve a superposition of components with different values of K,

$$\varphi_{\tau IM}(\omega)=\left(\frac{2I+1}{8\pi^2}\right)^{1/2}\sum_K c_{\tau I}(K)\mathscr{D}^I_{MK}(\omega) \tag{4-9}$$

The third rotational quantum number is denoted by τ, and the amplitudes $c_{\tau I}(K)$ depend on the relative magnitude of the moments of inertia, as will be discussed in Sec. 4-5.

4-2b Consequences of Axial Symmetry

If the system possesses axial symmetry, two consequences ensue:

(a) The projection I_3 on the symmetry axis is a constant of the motion.
(b) There are no collective rotations about the symmetry axis.

The first implication is well known from classical mechanics and expresses the invariance of the Hamiltonian with respect to rotations about the symmetry axis. More generally, I_3 is a constant of the motion if the 3 axis is a symmetry axis for the tensor of inertia.

Implication (b) is a feature of the quantal description expressing the impossibility of distinguishing orientations of the intrinsic frame that differ only by a rotation about the symmetry axis. This consequence of the axial symmetry is similar to the absence of collective rotations for a spherical system. It follows that the quantum number K represents the angular momentum of the intrinsic motion and has a fixed value for the rotational band based on a given intrinsic state. (In diatomic molecules, the angular momentum of the collective rotational motion is perpendicular to the symmetry axis because the nuclei can be treated as mass points and because the electrons do not rotate collectively in the axially symmetric binding field.)

The restriction on the rotational degrees of freedom resulting from axial symmetry corresponds to the constraint

$$I_3 = J_3 \tag{4-10}$$

where J_3 is the operator representing the component of intrinsic angular momentum. The condition (4-10) implies that the operations associated with rotations about the symmetry axis, which are generated by I_3, have prescribed values determined by the intrinsic structure.

Since the axial symmetry makes it impossible to distinguish orientations differing only in the value of the third Euler angle ψ, this variable is redundant. The constraint (4-10) ensures that the total nuclear wave function, which is a product of intrinsic and rotational wave functions (see Eq. (4-4)), is independent of the value of ψ. In fact, a rotation of the intrinsic frame through an angle $\Delta\psi$ about the 3 axis multiplies the intrinsic wave function by the factor $\exp\{-iJ_3\Delta\psi\}$ and the rotational wave function by $\exp\{iI_3\Delta\psi\}$; the total wave function is therefore invariant, for $J_3 = I_3$ $(=K)$. Instead of treating the Euler angle ψ as a redundant variable, one may constrain ψ to have a definite value, such as $\psi = 0$ or $\psi = -\phi$; see the comment on the helicity wave function in Appendix 3A, Vol. I, p. 361. If ψ were fixed, the normalization constants in Eqs. (4-7) and (4-8) would be multiplied by $(2\pi)^{1/2}$.

4-2c $\mathscr{R}$ Invariance

A further reduction in the rotational degrees of freedom follows if the intrinsic Hamiltonian is invariant with respect to a rotation of 180° about an axis perpendicular to the symmetry axis. Rotations about different axes perpendicular to the symmetry axis are equivalent; for definiteness, we choose a rotation $\mathscr{R} \equiv \mathscr{R}_2(\pi)$ with respect to the 2 axis. (For systems with axial symmetry but not spherical symmetry, $\mathscr{R}$ invariance is the only possible additional rotational invariance; in fact, invariance with respect to any other rotation would imply an infinity of symmetry axes and hence spherical symmetry.)

The $\mathscr{R}$ invariance implies that the rotation $\mathscr{R}$ is part of the intrinsic degrees of freedom, and is therefore not to be included in the rotational degrees of freedom. We can express this constraint by requiring that the operator $\mathscr{R}_e$, which performs the rotation $\mathscr{R}$ by acting on the collective orientation angles (external variables), is identical to the operator $\mathscr{R}_i$, which performs the same rotation by acting on the intrinsic variables

$$\mathscr{R}_e = \mathscr{R}_i \tag{4-11}$$

The condition (4-11) is analogous to the constraint (4-10) associated with the invariance under infinitesimal rotations about the symmetry axis.

Intrinsic states with $K=0$ can be labeled by the eigenvalue r of $\mathscr{R}_i$,

$$\begin{gathered}\mathscr{R}_i \Phi_{r,K=0}(q) = r\Phi_{r,K=0}(q) \\ r = \pm 1\end{gathered} \tag{4-12}$$

The eigenvalues of $\mathscr{R}_i$ are ± 1 because $\mathscr{R}^2 = \mathscr{R}_2(2\pi) = +1$ for a system with integer angular momentum.

The operation $\mathscr{R}_e$ acting on the rotational wave function (4-8) inverts the direction of the symmetry axis ($\theta \rightarrow \pi - \theta$, $\phi \rightarrow \phi + \pi$), and we obtain

$$\mathscr{R}_e Y_{IM}(\theta,\phi) = (-1)^I Y_{IM}(\theta,\phi) \tag{4-13}$$

The constraint $\mathscr{R}_i = \mathscr{R}_e$ therefore implies

$$(-1)^I = r \tag{4-14}$$

and the rotational spectrum contains states with only even values or only odd values of I,

$$\begin{gathered}\Psi_{r,K=0,IM} = (2\pi)^{-1/2} \Phi_{r,K=0}(q) Y_{IM}(\theta,\phi) \\ I = 0,2,4,\ldots \qquad r = +1 \\ I = 1,3,5,\ldots \qquad r = -1\end{gathered} \tag{4-15}$$

The constraint $\mathscr{R}_e = r$ halves the domain of independent orientation angles and removes every other I value from the rotational spectrum.

The intrinsic states with $K \neq 0$ are twofold degenerate, as a consequence of the $\mathscr{R}$ invariance. We shall use a notation such that K is taken to be positive, and the rotated states with negative eigenvalues of J_3 are denoted by $\bar{K}$,

$$\Phi_{\bar{K}}(q) \equiv \mathscr{R}_i^{-1} \Phi_K(q) \tag{4-16}$$

If the intrinsic states are expanded in components of total angular momen-.

tum J, the phase convention (4-16) implies (see Eq. (1A-47))

$$\Phi_K = \sum_J c_J \Phi_{JK}$$
$$\Phi_{\bar{K}} = \exp\{i\pi J_2\}\Phi_K = \sum_J (-1)^{J+K} c_J \Phi_{J,-K} \tag{4-17}$$

where $\Phi_{J,\pm K}$ are components of a J multiplet with $J_3 = \pm K$.

The effect of $\mathscr{R}_e$ on the rotational wave function is given by

$$\mathscr{R}_e \mathscr{D}^I_{MK}(\omega) = \exp\{-i\pi I_2\}\mathscr{D}^I_{MK}(\omega)$$
$$= (-1)^{I+K}\mathscr{D}^I_{M-K}(\omega) \tag{4-18}$$

In the derivation of Eq. (4-18), we have employed the relation (1A-47) and the fact that the matrix elements of I_2 have the opposite sign from those of I_y (see Eqs. (1A-93)). In order to fulfill the condition (4-11), the nuclear wave functions must therefore have the form

$$\Psi_{KIM} = 2^{-1/2}(1 + \mathscr{R}_i^{-1}\mathscr{R}_e)\left(\frac{2I+1}{8\pi^2}\right)^{1/2} \Phi_K(q)\mathscr{D}^I_{MK}(\omega)$$
$$= \left(\frac{2I+1}{16\pi^2}\right)^{1/2} \{\Phi_K(q)\mathscr{D}^I_{MK}(\omega) + (-1)^{I+K}\Phi_{\bar{K}}(q)\mathscr{D}^I_{M-K}(\omega)\}$$
$$I = K, K+1, \ldots \qquad (K>0) \tag{4-19}$$

(Note that $\mathscr{R}_i^2 = \mathscr{R}_e^2 = (-1)^{2I}$.) From the two intrinsic states Φ_K and $\Phi_{\bar{K}}$, we can thus form only a single rotational state for each value of I, as a consequence of the restrictions in the rotational degrees of freedom imposed by the $\mathscr{R}$ invariance of the deformation.

The wave function (4-19) is not of the simple product form (4-4), but involves a sum of two such terms, associated with the degenerate intrinsic states. The superposition of the two components represents an interweaving of intrinsic and rotational degrees of freedom, which gives rise to interference effects with no counterpart in a classical system. The matrix elements of an operator F between symmetrized states of the type (4-19) can be written

$$\langle K_2 I_2 M_2 | F | K_1 I_1 M_1 \rangle$$
$$= \langle K_2 I_2 M_2 | F | K_1 I_1 M_1 \rangle_{\text{unsym}} + (-1)^{I_1+K_1}\langle K_2 I_2 M_2 | F | \bar{K}_1 I_1 M_1 \rangle_{\text{unsym}} \tag{4-20}$$
$$(K_1 > 0,\ K_2 > 0)$$

in terms of matrix elements for unsymmetrized states

$$(\Psi_{KIM})_{\text{unsym}} = \left(\frac{2I+1}{8\pi^2}\right)^{1/2} \Phi_K(q)\mathscr{D}^I_{MK}(\omega)$$
$$(\Psi_{\bar{K}IM})_{\text{unsym}} = \left(\frac{2I+1}{8\pi^2}\right)^{1/2} \Phi_{\bar{K}}(q)\mathscr{D}^I_{M-K}(\omega) \tag{4-21}$$

In deriving the result (4-20), we have used the relation

$$\mathscr{R}_{\rm i}^{-1}\,\mathscr{R}_{\rm e} F\,\mathscr{R}_{\rm e}^{-1}\,\mathscr{R}_{\rm i} = F \tag{4-22}$$

expressing the condition that any physical operator must transform in the same manner under the equivalent operations $\mathscr{R}_{\rm e}$ and $\mathscr{R}_{\rm i}$.

The second term in Eq. (4-20) involves the phase factor

$$\sigma \equiv (-1)^{I+K} \tag{4-23}$$

which is referred to as the signature (following the terminology applied to Regge trajectories). The contribution of this term to the matrix element alternates in sign for successive values of I (if we assume the operator F itself to be a smooth function of orientation angles and angular momenta). The signature-dependent term in the matrix elements implies that the rotational bands with $K \neq 0$, in a system with axial symmetry and $\mathscr{R}$ invariance, tend to separate into two families distinguished from each other by the quantum number σ.

The occurrence of two interfering terms in the matrix element (4-20) is a specific quantal effect. For matrix elements between two states in the same band, the signature-dependent term contributes if the operator F can invert the sign of I_3 and thereby produce effects equivalent to a rotation $\mathscr{R}$ of the entire system.

Rotors with $\mathscr{R}$ symmetry were first encountered in the spectra of diatomic homonuclear molecules (see, for example, Herzberg, 1950; pp. 130ff.). For such a system, the quantum number r, for states with vanishing electronic angular momentum about the axis, can be expressed in the form $r = r_{\rm el}P(12)$, where $r_{\rm el}$ represents the r quantum number of the electronic wave function, while $P(12)$ represents the spatial exchange of the nuclei. The statistics for the identical nuclei implies the relation $P(12) = (-1)^S$, where S is the total nuclear spin. For the total rotational angular momentum L, exclusive of the nuclear spins, we therefore obtain, from Eq. (4-14), the condition

$$(-1)^L = r_{\rm el}(-1)^S \tag{4-24}$$

which expresses the restriction on the rotational motion associated with the statistics (Hund, 1927).

The conventional notation in molecular spectroscopy characterizes the electronic wave function as positive or negative with respect to reflection in a plane containing the molecular axis. The operation, denoted in the following by $\mathscr{S}$ with eigenvalue s, is the product of $\mathscr{R}$ and the parity operation $\mathscr{P}$ (see p. 16). The electronic state of the molecule is labeled by the quantum numbers $s_{\rm el}(=+\text{ or }-)$ and the parity $\pi_{\rm el}$, which is referred to as g (gerade; $\pi_{\rm el}=+1$) or u (ungerade, $\pi_{\rm el}=-1$). Thus, for example, an electronic ${}^1\Sigma_u^-$ state has $\pi_{\rm el}=s_{\rm el}=-1$ and hence $r_{\rm el}=\pi_{\rm el}s_{\rm el}=+1$; the upper left-hand index is the electronic spin multiplicity, while the letters $\Sigma, \Pi, \Delta, \ldots$ give the component Λ of the electronic orbital angular momentum with respect to the symmetry axis $(=0,1,2,\ldots)$.

The redundance in the degrees of freedom associated with the $\mathscr{R}$ invariance, as discussed in the present section, is associated with a lack of uniqueness of the variables (q,ω) when regarded as functions of the particle coordinates x. Thus, the two sets of variables (q,ω) and (q',ω'), related by

$$\begin{aligned} q' &= \mathscr{R}_{\rm i}^{-1} q\, \mathscr{R}_{\rm i} \\ \omega' &= \mathscr{R}_{\rm e}\omega\, \mathscr{R}_{\rm e}^{-1} \end{aligned} \tag{4-25}$$

refer to the same x. The condition $\mathscr{R}_{\rm i}^{-1}\mathscr{R}_{\rm e}=1$, therefore, expresses the requirement that wave functions and operators be single-valued functions of x. (The transformation to the variables (q,ω) and the associated invariance conditions can be exhibited in a simple form in a model describing the dynamical degrees of freedom in terms of the amplitudes for quadrupole deformation; see Appendix 6B.)

The lack of uniqueness in the orientation angles considered as a function of the particle variables is connected with the fact that these angles are taken to be symmetric functions of the coordinates (and momenta) of the identical particles. Such a description is especially appropriate in the characterization of collective deformations in a system like the nucleus, where the particles move throughout the volume of the system and are, therefore, continually exchanging positions. In some situations, one can employ an alternative approach that involves a labeling of the constituent particles. Thus, for example, in a two-body system, the orientation is usually chosen as the direction from particle 1 to particle 2. This procedure is the one employed in characterizing the orientation of a molecule and leads to angle variables that are single-valued functions of the coordinates. The consequences of the identity of the particles are expressed in a different form in the two approaches. When the orientation angles are symmetric functions of the variables of the identical particles, the exchange of two such particles is an intrinsic operator, and the intrinsic states carry the appropriate permutation symmetry. When the definition of the orientation angles involves a labeling of the particles, the exchange acts also on the orientation angles, and the requirements of permutation symmetry may lead to restrictions in the rotational spectrum.

4-2d $\mathscr{P}$ and $\mathscr{T}$ Symmetry

If the intrinsic Hamiltonian is invariant with respect to space reflection and time reversal, the operations $\mathscr{P}$ and $\mathscr{T}$ act on the intrinsic motion but do not affect the orientation angles.

Since $\mathscr{P}$ commutes with J_3, the intrinsic states can be labeled by the parity quantum number

$$\mathscr{P}\Phi_K(q) = \pi\Phi_K(q) \qquad \pi = \pm 1 \tag{4-26}$$

and all the states in the band have the parity π. For $K=0$ bands, the quantum numbers π and r are distinct; each may, independently of the other, take the value $+1$ or -1.

For a $\mathscr{T}$- and $\mathscr{R}$-invariant intrinsic Hamiltonian, we can choose phases for the intrinsic states such that $\mathscr{R}\mathscr{T}=1$ (compare Eq. (1-39)). From Eq. (4-16) then follow

$$\begin{aligned} \mathscr{T}\Phi_K(q) &= \Phi_{\overline{K}}(q) \\ \mathscr{T}\Phi_{\overline{K}}(q) &= (-1)^{2K}\Phi_K(q) \\ \mathscr{T}\Phi_{K=0}(q) &= r\Phi_{K=0}(q) \end{aligned} \tag{4-27}$$

where the phase factor $(-1)^{2K}$ is the value of $\mathscr{T}^2$ (see Eq. (1-41)). Since time reversal does not affect ω, the rotational wave function is transformed into its complex conjugate, and we obtain (see Eqs. (1A-38) and (4-19))

$$\begin{aligned} \mathscr{T}\Phi_K(q)\mathscr{D}^I_{MK}(\omega) &= \Phi_{\overline{K}}(q)(\mathscr{D}^I_{MK}(\omega))^* \\ &= (-1)^{M-K}\Phi_{\overline{K}}(q)\mathscr{D}^I_{-M-K}(\omega) \\ \mathscr{T}\Psi_{KIM} &= (-1)^{I+M}\Psi_{\overline{K}I,-M} \end{aligned} \tag{4-28}$$

in accordance with the general phase rule (1-40).

The transformation under time reversal, combined with Hermitian conjugation, gives the relation

$$\begin{aligned} \langle K_2|F(q,p)|K_1\rangle &= -c\langle \overline{K}_1|F(q,p)|\overline{K}_2\rangle \\ (\mathscr{T}F\mathscr{T}^{-1})^\dagger &= -cF \end{aligned} \tag{4-29}$$

for the intrinsic matrix elements of an operator depending on the internal variables. The phase factor c characterizes the transformation of the operator F under particle-hole conjugation (see Eq. (3B-21)).

For the diagonal matrix element in a $K=0$ band, the relation (4-29) yields

$$\langle K=0|F|K=0\rangle = 0 \qquad \text{for} \quad c=+1 \tag{4-30}$$

For bands with $K\neq 0$, we obtain the selection rule

$$\langle K|F|\overline{K}\rangle = 0 \qquad \text{for} \quad c(-1)^{2K} = +1 \tag{4-31}$$

for the signature-dependent term in the matrix element connecting two states in the same band (see Eq. (4-20)). (For an operator depending also on the rotational variables (ω, I_κ), the phase factor c to be used in the expressions above refers to the transformation of F in the space of the internal variables. Since ω is invariant while I_κ changes sign under time reversal combined with Hermitian conjugation, the phase factor characterizing the transformation of the total operator equals $+c$ or $-c$ depending on whether the operator is even or odd with respect to the inversion $I_\kappa \rightarrow -I_\kappa$.)

In the following sections (Secs. 4-2e and 4-2f), we shall consider rotational spectra associated with deformations that are not invariant with respect to $\mathscr{R}$, $\mathscr{P}$, and $\mathscr{T}$, while Sec. 4-2g deals with rotational motion resulting from a deformation that couples spin and isospin. The observed nuclear rotational spectra, at least in the great majority of cases, are found to have the full symmetry considered above, and some readers may therefore prefer to proceed directly to the discussion in Secs. 4-3 and 4-4.

4-2e Deformations Violating $\mathscr{P}$ or $\mathscr{T}$ Symmetry

In a system governed by overall $\mathscr{P}$ and $\mathscr{T}$ invariance, the occurrence of a deformation that violates either of these symmetries is associated with a two-valued collective degree of freedom, corresponding to the equivalence of configurations with opposite sign of the deformation. The spectrum, therefore, acquires a doublet structure.

An example of a parity-violating deformation would be provided by a pseudoscalar component in the one-body potential and density, proportional to $\mathbf{s}\cdot\mathbf{r}$. (The possibility of a pseudoscalar deformation in the nuclear potential was considered by Bleuler, 1966; see also Burr *et al.*, 1969). The potential $\mathbf{s}\cdot\mathbf{r}$ is rotationally invariant; in a nonspherical, but axially symmetric nucleus, the corresponding potential would involve two separate terms, proportional to s_3x_3 and $s_1x_1+s_2x_2$, respectively. Parity-violating potentials of the type considered also violate $\mathscr{T}$ symmetry, but conserve $\mathscr{P}\mathscr{T}$. The assumption of overall conservation of $\mathscr{P}$ as well as $\mathscr{T}$ implies that deformations of either sign have the same energy, and all the states of the system therefore occur in two modifications Ψ_+ and $\Psi_- = \mathscr{P}\Psi_+$. The two sets of states can be combined to form eigenstates of $\mathscr{P}$,

$$\Psi_\pi = 2^{-1/2}\begin{cases} (\Psi_+ + \Psi_-) & \pi = +1 \\ i(\Psi_+ - \Psi_-) & \pi = -1 \end{cases} \tag{4-32}$$

(We have assumed the phases of the states $\Psi_\pm$ to be chosen in such a manner that these states have the eigenvalue $+1$ for the operator $\mathscr{R}\mathscr{P}\mathscr{T}$; the states (4-32) then have the standard phasing, $\mathscr{R}\mathscr{T}=1$.) Hence, the spectrum consists of parity doublets. The doublets are connected by $\lambda\pi = 0-$ transitions of collective character, with matrix elements proportional to the pseudoscalar deformation.

The configurations with opposite sign of the deformation are separated by a potential barrier, but in a quantal system, an inversion of the deformation can take place by a "tunneling" motion. The frequency ω_t of this inversion gives the energy separation, $\Delta E = \hbar\omega_t$, between the parity doublets. The treatment of the tunneling effect requires a combination of the conjugate intrinsic Hamiltonians to form a $\mathscr{P}$- and $\mathscr{T}$-conserving Hamiltonian that includes the degree of freedom associated with the tunneling motion. In this manner, the collective degree of freedom associated with the doublet structure appears as a limiting feature of the vibrational spectrum, for a potential function with two minima separated by a barrier. The eigenstates are given by Eq. (4-32), in the approximation in which the overlap of the states Ψ_+ and Ψ_- can be neglected.

Similar considerations apply to deformations that violate other combinations of $\mathscr{P}$ and $\mathscr{T}$. Thus, a one-body potential proportional to $\mathbf{s}\cdot\mathbf{p}$ violates $\mathscr{P}$, but not $\mathscr{T}$, and gives rise to parity doublets, as in the case considered above. An example of a $\mathscr{T}$-, but not $\mathscr{P}$-violating deformation is provided by a potential proportional to $\mathbf{r}\cdot\mathbf{p}+\mathbf{p}\cdot\mathbf{r}$; such a deformation leads to a spectrum consisting of doublets having the same quantum numbers $I\pi$.

4-2f Combinations of Rotation and Reflection Symmetries

If a $\mathscr{P}$- or $\mathscr{T}$-violating deformation occurs in a nonspherical system, one obtains a doubling of all the states in the rotational band, as discussed in Sec. 4-2e. A connection between rotational motion and the degrees of freedom associated with $\mathscr{P}$- and $\mathscr{T}$-violating deformations occurs if the system, while violating $\mathscr{R}$ symmetry, is invariant with respect to a combination of $\mathscr{R}$ with $\mathscr{P}$ or $\mathscr{T}$ symmetry.

$\mathscr{S}$ invariance

An example is provided by an axially symmetric shape deformation containing components of odd multipole order. Such a deformation violates $\mathscr{R}$ and $\mathscr{P}$ symmetry, but conserves $\mathscr{R}\mathscr{P}$, as can be seen from the fact that this combined operation represents a reflection in a plane containing the symmetry axis. (Diatomic molecules with nuclei of different charge have deformations of this type; the possible occurrence of nuclei with stable deformations containing octupole components is discussed on p. 561.)

The invariance with respect to a reflection in a plane containing the symmetry axis is conveniently expressed in terms of the operation

$$\mathscr{S} = \mathscr{P}\mathscr{R}^{-1} \tag{4-33}$$

Such an invariance of the deformation implies that $\mathscr{S}$ is an intrinsic variable. The parity

$$\mathscr{P} = \mathscr{S}\mathscr{R} \tag{4-34}$$

is therefore an operator that acts on the intrinsic variables through $\mathscr{S}(=\mathscr{S}_i)$ and on the rotational variables through $\mathscr{R}$ $(=\mathscr{R}_e)$.

The intrinsic states with $K=0$ are eigenstates of $\mathscr{S}_i$ and $\mathscr{T}$ (assuming the intrinsic Hamiltonian to be invariant under $\mathscr{T}$, as in the case of shape deformations),

$$\begin{aligned} \mathscr{S}_i \Phi_{s,K=0}(q) &= s\Phi_{s,K=0}(q) \\ \mathscr{S}_i \mathscr{T} \Phi_{s,K=0}(q) &= \Phi_{s,K=0}(q) \end{aligned} \tag{4-35}$$

The last relation determines the phase of Φ. The effect of $\mathscr{R}_e$ on the rotational wave function is given by Eq. (4-13), and the relation (4-34) therefore yields

$$\pi = s(-1)^I \tag{4-36}$$

Hence, the band contains the states

$$\begin{gathered} \Psi_{s,K=0,IM} = (2\pi)^{-1/2} \Phi_{s,K=0}(q) Y_{IM}(\theta,\phi) \begin{cases} 1 & \pi = +1 \\ i & \pi = -1 \end{cases} \\ I\pi = 0+, 1-, 2+, \ldots \qquad s = +1 \\ I\pi = 0-, 1+, 2-, \ldots \qquad s = -1 \end{gathered} \tag{4-37}$$

The factor i for $\pi = -1$ ensures the standard transformation (4-28) under time reversal.

For $K \neq 0$, the intrinsic states are twofold degenerate, as a consequence of $\mathscr{S}$ (or $\mathscr{T}$) symmetry. We can choose the phases of the intrinsic states such that (compare Eqs. (4-16) and (4-27))

$$\begin{aligned}\Phi_{\bar{K}}(q) &= \mathscr{S}_i \, \Phi_K \\ &= \mathscr{T}\Phi_K\end{aligned} \qquad (K>0) \tag{4-38}$$

The relation (4-34) implies that the total nuclear states of definite parity involve the following combinations of Φ_K and $\Phi_{\bar{K}}$:

$$\begin{aligned}\Psi_{\pi KIM} &= \left(\frac{2I+1}{16\pi^2}\right)^{1/2} e^{i\alpha}(1+\pi\,\mathscr{S}_i\,\mathscr{R}_e)\Phi_K\mathscr{D}^I_{MK} \\ &= \left(\frac{2I+1}{16\pi^2}\right)^{1/2} \begin{cases} \Phi_K\mathscr{D}^I_{MK} + (-1)^{I+K}\Phi_{\bar{K}}\mathscr{D}^I_{M-K} & \pi=+1 \\ i\left(\Phi_K\mathscr{D}^I_{MK} - (-1)^{I+K}\Phi_{\bar{K}}\mathscr{D}^I_{M-K}\right) & \pi=-1 \end{cases}\end{aligned} \tag{4-39}$$

The phase α is chosen, in the last form of Eq. (4-39), so as to ensure the standard transformation under time reversal. (The wave functions (4-39) contain, as a special case, the helicity representation of the one-particle wave functions; the phase factor i corresponds to the phase i^l in Eq. (3A-5).)

The violation of $\mathscr{R}$ and $\mathscr{P}$ implies that there is no constraint on the rotational or parity degrees of freedom, separately. For each intrinsic state, the band contains states with all values of $I \geqslant |K|$, and both parity values occur. However, the $\mathscr{S}$ invariance implies a link between rotation and parity, which selects the set of states (4-37) and (4-39).

The spectrum associated with an $\mathscr{R}\mathscr{P}$-invariant but $\mathscr{R}$- and $\mathscr{P}$-violating deformation exhibits characteristic differences from that associated with a $\mathscr{P}$-violating, but $\mathscr{R}$-conserving, deformation. As mentioned earlier, a deformation of the latter type gives a parity doubling of all states and thus, for $K=0$, the rotational spectrum contains the states $I\pi = 0\pm,\ 2\pm, \ldots$, or $I\pi = 1\pm,\ 3\pm,\ldots$, depending on the value of r. Moreover, for $K \neq 0$, the signature-dependent terms (see Eq. (4-20)) are the same for the parity doublets, since the intrinsic states Φ_K and $\Phi_{\bar{K}}$ are of the type (4-32). In contrast, for the $\mathscr{R}\mathscr{P}$-invariant deformation, the signature-dependent terms have opposite sign for the two states (4-39) and, hence, contribute to the energy splitting of the parity doublets. (Such a splitting is similar to the Λ-type doubling in molecular spectra; see, for example, Herzberg, 1950, pp. 226ff.).

A further separation between the parity doublets arises from a tunneling motion, similar to that discussed on p. 15 for a $\mathscr{P}$-violating deformation. For the $\mathscr{R}\mathscr{P}$-invariant system, the tunneling leads to an inversion of the symmetry axis through continuous variations of the shape, for fixed orientation. This effect gives rise to an energy shift $\Delta E = \hbar\omega_t$ of the negative parity states relative to the positive parity states, which in first approximation is independent of I. The tunneling may be treated in terms of an extended, $\mathscr{R}$- and $\mathscr{P}$-invariant, intrinsic Hamiltonian, which includes the degree of freedom associated with the barrier penetration. The eigenstates of the extended intrinsic Hamiltonian consist of doublets with opposite parity, separated by the energy $\hbar\omega_t$. The total wave functions are given by the expressions (4-15) and (4-19), which reduce to the forms (4-37) and (4-39) if the intrinsic states are represented by symmetric and antisymmetric combinations of states referring to the two opposite directions of the $\mathscr{R}$- and $\mathscr{P}$-violating deformation (see Eq. (4-32)).

A corresponding tunneling, or inversion, effect plays an important role in the analysis of the spectra of polyatomic molecules (see, for example, Herzberg, 1945, p. 33 and pp. 220ff.). In a simple molecule, like NH_3, the tunneling frequency in the ground state is of the order of 10^{10} sec^{-1}, which is a factor 10^3 smaller than the frequency of vibrations in the tunneling direction and an order of magnitude smaller than the lowest rotational frequency. The inversion frequency decreases rapidly with the mass of the atoms involved and the complexity of the configurations. For most of the organic molecules with left-right asymmetry, the lifetime for transitions between the "isomers" exceeds the time scale of biological evolution.

Survey of symmetries

The considerations in the present and previous sections illustrate the general rules that govern the relationship between the collective degrees of freedom and the invariance of the deformation with respect to rotations and reflections. For an axially symmetric system, the invariance group of the intrinsic Hamiltonian is the product of rotations about the symmetry axis and the group G of discrete operations under which the deformation is invariant. The group may comprise up to eight elements formed out of combinations of $\mathscr{R}$, $\mathscr{P}$, and $\mathscr{T}$. In the case of maximum symmetry violation, for which G contains only the identity, the intrinsic states, which can be labeled by $K = I_3$, are nondegenerate, and the rotational spectrum has two states for each $I\pi$ with $I \geqslant |K|$ (that is $I\pi = (|K|\pm)^2$, $(|K|+1, \pm)^2, \ldots$). The invariance of the deformation reduces the number of states in the rotational spectrum by a factor equal to the number of elements in the group G. The two parity values occur with equal weight, except when G contains the element $\mathscr{P}$, in which case the entire collective spectrum has the same parity as the intrinsic state. If G includes an operation that inverts I_3, ($\mathscr{R}$, $\mathscr{T}$, $\mathscr{R}\mathscr{P}$, or $\mathscr{P}\mathscr{T}$), the intrinsic states are twofold degenerate (for $K \neq 0$), and the rotational states are formed by a combination of the conjugate intrinsic states.

Examples of symmetry combinations are illustrated in Fig. 4-2a. In the first case, the intrinsic Hamiltonian is assumed to be invariant under $\mathscr{R}$, $\mathscr{P}$, and $\mathscr{T}$; for

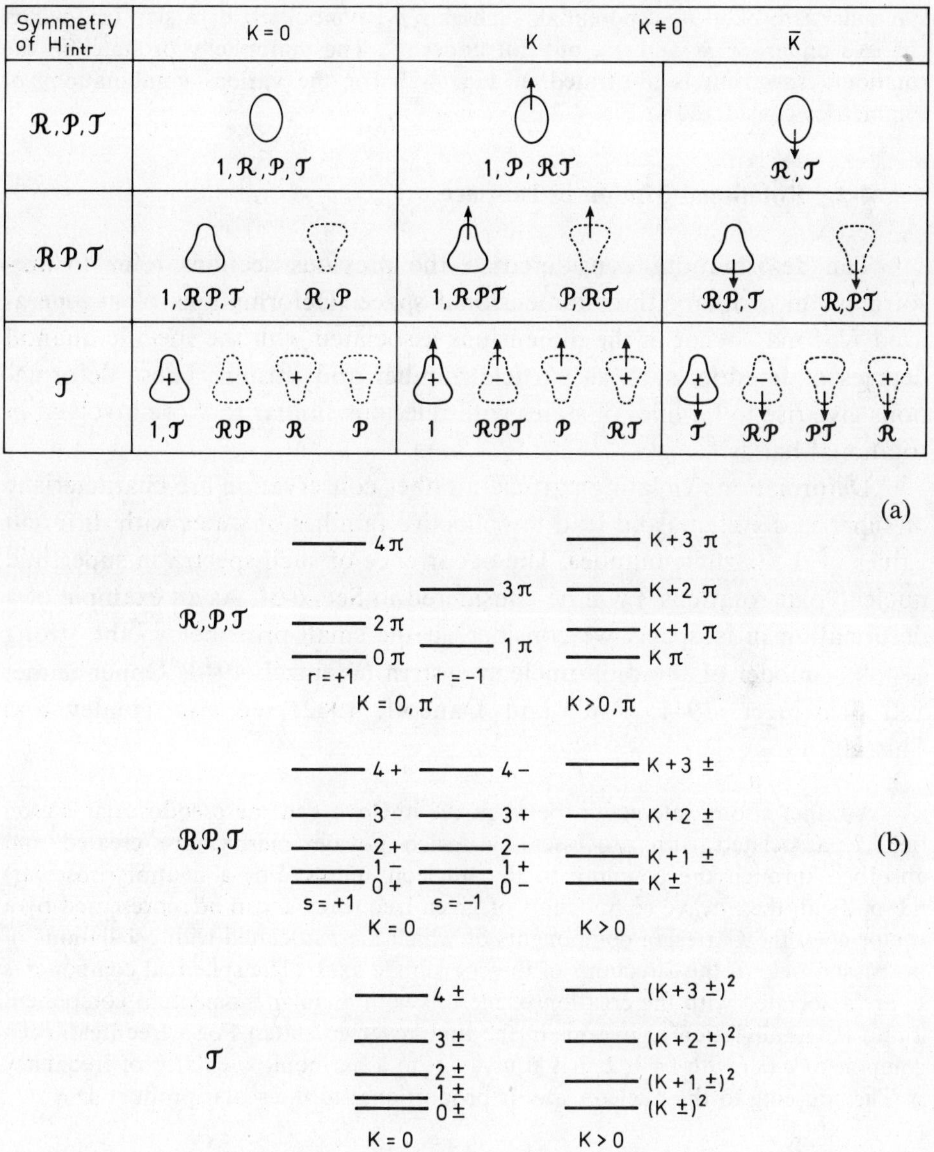

Symmetry of H_{intr}	K = 0	K ≠ 0: K	K ≠ 0: $\bar{K}$
$\mathscr{R}, \mathscr{P}, \mathscr{T}$	$1, \mathscr{R}, \mathscr{P}, \mathscr{T}$	$1, \mathscr{P}, \mathscr{R}\mathscr{T}$	$\mathscr{R}, \mathscr{T}$
$\mathscr{R}\mathscr{P}, \mathscr{T}$	$1, \mathscr{R}\mathscr{P}, \mathscr{T}$ $\mathscr{R}, \mathscr{P}$	$1, \mathscr{R}\mathscr{P}\mathscr{T}$ $\mathscr{P}, \mathscr{R}\mathscr{T}$	$\mathscr{R}\mathscr{P}, \mathscr{T}$ $\mathscr{R}, \mathscr{P}\mathscr{T}$
$\mathscr{T}$	$1, \mathscr{T}$ $\mathscr{R}\mathscr{P}$ $\mathscr{R}$ $\mathscr{P}$	1 $\mathscr{R}\mathscr{P}\mathscr{T}$ $\mathscr{P}$ $\mathscr{R}\mathscr{T}$	$\mathscr{T}$ $\mathscr{R}\mathscr{P}$ $\mathscr{P}\mathscr{T}$ $\mathscr{R}$

(a)

$\mathscr{R}, \mathscr{P}, \mathscr{T}$: K = 0, π — r = +1: 0π, 2π, 4π; r = −1: 1π, 3π. K > 0, π: Kπ, K+1 π, K+2 π, K+3 π

$\mathscr{R}\mathscr{P}, \mathscr{T}$: K = 0 — s = +1: 0+, 1−, 2+, 3−, 4+; s = −1: 0−, 1+, 2−, 3+, 4−. K > 0: K ±, K+1 ±, K+2 ±, K+3 ±

$\mathscr{T}$: K = 0: 0±, 1±, 2±, 3±, 4±. K > 0: $(K\pm)^2$, $(K+1\pm)^2$, $(K+2\pm)^2$, $(K+3\pm)^2$

(b)

Figure 4-2 Rotational spectra for axially symmetric systems with different combinations of $\mathscr{R}$, $\mathscr{P}$, and $\mathscr{T}$ invariance. The intrinsic states are illustrated in Fig. 4-2a and the associated rotational spectra in Fig. 4-2b; the quantum numbers $I\pi$ are given to the right of the energy levels.

$K \neq 0$, the intrinsic state is invariant under $\mathscr{R}\mathscr{T}$ and $\mathscr{P}$ and transforms into the degenerate state with opposite J_3 under $\mathscr{R}$ and $\mathscr{T}$. In the second case, the $\mathscr{R}$ symmetry is violated, but H_{int} is assumed to be invariant under $\mathscr{T}$ and $\mathscr{R}\mathscr{P}$ (or $\mathscr{S}$); under $\mathscr{R}$ or $\mathscr{P}$ separately, the states transform into eigenstates of a new equivalent Hamiltonian $\mathscr{R}^{-1}H_{\text{int}}\mathscr{R}$, as indicated by the dotted figures. In the third case, only $\mathscr{T}$ but neither $\mathscr{R}$, $\mathscr{R}\mathscr{P}$, nor $\mathscr{P}$ is a symmetry of H_{int}. The $\mathscr{R}\mathscr{P}$ symmetry is broken by a potential, such as $s_3 p_3$, symbolized by a sign ($\pm$) which changes under $\mathscr{R}\mathscr{P}$ and $\mathscr{P}$, but not under $\mathscr{T}$. The multiplicity of states in the rotational spectrum is illustrated in Fig. 4-2b for the various combinations of symmetries considered in Fig. 4-2a.

4-2g Rotational Motion in Isospace

The deformations considered in the previous sections refer to anisotropies in ordinary three-dimensional space. Deformations of a generalized type may occur in the dimensions associated with the specific quantal degrees of freedom such as particle number and isospin. These deformations give rise to families of states with relations similar to those involved in rotational bands.

Deformations violating particle-number conservation are characteristic of superfluid systems and lead to collective families of states with different values of the particle number. The occurrence of such spectra in superfluid nuclei ("pair rotations") will be considered in Sec. 6-3f. As an example of a deformation in isospace, we consider, in the small print below, the strong coupling model of the pion-nucleon system (Wentzel, 1940; Oppenheimer and Schwinger, 1941; Pauli and Dancoff, 1942; see also Henley and Thirring, 1962).

A rather strong interaction between the nucleon and the pseudoscalar meson field is associated with processes by which p-wave mesons are created and absorbed through the coupling to the nucleon spin **s**. For a neutral (isoscalar) meson field, the p-wave components of given frequency ω can be represented by a vector $\mathbf{q}(\omega)$, the Cartesian components of which are associated with oscillations of the meson field in the directions of the coordinate axes. (The spherical components q_μ are associated with the creation of mesons with angular momentum component μ and the annihilation of mesons in the time-reversed state.) For a free field, each component $q_i(\omega)$ with $i = 1, 2, 3$ is equivalent to a harmonic oscillator of frequency ω. The coupling to the nucleon spin is proportional to the scalar product $\mathbf{s} \cdot \mathbf{q}$,

$$H_c = \sum_{i=1}^{3} s_i \int q_i(\omega) F(\omega)\, d\omega \tag{4-40}$$

where the strength and form factor of the coupling are expressed by the function $F(\omega)$. The coupling (4-40) provides a force driving the meson field oscillators from their equilibrium position $q_i = 0$. A coupling of sufficient strength may lead to a static deformation that is large compared with the zero-point oscillations. The deformation of the p-wave field has the character of a vector and therefore possesses axial symmetry (about the direction of this vector). In the intrinsic system, the nucleon spin is in a state of $m'_s = 1/2$ (with suitable choice of direction along the symmetry axis), and only meson field components with $\mu' = 0$ are affected. Hence, the intrinsic state has $K\pi = 1/2+$. The deformation is $\mathscr{R}\mathscr{T}$ invariant (but violates $\mathscr{R}$ and $\mathscr{T}$ separately), and the rotational spectrum contains the states $I\pi = 1/2+, 3/2+, \ldots$ (see p. 18).

For the isovector pion field, each component $\mathbf{q}_i(\omega)$ is a vector in isospace, and the coupling takes the form

$$H_c = \sum_{i,j=1}^{3} s_i t_j \int q_{ij} F(\omega)\, d\omega \tag{4-41}$$

where t is the isospin of the nucleon. This coupling can produce a deformation in each of the two spaces similar to that associated with the coupling to the neutral field. However, the state of lowest energy involves a correlation in the two spaces such that the deformation is invariant with respect to rotations performed simultaneously in both spaces, with the same angle and about the same intrinsic axis $\kappa (= 1,2,3)$. This invariance implies that the intrinsic state is an eigenstate with eigenvalue 0 of the composite spin-isospin operator $\mathbf{u} = \mathbf{s} + \mathbf{t}$, defined in the intrinsic coordinate system. In such a state, $t_\kappa = -s_\kappa$, and hence, for each value of κ, the deformation $q_{\kappa\kappa}$ is as large as the q_{33} deformation in the uncorrelated state $m'_s = 1/2$, $m'_t = 1/2$; the correlation therefore gives an increase in the binding energy by a factor of three. (It is readily shown that the $u = 0$ state gives maximal binding; to this purpose, it is convenient to note that, by suitable (different) rotations in spin and isospace, it is possible to bring the coupling (4-41) to diagonal form involving only intrinsic components $q_{\kappa\kappa'}$, with $\kappa' = \kappa$. A deformation of similar symmetry is encountered in the description of superfluidity resulting from correlated pairs in ^3P states ($L = 1$, $S = 1$); Balian and Werthamer, 1963.) The intrinsic state with $u = 0$ does not possess axial symmetry in either of the two spaces, but is invariant with respect to arbitrary rotations performed simultaneously about any of the intrinsic axes κ in the two spaces. The rotational states therefore satisfy the condition $I_\kappa + T_\kappa = 0$ for $\kappa = 1,2,3$, and the spectrum consists of the series of states

$$\begin{gathered}\Psi_{IMTM_T} = \Phi_{\text{intr}} \frac{(2I+1)^{1/2}}{8\pi^2} \sum_K (-1)^{I-K} \mathscr{D}^I_{MK}(\omega)\, \mathscr{D}^{T=I}_{M_T -K}(\omega_\tau) \\ I = T = \tfrac{1}{2}, \tfrac{3}{2}, \ldots\end{gathered} \tag{4-42}$$

where ω and ω_τ represent the orientations in space and isospace, respectively.

It is possible that the $I = 3/2$, $T = 3/2$ nucleon isobar with a mass of 1236 MeV (see Fig. 1-11, Vol. I, p. 57) may correspond to the first excited state of such a generalized rotational band. This interpretation would imply relations between the properties of the nucleon and the isobar state similar to the intensity relations in nuclear rotational spectra (see Sec. 4-3), but there appears so far to be no test of these relations. However, the rather large excitation energies, which exceed the pion rest mass, may indicate that the coupling is not sufficiently strong for a quantitative treatment in terms of a static deformation. (For a discussion of the effective coupling strength, see Henley and Thirring, 1962, pp. 192 ff.) There is at present no evidence for the existence of higher members of the sequence (4-42), with $I=T \geqslant 5/2$. In contrast to the $I=T=1/2$ and $3/2$ members, the higher states cannot be constructed from a configuration of three quarks, each of which has spin and isospin 1/2 (see, for example, Vol. I, p. 41), and this circumstance might imply a discontinuity in the rotational band of a type similar to that encountered in the rotational spectra of light nuclei (see, for example, the discussion of ^{8}Be, pp. 101 ff.).

The experimental evidence concerning the excitation spectrum of the nucleon may suggest additional rotational relationships (see the trajectories of baryon states with fixed T in Fig. 1-13, Vol. I, p. 65). Several different bands based on a given intrinsic state are possible if a corresponding number of different deformations are present.

4-3 ENERGY SPECTRA AND INTENSITY RELATIONS FOR AXIALLY SYMMETRIC NUCLEI

The intimate connection between states in a rotational band is reflected in the relations between the matrix elements involving different members of a given band. The separation between rotational and intrinsic motion implies that, in first approximation, the matrix elements of the various operators can be expressed in terms of products of intrinsic and rotational factors. The intrinsic factor is the same for all the members of a band, and the rotational factor gives the dependence of the total matrix element on the rotational quantum numbers. For systems with axial symmetry, the rotational wave function is completely specified by the quantum numbers IKM (see Sec. 4-2a), and the dependence of the matrix elements on the rotational quantum numbers follows from the tensorial structure of the operators (Alaga *et al.*, 1955; Bohr, Fröman, and Mottelson, 1955; Satchler, 1955; Helmers, 1960). The relations are similar to the intensity rules that govern the radiative transitions in molecular band spectra (Hönl and London, 1925).

Departures from the leading-order intensity rules arise from the Coriolis and centrifugal forces acting in the rotating body-fixed system. These effects can be analyzed in terms of a coupling between rotational and intrinsic motion giving rise to a mixing of bands, as discussed in Sec. 4-4. Alternatively, the higher-order effects can be described in terms of renormalized, effective operators acting in the unperturbed basis; in such an approach, which provides a convenient basis for a phenomenological description, the coupling effects are represented by the dependence of the renormalized operators on the rotational angular momentum. For sufficiently small rotational frequencies, the Coriolis effects can be treated by perturbation theory, and the operators can be expanded in a power series of the rotational angular momentum. For axially symmetric systems, as discussed in the present section, the general form of such an expansion is greatly restricted by the symmetry of the rotational motion and the tensorial structure of the operators (Bohr and Mottelson, 1963). The particle-rotor model treated in Appendix 4A provides a simple explicit example illustrating many of these general relationships.

The renormalization of the operators associated with the rotational coupling effects may be expressed in terms of a canonical transformation that diagonalizes the Hamiltonian in the representation of the uncoupled motion. This transformation may also be viewed as a transformation to new intrinsic and rotational variables, in terms of which the perturbed wave functions retain the form (4-19). The explicit construction of the transformation is considered in Appendix 4A for the model consisting of a particle coupled to the motion of a rigid rotor.

4-3a Rotational Energies

For a slowly rotating system, the energy is given, to first approximation, by the intrinsic energy E_K associated with the state $\Phi_K(q)$, and is therefore the same for all the members of the band. The superimposed rotational motion gives an additional energy depending on the rotational angular momentum. Since the energy is a scalar with respect to rotations of the coordinate system, the effective Hamiltonian can only involve the components I_κ of the angular momentum with respect to the intrinsic axes.

Energy expansion for bands with $K=0$

For bands with vanishing intrinsic angular momentum K, the rotational energy has an especially simple form. Such bands involve only a single intrinsic state $\Phi_{K=0}$ (see Eq. (4-15)); the part of the energy operator acting in $K=0$ bands is therefore diagonal with respect to I_3 and can only depend on the combination $I_1^2+I_2^2$.

To leading order, the rotational Hamiltonian is of the form

$$(H_{\text{rot}})_{K=0}=h_0(q,p)(I_1^2+I_2^2) \tag{4-43}$$

where $h_0(q,p)$ is a function of the intrinsic variables. The expectation value of the operator (4-43) can be written

$$E_{\text{rot}} = \frac{\hbar^2}{2\mathscr{I}} I(I+1) \tag{4-44}$$

with the effective moment of inertia $\mathscr{I}$ defined by

$$\frac{\hbar^2}{2\mathscr{I}} \equiv \langle K|h_0|K\rangle \tag{4-45}$$

The form (4-44) is the familiar expression obtained by quantizing the classical Hamiltonian for a symmetric top.

More generally, the rotational energy can be expressed as a function of $I(I+1)$. For sufficiently small values of I, one can employ an expansion in powers of $I(I+1)$,

$$E_{\text{rot}}(I(I+1)) = AI(I+1) + BI^2(I+1)^2 + CI^3(I+1)^3 + DI^4(I+1)^4 + \cdots \tag{4-46}$$

where A is the intrinsic matrix element (4-45), while $B, C, D, \ldots$ are corresponding higher-order inertial parameters.

The deviations from the leading-order rotational energy (4-44) may be viewed as a dependence of the moment of inertia on the rotational angular momentum. The moment of inertia is defined as the ratio between the angular momentum $\hbar I_\kappa$ and the angular frequency ω_κ,

$$\mathscr{I}_\kappa = \frac{\hbar I_\kappa}{\omega_\kappa} \tag{4-47}$$

The frequency may be obtained from the canonical equation of motion for the angle conjugate to I_κ,

$$\omega_\kappa = \hbar^{-1} \frac{\partial H_{\text{rot}}}{\partial I_\kappa} = 2\hbar^{-1} I_\kappa \frac{\partial H_{\text{rot}}}{\partial (I_1^2 + I_2^2)} \tag{4-48}$$

since the rotational Hamiltonian, in a $K=0$ band, is a function of $I_1^2 + I_2^2$. For the moment of inertia ($\mathscr{I}_1 = \mathscr{I}_2 \equiv \mathscr{I}$), we thus obtain

$$\mathscr{I} = \frac{\hbar^2}{2} \left(\frac{\partial H_{\text{rot}}}{\partial (I_1^2 + I_2^2)} \right)^{-1} \tag{4-49}$$

and, hence, the expansion (4-46) for the energy in powers of $I(I+1)$ implies

$$\begin{aligned}\mathscr{I} &= \frac{\hbar^2}{2}\left(A+2BI(I+1)+3CI^2(I+1)^2+\cdots\right)^{-1} \\ &= \frac{\hbar^2}{2}\left(A^{-1}-2BA^{-2}I(I+1)+(4B^2A^{-3}-3CA^{-2})I^2(I+1)^2+\cdots\right)\end{aligned} \tag{4-50}$$

The energy and moment of inertia may also be expanded in powers of the rotational frequency rather than angular momentum (Harris, 1965). Such an expansion occurs naturally in the analysis of the rotational properties in terms of the response of the nucleonic motion to the rotating potential (cranking model; see pp. 75 ff.). The relation between the energy and the moment of inertia now takes the form (see Eqs. (4-47) and (4-49))

$$\begin{aligned}\frac{\partial H_{\text{rot}}}{\partial\omega^2} &= \frac{\partial H_{\text{rot}}}{\partial(I_1^2+I_2^2)}\frac{\partial(I_1^2+I_2^2)}{\partial\omega^2} = \frac{1}{2\mathscr{I}}\frac{\partial(\omega^2\mathscr{I}^2)}{\partial\omega^2} \\ &= \tfrac{1}{2}\mathscr{I}+\omega^2\frac{\partial\mathscr{I}}{\partial\omega^2}\end{aligned} \tag{4-51}$$

or, in terms of a power series expansion in $\omega^2(=\omega_1^2+\omega_2^2)$,

$$\begin{aligned}\mathscr{I} &= \alpha+\beta\omega^2+\gamma\omega^4+\delta\omega^6+\cdots \\ H_{\text{rot}} &= \tfrac{1}{2}\alpha\omega^2+\tfrac{3}{4}\beta\omega^4+\tfrac{5}{6}\gamma\omega^6+\tfrac{7}{8}\delta\omega^8+\cdots\end{aligned} \tag{4-52}$$

The value of ω^2 associated with the quantum state I can be obtained from the relation (see Eq. (4-47))

$$\hbar^2 I(I+1) = \omega^2\mathscr{I}^2 \tag{4-53}$$

If the moment of inertia is given by the expansion (4-52), the energy can be expressed in the form (4-46) with

$$\begin{aligned}\hbar^{-2}A &= \frac{1}{2\alpha} \\ \hbar^{-4}B &= -\frac{\beta}{4\alpha^4} \\ \hbar^{-6}C &= \frac{\beta^2}{2\alpha^7}-\frac{\gamma}{6\alpha^6} \\ \hbar^{-8}D &= -\frac{3\beta^3}{2\alpha^{10}}+\frac{\beta\gamma}{\alpha^9}-\frac{\delta}{8\alpha^8} \\ &\cdots\end{aligned} \tag{4-54}$$

The expansion coefficients $\alpha, \beta, \gamma, \ldots$ of Eq. (4-52) thus provide a parameterization of the rotational energy alternative to that given by Eq. (4-46) in terms of the coefficients $A, B, C, \ldots$. The nonlinear relation between the angular momentum and frequency implies that the two expansions in general have different radii of convergence.

Ground-state bands of even-even nuclei

The low-energy spectra of the even-even nuclei exhibit a striking simplicity, and in certain regions of N, Z values (see Fig. 4-3) consist of a sequence of states with $I\pi = 0+, 2+, 4+, \ldots$ and energies approximately following the relation (4-44), as shown in Fig. 4-4. Examples of such rotational spectra are discussed in more detail in connection with Fig. 4-7 (^{168}Er, p. 63), Fig. 4-9 (^{238}U, p. 67), Fig. 4-11 (^{172}Hf, p. 70), Fig. 4-13 (^{20}Ne, p. 97), Fig. 4-14 (^{8}Be, p. 100), Fig. 4-29 (^{166}Er, p. 159), and Fig. 4-31 (^{174}Hf, p. 168). The regions in which the rotational spectra occur correspond to ground-state configurations with many particles outside of closed shells; this pattern may be understood as a simple consequence of the nuclear shell structure. While the closed-shell configurations prefer the spherical symmetry, the orbits of the particles outside of closed shells are strongly anisotropic and drive the system away from spherical symmetry (Rainwater, 1950). General features of the shell-structure effects on the nuclear potential energy are discussed in Chapter 6, pp. 602 ff.

Metastable configurations with shapes very different from those of the ground state may occur as a result of shell structure in the strongly deformed potential. These states may be separated from those of lower deformation by appreciable barriers. Thus, the possibility of shape isomerism appears as a rather general property of the nucleus (see the discussion in Chapter 6, pp. 602 ff,). Evidence for such deformed excited configurations is provided by the rotational bands observed in the spectra of ^{16}O and ^{40}Ca, which are spherical in their ground-state (closed-shell) configuration[2], and by the discovery of the spontaneously fissioning isomers in the very heavy elements; see the discussion in Chapter 6, pp. 623 ff. (rotational band structure based on the fission isomer of ^{240}Pu has been observed by Specht et al., 1972).

---

[2] The fact that the first excited states in ^{16}O and ^{40}Ca have positive parity, while the low-lying particle–hole excitations are restricted to negative parity, implies that these states involve the excitation of a larger number of particles. It was suggested (Morinaga, 1956) that the excited positive parity states might be associated with collective quadrupole deformations. The existence of a rotational band structure in ^{16}O was convincingly established as a result of the ^{12}C($\alpha\alpha$) studies by Carter *et al.* (1964) and the observation of strongly enhanced $E2$-transition matrix elements (Gorodetzky *et al.*, 1963).

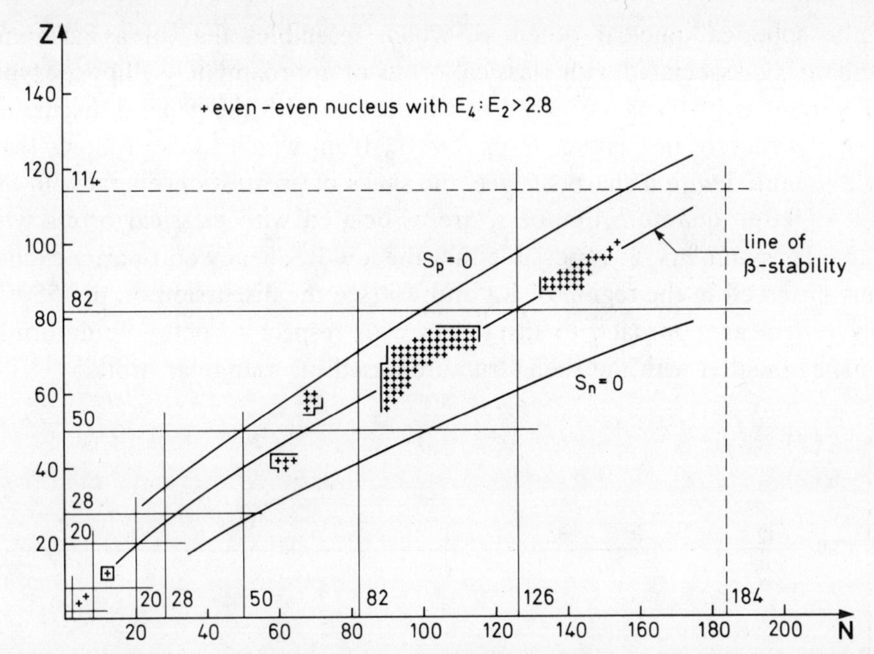

Figure 4-3 Regions of deformed nuclei. The crosses represent even-even nuclei, whose excitation spectra exhibit an approximate $I(I+1)$ dependence, indicating rotational structure. The nuclei included in the figure have been selected on the basis of the (rather arbitrary) criterion $E(I=4):E(I=2)>2.8$. The data are taken from the compilations by Sakai (1970 and 1972). The line of β stability and the estimated borders of instability with respect to proton and neutron emission are those shown in Fig. 2-18, Vol. I, p. 203.

The observed pattern of the rotational bands can be uniquely interpreted in terms of the symmetry of the nuclear deformation. First, the occurrence of only a single sequence of I values shows that we are dealing with an axially symmetric body constrained to rotate about axes perpendicular to the symmetry axis. (Rotational bands for systems without axial symmetry contain several states of the same I; see Sec. 4-5.) The approximate axial symmetry is also exhibited by the close agreement with the $I(I+1)$ rule for the energies, which is characteristic of a "symmetric rotor," a system for which the ellipsoid of inertia has spheroidal symmetry. Next, the absence of states with odd values of I reveals the $\mathscr{R}$ symmetry in the nuclear shape (see Fig. 4-2b). Finally, the absence of parity doublets shows the $\mathscr{P}$ invariance of the intrinsic motion, while the absence of doublets with the same $I\pi$ implies $\mathscr{T}$ invariance of the intrinsic motion (see p. 15). These symmetry features revealed by the rotational spectra can be interpreted in terms of a spheroidal deformation in the nuclear shape. The preference for such a shape can be understood from the fact that the one-particle motion in the spherical nuclear potential, which resembles that of a harmonic oscillator, is associated with classical orbits of approximately elliptical type. (This point is further considered in connection with the general discussion of shell structure in Chapter 6, pp. 589 ff., from which it also follows that, for a potential with a sharp surface, the shells of greatest degeneracy, in the limit of large quantum numbers, are associated with classical orbits with triangular symmetry. It is possible that the low-frequency odd-parity excitations observed in the region of Ra and Th (see the discussion on pp. 559 ff.) may reflect an approach to instability with respect to octupole deformations, associated with the shell structure based on triangular orbits.)

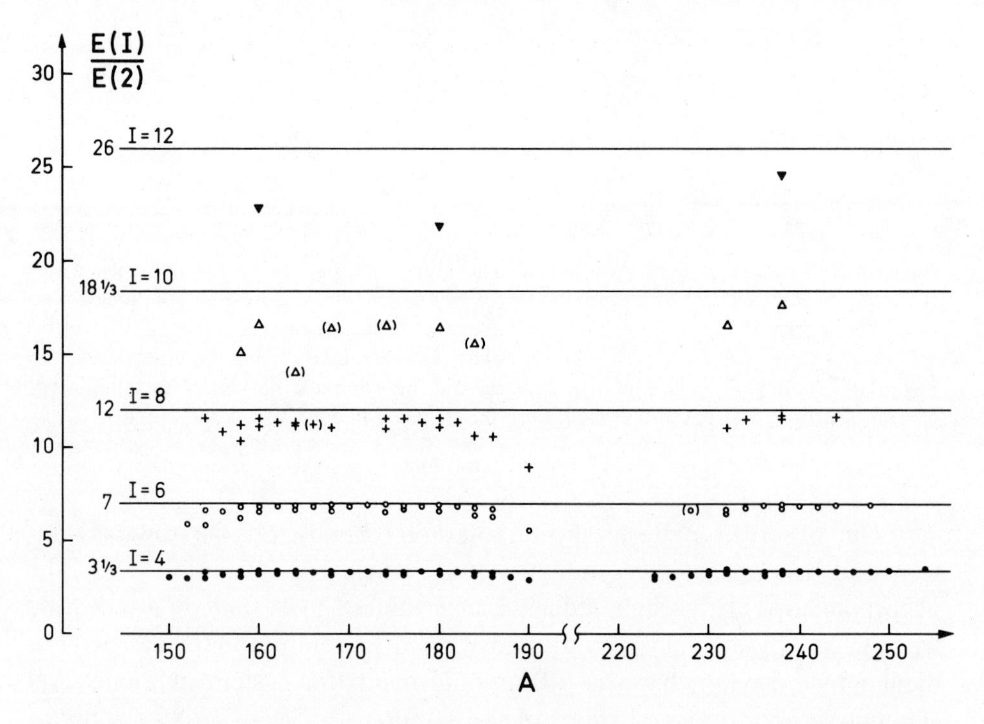

Figure 4-4 Energy ratios in the ground-state rotational band in even-even nuclei. The figure shows the observed energy ratios $E_I : E_2$ for the ground-state rotational bands in even-even nuclei in the regions $152 \lesssim A \lesssim 190$ and $224 \lesssim A$. For each element, only the β-stable isotopes are included. The data are taken from *Table of Isotopes* by Lederer *et al.*, 1967. The entries in parentheses represent tentative assignments. The horizontal lines give the theoretical energy ratios for the leading-order expression (4-44).

The observed bands have the quantum numbers $K\pi r = 0++$. This property of the ground states of even-even nuclei can be interpreted on the basis of an independent-particle description of the intrinsic structure. In an axially symmetric nuclear potential, the individual nucleonic orbits can be characterized by the projection Ω of the angular momentum j on the symmetry axis. Time reversal symmetry (or $\mathscr{R}$ symmetry) implies that the orbits occur in degenerate, conjugate pairs with opposite values of Ω, and the ground state of an even-even nucleus is therefore formed by adding particles pairwise in such conjugate orbits, with resulting $K=0$. For a $\mathscr{P}$-invariant potential, the conjugate orbits have the same parity, and thus each pair of particles has $\pi=+1$. Moreover, the conjugate single-particle states Ω and $\bar{\Omega}$ transform into each other under the rotation $\mathscr{R}_i$ (see Eq. (4-16)), and the antisymmetrized state representing a pair of particles in conjugate orbits therefore has $r=+1$,

$$\begin{aligned}\Phi(1,2) &= \frac{1}{\sqrt{2}}\left(\psi_{\Omega}(1)\psi_{\bar{\Omega}}(2)-\psi_{\bar{\Omega}}(1)\psi_{\Omega}(2)\right)\\ \mathscr{R}_i\Phi(1,2) &= \frac{1}{\sqrt{2}}\left(-\psi_{\bar{\Omega}}(1)\psi_{\Omega}(2)+\psi_{\Omega}(1)\psi_{\bar{\Omega}}(2)\right)=\Phi(1,2)\\ &\left(\psi_{\bar{\Omega}}=\mathscr{R}_i^{-1}\psi_{\Omega}=-\mathscr{R}_i\psi_{\Omega}\right)\end{aligned} \tag{4-55}$$

This result could also have been obtained from Eq. (4-17), since the antisymmetrized pair represents a superposition of even J values.

The absolute values of the rotational energies determine the moment of inertia by means of Eq. (4-44). The observed values are illustrated in Fig. 4-12 (p. 74). It is seen that the moments of inertia for the ground-state bands in even-even nuclei are smaller by a factor of 2 to 3 as compared with the values for rigid rotation.

The moments of inertia of a rotating system can be analyzed in terms of the effect of the Coriolis forces in perturbing the motion of the individual particles. For a system described by independent-particle motion in a self-consistent potential, the moment of inertia is found to be approximately equal to the rigid-body value. However, the pair correlations prevent the nucleons from following the rotational motion as effectively as would be the case for independent-particle motion, and reduce the moments below the values corresponding to rigid rotation. (See the discussion of the nuclear moments of inertia on pp. 75ff.)

In the most fully developed rotational bands, the energies are proportional to $I(I+1)$, with an accuracy of a few parts per thousand for the lowest values of I. The deviations have a systematic character and can be well represented by the next term in the expansion (4-46) (see, for example, Figs. 4-8, p. 64, and 4-9, p. 67). The value of B/A is of the order of 10^{-3} in the regions of largest stability of the deformed nuclear shape, but B/A increases as one approaches the configurations for which the deformed shape is no longer stable, as can be seen from Fig. 4-4. (The values of B/A for the well-developed nuclear rotational spectra are similar in magnitude to those for the ground-state band of the H_2 molecule; for heavier molecules, the deviations from the $I(I+1)$ energy dependence are usually much smaller (see, for example, Herzberg, 1950).)

The deviations from the $I(I+1)$ dependence of the rotational energy express the variation of the moment of inertia with the rotational frequency. Such a variation arises from the dependence of the moment of inertia on the collective parameters that characterize the nuclear shape and pair correlations. In turn, this dependence implies that the equilibrium values of the parameters have a term proportional to $I(I+1)$; this variation of the equilibrium with I gives rise to a decrease in the total energy proportional to $I^2(I+1)^2$.

If the nucleus rotated as a semirigid body (like a molecule), these effects would correspond to the classical centrifugal distortions and would in most cases be very small compared with the observed B terms. However, the moments of inertia are strongly affected by the nucleonic correlation effects, as revealed by the reduced values of the moments. This sensitivity may lead to a much stronger dependence of the moment of inertia on the collective parameters (see, for example, the dependence on the pair correlation parameter discussed on p. 82).

Direct information on these coupling effects can be obtained by measurements of the deformation parameters as functions of the rotational angular momentum (see, for example, the corrections to the leading-order $E2$-intensity rules discussed on p. 129). The dependence of the moment of inertia on the collective parameters also gives rise to a coupling between the rotational motion and the vibrational excitations associated with the oscillations in these parameters. These couplings imply appreciable corrections to the intensity relations for the transitions between the ground-state band and vibrational excitations; from an analysis of these intensities, one can infer the magnitude of the corresponding contribution to the B term in the rotational energy. In the cases that have been analyzed so far, the $\Delta K=2$ couplings contribute only a small fraction of the B coefficient in the ground-state band (see, for example, the analysis of the coupling to the $K=2$ band in ^{166}Er, p. 162). Somewhat larger contributions have been observed to result from couplings to low-lying $K=0$ excitations (see, for example, the analysis of the spectrum of ^{174}Hf, p. 172). At present, it is not yet clear to what extent the observed B terms can be attributed to the coupling to a few low-lying intrinsic excitations.

In many cases, the precision of the energy measurements makes possible a determination of several of the higher-order terms in the expansion in powers of $I(I+1)$. While the ratio of B/A might indicate a radius of convergence of $I \approx 30$, the magnitude of the higher-order terms implies a considerably poorer rate of convergence. If the energy is expressed as a power series in the rotational frequency, as in Eq. (4-52), rather than in the angular momentum, the rate of convergence is found to be significantly improved (Harris, 1965; Mariscotti *et al.,* 1969). The power series expansions in I and ω are compared on pp. 24 ff.; see also Fig. 4-11 (^{172}Hf), p. 70. The reason for the apparently greater simplicity of the expansion in ω^2 is not understood at present.

With increasing values of the angular momentum, the rotational motion may give rise to major distortions in the intrinsic structure that have the character of phase transitions with rather abrupt changes in the rotational spectrum. These effects will be discussed in connection with the more general problem of characterizing the elementary modes of excitation in the yrast region (see pp. 41 ff.).

Energy expansion for bands with $K \neq 0$

For $K \neq 0$, the nuclear states involve a combination of intrinsic wave functions with $I_3 = \pm K$ (see Eq. (4-19)). Thus, the rotational energy consists partly of terms with $\Delta K = 0$, which are of the same form as for $K=0$ bands (see Eq. (4-46)); in addition, the rotational energy may involve terms with $\Delta K = \pm 2K$ connecting the two components of the wave function. The leading-order terms with $\Delta K = \pm 2K$ have the form

$$(H_{\text{rot}})_{\Delta K = \pm 2K} = h_{2K}(q,p)(I_-)^{2K} + \mathscr{R}\text{-conj.} \qquad (4\text{-}56)$$

The second term is the Hermitian conjugate involving $(I_+)^{2K}$; this term can also be expressed as the conjugate term obtained by performing the transformation $\mathscr{R}_i^{-1}\mathscr{R}_e$. (Under $\mathscr{R}_e$, the operators $I_\pm$ transform into $-I_\mp$.) The $\mathscr{R}$ symmetry requires the operators and state vectors to be invariant under the transformation $\mathscr{R}_i^{-1}\mathscr{R}_e$ (see Sec. 4-2c); hence, the

$\mathscr{R}$-conjugate terms in operators such as (4-56) always contribute identical matrix elements when acting between $\mathscr{R}$-symmetrized states.

The expectation value of the additional rotational energy (4-56) is proportional to the signature of the level (see Eqs. (4-20) and (4-23)) and can be written

$$\Delta E_{\mathrm{rot}} = (-1)^{I+K} A_{2K} \frac{(I+K)!}{(I-K)!} \tag{4-57}$$

with

$$A_{2K} = \langle K | h_{2K} | \overline{K} \rangle \tag{4-58}$$

(The matrix elements of $I_{\pm}$ are given by Eq. (1A-93).)

For $K=1/2$, the additional energy term (4-56) is linear in I_{κ} and, hence, in the rotational frequencies ω_{κ}. This contribution to the energy represents the diagonal effect of the Coriolis forces acting in the rotating coordinate system. Thus, if the angular momentum $K=1/2$ can be associated with a single particle moving in the potential generated by the rotating nuclear core, we have (see the discussion of the particle-rotor model in Appendix 4A)

$$h_{+1} = -\frac{\hbar^2}{2\mathscr{I}} j_{+} \tag{4-59}$$

where $\mathbf{j}$ is the angular momentum of the particle. The energy parameter A_1 can then be expressed in the form

$$\begin{aligned} A_1 &\equiv \frac{\hbar^2}{2\mathscr{I}} a \\ a &= -\langle K=1/2 | j_{+} | \overline{K=1/2} \rangle \end{aligned} \tag{4-60}$$

defining the quantity a referred to as the decoupling parameter. (In Eq. (4-59), we have ignored the effect of the rotational motion on the nuclear potential (see pp. 278 ff.) as well as the difference between the moment of inertia $\mathscr{I}$ of the total system and the moment $\mathscr{I}_0$ associated with the nuclear core, with respect to which the particle is moving; see the discussion on pp. 280 ff.)

The expression (4-60) for a implies values that are typically of order unity or greater, and the A_1 term must therefore be included in the leading-order expression for the rotational energy spectrum[3]

$$
\begin{aligned}
E_{\text{rot}} &= AI(I+1) + A_1(-1)^{I+1/2}(I+1/2)\delta(K,1/2) \\
&= \frac{\hbar^2}{2\mathscr{I}}\left(I(I+1) + a(-1)^{I+1/2}(I+1/2)\delta(K,1/2)\right)
\end{aligned}
\tag{4-61}
$$

For $K=1/2$, the rotational energy (4-61) is, apart from a constant, proportional to $(I+(1+\sigma a)/2)^2$, where $\sigma=(-1)^{I+1/2}$ is the signature. Thus, the branches with different signatures have energy parabolas displaced in opposite directions along the I axis. When the decoupling parameter a is numerically larger than unity, these displacements lead to inversion of the normal spin sequence. In the special case of an intrinsic state in which the nucleonic spin has the component along the symmetry axis, $\Sigma=1/2$, while the remaining degrees of freedom are coupled to a state of $K=0$, the value of a given by Eq. (4-60) equals the r quantum number for the $K=0$ state; the energy spectrum (4-61) then has the doublet structure associated with a spin that is weakly coupled to a $K=0$ rotational band with angular momenta $0,2,4,\ldots$ (for $a=r=+1$) or $1,3,5,\ldots$ (for $a=r=-1$).

The coefficients A_{2K} represent perturbation effects of order $2K$ in the Coriolis interaction. For $K \geqslant 3/2$, the additional energy term (4-57) is, therefore, much smaller than the leading-order rotational energies (see, for example, the discussion of the A_3 term in the particle-rotor model, p. 204).

The general expression for the $\Delta K = \pm 2K$ parts of the rotational Hamiltonian is obtained by multiplying the operator (4-56) with a function of $I_1{}^2 + I_2{}^2$, and the associated energies can be expanded in a power series in $I(I+1)$, as in Eq. (4-46). For the total energy, we thus obtain for the first few values of K

$$
E(K,I) = E_K + AI(I+1) + BI^2(I+1)^2 + \cdots
$$
$$
+\begin{cases}
(-1)^{I+1/2}(I+\frac{1}{2})(A_1 + B_1 I(I+1) + \cdots) & K=\frac{1}{2} \\
(-1)^{I+1} I(I+1)(A_2 + B_2 I(I+1) + \cdots) & K=1 \\
(-1)^{I+3/2}(I-\frac{1}{2})(I+\frac{1}{2})(I+\frac{3}{2})(A_3 + B_3 I(I+1) + \cdots) & K=\frac{3}{2} \\
(-1)^{I}(I-1)I(I+1)(I+2)(A_4 + B_4 I(I+1) + \cdots) & K=2
\end{cases}
\tag{4-62}
$$

[3]The term in the rotational energy, linear in I, is known from molecular spectra as a modification of the Λ-type doubling caused by the spin-orbit coupling (Van Vleck, 1929). For the nuclear rotational spectra, the term was considered by Davidson and Feenberg, 1953, and by Bohr and Mottelson, 1953.

(An alternative form for the expansion may be obtained by replacing $I(I+1)$ by $I(I+1)-K^2$ (representing the expectation value of $I_1^2+I_2^2$). The treatment of the particle-rotor model indicates that such an expansion may be a somewhat more natural one. However, if $A \gg BK^2 \gg CK^4 \gg \cdots$, the difference between the two forms of the expansion is only slight.)

Rotational bands in odd-A nuclei

Extensive studies have been made of the low-lying bands in odd-A nuclei. Due to the greater density of states in these nuclei as compared with the even-even nuclei, the low-energy spectra appear more complex. However, in the regions of the deformed nuclei (see Fig. 4-3), it is found that the low-lying states can always be classified in a series of bands, each characterized by an intrinsic angular momentum K and a parity π. The observed values of $K\pi$ for the lowest bands correspond to the values of $\Omega\pi$ of the last odd particle moving in the deformed nuclear potential (see Chapter 5). At excitations of the order of 1 MeV, one also encounters bands involving vibrational excitations superposed on the quantum states of a single particle, as well as bands based on configurations with three unpaired particles.

The states contained in each band have $I=K, K+1, K+2, \ldots$; the energies follow the rotational expression (4-61) with an accuracy similar to the energies in the ground-state bands of even-even nuclei. Examples are illustrated in Fig. 4-15 (^{169}Tm, p. 103), Fig. 4-16 (^{177}Lu and ^{177}Hf, p. 106), Fig. 4-19 (^{239}Pu, p. 112); Figs. 5-7 (^{159}Tb, p. 254), Fig. 5-9 (^{175}Yb, p. 259), Fig. 5-12 (^{237}Np, p. 266), Fig. 5-14 (^{235}U, p. 274), and Fig. 5-15 (^{25}Mg, p. 284).

The dependence of the rotational energy on the signature is strikingly apparent in the $K=1/2$ bands, for which the A_1 term is usually comparable to the A term. (For an interpretation of the decoupling parameter on the basis of the one-particle orbits, see Sec. 5-3d.) The expected smaller signature-dependent effects have also been identified in bands with $K \geqslant 3/2$ (see, for example, the $K=3/2$ bands in Figs. 5-7 and 5-14, and the discussion of the $K=7/2$ band on p. 280).

The moments of inertia of the odd-A bands are found to be systematically larger than those of the ground-state bands in even-even nuclei, by amounts that are typically of the order of 20%, but that can be much larger in some cases (see Fig. 4-12). Such an effect can be understood in terms of the response of the last odd particle to the rotation of the potential; because of the presence of the pair correlations that reduce the inertial contribution from the particles in the core, the contribution of the unpaired particle becomes a significant fraction of the total. On this basis, it is possible to understand the observed strong dependence of the increment of the moment of inertia on the particular orbit of the odd particle (see pp. 310ff.). An additional effect on the moment of inertia results from the reduced pair correlation in the presence of the odd particle (see p. 311.) The B coefficients are even more sensitive to the presence of the odd nucleon and in some odd-A bands have positive sign (see, for example, Fig. 4-17 ($K\pi = 9/2+$ band in ^{177}Hf)).

Excited bands in even-even nuclei

Examples of rotational band structure associated with excited bands in even-even nuclei are shown in Figs. 4-7 (^{168}Er, p. 63), 4-13 (^{20}Ne, p. 97), 4-29 (^{166}Er, p. 159), 4-31 (^{174}Hf, p. 168), and in several illustrations in Chapter 6 (Figs. 6-31 and 6-44 (even- and odd-parity bands in Sm isotopes, pp. 534 and 577, respectively), Fig. 6-32 (Os isotopes, p. 536), Fig. 6-39 (^{234}U, p. 555)). A characteristic feature of these spectra is the absence of intrinsic excitations up to energies that are typically an order of magnitude larger than the energy spacing between the bands in odd-A nuclei.

Some of the observed excitations in the even-even nuclei can be interpreted as two-quasiparticle states resulting from the breaking of a pair; these excitations occur in the energy region above 2Δ, which represents the energy required to break a pair (see Fig. 2-5, Vol. I, p. 170). The two-quasiparticle bands have moments of inertia that—like those of odd-A nuclei—are systematically larger than the moments of the ground-state band (see, for example, the bands in ^{168}Er (Table 4-1, p. 65)).

The even-even spectra also exhibit a few excitations significantly below 2Δ, and it appears that these modes represent collective oscillations around the deformed equilibrium shape. (See the discussion in Sec. 6-3b.) The quadrupole vibrational excitations ($K\pi = 2+$ and $K\pi r = 0++$) exhibit moments of inertia that in most cases differ from those of the ground-state bands by less than 10% (see the examples quoted on p. 549 (γ vibration) and p. 551 (β vibration)); this feature may be understood from the fact that the moment of inertia is not expected to change appreciably for the rather small amplitudes of oscillation away from equilibrium. Examples of octupole vibrational excitations are provided by the $K\pi r = 0--$ bands that in many nuclei are observed among the lowest intrinsic excitations. The moments of inertia in these bands are typically as much as 50% larger than for the ground-state bands (see the example in Fig. 6-44 (^{152}Sm, p. 577)), and this effect may be attributed to the very large Coriolis coupling to the $K\pi = 1-$ octupole excitation (see the discussion on p. 578). There is no corresponding effect for the quadrupole oscillations, since the $K\pi = 1+$ quadrupole degree of freedom is represented by the rotational motion (see Fig. 6-3, p. 363).

Rotational bands in odd-odd nuclei

In spite of the high level density encountered in the low-energy spectra of odd-odd nuclei, it has been possible in a number of cases to establish rather detailed level schemes, and the available evidence is consistent with the expected rotational band structure (see the example in Fig. 4-20 (^{166}Ho, p. 120)). A classification of the intrinsic states has been possible in terms of the orbits of the odd neutron and proton. For each configuration (Ω_n, Ω_p), one obtains two bands with $K = |\Omega_n + \Omega_p|$ and $K = |\Omega_n - \Omega_p|$, respectively; in the special case of $\Omega_n = \Omega_p$, the $K = 0$ states separate into two bands with $r = +1$ and $r = -1$ with intrinsic configurations $2^{-1/2}(|\Omega_n \bar{\Omega}_p\rangle - r|\bar{\Omega}_n \Omega_p\rangle)$. The expectation value of the np interaction in these states gives rise to an I-independent term depending on r and thus to a separation of the two $K = 0$ bands.

The moments of inertia of the odd-odd configurations are systematically greater than those of the even-even nuclei and can be approximately described by the additive contributions $\delta\mathscr{I}_n$ and $\delta\mathscr{I}_p$ of the odd neutron and proton, as determined from the corresponding configurations in the odd-A nuclei (see the examples in Table 4-14, p. 121).

In the case of configurations with $\Omega_p = \Omega_n = 1/2$, which give rise to near-lying bands with $K = 0$ and $K = 1$, the effect of the Coriolis interaction may be especially large and leads to alternating energy shifts in the $K = 1$ band, corresponding to the A_2 term in Eq. (4-62). (For an analysis of configurations of this type, see Hansen, 1964, and Sheline *et al.*, 1966.)

Rotational structure in the fission channels

Rotational band structure similar to that encountered in the low-energy spectra is expected in the spectrum of fission channels at excitation energies not too far above the fission threshold (Bohr, 1956; the channel analysis of the fission process is discussed in Sec. 6-3c). Despite the fact that the fissioning nucleus has an excitation energy of about 5 MeV, the nucleus, in passing over the saddle point, is "cold", since the main part of the excitation energy is stored as potential energy of deformation. The quantized potential energy surfaces (the fission channels) are therefore widely separated as in the low-energy spectra.

The rotational band structure of the fission channels is determined by the symmetry of the saddle-point shape, as for the rotational bands based on the ground-state configurations. Evidence on the channel spectrum can be obtained from the measurement of fission yields (see, for example, pp. 123 ff.), and the angular distributions provide information on the distribution of the quantum number K in the channel states (see Eq. (4-178)). The observed angular distributions in the region of the fission threshold show striking anisotropies, which confirm the picture of fission proceeding through one or a few channels. However, the available evidence is in no case sufficiently detailed to uniquely determine the symmetry of the saddle-point shape. The evidence from photofission of ^{238}U is discussed in the example on pp. 123 ff. Evidence on the symmetry of the saddle-point shape may also be obtained from a statistical analysis of the level density of fission channels (see the discussion on pp. 617 ff.).

Rotational motion for configurations with many quasiparticles

Very little evidence is available about rotational motion for intrinsic configurations above about 1 or 2 MeV in heavy nuclei. With the excitation of increasing numbers of quasiparticles, one expects a weakening of the pair correlations, since more and more of the levels near the Fermi energy are occupied by unpaired particles and are thus not available for the pair correlation. Eventually, this effect is expected to lead to configurations in which the pair correlation parameter Δ vanishes. (The disappearance of Δ is similar to the phase transition from the superconducting to the normal state of the metal at the critical temperature.) With the decrease of Δ, the moment of inertia is expected to increase and to approach the rigid-body value for $\Delta = 0$ (see the discussion on pp. 77 ff.). A part of the systematic increase in the moment of inertia observed in the one- and two-quasiparticle states, as compared to that of the ground-state bands of even-even nuclei, can be recognized as the first step in this direction (see p. 311).

With increasing excitation, the density of levels rapidly increases, and the individual levels represent combinations of many different configurations containing different numbers of quasiparticles (see the discussion in Vol. I, pp. 156 ff.). However, if the Coriolis couplings can be neglected, the levels will continue to form rotational sequences; for systems of axial symmetry, the bands will be characterized by the quantum number K.

Evidence for strong admixing of different values of K at energies corresponding to the neutron binding can be derived from the statistical analysis of the resonances observed in slow-neutron capture. If K were a good quantum number for slow-neutron resonances, one would expect to observe three sets of uncorrelated strong resonances with $(K,I)=(I_0+1/2, I_0+1/2)$, $(I_0-1/2, I_0\pm 1/2)$, as well as additional weaker resonances (for $I_0>1/2$) associated with K-forbidden resonance processes. (The target with spin I_0 is assumed to have $K=I_0$.) It is found, however, that, in odd-A nuclei, the distribution of the energy spacing between neighboring resonances, as well as of the fluctuations in the observed neutron widths, can be accounted for by assuming complete randomness within each of the two sets of levels with $I=I_0 \pm 1/2$. (See, for example, Desjardins *et al.,* 1960.) This analysis therefore provides evidence against the existence of conserved quantum numbers beyond the total angular momentum and parity. The determination of the excitation energies at which the conservation of K breaks down and the manner in which this transition takes place remain to be explored.

Effect of rotational motion on nuclear level densities

Further information on nuclear rotational motion can be obtained from the study of nuclear level densities at excitation energies where statistical concepts can be applied. The occurrence of collective rotational motion implies an increase in the degrees of freedom available for low-energy excitations, which may lead to a significant increase in the total nuclear level density (Bjørnholm *et al.,* 1974). The magnitude of this effect depends on the symmetry of the deformation, which determines the degrees of freedom in the rotational motion. The nuclear deformation also implies an anisotropy in the effective moments of inertia for generating angular momentum with respect to different intrinsic axes. This anisotropy has consequences for the dependence of the level density on the angular momentum quantum numbers.

For a nucleus with axial symmetry, the level density, for given I, is obtained by summing over the intrinsic states with $|K| \leqslant I$ (Ericson, 1958). When many independent degrees of freedom can contribute to the intrinsic angular momentum K, the distribution in K is expected to be of the normal form, and we therefore obtain

$$\rho_{\text{intr}}(E,K)=(2\pi)^{-1/2}\sigma_K^{-1}\exp\left\{-\frac{K^2}{2\sigma_K^2}\right\}\rho_{\text{intr}}(E) \tag{4-63a}$$

$$\rho(E,I)=\tfrac{1}{2}\sum_{K=-I}^{I}\rho_{\text{intr}}(E-E_{\text{rot}}(K,I),K) \tag{4-63b}$$

$$\approx(8\pi)^{-1/2}\sigma_K^{-1}\rho_{\text{intr}}(E)$$

$$\times\sum_{K=-I}^{I}\exp\left\{-\frac{1}{T}\left(\frac{\hbar^2}{2\mathscr{I}_\perp}I(I+1)+\left(\frac{\hbar^2}{2\mathscr{I}_3}-\frac{\hbar^2}{2\mathscr{I}_\perp}\right)K^2\right)\right\} \tag{4-63c}$$

$$\rho(E,I) \approx (2I+1)(8\pi)^{-1/2}\sigma_K^{-1}\rho_{\mathrm{intr}}(E) \quad \text{for} \begin{cases} E_{\mathrm{rot}} \ll T \\ K \ll \sigma_K \end{cases} \tag{4-63d}$$

$$E_{\mathrm{rot}}(K,I) = \frac{\hbar^2}{2\mathscr{I}_\perp}(I(I+1)-K^2)$$

$$\sigma_K^{-2} = \frac{\hbar^2}{\mathscr{I}_3}\frac{1}{T}$$

In Eq. (4-63a), the total intrinsic level density is denoted by $\rho_{\mathrm{intr}}(E)$, while the parameter σ_K characterizes the root-mean-square average of K. In the expression for the total level density as a function of I, the factor $1/2$ results from the assumed $\mathscr{R}$ symmetry, which implies that the intrinsic states with $\pm K$ combine to form a single band with $I=|K|,|K|+1,\ldots$, while the bands with $K=0$ have only even or odd values of I (see Sec. 4-2c). In the expression (4-63c), the temperature, T, represents the inverse of the logarithmic derivative of $\rho_{\mathrm{intr}}(E)$; the rotational energy is characterized by the moment of inertia, $\mathscr{I}_\perp$, for collective rotation about an axis perpendicular to the symmetry axis, while σ_K is expressed in terms of an effective moment of inertia, $\mathscr{I}_3$, for generating angular momentum about the symmetry axis. The last expression (4-63d) for $\rho(E,I)$ is an approximation valid for small values of I. (The factorized form (4-63a) of $\rho_{\mathrm{intr}}(E,K)$ is an approximation valid when the excitation energy E is large compared with the energy $(E_{\mathrm{rot}})_3 = \hbar^2K^2/2\mathscr{I}_3$ required to generate the angular momentum K. If this condition is not satisfied, the statistical analysis leads to an intrinsic level density depending on $E-(E_{\mathrm{rot}})_3$, as in Eq. (2B-51).)

The angular distribution in the fission process can be expressed in terms of the rotational quantum numbers I and K, where K is the component along the axis of fission (see Eq. (4-178)). A uniform distribution in K, for given I, leads to an isotropic angular distribution. Thus, the anisotropy provides information on the difference between the moments of inertia, for rotations perpendicular and parallel to the fission direction. Experimental evidence on the angular distribution of fission induced by 40 MeV α particles is discussed in the example on pp. 617 ff.; the magnitude of the anisotropy leads to an estimate of the deformation at the saddle-point shape.

The expressions (4-63) may be compared with those that apply to a spherical nucleus, as discussed in Sec. 2B-6 for independent-particle motion. For a spherical system, the level density as a function of I is obtained by a decomposition of the total level density, $\rho(E)$, in terms of components $\rho(E,M)$ with specified values of the angular momentum M with respect to a fixed axis,

$$\rho(E,M)=(2\pi)^{-1/2}\sigma^{-1}\exp\left\{-\frac{M^2}{2\sigma^2}\right\}\rho(E) \tag{4-64a}$$

$$\begin{aligned}\rho(E,I)&=\rho(E,M=I)-\rho(E,M=I+1)\\&\approx(2I+1)(8\pi)^{-1/2}\sigma^{-3}\exp\left\{-\frac{I(I+1)}{2\sigma^2}\right\}\rho(E)\end{aligned} \tag{4-64b}$$

As for the K distribution in Eq. (4-63a), we have assumed a normal distribution in M, with the mean-square average value denoted by σ.

Comparing Eqs. (4-64b) and (4-63d), it is seen that, for small values of I, the level density for the deformed nucleus involves an extra factor σ^2 as compared with that of the spherical nucleus. This increase reflects the fact that only the fraction $(2\sigma^2)^{-1}$ of the $M=0$ states belongs to states with $I=0$, while half of the $K=0$ bands have an $I=0$ member. For larger values of I, the level density for the deformed nucleus contains a cutoff resulting from the rotational energy, and the total level density depends on the moment of inertia $\mathscr{I}_\perp$ for the collective rotation. Summing the expression (4-63c) over I, taking into account the $(2I+1)$-fold degeneracy of each level I, one obtains a total level density equal to the intrinsic density $\rho_{\text{intr}}(E)$ multiplied by the factor $(\hbar^{-2}\mathscr{I}_\perp T)$. The order of magnitude of this factor corresponds to the number of states in a rotational band, with $E_{\text{rot}} \lesssim T$.

In comparing the level densities of spherical and deformed systems, one must take into account that the occurrence of the collective rotational motion is associated with a corresponding reduction of the degrees of freedom in the intrinsic motion. However, the particle excitations of which the rotational motion is built have energies that, in heavy nuclei, are of the order of a few MeV (see the discussion on pp. 82ff.) and, for temperatures that are small compared with these energies, the removal of the "spurious" degrees of freedom is not expected to significantly affect the intrinsic level density.

The above discussion has assumed axial symmetry, in which case the collective motion is constrained to rotations about axes perpendicular to the symmetry axis. The full degrees of freedom for rotations in three dimensions come into play, if the deformation is not invariant under any element of the rotation group. In this case, the rotational bands contain $2I+1$ levels for each value of I (see the discussion in Sec. 4-5). The total level density, as a function of I, can be written

$$\rho(E,I)=\sum_{\tau=1}^{2I+1}\rho_{\text{intr}}(E-E_{\text{rot}}(\tau,I)) \tag{4-65a}$$

$$\approx(2I+1)\rho_{\text{intr}}(E)\quad\text{for}\quad E_{\text{rot}}(\tau,I)\ll T \tag{4-65b}$$

where the quantum number τ labels the different rotational levels with the same value of I in a given rotational band. The expression (4-65b) represents an approximation valid for small I, and is seen to be a factor of order σ_K larger than the corresponding expression (4-63d) appropriate to an axially symmetric nucleus. For larger values of I, the level density (4-65a) can be evaluated by employing the expression (4-269) for the rotational Hamiltonian and replacing the sum over τ by an integration over the surface of a sphere $I^2 = I_1^2 + I_2^2 + I_3^2$ in **I** space. (The volume element in this space is $\pi^{-1} dI_1 dI_2 dI_3$, which represents the element of phase space, in units of $(2\pi\hbar)^3$, integrated over the total solid angle $8\pi^2$ in the space of orientations; this volume element is seen to correspond to a number of quantum states per unit of I of $4I^2 \approx (2I+1)^2$.) The resulting I dependence of the level density makes possible, in principle, a determination of all three moments of inertia, $\mathscr{I}_\kappa$.

Restrictions in the rotational degrees of freedom may also arise from an invariance of the nuclear shape under finite rotations. Such a restriction has already been encountered in the $\mathscr{R}$ symmetry for the axially symmetric nuclei, which implies a reduction in the level density by a factor 2. For an axially symmetric shape, $\mathscr{R}$ symmetry is the only possible additional subgroup of rotations, but shapes without axial symmetry may be invariant under any of the point groups. If the shape is invariant under a group of g elements, the level density is reduced by a factor g^{-1} as compared with the values (4-65). This result can be simply obtained by noting that a loss of the symmetry considered would imply the possibility of distinguishing between g different regions of configuration space corresponding to the g different orientations of the symmetry-violating deformation. Hence, the total level density is increased by a factor g as compared with the symmetric case.

An example of symmetry under a point group is provided by the most general quadrupole deformation. Such a deformation violates axial symmetry but is invariant under the rotations $\mathscr{R}_\kappa$ of 180° with respect to each of the three principal axes. The associated point group, D_2, has four elements; hence, the invariance leads to a level density that is one quarter of the value (4-65).

Additional collective degrees of freedom result from deformations that violate space reflection, $\mathscr{P}$, or time reversal, $\mathscr{T}$. A deformation violating either of these symmetries leads to a doubling of the energy levels (see Sec. 4-2e) and thus to an increase in the level density by a factor 2.

The spectrum in the yrast region[4]

The study of nuclear structure in the region of the yrast line for large values of the angular momentum represents a new and challenging field of investigation. Despite the high excitation, the nucleus may be considered

---

[4]The lowest level of given I is referred to as an "yrast" level (Grover, 1967). The word appears to derive from the old Germanic *wór*, which developed into the English *weary* and the Nordic *ørr* (modern Swedish *yr*, modern Danish *ør*), meaning dizzy. The form *yrast* is the naturally constructed superlative of *yr*.

"cold," since the energy is almost entirely expended in generating the angular momentum. However, the elementary modes of excitation appropriate to the description of nuclear matter under these conditions may be modified in a major way by the large internal stresses associated with the centrifugal and Coriolis forces.

The available evidence on the yrast region comes partly from detailed spectroscopic studies that have identified individual levels up to $I \sim 20$. Additional, more indirect information concerning levels up to $I \sim 50$ has come from the analysis of γ cascades following compound nucleus reactions initiated by heavy ions; these studies indicate that, even for such high values of the angular momentum, the yrast region involves families of collective excitations that may be of rotational character (see the discussion of the example on pp. 68 ff.).

At the present time, the exploration of the yrast region is in a very tentative stage; in the following, we shall indicate a number of structural effects that may be significant.

1. The strong dependence of the moment of inertia on the pair correlations implies that these are weakened with increasing angular momentum. One may envisage a rather abrupt phase transition from the pair-correlated (or superfluid) phase to that of a normal Fermi gas (Mottelson and Valatin, 1960). The observed rotational spectra for nuclei with $A \sim 160$ exhibit striking changes for $I \sim 20$ that may possibly be associated with such an effect (see the example discussed on pp. 71 ff. and the more extensive data presented by Johnson *et al.*, 1972).

2. The axial symmetry of the nucleus represents a degenerate situation with two equal moments of inertia, and a deviation from this symmetry will in most cases increase one of these moments at the expense of the other. The possibility of aligning the rotational angular momentum along the axis with the largest moment of inertia thus implies a tendency for the system to break away from axial symmetry, with increasing angular momentum. Tentative estimates of this effect are considered in connection with the analysis of the coupling of the rotational motion to the collective oscillations away from axial symmetry on pp. 166 ff.; see also the comments under point 5. (Rotational spectra of systems that deviate from axial symmetry will be discussed in Sec. 4-5.)

3. The increase in the moment of inertia, with increasing elongation of the nucleus, implies that the rotational angular momentum may lead to instability with respect to fission (see the estimate in Sec. 6A-2b based on the liquid-drop model).

4. The Coriolis force produces a major modification in the single-particle orbits if the rotational frequency becomes comparable with the difference in the frequencies of the particle motion parallel and perpendicular to the symmetry axis. These modifications imply a tendency for the angular momenta of the individual nucleons to become aligned in the direction of the axis of rotation, with a resulting reduction in the magnitude of the deformation with respect to this axis. In some situations, this effect may lead to a gradual change in the shape of the nucleus, resulting ultimately in axial symmetry in the direction of rotation, in which case the rotational band terminates at a finite value of the angular momentum (see the discussion on pp. 84 ff.). The critical values I_{max} are of the order of the particle number (see Table 4-3, p. 87); hence, this phenomenon is of special importance for the light nuclei. Tentative evidence for such an effect, occurring in ^{20}Ne, is considered on p. 98; see also the discussion of ^{8}Be on pp. 101 ff. In other situations, the Coriolis force may act especially strongly on selected orbits (in particular the orbits with large j and small Ω in heavy nuclei; see p. 311), in which case there may be a transition to a coupling scheme with the angular momentum of these particles aligned along the axis of rotation long before the total angular momentum approaches I_{max} (Stephens and Simon, 1972).

5. The nucleus may accommodate large values of the angular momentum either by collective rotations or by aligning the angular momentum of individual nucleons in the direction of a symmetry axis. The energies required by the second mechanism may be comparable to the rotational energies (see the statistical estimate on p. 81), and such a coupling scheme may, therefore, compete effectively in the yrast region. Isomeric states with large K values have been observed on the yrast line in special cases (see the examples of the $K\pi = 8-$ and $K\pi = 16+$ isomers in ^{178}Hf discussed on p. 73). The studies of the nuclear levels with still higher spin that have been populated in heavy ion reactions have so far provided no indication for the occurrence of isomerism. This evidence may suggest that, in the yrast regions explored, the angular momentum is not carried by individual nucleons, possibly as a result of a departure from axial symmetry in the nuclear shape. For still larger values of the angular momentum, of the order of magnitude of the critical values for instability with respect to fission (see point 3), one may envisage a regime in which the centrifugal forces, which favor oblate deformations (as for a classical rotating fluid), dominate over the effects of shell structure and produce an oblate nuclear shape with the angular momentum aligned in the direction of the symmetry axis. In such a situation, the yrast levels would not form regular sequences

and the radiative transition matrix elements would be those associated with changes of orbits of individual particles.

6. As a consequence of the rotational distortions in the nuclear shape, collective rotational motion may occur in the yrast region also for nuclei that do not exhibit rotations for lower values of the angular momentum (see the vibrational model discussed in Sec. 6B-3, and the evidence presented in Fig. 6-33, p. 537).

4-3b $E2$-Matrix Elements within Band

The deformation that underlies the separation between rotational and intrinsic motion (see Sec. 4-2) can be determined from the study of appropriate matrix elements connecting the members of a band. The occurrence of a stable deformation implies that these rotational matrix elements are large compared to the corresponding matrix elements associated with fluctuations in the deformation parameter (vibrational or single-particle transitions). The observed nuclear shape deformations are mainly of quadrupole type, and the most detailed evidence on these deformations is provided by the measurement of $E2$-matrix elements within a rotational band.

Leading-order intensity relations

An axially symmetric quadrupole deformation can be characterized by the intrinsic electric quadrupole moment (compare Eq. (3-26))

$$\begin{aligned} eQ_0 &\equiv \langle K | \int \rho_e(\mathbf{r}') r^2 (3\cos^2\vartheta' - 1)\, d\tau' | K \rangle \\ &= \left(\frac{16\pi}{5}\right)^{1/2} \langle K | \mathcal{M}(E2, \nu=0) | K \rangle \end{aligned} \tag{4-66}$$

The primed coordinates refer to the intrinsic (body-fixed) system, and $\mathcal{M}(E2,\nu)$ denotes the components of the electric quadrupole tensor (see Eq. (3-29)) relative to the intrinsic system.[5]

The $E2$ moments $\mathcal{M}(E2,\mu)$ referring to the axes of the fixed system are obtained from the intrinsic moments by the standard transformation of

[5]Components of tensor operators such as $\mathcal{M}(E2,\nu)$, referring to the intrinsic system, are often indicated by a prime, $\mathcal{M}'(E2,\nu)$; see, for example, Sec. 1A-6. However, in the present context, the appearance of the tensor label ν is sufficient to identify an operator as referring to the intrinsic coordinate system, and the prime has therefore been omitted.

tensor operators (see Eq. (1A-52)),

$$\begin{aligned}\mathscr{M}(E2,\mu) &= \sum_{\nu} \mathscr{M}(E2,\nu)\mathscr{D}^2_{\mu\nu}(\omega) \\ &= \left(\frac{5}{16\pi}\right)^{1/2} eQ_0\, \mathscr{D}^2_{\mu 0}(\omega) \\ &= \frac{e}{2} Q_0 Y_{2\mu}(\theta,\phi)\end{aligned} \tag{4-67}$$

where $\omega=(\phi,\theta,\psi)$ are the orientation angles for the body-fixed system. (In the last expression in Eq. (4-67), we have employed the relation (1A-42).) The moment (4-67) includes only the collective part associated with the average deformation of the intrinsic state.

The collective $E2$ moment (4-67) connects states (with $\Delta I \leqslant 2$) belonging to the same rotational band. The $E2$-matrix elements between two such states, with wave functions (4-15) or (4-19), can be evaluated by employing the relations (1A-41) and (1A-43) for the $\mathscr{D}$ functions, and one obtains the reduced matrix elements and transition probabilities (see Eqs. (1A-61) and (1A-67))

$$\langle KI_2 \| \mathscr{M}(E2) \| KI_1 \rangle = (2I_1+1)^{1/2} \langle I_1 K 2 0 | I_2 K \rangle \left(\frac{5}{16\pi}\right)^{1/2} eQ_0 \tag{4-68a}$$

$$B(E2;\, KI_1 \rightarrow KI_2) = \frac{5}{16\pi} e^2 Q_0^2 \langle I_1 K 2 0 | I_2 K \rangle^2 \tag{4-68b}$$

The vector addition coefficient $\langle I_1 K 2 0 | I_2 K \rangle$ represents the coupling of the angular momenta in the intrinsic frame.

The diagonal matrix elements $(I_1 = I_2)$ give the static moments (see Eq. (3-30)), and Eq. (4-68a) yields

$$\begin{aligned}Q &= \langle IK20|IK\rangle\langle II20|II\rangle Q_0 \\ &= \frac{3K^2 - I(I+1)}{(I+1)(2I+3)} Q_0\end{aligned} \tag{4-69}$$

The ratio between Q and Q_0 is the expectation value of $P_2(\cos\theta)$ in the substate $M=I$ and reflects the averaging of the charge eccentricity associated with the rotational motion. Thus, $|Q|$ is always smaller than $|Q_0|$; for the particular case of $I=K$ (which usually applies to the ground state), one obtains

$$Q = \frac{I}{I+1} \frac{2I-1}{2I+3} Q_0 \tag{4-70}$$

For $I=0$ or $1/2$, the nuclear axis points with equal probability in all directions; hence, the intrinsic deformation does not give rise to an external moment (as follows from general requirements of rotational invariance). In the classical limit of $I \to \infty$ (and $K=I$), the zero-point fluctuations associated with the rotation become negligible, and Q approaches Q_0. For the states of very large I in a given band (with fixed K), Eq. (4-69) gives $Q \approx -Q_0/2$, corresponding to the fact that the total angular momentum becomes perpendicular to the nuclear symmetry axis ($P_2(\cos \frac{1}{2}\pi) = -\frac{1}{2}$).

The $E2$ transitions are governed by the selection rule $\Delta I \leqslant 2$. For $I \gg K$, the vector addition coefficients in Eq. (4-68) have the approximate values

$$\langle I_1 K 2 0 | I_2 K \rangle \approx \begin{cases} \left(\frac{3}{8}\right)^{1/2} & I_2 = I_1 \pm 2 \\ \pm\left(\frac{3}{2}\right)^{1/2}(K/I) & I_2 = I_1 \pm 1 \\ -\frac{1}{2} & I_2 = I_1 \end{cases} \tag{4-71}$$

Hence, the large matrix elements have $\Delta I = 2$ or 0 and connect states with the same signature (see Eq. (4-23)), while the transitions between states of different signature ($\Delta I = 1$) are weaker by a factor of order $(K/I)^2$. This selection rule corresponds to the fact that, in the classical limit, a rotating body produces a quadrupole field with frequencies 0 and $2\omega_{\text{rot}}$.

In the approximation used in the present section, in which the $E2$ moments are assumed to be associated with a static intrinsic deformation with axial symmetry, all the $E2$-matrix elements within a given band are expressed in terms of a single parameter Q_0, by relations of a geometrical character. The empirical evidence testing these relations is illustrated in Fig. 4-24 (p. 129); see also the examples discussed in connection with Tables 4-4 (^{20}Ne, p. 98), 4-6 (^{169}Tm, p. 104), 4-8 (^{177}Lu and ^{177}Hf, p. 108), and 4-10 (^{239}Pu, p. 113). For the well-developed rotational spectra, the ratios of the $E2$-matrix elements for the lowest states in the band are found to obey the leading-order intensity rule to within the experimental accuracy of a few percent. In the case of ^{20}Ne, the measured $E2$-matrix elements, for the states with $I=6$ and 8, exhibit major deviations from the leading-order intensity relations. These deviations can be understood in terms of the alignment of the particle angular momenta that may lead to a termination of the band and a vanishing $E2$-transition moment (see point 4 on p. 43).

Deformation parameters

The strength of the $E2$ transitions between members of a rotational band, which, in heavy nuclei, is typically of the order of a hundred single-particle units (see Fig. 4-5), strikingly exhibits the collective character of the quadrupole deformation in the rotating nuclei.

The magnitude of the observed matrix elements determines the intrinsic moment Q_0. Assuming a collective deformation of the nucleus as a whole, one may express Q_0 in terms of a distortion parameter δ, which is conveniently defined by the relation

$$Q_0 = \tfrac{4}{3}\langle \sum_{k=1}^{Z} r_k^2 \rangle \delta \tag{4-72}$$

where the sum is extended over all the protons in the nucleus. The definition (4-72) implies that, for a uniformly charged spheroidal nucleus with a sharp surface, the parameter δ is given by

$$\delta = \frac{3}{2}\frac{(R_3)^2 - (R_\perp)^2}{(R_3)^2 + 2(R_\perp)^2}$$

$$\approx \frac{\Delta R}{R} + \frac{1}{6}\left(\frac{\Delta R}{R}\right)^2 + \cdots \tag{4-73}$$

$$\Delta R = R_3 - R_\perp \qquad R = \left(\tfrac{5}{3}\langle r^2 \rangle\right)^{1/2} \approx \tfrac{1}{3}(R_3 + 2R_\perp)$$

in terms of thc difference ΔR between the radii parallel and perpendicular to the symmetry axis and the mean radius R.

The quadrupole deformations may also be characterized by the parameter β_2, associated with the expansion of the density distribution in spherical harmonics. The relationship between the two parameters δ and β_2 is given by Eq. (4-191); to leading order, we have $\delta \approx 0.95\beta_2$, in the limit of a sharp surface. The advantage of the deformation parameter δ is its rather direct relationship to the experimentally determined quadrupole moment, and this parameter is therefore usually employed in the present chapter, as well as in Chapter 5.

The measured deformations of heavy nuclei are shown in Fig. 4-25, p. 133. It is seen that typical values of δ are of order 0.2 to 0.3 and that all the strongly deformed nuclei have positive Q_0 values (prolate shape). For the lighter nuclei, the deformation parameters are somewhat larger (see Table 4-15, p. 134). The odd-A and odd-odd nuclei are found to have shapes similar to neighboring even-even nuclei (see Fig. 4-25).

In a few cases, it has been possible to obtain evidence concerning the quadrupole deformation of excited bands. Usually, these bands are found to have deformations similar to those of the ground-state bands (see, for example, the determination of Q_0 $(K\pi = 3+)$ in ^{172}Yb (Forker and Wagner, 1969), and the discussion of indirect evidence concerning excited bands in

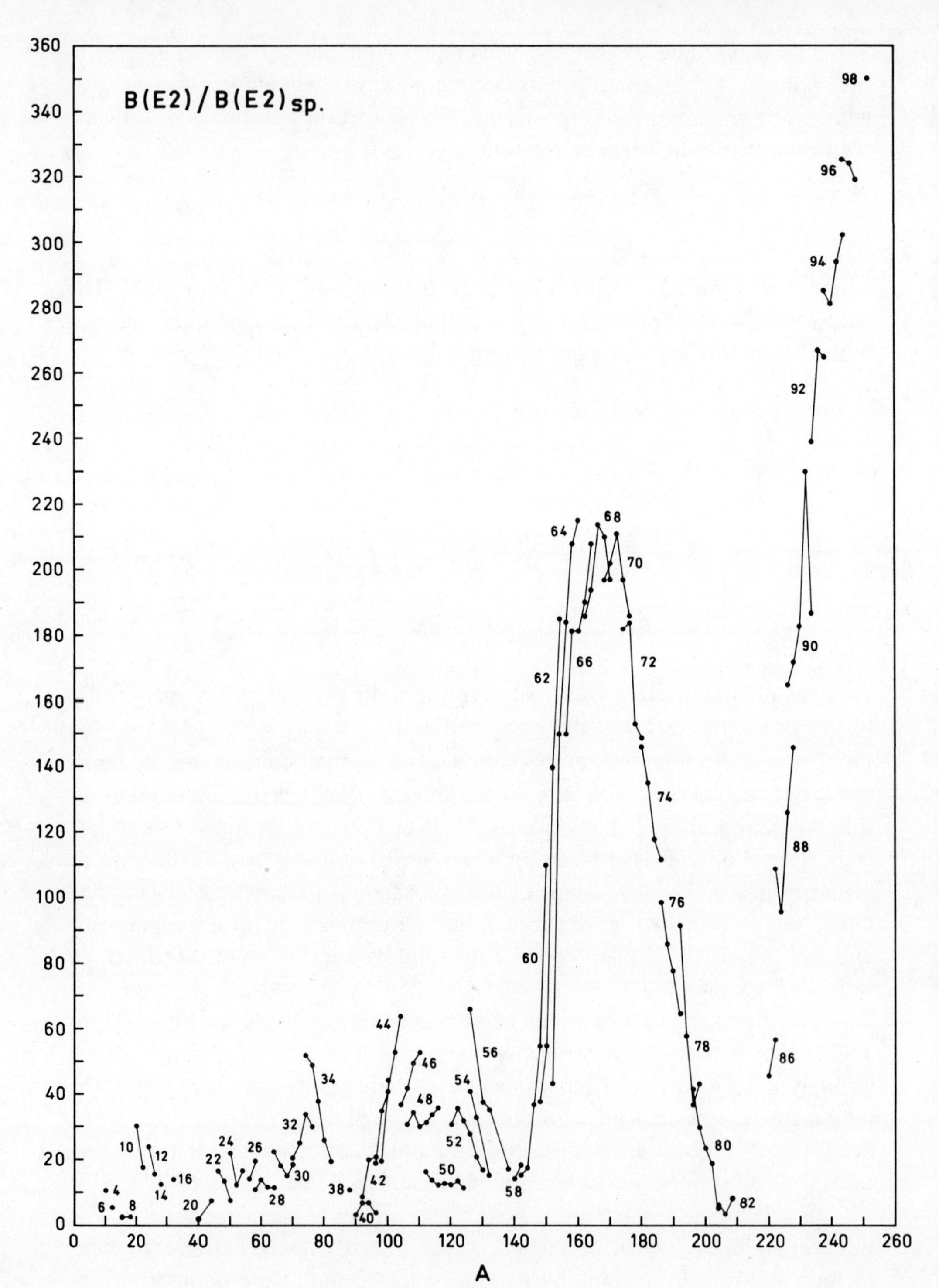
B(E2)/B(E2)sp.
A
0
20
40
60
80
100
120
140
160
180
200
220
240
260
280
300
320
340
360
4
6
8
10
12
14
16
20
22
24
26
28
30
32
34
38
40
42
44
46
48
50
52
54
56
58
60
62
64
66
68
70
72
74
76
78
80
82
86
88
90
92
94
96
98

^{175}Lu (p. 156) and ^{166}Er (p. 162)). The occurrence of shape isomerism involving major differences in shape among low-lying excitations has been referred to on p. 26.

Theoretical estimates of nuclear equilibrium deformations can be obtained by considering independent-particle motion in deformed potentials. The equilibrium shape can be characterized by the minimum in the total nuclear energy or by the condition of self-consistency between the density and the potential (see pp. 135 ff., and the more general discussion of the shell-structure effects on the nuclear potential energy surface, pp. 602 ff.). The magnitude of δ is found to be of the order of the ratio between the number of particles outside of closed shells and the total number of nucleons and, thus, typically of order $\delta \sim A^{-1/3}$. (See the discussion on pp. 135 ff. and the comments on p. 605.) With the approach to closed-shell configurations, the magnitude of the equilibrium deformation decreases. For configurations with relatively few particles outside of closed shells, the effect of deformation on the single-particle motion is small compared with the effect of pair correlations, and a transition takes place to a spherical coupling scheme with a vibrational quadrupole excitation spectrum (see the discussion in Chapter 6, pp. 520 ff.).

While the main nuclear deformations are found to be of quadrupole type, higher multipole components are to be expected and have been experimentally observed. Evidence on $\lambda=4$ moments is discussed on pp. 139 ff; in the rare-earth region, the $\lambda=4$ deformations appear to be typically about a factor of 5 smaller than the quadrupole deformations (see Table 4-16, p. 140).

Figure 4-5 $E2$-transition probabilities connecting the ground state with the first excited 2+ state in even-even nuclei. The $B(E2; 0+ \rightarrow 2+)$ values are plotted in units of the "single-particle" value

$$B_{\text{sp}}(E2) = \frac{5}{4\pi} e^2 \left(\tfrac{3}{5} R^2\right)^2 = 0.30 A^{4/3} e^2 \,\text{fm}^4 \qquad (R = 1.2 A^{1/3} \,\text{fm})$$

which represents the transition probability for a two-proton excitation $(j^2)_{J=0} \rightarrow (j^2)_{J=2}$, in the limit of large j, with a radial matrix element taken to be $\langle j | r^2 | j \rangle = \frac{3}{5} R^2$ (see, for example, Eqs. (1A-72a) and (3C-33)). The quantity $B_{\text{sp}}(E2)$ also represents the transition probability for a single proton from $l=0$ to $l=2$ (see Eq. (3C-34)). The unit $B_W(E2)$ (see Eq. (3C-38)) equals $\frac{1}{5} B_{\text{sp}}(E2)$ and would be the appropriate single-particle unit for the transition $2 \rightarrow 0$.

The empirical $B(E2)$ values are taken from the compilation by Stelson and Grodzins (1965). For the doubly closed-shell nuclei, the data are taken from: ^{16}O (Skorka *et al.*, 1966); ^{40}Ca (J. R. MacDonald, D. F. H. Start, R. Anderson, A. G. Robertson, and M. A. Grace, *Nuclear Phys.* **A108**, 6, 1968); ^{208}Pb (J. F. Ziegler and G. A. Peterson, *Phys. Rev.* **165**, 1337, 1968).

Generalized intensity relations

In the above discussion, we have considered the $E2$-matrix elements associated with an axially symmetric deformation with a quadrupole moment that is independent of the rotational angular momentum. Corrections to the leading-order relations arise from the perturbations of the intrinsic structure produced by the Coriolis and centrifugal forces, as well as from nonaxially symmetric components in the nuclear charge distribution.

A systematic analysis of the generalized intensity relations may be obtained by expanding the intrinsic moments $\mathscr{M}(E2,\nu)$ (see Eq. (4-67)) in powers of the rotational angular momentum. The procedure corresponds to that employed in the analysis of the rotational energy in Sec. 4-3a with the modifications resulting from the tensor character of the $E2$ moment. In the small print below, we give a rather detailed discussion of the structure of the $E2$-matrix elements within a rotational band, as a prototype of the general analysis of tensorial matrix elements in the rotational coupling scheme.

In zeroth order, neglecting the I dependence of $\mathscr{M}(E2,\nu)$, the matrix elements within a band may receive contributions from the components with $\nu=0$ and $\nu=\pm 2K$. The $\nu=0$ components give the matrix element (4-68), while the components with $\nu=\pm 2K$ give a term proportional to the signature. Since the I-independent part of $\mathscr{M}(E2,\nu)$ is invariant under time reversal combined with Hermitian conjugation ($c=-1$), the selection rule (4-31) implies that the signature-dependent term vanishes for half integer K. Hence, this term occurs only for $K=1$ bands, and in the I-independent approximation, the $E2$-matrix elements have the form

$$\langle KI_2 \| \mathscr{M}(E2) \| KI_1 \rangle = (2I_1+1)^{1/2} \Big(\langle I_1 K 2 0 | I_2 K \rangle \langle K | \mathscr{M}(E2,\nu=0) | K \rangle + (-1)^{I_1+1} \langle I_1 -1 2 2 | I_2 1 \rangle \langle K=1 | \mathscr{M}(E2,\nu=2) | \overline{K=1} \rangle \delta(K,1) \Big) \tag{4-74}$$

where the first term is identical to the matrix element (4-68). The contribution of the second term will usually be very small compared to the contribution from the collective $\nu=0$ moment. The term vanishes for a two-particle configuration (Ω_1,Ω_2), but may receive contributions from linear combinations of such configurations; for a vibrational excitation with $K=1$, the term represents the small $E2$ moment with $\nu=2$ that may arise as a second-order effect in the vibrational amplitude.

With the inclusion of terms linear in $I_\pm$, the effective moments can be expressed in the form

$$\mathscr{M}(E2,\mu)=\sum_{\nu=-2}^{2}\left(m_{\nu,\nu}\mathscr{D}^2_{\mu\nu}+\tfrac{1}{2}m_{\nu+1,\nu}\{I_-,\mathscr{D}^2_{\mu\nu}\}+\tfrac{1}{2}m_{\nu-1,\nu}\{I_+,\mathscr{D}^2_{\mu\nu}\}\right) \tag{4-75a}$$

$$\begin{gathered}\mathscr{R}_e\{I_-,\mathscr{D}^2_{\mu\nu}\}\,\mathscr{R}_e^{-1}=(-1)^{\nu+1}\{I_+,\mathscr{D}^2_{\mu-\nu}\}\\ \mathscr{R}_i^{-1}m_{\nu+1,\nu}\,\mathscr{R}_i=(-1)^{\nu+1}m_{-\nu-1,-\nu}\end{gathered} \tag{4-75b}$$

with the notation $m_{\Delta K,\nu}$ for the intrinsic operators indicating the tensor component ν and the change in K induced by the operator. As indicated by Eq. (4-75b), the terms in $\mathscr{M}(E2,\mu)$ involving I_+ can also be expressed as the $\mathscr{R}$ conjugate of the terms involving I_- (see p. 31). The intrinsic operators $m_{\Delta K,\nu}$ satisfy the symmetries

$$(m_{\Delta K,\nu})^{\dagger}=(-1)^{\nu}m_{-\Delta K,-\nu} \tag{4-76a}$$

$$\mathscr{R}_i\mathscr{T}m_{\Delta K,\nu}(\mathscr{R}_i\mathscr{T})^{-1}=m_{\Delta K,\nu} \tag{4-76b}$$

that follow from the transformation of $\mathscr{M}(E2,\mu)$ under Hermitian conjugation and time reversal. The operators $I_{\pm}$ do not commute with the $\mathscr{D}$ functions, and in Eq. (4-75), we have chosen to express the $I_{\pm}$ dependence in terms of symmetrized products. Since the commutator of $I_{\pm}$ and $\mathscr{D}^{\lambda}_{\mu\nu}$ is proportional to $\mathscr{D}^{\lambda}_{\mu,\nu\mp1}$ (see Eq. (1A-91)), a difference in the ordering of the operators in Eq. (4-75) is equivalent to a redefinition of the operators $m_{\nu,\nu}$ occurring in the I-independent terms.

Matrix elements within a band may receive contributions from the terms in Eq. (4-75) with $\Delta K=0$ and $\pm 2K$, and one obtains

$$\begin{aligned}\langle KI_2\|\,&\mathscr{M}(E2)\|KI_1\rangle\\ =&(2I_1+1)^{1/2}\big(\langle I_1K20|I_2K\rangle\langle K|m_{0,0}|K\rangle\\ &+(-1)^{I_1+1}\langle I_1-122|I_21\rangle\langle K=1|m_{2,2}|\overline{K=1}\rangle\delta(K,1)\\ &+\big(\langle I_1\,K+1\,2-1|I_2K\rangle(I_1-K)^{1/2}(I_1+K+1)^{1/2}\\ &-\langle I_1\,K-1\,21|I_2K\rangle(I_1+K)^{1/2}(I_1-K+1)^{1/2}\big)\langle K|m_{0,-1}|K\rangle\\ &+\tfrac{1}{2}(-1)^{I_1+K}\big(\langle I_1\,-K\,2\,2K-1|I_2K-1\rangle(I_2+K)^{1/2}(I_2-K+1)^{1/2}\\ &+\langle I_1\,-K+1\,2\,2K-1|I_2K\rangle(I_1+K)^{1/2}(I_1-K+1)^{1/2}\big)\langle K|m_{2K,2K-1}|\overline{K}\rangle\\ &+\tfrac{1}{2}(-1)^{I_1+K}\big(\langle I_1\,-K\,2\,2K+1|I_2K+1\rangle(I_2-K)^{1/2}(I_2+K+1)^{1/2}\\ &+\langle I_1\,-K-1\,2\,2K+1|I_2K\rangle(I_1-K)^{1/2}(I_1+K+1)^{1/2}\big)\,\langle K|m_{2K,2K+1}|\overline{K}\rangle\big)\end{aligned} \tag{4-77}$$

In the derivation of this expression, we have employed the relation $\langle K|m_{0,1}|K\rangle = -\langle K|m_{0,-1}|K\rangle$, which follows from the symmetry relations (4-76). The terms involving products of vector addition coefficients and matrix elements of $I_{\pm}$ can be expressed in various alternative ways; the form (4-77) is obtained from the relation

$$\{I_-,\mathscr{D}^2_{\mu-1}\}-\{I_+,\mathscr{D}^2_{\mu 1}\}=2(\mathscr{D}^2_{\mu-1}I_- - \mathscr{D}^2_{\mu 1}I_+) \tag{4-78}$$

The intrinsic moments $m_{\Delta K=\nu\pm1,\nu}$ are odd with respect to time reversal followed by Hermitian conjugation ($c=+1$); hence, the terms in the effective moment (4-75a) that are linear in $I_{\pm}$ do not contribute to the matrix elements in a $K=0$ band (see Eq. (4-30)). Moreover, the matrix elements $\langle K|m_{2K,2K\pm1}|\bar{K}\rangle$ vanish for integer K (see Eq. (4-31)). Estimates of the I-dependent terms in the effective moment (4-75a) arising from band mixing caused by the Coriolis interaction are considered on pp. 130ff.

To second order in the rotational angular momentum, the effective $E2$ moment may contain terms proportional to I_+^2 and I_-^2 ($\nu=\Delta K\pm2$) and, in addition, terms with $\nu=\Delta K$ involving intrinsic moments proportional to $\frac{1}{2}\{I_+,I_-\}=I_1^2+I_2^2$. We shall confine ourselves to the part of the $E2$ operator that contributes to $\Delta K=0$ transitions,

$$\begin{aligned}\mathscr{M}(E2,\mu)_{\Delta K=0}&=m_{0,0}\mathscr{D}^2_{\mu 0}+\tfrac{1}{4}m'_{0,0}\{\{I_+,I_-\},\mathscr{D}^2_{\mu 0}\}\\&\quad+\left(\tfrac{1}{2}m_{0,-1}\{I_-,\mathscr{D}^2_{\mu-1}\}+\mathscr{R}\text{-conj.}\right)+\left(\tfrac{1}{2}m_{0,-2}\{I_-^2,\mathscr{D}^2_{\mu-2}\}+\mathscr{R}\text{-conj.}\right)\end{aligned} \tag{4-79}$$

including terms up to second order in $I_{\pm}$. The intrinsic operators satisfy the symmetry relations (4-76); the $\nu=\pm2$ terms can be conveniently evaluated by using the identity

$$\begin{aligned}\left[\mathbf{I}^2,\left[\mathbf{I}^2,\mathscr{D}^\lambda_{\mu 0}\right]\right]&=\tfrac{1}{2}(\lambda(\lambda+1)(\lambda(\lambda+1)-2))^{1/2}\left(\{I_+^2,\mathscr{D}^\lambda_{\mu 2}\}+\{I_-^2,\mathscr{D}^\lambda_{\mu-2}\}\right)\\&\quad+\tfrac{1}{2}\lambda(\lambda+1)\{\{I_+,I_-\},\mathscr{D}^\lambda_{\mu 0}\}+\tfrac{1}{2}(\lambda(\lambda+1))^{1/2}\{I_3,\{I_+,\mathscr{D}^\lambda_{\mu 1}\}-\{I_-,\mathscr{D}^\lambda_{\mu-1}\}\}\\&\quad-\lambda(\lambda+1)(\lambda(\lambda+1)-2)\mathscr{D}^\lambda_{\mu 0}\end{aligned} \tag{4-80}$$

which may be derived from Eq. (4A-34).

For a $K=0$ band, the $\nu=\pm1$ terms in the moment (4-79) do not contribute (as discussed above), and the $E2$-matrix elements have the form

$$\begin{aligned}\langle K=0,I_2\|\mathscr{M}(E2)\|K=0,I_1\rangle&=(2I_1+1)^{1/2}\langle I_1 0 2 0|I_2 0\rangle\\&\quad\times\left(M_1+M_2(I_2(I_2+1)+I_1(I_1+1))+M_3(I_2(I_2+1)-I_1(I_1+1))^2\right)\end{aligned} \tag{4-81}$$

with the intrinsic matrix elements

$$
\begin{aligned}
M_1 &= \langle K=0|m_{0,0}+2\sqrt{6}\, m_{0,2}|K=0\rangle \\
M_2 &= \langle K=0|\frac{1}{2}m'_{0,0}-\frac{\sqrt{6}}{2}m_{0,2}|K=0\rangle \\
M_3 &= \langle K=0|\frac{1}{2\sqrt{6}}m_{0,2}|K=0\rangle
\end{aligned}
\tag{4-82}
$$

The terms of higher order in the rotational angular momentum can be obtained by multiplying the terms in Eq. (4-79) by functions of $\mathbf{I}^2$ (as in the analysis of the higher-order terms in the rotational energy; see p. 24). Thus, in a $K=0$ band, the I dependence of the effective moment can be expressed in terms of $I_1(I_1+1)$ and $I_2(I_2+1)$; the symmetry of the matrix element with respect to initial and final states implies that the $E2$-matrix element can always be expressed in the form (4-81) with an intrinsic moment (the factor in curly brackets) depending on the combinations $I_1(I_1+1)+I_2(I_2+1)$ and $(I_1(I_1+1)-I_2(I_2+1))^2$.

The corrections to the leading-order intensity relations, associated with the matrix elements M_2 and M_3 in Eq. (4-81), represent the distortions in the nuclear shape produced by the rotational motion. The various distortion effects mentioned above (p. 50) are characterized by different ratios of M_2 to M_3. Thus, for example, a centrifugal stretching, by which the shape remains axially symmetric while the eccentricity increases with I, is described by the $m'_{0,0}$ term in Eq. (4-79) and thus gives an intensity relation with $M_2:M_1>0$ and $M_3=0$; such an effect is connected with the coupling between the ground-state band and excited $K=0$ bands representing axially symmetric oscillations (β vibrations) in the nuclear shape (see Eq. (4-259)).

The distortion effects associated with coupling to $K=2$ bands give corrections to the $E2$-matrix elements with $M_2=-6M_3$ (see Eq. (4-215)); the coupling between the bands involves the difference between the moments of inertia with respect to the intrinsic 1 and 2 axes associated with departures from axial symmetry (see Eq. (4-285) for an asymmetric rotor and Eq. (6-295) for the coupling to γ-vibrational excitations). The coupling thus expresses the favoring of rotations about the axis with maximum possible moment of inertia and is connected with the classical centrifugal effect that tends to destroy the axial symmetry of an object that rotates about an axis perpendicular to the symmetry axis (see point 2 on p. 42).

In addition to the contributions from the coupling to the $K=0$ and $K=2$ excitations, the $E2$-intensity relations in the ground-state band may be affected by the coupling to $K=1$ excitations acting in second order. The effective moment can be obtained from the term in Eq. (4-202) that is of second order in the perturbation operator S given by Eq. (4-201). The resulting contribution to the matrix element (4-81) has $M_3=0$ and M_2 proportional to and with the same sign as $Q_0(K=0) - Q_0(K=1)$.

It should be emphasized that the total I dependence of the intrinsic quadrupole moment cannot in general be obtained from the coupling to observed intrinsic excitations, since additional effects may be associated with the coupling to the "spurious" $K=1$ mode of excitation that describes the degree of freedom of the rotational motion. The model of particle motion in a rotating harmonic oscillator potential (as discussed on pp. 84ff.) represents an extreme case in which the entire I dependence is connected with the spurious mode.

A determination of both the matrix elements M_2 and M_3 requires measurements of static moments as well as transition moments. The static moments involve only the parameter M_2, while for the transition $I \rightarrow I+2$, the I-dependent part of the intrinsic moment is proportional to $(M_2+8M_3)I(I+3)$. The accuracy of the present measurements of $E2$-matrix elements in the ground-state bands of even-even nuclei is in most cases barely sufficient to detect deviations from the leading-order intensity relations (see Fig. 4-24, p. 129).

4-3c *M*1-Matrix Elements within Band

$K=0$ bands. Rotational magnetic moment

In a $K=0$ band, the magnetic moment produced by the intrinsic motion vanishes, as a result of time-reversal symmetry, and the leading-order $M1$ moment is proportional to the rotational angular momentum. To this order, the static moments are given by

$$\mu = g_R I \tag{4-83}$$

where the parameter g_R is the effective g factor for the rotational motion. The higher-order terms can be included by treating g_R as a function of $I(I+1)$. No $M1$ transitions occur within the band, since successive states have $\Delta I=2$.

The g_R factors for even-even nuclei, derived from the observed precession of the $2+$ first rotational states in a magnetic field, are plotted in Fig. 4-6. The measured g_R values are comparable to, though in most cases somewhat smaller than, the value Z/A, which would apply to a uniform rotation of a charged body.

The rotational angular momentum is primarily associated with orbital motion of the nucleons. (The relative contribution of the spins is even smaller than the ratio of spin to orbital angular momentum quantum numbers of the individual nucleons, because the splittings of the single-particle levels produced by the spin-orbit force is on the average smaller than the splittings caused by the deformation; see Chapter 5, p. 217). The fact that the g_R values are somewhat smaller than Z/A thus provides an indication that the protons contribute less effectively than the neutrons to the rotational motion. Such an effect results from the pair correlations,

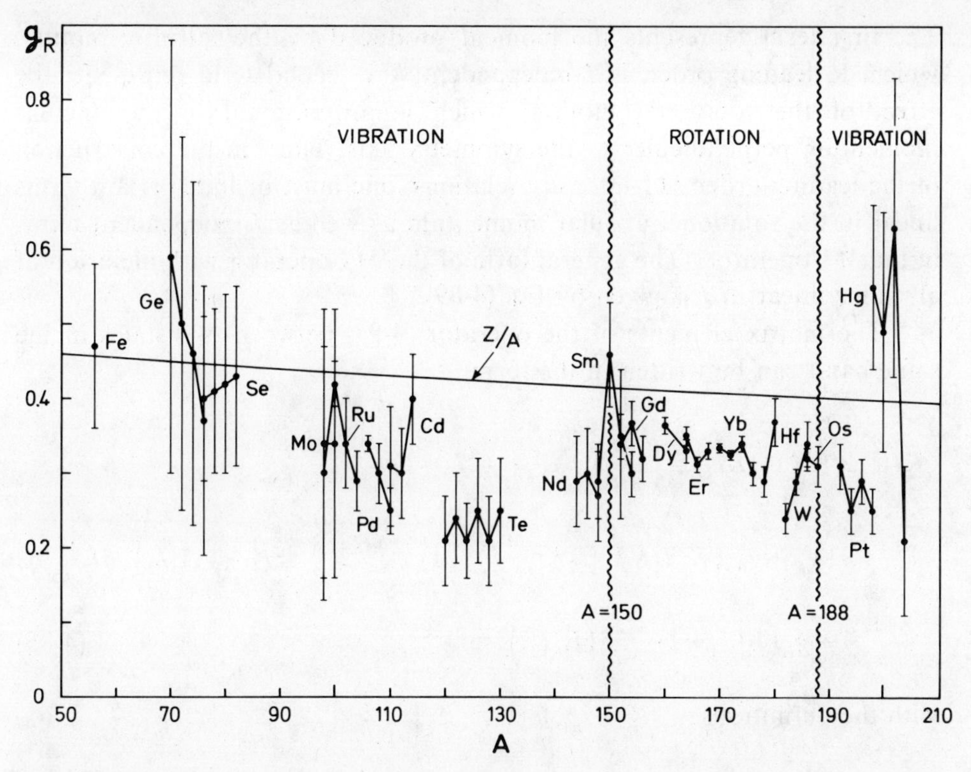

Figure 4-6 g factors for first excited 2+ states in even-even nuclei. The figure is based on the experimental data given by G. M. Heestand, R. R. Borchers, B. Herskind, L. Grodzins, R. Kalish, and D. E. Murnick, *Nuclear Phys.* **A133**, 310 (1969) and on the review by Grodzins (1968). We are indebted to L. Grodzins, B. Herskind, and S. Ogaza for help in the preparation of the figure.

which are somewhat larger for protons than for neutrons (Nilsson and Prior, 1961; see also the comment on p. 83).

Some information is available on the magnetic moments of higher members of the ground-state rotational bands of even-even nuclei; the data on states up to $I=8$ appear to be consistent with the relation (4-83), within the experimental accuracies (see, for example, Kugel *et al.*, 1971).

Bands with $K\neq 0$. Intrinsic magnetic moment

In bands with $K\neq 0$, the intrinsic nucleonic motion generates a magnetic moment, in addition to that produced by the rotational motion, and the $M1$ operator takes the form (see Eq. (3-39))

$$\mathscr{M}(M1,\mu)=\sum_{\nu}\mathscr{M}(M1,\nu)\mathscr{D}^{1}_{\mu\nu}+\left(\frac{3}{4\pi}\right)^{1/2}\frac{e\hbar}{2Mc}g_R\left(I_\mu-I_3\mathscr{D}^{1}_{\mu 0}\right) \quad (4\text{-}84)$$

The first term represents the moment produced by the intrinsic motion, which, to leading order, is I independent; the second term represents the effect of the rotational motion, which is proportional to the angular momentum perpendicular to the symmetry axis. Thus, in the construction of the leading-order $M1$-intensity relations, one must include certain terms linear in the rotational angular momentum as well as I-independent terms in the $M1$ operator. (The general form of the $M1$ operator with inclusion of all terms linear in I is given by Eq. (4-89).)

The matrix elements of the operator (4-84) between two states in the same band can be written in the form

$$\begin{aligned}\langle KI_2 \| \mathscr{M}(M1) \| KI_1 \rangle = & \left(\frac{3}{4\pi}\right)^{1/2} \frac{e\hbar}{2Mc} (2I_1+1)^{1/2} \\ & \times \Big((g_K - g_R)\big(K\langle I_1 K 1 0 | I_2 K\rangle - b(-1)^{I_1+1/2} 2^{-1/2} \langle I_1 -\tfrac{1}{2} 1 1 | I_2 \tfrac{1}{2}\rangle \delta(K,\tfrac{1}{2})\big) \\ & + g_R \big(I_1(I_1+1)\big)^{1/2} \delta(I_1, I_2) \Big) \end{aligned} \tag{4-85}$$

with the definitions

$$\begin{aligned} \left(\frac{3}{4\pi}\right)^{1/2} \frac{e\hbar}{2Mc} g_K K &\equiv \langle K | \mathscr{M}(M1, \nu=0) | K \rangle \\ \left(\frac{3}{4\pi}\right)^{1/2} \frac{e\hbar}{2Mc} (g_K - g_R) b &\equiv -2^{1/2} \langle K = \tfrac{1}{2} | \mathscr{M}(M1, \nu=1) | \overline{K = \tfrac{1}{2}} \rangle \end{aligned} \tag{4-86}$$

for the intrinsic g factor, g_K, and the quantity b, referred to as the magnetic decoupling parameter.

For a band with $K > 1/2$, the static moments and $M1$-transition probabilities can be written

$$\begin{aligned} \mu &= g_R I + (g_K - g_R) \frac{K^2}{I+1} \\ B(M1; KI_1 \rightarrow K, I_2 = I_1 \pm 1) &= \frac{3}{4\pi} \left(\frac{e\hbar}{2Mc}\right)^2 (g_K - g_R)^2 K^2 \langle I_1 K 1 0 | I_2 K \rangle^2 \end{aligned} \tag{4-87}$$

The expression for μ has the simple classical limits $\mu \rightarrow g_R I$ for $I \rightarrow \infty$, K fixed, and $\mu \rightarrow g_K I$ for $I = K \rightarrow \infty$. The transition probability vanishes for $g_K = g_R$, since then the magnetic moment is proportional to the total angular momentum and hence is a constant of the motion.

For $K=1/2$, the $M1$-matrix elements (4-85) contain an additional contribution from the $\Delta K=1$ term in the intrinsic moment,

$$\mu(K=\tfrac{1}{2},I)=g_R I+\frac{g_K-g_R}{4(I+1)}\left(1+(2I+1)(-1)^{I+1/2}b\right)$$

$$B(M1;K=\tfrac{1}{2},I_1\rightarrow K=\tfrac{1}{2},I_2=I_1\pm 1) \tag{4-88}$$

$$=\frac{3}{16\pi}\left(\frac{e\hbar}{2Mc}\right)^2(g_K-g_R)^2\left(1+(-1)^{I_>+1/2}b\right)^2\langle I_1\tfrac{1}{2}10|I_2\tfrac{1}{2}\rangle^2$$

where $I_>$ denotes the greater of I_1 and I_2.

The intensity relations for rotational $M1$-matrix elements have been subjected to extensive tests. In particular, it has been found that the $M1$-transition probabilities in bands with $K>1/2$ follow the leading-order expression (4-87) involving the single parameter (g_K-g_R), to within the accuracy of the experiments, which in the best cases amounts to a few percent (Boehm *et al.*, 1966; Haverfield *et al.*, 1967; Seaman *et al.*, 1967. Examples are shown in Tables 4-8, p. 108, and 4-10, p. 110). For bands with $K=1/2$, the $M1$-transition amplitudes involve two parameters (g_K-g_R) and b, and the static moments the additional parameter g_R (see Eq. (4-88)). For a test of these relations, see Table 4-7, p. 104.)

The observed g_R values for odd-A nuclei are shown in Table 5-14. The values deviate somewhat from those of neighboring even-even nuclei; such a difference arises from the renormalization of g_R associated with the Coriolis effect on the last odd particle (see Eq. (4A-35)). The g_K and b values reflect the properties of the orbit of the last odd particle and will be considered in Chapter 5, pp. 302ff.

The most general form of the $M1$ operator with inclusion of terms linear in the rotational angular momentum can be written

$$\mathscr{M}(M1,\mu)=\sum_{\nu}\left(m_{\nu,\nu}\mathscr{D}^1_{\mu\nu}+\left(\tfrac{1}{2}m_{\Delta K=\nu+1,\nu}\{I_-,\mathscr{D}^1_{\mu\nu}\}+\mathscr{R}\text{-conj.}\right)\right) \tag{4-89}$$

where the intrinsic operators $m_{\Delta K,\nu}$ satisfy the symmetry relations (4-76). The I-independent intrinsic operators $m_{\nu,\nu}$ are equal to the moments $\mathscr{M}(M1,\nu)$ in Eq. (4-84) and, from the identity (4A-33), it follows that the $\Delta K=0$, $\nu=-1$ term in Eq. (4-89), acting within a band, corresponds to the collective rotational magnetic moment in Eq. (4-84). The additional I-dependent terms in Eq. (4-89) that do not occur in expression (4-84) give a signature-dependent contribution $(\Delta K=2,\nu=1)$ to the $M1$-matrix elements in a $K=1$ band; for a $K=1/2$ band, the $\Delta K=1,\nu=0$ term vanishes as a consequence of the selection rule (4-31). While the $\Delta K=0,\nu=-1$

term represents a collective effect of the rotational motion, the additional I-dependent terms in Eq. (4-89) are associated with the effect of the Coriolis interaction on the one or few particles that contribute the angular momentum K (see, for example, the analysis of the $M1$ moments in the particle-rotor model on pp. 208ff.); these terms are therefore expected to be small compared with the leading-order terms included in Eq. (4-84), except when special degeneracies are present. (Such a degeneracy may occur, for example, in two-particle configurations with $\Omega_1 = 1/2$ and $\Omega_2 = 1/2$, which give rise to near-lying bands with $K=1$ and 0, coupled by the Coriolis interaction; see p. 122.)

4-3d General Structure of Matrix Elements

The examples considered above illustrate the general procedure by which one may determine the form of matrix elements in the rotational coupling scheme. For any given tensor operator, such as a multipole moment for γ or β transitions, or a particle transfer moment, a transformation to the intrinsic coordinate system gives (see Eq. (1A-98))

$$\mathscr{M}(\lambda\mu) = \sum_{\nu} \mathscr{M}(\lambda\nu; I_{\pm}) \mathscr{D}^{\lambda}_{\mu\nu}(\omega) \tag{4-90}$$

The intrinsic moments $\mathscr{M}(\lambda\nu)$ are scalars (independent of the orientation ω), but may depend on the components $I_{\pm}$ of the rotational angular momenta, as well as on the intrinsic variables. The angular momentum dependence of the intrinsic moments represents the dynamical effects of the rotational motion.

Intensity relations for I-independent moments

If the dependence of the intrinsic moments on $I_{\pm}$ can be neglected, the matrix element between two states (4-19) is given by (see Eq. (1A-43))

$$\begin{aligned} &\langle K_2 I_2 \| \mathscr{M}(\lambda) \| K_1 I_1 \rangle \\ &= (2I_1+1)^{1/2} \big(\langle I_1 K_1 \lambda\, K_2 - K_1 | I_2 K_2 \rangle \langle K_2 | \mathscr{M}(\lambda, \nu = K_2 - K_1) | K_1 \rangle \\ &\quad + (-1)^{I_1 + K_1} \langle I_1 - K_1 \lambda\, K_1 + K_2 | I_2 K_2 \rangle \langle K_2 | \mathscr{M}(\lambda, \nu = K_1 + K_2) | \overline{K}_1 \rangle \big) \\ &\qquad (K_1 \neq 0, K_2 \neq 0) \end{aligned} \tag{4-91}$$

For matrix elements within a band, the signature-dependent term obeys the selection rule (4-31).

If one of the bands, or both, has $K=0$, the two terms in Eq. (4-91) become equal, and we obtain (see the normalization (4-15) of states with $K=0$)

$$\langle K_2 I_2 \| \mathcal{M}(\lambda) \| K_1 = 0, I_1 \rangle \tag{4-92}$$
$$= (2I_1+1)^{1/2} \langle I_1 0 \lambda K_2 | I_2 K_2 \rangle \langle K_2 | \mathcal{M}(\lambda, \nu = K_2) | K_1 = 0 \rangle \begin{cases} \sqrt{2} & K_2 \neq 0 \\ 1 & K_2 = 0 \end{cases}$$

The intrinsic matrix element within a $K=0$ band vanishes if $\mathcal{M}(\lambda, \nu=0)$ is odd under time reversal combined with Hermitian conjugation (see Eq. (4-30)).

The I dependence of the matrix elements (4-91) and (4-92) is of a simple geometrical nature, reflecting the different distributions of orientation angles for the different members of a rotational band.

K-forbidden transitions

For transitions with $|K_1 - K_2| > \lambda$, the matrix elements (4-91) and (4-92) vanish since, in the above approximation, it is not possible to conserve the component of angular momentum along the symmetry axis. Transitions of this type are referred to as K forbidden and

$$n \equiv |K_1 - K_2| - \lambda \tag{4-93}$$

denotes the order of K forbiddenness. For such transitions, the leading-order contributions arise from the terms in the intrinsic moments proportional to $(I_\pm)^n$,

$$\mathcal{M}(\lambda\mu) = m_{\Delta K = \lambda + n, \nu = \lambda} \mathcal{D}^{\lambda}_{\mu\lambda} (I_-)^n + \mathcal{R}\text{-conj.} \tag{4-94}$$

where the operator m depends only on the intrinsic variables. The resulting matrix element has the form

$$\langle K_2 I_2 \| \mathcal{M}(\lambda) \| K_1 I_1 \rangle = (2I_1+1)^{1/2} \langle I_1 \; K_2 - \lambda \; \lambda\lambda | I_2 K_2 \rangle$$
$$\times \left(\frac{(I_1 - K_1)!(I_1 + K_1 + n)!}{(I_1 - K_1 - n)!(I_1 + K_1)!} \right)^{1/2} \langle K_2 | m_{\Delta K = \lambda + n, \nu = \lambda} | K_1 \rangle \begin{cases} \sqrt{2} & K_1 = 0 \\ 1 & K_1 \neq 0 \end{cases}$$
$$(K_2 = K_1 + \lambda + n) \tag{4-95}$$

assuming $K_2 > K_1$.

Higher-order corrections

Corrections to the leading-order intensity relations are obtained by including higher-order terms in the intrinsic moments, and the general form of the expansion can be found by similar considerations as employed in the above discussion of the energy and matrix elements within a rotational band. For transitions between different bands, the first-order correction terms have an especially simple form when $K_2 - K_1 \geqslant \lambda$. In this case, the effective moment with $\Delta K = K_2 - K_1 = n + \lambda$ involves only a single term of order $(I_{\pm})^{n+1}$,

$$\begin{aligned}
\mathscr{M}(\lambda\mu)_{\Delta K = K_2 - K_1} &= m_{\Delta K = n+\lambda, \nu=\lambda}\, \mathscr{D}^{\lambda}_{\mu\lambda}(I_-)^n + \tfrac{1}{2} m_{\Delta K, \nu = \lambda - 1}\left\{(I_-)^{n+1}, \mathscr{D}^{\lambda}_{\mu,\lambda-1}\right\} + \mathscr{R}\text{-conj.} \\
&= \left(m_{\Delta K,\lambda} + (2\lambda)^{1/2}\left(\tfrac{1}{2}(n+\lambda) - I_3\right) m_{\Delta K, \lambda-1}\right)\mathscr{D}^{\lambda}_{\mu\lambda}(I_-)^n \\
&\quad + (2\lambda)^{-1/2} m_{\Delta K,\lambda-1}\left[\mathbf{I}^2, \mathscr{D}^{\lambda}_{\mu\lambda}\right](I_-)^n + \mathscr{R}\text{-conj.}
\end{aligned} \tag{4-96}$$

where the last form has been obtained by means of the identity (4A-34). For $K_1 = 1/2$, an additional contribution may arise from a term of the form

$$\mathscr{M}(\lambda\mu)_{\Delta K = K_2 + 1/2} = m_{\Delta K = K_2 + 1/2, \nu=\lambda}\, \mathscr{D}^{\lambda}_{\mu\lambda}(I_-)^{n+1} + \mathscr{R}\text{-conj.} \tag{4-97}$$

The total matrix element can, therefore, be written (Mikhailov, 1966)

$$\begin{aligned}
\langle K_2 I_2 \| \mathscr{M}(\lambda) \| K_1 I_1 \rangle &= (2I_1+1)^{1/2} \langle I_1 K_2 - \lambda\, \lambda\lambda | I_2 K_2 \rangle \\
&\times \left(\frac{(I_1 - K_1)!(I_1 + K_1 + n)!}{(I_1 - K_1 - n)!(I_1 + K_1)!} \right)^{1/2} \left(M_1 + M_2 (I_2(I_2+1) - I_1(I_1+1)) \right. \\
&\left. + M_3 (-1)^{I_1 + 1/2}(I_1 + 1/2)\delta(K_1, 1/2) \right) \begin{cases} \sqrt{2} & K_1 = 0 \\ 1 & K_1 \neq 0 \end{cases}
\end{aligned} \tag{4-98}$$

where M_1, M_2, and M_3 are intrinsic matrix elements. For $K_2 - K_1 < \lambda$, the first-order corrections may involve two different terms with $\Delta K = K_2 - K_1$, $\nu = \Delta K \pm 1$ and, if $K_2 + K_1 \leqslant \lambda + 1$, additional terms with $\Delta K = K_2 + K_1$.

Empirical data

The leading-order intensity relations are expected to represent the main part of the transition amplitudes, provided

(a) there is a good separation between intrinsic and rotational motion as evidenced, for example, by a rapid convergence of the energy expansion;

(b) the matrix element considered is not sensitive to small perturbations. A sensitivity may be expected if the leading-order matrix elements are very small, due to selection rules in the intrinsic motion or to accidental cancellations.

A special situation occurs for operators, such as the magnetic multipole moments, which depend explicitly on the velocity, or angular momentum, of the nucleons. For such operators, the terms proportional to the rotational angular momentum may be comparable with the terms depending only on the intrinsic motion of the nucleons, and both types of terms must be included in the leading-order intensity relations (see, for example, the $M1$ moment (4-84)).

As discussed in Secs. 4-3b, c, the leading-order intensity relations describe the $E2$ and $M1$ transitions within bands with considerable accuracy. These relations have also been found to apply to a variety of transitions between bands. (See the evidence on favored α transitions in Table 4-13, p. 118, and Fig. 5-13, p. 268, and on β transitions in Fig. 4-26, p. 141, and Table 4-17, p. 143. For a test of the intensity relations for nucleon transfer reactions, see, for example, Dehnhard and Mayer-Böricke, 1967.)

Examples of transitions that are sensitive to small admixtures in the wave functions are provided by the $E1$ transitions between low-lying bands in odd-A nuclei; these transitions are strongly hindered by selection rules in the quantum numbers of the single-particle motion. The relative $B(E1)$ values violate the leading-order intensity relations, but are found to be rather well reproduced by the generalized relations obtained by including linear terms in $I_\pm$ in the transition moments (Grin and Pavlichenkow, 1965). As an example, the analysis of $E1$ transitions with $\Delta K=1$, on the basis of Eq. (4-98) with $n=0$, is illustrated in Fig. 4-18, p. 110. Low-energy $E1$ transitions with larger matrix elements occur in the decay of the $K\pi r=0--$ bands in even-even nuclei, and are found to approximately obey the leading-order intensity relations (Stephens *et al.*, 1955; see also the example in Table 4-12, p. 114, of $E1$ transitions in an odd-A nucleus, which follow the leading-order rules and which may be of related character).

$E2$ transitions between bands are expected to be sensitive to a small admixing of the two bands, since the admixed components contribute with the collective matrix elements. An example of $E2$ transitions with $\Delta K=2$ exhibiting major effects of the M_2 term in Eq. (4-98) is illustrated in Fig. 4-30, p. 160. (For $E2$ transitions with $\Delta K=1$, the band mixing does not affect the relative transition rates to leading order; see p. 148 and the example shown in Table 4-19, p. 155.)

For the K-forbidden transitions, the relative intensities are found to obey the leading-order intensity relation (4-95) with an accuracy similar to that of the K-allowed transitions. (See the examples of $M1$ transitions with $n=1$ in Table 4-11, p. 113, and of $E2$ transitions with $n=4$ in Table 4-18, p. 144.)

The absolute transition rates for the K-forbidden transitions between low-lying states are observed to be systematically hindered by large factors, which are typically of magnitude 10^2 for each order of K forbiddenness. (See the examples discussed on p. 107 (^{177}Lu and ^{177}Hf), p. 142 (^{176}Lu), and p. 144 (^{244}Cm).) This retardation provides important evidence for the validity of the quantum number K in the low-energy part of the spectrum. The K-forbidden matrix elements involve Coriolis coupling effects of order n, and the observed rates imply amplitudes for band mixing of order 10^{-n}.

▼

ILLUSTRATIVE EXAMPLES TO SECTION 4-3

Spectrum of ^{168}Er as obtained from study of ($n\gamma$) reaction (Figs. 4-7 and 4-8, Tables 4-1 and 4-2)

The capture of a neutron by a heavy nucleus is in general followed by a very complicated cascade of γ rays involving of the order of 10^6 different transitions between something of the order of 10^4 different levels. The great resolving power of crystal diffraction spectrometers ($\Delta E/E \sim 10^{-5}$ in favorable cases; see, for example, the review article by Knowles, 1965) and the relative simplicity of the low-energy nuclear excitation spectra have made it possible in many cases to disentangle the complex γ spectra and construct partial nuclear level schemes by looking for transition energies equal to the sum of energies for other transitions, a method familiar from the study of atomic spectra (Ritz combination principle). The spectrum of ^{168}Er shown in Fig. 4-7 has been constructed in this manner from the measured γ-transition energies following the capture of thermal neutrons in ^{167}Er.

▲

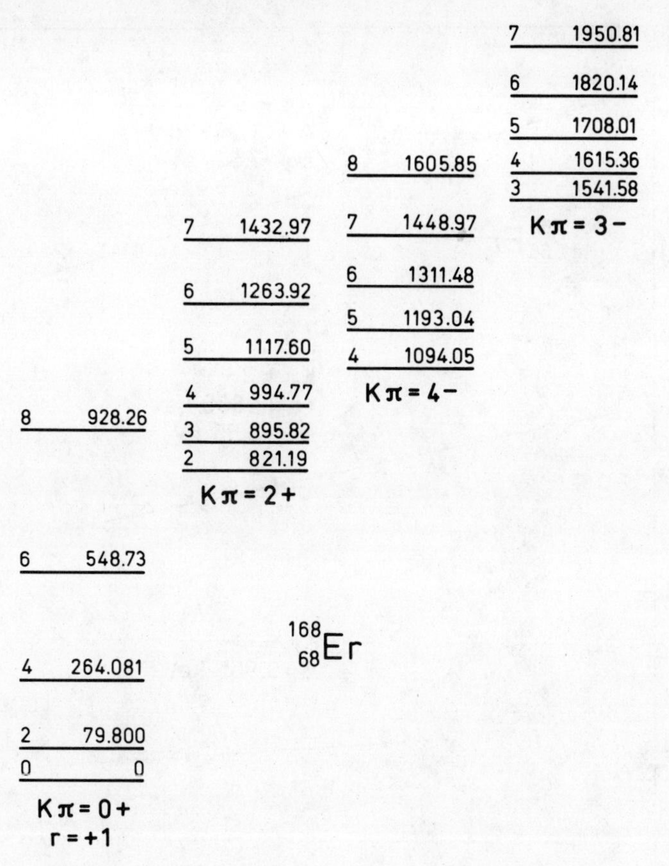

Figure 4-7 Spectrum of ^{168}Er. The data are taken from H. R. Koch, *Z. Physik* **192**, 142 (1966) and thesis submitted to Technische Hochschule, München (1965).

▼ It is seen that, besides the usual ground-state rotational band of even-even nuclei ($K\pi = 0+$, $r = +1$), three other bands have been identified in the spectrum, having $K\pi = 2+$, $4-$, and $3-$, respectively. From the accurately measured transition energies, one can obtain quantitative determinations of the higher terms in the expansion (4-62). Figure 4-8 shows the energy $E(I) - E(I = K)$ divided by $I(I+1) - K(K+1)$ as a function of $I(I+1)$. In such a plot, a simple $I(I+1)$ energy dependence gives a horizontal straight line, and the inclusion of the $I^2(I+1)^2$ term gives a sloping straight line. It is seen that, in the present case, these two leading terms in the expansion describe the observed energies with an accuracy of order 0.5% up to the highest states measured ($I = 8$). In the $K\pi = 0+$ and $K\pi = 2+$ bands, the remaining discrepancies exhibit a smooth variation with I and can be fitted by the C and D coefficients given in Table 4-1. ▲

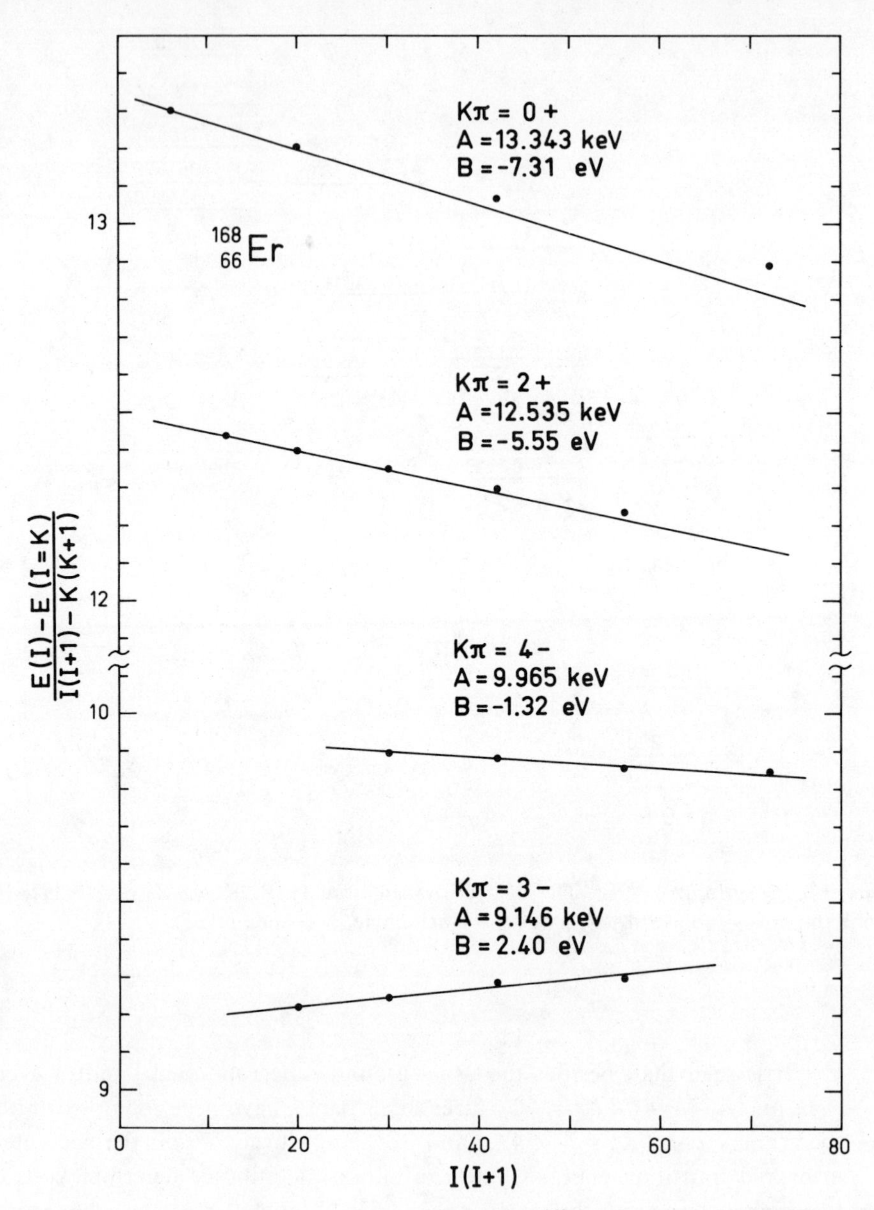

Figure 4-8 Analysis of energies in rotational bands in ^{168}Er. The data are taken from Fig. 4-7. The straight lines represent fits to the rotational spectra that deviate slightly from the parameters in Table 4-1.

▼ The energies in the odd-parity bands show deviations from a smooth variation with I that are of order 0.3 keV, which is several times the ▲ experimental uncertainty. These variations may reflect a dependence on the

▼ signature that is a general feature of the rotational energies for bands with $K \neq 0$ (see Eq. (4-62); the energies of the $K\pi = 2+$ band in ^{168}Er also provide
▲ evidence for a small signature dependence (see Table 4-1)).

$K\pi$	$E(I=K)$ (MeV)	A (keV)	B (eV)	C (meV)	D (μeV)	A_4 (eV)
0+	0	13.343	−7.31	18.6	−56	
2+	0.821	12.535	−5.55	14.3		0.017
4−	1.094	9.965	−1.32			
3−	1.542	9.146	2.40			

Table 4-1 Rotational energy coefficients for bands in ^{168}Er. The table is based on the experimental data in Fig. 4-7; the expansion parameters are determined from the energies of the lowest members in each band.

▼ The very high precision of the energy measurements of the rotational spectra in ^{168}Er observed in the (nγ) process makes possible a comparison of different parameterizations of the rotational energy. Thus, we may consider an expansion in the rotational frequency as an alternative to the expansion of the energy in powers of $I(I+1)$ (see p. 25). If the expansion in ω^2 is more rapidly convergent than that in $I(I+1)$, we can exploit this fact to obtain relations between the higher coefficients in the latter expansion. Including only two terms in the expansion (4-52) implies the relations

$$\frac{C}{A} = 4\left(\frac{B}{A}\right)^2$$
$$\frac{D}{A} = 24\left(\frac{B}{A}\right)^3 \qquad (4\text{-}99)$$

In Table 4-2, these relations are compared with the observed data for ^{168}Er and other nuclei for which sufficiently accurate energy measurements are available; it is seen that the relations (4-99) are obeyed in most cases to better than 20%. (See also the similar though somewhat less accurate evidence on even-even actinide nuclei; Schmorak *et al.*, 1972). Another way to exhibit the rate of convergence of the expansion in ω^2 is to plot the moment of inertia as a function of ω^2 (see the example in Fig. 4-11b, p. 70). The present analysis appears to suggest that the expansion of the energy of the ground-state bands of even-even nuclei in powers of $I(I+1)$ has a radius of convergence that is considerably smaller than for the corresponding expansion in ω^2.

The (nγ) measurements also provide a large body of data on γ-ray branching ratios, which can be interpreted in terms of the intensity rules
▲ implied by the rotational description (Koch, *loc. cit.*, Fig. 4-7). This analysis

▼ further confirms the significance of the expansion in powers of the rotational angular momentum, and at the same time provides information on the mixing of bands produced by the Coriolis interaction. The mixing of the $K\pi=0+$ and $K\pi=2+$ bands can be analyzed along the same lines as the corresponding effect in ^{166}Er discussed on pp. 160ff.; the mixing of the $K\pi=4-$ and $3-$ ▲ bands with $\Delta K=1$ is similar to that in ^{175}Lu discussed on pp. 154ff.

Nucleus	A (keV)	$-\frac{B}{A}\times 10^3$	$\frac{C}{A}\times 10^6$	$4(\frac{B}{A})^2\times 10^6$	$-\frac{D}{A}\times 10^9$	$-24(\frac{B}{A})^3\times 10^9$
^{156}Gd	15.0332	2.362	14.3	22.3		
^{158}Gd	13.3353	1.069	4.10	4.56		
^{162}Dy	13.5174	0.934	3.74	3.49	16	20
^{164}Dy	12.2858	0.738	1.44	2.18		
^{168}Er	13.3431	0.547	1.39	1.20	4.2	4.0
^{178}Hf	15.6203	1.005	4.14	4.04	22	24

Table 4-2 Test of relations implied by linear relationship between $\mathscr{I}$ and ω^2. The rotational energy expansion coefficients A, B, C, and D are calculated from the energies of the $I=0$, 2, 4, 6, and 8 members of the ground-state rotational band. The data are taken from the compilation of (nγ) spectra by Groshev *et al.*, 1968 and 1969, and from a private communication by H. R. Koch, 1970 (^{156}Gd and ^{158}Gd).

▼ The excitation mode responsible for the $K\pi=2+$ band is found to be a systematic feature of almost all deformed nuclei and furthermore to be characterized by rather large $E2$-matrix elements to the ground-state band (see Fig. 6-38). These features suggest an interpretation of this excitation mode in terms of a collective quadrupole vibration away from axial symmetry (γ vibration; see pp. 549ff. and the analysis on pp. 164ff. of the corresponding excitation in ^{166}Er, which also considers a possible interpretation in terms of a rotation of a nucleus of triaxial shape).

The $K\pi=4-$ and $3-$ bands may be associated with intrinsic excitations in which a pair is broken. The properties of the bands and their coupling suggest that we are dealing with a configuration involving two neutrons in the single-particle orbits [633 7/2] and [521 1/2], and having $K=\Omega_1\pm\Omega_2$. (See Table 5-13, p. 300.)

Rotational spectrum of ^{238}U studied by heavy ion Coulomb excitation (Fig. 4-9)

The cross section for the Coulomb excitation process is proportional to the electric multipole transition matrix element. This reaction is therefore very effective in inducing rotational excitations and played a decisive role in ▲ the discovery and early exploration of the rotational relationships. (See the

▼ summary of the development given by Alder *et al.*, 1956, and by Biedenharn and Brussaard, 1965.) By employing heavy ions, it has been possible to produce multiple $E2$ excitations, which lead to high angular momentum members of the rotational band built on the ground state of the target nucleus. Figure 4-9 shows the γ-ray spectrum resulting from the bombardment of ^{238}U by 182 MeV argon ions. The strong γ transitions are seen to form a sequence corresponding to the decay of the successive members of the ground-state rotational band ^{238}U ($K\pi r = 0++$). The deviations from the $I(I+1)$ dependence of the energy increase regularly and reach about 10% for the highest observed ($I=14$); the two-parameter fit, which includes a term proportional to $I^2(I+1)^2$, reproduces the observed spectrum within the accuracy of the measurements. ▲

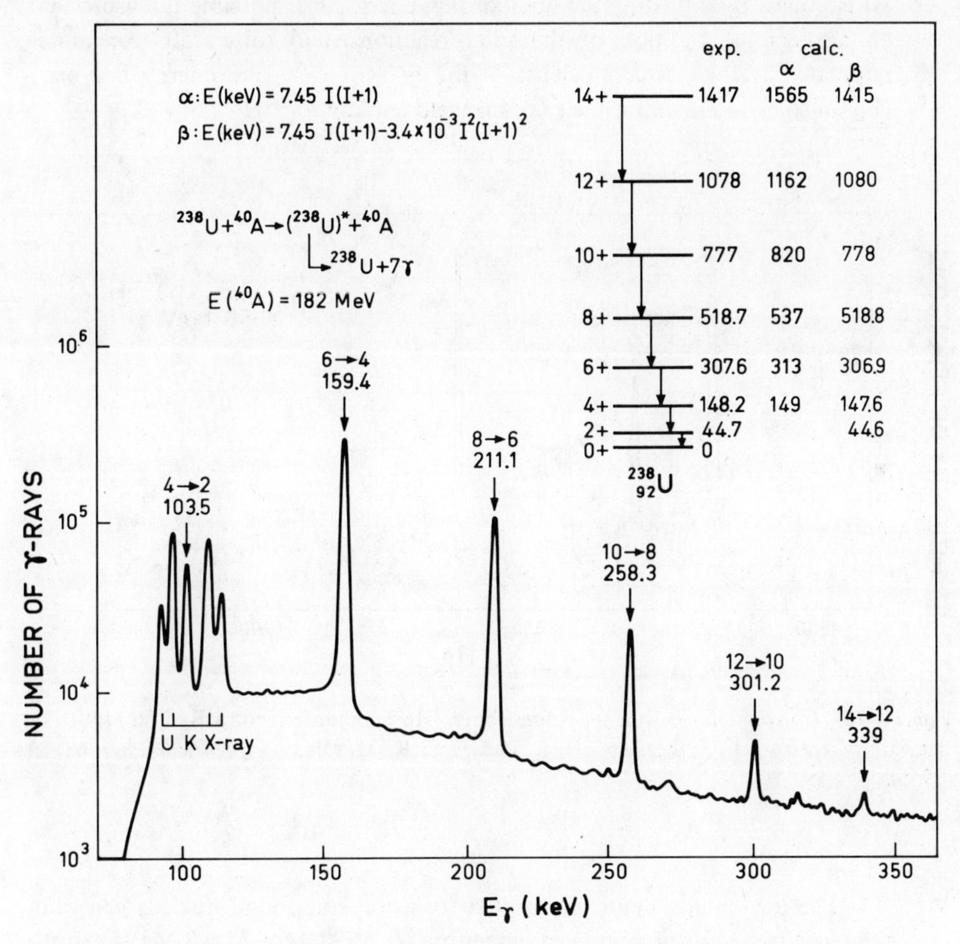

Figure 4-9 Coulomb excitation of ground-state rotational band in ^{238}U. The data are taken from R. M. Diamond and F. S. Stephens, *Arkiv Fysik* **36**, 221 (1967) and private communication.

▼ *Ground-state rotational band in ^{172}Hf as studied in the ^{165}Ho (^{11}B, 4n) reaction (Figs. 4-10 and 4-11)*

The most effective method discovered so far for populating the high angular momentum states in rotational bands is to form a compound nucleus by heavy ion bombardment. (This method was pioneered by Morinaga and Gugelot, 1963.) Figure 4-10 shows the conversion electron spectrum from the reactions produced by bombarding ^{165}Ho with ^{11}B ions of 56 MeV. The compound nucleus formed in this collision is ^{176}Hf at about 50 MeV of excitation and involves states with angular momentum up to $I \sim 25$ (see Eq. (6-547)). The highly excited compound nucleus decays mainly by successive emission of low-energy neutrons ($E_n \sim 2$ MeV) followed by a cascade of γ rays leading eventually to the ground states of a number of different lighter Hf isotopes. By adjusting the bombarding energy, it is possible to ensure that an appreciable fraction of the total reaction yield (often 50% or more) populates a single final nucleus; in the present case, the energy has been
▲ chosen so as to maximize the (^{11}B, 4n) yield leading to ^{172}Hf.

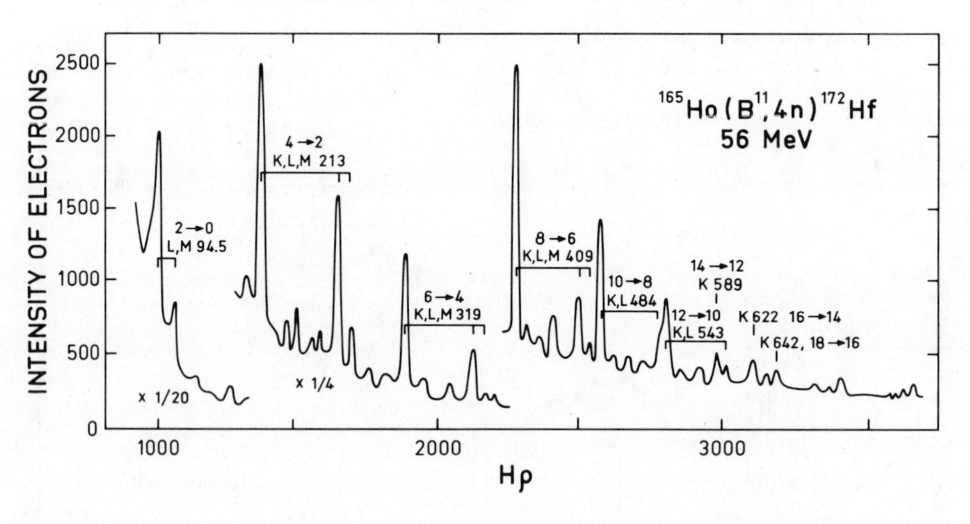

Figure 4-10 Conversion electron spectrum from ^{165}Ho bombarded with ^{11}B of 56 MeV. The data are taken from F. S. Stephens, N. L. Lark, and R. M. Diamond, *Nuclear Phys.* **63**, 82 (1965).

▼ The low-energy neutrons emitted from the compound nucleus can only remove a few units of angular momemtum ($k_n R \approx 2.4$ for $E_n = 2$ MeV; expres-
▲ sions for the centrifugal barrier penetration are given in Table 3F-1, Vol. I, p.

439), and thus the final γ-ray cascade in ^{172}Hf must involve states of quite high angular momentum. Since the ground-state rotational band provides the lowest states of given angular momentum up to rather large values of I, the excited states of this band are populated in an appreciable fraction of all cascades. Indeed, the conversion electron spectrum given in Fig. 4-10 shows that the transitions between the members of the ground-state rotational band can be clearly resolved from the background generated by the innumerable other unresolved transitions in the cascade.

Expansions of rotational energies for low values of I

By adding the energies of the different transitions, we obtain the energy for the ground-state band as a function of angular momentum, which is plotted in Fig. 4-11a. This plot shows that the main part of the energy variation in the band is described by the leading-order term proportional to $I(I+1)$, which provides an accuracy of order 30% up to the highest states populated ($I=18$). For not too large values of I, the deviations can be approximately described by a term proportional to $I^2(I+1)^2$; the coefficients A and B in this expansion (see Eq. (4-46)) have been obtained by requiring a fit to the observed energies of the lowest members of the band. (There is considerable uncertainty in the determination of the third-order coefficient, C, as a result of the experimental error in the energies of the lowest states in the band.) The ratio of successive terms in the expansion (4-46) is somewhat larger than in the examples discussed in Figs. 4-8 and 4-9 (in the present case, $B/A = -1.4\times10^{-3}$ compared with $B/A = -0.54\times10^{-3}$ in ^{168}Er).

For the neighboring nucleus ^{174}Hf, more accurate values of the rotational energies in the ground-state band are available (see Fig. 4-31, p. 168), and the expansion coefficients determined up to the seventh power in $I(I+1)$ are exhibited in Table 4-20, p. 169. It is seen from this table that the coefficients A, B, and C are rather accurately determined, but that even with the available more precise data, the higher coefficients have doubtful significance due to the poor convergence of the series for $I \gtrsim 10$.

Though the expansion of the energy in powers of $I(I+1)$ becomes rather poorly convergent in ^{172}Hf for $I \gtrsim 10$, the function $E(I)$ continues to depend smoothly on I. This feature has motivated the search for other parameterizations of the rotational energy. Thus, Fig. 4-11b shows the same data as Fig. 4-11a, plotted in terms of the moment of inertia $\mathscr{I}$ as a function of the rotational frequency ω_{rot} (see also the discussion on pp. 65 ff. (^{168}Er)).

In order to obtain the rotational frequency, we must calculate the derivative of E_{rot} with respect to I (see Eq. (4-48)), and for this purpose we have employed an approximate description in which E_{rot} is fitted with a linear function of $I(I+1)$ over the energy intervals between successive members of

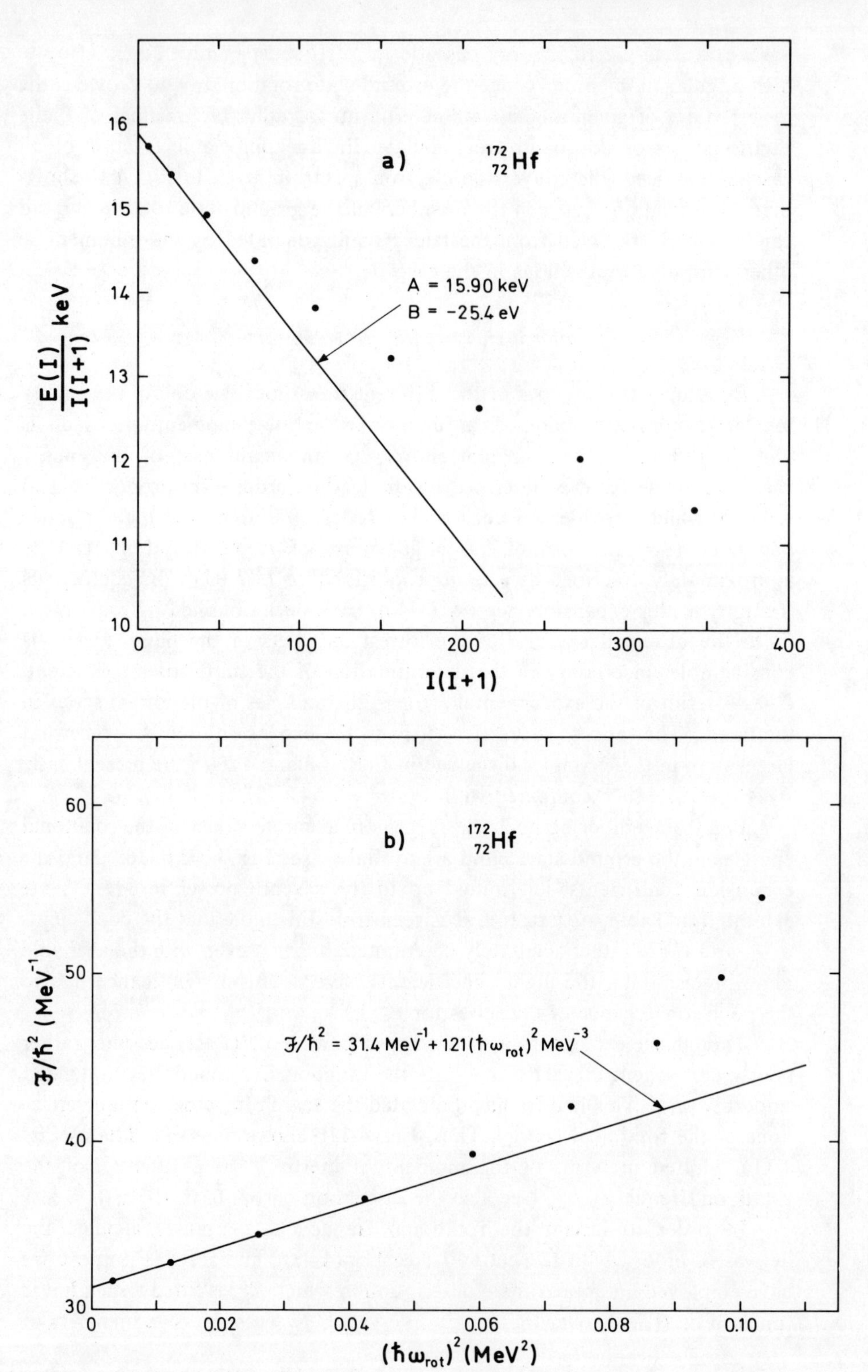

a)
$^{172}_{72}$Hf
A = 15.90 keV
B = −25.4 eV
$\frac{E(I)}{I(I+1)}$ keV
I(I+1)
b)
$^{172}_{72}$Hf
$\mathcal{J}/\hbar^2 = 31.4\ \mathrm{MeV}^{-1} + 121(\hbar\omega_{rot})^2\ \mathrm{MeV}^{-3}$
$\mathcal{J}/\hbar^2$ (MeV^{-1})
$(\hbar\omega_{rot})^2$ (MeV^2)

▼ the rotational band. Thus, we obtain

$$(\hbar\omega_{\rm rot})^2 = 4I(I+1)\left(\frac{\partial E_{\rm rot}}{\partial I(I+1)}\right)^2$$

$$\approx 2(I_1(I_1+1)+I_2(I_2+1))\left(\frac{E(I_2)-E(I_1)}{I_2(I_2+1)-I_1(I_1+1)}\right)^2 \tag{4-100}$$

and

$$\mathscr{I}(\omega_{\rm rot}) = \frac{\hbar^2}{2}\left(\frac{\partial E_{\rm rot}}{\partial I(I+1)}\right)^{-1}$$

$$\approx \frac{\hbar^2}{2}\frac{I_2(I_2+1)-I_1(I_1+1)}{E(I_2)-E(I_1)} \tag{4-101}$$

It is seen that, in the case of ^{172}Hf, the function $\mathscr{I}(\omega_{\rm rot}^2)$ is described as a simple linear function of $\omega_{\rm rot}^2$ over a larger part of the spectrum than the corresponding two-term expansion of the energy in powers of $I(I+1)$. This conclusion is the same as that derived from Table 4-2, p. 66, which refers to a number of other even-even nuclei, for which very accurate measurements of energies of the lowest members of the rotational bands are available.

Although, for small values of I, the expansions in $I(I+1)$ and $\omega_{\rm rot}^2$ both are of general validity, the radii of convergence may be very different. Thus, if the moment of inertia is linear in $\omega_{\rm rot}^2$, the relation (4-53) implies that the $I(I+1)$ expansion has a radius of convergence

$$I(I+1) = \frac{4}{27}\left|\frac{A}{B}\right| \tag{4-102}$$

For ^{172}Hf, with $|A/B| \approx 600$, the critical value of I implied by Eq. (4-102) is of order 10, which is consistent with the behavior shown in Fig. 4-11a. Although the evidence discussed in the present section provides strong indication for the greater simplicity of the expansion in the rotational frequency, the significance of this fact is not yet apparent.

Evidence for phase transition

For the highest rotational states observed in the ground-state band of ^{172}Hf, the energies, as analyzed in Fig. 4-11b, show a rapid increase in the ▲ moment of inertia, indicating a major modification in the intrinsic structure

Figure 4-11 Rotational energies for the ground-state band in ^{172}Hf. The experimental data are taken from Fig. 4-10. In Fig. 4-11a, the energy, in units of $I(I+1)$, is plotted against $I(I+1)$. In Fig. 4-11b, the moment of inertia $\mathscr{I}$ is plotted as a function of the square of the rotational frequency. For comparison, the rigid moment of inertia for ^{172}Hf is $\mathscr{I}_{\rm rig} \approx 80\hbar^2$ MeV^{-1}, assuming $\delta = 0.3$ and $\langle r^2\rangle = \frac{3}{5}(1.2\times A^{1/3})^2$ fm^2.

▼ occurring for $I \approx 20$. The fact that the moment of inertia is almost doubled and is approaching the value for rigid rotation suggests that the transition is associated with a considerable reduction in the pair correlation (see Eq. (4-128) for the effect of the pair correlation on the moment of inertia). Such a destruction of the pair correlations by the rotational motion is expected as a general feature of superfluid systems (see p. 37), and the critical value of the angular momentum may be estimated by comparing the decrease in rotational energy associated with the transition to the uncorrelated system with the energy gain associated with the pair correlations. The latter energy $\mathscr{E}_{\text{pair}}$ is of order (Δ^2/d), where d is the average spacing between the twofold degenerate single-particle levels, since the number of pairs participating in the correlation effect is of order Δ/d. Thus, for a spectrum of uniform levels with spacing d, the energy gain associated with the pairing is $\frac{1}{2}(\Delta^2/d)$ for the neutrons and protons separately (see Eq. (6-618)). For nuclei in the region of Hf, we have $\Delta \approx 0.9$ MeV (see, for example, Fig. 2-5, Vol. I, p. 170) and $d \approx 0.4$ MeV (see, for example, Figs. 5-2 and 5-3), and hence $\mathscr{E}_{\text{pair}} \approx 2$ MeV. The difference in rotational energy for the superfluid and normal systems is $\Delta\mathscr{E}_{\text{rot}} \approx 10 I^2$ keV (see Fig. 4-12), which implies $I_{\text{crit}} \sim 15$, as an order of magnitude estimate. Since the pair correlations involve the neutrons and protons separately, the phase transition to the normal system may take place in two distinct steps. However, it must be emphasized that also other effects lead to major modifications of the rotational motion in the angular momentum range considered (see pp. 42ff.).

The yrast region for $I > 20$

Even though the resolved transitions in Fig. 4-10 only extend to states with $I = 18$, experiments of this type reveal important features of the spectra for higher angular momentum, in particular concerning the states with energies close to the minimum value for given I (yrast region). Relevant features of the available data are

(i) The cross section for populating the ground-state rotational band indicates that these processes involve the descendants of compound nuclei with angular momenta that in some experiments are as great as 50 units (Stephens *et al.*, 1968).

(ii) The probability for populating the states in the ground-state rotational band decreases strongly in the range of I from 10 to 20 (Stephens *et al.*, 1965; Stephens *et al.*, 1968).

(iii) The γ spectra do not exhibit any single strong lines besides those of the ground-state rotational band. It appears that the missing angular momentum is carried away by a large number of γ rays, which combine to form an unresolved background. ▲

▼ (iv) The total delay time for the unresolved transitions preceding the ground-state cascade is, for the main part of the intensity, of the order of 10^{-11} sec (Diamond *et al.*, 1969a).

The picture of the γ cascade that appears to emerge from this evidence starts from the production of a compound nucleus which, after neutron evaporation, has an energy above the yrast line ranging up to values of the order of the neutron separation energy. The first few γ transitions are expected to be similar to those observed in neutron capture processes and to lead to a cooling of the nucleus. However, the absence of strong lines indicates that, in most of the cascades, the nuclei do not completely cool to the yrast line, but proceed along a large number of paths somewhat above this line. The slope of the yrast line is expected to be given approximately by the rigid-body value for the moment of inertia (as is obtained both for collective rotations with $\Delta = 0$ and in a statistical analysis; see pp. 77 ff.). Thus, the transition energy for $E2$ transitions with $\Delta I = 2$ and $I \approx 30$, $A \approx 160$ is of the order of 1 MeV. The evidence quoted above implies lifetimes for these transitions $\lesssim 10^{-12}$ sec, corresponding to transition rates at least an order of magnitude larger than for single-particle $E2$ transitions (see Eqs. (3C-18) and (3C-38)). Such an enhancement, together with the relative weakness of the transitions leading toward the yrast line, suggests that the main transitions proceed along trajectories of collective families. These sequences may have many of the properties of rotational bands (Williamson *et al.*, 1968; Newton *et al.*, 1970). The absence of delays in the great majority of the γ cascades implies that the collective families do not have band heads, as would be associated with states of large K in axially symmetric nuclei.

For lower values of the angular momentum, bands with large K values, giving rise to isomeric states, have been observed on the yrast line, or very close to it, in a number of even-even nuclei (see, for example, Borggreen *et al.*, 1967). A striking example is provided by the spectrum of ^{178}Hf (Helmer and Reich, 1968); in this nucleus, a $K\pi = 8-$ band begins at 1.148 MeV, which is 89 keV larger than the energy of the $I = 8$ member of the ground-state band; however, because of the larger moment of inertia of the $K = 8$ band, it crosses the (extrapolated) ground-state band between $I = 10$ and 12. Another intrinsic configuration, which appears to represent a four-quasiparticle state with $K\pi = 16+$, has been found at an excitation energy of about 2.5 MeV, which is lower than the extrapolated position of the $K = 8$, $I = 16$ level by about 1 MeV.

The occurrence of states with $K \approx I$ in the yrast region may be understood from the fact that the energy expended in generating angular momentum through the alignment of the orbits of a few individual nucleons can be ▲ characterized by a moment of inertia similar to that for collective rotation

▼ (see p. 81). The absence of isomers with still higher spin, which is suggested by the evidence discussed above, might indicate a change to a nuclear coupling scheme associated with nonaxially symmetric shapes.

Moments of inertia for ground-state bands of nuclei with $150 \leqslant A \leqslant 188$ (Fig. 4-12)

To leading order in the expansion in powers of the angular momentum, the rotational energies involve the two parameters A and A_1, which can also be expressed in terms of the effective moments of inertia $\mathscr{I}$ and the decoupling parameter a (see Eqs. (4-61)). The available information on the moments of inertia for ground-state bands of nuclei with $150 \leqslant A \leqslant 188$ is illustrated in Fig. 4-12. For the bands with $K \neq 1/2$, the moments of inertia in the figure have been determined from the energy separation of the two lowest band members ($I = K$ and $K+1$); for the $K = 1/2$ bands, the determination of $\mathscr{I}$ is ▲ based on the observed positions of the three lowest band members.

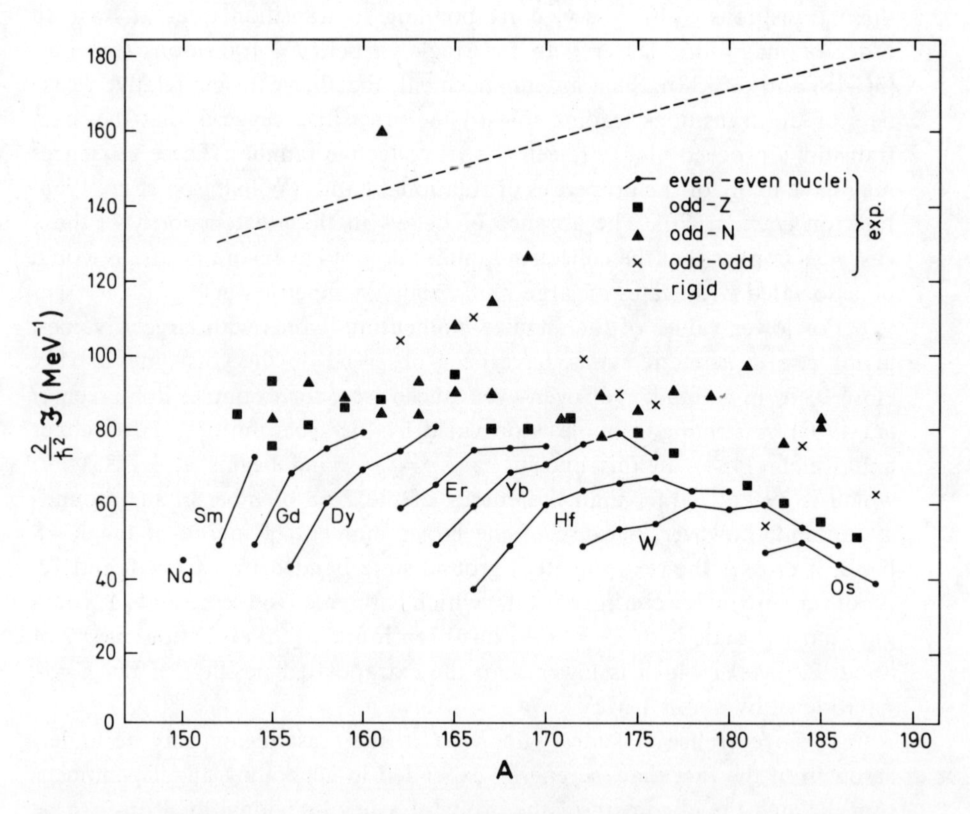

Figure 4-12 Systematics of moments of inertia for nuclei with $150 \leqslant A \leqslant 188$. The moments of inertia are obtained from the empirical energy levels in *Table of Isotopes* by Lederer *et al.*, 1967.

▼ *Comparison with rigid rotation and irrotational flow*

In Fig. 4-12, the empirical moments of inertia are compared with the moments that would be associated with a rigid rotation of the deformed nucleus,

$$\mathscr{I}_{\text{rig}} = AM\langle x_2^2 + x_3^2 \rangle \tag{4-103}$$

Assuming axial symmetry about the intrinsic 3 axis (and the same density distribution for neutrons and protons), the moment (4-103) can be expressed in the form

$$\begin{aligned}\mathscr{I}_{\text{rig}} &= \tfrac{2}{3}AM\langle r^2 \rangle(1 + \tfrac{1}{3}\delta) \\ &= \tfrac{2}{5}AMR^2(1 + \tfrac{1}{3}\delta)\end{aligned} \tag{4-104}$$

where δ is the deformation parameter determined from the electric quadrupole moment (see Eq. (4-72)), while R is the mean radius defined by

$$R^2 = \tfrac{5}{3}\langle r^2 \rangle \tag{4-105}$$

and taken to have the value $R = 1.2A^{1/3}$ fm. For a nucleus of spheroidal shape, the parameter δ, for small eccentricities, equals $\Delta R/R$, where ΔR is the difference between the semi-axes of the spheroid (see Eq. (4-73)). The values of δ employed in the evaluation of $\mathscr{I}_{\text{rig}}$ in Fig. 4-12 are taken from Fig. 4-25, p. 133.

From Fig. 4-12, it is seen that the observed moments are smaller than $\mathscr{I}_{\text{rig}}$ by a factor of order 2. The observed moments show systematic variations that are especially apparent for the even-even nuclei. Thus, as one approaches the limits of the deformed regions, the moments decrease, and this variation of $\mathscr{I}$ is associated with a corresponding variation in δ (see Fig. 4-25). This correlation is characteristic of another classical model, the rotation of a drop of irrotational fluid. The flow pattern for such a rotational motion is illustrated in Fig. 6A-2, p. 675, and the moment of inertia is (see Eq. (6A-81))

$$\mathscr{I}_{\text{irrot}} \approx \mathscr{I}_{\text{rig}}\,\delta^2 \approx \mathscr{I}_{\text{rig}}\left(\frac{\Delta R}{R}\right)^2 \tag{4-106}$$

to leading order in δ. The observed values of $\mathscr{I}$, however, are about a factor of 5 greater than $\mathscr{I}_{\text{irrot}}$. (In the region illustrated in Fig. 4-12, the deformation δ is typically of order 0.2–0.3.)

Analysis in terms of nucleonic response to rotation of the nuclear field (cranking model)

The nuclear moments of inertia can be analyzed by considering the motion of the nucleons in the rotating nuclear potential. The Coriolis and centrifugal forces acting in the rotating body-fixed system give rise to an increase in the energy of the nucleonic motion, which can be identified with
▲ the rotational energy.

▼ For a single nucleon, the motion with respect to the potential rotating with frequency ω_{rot} is described by the Hamiltonian

$$H = H_0 - \hbar\boldsymbol{\omega}_{rot} \cdot \mathbf{j} \tag{4-107}$$

where H_0 describes the motion in the absence of the rotation. The Hamiltonian (4-107) may be seen to be equivalent to the general relation between the time derivative of operators evaluated in the rotating system and in the fixed system (labeled by the subscript 0)

$$\frac{dF}{dt} = \frac{i}{\hbar}[H,F] = \left(\frac{dF}{dt}\right)_0 - i\boldsymbol{\omega}_{rot} \cdot [\mathbf{j},F] \tag{4-108}$$

In this expression, F is an operator depending on the particle variables, and the commutator of $\mathbf{j}$ with F gives the change in F associated with a rotation of the coordinate system (see Vol. I, p. 10). It is readily verified that the equations of motion derived from the Hamiltonian (4-107) include the Coriolis and centrifugal forces; indeed, the orbital part $-\boldsymbol{\omega}_{rot} \cdot (\mathbf{r} \times \mathbf{p})$ of the rotational perturbation term in Eq. (4-107) is the familiar expression from classical Hamiltonian mechanics. (It must be emphasized that the above derivation of the Hamiltonian in the rotating system assumes that the potential in which the particles move is not affected by the rotational motion. The occurrence of terms in the potential that are linear in the rotational frequency is considered on pp. 80 and 278.)

If the effect of the rotation can be considered as a small perturbation of the nucleonic motion, the resulting shift in the one-particle energies is given by

$$\delta\varepsilon(\nu) = \hbar^2\omega_{rot}^2 \sum_{\nu'} \frac{\langle \nu' | j_1 | \nu \rangle^2}{\varepsilon(\nu') - \varepsilon(\nu)} \tag{4-109}$$

where j_1 is the component of $\mathbf{j}$ along the axis of rotation. The unperturbed nucleonic state is labeled by ν, while the states coupled to ν by the Coriolis force are labeled by ν'. (In the derivation of Eq. (4-109), one must take into account that the energy of the system is represented by H_0, while the Hamiltonian H is the time-displacement operator describing the evolution of the system in the rotating (time-dependent) coordinate frame; the difference between H and H_0 gives rise to a change of sign of $\delta\varepsilon$.)

The energy shift (4-109) is proportional to the square of the rotational frequency and can therefore be interpreted as a contribution to the moment of inertia. Adding the contributions from all the particles, we obtain the total moment of inertia for a given state 0 (Inglis, 1954)

$$\mathscr{I}_1 = 2\hbar^2 \sum_{i \neq 0} \frac{\langle i | J_1 | 0 \rangle^2}{E_i - E_0} \tag{4-110}$$

$$J_1 = \sum_{k=1}^{A} (j_1)_k$$

▲

▼ where i labels the many-particle states of the A nucleons in the deformed potential, with excitation energies E_i. The moment of inertia (4-110) can also be obtained from the expectation value of J_1 in the perturbed state and the relation (4-47). The expression (4-110) is referred to as the "cranking formula", since the frequency of the rotating potential is treated as an externally prescribed quantity; however, a result equivalent to Eq. (4-109) is obtained in the particle-rotor model, in which the frequency of the rotating potential is treated as a dynamical variable (see Eq. (4A-15)). An alternative derivation of the expression (4-110) can be based on the treatment of the residual interactions that are necessary to restore rotational invariance to the Hamiltonian based on the deformed potential (see pp. 446 ff.).

Moments of inertia from independent-particle motion

Some of the main features of the moment of inertia (4-110) for independent-particle motion may be exhibited by considering a system of particles moving in an anisotropic harmonic oscillator potential,

$$V = \sum_{\kappa=1}^{3} \tfrac{1}{2} M \omega_\kappa^2 x_\kappa^2 \tag{4-111}$$

The one-particle states can be labeled by the oscillator quantum numbers n_κ, and the total energy is

$$\mathscr{E} = \sum_{\kappa=1}^{3} \hbar\omega_\kappa \sum_{k=1}^{A} (n_\kappa + \tfrac{1}{2})_k \tag{4-112}$$

For a given configuration with specified sets of quantum numbers $(n_\kappa)_k$, the equilibrium shape can be obtained from a self-consistency argument. The equipotential surfaces for the deformed potential (4-111) are ellipsoids with axes proportional to ω_κ^{-1}. The anisotropy of the density distribution can be characterized by the mean square values

$$\langle \sum_{k=1}^{A} (x_\kappa^2)_k \rangle = \frac{\hbar}{M\omega_\kappa} \Sigma_\kappa \tag{4-113}$$

where we have employed the notation

$$\Sigma_\kappa \equiv \sum_{k=1}^{A} (n_\kappa + \tfrac{1}{2})_k \tag{4-114}$$

Hence, the deformation of the density distribution will be equal to that of the potential if ω_κ is proportional to Σ_κ^{-1},

$$\omega_1 \Sigma_1 = \omega_2 \Sigma_2 = \omega_3 \Sigma_3 \tag{4-115}$$

For a spin-independent potential, only the orbital momentum contri-
▲ butes to the moment of inertia (4-110), and the excitations produced by the

▼ Coriolis coupling involve two types of terms corresponding to $\Delta N=0$ and 2, for the change of the total oscillator quantum number ($N=n_1+n_2+n_3$). The $\Delta N=0$ terms have energy denominators $\pm\hbar(\omega_2-\omega_3)$, which are proportional to the deformation, while the energy denominators for the $\Delta N=2$ terms involve the sum of the frequencies. For independent-particle motion, the moment of inertia (4-110) can be expressed as a sum of the contributions from the individual particles, since the terms forbidden by the exclusion principle cancel pairwise; employing the expression (4-130) for the particle angular momentum in terms of the harmonic oscillator variables, one obtains

$$\mathscr{I}_1=\frac{\hbar}{2\omega_2\omega_3}\left(\frac{(\omega_2+\omega_3)^2}{\omega_2-\omega_3}(\Sigma_3-\Sigma_2)+\frac{(\omega_2-\omega_3)^2}{\omega_2+\omega_3}(\Sigma_2+\Sigma_3)\right) \tag{4-116}$$

For the equilibrium deformation satisfying Eq. (4-115), it is seen from Eq. (4-113) that the expression (4-116) yields a moment of inertia equal to that for a rigid rotation of a system with the same density distribution (Bohr and Mottelson, 1955),

$$\mathscr{I}_1=\langle\sum_{k=1}^{A}M(x_2^2+x_3^2)_k\rangle=\mathscr{I}_{\text{rig}} \tag{4-117}$$

This result is independent of the configuration of the particles in the oscillator potential.

The equilibrium condition is essential for the relation (4-117). For example, if one considers a configuration with $\Sigma_1=\Sigma_2=\Sigma_3$, as for closed shells, and imposes a deformation by an external force, only the second term in Eq. (4-116) contributes, and the moment of inertia has the value

$$\mathscr{I}_1=M\frac{\langle\sum_k(x_2^2-x_3^2)_k\rangle^2}{\langle\sum_k(x_2^2+x_3^2)_k\rangle} \tag{4-118}$$

which is smaller than the rigid-body value by the square of the deformation parameter. In fact, the moment (4-118) is that associated with irrotational flow (see Eq. (6A-81)).

The actual nuclear potential differs significantly from that of a harmonic oscillator in the radial dependence and due to the presence of the spin-orbit coupling. However, for the potentials employed in Chapter 5 (see pp. 218 ff.), the numerical evaluation of the expression (4-110) gives values close to $\mathscr{I}_{\text{rig}}$, provided the eccentricity of the potential is adjusted to the equilibrium value. Still, it should be emphasized that the exact relationship to the rigid-body moment of inertia has been established only for the harmonic oscillator potential and that the accuracy and generality of this result in other cases remain to be further explored. ▲

▼ The significance of the rigid-body value for the moment of inertia may be further illuminated by means of the semiclassical limit provided by the Fermi gas approximation. The velocity distribution in the rotating coordinate system is determined by the Hamiltonian (4-107), which can be written in the form (assuming a spin- and velocity-independent potential V)

$$H=\frac{1}{2}M\mathbf{v}^2+V(\mathbf{r})-\frac{1}{2}M(\boldsymbol{\omega}_{\text{rot}}\times\mathbf{r})^2 \tag{4-119}$$

where the velocity

$$\mathbf{v}=\frac{\mathbf{p}}{M}-(\boldsymbol{\omega}_{\text{rot}}\times\mathbf{r}) \tag{4-120}$$

is that of the particle relative to the rotating frame. Since the Hamiltonian (4-119) involves only the square of the velocity, the distribution $\rho(\mathbf{r},\mathbf{v})$ remains isotropic in velocity space at each point $\mathbf{r}$. Hence, in the rotating frame, there is no net current and, in the fixed coordinate system, the flow pattern is the same as for a rotating rigid body. The argument used here is the same as that involved in demonstrating the absence of diamagnetism in a classical electron gas (Bohr, 1911; Van Leuwen, 1921; see also Van Vleck, 1932, pp. 94ff. For an electron in a constant magnetic field $\mathbf{H}$, the Hamiltonian is given by the first two terms of Eq. (4-119) with ω_{rot} replaced by $(e/2Mc)\mathbf{H}$; the last term in the Hamiltonian (4-119) is the centrifugal potential, which has no counterpart in the magnetic interaction. However, the centrifugal potential is velocity independent and therefore does not affect the above arguments.)

The role of the isotropy of the velocity distribution sheds light on the significance of the equilibrium deformation in the derivation of the result (4-117). In fact, for the harmonic oscillator potential, we have

$$\left\langle \sum_{k=1}^{A} (v_\kappa^2)_k \right\rangle = \frac{\hbar}{M}\omega_\kappa\Sigma_\kappa \tag{4-121}$$

and the overall velocity distribution is therefore isotropic up to the second moments, when the self-consistency conditions (4-115) are fulfilled. Indeed, the energy gain by the deformation of the potential may be viewed as a lowering of the kinetic energy through the establishment of an isotropic velocity distribution. If the velocity distribution is not isotropic, the Coriolis forces may introduce a flow in the rotating coordinate system proportional to the rotational frequency; for example, for a deformed closed-shell configuration, the flow is that illustrated in Fig. 6A-2 for an irrotational fluid.

The above arguments for the rigid-body value for the moment of inertia are not affected by the occurrence of a velocity dependence in the one-body potential (effective mass term, see Vol. I, p. 147). Though the rotational motion does induce extra velocities in the motion of the particles, the potential energy is determined by the relative velocities of the particles at ▲ each point, which is not affected by the collective rotational motion. In the

▼ evaluation of the expression (4-110), an effective mass term in the potential leads to a moment $(M^*/M)\,\mathscr{I}_{\text{rig}}$, but this spurious effect is compensated by an additional term in the rotating potential, which is required to restore the local Galilean invariance. This term is obtained by substituting, in the one-particle potential, $\mathbf{p}\rightarrow\mathbf{p}-M\mathbf{u}$ (see Eq. (1-16)), where $\mathbf{u}=\boldsymbol{\omega}_{\text{rot}}\times\mathbf{r}$ is the collective flow. The additional term implies that the Coriolis coupling is multiplied by the factor (M/M^*), and the resulting moment of inertia is again equal to $\mathscr{I}_{\text{rig}}$, as follows directly from the evaluation of $\langle J_1\rangle$ in the perturbed state. (The compensation of the effective mass in the rotational moment was exhibited by Migdal, 1959; a similar compensation of the effective mass term in the description of the translational mass is discussed on p. 445.)

Moment of inertia associated with alignment of single-particle orbits along an axis of symmetry

For an axis of symmetry, the moment of inertia (4-110) vanishes, corresponding to the fact that the nucleus cannot perform collective rotations about such an axis. However, the individual nucleons carry angular momentum with respect to such an axis, and a net angular momentum can be generated by aligning that of the individual particles. The average energy of the lowest state of given angular momentum along a symmetry axis can be obtained from a statistical analysis. This state is obtained by occupying all the lowest particle orbitals up to a Fermi energy $\varepsilon_{\text{F}}(m)$, considered as a function of the one-particle angular momentum m along the given axis. (For a deformed but axially symmetric potential, the quantum number m corresponds to Ω and M to K.) The function $\varepsilon_{\text{F}}(m)$ is to be determined by minimizing the sum of one-particle energies subject to the two constraints that the particle number and total angular momentum component M are specified.

Thus, we consider variations of the quantity

$$\begin{aligned}\mathscr{E}' &= \mathscr{E}-\lambda A-\mu M\\ &= \sum_m \int^{\varepsilon_{\text{F}}(m)} g(\varepsilon,m)(\varepsilon-\lambda-\mu m)\,d\varepsilon \qquad (4\text{-}122)\end{aligned}$$

involving the two Lagrange multipliers λ and μ. The one-particle spectrum, for a given m, is described by the level density $g(\varepsilon,m)$. The condition that $\mathscr{E}'$ be stationary with respect to arbitrary variations of $\varepsilon_{\text{F}}(m)$ yields

$$\varepsilon_{\text{F}}(m)=\lambda+\mu m \qquad (4\text{-}123)$$

▲ and the Lagrange multipliers λ and μ, which are seen to represent the

▼ derivatives of $\mathscr{E}$ with respect to A and M, respectively, are determined by the relations

$$A=\sum_m \int^{\varepsilon_F(m)} g(\varepsilon,m)\,d\varepsilon \approx \int^{\lambda} g(\varepsilon)\,d\varepsilon$$

$$g(\varepsilon)\equiv \sum_m g(\varepsilon,m)$$

$$M=\sum_m \int^{\varepsilon_F(m)} g(\varepsilon,m)m\,d\varepsilon \approx \mu \sum_m m^2 g(\varepsilon=\lambda,m)$$

$$=\mu\langle m^2\rangle g_0 \tag{4-124}$$

$$g_0\equiv g(\varepsilon=\lambda)$$

where $\langle m^2\rangle$ denotes the mean value of m^2 for the orbits near the Fermi energy. In Eq. (4-124), we have assumed $g(\varepsilon,-m)=g(\varepsilon,m)$ and have ignored the variation of $g(\varepsilon,m)$ with ε in the energy interval μm around the Fermi surface. In the same approximation, we obtain for the energy

$$\mathscr{E}=\sum_m \int^{\varepsilon_F(m)} g(\varepsilon,m)\varepsilon\,d\varepsilon$$

$$\approx \mathscr{E}(M=0)+\tfrac{1}{2}\mu^2\langle m^2\rangle g_0$$

$$=\mathscr{E}(M=0)+\frac{\hbar^2}{2\mathscr{I}}M^2 \tag{4-125}$$

where the effective moment of inertia is given by

$$\mathscr{I}=\hbar^2\langle m^2\rangle g_0 \tag{4-126}$$

An evaluation of $\langle m^2\rangle g_0$ on the basis of the Fermi gas approximation yields the rigid-body value for the moment of inertia (4-126); see Eqs. (2B-59) to (2B-61). This result is intimately related to the above derivation of the moment of inertia on the basis of the cranking model. Indeed, the quantity $\mathscr{E}'$ in Eq. (4-122) may be recognized as the expectation value of the Hamiltonian describing motion in a potential rotating with frequency $\mu\hbar^{-1}$. Thus, in both cases, we are considering the perturbations produced by the Coriolis and centrifugal interactions, and the rigid-body value for the moment of inertia results from the fact that the velocity distribution in the rotating coordinate system remains isotropic.

While the moment of inertia (4-110) for the collective rotational motion is associated with small coherent perturbations in the motion of a large ▲ number of particles, the moment of inertia (4-126) for rotation about a symmetry axis represents large changes in the quantum numbers of a few individual nucleons at the Fermi surface. This type of rotational motion, therefore, does not involve simple generic relations between the states with successive values of the angular momentum, such as for the collective rotations of deformed systems.

Effect of pair correlations

The fact that the moments of inertia are found to be appreciably smaller than $\mathscr{I}_{\text{rig}}$ must be attributed to correlations in the intrinsic nucleonic motion. The main effect appears to be associated with the pair correlations.

In the pair-correlated (or superfluid) nucleus, the excitations can be described in terms of quasiparticles (see the discussion in Chapter 6, pp. 647 ff.). For an even-even nucleus, the ground state is the quasiparticle vacuum ($\text{v}=0$) and the excited states in Eq. (4-110) are two-quasiparticle states ($\text{v}=2$). The matrix elements connecting the $\text{v}=0$ and the $\text{v}=2$ states can be obtained from Eq. (6-610b), and the moment of inertia is therefore given by (Belyaev, 1959; Migdal, 1959)

$$\mathscr{I}_\kappa(\text{v}=0)=2\hbar^2\sum_{\nu_1,\nu_2}\frac{|\langle\nu_2|j_\kappa|\nu_1\rangle|^2}{E(\nu_1)+E(\nu_2)}\left(u(\nu_1)v(\nu_2)-v(\nu_1)u(\nu_2)\right)^2 \tag{4-127}$$

where the sum extends over all two-quasiparticle states, ($\text{v}=2$, $\nu_1\nu_2$). The quasiparticle energies E and the amplitudes u and v can be determined from single-particle energies ε and the energy gap parameter Δ by means of Eqs. (6-601) and (6-602); see also Eq. (6-611) for λ. It is seen that the pair correlations reduce the moment of inertia partly by increasing the energy denominators in Eq. (4-127), partly by reducing the matrix elements in the numerators.

A qualitative estimate of the effect of the pair correlations can be obtained by noting that the most important single-particle transitions contributing to the moment of inertia (4-127) have the common excitation energy $\varepsilon_2-\varepsilon_1=\hbar(\omega_2-\omega_3)=\delta\hbar\omega_0$ corresponding to the shift of an oscillator quantum from the direction of the 3 to that of the 2 axis. Averaging the position of the chemical potential with respect to the orbits that contribute, one obtains

$$\begin{aligned}\mathscr{I}&=\tfrac{1}{2}\mathscr{I}_{\text{rig}}\int_{-\infty}^{+\infty}\frac{\left(u(\varepsilon+\delta\hbar\omega_0)v(\varepsilon)-u(\varepsilon)v(\varepsilon+\delta\hbar\omega_0)\right)^2}{E(\varepsilon)+E(\varepsilon+\delta\hbar\omega_0)}\,d\varepsilon\\&=\mathscr{I}_{\text{rig}}\left(1-g\left(\frac{\delta\hbar\omega_0}{2\Delta}\right)\right)\end{aligned} \tag{4-128}$$

$$g(x)\equiv\frac{\ln\left(x+(1+x^2)^{1/2}\right)}{x(1+x^2)^{1/2}}$$

For a nucleus with $A=160$ and $\delta\approx 0.3$, corresponding to the most strongly deformed nuclei in the region illustrated in Fig. 4-12, we obtain $\hbar\omega_0\delta\approx 2.3$ MeV (see Eq. (2-131)). With $\Delta\approx 0.9$ MeV (see Fig. 2-5, Vol. I, p. 170), the expression (4-128) implies a reduction of the moment by a factor of about 2 relative to the rigid-body value, in accordance with the empirical values in Fig. 4-12. Since Δ_{p} is somewhat larger than Δ_{n} (see Fig. 2-5), the proton contribution to the moment of inertia is expected to be more strongly decreased by the pairing than is that of the neutrons, in agreement with the evidence from the g_R factors (see p. 54).

The more detailed evaluation of expression (4-127), on the basis of the available evidence on the single-particle spectrum and the values of Δ obtained from the odd-even mass differences, yields moments of inertia that reproduce rather well the main trends of the empirical moments, though the estimated values appear to be systematically 10–20% too small (Griffin and Rich, 1960; Nilsson and Prior, 1961).

In a rotating paired system, the pair field acquires a component that is proportional to the rotational frequency and hence contributes to the moment of inertia (Migdal, 1959; the structure of this additional term in the pair potential is discussed on p. 278). For $\Delta\ll\hbar\omega_0$, as in the nucleus, this contribution is relatively small; it may account for a significant part of the remaining 10–20% in the estimate of the moment of inertia. The additional contribution to $\mathscr{I}$ becomes the dominant one in the limit $\Delta\gg\hbar\omega_0$, where it ensures that the moment of inertia approaches that of irrotational flow, as expected for a superfluid with dimensions large compared with the coherence length (Migdal, *loc. cit.*; see also the discussion of superflow on pp. 397 ff.).

The strong dependence of the moment of inertia on the pair correlation parameter Δ implies a significant coupling between Δ and the rotational motion. Since $\mathscr{I}$ increases with decreasing Δ, it must be expected that Δ will be a decreasing function of I. For sufficiently large I, the coupling to the rotation is expected to destroy the pair correlations, an effect analogous to the destruction of superconductivity by a magnetic field (Mottelson and Valatin, 1960). Empirical evidence that may indicate such an effect is discussed on p. 72.

For odd-A and odd-odd nuclei, the moments of inertia are found to be significantly larger than for the ground-state bands of neighboring even-even nuclei. Such an increase is a simple consequence of the pair correlations, as discussed in Chapter 5, pp. 251 ff.

Evidence on the moment of inertia for very large deformations has been obtained from the observation of transitions between rotational states based on the fission isomers. In ^{240}Pu, the rotational constant is found to be $\hbar^2/2\mathscr{I}=3.3$ keV (Specht *et al.*, 1972), which is about a factor of 2 smaller than the observed value in the ground-state band of this nucleus. Part of this difference may be attributed to the increase of the rigid-body value as a result

▼ of the deformation; for a prolate spheroidal shape with a ratio of axes of 2 : 1, as expected from the shell-structure interpretation of the fission isomer (see the discussion on pp. 634ff.), the rigid-body moment of inertia is larger than for the spherical shape by a factor of $(5/4)2^{1/3} \approx 1.57$, corresponding to $\hbar^2/2\mathscr{I}_{\text{rig}} = 2.5$ keV. Thus, the observed moment is about 75% of the rigid-body value. The fact that the moment is appreciably closer to the rigid-body value than is the case for the ground-state band is in agreement with the above analysis of the influence of the pair correlations, since the relevant parameter $(\delta\hbar\omega_0/2\Delta)$ is considerably larger in the more strongly deformed shape isomer. However, the assumption underlying Eq. (4-128), that the main contributions to the moment of inertia are associated with intrinsic excitations with energies approximately equal to $\delta\hbar\omega_0$, is no longer valid for δ of order unity.

Significance of finite particle number for nuclear rotations, illustrated by harmonic oscillator model (Table 4-3)

In the present example, we consider features of the nuclear rotational motion that are connected with the fluctuations in the nuclear orientation resulting from the finite particle number. In some situations, the bands may terminate at a finite value of the angular momentum. Such a termination is characterized by major deviations from the leading-order $E2$-intensity relations. A related problem concerns the constraint imposed on the intrinsic motion due to the inclusion of some of the particle degrees of freedom in the rotational motion. These problems may be illustrated by an analysis of particle motion in a harmonic oscillator potential. The simplicity of this model makes possible the explicit construction of the many-particle wave function of the rotating system, which can also be recognized as resulting from the diagonalization of appropriate two-body interactions representing the deformed field.[5]

Alignment of angular momenta produced by rotation

We first consider the motion of nucleons in an anisotropic oscillator potential that is externally rotated, as in the cranking model. For a harmonic oscillator potential, it is convenient to consider the motion in terms of the quanta of oscillation. In the absence of rotation, the eigenmodes correspond to oscillations in the directions of the principal axes

$$H_0 = \sum_{\kappa=1}^{3} \left(c_\kappa^\dagger c_\kappa + \tfrac{1}{2}\right)\hbar\omega_\kappa \tag{4-129}$$

▲ where $c_\kappa^\dagger$ and c_κ create and destroy a quantum in the direction κ. The

---

[5]These features were obtained by Elliott (1958) on the basis of SU_3 classification of particle motion in a harmonic oscillator potential.

▼ eigenvalues of the Hamiltonian (4-129) are given by Eq. (4-112), in terms of the quantum numbers n_κ. (The relationship between the variables $(c_\kappa^\dagger, c_\kappa)$ and the coordinates and momenta of a particle is considered in the example in Chapter 5, pp. 231 ff.)

We shall ignore the spin-orbit coupling, and thus the Coriolis term in the Hamiltonian is proportional to the orbital angular momentum of the particle. For a rotation about the 1 axis, we have (see Eqs. (5-23) and (5-26))

$$H_{\text{Cor}} = -\hbar\omega_{\text{rot}} l_1 = -\omega_{\text{rot}}(x_2 p_3 - x_3 p_2)$$

$$= -\hbar\omega_{\text{rot}}\left(\frac{(\omega_2+\omega_3)}{2(\omega_2\omega_3)^{1/2}}(c_3^\dagger c_2 + c_2^\dagger c_3) - \frac{(\omega_2-\omega_3)}{2(\omega_2\omega_3)^{1/2}}(c_3^\dagger c_2^\dagger + c_2 c_3)\right) \quad (4\text{-}130)$$

The first term in Eq. (4-130) involves shifts of quanta from the 3 direction to the 2 direction and *vice versa* ($\Delta N = 0$), while the second term excites or annihilates two quanta ($\Delta N = 2$). The latter effects involve the large energy denominator $\hbar(\omega_2+\omega_3)$, and the perturbations in the wave function produced by these terms are therefore smaller than the effects of the $\Delta N = 0$ terms, by a factor $(\omega_2-\omega_3)^2/(\omega_2+\omega_3)^2$, which is of the order of the square of the deformation parameter. In the present treatment of the rotational perturbations, we shall consistently ignore effects of this order. (For a solution with inclusion of the $\Delta N = 2$ terms, see Valatin, 1956.)

The Hamiltonian with the inclusion of the $\Delta N = 0$ part of the Coriolis coupling can be diagonalized by a linear transformation

$$\begin{aligned} c_2 &= a c_\alpha + b c_\beta \\ c_3 &= -b c_\alpha + a c_\beta \end{aligned} \qquad (a^2 + b^2 = 1) \quad (4\text{-}131)$$

to new oscillator variables, labeled α and β. With the (real) coefficients a and b satisfying the relations

$$2ab = p(1+p^2)^{-1/2} \qquad a^2 - b^2 = (1+p^2)^{-1/2}$$

$$p \equiv \frac{2\omega_{\text{rot}}}{\omega_2 - \omega_3} \quad (4\text{-}132)$$

we obtain

$$H = H_0 - \hbar\omega_{\text{rot}} l_1$$

$$= (c_1^\dagger c_1 + \tfrac{1}{2})\hbar\omega_1 + (c_\alpha^\dagger c_\alpha + \tfrac{1}{2})\hbar\omega_\alpha + (c_\beta^\dagger c_\beta + \tfrac{1}{2})\hbar\omega_\beta \quad (4\text{-}133)$$

and

$$\omega_\alpha = \tfrac{1}{2}(\omega_2+\omega_3) + \tfrac{1}{2}(\omega_2-\omega_3)(1+p^2)^{+1/2}$$

$$\omega_\beta = \tfrac{1}{2}(\omega_2+\omega_3) - \tfrac{1}{2}(\omega_2-\omega_3)(1+p^2)^{+1/2} \quad (4\text{-}134)$$

▲

▼ The eigenmodes of oscillation in the rotating deformed potential have a nonvanishing value of the orbital angular momentum along the axis of rotation,

$$\langle l_1 \rangle = 2ab(n_\beta - n_\alpha) \tag{4-135}$$

In the limit $\omega_{rot} \gg \omega_2 - \omega_3$ (assuming $\omega_2 > \omega_3$), the quanta α and β become eigenstates of the angular momentum in the 1 direction, with eigenvalues $-\hbar$ and $+\hbar$, respectively. In this limit, the frequencies ω_α and ω_β approach the values $\frac{1}{2}(\omega_2 + \omega_3) \pm \omega_{rot}$.

If the rotational frequency is gradually increased, the number of quanta in the corresponding modes remains constant (adiabatic condition). Thus, a particle initially in the state n_1, n_2, n_3 (for $\omega_{rot} = 0$) occupies the state n_1, $n_\alpha = n_2$, $n_\beta = n_3$ in the rotating potential. For a system of particles initially in a state specified by the quantities

$$\Sigma_\kappa = \sum_{k=1}^{A} (n_\kappa + \tfrac{1}{2})_k \tag{4-136}$$

the total angular momentum in the rotating potential has the expectation value

$$\begin{aligned} \langle I_1 \rangle &= \langle \sum_k (l_1)_k \rangle = 2ab(\Sigma_3 - \Sigma_2) \\ &= 2abI_{max} \end{aligned} \tag{4-137}$$

where

$$I_{max} = \Sigma_3 - \Sigma_2 \tag{4-138}$$

is the maximum value of I_1, obtained in the limit $\omega_{rot} \gg \omega_2 - \omega_3$.

The perturbations produced by the Coriolis interaction imply that the shape of the system changes with increasing rotational frequency. In fact, the shift of quanta from the 3 to the 2 direction leads to a decrease in the deformation with respect to the 1 axis and to a corresponding modification in the self-consistent potential. The density distribution of the perturbed states retains the symmetry with respect to the inversions $\mathscr{S}_2$ and $\mathscr{S}_3$ of the 2 and 3 axes, characteristic of the initial state with specified n_2 and n_3; in fact, the Coriolis coupling, though violating $\mathscr{S}_2$ and $\mathscr{S}_3$ as well as time reversal $\mathscr{T}$, is invariant under the transformations $\mathscr{S}_2\mathscr{T}$ and $\mathscr{S}_3\mathscr{T}$, which ensure the inversion symmetries $\mathscr{S}_2$ and $\mathscr{S}_3$ of the density $\rho(\mathbf{r})$. For the static system, the consistency of density and potential is expressed by the relations (4-115), and for the rotating potential, the corresponding condition yields

$$\omega_1 \Sigma_1 = \omega_2(a^2\Sigma_2 + b^2\Sigma_3) = \omega_3(b^2\Sigma_2 + a^2\Sigma_3) \tag{4-139}$$ ▲

▼ and, hence (see Eqs. (4-132), (4-137), and (4-138)),

$$\frac{\omega_2-\omega_3}{\omega_2+\omega_3}=(1+p^2)^{-1/2}\frac{\Sigma_3-\Sigma_2}{\Sigma_3+\Sigma_2}$$

$$=\left(1-\frac{\langle I_1\rangle^2}{I_{\max}^2}\right)^{1/2}\frac{\Sigma_3-\Sigma_2}{\Sigma_3+\Sigma_2} \qquad \text{(4-140a)}$$

$$(\omega_2-\omega_3)^2=(\omega_2-\omega_3)^2_{\omega_{\text{rot}}=0}-4\omega_{\text{rot}}^2 \qquad \text{(4-140b)}$$

These relations exhibit the gradual disappearance of the deformation with respect to the 1 axis, as the rotational frequency approaches the value corresponding to $\langle I_1\rangle=I_{\max}$.

For the self-consistent potential, the moment of inertia, representing the ratio between the angular momentum (4-137) and the rotational frequency, has the constant value (see Eqs. (4-132), (4-140), and (4-113)),

$$\mathscr{I}=\frac{\hbar\langle I_1\rangle}{\omega_{\text{rot}}}=2\hbar\frac{\Sigma_2+\Sigma_3}{\omega_2+\omega_3}$$

$$\approx\hbar\left(\frac{\Sigma_2}{\omega_2}+\frac{\Sigma_3}{\omega_3}\right)=\mathscr{I}_{\text{rig}} \qquad \text{(4-141)}$$

For a configuration with axial symmetry ($\Sigma_1=\Sigma_2$), the rotational motion is constrained to axes perpendicular to the symmetry axis, and the rotational energy is proportional to $I(I+1)$ all the way to the limit $I_{\max}$, at which the band abruptly terminates. For configurations with triaxial shape, one must consider rotations about all three principal axes; however, since the moments of inertia $\mathscr{I}_\kappa$ are equal, apart from terms involving the deformation, the rotational energy remains proportional to $I(I+1)$, to this accuracy.

Estimates of $I_{\max}$ for the ground-state configurations of several representative nuclei are given in Table 4-3. The estimates are obtained by ▲

Nucleus	$I_{\max}$
^{8}Be	4
^{20}Ne	8
^{164}Er	≈100
^{238}U	≈140

Table 4-3 Maximum values of the rotational angular momentum corresponding to alignment of the angular momenta of the individual particles.

assigning the ground-state configurations implied by the level order in Figs. 5-1 to 5-5 for prolate deformations with values of δ given in Fig. 4-25, p. 133. The values of Σ_κ are calculated from the asymptotic quantum numbers of the occupied orbits, and I_{max} is obtained from Eq. (4-138). The values of I_{max} in the middle of shells are of order A, since the difference between Σ_3 and Σ_2 is contributed by the particles outside of closed shells; there are $\sim A^{2/3}$ such particles, each carrying $\sim A^{1/3}$ quanta (see, for example, Eqs. (2-151) and (2-158)).

In the case of ^{8}Be and ^{20}Ne, the states of the ground-state band up to $I=I_{max}$ have been observed, but the evidence is inconclusive concerning the structure of the spectrum in the channels with $I>I_{max}$ (see the discussion on pp. 101 ff. and 97 ff.). In heavier nuclei, the estimated values of I_{max} are much larger than the domain of I values so far studied. It must be emphasized that, for the very high angular momenta, it may be important to consider the centrifugal distortion effects that result from the dependence of the moment of inertia on the deformation. Such effects imply an increase in the deformation as a function of the total angular momentum and may eventually lead to fission. (For the liquid-drop model, these effects are discussed in Sec. 6A-2b.)

E2-matrix elements with inclusion of Coriolis distortion

The shape variations produced by the rotational motion imply corrections to the $E2$-matrix elements within the band. (We shall assume that the electric moments are proportional to those of the mass distribution, as for $T=0$ states.)

If the initial equilibrium shape, in the absence of rotation, is prolate about the 3 axis ($\Sigma_1=\Sigma_2<\Sigma_3$; $\omega_1=\omega_2>\omega_3$), the final shape, for $I=I_{max}$, is oblate ($\omega_1>\omega_2=\omega_3$) and is reached through a succession of triaxial shapes ($\omega_1>\omega_2>\omega_3$). The static quadrupole moment, which is given by the diagonal matrix element of $\mathcal{M}(E2,\mu=0)$ in the state $M=I$, involves the density distribution averaged with respect to the axis of rotation. Since the expectation value of $x_2^2+x_3^2$ is not affected by the rotational perturbations, in the approximation considered, there are no corrections to the static moments. The transition moments, however, are associated with the time-dependent part of the density distribution resulting from the rotational motion and are proportional to (see Eqs. (4-112), (4-132), and (4-140))

$$\langle \sum_k (x_2^2-x_3^2)_k \rangle = \frac{\hbar}{M}\left(\frac{a^2\Sigma_2+b^2\Sigma_3}{\omega_2} - \frac{b^2\Sigma_2+a^2\Sigma_3}{\omega_3} \right)$$

$$= \left(1-\frac{\langle I_1\rangle^2}{I_{max}^2}\right)^{1/2} \langle \sum_k (x_2^2-x_3^2)_k \rangle_{\omega_{rot}=0} \tag{4-142}$$

▼ When I approaches I_{max}, the transition probability vanishes, since a rotation about a symmetry axis does not produce a time-dependent field.

The above discussion of the $E2$ moments is based on a semiclassical approximation, valid for $I \gg 1$. The quantal form for the $E2$-matrix elements in a $K=0$ band can be expressed in terms of an intrinsic moment that is a function of the combinations $I_1(I_1+1)+I_2(I_2+1)$ and $(I_1(I_1+1)-I_2(I_2+1))^2$ (see p. 53). Since, in the present model, the static moments are not affected by the rotational perturbations, the intrinsic moment depends only on the combination $(I_1(I_1+1)-I_2(I_2+1))^2$; hence the quantity $\langle I_{\kappa=1}\rangle^2$ in Eq. (4-142) is to be replaced by a multiple of this combination. The derivation of the proportionality factor I_{max}^{-2} is valid only to leading order in I_{max}. Since the matrix element vanishes identically for the transition $I=I_{max}\rightarrow I+2$, one is led to the following expression for the I dependence of the matrix element

$$\langle K=0,I_2\| \mathscr{M}(E2)\|K=0,I_1\rangle$$
$$=(2I_1+1)^{1/2}\langle I_1 0 2 0|I_2 0\rangle M_1\left(1-\left(\frac{I_1(I_1+1)-I_2(I_2+1)}{2(2I_{max}+3)}\right)^2\right)^{1/2} \tag{4-143}$$

The parameter M_1 in Eq. (4-143) may be obtained by considering the static moment in the fully aligned state with $M=I=I_{max}$, which can be described, in the fixed coordinate system, in terms of a distribution of quanta, with $\Sigma_x=\Sigma_y=\Sigma_z+\frac{1}{2}I_{max}$. We therefore have (see Eqs. (4-113), (4-115), and (4-138))

$$\begin{aligned} Q(I=I_{max}) &= \frac{Z}{A}\langle I=I_{max}, M=I|\sum_k (2z^2-x^2-y^2)_k|I=I_{max}, M=I\rangle \\ &= \frac{Z}{A}\left(2\frac{\hbar}{M\omega_z}\Sigma_z-\frac{\hbar}{M\omega_x}\Sigma_x-\frac{\hbar}{M\omega_y}\Sigma_y\right) \\ &\approx -2\frac{Z}{A}\frac{\hbar}{M\omega_0}I_{max} \end{aligned} \tag{4-144}$$

where ω_0 is the mean oscillator frequency. Hence, one finds (see Eqs. (4-68) and (4-69))

$$M_1=\left(\frac{5}{16\pi}\right)^{1/2} 2\frac{Ze}{A}\frac{\hbar}{M\omega_0}(2I_{max}+3) \tag{4-145}$$

To leading order in I_{max}, the quantity M_1 coincides with the intrinsic moment for the particle motion in the static potential (see Eqs. (4-113) and (4-138)). Since I_{max} is of order A (see p. 88), the constant term in Eq. (4-145) is of higher order than the correction terms of order δ^2 that have been consistently neglected in the above discussion. ▲

▼ The quadrupole matrix elements derived above are seen to obtain half their value from the anisotropy of the configuration (differences in Σ_κ) and half the value from the anisotropy of the potential (differences in ω_κ); see, for example, Eqs. (4-115) and (4-144). The part associated with the deformation of the potential can also be described as a small admixture of components with $\Delta N=2$ into the wave functions of the spherical potential. As discussed in Chapter 6, such admixtures can be taken into account in terms of a renormalization of the effective operators for the particles outside of closed shells (the doubling of the effective quadrupole moment corresponds to the static polarizability, $\chi(\tau=0,\lambda=2)=1$, obtained from Eq. (6-370)).

Rotating states expressed as angular momentum projection of intrinsic states

In the harmonic oscillator model, the perturbations of the particle motion produced by the Coriolis interaction have an especially simple character due to the fact that the excitations all have the same frequency, if we neglect the $\Delta N=2$ terms. Thus, for rotation about the 1 axis, the $\Delta N=0$ excitations have the single frequency $\omega_2-\omega_3$, and the perturbation of the wave function produced by the Coriolis interaction can therefore be obtained by operating with the component I_1 of the total angular momentum (and powers of this operator). Since the components I_κ are the generators of infinitesimal rotations, the rotating state can be expressed as a superposition of states corresponding to the intrinsic configuration oriented in different directions.

The weighting of the different orientations ω can be obtained by exploiting the rotational invariance of the total Hamiltonian. Thus, for an axially symmetric configuration with $I_3=K$, the intrinsic state $|K;\omega\rangle$, with orientation ω, can be expanded in components with different IM (see Eqs. (1A-94) and (1A-39))

$$\begin{aligned}|K;\omega\rangle &= \sum_I c_I\left(\frac{2I+1}{8\pi^2}\right)^{1/2}|KI,M'=K\rangle_{\mathscr{K}'}\\ &= \sum_{IM} c_I\left(\frac{2I+1}{8\pi^2}\right)^{1/2}(\mathscr{D}^I_{MK}(\omega))^*|KIM\rangle_{\mathscr{K}}\end{aligned}\tag{4-146}$$

where $\mathscr{K}'$ is the coordinate system with orientation ω with respect to the fixed frame $\mathscr{K}$, and where c_I are suitably normalized expansion coefficients, which are real, assuming the states $|K;\omega\rangle$ as well as $|KIM'\rangle_{\mathscr{K}'}$ to have the standard phasing. Inverting the relation (4-146), we obtain (omitting the label $\mathscr{K}$ for the state vector)

$$|KIM\rangle = c_I^{-1}\left(\frac{2I+1}{8\pi^2}\right)^{1/2}\int d\omega\,\mathscr{D}^I_{MK}(\omega)|K;\omega\rangle \tag{4-147}$$

▲

▼ Thus, the eigenstates $|KIM\rangle$ are expressed as the angular momentum projection from the aligned intrinsic state, and the relation (4-146) exhibits the intrinsic state as a wave packet constructed by superposition of the different members of the rotational band.[6]

The expansion coefficients c_I can be obtained from the normalization integral

$$c_I^2 = \frac{2I+1}{8\pi^2}\int d\omega_2 \int d\omega_1 (\mathscr{D}^I_{MK}(\omega_2))^* \mathscr{D}^I_{MK}(\omega_1)\langle K;\, \omega_2 | K;\, \omega_1\rangle$$

$$= \int d\omega (\mathscr{D}^I_{KK}(\omega))^* \langle K;\, \omega | K;\, \omega = 0\rangle \tag{4-148}$$

The last relation follows from the fact that the overlap factor $\langle K;\, \omega_2 | K;\, \omega_1\rangle$ depends only on the relative orientation ω of ω_2 with respect to ω_1 (see also Eq. (1A-45)).

We illustrate the evaluation of the projection integrals by considering configurations obtained by filling the particles in the lowest available orbits in an axially symmetric (and prolate) potential; in such a state, the quanta are aligned along the 3 axis to the maximal possible extent consistent with the exclusion principle. We shall also assume the aligned state to involve filled subshells in the oscillator potential and thus to have $K=0$; for other configurations, the self-consistent deformation deviates from axial symmetry. The aligned state may be described in terms of quanta belonging to an isotropic oscillator potential, since the effect of anisotropy of the potential can be included in terms of a renormalization of the effective moments, as discussed on p. 90.

In the aligned $K=0$ state, the quanta in the 1 and 2 directions, together with a corresponding number of quanta in the 3 direction, form a spherically symmetric state, and the anisotropy of the aligned state is determined by the $\Sigma_3 - \Sigma_2$ extra quanta in the 3 direction. (A formal proof of this equivalence can be based on the transformation properties of the states under the SU_3 symmetry, as discussed in the small print on p. 96.) For a single quantum, the overlap factor $\langle K=0; \omega | K=0; \omega=0\rangle$ is equal to $\cos\theta\,(=\mathscr{D}^1_{00}(\omega))$; for the aligned state, we therefore obtain

$$\langle K=0;\, \omega=\phi\theta\psi | K=0;\, \omega=0\rangle = \cos^\lambda\theta \tag{4-149}$$

$$\lambda \equiv \Sigma_3 - \Sigma_2$$

▲

---

[6]The use of wave functions of the type (4-147) to describe collective motion in terms of the nucleonic coordinates was first suggested by Hill and Wheeler (1953). The rotational properties implied by such wave functions were further investigated by Peierls and Yoccoz (1957), Yoccoz (1957), Villars (1957), and Elliott (1958). Projected wave functions that restore the full symmetry of the Hamiltonian have also been employed in the study of atoms and molecules by means of the "unrestricted" Hartree-Fock approximation (see, for example, Löwdin (1966) and references quoted there).

▼ An evaluation of the normalization integral (4-148) thus yields for the coefficients c_I (see, for example, Abramowitz and Stegum, 1964, p. 338),

$$c_I = \left(\frac{8\pi^2 \lambda!}{(\lambda - I)!!(\lambda + I + 1)!!} \right)^{1/2} \tag{4-150}$$

and it is seen that the rotational band contains the members

$$I = \begin{cases} 0, 2, \ldots, \lambda & \lambda \text{ even} \\ 1, 3, \ldots, \lambda & \lambda \text{ odd} \end{cases} \tag{4-151}$$

corresponding to a $K=0$ sequence with the $\mathscr{R}$-symmetry quantum number $r=(-1)^\lambda$ (each quantum in the 3 direction transforms with a phase factor -1 under the rotation $\mathscr{R} = \mathscr{R}_2(\pi)$). The band terminates at the value of $I_{\max}$ derived above (see Eq. (4-138)).

In the limit of $\lambda \gg 1$, the overlap factor (4-149) approaches a δ function

$$\langle K=0; \omega_2 | K=0; \omega_1 \rangle \underset{\lambda \gg 1}{\approx} \frac{4\pi^2}{\lambda} \left(\delta(\omega_2 - \omega_1) + (-1)^\lambda \delta(\omega_2 - \mathscr{R}\omega_1) \right) \tag{4-152}$$

where $\mathscr{R}\omega_1$ is the orientation obtained by rotating the coordinate system with orientation ω_1 by 180° about its 2 axis. In this limit, the values of c_I become independent of I, for $I \ll \lambda$;

$$c_I \underset{I \ll \lambda}{\approx} \left(\frac{8\pi^2}{\lambda} \right)^{1/2} \tag{4-153}$$

In the above analysis, the rotational degree of freedom is expressed in terms of the superposition of degenerate intrinsic excitations produced by the operators I_κ. These particular combinations of excitations are therefore to be regarded as spurious in the sense that they are not to be included as separate degrees of freedom in the intrinsic spectrum. (The removal of the spurious excitations is illustrated by the example (A = 19) considered on pp. 288 ff.) In the more general situation, where the excitations produced by the rotation operator have nondegenerate frequencies, the problem of isolating the spurious intrinsic excitations is part of the dynamical problem of constructing the rotating state in terms of the particle degrees of freedom (see the discussion in Chapter 6, pp. 446 ff.).

The representation (4-147) for the state of the rotating nucleus was motivated above on the basis of the special properties of the harmonic oscillator model, but more generally one may consider the projected state (4-147) as the zeroth approximation to the wave function expressed in terms of the coordinates of the individual particles.

It can be simply verified that the evaluation of matrix elements for the projected wave function (4-147) gives the leading-order rotational intensity relations (I-independent intrinsic moments; see Sec. 4-3), if one neglects the zero-point fluctuations in the orientation of the aligned state, corresponding to the δ-function approxi- ▲

▼ mation (4-152) for the overlap factor. The finite zero-point fluctuations imply corrections to the leading-order intensity relations, as in the quadrupole matrix elements considered below; for not too large values of I, these higher-order terms can be expressed as a power series in the rotational angular momentum I of the form derived in Sec. 4-3 on the basis of symmetry considerations.

However, in the analysis of the I-dependent terms, it must be recognized that, in general, there will be additional contributions arising from the effects of the rotation on the intrinsic motion. Only in the special case in which the excitations produced by the Coriolis interaction all have the same frequency, are the rotational perturbations systematically included in the wave functions projected from I-independent intrinsic states. (The necessity of including I-dependent terms in the intrinsic wave function in order to describe the higher-order rotational properties, such as the moment of inertia, has been discussed from somewhat different points of view by Bohr and Mottelson, 1958, and by Peierls and Thouless, 1962.)

Quadrupole operators as generators of rotational band (SU_3 symmetry)

The structure of the projected states (4-147) can be characterized by the matrix elements of the quadrupole operators. In the analysis of matrix elements between projected states, it is convenient to employ the set of operators

$$E(\kappa',\kappa) \equiv \sum_{k=1}^{A} (c_{\kappa'}^{\dagger} c_{\kappa})_k \tag{4-154a}$$

$$[E(\kappa_b,\kappa_a), E(\kappa_d,\kappa_c)] = \delta(a,d)E(\kappa_b,\kappa_c) - \delta(b,c)E(\kappa_d,\kappa_a) \tag{4-154b}$$

that shift an oscillator quantum from one to another of the Cartesian axes, $\kappa = 1,2,3$. The commutation relations (4-154b) for the shift operators follow directly from those of the boson operators (see Eq. (6-2)). In terms of these operators, the isoscalar quadrupole moments, referred to the same set of axes, are given by (see Eq. (5-24))

$$\mathscr{M}(2,\nu)_{\Delta N=0} = \left(\frac{5}{16\pi}\right)^{1/2} \frac{\hbar}{M\omega_0} \times \begin{cases} 2E(3,3) - E(1,1) - E(2,2) & \nu = 0 \\ (\frac{3}{2})^{1/2}(E(2,3) - E(3,2) \mp (E(3,1) + E(1,3))) & \nu = \pm 1 \\ (\frac{3}{2})^{1/2}(E(1,1) - E(2,2) \mp (E(2,1) - E(1,2))) & \nu = \pm 2 \end{cases} \tag{4-155}$$

where the $\Delta N = 2$ terms have been neglected, as indicated by the subscript $\Delta N = 0$. The moment (4-155) is expressed in terms of quanta referring to particle motion in a spherical oscillator potential with frequency ω_0. The ▲ angular momentum operators I_κ can also be expressed in terms of the shift

▼ operators (see Eq. (5-26)),

$$I_\kappa = \begin{cases} E(2,3)+E(3,2) & \kappa=1 \\ i(E(1,3)-E(3,1)) & \kappa=2 \\ -(E(1,2)+E(2,1)) & \kappa=3 \end{cases} \tag{4-156}$$

The axially symmetric states of maximum alignment considered in the previous section are characterized by the fact that they are annihilated by the shift operators $E(3,1)$ and $E(3,2)$, as well as by $E(1,2)$ and $E(2,1)$; moreover, they are eigenstates of the operators $E(\kappa,\kappa)$. Thus, the quadrupole moment (4-155) acting on such a state yields

$$\mathscr{M}(2,\nu)_{\Delta N=0}|K=0;\omega\rangle = \left(\frac{5}{16\pi}\right)^{1/2}\frac{\hbar}{M\omega_0}\begin{cases} 2\lambda\,|K=0;\omega\rangle & \nu=0 \\ (\tfrac{3}{2})^{1/2}I_\pm\,|K=0;\omega\rangle & \nu=\pm 1 \\ 0 & \nu=\pm 2 \end{cases} \tag{4-157}$$

The moment $\mathscr{M}(2,\mu)$ acting on the projected state (4-147) is obtained by the standard transformation to the intrinsic coordinate system (see Eq. (4-67))

$$\mathscr{M}(2,\mu)_{\Delta N=0}|K=0,IM\rangle = c_I^{-1}\left(\frac{2I+1}{8\pi^2}\right)^{1/2}\left(\frac{5}{16\pi}\right)^{1/2}\frac{\hbar}{M\omega_0}$$

$$\times\int d\omega\,\mathscr{D}^I_{M0}(\omega)\big(2\lambda\mathscr{D}^2_{\mu0}(\omega)+(\tfrac{3}{2})^{1/2}(\mathscr{D}^2_{\mu1}I_+ + \mathscr{D}^2_{\mu-1}I_-)\big)|K=0;\omega\rangle \tag{4-158}$$

The operators $I_\pm$ that rotate the intrinsic state can also be viewed as operators that generate the inverse rotations of the orientation angles ω. Thus, the integral (4-158) represents a superposition of states in the rotational band based on the intrinsic state $|K=0\rangle$, and the reduced matrix elements of the quadrupole moment are found to be (see Eq. (1A-91) and (4A-34))

$$\langle K=0,I_2\|\,\mathscr{M}(2)\|K=0,I_1\rangle$$

$$=\left(\frac{5}{16\pi}\right)^{1/2}\frac{\hbar}{M\omega_0}(2I_1+1)^{1/2}\langle I_1 0 2 0|I_2 0\rangle\frac{c_{I_2}}{c_{I_1}}\big(2\lambda+3+\tfrac{1}{2}(I_2(I_2+1)-I_1(I_1+1))\big)$$

$$=\left(\frac{5}{16\pi}\right)^{1/2}\frac{\hbar}{M\omega_0}(2\lambda+3)(2I_1+1)^{1/2}\langle I_1 0 2 0|I_2 0\rangle$$

$$\times\begin{cases} 1 & I_2=I_1 \\ \left(1-\left(\dfrac{2I_1+3}{2\lambda+3}\right)^2\right)^{1/2} & I_2=I_1+2 \end{cases} \tag{4-159}$$

▲

▼ In the last form, we have employed the relation (4-150) for the coefficients c_I. The result (4-159) corresponds to Eqs. (4-143) and (4-145), but is smaller by a factor 2 reflecting the fact that the wave functions in the matrix element (4-159) refer to a spherical oscillator; the effect of the self-consistent deformation of the oscillator potential can be included as a renormalization of the effective quadrupole moment by a factor 2 (see the discussion on p. 90).

The fact that the quadrupole operators induce transitions only within a single band exhibits these operators as generators of the rotational motion. As a consequence, the states $|K=0,IM\rangle$ are eigenstates of the rotational invariants that can be constructed as products of the quadrupole operators $\mathscr{M}(2,\mu)_{\Delta N=0}$. In particular, the quadratic invariant

$$H' = \tfrac{1}{2}\kappa \sum_{\mu} \mathscr{M}^{\dagger}(2,\mu)_{\Delta N=0}\, \mathscr{M}(2,\mu)_{\Delta N=0} \tag{4-160}$$

can be viewed as an effective two-body force that produces the superposition of configurations corresponding to rotational motion, when acting between particles moving in a spherically symmetric harmonic oscillator potential. As discussed in Chapter 6 (see, for example, pp. 340ff.), the interaction (4-160) can also be viewed as the effect on the motion of each particle of the deformed field produced by the total nuclear quadrupole moment. The value of the coupling constant, κ, that is required to give a deformed field equal to the self-consistent value (4-115) is given by Eq. (6-78), aside from a renormalization factor of 2 that results from the neglect of the $\Delta N=2$ terms in the interaction (4-160); see p. 521 in Chapter 6.

The eigenvalue of H' in the states $|K=0,IM\rangle$ can be obtained by noting the connection between this interaction and the quadratic expression formed from the shift operators (see Eqs. (4-155) and (4-156))

$$\sum_{\kappa,\kappa'} E(\kappa',\kappa)E(\kappa,\kappa') - \tfrac{1}{3}\Big(\sum_{\kappa} E(\kappa,\kappa)\Big)^2$$

$$= \frac{8\pi}{15}\left(\frac{M\omega_0}{\hbar}\right)^2 \sum_{\mu} \mathscr{M}^{\dagger}(2,\mu)_{\Delta N=0}\, \mathscr{M}(2,\mu)_{\Delta N=0} + \tfrac{1}{2}\mathbf{I}^2 \tag{4-161}$$

This expression commutes with all the shift operators (see Eq. (4-154b)) and is therefore a constant for a given band. Hence, apart from a constant, the eigenvalue of H' is proportional to $I(I+1)$, and the coefficient corresponds to the rigid moment of inertia, if the coupling constant κ is chosen to have the self-consistent value referred to above. This result is equivalent to that obtained (see Eq. (4-141)) by treating the interactions of the particles in terms of the rotating deformed field.

The many simple relationships derived above for the harmonic oscillator model can be recognized as consequences of an underlying symmetry of this model. In fact,

▲ the shift operators obey commutation relations (4-154b) that characterize the genera-

▼ tors of unitary transformations in three dimensions, U_3 (for an elementary discussion of unitary symmetry, see Sec. 1C-3). The generators that are orthogonal to the total oscillator quantum number, $E(1,1)+E(2,2)+E(3,3)$, define the group SU_3, and the rotational bands obtained above carry irreducible representations of this group. The invariant (4-161) is recognized as the Casimir operator $G_2(SU_3)$; see Eq. (1C-50).

The representations of U_3 are labeled by the quantum numbers $[f_1 f_2 f_3]$, which at the same time characterize the permutation symmetry of the quanta. Under SU_3, the corresponding representations are characterized by the quantum numbers $\lambda = f_1 - f_2$ and $\mu = f_2 - f_3$. The states belonging to the representation $[f_1 f_2 f_3]$ have the property that, when acted on by the shift operators, it is possible to align f_1 quanta (and no more) in any given direction (such as along the 3 axis) and subsequently f_2 quanta (but no more) in one of the other directions. Thus, the axially symmetric aligned state $|K=0;\omega\rangle$ considered above belongs to the representation $(\lambda,\mu=0)$. The representations with $\mu\neq 0$ (and $\lambda\neq 0$) correspond to configurations without axial symmetry, and the associated band structure may contain several states with the same value of I, as for the spectrum of the asymmetric rotor. (The angular momentum values contained in the representation $(\lambda\mu)$ are given in Eqs. (1C-59) and (1C-60).)

The coefficients c_I describe the transformation from a basis in which the operators $E(\kappa,\kappa)$ are diagonal to a basis employing the quantum numbers IM. These coefficients are determined by the group structure and are thus the same for different configurations that have the same symmetry $(\lambda\mu)$.

Rotational bands in ^{20}Ne (Fig. 4-13 and Table 4-4)

In the region beyond $A=16$ and extending up to $A\approx 28$, rotational band structure has been recognized as a systematic feature in the nuclear spectra. As an example, Fig. 4-13 shows sequences of states in the low-energy spectrum of ^{20}Ne. The study of the reactions ^{16}O$(\alpha\alpha)$ and ^{12}C$(^{12}$C $\alpha)^{20}$Ne has played an especially important role in establishing the high angular momentum states in this spectrum. (Other examples of rotational spectra for nuclei in this region are discussed in Chapter 5, pp. 283ff.)

The levels shown in Fig. 4-13 can be arranged into three rotational bands. In addition to the states shown in this figure, a number of even-parity states, beginning at about 6.7 MeV, have been observed and tentatively classified in terms of rotational bands (see, for example, Kuehner and Almqvist, 1967; Nagatani *et al.*, 1971). In these light nuclei, the rotational energy expansion does not have the high quantitative accuracy characteristic of the bands in heavier nuclei, but the first-order term, proportional to $I(I+1)$, does describe the main trend in the energies. The energies in the ground-state band exhibit a small oscillatory component lowering the 4+ and 8+ states relative to the 2+ and 6+ members; the significance of this energy alternation in terms of the collective description is not yet apparent.

Further evidence for the rotational interpretation is obtained from the measured γ-transition intensities (Smulders *et al.*, 1967). Thus, the decay of the 4− level at 7.02 MeV to the 3− and 2− states at 5.63 MeV and 4.97 MeV, respectively, is found to involve strongly enhanced $E2$ transitions of ▲ strength $B(E2;4-\rightarrow 3-)=26B_W(E2)$ and $B(E2;4-\rightarrow 2-)=11B_W(E2)$

▼ (with $B_W(E2)=3.2e^2\,\text{fm}^4$; see Eq. (3C-38)). The ratio of the transition probabilities (2.4 ± 0.2) is in good agreement with the intensity relation for $E2$ transitions in a $K=2$ band $(B(E2;I=4\rightarrow I=3):B(E2;I=4\rightarrow I=2) = \langle 4220|32\rangle^2:\langle 4220|22\rangle^2=2.24$; see Eq. (4-68)). The absolute transition strength yields the intrinsic quadrupole moment $Q_0=56\,\text{fm}^2$, which equals that of the ground-state band within the experimental uncertainties (see Table
▲ 4-4).

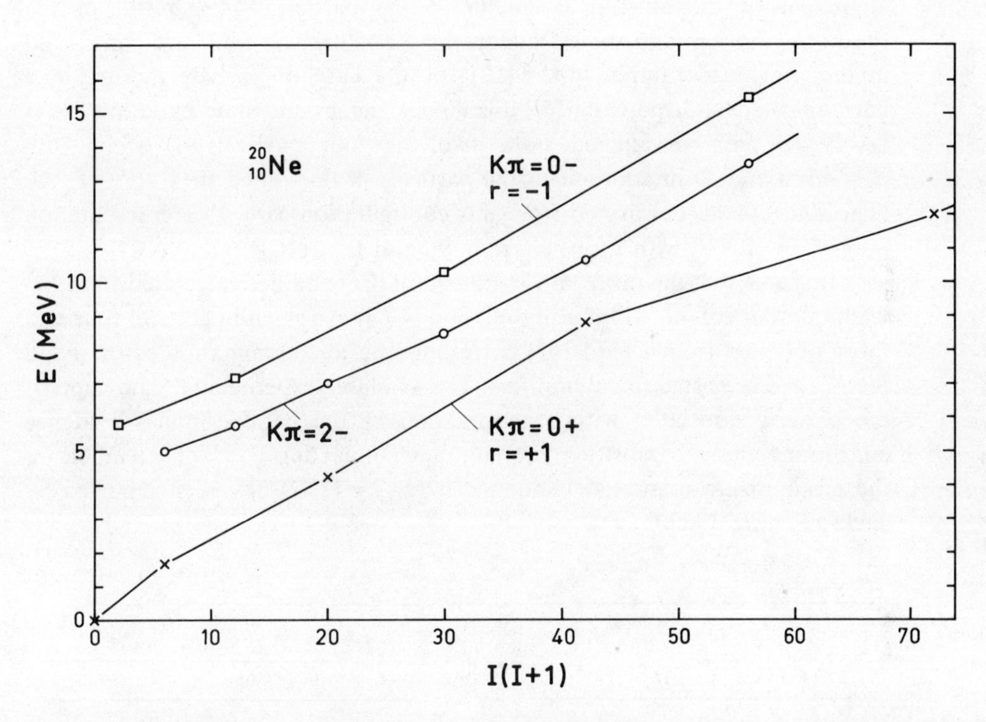

Figure 4-13 Energy levels of ^{20}Ne. The data are taken from J. A. Kuehner and E. Almqvist, *Can. J. Phys.* **45**, 1605 (1967) and from W. E. Hunt, M. K. Mehta, and R. H. Davis, *Phys. Rev.* **160**, 782 (1967).

▼ The ground-state rotational band in ^{20}Ne has also been discussed in terms of a configuration involving four particles in s or d orbits outside a closed-shell core corresponding to ^{16}O; indeed, the comparison between the shell-model description and the rotational interpretation of the spectra in this region was an important step in the development of a microscopic theory of nuclear rotations (Paul, 1957; Elliott, 1958). The shell-model configuration $(\text{sd})^4$ gives a maximum total angular momentum I of 8 units, but there is so far no evidence as to whether the band in ^{20}Ne breaks off at this value of I, or
▲ whether it continues to still higher angular momenta.

▼ An approach to the endpoint of the band is expected to manifest itself in large rotational perturbations of the nuclear shape resulting in major deviations from the leading-order intensity relation for $E2$ transitions within the band. In fact, the endpoint of the band corresponds to an alignment of the angular momenta of all the particles outside closed shells and is therefore associated with a transition to a coupling scheme corresponding to an oblate nucleus rotating about the new symmetry axis ($K = I_1 = I$). For such a rotational motion, the $E2$-transition probability vanishes. The gradual alignment of the angular momenta of the particles, as a result of the rotational motion, and the effects on the $E2$-matrix elements are considered in the previous example (pp. 84ff.) for the case of particle motion in a harmonic oscillator potential. In this model, the ground-state band in ^{20}Ne is associated with an aligned state involving four nucleons in orbits with $n_1 = n_2 = 0$, $n_3 = 2$, in addition to the particles in the closed-shell core of ^{16}O. (The states (4-147) projected from this configuration carry the representation $(\lambda\mu) = (80)$ of SU_3 with $I_{max} = 8$ (see p. 96 and Eqs. (1C-59) and (1C-60).)

In Table 4-4, the ratios of $E2$-transition probabilities calculated from the leading-order rotational intensity relation (4-68) are given in column four and those obtained from Eq. (4-143) corresponding to a termination of the band for $I_{max} = 8$ are listed in column five. The available experimental data support the pattern associated with a termination of the band. (Similar evidence concerning the $E2$-transition probabilities connecting the high members of the ground-state rotational band in ^{19}F ($I_{max} = 13/2$) has been reported by ▲ Jackson *et al.*, 1969.)

Transition	$B(E2)$ in e^2fm^4	$B(E2)$: $B(E2; 2\to0)$ Experiment	Q_0 const.	$I_{max} = 8$
$2\to0$	57 ± 8	1	1	1
$4\to2$	70 ± 7	1.2 ± 0.15	1.43	1.26
$6\to4$	66 ± 8	1.15 ± 0.15	1.57	1.06
$8\to6$	24 ± 8	0.4 ± 0.15	1.65	0.64
$10\to8$			1.69	

Table 4-4 Rotational $E2$-transition probabilities in ^{20}Ne. The experimental data are taken from T. K. Alexander, O. Häusser, A. B. McDonald, A. J. Ferguson, W. T. Diamond, and A. E. Litherland, *Nuclear Phys.* **A179**, 477 (1972) and references quoted there. Evidence on the static quadrupole moment of the 2+ state has been obtained from Coulomb excitation measurements; the value $Q = -23 \pm 8$ fm^2 (D. Schwalm, A. Bamberger, P. G. Bizzeti, B. Povh, G. A. P. Engelbertink, J. W. Olness, and E. K. Warburton, *Nuclear Phys.* **A192**, 449, 1972) yields $Q_0 = 80 \pm 25$ fm^2, to be compared with the value $|Q_0| = 54 \pm 4$ fm^2 obtained from the transition probability $B(E2; 2\to0)$ in Table 4-4.

▼ The ground-state band of ^{20}Ne may also be viewed in terms of an α-particle cluster moving with respect to the ^{16}O core (Wildermuth and Kanellopoulos, 1958; for the analysis of light nuclei in terms of α clusters and the relationship to the shell-model interpretation, see the reviews by Neudatchin and Smirnow, 1969, and Ikeda *et al.*, 1972). Within the framework of the harmonic oscillator model, such a cluster representation is equivalent to the representation considered above, in terms of the intrinsic state with the maximal alignment of the orbits of the individual particles. This equivalence is an example of a general class of relations that follow from the invariance of the harmonic oscillator Hamiltonian with respect to orthogonal transformations among the particle coordinates.

A transformation of this type leads to a new set of independent harmonic oscillators all having the frequency ω_0, and the eigenstates can be labeled by the number of quanta in each mode. A familiar example is provided by the transformation to center-of-mass and relative coordinates (see footnote 19 on p. 474). In the cluster representations, the A particles are divided into subunits each containing $A_1, A_2, \ldots$ particles, and the coordinates in such a representation describe internal motion within each cluster, relative motion of the center-of-masses of the clusters, and center-of-mass motion for the total system.

The generators (4-154) of U_3 symmetry are unaffected by a transformation to cluster coordinates; in fact, such a transformation acts among the x_1, x_2, and x_3 coordinates separately and therefore has no effect on the distribution of the oscillator quanta in the three different spatial directions. Thus, cluster states with specified numbers of oscillation quanta associated with the A different particle degrees of freedom carry representations of U_3. The different representations of the motion in the harmonic oscillator model, in terms of clusters or independent particles, provide alternative sets of basis states for describing the (antisymmetrized) many-particle states with specified number N of oscillator quanta, SU_3 symmetry $(\lambda\mu)$, and orbital permutation symmetry $[f]$.

In the case of ^{20}Ne, the lowest configuration $(1s)^4(1p)^{12}(2s,1d)^4$ has the total number of oscillator quanta $N=20$ and gives rise to only a single set of states with the symmetry quantum numbers $(\lambda\mu)=(80)$ and $[f]=[44444]$. (This result follows immediately from the fact that the $(sd)^4$ configuration gives only a single state with $L=8$.) The states of the $(\lambda\mu)=(80)$ band, which were obtained above by the projection (4-147) from the aligned intrinsic state, can also be generated by antisymmetrization of wave functions describing an α particle-like cluster ($A_1=4, N_1=0$) and an ^{16}O cluster ($A_2=16, N_2=12$) in relative motion with 8 quanta. The two representations provide alternative ways of viewing the same total wave function.

In the cluster description of ^{20}Ne, the rotational perturbations discussed above correspond to the effect of the Coriolis force in changing the motion of the α particle from a pendulating orbit ($L=0$) with a prolate intrinsic density distribution to a circular orbit ($L=L_{max}=8$) with an oblate density distribution. The relation (4-159) for the quadrupole matrix elements is seen to be the same as for a single particle in oscillator states with $N=\lambda$; in fact, this relation is completely specified by the SU_3 quantum numbers.

Spectrum of ^{8}Be seen in $\alpha\alpha$ scattering (Fig. 4-14 and Table 4-5)

Due to the exceptionally large binding energy of ^{4}He, all the states of ^{8}Be are unstable to particle emission. The lowest states in ^{8}Be have been studied ▲ by means of $\alpha\alpha$ elastic scattering. The ground state ($I\pi=0+$) gives rise to a

▼ sharp resonance, but for the excited states ($I\pi = 2+$ and $4+$) the α widths are large and the resonance parameters are determined from a phase-shift analysis (see Fig. 4-14). The resonance parameters, given in Table 4-5, have been obtained by fitting the nuclear scattering amplitude to the standard resonance expression (Eqs. (3F-10) and (3F-12)), assuming the nonresonant ▲ amplitude to be represented by hard-sphere scattering.

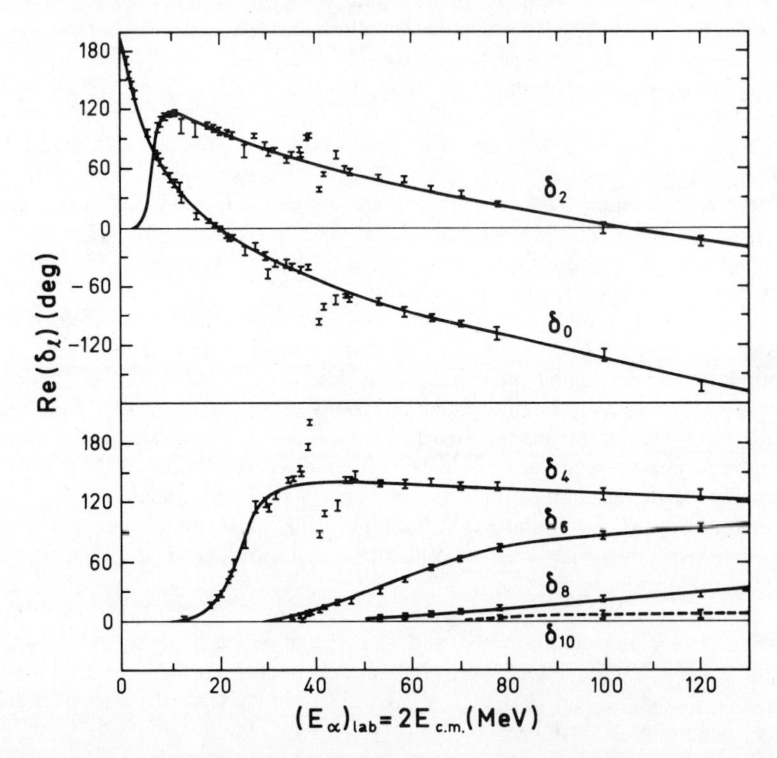

Figure 4-14 Phase shifts in $\alpha\alpha$ scattering. The figure showing the real part of the $\alpha\alpha$ phase shift is taken from the analysis by P. Darriulat, G. Igo, H. G. Pugh, and H. D. Holmgren, *Phys. Rev.* **137**, B315 (1965), which also includes the results of previous investigations. The s-wave phase shift rises from approximately 0° to approximately 180° over a very narrow energy interval ($\Gamma_{res} = 7$eV) around the resonance at $E_\alpha = 184$ keV.

▼ The spins and parities of the three lowest states in ^{8}Be are seen to correspond with those expected for the ground-state band of an even-even nucleus ($K\pi r = 0++$), and the energies are in rough agreement with the leading-order expression proportional to $I(I+1)$.

The phase shifts in Fig. 4-14 also provide evidence for attractive interactions in the channels with higher angular momentum, and attempts have been made to interpret these phase shifts in terms of broad resonances ($E_{res}(I=6) \approx (30+10i)$ MeV, $E_{res}(I=8) \approx (60+35i)$ MeV; Darriulat *et al.*, *loc.cit.*, Fig.
▲

▼ 4-14). However, the detailed analysis of the high-energy scattering is difficult due to strong inelastic processes, starting at the threshold for $\alpha+\alpha \rightarrow {}^7\text{Li}+\text{p}$ at $(E_\alpha)_{\text{lab}}=34.7$ MeV. The rapid fluctuations in the phase shifts in Fig. 4-14 around $E_\alpha=40$ MeV are not well understood. ▲

$I\pi$	E_{cm} (MeV)	Γ (MeV)
0+	0.092	6.8×10^{-6}
2+	2.9	1.5
4+	11.4	6.7

Table 4-5 Parameters for the lowest resonances in $\alpha\alpha$ scattering. The parameters for the 0+ resonance are from J. Benn, E. B. Dally, H. H. Muller, R. E. Pixley, H. H. Staub, and H. Winkler, *Phys. Letters* **20**, 43 (1966). The parameters for the 2+ and 4+ levels are taken from the summary of the phase shift analyses given by Lauritsen and Ajzenberg, 1966.

▼ The possible occurrence of $I=6$ and 8 members in the ^{8}Be rotational band acquires special significance in connection with the analysis of the rotational motion in terms of the one-particle configurations. The lowest shell-model configuration for ^{8}Be is $(1s)^4(1p)^4$, and the interactions between the p particles favor the states with totally symmetric orbital wave function. This set of states has $L=0,2,4$ and $S=0, T=0$, as can be seen from a counting of the m components or by the general methods described in Appendix 1C (see the states with $n=4$ and permutation symmetry $[f]=[4]$ in orbital space (Table 1C-4, Vol. I, p. 132) and $[f]=[1111]$ in isospin-spin space (Table 1C-5, Vol. I, p. 134)). These states can be viewed as the rotational band associated with the intrinsic state formed by aligning the p orbits in a given direction; thus, the states can be obtained by angular momentum projection (see Eq. (4-147)) from the aligned configuration $(l=1, m=0)^4$.

One can also analyze the states of the configuration $(1s)^4(1p)^4$ in terms of the relative motion of two α-particle-like clusters (Perring and Skyrme, 1956). If the nucleons are assumed to move in a harmonic oscillator potential, the Hamiltonian is invariant with respect to orthogonal transformations among the coordinates of the $A=8$ particles. Thus, one can transform to a set of coordinates describing intrinsic motion within each of the α clusters, in addition to relative motion of the two α clusters and center-of-mass motion for the total system. Each of these degrees of freedom is associated with an oscillator with the same frequency ω_0. If the α clusters are in the ground state, ▲

▼ the lowest state of relative motion, from which one can construct a total wave function that is antisymmetric in all eight nucleons, possesses four quanta ($N=4$), corresponding to the number of quanta of the lowest shell-model configuration. As for the states of a single particle in an oscillator potential, the relative motion of the α clusters with $N=4$ gives rise to states with $L=0$, 2, and 4. The states generated in this manner are identical to the states formed from the shell-model configuration $(1s)^4(1p)^4$, since these states are uniquely characterized by the total number of quanta and the orbital permutation symmetry.

States of ^8Be with $L>4$ require excited configurations. For nucleonic motion in an oscillator potential, the lowest states of $L\pi=6+$, $8+$, etc., possess $N=L$ quanta and can be generated by considering the relative motion of two α clusters with this number of quanta. These states form a vibrational sequence as distinct from the rotational relation of the $L=0,2,4$ states, all having $N=4$. However, it is possible that additional interactions not included in this simple picture may blur the sharp distinction between the two types of excitation and produce a sequence of states varying more smoothly with angular momentum.

Ground-state rotational band of ^{169}Tm (Fig. 4-15; Tables 4-6 and 4-7)

The ground state of ^{169}Tm has $I\pi=1/2+$ and forms the head of a rotational band with $K=1/2$. The properties of this band have been studied through a variety of different types of experiments, including Coulomb excitation, analysis of the radiation following the electron-capture decay of ^{169}Yb, and Mössbauer measurements to determine the static moments of excited states.

Energy levels and transition matrix elements

The energy levels of the ground-state band of ^{169}Tm are shown in Fig. 4-15. The observed energies are fit by the expansion (4-62) including up to the fourth-order term. The rate of convergence is similar to that found in the even-even nuclei.

The $E2$-matrix elements within the band are analyzed in Table 4-6. The static and transition moments are expressed in terms of the intrinsic quadrupole moment Q_0, using expressions (4-69) and (4-68). The constancy of the derived values of Q_0 confirms these $E2$-intensity rules within the experimental accuracy of about 10%.

The leading-order $M1$-matrix elements within a $K=1/2$ band depend on three parameters g_K, g_R, and b (see Eq. (4-88)). With the values $g_K=-1.57$, $g_R=0.406$, $b=-0.16$, it is possible to fit all the observed $M1$-matrix elements in the band of ^{169}Tm (see Table 4-7). ▲

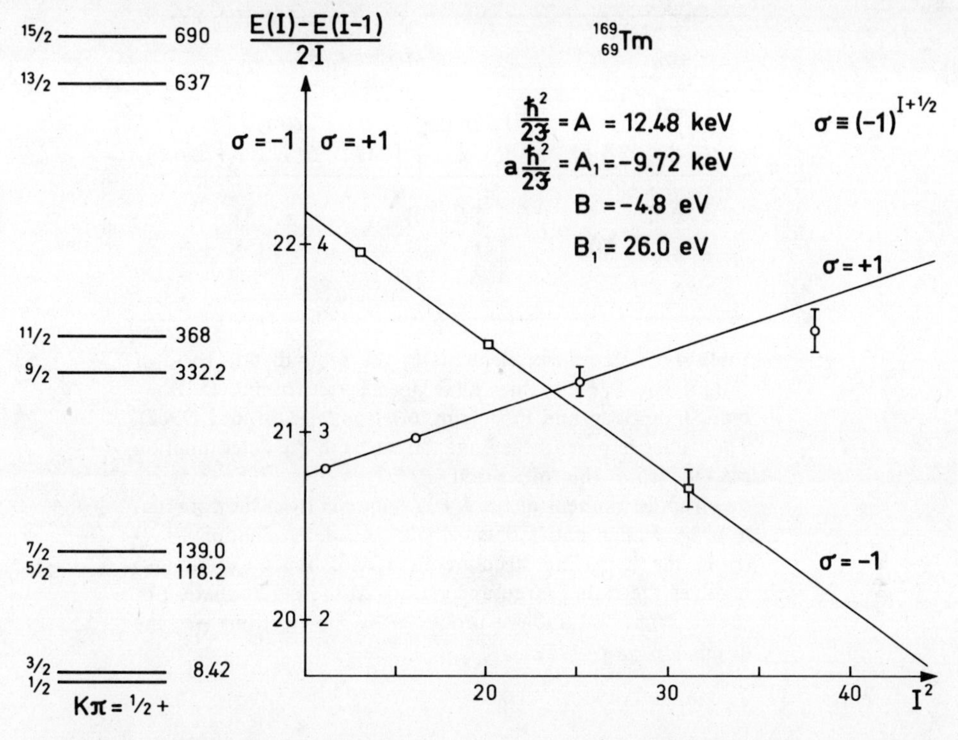

Figure 4-15 Ground-state rotational band of ^{169}Tm. The figure is based on the measurements by P. Alexander and F. Boehm, *Nuclear Phys.* **46**, 108 (1963) and R. M. Diamond, B. Elbek, and F. S. Stephens, *Nuclear Phys.* **43**, 560 (1963). All energies are in keV, except where other units are explicitly given.

▼ *Interpretation of the intrinsic configuration*

The intrinsic configuration responsible for the $K\pi = 1/2+$ ground-state band in ^{169}Tm can be identified with an orbit of the last odd proton moving in a nonspherical average potential and characterized by the quantum numbers $[411\frac{1}{2}]$ (see Table 5-12, p. 296). From the wave function for this one-particle orbital given in Table 5-2 (p. 229) the decoupling parameter can be estimated from Eq. (5-46), which yields $a = -0.9$, to be compared with the empirical value of -0.78 (see Fig. 4-15). The intrinsic magnetic parameters can be estimated from Eqs. (5-86) and (5-87), which yield $g_K = -1.2$ and $b = -0.1$, while the values deduced above from the empirical moments are $g_K = -1.57$ and $b = -0.16$. The estimate of the magnetic properties is based on an effective g_s factor of 0.7 $(g_s)_{\text{free}}$ (see Eq. (5-89)), as derived from the systematics of the one-particle moments in deformed nuclei (see Table 5-14, p. 303). It must be emphasized that, in general, the g factors in the effective ▲

	$Q_0^2\,(10^{-48}\,\text{cm}^4)$		
I	from $Q(I)$	from $B(E2; I+1 \rightarrow I)$	from $B(E2; I+2 \rightarrow I)$
1/2		56 (10)	57 (3)
3/2	40	60 (7)	63 (5)
5/2		56 (13)	

Table 4-6 $E2$-matrix elements in the ground-state band of ^{169}Tm. The $B(E2)$ values have been taken from J. D. Bowman, J. de Boer, and F. Boehm, *Nuclear Phys.* **61**, 682 (1965). The values in parentheses give the errors in the determination of Q_0^2 from the measured $B(E2)$ values. The electric quadrupole moment of the $I=3/2$ state is from the compilation by Fuller and Cohen, 1969; the determination of Q (from the hyperfine structure in a Mössbauer experiment) involves the rather large uncertainty in the determination of the electric field gradient produced by the electrons (see the discussion on p. 132).

I	μ_{exp}	μ_{calc}	$B(M1; I+1 \rightarrow I)_{exp}$	$B(M1; I+1 \rightarrow I)_{calc}$
1/2	−0.229 (3)	−0.233	0.047 (10)	0.055
3/2	+0.534 (15)	0.538	0.150 (8)	0.125
5/2	+0.73 (5)	0.74	0.055 (15)	0.071
7/2	+1.32 (6)	1.45	0.144 (7)*	0.14
9/2		1.59		0.075
11/2		2.30	0.143 (12)*	0.14

Table 4-7 $M1$-matrix elements in the ground-state rotational band of ^{169}Tm. The experimental data and the theoretical analysis are taken from Bowman *et al.*, *loc.cit.*, Table 4-6, from E. N. Kaufmann, J. D. Bowman, and S. K. Bhattacherjee, *Nuclear Phys.* **A119**, 417 (1968), and from C. Günther, H. Hübel, A. Kluge, K. Krien, and H. Toschinski, *Nuclear Phys.* **A123**, 386 (1969). The magnetic moments are given in units of $(e\hbar/2Mc)$, while the $B(M1)$ values are in units of $(e\hbar/2Mc)^2$. For the transitions marked with an asterisk, only the relative intensity of the transitions with $\Delta I=1$ and $\Delta I=2$ has been determined; the value of $B(M1)$ in the table has been obtained by assuming the $E2$-intensity relations with $Q_0^2=58\times10^{-48}$ cm^4 (see Table 4-6).

▼ $\Delta K=1$ and $\Delta K=0$ one-particle operators may be different (see p. 304). In addition, the Coriolis coupling may contribute to the renormalization of the g_K and g_R factors. (See Eq. (4A-35) and the discussion of g_R (^{159}Tb) on p. 255); for estimates of these effects for ^{169}Tm, see Günther *et al.*, 1969.)

Rotational bands populated in decay of $^{177}Lu^m$ (Figs. 4-16 to 4-18; Tables 4-8 and 4-9)

The occurrence of the high-spin isomeric state in ^{177}Lu($I\pi=23/2-$) provides the opportunity for a detailed study of long sequences of rotational states (Jørgensen *et al.*, 1962). The decay scheme of this isomeric state is shown in Fig. 4-16. It is seen that the decay proceeds partly by an $E3$ transition populating the $K\pi=7/2+$ ground-state band in ^{177}Lu and partly by a first forbidden β transition to an $I\pi=23/2+$ isomer in ^{177}Hf, which subsequently decays through two different rotational bands with $K\pi=7/2-$ and $K\pi=9/2+$, respectively.

Energy spectrum

The energies of the three rotational sequences are analyzed in Fig. 4-17. It is seen that the $K\pi=7/2+$ band in ^{177}Lu is fit within the experimental errors by the rotational energy expansion including two terms. For the $7/2-$ band in ^{177}Hf, small deviations from the two-term expansion are observed for the high-spin members and give evidence for a third term with C of order 10 meV.

For the $9/2+$ band in ^{177}Hf, the high-spin members exhibit deviations from the two-term expansion with a more complex structure. These deviations may be connected with the especially strong Coriolis interaction that affects the $K\pi=9/2+$ band. The neutron moves in an orbit with quantum numbers [624 9/2], which contains large components with $j=13/2$ (see Fig. 5-3) and for which, therefore, the Coriolis matrix elements are especially large (see, for example, Eq. (4-197)). Acting in second order, the Coriolis coupling manifests itself in the large moment of inertia for this band (see Eq. (4-205)), which is observed to be about 50% larger than for the other two bands considered. In higher orders, the Coriolis coupling produces terms in the rotational energy of correspondingly higher power in I. Thus, a positive sign of the B coefficient as observed for the $9/2+$ band can arise from a fourth-order effect involving the last odd particle that more than compensates the effect of all the other particles, which give a negative B term, as in the even-even nuclei. In ninth order, the Coriolis coupling generates a signature-dependent term of the form (4-57), which may be responsible for the irregularities in the observed energies. However, no quantitative interpretation has so far been given of the anomalies in the $K\pi=9/2+$ band of ^{177}Hf. (See also the discussion of the rotational energies for the [743 7/2] band in ^{235}U, pp. 280ff.) ▲

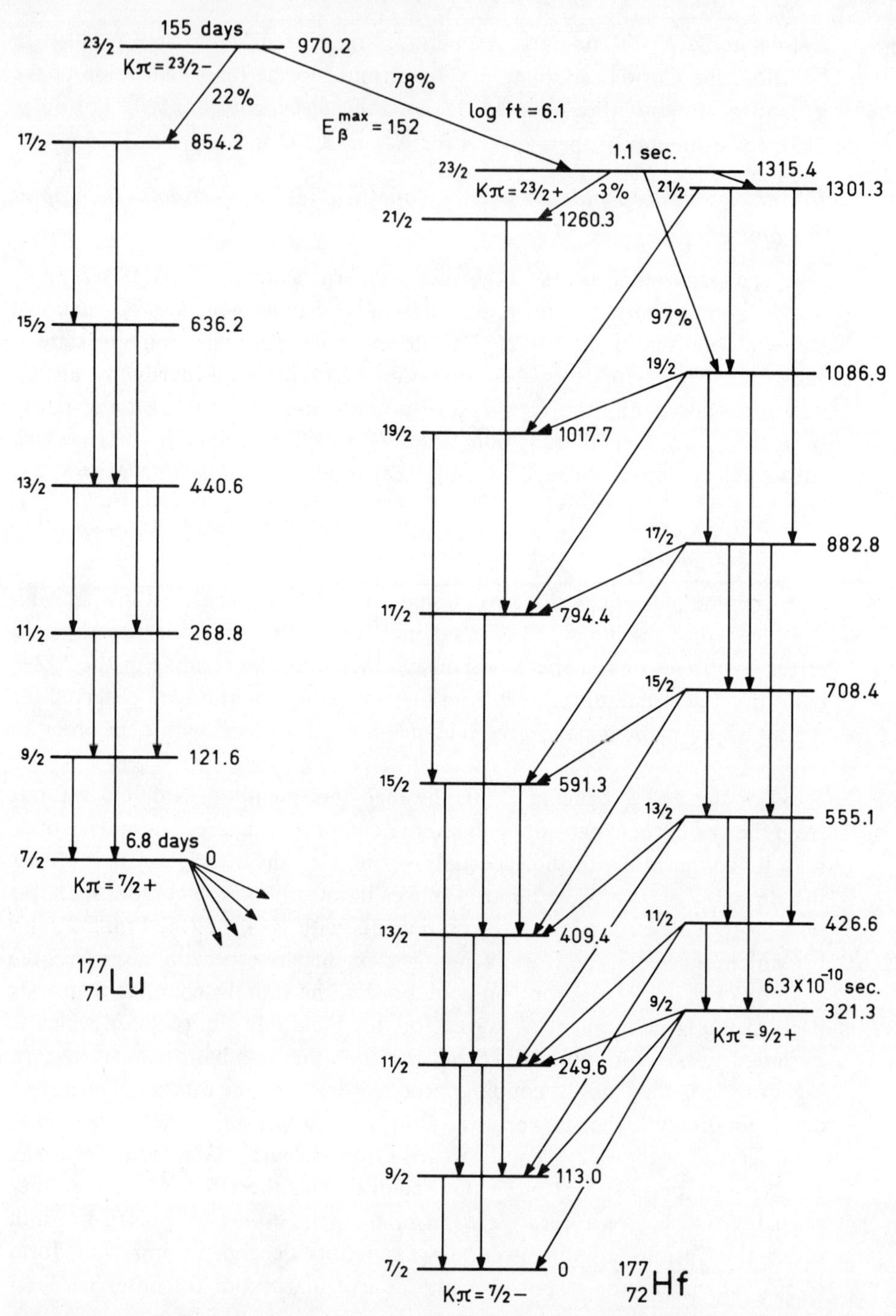

Figure 4-16 Decay scheme for $^{177}Lu^m$. The data are taken from A. J. Haverfield, F. M. Bernthal, and J. M. Hollander, *Nuclear Phys.* **A94**, 337 (1967), which also includes results of previous investigations. Additional transitions reported by F. Bernthal (private communication) have also been included. The energies in the figure are in keV, and the decay times represent half lives.

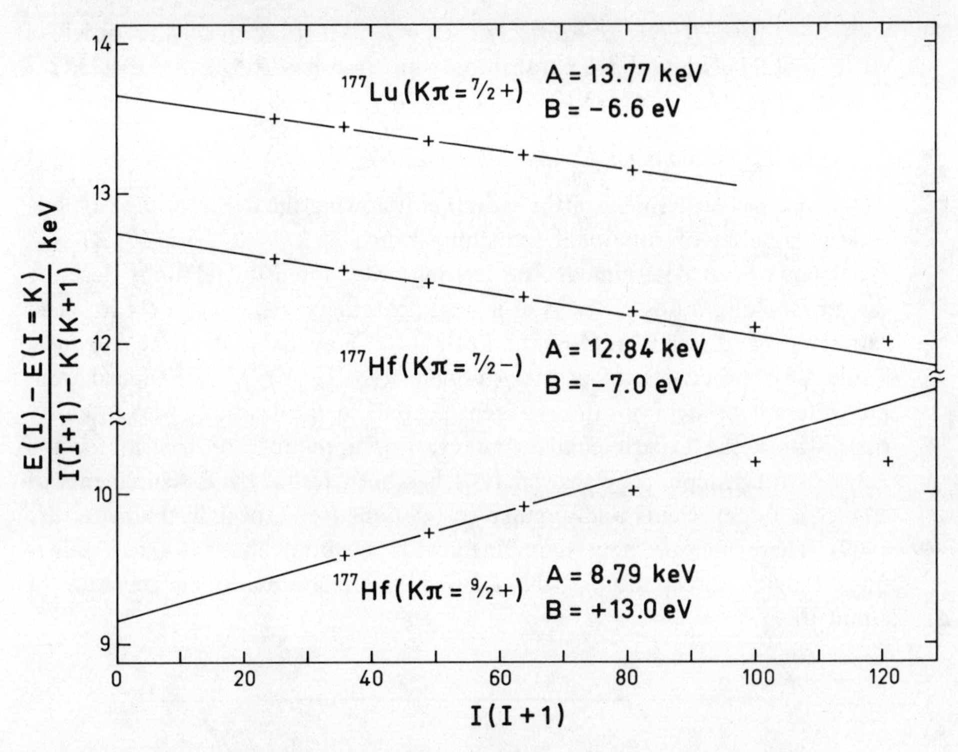

Figure 4-17 Test of rotational energy ratios for ^{177}Lu and ^{177}Hf (see Fig. 4-16).

▼ *Intrinsic configurations*

The intrinsic configurations in Fig. 4-16 can be simply interpreted in terms of the one-particle orbits discussed in Chapter 5. The ground-state band in ^{177}Lu is given the classification [404 7/2] (see Table 5-12, p. 296) as for the other odd Lu isotopes, and the two low-lying bands in ^{177}Hf are associated with the neutron orbits [514 7/2] and [624 9/2] (see Table 5-13, p. 300). The high-spin isomers can be interpreted in terms of the three-particle configurations ($[404\ 7/2]_p$, $[514\ 7/2]_n$, $[624\ 9/2]_n$) for ^{177}Lu ($K\pi = 23/2-$), and ($[404\ 7/2]_p$, $[514\ 9/2]_p$, $[514\ 7/2]_n$) for ^{177}Hf ($K\pi = 23/2+$).

The two isomeric states are connected by a one-particle β transition ($[624\ 9/2]_n \rightarrow [514\ 9/2]_p$) with $\Delta N = \Delta n_z = 1$, $\Delta\Lambda = \Delta\Sigma = 0$. Such a transition is classified as first forbidden unhindered. The reduced transition rate is $\log ft = 6.1$, which is similar to, though somewhat faster than, the transition between the same one-particle states as observed in ^{177}Yb $(9/2+) \rightarrow ^{177}$Lu $(9/2-)$, with $\log ft = 6.8$, and ^{181}W $(9/2+) \rightarrow ^{181}$Ta $(9/2-)$, with $\log ft = 6.7$.

The $E3$ γ decay of the ^{177}Lu isomer is highly K forbidden ($n = 5$). The observed decay rate gives a value of $B(E3; I = K = 23/2 \rightarrow K = 7/2, I = 17/2)$ that is of order 10^{-9} of the single-particle unit $B_W(E3)$, given by Eq. ▲ (3C-38). For the decay of the ^{177}Hf isomer, one finds that the $E1$ transition

▼ with $n=7$ has $B(E1;\ I=K=23/2\rightarrow K=7/2,\ I=21/2)$ of the order of $10^{-13}\ B_W(E1)$, while the $E2$ transition with $n=5$ has $B(E2;\ I=K=23/2\rightarrow K=9/2,\ I=19/2)$ of order $10^{-8}\ B_W(E2)$.

Intensity relations for E2 and M1 transitions

From a measurement of the radiation following the decay of the isomers, a large number of rotational branching ratios $T(KI\rightarrow K,\ I-1)/T(KI\rightarrow K,\ I-2)$ have been determined. The leading-order intensity relations describe these branching ratios in terms of a single parameter $(g_K-g_R)^2/Q_0^2$ for each rotational band (see Eqs. (4-68) and (4-87)). The available data are shown in Table 4-8. The constancy of the parameter $(g_K-g_R)^2/Q_0^2$ as deduced from the different branching ratios is seen to confirm the leading-order intensity rules within the experimental accuracy. For a number of transitions, the $E2/M1$ ratio implied by this analysis has been tested by measurements of conversion coefficients and angular correlations (see especially Hübel *et al.*, 1969). These measurements combined with the branching ratio determinations provide direct tests of the $E2$-intensity relations, to an accuracy of ▲ about 10%.

	I	$(g_K-g_R)^2/Q_0^2$ 10^{-7} fm^{-4}
^{177}Lu	11/2	2.6 ± 0.4
$K\pi=7/2+$	13/2	2.4 ± 0.3
	15/2	2.6 ± 0.3
	17/2	2.5 ± 0.4
^{177}Hf	11/2	0.04 ± 0.6
$K\pi=7/2-$	13/2	0.26 ± 0.12
	15/2	0.24 ± 0.14
^{177}Hf	13/2	2.7 ± 0.3
$K\pi=9/2+$	15/2	2.8 ± 0.3
	17/2	2.9 ± 0.3
	19/2	3.0 ± 0.3
	21/2	2.7 ± 0.3

Table 4-8 Test of $E2$- and $M1$-intensity relations in decay of rotational bands of ^{177}Lu and ^{177}Hf. The ratio $(g_K-g_R)^2/Q_0^2$ is obtained from relative γ intensities $T(E2+M1;\ KI\rightarrow K,I-1)$: $T(E2;\ KI\rightarrow K,I-2)$. The experimental data and the theoretical analysis are taken from Haverfield *et al.*, *loc.cit.*, Fig. 4-16.

▼ Besides the data listed in Table 4-8, an additional test of the intensity rules may be obtained from the measured magnetic moments in the $I=7/2$ ($\mu=0.784\pm0.001$) and $I=9/2$ ($\mu=1.04\pm0.06$) members of the $K\pi=7/2-$ band in ^{177}Hf. (See Büttgenbach *et al.*, 1973, and Lindgren, 1965.) The relation (4-87) for the static moments yields $(g_K-g_R)=-0.02\pm0.05$, and combining this value with the Q_0 value of 674 fm^2, obtained from Coulomb excitation of the $7/2\rightarrow9/2$ transition (Hansen *et al.*, 1961), one thus finds $(g_K-g_R)^2/Q_0^2=(0.0\pm0.1)\times10^{-7}$ fm^{-4}.

E1 transitions

The evidence on $E1$-matrix elements connecting the $K\pi=9/2+$ and $7/2-$ bands in ^{177}Hf is collected in Table 4-9. The absolute magnitude of the $B(E1)$ values is determined partly from the lifetime of the 321 keV level, partly from the branching ratios between the $E1$ transitions (interband) and the competing rotational $E2$ transitions (within the $K\pi=9/2+$ band). The absolute rate of the latter transitions has been assumed to follow the leading-order intensity relation with $Q_0=855$ fm^2 (see below). ▲

I_i	I_f	E_γ keV	$B(E1)$ $10^{-5}e^2$fm^2
9/2	7/2	321.3	0.036
	9/2	208.3	6.9
	11/2	71.7	2.3
11/2	9/2	313.7	0.70
	11/2	177.0	10
13/2	11/2	305.5	2.4
	13/2	145.8	13
15/2	13/2	299.0	4.4
	15/2	117.2	9.8
17/2	15/2	291.4	8.4
	17/2	88.4	13
19/2	17/2	292.5	11
	19/2	69.2	8.4
21/2	19/2	283.4	14

Table 4-9 $E1$-transition rates in ^{177}Hf. The data are taken from Haverfield *et al.*, *loc.cit.*, Fig. 4-16, and F. Bernthal (private communication; see also UCRL-18651, 1969). The $B(E1)$ values for the states with $I_i>9/2$ differ somewhat from those given by these references, on account of the different assumptions concerning the absolute $E2$ rates (see the discussion in the text).

▼ The leading-order intensity relation for the $E1$-matrix elements gives $B(E1; I_i \to I_f) = \text{const.}\langle I_i\, 9/2\, 1-1 | I_f\, 7/2\rangle^2$ and would thus imply a constant ordinate for the points plotted in Fig. 4-18. The observed values exhibit major deviations from constancy, but are seen to be consistent with the generalized intensity relation (see Eq. (4-98))

$$B(E1; KI_i \to K-1, I_f)$$
$$= \langle I_i K 1 -1 | I_f\, K-1\rangle^2 \left(M_1 + M_2(I_f(I_f+1) - I_i(I_i+1))\right)^2 \qquad (4\text{-}162)$$

obtained by including the leading-order I-dependent term in the intrinsic moment. The possibility of improving the fit to the experimental data by inclusion of quadratic I-dependent terms in the intrinsic moments has been discussed by Grin (1967) and by Bernthal and Rasmussen (1967), but the available data, as analyzed in Fig. 4-18, do not seem to require such ▲ higher-order terms.

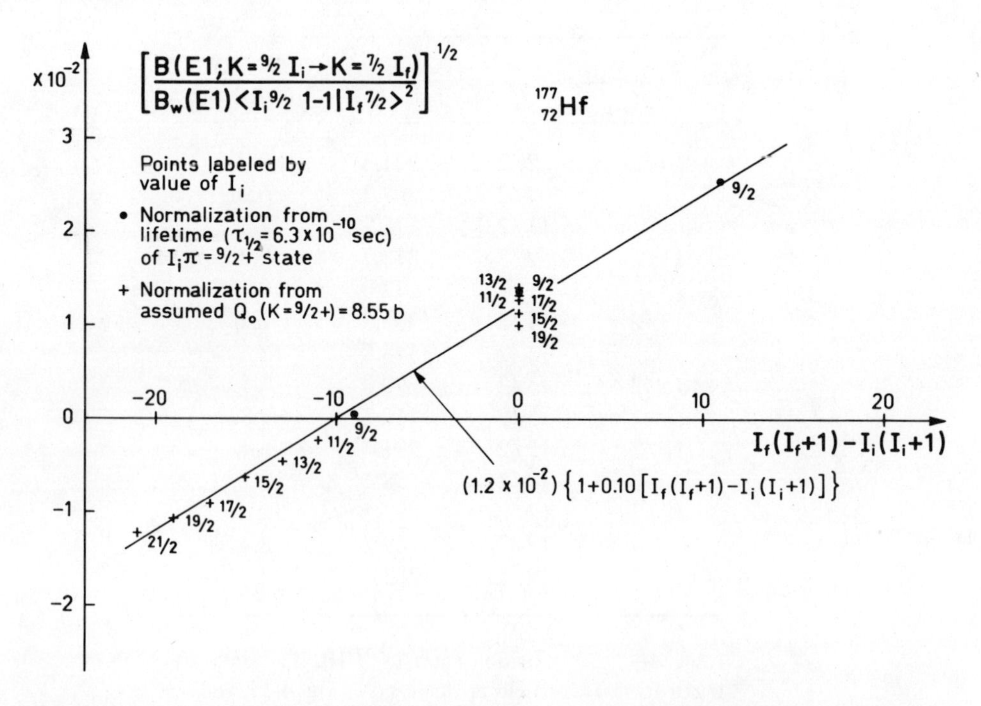

Figure 4-18 $E1$ amplitudes in ^{177}Hf. The experimental $B(E1)$ values are taken from Table 4-9 and are measured in units of $B_W(E1) = 2.0e^2\text{fm}^2$ (see Eq. (3C-38)). Similar analyses of the $E1$ transitions in ^{177}Hf have been given by Yu. T. Grin and I. M. Pavlichenkow, *Phys. Letters* **9**, 249 (1964) and by M. N. Vergnes and J. O. Rasmussen, *Nuclear Phys.* **62**, 233 (1965). See also Haverfield *et al., loc. cit.*, Fig. 4-16.

▼ The experimental data testing the $E1$-intensity relations may be divided into three categories. First, the relative intensities of $E1$ transitions from the same initial level can be compared without making any further assumptions, and therefore provide especially direct tests. Second, the analysis of the relative rates of the transitions from the different levels with $I_i > K = 9/2$ involves the $E2$-intensity rule for the transitions within the $K = 9/2$ band. The evidence presented in Table 4-9 appears to confirm this rule to an accuracy of about 10%. Finally, the comparison between the $E1$ transitions from the $I_i = 9/2$ level and those from all the other levels requires a knowledge of the absolute $E2$ rates, which have not been measured. The assumed value of Q_0 has been chosen so as to obtain the optimum fit to the $E1$-intensity relation (4-162). The value of Q_0 required to obtain this agreement is about 30% larger than the values ($Q_0 = 674$ fm^2; Hansen *et al.*, *loc.cit.*) observed for the ground-state band of ^{177}Hf and for the [624 9/2] ground-state configuration in ^{179}Hf ($Q_0 = 685$ fm^2; Hansen *et al.*, *loc.cit.*). Such a large value of Q_0 is rather surprising and may suggest either an error in the lifetime of the 321 keV level of ^{177}Hf or an irregularity in the $E1$ intensities that goes beyond the present analysis.

The fit in Fig. 4-18 implies comparable contributions from the M_1 and M_2 terms in the $E1$-matrix element (4-162). The important role of the I-dependent term is connected with the smallness of the leading term, which in turn can be understood on the basis of the asymptotic selection rules. These rules forbid the [624 9/2]→[514 7/2] transition because of the spin flip involved (see Table 5-3, p. 234). The M_2 term may be attributed to the Coriolis coupling, which admixes bands with $\Delta K = \pm 1$ (see the discussion on pp. 147ff. as well as Eq. (4A-31)). Among the significant coupling effects are the admixing of the orbit [624 7/2] into the [624 9/2] band and the orbit [514 9/2] into the [514 7/2] band. These admixed components can then undergo unhindered $E1$ transitions. Additional significant contributions may arise from the admixture of the octupole vibrational excitations, which are known to give rise to rather strong $E1$ transitions in the low-energy spectrum. (For an estimate of these contributions as well as of the one-particle effects, see Bernthal and Rasmussen, 1967.) Since the first-order corrections involve unhindered $E1$-matrix elements, the subsequent terms in the expansion should exhibit the usual rate of convergence.

Rotational bands in ^{239}Pu (Fig. 4-19; Tables 4-10 to 4-13)

The spectra of the heaviest nuclei show the systematic occurrence of rotational band structure for $A \gtrsim 224$; in this region of the elements, α radioactivity provides an important additional tool for the study of the rotational coupling scheme. ▲

Energy spectrum

The spectrum of ^{239}Pu has been studied in the α decay of ^{243}Cm, in the β decay of ^{239}Np, and the electron capture decay of ^{239}Am, as well as by the Coulomb excitation process. As shown in Fig. 4-19, the known levels can be arranged into a number of rotational bands. The primary evidence for this interpretation of the spectrum comes from the spins and parities of the levels and from the quantitative agreement of the observed energies with the values (in parentheses) calculated from the first-order expression (4-61). A number of additional properties of the levels confirm the assignments and provide further tests of the rotational coupling scheme.

E2 and M1 intensities in rotational transitions

The evidence on $E2$- and $M1$-intensity ratios for transitions within the ground-state rotational band ($K\pi = 1/2+$) is compared in Table 4-10 with the leading-order intensity rules given by Eqs. (4-68) and (4-88). (For the two $M1$

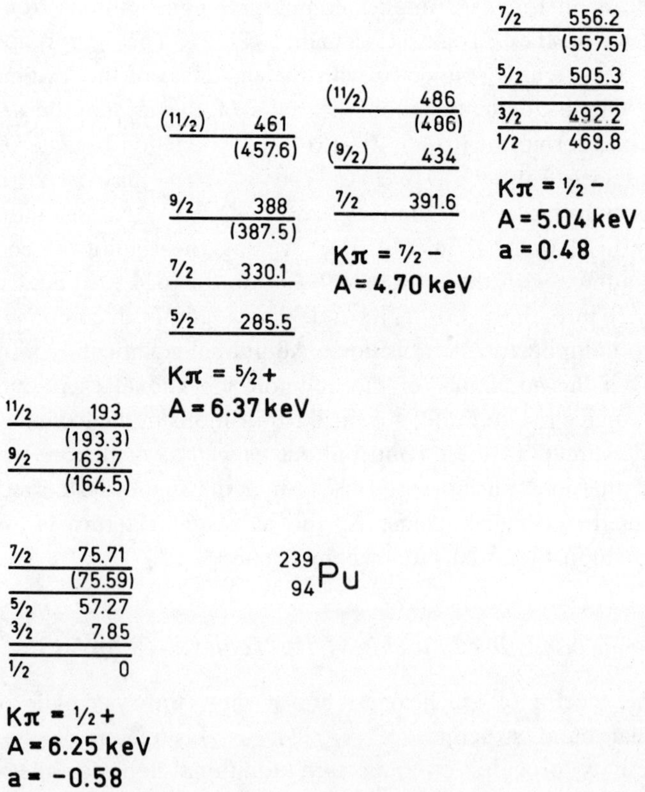

11/2 193 (193.3)
9/2 163.7 (164.5)
7/2 75.71 (75.59)
5/2 57.27
3/2 7.85
1/2 0
$K\pi = 1/2+$
$A = 6.25$ keV
$a = -0.58$

(11/2) 461 (457.6)
9/2 388 (387.5)
7/2 330.1
5/2 285.5
$K\pi = 5/2+$
$A = 6.37$ keV

(11/2) 486 (486)
(9/2) 434
7/2 391.6
$K\pi = 7/2-$
$A = 4.70$ keV

7/2 556.2 (557.5)
5/2 505.3
3/2 492.2
1/2 469.8
$K\pi = 1/2-$
$A = 5.04$ keV
$a = 0.48$

$^{239}_{94}$Pu

Figure 4-19 Spectrum of ^{239}Pu. The level scheme is taken from *Table of Isotopes* by Lederer *et al.*, 1967.

▼ transitions considered in Table 4-10, the term involving the magnetic decoupling parameter b cancels in the ratio of the $B(M1)$ values.) The ratios of the reduced transition probabilities are determined from relative intensities and $M1:E2$ mixing ratios. In the derivation of the $B(M1)$ ratio, the rotational
▲ $E2$-intensity rule was assumed.

	Experiment	Theory
$\dfrac{B(E2;9/2\rightarrow5/2)}{B(E2;9/2\rightarrow7/2)}$	14 ± 3	16.5
$\dfrac{B(E2;5/2\rightarrow1/2)}{B(E2;5/2\rightarrow3/2)}$	3.4 ± 0.2	3.5
$\dfrac{B(M1;9/2\rightarrow7/2)}{B(M1;5/2\rightarrow3/2)}$	1.1	1.1

Table 4-10 $E2$-and $M1$-intensity ratios for transitions within the ground-state rotational band of ^{239}Pu, with $K\pi=1/2+$. The experimental data are taken from G. T. Ewan, J. S. Geiger, R. L. Graham, and D. R. MacKenzie, *Phys. Rev.* **116**, 950 (1959).

▼ *Forbidden M1 transitions*

The $M1$ transitions from the $K\pi=5/2+$ band to the ground-state band are K forbidden and the measured half life of the 285 keV level, $\tau_{1/2}=1.1\times10^{-9}$ sec. implies a hindrance factor of the order of 10^3 as compared with the single-particle estimate (see Eq. (3C-38)). The relative intensities of these $M1$ transitions are given in Table 4-11 and are compared with the leading-order
▲ intensity rules for K-forbidden transitions (4-95).

$K_i\pi=5/2+$	$K_f\pi=1/2+$		$\dfrac{B(M1;K_iI_1\rightarrow K_fI_2)}{B(M1;K_iI_1\rightarrow K_fI_3)}$	
I_1	I_2	I_3	Experiment	Theory
5/2	3/2	5/2	0.6	0.87
	7/2	5/2	0.4	0.31
7/2	5/2	7/2	0.6	0.75
	9/2	7/2	0.4	0.35

Table 4-11 $M1$-intensity ratios for K-forbidden transitions in ^{239}Pu. The experimental data are taken from the reference quoted in Table 4-10.

▼ *E2 transitions between the $K\pi = 1/2-$ and $K\pi = 1/2+$ bands*

A number of $E1$ transitions between the $K\pi = 1/2-$ and $K\pi = 1/2+$ bands in ^{239}Pu have been obtained and are compared with the leading-order intensity relations in Table 4-12. In general, the leading-order $E1$-intensity relations between two $K = 1/2$ bands involve two different intrinsic matrix elements (see Eq. (4-91))

$$B(E1;\, K\pi = 1/2-, I_1 \rightarrow K\pi = 1/2+, I_2)$$

$$= |\langle I_1 \tfrac{1}{2} 1 0 | I_2 \tfrac{1}{2}\rangle M_1 + (-1)^{I_1+\frac{1}{2}} \langle I_1 -\tfrac{1}{2} 1 1 | I_2 \tfrac{1}{2}\rangle M_2|^2 \qquad \text{(4-163)}$$

$$M_1 = \langle K\pi = 1/2+ |\, \mathscr{M}(E1, \nu = 0) | K\pi = 1/2- \rangle$$

$$M_2 = \langle K\pi = 1/2+ |\, \mathscr{M}(E1, \nu = 1) | \overline{K\pi = 1/2-} \rangle$$

The observed relative intensities in Table 4-12 are fitted, within the experimental error, by the relation (4-163) with $M_2 = 0$ and place an upper limit of ▲ about 0.1 on the ratio M_2/M_1.

$K_i\pi = 1/2-$	$K_f\pi = 1/2+$		$\frac{B(E1;\, K_i I_1 \rightarrow K_f I_2)}{B(E1;\, K_i I_1 \rightarrow K_f I_3)}$	
I_1	I_2	I_3	Experiment	Theory ($M_2 = 0$)
1/2	1/2	3/2	0.6	0.5
3/2	1/2	5/2	0.34	0.55
	3/2	5/2	0.06	0.11
5/2	3/2	7/2	0.5	0.7
	5/2	7/2	0.06	0.05
7/2	5/2	9/2	~0.5	0.77
	7/2	9/2	<0.05	0.03

Table 4-12 $E1$-intensity relations for transitions in ^{239}Pu. The experimental data are taken from D. W. Davies and J. M. Hollander, *Nuclear Phys.* **68**, 161 (1965).

▼ In most experimental tests of the $E1$-intensity relations, it has been found necessary to include rather large I-dependent terms in the intrinsic moment, and this failure of the leading-order description has been attributed to the strongly hindered character of the $E1$ transitions usually observed in the low-energy odd-A spectra. (As an example, see the $E1$ transitions in the ▲ spectrum of ^{177}Hf discussed on pp. 109ff.) The $E1$ transitions in Table 4-12

▼ are an exception to this rule, and this may suggest that, in the present case, the $E1$ transitions are not so strongly hindered (see the discussion below of the excitation mode associated with the $K\pi = 1/2-$ band).

Alpha transitions

A convenient unit for expressing the strength of α-decay transitions is provided by the transition rate T_0 to the ground states of even-even nuclei, which is observed to vary smoothly as a function of the charge and mass numbers of the nucleus. The empirical values of T_0 can be approximately represented by the Geiger-Nuttal relation

$$\log_{10} T_0(E_\alpha, Z) = -\frac{K_1(Z)}{\sqrt{E_\alpha}} + K_2(Z) \tag{4-164}$$

where Z is the charge number of the parent nucleus and E_α the kinetic energy of the α particle. The form of the relation (4-164), as well as the order of magnitude of the coefficients K_1 and K_2, can be obtained by considering the penetration of the α particle through the Coulomb barrier (see, for example, Hyde *et al.*, 1964). An empirical fit of the observed decay rates for nuclei with $Z \geqslant 84$ (Fröman, 1957) yields, for the coefficients in Eq. (4-164),

$$\begin{aligned} K_1 &= 139.8 + 1.83(Z-90) + 0.012(Z-90)^2 \\ K_2 &= 52.1 + 0.30(Z-90) + 0.001(Z-90)^2 \end{aligned} \tag{4-165}$$

if T_0 is expressed in sec^{-1} and E_α in MeV. Expressing the α-transition rates to the excited states of the ground-state bands of even-even nuclei in units of T_0, we obtain the reduced intensities C_L

$$T_L = C_L(Z,A)\, T_0(E_\alpha, Z) \tag{4-166}$$

(For a compilation of the coefficients C_L see, for example, Hyde *et al.*, 1964.)

The transitions to the ground-state bands in even-even nuclei proceed through the formation of the α particle from nucleon pairs in conjugate orbits related by time reversal (Ω_1 and $\Omega_2 = \bar{\Omega}_1$). This process is greatly enhanced by the pair correlation effects (see the discussion in Chapter 5, pp. 270ff.) and is therefore strongly favored as compared with transitions that involve pairs of nucleons with $\Omega_2 \neq \bar{\Omega}_1$.

In deformed odd-A nuclei, most of the α transitions are strongly hindered relative to the even-even ground-band transitions. The hindrance factor describing the rate relative to that given by Eq. (4-166) is typically of order 10^2 to 10^3 (see, for example, Fig. 5-13, p. 268). In each odd-A decay, however,

▲ there is a single band in the daughter that is populated with a hindrance

▼ factor of order unity. This group, the "favored transitions," is interpreted as representing transitions in which the α particle is formed from nucleons in paired orbits, as in the ground-state transitions of even-even nuclei (Perlman *et al.*, 1950; Bohr *et al.*, 1955). Such an interpretation implies $\Delta K=0$ and $\pi_i=\pi_f$ for these transitions; whenever independent evidence is available, these selection rules are confirmed.

In the favored transitions in even-even nuclei, the full intensity C_L goes into a single α group (to the state with $I=L$), but in odd-A nuclei this intensity may be distributed over a number of transitions. This distribution is determined in first approximation by the leading-order intensity relation applying to I-independent transition operators (Bohr *et al.*, 1955). Thus, if we can neglect the effect of the rotational motion during the emission process, the α particles are formed in a state of $\Omega_\alpha=0$ relative to the nuclear axis and with an amplitude that is independent of the rotational angular momentum. In this approximation, the transition rates for emission of α particles with angular momentum L follow the leading-order intensity relation for a tensor operator of rank L and intrinsic component $\nu=\Delta K=0$ (see Eq. (4-91))

$$T(L;\, KI_1 \to KI_2) = \langle I_1 K L 0 | I_2 K\rangle^2 C_L T_0 \tag{4-167}$$

It should be emphasized that the derivation of Eq. (4-167) involves additional assumptions beyond those considered in connection with the intensity rules governing the emission of weakly interacting particles (β and γ transitions). In the case of α decay, the escape of the α particle from the nuclear surface involves the penetration of a nonspherical potential, which may give rise to subsequent exchange of angular momentum between the α particle and the daughter nucleus. The validity of the I-independent intensity relation (4-167) therefore requires that the rotational motion can be neglected during the passage of the nonspherical part of the barrier. This requirement corresponds to the smallness of the parameter

$$x = \frac{\tau_{\text{pas}}}{\tau_{\text{rot}}} \tag{4-168}$$

which measures the time required by the α particle to pass through the region of nonspherical couplings, in units of the rotational period. Since the (imaginary) velocity of the α particle just outside the nuclear surface R is of magnitude

$$|v_\alpha| \approx \left(\frac{2}{M_\alpha}\left(\frac{2Ze^2}{R} - E_\alpha \right) \right)^{1/2} \tag{4-169}$$

we obtain

$$\tau_{\text{pas}} \approx \frac{R}{|v_\alpha|} \approx 3\times 10^{-22}\ \text{sec} \tag{4-170}$$

for a typical heavy element α decay. The characteristic time for rotation is

$$\tau_{\text{rot}} \approx \frac{\mathscr{I}}{\hbar I} \approx 5\times 10^{-20} I^{-1}\ \text{sec} \tag{4-171}$$

▲ for a heavy nucleus. Hence, $x \ll 1$ for values of I that are not too large.

▼ A more detailed treatment of the penetration problem can be based on the coupled equations describing the α waves in the different channels $(LI_2)I_1$ with the same total I_1. (See the similar problem of nucleon scattering on deformed nuclei discussed in Appendix 5A.) Numerical solutions of the coupled barrier penetration equations have confirmed that the relation (4-167) provides a good approximation, except for partial waves with small values of C_L (see, for example, Chasman and Rasmussen, 1959).

In the α decay of ^{243}Cm to ^{239}Pu, the groups populating the $K\pi=5/2+$ band have absolute intensities within a factor of 2 of the values given by Eq. (4-167), while all other transitions are hindered by factors of 10^2 or more. In Table 4-13, the relative intensities of the α groups to the different members of the $K\pi=5/2+$ band are compared with the estimate (4-167) and are seen to be rather well described by assuming the same values of C_L as observed in neighboring even-even nuclei. The absolute rates of the transitions in Table 4-13 imply C_L coefficients that are smaller than for the even-even nuclei by a factor of about 0.5. (This value is based on a direct comparison with the observed decay rates of the neighboring even Cm isotopes and a dependence on the transition energy as given by Eqs. (4-164) and (4-165).) The hindrance in the odd-A favored transitions is discussed in Sec. 5-3, pp. 271ff.; the fine-structure intensities for unfavored transitions are discussed in connection with the example illustrated in Fig. 5-13, p. 268.

Intrinsic configurations

The rotational classification and intensity rules discussed above are independent of any detailed assumption about the structure of the intrinsic states $K\pi$, but, as discussed in Chapter 5, it is possible to interpret many of the properties of the low-lying bands in odd-A nuclei on the basis of independent-particle motion in a spheroidal potential. For $^{239}_{94}\mathrm{Pu}_{145}$ with deformation parameter $\delta\approx0.25$ (see the systematics in Fig. 4-25), it is seen from Fig. 5-5, p. 225, that the two lowest orbits should be [631 1/2] and [622 5/2], in good agreement with the observed $K\pi$ values for the lowest intrinsic states; the $K\pi=7/2-$ band can be associated with the orbit [743 7/2], which is also expected from Fig. 5-5. (For a discussion of the magnetic moment, decoupling parameter, moments of inertia, and transition rates on the basis of these configuration assignments, see Mottelson and Nilsson, 1959.) These configuration assignments have also been employed in a more detailed analysis of the α-formation process, which gives an interpretation of the main features of the intensities of the unfavored α transitions (Poggenburg, 1965).

The $K\pi=1/2-$ excitation in ^{239}Pu appears to represent an octupole vibrational excitation ($\nu\pi=0-$; $r=-1$) built on the $K\pi=1/2+$ ground-state configuration. Such an interpretation is strongly suggested by the relatively large cross section for the excitation of this band in the (dd′) reaction, which is similar to the cross section for exciting the $K\pi=0-$ band in ^{238}Pu (Grotdal ▲ *et al.*, 1973). For a $\nu=0$ octupole mode of excitation, the decoupling para-

I_f	$C_L\langle 5/2\ 5/2\ L0 \vert I_f\ 5/2\rangle^2$					$T_0(E_\alpha)$ (relative values)	Relative intensities	
	$L=0$	$L=2$	$L=4$	$L=6$	Sum		Calculated	Experimental
5/2	1	0.20	0.00		1.20	1	1	1
7/2		0.26	0.00		0.26	0.57	0.12	0.16
9/2		0.09			0.09	0.28	0.021	0.022
11/2			0.0007	0.0006	0.0013	0.11	0.00012	≈ 0.0003

Table 4-13 Relative intensities of favored α transitions in the decay of ^{243}Cm. The relative α intensities are taken from *Table of Isotopes* by Lederer *et al.*, 1967. (The identification of the branch to the 11/2 state appears to be uncertain.) The parameters C_L are the mean values derived from the decay of the neighboring even-even nuclei ^{242}Cm and ^{244}Cm($C_0=1$, $C_2=0.55$, $C_4=0.0022$, and $C_6=0.0025$; Lederer *et al.*, 1967). The quantity $T_0(E_\alpha)$ is calculated from Eqs. (4-164) and (4-165).

▼ meter (4-60) has the value (see Eq. (6-89))

$$a(K\pi=1/2-)=-a(K\pi=1/2+) \tag{4-172}$$

since the octupole excitation has the eigenvalue -1 for the $\mathscr{T}$ (or $\mathscr{R}$) operator. The relation (4-172) is in rather good agreement with the observed values. In addition, the $E1$ transitions to the ground state should have a vanishing $\Delta K=1$ matrix element, M_2 (see Eq. (4-163)), in agreement with the data in Table 4-12 and, as in the case of the low-lying $K\pi r=0--$ octupole excitations in even-even nuclei, should obey the leading-order intensity rules.

Rotational bands in the odd-odd nucleus ^{166}Ho (Fig. 4-20 and Table 4-14)

The low-energy spectra of odd-odd nuclei have a very high level density, and the unraveling of these spectra has posed challenging problems. By combining the results of a variety of different nuclear measurements, it has been possible, in a number of cases, to establish the main features of the low-energy spectra. As an example, the term scheme of ^{166}Ho is shown in Fig. 4-20.

Rotational bands

It is seen that the observed levels in ^{166}Ho can be arranged into rotational sequences characterized by the quantum numbers K and π (and r). The numbers given in parentheses are the predicted energies obtained by using the observed energies of the first two members of the band and assuming a rotational energy proportional to $I(I+1)$. This comparison, as well as the derived values of A and B listed under each band, show that the rate of convergence of the energy expansion (4-62) is similar in this odd-odd nucleus to that found in neighboring odd-A and even-even nuclei. A number of relative γ-ray intensities have been measured for transitions within the different rotational bands and have been found to be in agreement with the leading-order rotational intensity rules (Motz *et al.*, 1967).

Intrinsic configurations

One may attempt to describe the intrinsic states of an odd-odd nucleus in terms of the configurations Ω_n and Ω_p of the last odd neutron and proton. In the spectrum of ^{166}Ho, the three lowest bands may be associated with the configuration $[633\ 7/2]_n[523\ 7/2]_p$, which is expected from the observed systematics of the intrinsic states in odd-A nuclei (see Tables 5-12, p. 296, and 5-13, p. 300). The $K\pi=3+$ and $K\pi=4+$ bands may be associated with the configuration $[521\ 1/2]_n[523\ 7/2]_p$, while the $K\pi=1+$ band appears to have the configuration $[523\ 5/2]_n[523\ 7/2]_p$. An additional low-lying $K\pi=1+$ band is expected from the configuration $[512\ 5/2]_n[523\ 7/2]_p$.

The most direct and quantitative test of the configuration assignments is ▲ provided by the measured (dp) intensities for populating the different members of each rotational band. If the odd-proton target nucleus is described as an even-even core plus a proton in the configuration Ω_p, while the final state in the odd-odd nucleus has quantum numbers I_2 and $K_2 = |\Omega_p \pm \Omega_n|$, the leading-order (dp) intensity pattern can be expressed in the form

$$d\sigma(\Omega_p, I_1 \rightarrow (\Omega_p, \Omega_n), K_2 = |\Omega_p \pm \Omega_n|, I_2)$$
$$= \sum_j \tfrac{1}{2} \langle I_1 \Omega_p j \pm \Omega_n | I_2\ \Omega_p \pm \Omega_n \rangle^2 d\sigma(0 \rightarrow \Omega_n, I_n = j) \qquad (4\text{-}173)$$

where $d\sigma(0 \rightarrow \Omega_n, I_n)$ is the cross section for populating the state $\Omega_n I_n$ in a neighboring odd-A nucleus by a (dp) process on an even-even target. (For an

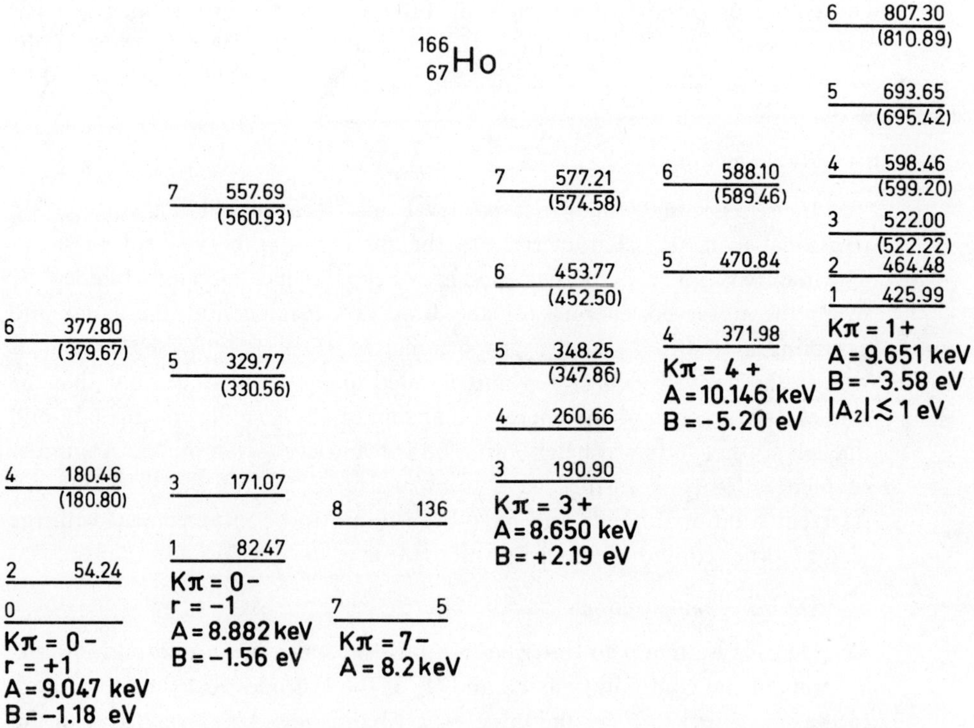

Figure 4-20 Rotational band structure in ^{166}Ho. The level scheme of ^{166}Ho has been established through a combination of high-precision measurements involving the β decay of ^{166}Dy, the ^{165}Ho (dp) reaction, and the neutron-capture reaction ^{165}Ho (nγ). The data in the figure are taken from the cooperative investigation by H. T. Motz, E. T. Jurney, O. W. B. Schult, R. H. Koch, U. Gruber, B. P. Maier, H. Baader, G. L. Struble, J. Kern, R. K. Sheline, T. von Egidy, Th. Elze, E. Bieber, and A. Bäcklin, *Phys. Rev.* **155**, 1265 (1967). (Higher-lying bands have been identified by Bollinger and Thomas, 1970.) The coefficients in the rotational expansion have been obtained from the lowest members of each rotational band, and the energies in parentheses have been calculated from these parameters.

▼ example of the analysis of (dp) intensities leading to an odd-A nucleus, see Chapter 5, pp. 258 ff.) The ground-state configuration of ^{165}Ho is [523 7/2], and thus all the bands of Fig. 4-20 can be populated by direct neutron transfer. The measured cross sections have been found to be consistent with the relation (4-173) (Struble *et al.*, 1965).

The observed rotational energy parameters A for the different bands in Fig. 4-20 are systematically smaller than those observed in the ground-state bands of neighboring even-even nuclei. (An average of these gives a value of about 12.6 keV.) As discussed in Chapter 5, the extra increment of the moment of inertia in an odd-A nucleus is observed to be a characteristic property that can be related to the configuration of the last odd nucleon (see Table 5-17, p. 312). If one neglects the interaction between the last odd neutron and the last odd proton, the increment to the moment of inertia of an odd-odd nucleus is given by the sum of the contributions from the odd particles (Peker, 1960; see also the discussion in Appendix 4A, p. 206)

$$\delta\mathscr{I}(\Omega_n,\Omega_p)\approx\delta\mathscr{I}(\Omega_n)+\delta\mathscr{I}(\Omega_p) \tag{4-174}$$

This relation is compared in Table 4-14 with the observed moments of inertia of the low-lying bands of ^{166}Ho. In view of the difficulty in obtaining quantitative values of the odd-particle contributions from the odd-A spectra (see Table 5-17), the qualitative agreement with the observed values of $\delta\mathscr{I}$ in Table 4-14 appears satisfactory. For a summary of tests of the relation (4-174) ▲ in odd-odd nuclei, see Scharff-Goldhaber and Takahashi (1967).

Neutron	Proton	$\delta\mathscr{I}_n$	$\delta\mathscr{I}_p$	$\delta\mathscr{I}_{calc}$	$\delta\mathscr{I}_{obs}$
[633 7/2]	[523 7/2]	32	17	49	38
[521 1/2]	[523 7/2]	13	17	30	28
[523 5/2]	[523 7/2]	19	17	36	25

Table 4-14 Moments of inertia of rotational bands in ^{166}Ho. All moments in the table are in units of $\frac{1}{2}\hbar^2\,\text{MeV}^{-1}$. The contributions of the different one-particle configurations to $\delta\mathscr{I}$ are taken from the observed rotational energies of neighboring odd-A nuclei (see Table 5-17 pp. 312 ff.). Thus, the value $\delta\mathscr{I}_p$ is that of the proton orbit [523 7/2] in ^{165}Ho ($\delta\mathscr{I}_p=\mathscr{I}(^{165}_{67}\text{Ho})-\frac{1}{2}(\mathscr{I}(^{164}_{66}\text{Dy})+\mathscr{I}(^{166}_{68}\text{Er}))$; the values of $\delta\mathscr{I}_n$ for the neutron orbits [633 7/2] and [521 1/2] represent averages for ^{165}Dy and ^{167}Er and, for the orbit [523 5/2], the average for ^{163}Dy and ^{165}Er. The increments in ^{166}Ho correspond to the differences between the observed moments and the average $(2\mathscr{I}/)\hbar^2 = 78\ \text{MeV}^{-1}$) of the neighboring even-even nuclei; the values are the averages of the increments observed in the two rotational bands ($K=|\Omega_p\pm\Omega_n|$) associated with each configuration (see Fig. 4-20).

▼ The rather large difference in the moments of inertia of the $K\pi=3+$ and $K\pi=4+$ bands (both associated with the configuration $[521\ 1/2]_n$ $[523\ 7/2]_p$) can be related to the Coriolis coupling of these two bands. Using the coupling term (4-197),

$$\langle\Omega_n=1/2,\Omega_p,K=\Omega_p+1/2,I|H_c|\overline{\Omega_n=1/2},\Omega_p,K=\Omega_p-1/2,I\rangle$$

$$=-A\left((I-\Omega_p+\tfrac{1}{2})(I+\Omega_p+\tfrac{1}{2})\right)^{1/2}\langle\Omega_n=1/2|j_+|\overline{\Omega_n=1/2}\rangle \qquad (4\text{-}175)$$

where A is the rotational energy constant, in the absence of the coupling (4-175). The matrix element of j_+ appearing in Eq. (4-175) is the same as that which determines the decoupling parameter of the rotational band associated with the orbit Ω_n (see Eq. (4-60)). From a second-order perturbation calculation, we thus obtain

$$A(K=\Omega_p+1/2)-A(K=\Omega_p-1/2)=\frac{2(Aa)^2}{E(K=\Omega_p+1/2)-E(K=\Omega_p-1/2)} \qquad (4\text{-}176)$$

In the approximation considered, we may take $A=\frac{1}{2}(A(K=\Omega_p+1/2)+A(K=\Omega_p-1/2))$. Inserting the observed values (see Fig. 4-20) of $A(K=3)$, $A(K=4)$, and ΔE (taken from the difference of the $I=4$ states), we obtain the value $a=0.97$, which is to be compared with the average value $a=0.6$ observed for the [521 1/2] orbit in neighboring odd-N nuclei (see Table 5-16, p. 309). Thus, the estimated coupling appears to account for only half the observed difference in the moments of inertia.

The rotational energy for $K=1$ bands contains a signature-dependent term of the type (see Eq. (4-62))

$$\delta E=A_2(-1)^{I+1}I(I+1) \qquad (4\text{-}177)$$

The measured energies of the rotational band beginning at 426 keV set a limit on A_2 that is at least four orders of magnitude smaller than the parameter A. A relatively small value for the A_2 term is expected on the basis of the configuration assignment given above. In fact, the term arises from second-order effects of the Coriolis coupling to $K=0$ bands and therefore vanishes for two-particle configurations, except those with $(\Omega_n,\Omega_p)=(1/2,\ 1/2)$.

Direct support for the configuration assignment of the $K\pi=1+$ band comes from the observation of a rather large GT matrix element for β decay of ^{166}Dy populating the $I\pi=1+$ level ($\log ft\approx4.9$; Motz *et al., loc. cit.*, Fig. 4-20). Such an "allowed unhindered" matrix element identifies the transition as occurring between spin-orbit partners. (For a discussion of the classification of the allowed β decays in terms of the intrinsic one-particle states in

▲ deformed nuclei, see Chapter 5, pp. 306ff.)

▼ *Fission channels observed in photofission of ^{238}U (Figs. 4-21 to 4-23)*

Evidence on rotational quantum numbers in intermediate stages of the fission process has been obtained from the analysis of angular distributions of the fission fragments. Quite generally, these angular distributions can be characterized by the probability distribution of the quantum number K representing the component of the angular momentum in the direction of the emitted fragments (see the helicity wave function (3A-5)). For fission proceeding through a compound nucleus, we can ignore the interference between channels of different total angular momentum, and the differential cross section for emission of fragments in the angle θ relative to the incident beam has the form (Bohr, 1956; Strutinsky, 1956; Wheeler, 1963)

$$d\sigma_f(\theta) = \sum_I \sum_{M=-I}^{I} \sigma(IM) \sum_{K=0}^{I} \frac{\Gamma_f(IK)}{\Gamma(I)} \frac{2I+1}{4} \left(|\mathscr{D}^I_{MK}(\theta)|^2 + |\mathscr{D}^I_{M-K}(\theta)|^2\right) \sin\theta \, d\theta \tag{4-178}$$

where $\sigma(IM)$ is the cross section for forming the compound nucleus with the total angular momentum I and component M in the beam direction. The partial width for fission with specified K is denoted by $\Gamma_f(IK)$, and $\Gamma(I)$ is the total width for given I.

Angular anisotropies in fission processes were first observed in the photofission of ^{232}Th and ^{238}U near threshold (Winhold, Demos, and Halpern, 1952). The photoabsorption in an even-even nucleus leads to states with $M = \pm 1$ (since the photon has helicity ± 1) and $I = 1$ or 2, corresponding to dipole or quadrupole absorption. The angular distribution (4-178) can thus be written in the form

$$W(\theta) = a + b \sin^2\theta + c \sin^2 2\theta \tag{4-179}$$

with

$$\begin{aligned}
a &= \tfrac{3}{4}P(11) + \tfrac{5}{4}P(21) \\
b &= \tfrac{3}{4}P(10) - \tfrac{3}{8}P(11) - \tfrac{5}{8}P(21) + \tfrac{5}{8}P(22) \\
c &= \tfrac{15}{16}P(20) - \tfrac{5}{8}P(21) + \tfrac{5}{32}P(22)
\end{aligned} \tag{4-180}$$

$$P(IK) \equiv \frac{\Gamma_f(IK)}{\Gamma(I)} \sum_M \sigma(IM)$$

Experimental data on the angular distribution of photofission in even-even nuclei are found to be in good agreement with the expression (4-179); see the example in Fig. 4-21, which refers to the photofission of ^{238}U in the ▲

▼ region just below threshold. (For the excitation function of ^{238}U in the threshold region, see Fig. 4-23.) The energy dependence of the ^{238}U(γf) ▲ anisotropy is illustrated in Fig. 4-22.

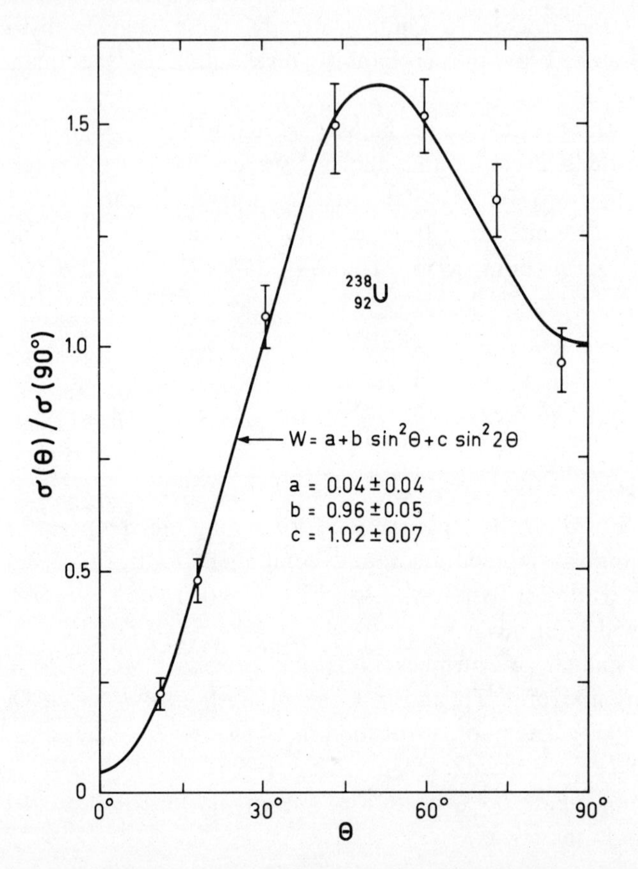

Figure 4-21 Angular distribution of fission fragments from photofission of ^{238}U. The data are taken from A. S. Soldatov, G. N. Smirenkin, S. P. Kapitza, and Y. M. Tsipeniuk, *Phys. Letters* **14**, 217 (1965). The fission processes result from the continuous bremsstrahlung spectrum produced by 5.2 MeV electrons. On account of the rapid variation of the fission cross section for the photon energies involved (see Fig. 4-23), the γ quanta contributing effectively to the observed fission processes have energies within a rather narrow interval, estimated to extend approximately 100 keV below the end point.

▼ The large anisotropies observed in the threshold region imply that the K quantum number characterizing the fission process is far from randomly distributed. This feature implies that the K distribution is determined by a stage in the fission process where the nucleus is "cold" and where, therefore, only few channels are available (Bohr, 1956). Such a stage occurs when the nucleus passes the saddle-point configuration, where almost the entire excitation energy is expended in the deformation. ▲

▼ The spectrum of channels is intimately connected with the symmetry of the sadddle-point shape. If this shape has axial symmetry and $\mathscr{R}$ invariance, the spectrum is similar to that observed near the nuclear ground states; in particular, the lowest channels form a band with $K\pi r = 0++$. The contribution of the $I\pi = 2+$ member of this band would account for the strong increase in the c coefficient for photofission of ^{238}U at the lowest energies ▲ shown in Fig. 4-22.

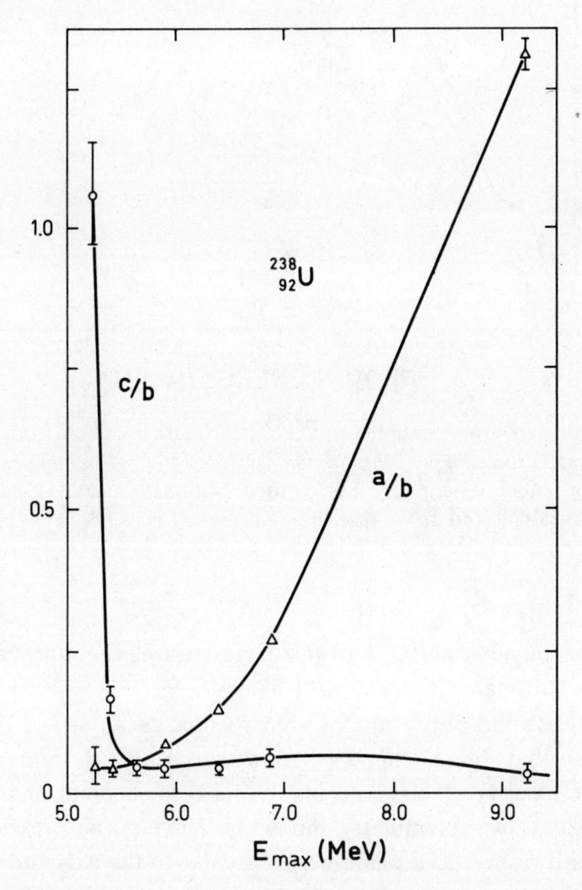

Figure 4-22 Energy dependence of anisotropy parameters in photofission of ^{238}U. The data are taken from Soldatov *et al., loc. cit.,* Fig. 4-21. Each point refers to the angular distribution obtained for a continuous bremsstrahlung spectrum with maximum energy E_{max}.

▼ The $E2$-photoabsorption process is intrinsically much weaker than the $E1$ process and therefore, at energies above that of the lowest $I\pi = 1-$ fission channel, one expects a strong dominance of dipole photofission. Indeed, in ▲ the energy region above 5.5 MeV in Fig. 4-22, the coefficient b is much larger

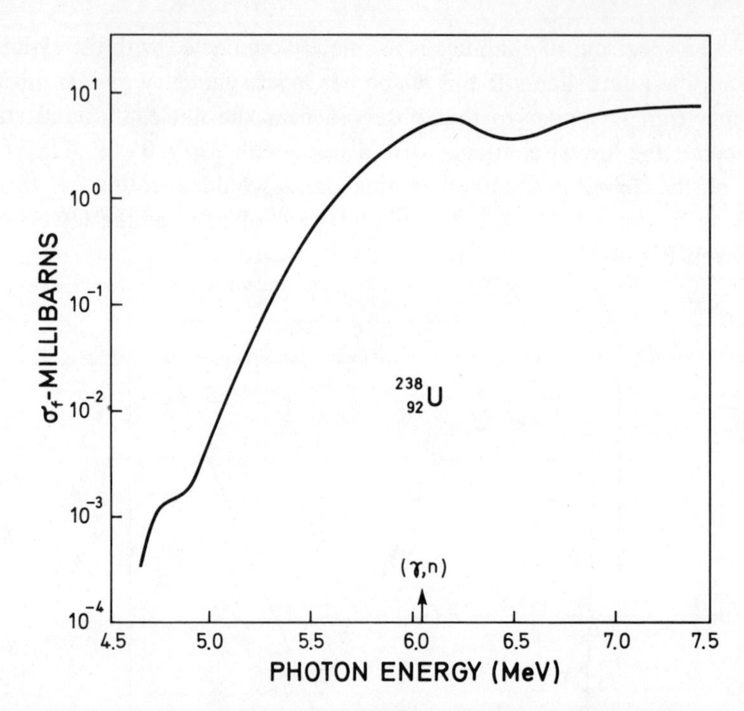

Figure 4-23 Photofission cross section for ^{238}U. The figure is taken from L. Katz, A. P. Baerg, and F. Brown, *Proceedings of the Second United Nations International Conference on the Peaceful Uses of Atomic Energy* **15**, 188, United Nations, Geneva, 1958, and the more recent data by A. M. Khan and J. W. Knowles, *Nuclear Phys.* **A179**, 333 (1972).

▼ than the coefficients a and c, and this characterizes the dominant process as dipole fission through a channel with $K\pi r = 0--$.

The lowest odd-parity modes are expected to be associated with shape deformations that break the $\mathscr{P}$ symmetry. A $K=0$ mode of this type preserves the axial symmetry and involves a displacement of mass along the symmetry axis (mass asymmetry mode); a $K=1$ mode preserves reflection symmetry with respect to a plane perpendicular to the axis and gives rise to a bending of the system. The photofission evidence implies that the $K\pi = 0-$ mode is appreciably lower than the $K\pi = 1-$ mode, and provides an indication that the saddle-point configuration is relatively soft, though not unstable, with respect to the mass asymmetry mode. If the system possessed a static deformation of this type, the $I\pi = 1-$ and $I\pi = 2+$ states would belong to the ground-state band (see Fig. 4-2b), and the dipole photofission would be expected to dominate over the entire threshold region. (The evidence from the photofission anisotropies in even Pu isotopes is similar to that in ^{238}U, while in ^{232}Th the magnitude of the c term is much smaller, which might imply that the saddle configuration in this nucleus violates $\mathscr{P}$ symmetry; see Rabotnov *et al.*, 1970.) ▲

▼ At higher energies, the photofission becomes approximately isotropic, which can be understood in terms of dipole fission with $P(11) = 2P(10)$, as expected for a statistical distribution. (In such a distribution, the intrinsic states with $K \neq 0$, which represent angular momentum projections $I_3 = \pm K$, have double weight, as compared with the $K=0$ states. In the enumeration of rotational states, the $\mathscr{R}$ symmetry implies that there is only a single state (KI) associated with the pair of intrinsic states with $I_3 = \pm K \neq 0$, while for $K=0$, a corresponding reduction in the number of rotational states results from the condition $r = (-1)^I$; see Sec. 4-2c.)

If the saddle-point shape deviates from axial symmetry, the spectrum of channels will be significantly different from that considered above (see Sec. 4-5). In the absence of invariance with respect to any finite rotation, the rotational bands contain $2I+1$ states for each value of I, and all values of K occur with equal weight. The angular distribution for photofission would therefore be isotropic, since the rotational energy spacings for $I=1$ and $I=2$ are small compared with the energies characterizing the opening of a single channel. If the shape retains invariance with respect to the rotations $\mathscr{R}_\kappa(\pi)$ of $180°$ with respect to the three principal axes, as for an ellipsoidal shape, the bands are labeled by the quantum numbers $(r_1 r_2 r_3)$; see pp. 177 ff. The lowest band is expected to have $(r_1 r_2 r_3) = (+++)$ and contains the set of states $I = 0, 2^2, 3, \ldots$ (see Eq. 4-277). The $I=2$ states span a space comprising $K=0$ and $K=2$ components with equal weight, which would give a quadrupole photofission with $a=0$ and $c/b = 7/4$ (see Eq. (4-180)). Such an interpretation is not ruled out by the data in Fig. 4-22. An ellipsoidal deformation does not couple $K=0$ and $K=1$ (symmetry with respect to the operation $\mathscr{R}_3(\pi)$) and therefore has no effect on the dipole fission.

Further evidence on the K distribution in the channel spectrum has been obtained from the measurements of angular distributions in various direct nuclear reactions that populate the compound nucleus in the energy region around the fission threshold ((d, pf), see, for example, Britt *et al.*, 1965; (t, pf), see, for example, Eccleshall and Yates, 1965; $(\alpha, \alpha' \mathrm{f})$, see Wilkins *et al.*, 1964; Britt and Plasil, 1966). Especially striking angular distributions are observed in the $(\alpha\alpha')$ process, which excites states of high angular momentum ($I \sim 8$ or greater, for $E_\alpha \approx 40$ MeV). These angular distributions are strongly peaked in the direction of the momentum transfer; from the width of the peak it can be concluded that the values of K (and M) are very much smaller than I. Although it is difficult to give an unambiguous interpretation, due to the lack of knowledge of the reaction mechanism and therefore of the relative magnitude of the partial cross sections $\sigma(IM)$ in Eq. (4-178), the data appear to be consistent with fission proceeding predominantly through $K=0$ channels, for energies within about 1 MeV of the threshold (Wilkins *et al.*, 1964). For excitation energies of a few MeV above threshold, the angular anisotropies decrease due to the participation of channels with higher values of K, as in ▲ photofission. For fission reactions involving still higher excitation energies

▼ (and large angular momenta), the angular distributions can be interpreted in terms of a statistical distribution of channels with different K and yield evidence on the moments of inertia parallel and perpendicular to the symmetry axis (see the example discussed in Chapter 6, pp. 617ff.).

The interpretation of the K distribution of the fission fragments in terms of the channel spectrum at saddle point involves the assumption that the K quantum number remains constant during the passage from saddle point to scission, as is expected if this stage of the process proceeds rapidly compared with the rotational period. The large anisotropies observed in the threshold region for the photofission of ^{238}U are consistent with this assumption. The passage from saddle point to scission may be significantly prolonged if the potential energy function contains a secondary minimum in this region; the system may then become trapped in this minimum, in which case the angular distribution of the fissioning nuclei will be determined by the channel spectrum at the outer barrier that temporarily retains the nuclei in the second minimum; see the discussion in Chapter 6, pp. 634.

Intensity relations for E2-matrix elements within a rotational band (Fig. 4-24)

In first approximation, the $E2$-matrix elements within a rotational band are determined by the single parameter Q_0 (see Eq. (4-68)), and ratios of matrix elements are therefore completely specified by the rotational quantum numbers KIM.

E2-transition probabilities

A favorable opportunity for testing the $E2$- rotational intensity relation is provided by the Coulomb excitation process. In an odd-A nucleus with ground-state spin I_0 $(=K)$, the first-order Coulomb excitation can populate the two first excited states with $I=I_0+1$ and I_0+2, and the ratio of the $B(E2)$ values can be directly determined by comparing the intensities of the corresponding groups of inelastically scattered projectiles. The relative $B(E2)$ values obtained in this manner are shown in Fig. 4-24. (Additional evidence for the validity of the $E2$-intensity rules in the bands of odd-A nuclei is given in Tables 4-6, p. 104, 4-8, p. 108, and 4-10, p. 113.)

Tests of the $E2$-intensity relation for the ground-state band of even-even nuclei are provided by multiple Coulomb excitation measurements as well as by lifetime determinations. Examples of measured $B(E2)$ ratios for the $6 \rightarrow 4$, $4 \rightarrow 2$, and $2 \rightarrow 0$ transitions are shown in Fig. 4-24.

The generalized $E2$-intensity relations, which include the effect of rotational perturbations, are discussed on pp. 50ff. For $K=0$ bands, the lowest-

▲ order corrections are quadratic in I and are of the form (4-81); for the

▼ transitions $I+2 \rightarrow I$, the $B(E2)$ values can be written

$$B(E2; I+2 \rightarrow I) = \frac{5}{16\pi} e^2 Q_0^2 \langle I+2020|I0\rangle^2 (1+2\alpha I(I+3))^2 \tag{4-181}$$

with

$$\left(\frac{5}{16\pi}\right)^{1/2} eQ_0 = M_1 + 6M_2 + 36M_3$$
$$\alpha \approx \frac{M_2 + 8M_3}{M_1} \tag{4-182}$$

The corrections to the leading-order intensity relations therefore involve only a single parameter, α. The data in Fig. 4-24 are consistent with the leading-order relation and imply $|\alpha| \lesssim 10^{-3}$, except for the case of ^{152}Sm, for which ▲ $\alpha \approx (1 \pm 0.5) \times 10^{-3}$.

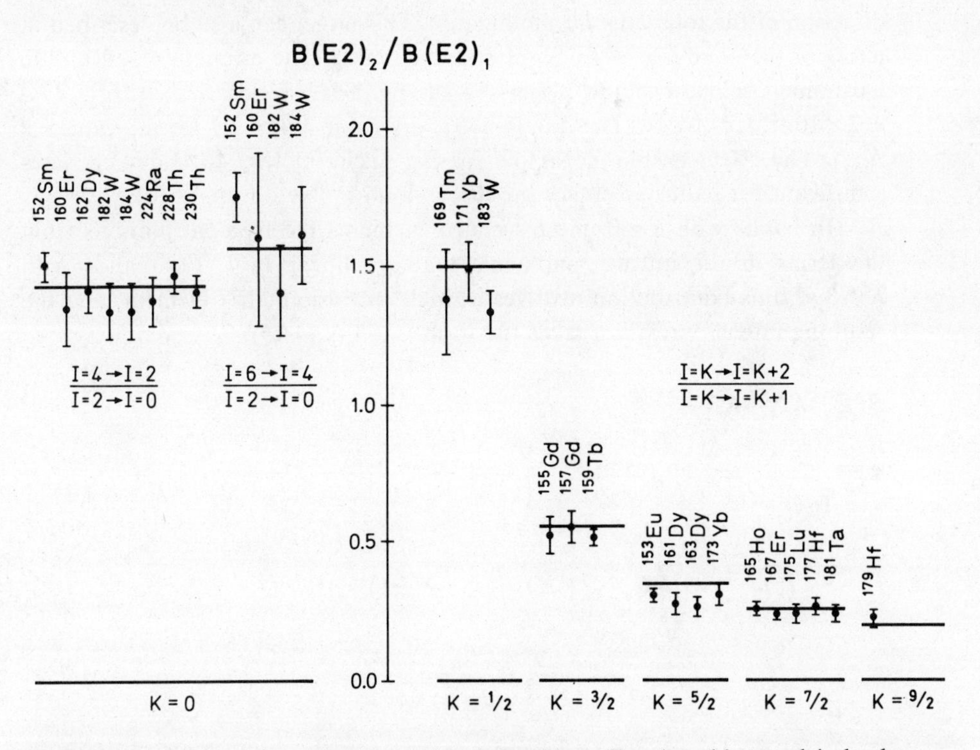

Figure 4-24 $E2$-intensity ratios in rotational transitions. For the odd-A nuclei, the data are taken from B. Elbek, *loc. cit.*, Fig. 4-25. For ^{169}Tm, see Table 4-6. For the even-even nuclei, the data are from A. C. Li and A. Schwarzschild, *Phys. Rev.* **129**, 2664 (1963); W. R. Neal and H. W. Kraner, *Phys. Rev.* **137**, B1164 (1965); R. M. Diamond, F. S. Stephens, R. Nordhagen, and K. Nakai, *Contributions to International Conference on Properties of Nuclear States*, p. 7, Montreal (1969); R. M. Diamond, F. S. Stephens, W. H. Kelly, and D. Ward, *Phys. Rev. Letters* **22**, 546 (1969); W. T. Milner, K. K. McGowan, R. L. Robinson, P. H. Stelson, and R. O. Sayer, *Nuclear Phys.* **A177**, 1 (1971). Evidence on $E2$-intensity relations in the ground-state band of ^{20}Ne is shown in Table 4-4.

▼ Part of the distortion effect described by the parameter α may be associated with the coupling between the ground-state band and the vibrational modes representing quadrupole oscillations about the equilibrium shape. This coupling can be obtained from an analysis of the $E2$ intensities for the decay of the vibrational bands. Such analyses are illustrated by the discussion of the $K=2$ vibrational band (γ vibration) in ^{166}Er (see pp. 158 ff.) and of the $K=0$ band (β vibration) in ^{174}Hf (see pp. 168 ff.). In these cases, the contributions to α amount to a few times 10^{-4} and thus imply corrections that are beyond the accuracy of present experiments. However, significantly larger effects are expected in the regions of transition between spherical and deformed nuclei, where the amplitudes of the oscillations around the equilibrium shape may become large.

An additional contribution to α arises from the tendency of the rotation to align the angular momenta of the individual particles with respect to the direction of the total angular momentum. This effect can also be described in terms of the coupling to the $K=1$ spurious mode. The estimate based on the harmonic oscillator model discussed on pp. 84 ff. yields $\alpha \approx -(2I_{max})^{-2} \approx -2\times 10^{-5}$, for $A\approx 160$ (see Eq. (4-143) and Table 4-3, p. 27 for the values of I_{max}). The effect is thus negligible for the nuclei in Fig. 4-24, but becomes significant for light nuclei (see the discussion of ^{20}Ne on pp. 96 ff.).

In bands with $K\neq 0$, the $E2$-matrix elements receive contributions from the terms in the intrinsic moments that are linear in I. For bands with $K>3/2$, this contribution involves a single intrinsic matrix element, and the resulting intensity relation takes the form (see Eq. (4-75))

$$\begin{aligned} B(E2; KI_1 \to KI_2) &= \frac{5}{16\pi} e^2 Q_0^2 \Big(\langle I_1 K 2 0 | I_2 K\rangle + \zeta \big(\langle I_1\, K+1\, 2\, -1 | I_2 K\rangle (I_1 - K)^{1/2} (I_1 + K + 1)^{1/2} \\ &\quad - \langle I_1\, K-1\, 2\, 1 | I_2 K\rangle (I_1 + K)^{1/2} (I_1 - K + 1)^{1/2} \big) \Big)^2 \end{aligned} \tag{4-183}$$

$$\zeta = \frac{\langle K | m_{0,-1} | K\rangle}{\langle K | m_{0,0} | K\rangle}$$

$$\left(\frac{5}{16\pi}\right)^{1/2} eQ_0 = \langle K | m_{0,0} | K\rangle$$

For $K=1/2$ and $3/2$, the corrections to the leading-order intensity matrix element involve an additional term proportional to the signature (see Eq. (4-77)).

The term proportional to ζ in Eq. (4-183) can be attributed to the ▲ $\Delta K=1$ band mixing due to the Coriolis coupling. The effect of this coupling

▼ can be obtained from Eqs. (4-202) and (4-203),

$$(\delta\,\mathscr{M}(E2,\mu))_{\Delta K=0}$$

$$= i[S, \mathscr{M}(E2,\mu)] - \tfrac{1}{2}[S,[S, \mathscr{M}(E2,\mu)]]$$

$$= \left(\tfrac{1}{2}\sqrt{6}\,\{\varepsilon_{+1}, \mathscr{M}_{-1}\} + \tfrac{3}{2}\{\varepsilon_{+1},\varepsilon_{-1}\}\,\mathscr{M}_0\right)\mathscr{D}^2_{\mu 0}$$

$$+ \left(\tfrac{1}{2}[\varepsilon_{+1}, \mathscr{M}_{-1}] + \tfrac{1}{4}\sqrt{6}\,[\varepsilon_{+1},\varepsilon_{-1}]\,\mathscr{M}_0\right)\left\{I_-, \mathscr{D}^2_{\mu-1}\right\} + \mathscr{R}\text{-conj.} \quad (4\text{-}184)$$

in terms of the amplitudes $\varepsilon_{\pm 1}$ of the admixed bands. We have assumed the unperturbed intrinsic moments ($\mathscr{M}_\nu \equiv \mathscr{M}(E2,\nu)$) to be I independent, and have included the second-order terms involving the collective moment $\mathscr{M}_0 = (5/16\pi)^{1/2} eQ_0$, with Q_0 treated as a constant. The terms proportional to $\mathscr{D}^2_{\mu 0}$ give a small renormalization of the intrinsic moment Q_0, while the terms linear in $I_\pm$ are of the same form as in Eq. (4-203) and yield

$$\zeta = \zeta_1 + \zeta_2$$

$$\zeta_1 = \left(\left(\frac{5}{16\pi}\right)^{1/2} eQ_0\right)^{-1} \langle K\|[\varepsilon_{+1}, \mathscr{M}_{-1}]\|K\rangle \quad (4\text{-}185)$$

$$\zeta_2 = \frac{\sqrt{6}}{2}\langle K\|[\varepsilon_{+1},\varepsilon_{-1}]\|K\rangle$$

$$= \frac{\sqrt{6}}{2}\left(\sum_i \langle i,K+1|\varepsilon_{+1}|K\rangle^2 - \sum_i \langle i,K-1|\varepsilon_{-1}|K\rangle^2\right)$$

The expectation values of the commutators can be expressed in terms of sums over the intermediate states with $K \pm 1$, as shown for the second-order contribution ζ_2. (Note that $\varepsilon_{-1} = -(\varepsilon_{+1})^\dagger$.)

Since ζ vanishes for $K=0$ bands, one may attempt to estimate the value of ζ for the low-lying bands in odd-A nuclei in terms of the contribution from the last odd particle. In a number of cases, it has been possible to obtain experimental estimates of the $\Delta K=1$ band mixing amplitudes, and it appears that, in the $E2$ transitions between the bands, the term proportional to $\varepsilon_{+1}Q_0$ is usually appreciably larger than the direct term, which is proportional to $\mathscr{M}(E2,\nu=1)$ (see, for example, the discussion of ^{175}Lu, p. 155, and of ^{235}U, p. 275). Thus, we may expect the main contribution to ζ to arise from the second-order term ζ_2.

The magnitude of ζ can, of course, become very large for accidentally near-lying bands, but for the ground-state bands, such degeneracies do not ▲ occur. Typical of the largest known ground-state admixtures are the ampli-

▼ tudes $\langle[752\ 5/2]\|\varepsilon_{-1}\|[743\ 7/2]\rangle \approx 3\times 10^{-2}$ as observed in ^{235}U (see p. 277) and $\langle[521\ 3/2]\|\varepsilon_{-1}\|[523\ 5/2]\rangle \approx 2\times 10^{-2}$ as observed in ^{163}Dy (see Tveter and Herskind, 1969). These admixtures imply $\zeta \approx \zeta_2 \approx -10^{-3}$, which in turn leads to an increase in the ratio of $B(E2; I_0 \rightarrow I_0+2) : B(E2; I_0 \rightarrow I_0+1)$ by about 1%. It must be emphasized that at present the evidence on the $\Delta K=1$ couplings is very incomplete, and in the cases considered, there will be additional contributions to ζ, positive as well as negative, from other $\Delta K=1$ excitations.

The deviations indicated by the measured cross sections for the Dy isotopes (see Fig. 4-24) are an order of magnitude larger than the effect estimated above; the deviations are only a few times the standard error, but if confirmed by more accurate measurements, would appear to be difficult to account for in terms of the couplings considered.

Static E2 moments

The measured static quadrupole moments provide still another source of evidence on $E2$-matrix elements within a rotational band. Indeed, the discovery of the very large static quadrupole moments deduced from the analysis of the atomic hyperfine structure gave the first indication of collective nuclear deformation effects (see footnote 1, p. 2) and also established the prolate character of the nuclear deformations.

The magnitude of the static quadrupole moments may be compared with the transition moments, in order to provide additional tests of the $E2$-intensity relations (see Eqs. (4-70) and (4-68)). However, the extensive body of data from atomic and molecular hyperfine structure measurements cannot be directly employed in such tests, due to the uncertainties in the knowledge of the electric field gradient at the nucleus that is produced by the bound electrons. The ratios of quadrupole moments for different isotopes of the same element are not subject to this uncertainty, and the values obtained are found to be consistent with the ratios of the transition moments to within the experimental accuracy, which is typically of the order of a few percent (see, for example, the ratio of the moments of Lu (Spalding and Smith, 1962), Gd (Stevens *et al.*, 1967), and W (Persson *et al.*, 1968).

Quantitative determinations of static quadrupole moments can be obtained from high-resolution studies of the energy levels of μ-mesic atoms, in states with large orbital angular momentum; preliminary experiments of this type have yielded $Q(^{175}\text{Lu}) \approx 350$ fm^2 (Leisi *et al.*, 1973), in good agreement with the intrinsic moment $Q_0 \approx 750$ fm^2, as obtained from the data quoted in Fig. 4-25. A further source of information on static nuclear quadrupole moments is provided by the multiple Coulomb excitation process

▲ (see, for example, the moment of the 2+ state of ^{188}Os, shown in Fig. 6-32).

▼ *Determination of nuclear deformations (Fig. 4-25; Tables 4-15 and 4-16)*

Quadrupole deformations from electromagnetic moments

The evidence on the magnitude of the quadrupole distortion in deformed nuclei, as determined from $E2$-matrix elements by means of Eqs. (4-68) and (4-72), is summarized in Fig. 4-25 and Table 4-15.

Most of the data have been obtained from Coulomb excitation cross sections or lifetime determinations. These measurements determine only the square of the intrinsic quadrupole moment, but the sign can be obtained from the measurement of static quadrupole moments. For the regions of nuclei shown in Fig. 4-25, the available evidence on static moments is consistent with a prolate shape in all cases. ▲

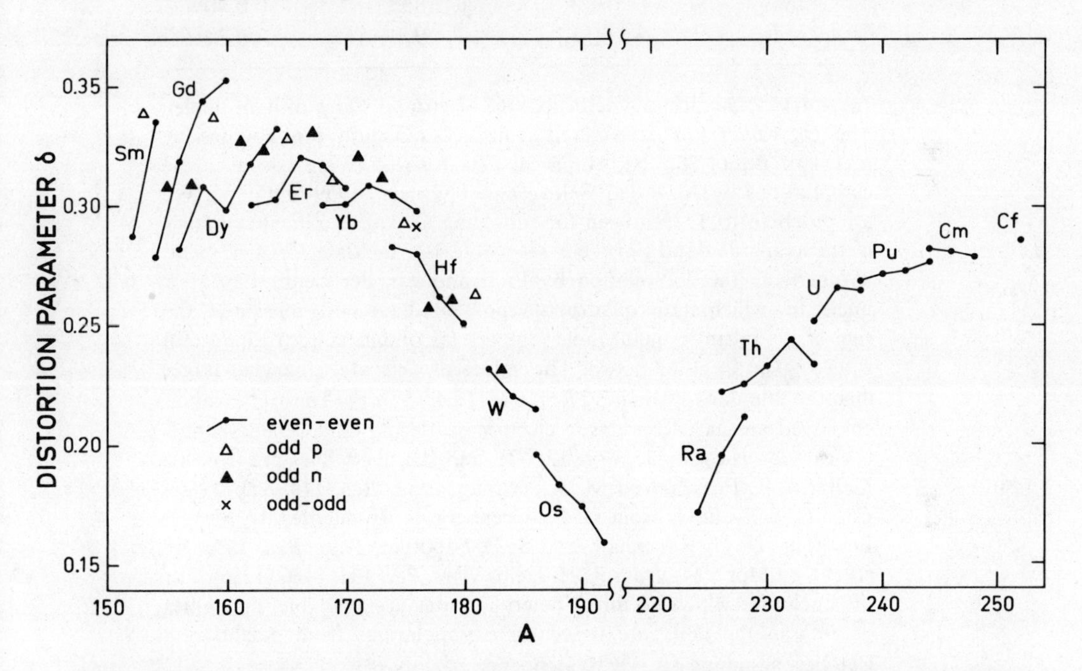

Figure 4-25 Quadrupole deformations of nuclear ground states ($A > 150$). The data for the region $150 < A < 190$ are taken from B. Elbek, *Determination of Nuclear Transition Probabilities by Coulomb Excitation*, Munksgaard, Copenhagen (1963), which summarizes the results of the series of experiments: B. Elbek, K. O. Nielsen, and M. C. Olesen, *Phys. Rev.* **108**, 406 (1957); V. Ramšak, M. C. Olesen, and B. Elbek, *Nuclear Phys.* **6**, 451 (1958); M. C. Olesen, and B. Elbek, *Nuclear Phys.* **15**, 134 (1960); B. Elbek, M. C. Olesen, and O. Skilbreid, *Nuclear Phys.* **10**, 294 (1959); **19**, 523 (1960); O. Hansen, M. C. Olesen, O. Skilbreid, and B. Elbek, *Nuclear Phys.* **25**, 634 (1961). For the heavy-element region, the data are taken from the lifetime determinations by R. E. Bell, S. Bjørnholm, and J. C. Severiens, *Mat. Fys. Medd. Dan. Vid. Selsk.* **32**, no. 12 (1960) and from the Coulomb excitation studies by J. L. C. Ford, Jr., P. H. Stelson, C. E. Bemis, Jr., F. K. McGowan, R. L. Robinson, and W. T. Milner, *Phys. Rev. Letters* **27**, 1232 (1971).

Nucleus	$\lvert Q_0 \rvert$ fm^2	$\langle r^2 \rangle$ fm^2	δ
^{7}Li	19+	5.4	0.88
^{9}Be	32+	5.1	1.2
^{10}Be	26	5.1	0.95
^{11}B	20	5.0	0.60
^{12}C	22	6.25	0.44
^{13}N	16	6.5	0.26
^{19}F	33	7.6	0.36
^{19}Ne	43	7.6	0.43
^{20}Ne	54+	7.9	0.51
^{24}Mg	57+	8.9	0.40
^{25}Mg	50+	9.1	0.34

Table 4-15 Quadrupole deformations of ground states of light nuclei. The Q_0 values are determined from $E2$-transition probabilities by means of Eq. (4-68). The empirical data for $A \leqslant 19$ are based on the compilations by Ajzenberg-Selove and Lauritsen, 1959 and 1966. We are indebted to T. Lauritsen for supplying us with a critical summary of the available data. For ^{20}Ne, see Table 4-4; the data for $A = 24$ and 25 are from the compilation by Endt and van der Leun, 1967. For nuclei for which static quadrupole moments have been measured, the sign of the intrinsic quadrupole moment (as obtained from Eq. (4-69)) is also given in column two. The values of $\langle r^2 \rangle$ are somewhat larger than the standard estimate $\frac{3}{5}R^2 = 0.6(1.2A^{1/3}\text{ fm})^2$. The $\langle r^2 \rangle$ values employed are based on elastic electron scattering data, whenever this information is available. For ^{7}Li, ^{9}Be, and ^{11}B, the values are from R. E. Rand, R. Frosch, and M. R. Yearian, *Phys. Rev.* **144**, 859 (1966). For ^{12}C, the value is from H. F. Ehrenberg, R. Hofstadter, U. Meyer-Berkhout, D. G. Ravenhall, and S. E. Sobottka, *Phys. Rev.* **113**, 666 (1959), and for ^{24}Mg from R. H. Helm, *Phys. Rev.* **104**, 1466 (1956). For the nuclei for which no direct determinations are available, the values of $\langle r^2 \rangle$ in the table are based on extrapolations from neighboring isotopes, assuming an $A^{2/3}$ dependence.

▼ Additional information on the quadrupole deformation of the nuclear charge density is provided by the study of μ-mesic spectra and electron scattering. The nonspherical components of the Coulomb field of a deformed nucleus imply that, in heavy atoms, the motion of the μ meson in states close to the nucleus is strongly coupled to the nuclear rotation, and the energy levels of the μ-mesic atom therefore yield information on a number of different $E2$-matrix elements within the ground-state rotational band (Wilets, 1954; Jacobsohn, 1954). The present data yield values of the intrinsic deformation consistent with those given in Fig. 4-25, and of comparable accuracy ▲ (see, for example, the reviews by Devons and Duerdoth, 1969, and by Wu

▼ and Wilets, 1969; see also Chen, 1969). Since the radius of the lower muonic orbits are comparable with the nuclear radius, the muonic spectra are sensitive to the radial distribution of the nuclear quadrupole moment; the evidence is consistent with the simple form factor obtained by a uniform deformation of the surfaces of constant density (see Eq. (4-189)). Detailed evidence on the $E2$-transition moments within rotational bands is also becoming available from inelastic electron scattering (Bertozzi *et al.*, 1972).

Estimates based on shell structure

An estimate of the nuclear eccentricity may be based on the aligned coupling scheme, in which each nucleon moves independently in the average deformed field produced by the rest of the nucleons. The equilibrium deformation can be determined from a self-consistency argument, requiring that the eccentricity of the potential and of the density should be equal at equilibrium. A simple estimate of the nuclear quadrupole deformations can be obtained by considering independent-particle motion in an anisotropic oscillator potential (see p. 77). The condition of self-consistency implies that the frequencies ω_κ in the directions of the three principal axes are inversely proportional to the quantities Σ_κ, which measure the total number of quanta in the corresponding directions (see Eq. (4-115)). Evaluating the deformation parameter δ given by Eq. (4-72), one obtains (see Eq. (4-113))

$$\begin{aligned}\delta &= \frac{3}{4}\frac{\langle \sum_{k=1}^{A}(2x_3^2 - x_1^2 - x_2^2)_k\rangle}{\langle \sum_{k=1}^{A}(x_1^2 + x_2^2 + x_3^2)_k\rangle}\\ &= \frac{3}{4}\frac{2\omega_3^{-1}\Sigma_3 - \omega_1^{-1}\Sigma_1 - \omega_2^{-1}\Sigma_2}{\omega_1^{-1}\Sigma_1 + \omega_2^{-1}\Sigma_2 + \omega_3^{-1}\Sigma_3}\\ &= \frac{3}{4}\frac{2\Sigma_3^2 - \Sigma_1^2 - \Sigma_2^2}{\Sigma_1^2 + \Sigma_2^2 + \Sigma_3^2}\end{aligned} \tag{4-186}$$

The eccentricity parameter δ depends on the differences between the numbers of oscillator quanta in the different directions, as described by Σ_κ, and for axially symmetric deformations with $\delta \ll 1$, the eccentricity can be expressed in the form

$$\begin{aligned}\delta &\approx \frac{3(\Sigma_3 - \Sigma_2)}{\Sigma_1 + \Sigma_2 + \Sigma_3}\\ &\approx \pm 4(\tfrac{2}{3})^{1/3} I_{\max} A^{-4/3}\end{aligned} \tag{4-187}$$

▲ where, in the last line, we have employed the estimates (2-157) and (2-158) for

▼ $\Sigma(N+3/2)$ and have expressed the anisotropy of the quanta in terms of the quantity I_{max} $(=|\Sigma_3-\Sigma_2|)$, which gives the maximum value of I contained in the aligned state (see Eq. (4-138)).

The values of δ given by Eq. (4-187) are of the order of the ratio of the particles outside of closed shells to the total number of particles and thus, in the middle of shells, the lowest configurations have eccentricities of order $A^{-1/3}$. An evaluation of Σ_κ on the basis of the observed order of filling of the single-particle orbits (as discussed in Chapter 5) leads to values of δ in qualitative agreement with the data in Fig. 4-25 and Table 4-15; see, for example, the values of δ implied by the estimated values of I_{max}, given in Table 4-3. This simple estimate of the equilibrium deformation neglects the deviations of the one-particle wave functions from those of the harmonic oscillator potential, as well as the effect of the pair correlations. The latter favor the spherical shape (with its higher degeneracy of the one-particle levels) and, in particular, are responsible for the transition to a spherical equilibrium shape, with the approach to closed-shell configurations (see the discussion in Chapter 6, pp. 520ff.).

It is a remarkable fact that all the strongly deformed nuclei that have so far been identified have a prolate shape. For a harmonic oscillator potential, in the limit of large quantum numbers, there is a symmetry between the prolate shapes occurring at the beginning of the shell and the oblate deformations at the end of the shell. (It is seen from Eq. (4-186) that, in this limit, configurations corresponding to the interchange of particles and holes give values of δ of the same magnitude.) The deviations from the harmonic oscillator potential imply that the single-particle spectrum in the spherical potential is unsymmetrical with respect to particles and holes in major shells (see, for example, Fig. 5-1 and the comments in Chapter 5, p. 298), and detailed estimates appear to be consistent with the observed prolate shapes for the nuclei in Fig. 4-25 (as well as the nuclei at the beginning of the (sd) shell). However, it remains an interesting problem to identify the structural features that are responsible for the systematic favoring of the prolate deformations.

Another important feature in the qualitative characterization of the nuclear deformations is the preference for axial symmetry. For a harmonic oscillator potential, an axially symmetric quadrupole deformation leaves the single-particle states degenerate with respect to the distribution of quanta in the x_1x_2 plane and in general, therefore, this potential would lead to ellipsoidal deformations with three different axes. Since the degeneracy of the orbits in the axially symmetric potential is of order $A^{1/3}$, the deviations from axial symmetry are only of order $A^{-2/3}$, corresponding to the fractional number of particles that can occupy a subshell. Effects opposing deviations from axial symmetry are provided by the spin-orbit coupling (since the orbital moment is quenched by the asymmetric potential; see Sec. 5-1c, p. 218) and the pair correlations (which favor maximum degeneracy of the ▲

▼ single-particle orbits); there is at present no clear-cut evidence for the occurrence of stable nuclear deformations without axial symmetry (see, for example, the discussion of the spectra of ^{166}Er (p. 166) and ^{24}Mg (p. 287)). It is possible, however, that such shapes may be encountered for states with high angular momentum (see the discussion on pp. 42ff.).

Deformation of nuclear fields

The analysis of the electromagnetic interaction effects yields information on the shape of the nuclear charge distribution; evidence on the deformation of the nuclear field may be obtained from the scattering or bound states of strongly interacting particles. The available data are rather preliminary, but consistent with values of δ similar to those obtained from the $E2$ measurements. (See, for example, the data in Table 4-16, obtained from ($\alpha\alpha'$) scattering. For the analysis of (pp′) scattering, see, for example, Table 6-2, p. 354. Evidence from the scattering of neutrons and protons on oriented nuclei has been reported by Marshak *et al.*, 1968, and Fisher *et al.*, 1971, respectively. The extensive evidence on the effect of the deformation of the nuclear potential on the bound states of neutrons and protons is discussed in Chapter 5. Evidence on the deformation of the potential acting on π mesons has been obtained from the analysis of π-mesic atoms (Leisi *et al.*, 1973).)

The Coulomb interaction and the presence of a neutron excess imply that the deformation of the neutron and proton density distributions may be somewhat different and thus give rise to an isovector potential with a deformation different from that of the isoscalar potential. The deformation of the isovector field could be obtained by comparing the cross sections for excitation of rotational states by inelastic scattering of neutrons and protons, of π^+ and π^-, or by observing the isobaric analogs of the rotational excitations in the direct (pn) process (see the discussion of the isobaric structure of vibrations in Chapter 6, pp. 375ff.).

Another measure of the deformation of the nuclear shape is provided by the splitting of the dipole resonance in deformed nuclei. The evidence from these measurements is summarized in Table 6-7 and gives deformation parameters that are consistent with those in Fig. 4-25 to within the accuracy of the determinations.

Parameters for describing shapes with higher multipole deformation

A systematic characterization of the different components in the nuclear shape deformations can be based on a macroscopic analysis of these deformations, as occurring in a system with a surface diffuseness small compared ▲ with the radius.[7] In such a system, the density $\rho_0(r)$ for spherical shape can be

[7] The macroscopic analysis of properties that follow from the existence of a surface region thin compared with the nuclear radius has been especially considered by Myers and Swiatecki (1969) and Myers (1973); such systems are referred to as "leptodermous."

▼ expressed in the form $\rho_0 f((r-R_0)/a_0)$, where ρ_0 is the central density in a heavy nucleus, while the form factor f involves the radius R_0 and the diffuseness parameter a_0 (as, for example, in the Woods-Saxon form factor (2-62)). A deformation with a wavelength large compared with the diffuseness can be described as a bodily displacement of the surface in each local region. The resulting density distribution is obtained by replacing R_0 by the angular-dependent radius $R(\vartheta,\varphi)$ and at the same time correcting the diffuseness parameter so that the magnitude of the normal derivative of the density at each point on an equidensity surface is unaffected by the deformation. Characterizing the deformed surface by the multipole parameters $\alpha_{\lambda\mu}$, we have

$$\rho(\mathbf{r}) = \rho_0 f((r - R(\vartheta,\varphi))/a(\vartheta,\varphi)) \tag{4-188a}$$

$$R(\vartheta,\varphi) = R_0\left(1 + \sum_{\lambda \geq 2,\mu} \alpha_{\lambda\mu} Y^*_{\lambda\mu}(\vartheta,\varphi) - \frac{c}{4\pi}\sum_{\lambda\geq 2,\mu} |\alpha_{\lambda\mu}|^2\right) \tag{4-188b}$$

$$a(\vartheta,\varphi) = a_0\left(1 + \tfrac{1}{2}(\nabla R)^2_{r=R(\vartheta,\varphi)}\right)$$

$$= a_0\left(1 + \tfrac{1}{2}R_0^2\left(\sum_{\lambda\geq 2,\mu} \alpha_{\lambda\mu}\nabla Y^*_{\lambda\mu}\right)^2_{r=R(\vartheta,\varphi)}\right) \tag{4-188c}$$

including the leading-order correction to the diffuseness a. The last term in R has been included in order to conserve the particle number, to order α^2; the coefficient c is unity for a sharp surface and can more generally be expressed in terms of radial moments (see Eq. (4-189c)). Deformations with $\lambda = 1$ are not included in the expansions in Eq. (4-188), since such a deformation is equivalent to a displacement of the center of the sphere. The expansion of the deformed density distribution in powers of $\alpha_{\lambda\mu}$ yields

$$\rho(\mathbf{r}) = \rho_0(r) - R_0\frac{\partial\rho_0}{\partial r}\left(\sum \alpha_{\lambda\mu}Y^*_{\lambda\mu} - \frac{c}{4\pi}\sum|\alpha_{\lambda\mu}|^2 + \tfrac{1}{2}(r-R_0)R_0\left(\sum\alpha_{\lambda\mu}\nabla Y^*_{\lambda\mu}\right)^2\right)$$

$$+\tfrac{1}{2}R_0^2\frac{\partial^2\rho_0}{\partial r^2}\left(\sum\alpha_{\lambda\mu}Y^*_{\lambda\mu}\right)^2 + \cdots \tag{4-189a}$$

$$\rho_0(r) \equiv \rho(\mathbf{r}, \alpha_{\lambda\mu}=0) = \rho_0 f((r-R)/a_0)$$

$$\rho_0 R_0 = \int_0^\infty \rho_0(r)\,dr \tag{4-189b}$$

$$c = \tfrac{1}{2}\frac{\langle r^{-2}\rangle}{\langle r^{-1}\rangle}R_0 \tag{4-189c}$$

▲

$$\langle r^n \rangle \equiv \frac{\int_0^\infty \rho_0(r) r^{n+2} dr}{\int_0^\infty \rho_0(r) r^2 dr} \tag{4-189d}$$

In the evaluation of the coefficient c from the normalization condition, it has been assumed that the radius R_0 is defined by the relation (4-189b), which, in particular, is satisfied for the radius parameter in the Woods-Saxon form factor (2-62), ignoring terms of order $\exp\{-R_0/a\}$.

For the density distribution (4-189a), the multipole moments can be expressed as power series in $\alpha_{\lambda\mu}$ with coefficients involving the radial moments $\langle r^n \rangle$ of ρ_0 (see Vol. I, p. 160, for the evaluation of these moments for the Woods-Saxon form factor). Thus, for example, the electric quadrupole and hexadecapole moments for a charge distribution of the form (4-189a) are given by

$$\mathscr{M}(E2,\mu) = \frac{3}{4\pi} ZeR_0^2 \left(\frac{4}{3} \frac{\langle r \rangle}{R_0} \alpha_{2\mu} - \left(\frac{10}{7\pi} \right)^{1/2} (\alpha_2 \alpha_2)_{2\mu} \left(1 + \frac{3}{4} \left(1 - \frac{2}{3} \langle r^{-1} \rangle R_0 \right) \right) \right)$$

$$\mathscr{M}(E4,\mu) = \frac{3}{4\pi} ZeR_0^4 \left(2 \frac{\langle r^3 \rangle}{R_0^3} \alpha_{4\mu} \right. \tag{4-190}$$

$$\left. + \frac{5}{3} \frac{\langle r^2 \rangle}{R_0^2} \left(\frac{45}{14\pi} \right)^{1/2} (\alpha_2 \alpha_2)_{4\mu} \left(1 - \frac{2}{3} \left(1 - \frac{4}{5} \frac{\langle r \rangle R_0}{\langle r^2 \rangle} \right) \right) \right)$$

with inclusion of the second-order terms involving $\alpha_{2\mu}$.

For an axially symmetric deformation, we have $\alpha_{\lambda\nu} = \beta_\lambda \delta(\nu,0)$ for the components with respect to the intrinsic coordinate system. The relation between the quadrupole deformation parameters δ and β_2 is obtained by comparing Eqs. (4-72) and (4-190)

$$\delta = \left(\frac{45}{16\pi} \right)^{1/2} \left(\frac{4}{5} \frac{\langle r \rangle R_0}{\langle r^2 \rangle} \beta_2 + \frac{3}{5} \frac{R_0^2}{\langle r^2 \rangle} \left(\frac{20}{49\pi} \right)^{1/2} \beta_2^2 + \cdots \right)$$

$$\approx 0.945 \beta_2 \left(1 - \frac{4}{3} \pi^2 \left(\frac{a_0}{R_0} \right)^2 \right) + 0.34 \beta_2^2 \tag{4-191}$$

where the diffuseness correction has been included only in the linear term and has been evaluated for the Woods-Saxon form factor (see Eq. (2-65)).

Evidence on hexadecapole moments ($\lambda = 4$) has been obtained from the excitation of the 4+ states in the ground-state bands of even-even nuclei by Coulomb excitation, inelastic electron scattering, as well as from the inelastic scattering of nuclear projectiles. As an example, Table 4-16 shows deforma-

▼ tion parameters obtained from an analysis of 50 MeV α particles, which gave the first determinations of the β_4 deformations. These experiments determine the deformation parameters $(\beta_\lambda)_{pot}$ of the effective potential acting on the α particle; the deformation parameter β_λ of the density has been obtained by assuming that the quantity $R_0\beta_\lambda$ is the same for the potential and the density (which in this case implies $\beta_\lambda \approx 1.2(\beta_\lambda)_{pot}$, since the mean radius of the potential is about 20% larger than that of the density; see the discussion in Chapter 6, p. 354. The deformation parameters in Table 4-16 agree rather well with those determined from electromagnetic excitation (see Erb *et al.*, 1972, ▲ and Bertozzi *et al.*, 1972).

	^{152}Sm	^{154}Sm	^{158}Gd	^{166}Er	^{174}Yb	^{176}Yb	^{178}Hf
β_2	0.246	0.270	0.282	0.276	0.276	0.276	0.246
β_4	0.048	0.054	0.036	0.0	−0.048	−0.054	−0.072

Table 4-16 Deformations of rare-earth nuclei obtained from $(\alpha\alpha')$ scattering. The parameters β_2 and β_4 for quadrupole and hexadecapole deformations are taken from the measurements and analysis of $(\alpha\alpha')$ scattering by D. L. Hendrie, N. K. Glendenning, B. G. Harvey, O. N. Jarvis, H. H. Duhm, J. Saudinos, and J. Mahoney, *Phys. Letters* **26B**, 127 (1968a).

▼ A qualitative estimate of the trend of the hexadecapole deformations may be obtained in terms of the moments of the orbits near the Fermi surface (Hendrie *et al.*, 1968, 1968a). For particles moving in an axially symmetric harmonic oscillator potential, the orbits may be characterized by the quantum numbers N, n_3, and Λ (see p. 216), and one obtains

$$\langle Nn_3\Lambda|r^4Y_{40}|Nn_3\Lambda\rangle$$
$$=\left(\frac{9}{4\pi}\right)^{1/2}\frac{3}{4}(27n_3^2-22n_3N+3N^2-\Lambda^2-6n_3-2N)\left(\frac{\hbar}{2M\omega_0}\right)^2 \quad (4\text{-}192)$$

In this expression, we have ignored the anisotropy in the oscillator potential; for an individual orbit, the effect of the anisotropy is small (of order δ), but the cumulative effect contributes a significant hexadecapole moment of order $A\delta^2$, which corresponds to the second term in the relation (4-190) for the $\lambda=4$ moment.

For a prolate potential, the first orbits to be filled in a given major shell are those with $n_3=N$, and these orbits have large positive hexadecapole moments. With decreasing values of n_3 (for fixed N), the matrix elements ▲ (4-192) decrease, becoming negative for intermediate values of n_3. Typical

▼ orbits that are being filled for the nuclei listed in Table 4-16 have $N=5$, $n_3=2$ for the neutrons and $N=4$, $n_3=1$ for the protons (see Tables 5-12 and 5-13, pp. 266 ff.); the negative $\lambda=4$ moments of these orbits account for the observed decrease of β_4 with increasing A. A similar trend has been observed in the first half of the (sd) shell ($\beta_4(^{20}\text{Ne})\approx+0.2$; $\beta_4(^{24}\text{Mg})\approx-0.05$; see Horikawa *et al.*, 1971, and references quoted there). The first orbit to fill in the (sd) shell has $(Nn_3\Lambda)=(220)$ (see Fig. 5-1, p. 220) and gives rise to the large positive value of β_4 for ^{20}Ne; the next orbit has $(Nn_3\Lambda)=(211)$, with an almost equal and opposite sign of $\langle r^4 Y_{40}\rangle$ and thus accounts for the smallness of β_4 for ^{24}Mg. Early evidence for $\lambda=4$ deformations in the actinide nuclei was obtained from the analysis of the intensities of α-decay fine-structure components (Fröman, 1957); for more detailed determinations based on Coulomb excitation, see Bemis *et al.*, 1973.

The β decay of ^{176}Lu (Fig. 4-26)

Two states of ^{176}Lu decay by β emission to different members of the ▲ ^{176}Hf ground-state rotational band (see Fig. 4-26).

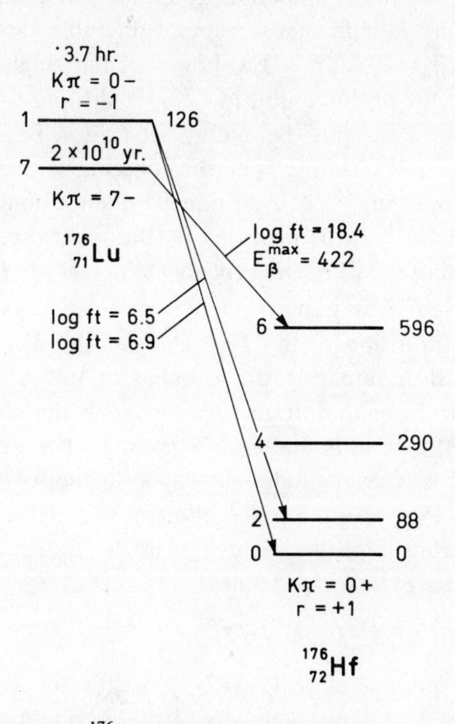

Figure 4-26 Decay scheme for ^{176}Lu. The data are taken from *Table of Isotopes* by Lederer *et al.*, 1967, except for the excitation energy of the ^{176}Lu isomer, which is from M. M. Minor, R. K. Sheline, E. B. Shera, and E. T. Jurney, *Phys. Rev.* **187**, 1516 (1969). All energies are in keV.

▼ The short-lived isomer ($\tau_{1/2}=3.7$ hr, $I\pi=1-$) decays by first forbidden transitions to the $I\pi=0+$ and $2+$ states of ^{176}Hf. The ground-state decay must be $\lambda=1$, while the transition to the $2+$ level could be a mixture of $\lambda=1$ and $\lambda=2$. However, the calculated *ft* value for a first forbidden single-particle transition of $\lambda=2$ type is of order $\xi^2(\approx 10^2)$ larger than the *ft* value for a similar transition of $\lambda=1$ type (see Vol. I, pp. 412 ff.); the observed *ft* value for the transition in question corresponds to that of an unhindered $\lambda=1$ first forbidden transition and, hence, it is not expected that $\lambda=2$ can contribute significantly to this transition. Thus, the leading-order intensity relation (4-91) gives

$$\frac{ft(1\rightarrow 2)}{ft(1\rightarrow 0)}=\frac{\langle 1K1-K|00\rangle^2}{\langle 1K1-K|20\rangle^2}=\begin{cases}1/2 & (K=0)\\ 2 & (K=1)\end{cases} \tag{4-193}$$

in terms of the K quantum number of the ^{176}Lu isomer. The observed value of 0.5 ± 0.1 for the ratio of the *ft* values (Hansen, 1964) clearly decides in favor of $K=0$ and at the same time provides a test of the intensity rules in β decay.

The interpretation of ^{176}Lu ($I\pi=1-$) as $K\pi=0-$, $r=-1$ receives strong support from the observation of a negative quadrupole moment for this state ($Q=-230$ fm^2; see the compilation by Fuller and Cohen, 1965). The lowest intrinsic states in ^{176}Lu in many respects resemble those of ^{166}Ho (see Fig. 4-20) with $\Omega_p=\Omega_n=7/2$. In ^{176}Lu, however, the single-particle assignments are [404 7/2] for the proton and [514 7/2] for the neutron (see Tables 5–12, p. 266, and 5-13, p. 268), and the energy separation between the three bands ($K=7$; $K=0$, $r=\pm 1$) resulting from the coupling of these two orbits is rather different from those in ^{166}Ho. Additional evidence confirming the classification of the levels in ^{176}Lu is provided by the spectroscopic studies based on the evidence from the ^{175}Lu(nγ) process (Minor *et al.*, *loc. cit.*, Fig. 4-26).

The long-lived ^{176}Lu ground state ($I\pi=7-$) decays by a highly hindered first forbidden transition to the $I=6$ state of the $K\pi r=0++$ ground-state band in ^{176}Hf. By comparing the *ft* value of this transition with that for unhindered first forbidden transition (such as the decay of the $I\pi=1-$ isomer of ^{176}Lu), the hindrance may be estimated to be of the order of 10^{12}. This hindrance can be attributed to the K forbiddenness of the transition. Assuming $\lambda=1$, the transition is K forbidden to order $n=6$, and the observed hindrance therefore implies a reduction in the transition intensity of about 10^2 for each order of K forbiddenness.

The β decay of ^{172}Tm (Table 4-17)

The ^{172}Tm ground state ($I\pi=2-$) decays by $\lambda=2$ first forbidden β transitions to the first three members of the ground-state rotational band of ^{172}Yb. Assuming the first-order intensity rules, we expect the $f_1 t$ values to be proportional to $\langle 222-2|I_f 0\rangle^{-2}$ (see Eq. (4-92)). The comparison of the measured intensity ratios with this expression is given in Table 4-17. ▲

I_f	$\log f_1 t$	$\dfrac{f_1 t(2\to I_f)}{f_1 t(2\to 0)}$	$\dfrac{\langle 2\,-2\,2\,2\vert 0\,0\rangle^2}{\langle 2\,-2\,2\,2\vert I_f 0\rangle^2}$
0	8.63	1	1
2	8.43	0.63 ± 0.19	0.7
4	9.77	13.7 ± 2	14

Table 4-17 Intensity ratios for first forbidden transitions in the decay of ^{172}Tm. The data are taken from P. G. Hansen, H. L. Nielsen, K. Wilsky, Y. K. Agarwal, C. V. K. Baba, and S. K. Bhattacherjee, *Nuclear Phys.* **76**, 257 (1966).

▼ The transition to the 2+ state in ^{172}Yb may contain admixtures of $\lambda=0$ and $\lambda=1$, but these components, which are favored by the leptonic matrix elements, are K forbidden. Information on the contribution of $\lambda=0$ and $\lambda=1$ components in the $2-\to 2+$ transition can be obtained from β-γ directional and polarization correlations and from measurements of the spectral shape of the emitted β particles. Transitions with $\lambda=2$ have a spectral shape differing significantly from that of allowed β decays, while $\lambda=0$ and $\lambda=1$ transitions have approximately the allowed shape (in the approximation $\xi\gg 1$; see Vol. I, p. 413). The observed spectral shape for the $2-\to 2+$ β branch gives an upper limit of 10% for the transition strength contributed by $\lambda=0$ or $\lambda=1$ components (Hansen *et al.*, *loc. cit.*, Table 4-17). Comparison of this limit with typical ft values observed for unhindered transitions of this type ($ft \sim 10^6$–10^7) implies a reduction of the transition probability by a factor of order 10^3 for this first-order K-forbidden transition.

K-forbidden E2 transitions in ^{244}Cm (Fig. 4-27 and Table 4-18)

The decay of the 1042 keV isomer in ^{244}Cm (see Fig. 4-27) provides an opportunity for testing the $E2$-intensity relation for K-forbidden transitions. The isomeric state has $K=6$ and the transitions to the ground-state band thus have $n=4$. For such transitions, the leading-order $E2$-intensity relation has the form (see Eq. (4-95))

$$B(E2;K_{\mathrm{i}}=6,I_{\mathrm{i}}\to K_{\mathrm{f}}=0,I_{\mathrm{f}})=\langle I_{\mathrm{i}}6\,2\,-2|I_{\mathrm{f}}4\rangle^2\frac{(I_{\mathrm{f}}+4)!}{(I_{\mathrm{f}}-4)!}2\langle K=6|m_{6,2}|K=0\rangle^2 \tag{4-194}$$

It is seen from Table 4-18 that the observed $B(E2)$ values are in agreement
▲ with this relation.

	Experiment	Theory
$\dfrac{B(E2;\, 6\to6)}{B(E2;\, 6\to4)}$	3.6	3.7
$\dfrac{B(E2;\, 6\to8)}{B(E2;\, 6\to4)}$	0.34	0.39

Table 4-18 Ratios of $B(E2)$ values for K-forbidden $E2$ transitions in ^{244}Cm (see references to Fig. 4-27).

▼ The absolute decay rates implied by the lifetime of the isomer show that the $E2$ transitions are hindered by factors of about 10^{10} and that the $M1$ transition ($KI=66\to KI=06$; see Fig. 4-27) is hindered by about 10^{12} as compared with single-particle estimates (Eq. (3C-38)). For both multipoles, the hindrance is, therefore, about a factor of 300 for each order of K forbiddenness. ▲

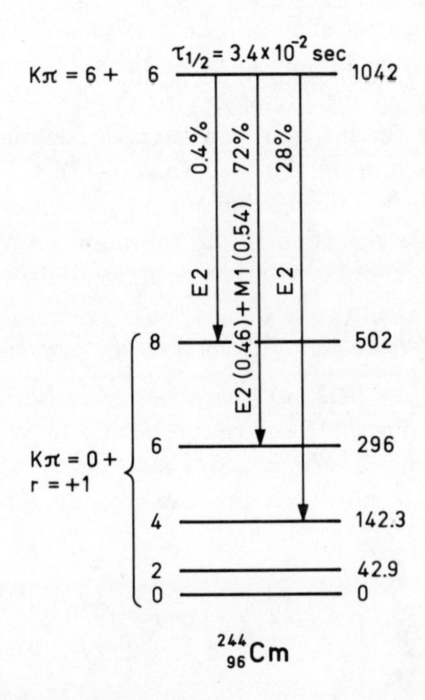

Figure 4-27 Decay scheme for ^{244}Cm^m. The data are taken from S. E. Vandenbosch and P. Day, *Nuclear Phys.* **30**, 177 (1962) and P. G. Hansen, K. Wilsky, C. V. K. Baba, and S. E. Vandenbosch, *Nuclear Phys.* **45**, 410 (1963).

4-4 COUPLING BETWEEN ROTATIONAL AND INTRINSIC MOTION FOR AXIALLY SYMMETRIC NUCLEI

The phenomenological analysis of matrix elements in the rotational coupling scheme, considered in Sec. 4-3, is based on the symmetry of the deformation and on the adiabatic condition that makes it possible to represent the matrix elements in terms of an expansion in powers of the rotational angular momentum. In such a description, the coupling between rotational and intrinsic motion resulting from the Coriolis and centrifugal forces does not appear explicitly, but manifests itself in the I dependence of the effective operators. These coupling effects can also be viewed in terms of the admixing of different rotational bands. In the present section, we shall consider such band mixing involving the low-lying intrinsic excitations. By further development of this approach, one may attempt to derive all the I-dependent nuclear properties on the basis of the interaction between rotation and intrinsic motion. (The cranking model considered on pp. 75 ff. is an example of such an extension.)

Coriolis interaction

The Hamiltonian describing the coupling between a set of bands, labeled by $i=1,2,\ldots,n$, can be written

$$H=H_0+H_c \tag{4-195}$$

where H_0 is diagonal in i, while H_c connects states belonging to different bands (but having the same value of I). In the representation (4-195), the Hamiltonian is assumed to be diagonalized with respect to all modes of excitation other than those explicitly under consideration, and the operators therefore include the I-dependent terms associated with all couplings other than those described by H_c.

The coupling H_c expresses the nondiagonal effects of the rotation within the space of states i, and the lowest-order coupling term is proportional to the rotational angular momentum

$$H_c=h_{+1}I_- + \mathscr{R}\text{-conj.} \tag{4-196}$$

where h_{+1} $(=h_{\Delta K=+1})$ is an intrinsic operator (a matrix in the space i). The interaction (4-196) gives rise to a coupling between bands with $\Delta K=1$, as well as between two bands with $K=1/2$. The diagonal effect of the interaction (4-196), in a $K=1/2$ band, gives rise to a signature-dependent term in the rotational energy (see the decoupling term in Eq. (4-61)).

If the intrinsic degrees of freedom i can be described in terms of one or a few particles moving in an average potential produced by the rest of the nucleons, and if it is assumed that this potential is independent of the rotational motion, the coupling H_c is the Coriolis interaction (see the discussion of the particle-rotor model in Appendix 4A), and we have

$$h_{+1} = -\frac{\hbar^2}{2\mathscr{I}_0} J_+ \tag{4-197}$$

where $\mathscr{I}_0$ is the moment of inertia of the residual nucleus, while $J_+ = J_1 + iJ_2$ is the component of the total angular momentum of the particles considered.

First-order effects of Coriolis coupling

The coupling (4-196) between two bands with $\Delta K=1$ gives rise to a mixing of these bands,

$$\begin{aligned} |\widehat{KI}\rangle &\approx |KI\rangle - c(I)|K+1,I\rangle \\ |\widehat{K+1},I\rangle &\approx |K+1,I\rangle + c(I)|KI\rangle \end{aligned} \tag{4-198}$$

where the caret denotes the coupled (or renormalized) states; the first-order amplitude of admixture is

$$c(I) = \langle K+1|\varepsilon_{+1}|K\rangle (I-K)^{1/2}(I+K+1)^{1/2} \begin{cases} \sqrt{2} & K=0 \\ 1 & K\neq 0 \end{cases} \tag{4-199}$$

$$\langle K_2|\varepsilon_{+1}|K_1\rangle \equiv \frac{\langle K_2|h_{+1}|K_1\rangle}{E(K_2)-E(K_1)} \qquad (K_2 = K_1+1)$$

In the energy denominator of the last expression, we ignore rotational energies as compared with intrinsic energies. The above expressions also apply to the mixing of two $K=1/2$ bands, if the intrinsic state $|K_1\rangle$ is replaced by the time-reversed $|\overline{K}_1\rangle$ and if the signature factor $(-1)^{I_1+K_1}$ is added to $c(I)$.

The perturbation treatment of the Coriolis coupling is based on the smallness of the quantity ε_{+1}. Since the intrinsic excitation energies in heavy nuclei are of the order of 1 MeV, the expression (4-197) for h_{+1} together with Eq. (4-199) yields values for the matrix elements of ε_{+1} that are typically of the order of a few percent. (Empirical values of ε_{+1} are derived from the analysis of the spectra of ^{175}Lu (p. 154) and ^{235}U (p. 273).)

The band mixing (4-198) leads to modifications in the various matrix elements involving the coupled bands (Kerman, 1956). Compact expressions for these effects are obtained by describing the band mixing in terms of a canonical transformation

$$\begin{aligned} |\hat{K}I\rangle &= \exp\{-iS\}|KI\rangle \\ &\approx |KI\rangle - iS|KI\rangle \end{aligned} \tag{4-200}$$

where the operator S is linear in the matrix ε_{+1} defined by Eq. (4-199)

$$S = -i(\varepsilon_{+1}I_- + \mathscr{R}\text{-conj.}) \tag{4-201}$$

The transformation $\exp\{iS\}$ diagonalizes the Hamiltonian (4-195), to leading order in the coupling (see the discussion in Appendix 4A, pp. 206 ff.).

The matrix elements of an operator $\mathscr{M}$ between the perturbed states can be obtained by evaluating the transformed operator

$$\begin{aligned} \exp\{iS\}\,\mathscr{M}\exp\{-iS\} &= \mathscr{M} + i[S, \mathscr{M}] - \tfrac{1}{2}[S,[S, \mathscr{M}]] + \cdots \\ &= \mathscr{M} + \delta\mathscr{M} \end{aligned} \tag{4-202}$$

in the unperturbed basis. For a tensor operator $\mathscr{M}(\lambda, \mu)$, the transformation (4-90) to intrinsic axes yields, to first order in S,

$$\begin{aligned} \delta\mathscr{M}(\lambda\mu) &\approx i[S, \mathscr{M}(\lambda\mu)] \\ &= \Big[(\varepsilon_{+1}I_- + \mathscr{R}\text{-conj.}), \sum_\nu \mathscr{M}(\lambda\nu)\mathscr{D}^\lambda_{\mu\nu}\Big] \\ &= \sum_\nu \tfrac{1}{2}\{\varepsilon_{+1}, \mathscr{M}(\lambda\nu)\}[I_-, \mathscr{D}^\lambda_{\mu\nu}] + \tfrac{1}{2}[\varepsilon_{+1}, \mathscr{M}(\lambda\nu)]\{I_-, \mathscr{D}^\lambda_{\mu\nu}\} + \mathscr{R}\text{-conj.} \end{aligned} \tag{4-203}$$

where the intrinsic moments $\mathscr{M}(\lambda\nu)$ have been assumed to be I independent. We have employed the identity (4A-26) for commutators of products and have used the conventional notation for anticommutators ($\{a,b\} = ab + ba$).

In the first term of Eq. (4-203), the commutator of I_- and $\mathscr{D}^\lambda_{\mu\nu}$ is proportional to $\mathscr{D}^\lambda_{\mu,\nu+1}$ (see Eq. (1A-91)). This term, therefore, gives rise to an I-independent renormalization of the intrinsic moments, and does not lead to a modification of the leading-order intensity relations. The second term in Eq. (4-203) introduces effective moments that are linear in $I_\pm$ and leads to generalized intensity relations of the type discussed in Sec. 4-3.

For dipole moments, the various terms occurring in Eq. (4-203) are discussed in Appendix 4A (see pp. 208 ff.). The coupling partly leads to a renormalization of the leading-order terms, such as the parameters g_K, g_R, and b that characterize $M1$ transitions within a band and partly introduces new terms linear in $I_{\pm}$, which contribute to matrix elements with $\Delta K=0$, ± 1, and ± 2. (Coriolis effects in $M1$ transitions between bands in ^{175}Lu are considered on pp. 156ff. The I-dependent terms in the $E1$-transition moments observed in ^{177}Hf may also be attributed to the effect of first-order Coriolis couplings; see p. 110.)

The coupling between two bands has an especially large effect on the $E2$ transitions between the bands, since the admixed components in the wave function contribute with the collective matrix element proportional to Q_0. The resulting contribution to the $E2$-transition moment is contained in the terms in Eq. (4-203) involving $\mathscr{M}(E2,\nu=0)=(5/(16\pi))^{1/2}eQ_0$. If we neglect the difference between the Q_0 values for the two bands, the commutator of ε_{+1} and $\mathscr{M}(E2,\nu=0)$ vanishes, and the induced moment $\delta\, \mathscr{M}(E2)$ gives an I-independent renormalization of the intrinsic transition moment $\mathscr{M}(E2,\nu=1)$, without any effect on the relative transition probabilities. For low-frequency $E2$ transitions ($\Delta E \lesssim 1$ MeV), the induced moment is large compared with the unperturbed moment and implies a strong enhancement of the absolute transition rates. (See the estimate in Chapter 5, pp. 281ff., and the example discussed on pp. 154ff.)

For $E2$ transitions within a band, the second term in Eq. (4-203) gives rise to deviations from the leading-order intensity relations (provided $K\neq 0$), but the effects are expected to be small, since they involve the intrinsic transition moments $\mathscr{M}(E2,\nu=1)$ compared with the collective moments $\mathscr{M}(E2,\nu=0)$ in the leading-order term. (See the example discussed on pp. 130ff.)

In a number of cases, it has been possible to estimate the band-mixing amplitude $\langle K+1|\varepsilon_{+1}|K\rangle$ on the basis of several different observed contributions to the transitions between the bands, and thus to test the consistency of the analysis. (See the example (^{175}Lu) discussed on pp. 154ff. The interpretation of the observed matrix elements of h_{+1} in terms of the intrinsic one-particle configurations is discussed in Sec. 5-3d.)

Second-order contributions to rotational energy

The coupling (4-196) produces a repulsion between the interacting bands resulting in the energy shifts

$$\begin{aligned}\delta E(K_1 I) &= -\,\delta E(K_2=K_1+1,I)\\ &= -\frac{\langle K_2|h_{+1}(q)|K_1\rangle^2}{E(K_2)-E(K_1)}\left(I(I+1)-K_1(K_1+1)\right)\end{aligned} \qquad (4\text{-}204)$$

The term proportional to $I(I+1)$ amounts to a renormalization of the A coefficient in the rotational energy (see Eq. (4-46)), and corresponds to a contribution to the moment of inertia

$$\delta \mathscr{I}(K_1) \approx -\delta \mathscr{I}(K_2) \approx 2\hbar^2 \frac{\langle K_2|J_1|K_1\rangle^2}{E(K_2)-E(K_1)} \tag{4-205}$$

In the derivation of Eq. (4-205), we have assumed $\delta\mathscr{I} \ll \mathscr{I}_0$ and have employed the relation (4-197) for h_{+1} (note that $J_1 = \frac{1}{2}(J_+ + J_-)$). The contribution of the term (4-205) to the difference of the moments of inertia of even-even and odd-A nuclei is discussed in Sec. 5-3d, pp. 251 ff.

The expression (4-205) is the same as that obtained by considering particles moving in an external potential rotated with a fixed frequency (the cranking model; see pp. 75 ff.). The above derivation applies to a single degree of freedom; if several degrees of freedom together contribute a significant fraction of the total moment of inertia, the results (4-204) and (4-205) differ. The analysis based on the cranking model corresponds to adding the contributions (4-205) to obtain the resultant moment $\mathscr{I}$ (see Eq. (4-110)), and this is expected to be the correct answer as long as no single degree of freedom contributes a significant fraction of the total. This result may be obtained by treating the coupling between the intrinsic degrees of freedom produced by the "recoil term", as illustrated by the analysis of the particle-rotor model in Appendix 4A, pp. 205 ff. (The early discussion of Casimir, 1931, established the result for the inertial effect of a particle bound harmonically to a rotor.)

Effective couplings with $\Delta K = 2$

When the spectrum involves elementary modes of excitation with $\Delta K = 2$, significant rotational coupling effects may be described by an effective coupling of the form

$$H_c = h_{+2} I_-^2 + \mathscr{R}\text{-conj.} \tag{4-206}$$

which produces a direct mixing of the bands. Coupling effects that are of second order in the rotational angular momentum arise from the Coriolis coupling (4-196) acting in second order through intermediate states with $\Delta K = 1$. These effects can be represented by the direct coupling (4-206), provided the energies of the $\Delta K = 1$ excitations are large compared with the energy of the $\Delta K = 2$ mode.

A term such as (4-206) that is of second order in I may be considered as part of the rotational energy, expressing the dependence of the moment of inertia on the intrinsic variable associated with the $\Delta K=2$ excitation. (See expression (6-296) for the coupling to γ vibrations and Eq. (4-285) for the asymmetric rotor.)

The effect of the coupling (4-206) can be treated by a canonical transformation

$$S = -i(\varepsilon_{+2}I_-^2 + \mathscr{R}\text{-conj.}) \tag{4-207}$$

with

$$\langle K_2|\varepsilon_{+2}|K_1\rangle = \frac{\langle K_2|h_{+2}|K_1\rangle}{E(K_2)-E(K_1)} \qquad (K_2 = K_1+2) \tag{4-208}$$

(See the analogous equations (4-201) and (4-199) for the $\Delta K=1$ coupling.)

The $E2$ transitions between the coupled bands provide a sensitive indicator of the coupling, since the admixed amplitudes contribute with the large matrix elements characteristic of transitions within a band. The associated additional moment is (see the relations (1A-91) and (4A-34) for the $\mathscr{D}$ functions)

$$\begin{aligned}
\delta\mathscr{M}(E2,\mu)_{\Delta K=2} &= [\varepsilon_{+2}I_-^2, \mathscr{M}(E2,\mu)] + \mathscr{R}\text{-conj.} \\
&= \varepsilon_{+2}\left(\frac{5}{16\pi}\right)^{1/2} eQ_0[I_-^2, \mathscr{D}_{\mu 0}^2] + \mathscr{R}\text{-conj.} \\
&= \varepsilon_{+2}\left(\frac{15}{8\pi}\right)^{1/2} eQ_0\{I_-, \mathscr{D}_{\mu 1}^2\} + \mathscr{R}\text{-conj.} \\
&= \varepsilon_{+2}\left(\frac{15}{8\pi}\right)^{1/2} eQ_0\left([\mathbf{I}^2, \mathscr{D}_{\mu 2}^2] - 2\{I_3, \mathscr{D}_{\mu 2}^2\}\right) + \mathscr{R}\text{-conj.}
\end{aligned} \tag{4-209}$$

We have used the expression (4-67) for the collective part of the $E2$ operator, assuming Q_0 to have the same value in the two bands. A difference in the Q_0 values for the admixed bands would give terms of higher order in I (see Eq. (4-235)).

With the inclusion of the induced moment (4-209). the total transition matrix element becomes

$$\begin{aligned}
\langle K_2 = K_1+2, I_2\| \mathscr{M}(E2)\| K_1 I_1\rangle &= (2I_1+1)^{1/2}\langle I_1 K_1 2 2|I_2 K_2\rangle \\
&\times \left(M_1 + M_2(I_2(I_2+1) - I_1(I_1+1))\right)\begin{cases} \sqrt{2} & K_1 = 0 \\ 1 & K_1 \neq 0 \end{cases}
\end{aligned} \tag{4-210}$$

with

$$M_1 = \langle K_2 | \mathscr{M}(E2, \nu=2) | K_1 \rangle - 4(K_1+1) M_2$$

$$M_2 = \left(\frac{15}{8\pi} \right)^{1/2} eQ_0 \langle K_2 | \varepsilon_{+2} | K_1 \rangle \qquad (4\text{-}211)$$

We have here assumed that the $E2$-transition moment, in the absence of the coupling H_c, is I independent. However, additional coupling effects, such as the first-order Coriolis coupling acting through intermediate states with $K = K_1 + 1$, may contribute I-dependent terms to the $E2$ moment (see Eq. (4-203)). In the present analysis, these terms are included in the effective operators in the representation that diagonalizes H_0. Since the effective moment for $E2$ transitions with $\Delta K = 2$, with the inclusion of terms linear in $I_\pm$, has the unique form (4-96) and leads to a generalized intensity relation of the type (4-210), it is not possible on the basis of the I dependence of the observed matrix elements to distinguish these different coupling effects.

The observed $E2$ transitions between the ground-state bands and the low-lying $K\pi = 2+$ bands in even-even nuclei show large deviations from the leading-order intensity relations. An example testing the generalized relation (4-210) is discussed on pp. 158ff.

In second order, the coupling (4-206) produces a repulsion between the levels in the interacting bands. The level shifts are of fourth order in I and contribute to the B term in the rotational energy by the amount

$$\delta B(K_1) = -\delta B(K_2)$$
$$= -(E(K_2) - E(K_1)) \langle K_2 | \varepsilon_{+2} | K_1 \rangle^2 \qquad (K_1 \neq 0 \neq K_2) \qquad (4\text{-}212)$$

If one of the bands has $K=0$ (and the symmetry r), the coupling only affects levels with $(-1)^I = r$. For the $K=2$ band, we then obtain a contribution to the A_4 term as well as to the B term (see Eq. (4-62)),

$$\delta B(K=2) = r\,\delta A_4(K=2) = -\tfrac{1}{2}\delta B(K=0)$$
$$= (E(K=2) - E(K=0)) \langle K=2 | \varepsilon_{+2} | K=0 \rangle^2 \qquad (4\text{-}213)$$

The perturbation of the $E2$ transitions within the bands, arising from the $\Delta K=2$ coupling, can be expressed in terms of the additional moment

$$\begin{aligned}\delta\mathscr{M}(E2,\mu)_{\Delta K=0} &= \left[\varepsilon_{+2}I_-^2,\ \mathscr{M}(E2,\nu=-2)\mathscr{D}^2_{\mu-2}\right]+\mathscr{R}\text{-conj.}\\ &= \tfrac{1}{2}\{\varepsilon_{+2},\ \mathscr{M}(E2,\nu=-2)\}\left[I_-^2,\mathscr{D}^2_{\mu-2}\right]\\ &\quad+\tfrac{1}{2}[\varepsilon_{+2},\ \mathscr{M}(E2,\nu=-2)]\{I_-^2,\mathscr{D}^2_{\mu-2}\}+\mathscr{R}\text{-conj.}\end{aligned} \tag{4-214}$$

For a $K=0$ band, the expectation value of the first term vanishes, since ε_{+2} is odd and $\mathscr{M}(E2,\nu=-2)$ is even under time reversal followed by Hermitian conjugation. From the identity (4-80) it therefore follows that the reduced $E2$-matrix element is of the form (4-81) with

$$\begin{aligned}M_1 &= \left(\frac{5}{16\pi}\right)^{1/2} eQ_0+24M_3\\ M_2 &= -6M_3\\ M_3 &= \frac{1}{2\sqrt{6}}\langle K=0|[\varepsilon_{+2},\ \mathscr{M}(E2,\nu=-2)]|K=0\rangle\\ &= -\frac{1}{\sqrt{6}}\langle K=2|\varepsilon_{+2}|K=0\rangle\langle K=2|\ \mathscr{M}(E2,\nu=2)|K=0\rangle\end{aligned} \tag{4-215}$$

The $\Delta K=2$ coupling also gives rise to $M1$ transitions between the bands, which are forbidden by the leading-order intensity rules. This effect, as well as additional consequences of the coupling between the ground-state bands and the $K=2$ γ-vibrational bands in even-even nuclei, are discussed in connection with the example considered on p. 158.

Effective couplings with $\Delta K=0$

The coupling between two bands with the same value of K can be represented by an interaction of the form

$$H_c=\tfrac{1}{2}h_0(I_+I_-+I_-I_+)=h_0(\mathbf{I}^2-I_3^2) \tag{4-216}$$

The intrinsic operator h_0 can be viewed as representing the dependence of the moment of inertia on the internal variables associated with the $\Delta K=0$ mode of excitation. For example, the coupling (4-216) may describe the centrifugal distortion of the shape (see Eq. (6-296)) or the effect of the rotation on the pair correlations (see the qualitative estimate (4-128) of the dependence of $\mathscr{I}$ on Δ). The conditions for representing the second-order Coriolis effects in terms of the direct coupling (4-216) are the same as those discussed above for the effective $\Delta K=2$ coupling.

The effect of the interaction (4-216) on the $E2$ transitions between the coupled bands is represented by the additional moment

$$\delta\mathscr{M}(E2,\mu)=\varepsilon_0\left(\frac{5}{16\pi}\right)^{1/2} eQ_0\left[\mathbf{I}^2,\mathscr{D}^2_{\mu 0}\right] \tag{4-217}$$

with

$$\langle\alpha_2 K|\varepsilon_0|\alpha_1 K\rangle=\frac{\langle\alpha_2 K|h_0|\alpha_1 K\rangle}{E(\alpha_2 K)-E(\alpha_1 K)} \tag{4-218}$$

The admixed bands are distinguished by the quantum number α, and we have assumed equal Q_0 values for these two bands. The total transition amplitude can be written

$$\langle\alpha_2 K I_2\|\mathscr{M}(E2)\|\alpha_1 K I_1\rangle$$
$$=(2I_1+1)^{1/2}\langle I_1 K 2 0|I_2 K\rangle\big(M_1+M_2(I_2(I_2+1)-I_1(I_1+1))\big) \tag{4-219}$$

with

$$M_1=\langle\alpha_2 K|\mathscr{M}(E2,\nu=0)|\alpha_1 K\rangle$$
$$M_2=\left(\frac{5}{16\pi}\right)^{1/2} eQ_0\langle\alpha_2 K|\varepsilon_0|\alpha_1 K\rangle \tag{4-220}$$

The matrix element (4-219) has the unique form obtained by including terms linear in $I_\pm$ in the intrinsic moment (see Eq. (4-75)). Additional contributions to M_2 may arise from coupling effects not included in H_c (see the comments following Eq. (4-210)).

A test of the generalized intensity relation (4-219) is discussed in the example on pp. 168 ff. Various additional consequences of the coupling (4-216), including energy shifts and the effect on $E0$- and $E2$-matrix elements in the coupled bands, are also considered in connection with this example.

▼

ILLUSTRATIVE EXAMPLES TO SECTION 4-4

Effects of Coriolis coupling on M1 and E2 transitions in ^{175}Lu (Fig. 4-28 and Table 4-19)

The ground-state band of ^{175}Lu has $K=7/2$, and a $K=5/2$ excitation occurs at 343 keV (see Fig. 4-28). The excited band is populated in the β decay of ^{175}Hf, and the measured $B(M1)$ and $B(E2)$ values for transitions between the bands are listed in Table 4-19. From these data, it is possible to obtain information about the Coriolis coupling of the two bands (Hansen *et al., loc. cit.*, Fig. 4-28).

E2 transitions

The $E2$-transition probabilities are found to satisfy the leading-order intensity relation (4-68). However, the absolute rates, which are of the order of a single-particle unit $B_W(E2)$, provide evidence for a coupling between the bands. The $K=7/2$ and $K=5/2$ bands can be associated with the orbits

▲

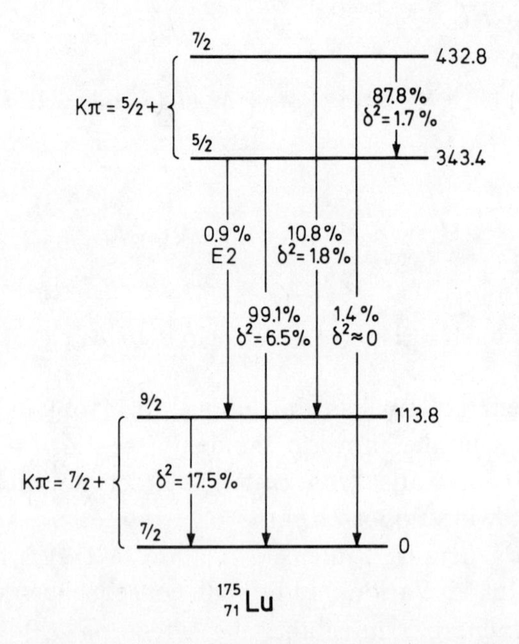

Figure 4-28 $E2$ and $M1$ transitions in ^{175}Lu. The experimental data are from P. G. Hansen, P. Hornshøj, and K. H. Johansen, *Nuclear Phys.* **A126**, 464 (1969). The figure lists the branching ratios in the decay of the various levels; for the mixed $M1+E2$ transitions, the quantity δ^2 is the ratio between the intensities of the $E2$ and $M1$ γ transitions.

▼ [404 7/2] and [402 5/2] of the odd proton (see Table 5-12, p. 297). The $E2$ transition between these orbits violates the $E2$, $\Delta K=1$ selection rules for the asymptotic quantum numbers ($\Delta n_3=1$, $\Delta\Lambda=1$, $\Delta\Sigma=0$) and is therefore hindered; an estimate of the matrix element $\langle\Omega=7/2|\ \mathscr{M}(E2,\nu=1)|\Omega=5/2\rangle$ based on the proton wave functions in Table 5-2, p. 228, yields $B(E2)$ values that are of order 10^{-3} of the observed ones, and we may thus conclude that the transition between the single-particle orbits can account for only a small ▲ fraction of the observed rate.

K_i	I_i	K_f	I_f	$B(E2)$ $e^2\text{fm}^4$	$m(E2)$ $e\,\text{fm}^2$	$B(M1)$ $(e\hbar/2Mc)^2$	$m(M1)$ $e\hbar/2Mc$
5/2	5/2	7/2	7/2	21 ± 5	20 ± 2	$(2.6\pm0.2)\times10^{-3}$	$-(10.4\pm0.5)\times10^{-2}$
		7/2	9/2	22 ± 3	22 ± 1		
5/2	7/2	7/2	7/2			$(3.6\pm0.6)\times10^{-3}$	$\pm(26\pm2)\times10^{-2}$
		7/2	9/2	2.8 ± 1.0	18 ± 3	$(1.1\pm0.3)\times10^{-3}$	$-(7.8\pm0.9)\times10^{-2}$
		5/2	5/2		750*	$(6.5\pm1.1)\times10^{-1}$	$+3.6\pm0.3$
7/2	9/2	7/2	7/2	1.90×10^4	750	$(8.0\pm0.8)\times10^{-2}$	$+1.38\pm0.07$

*value assumed

Table 4-19 $E2$ and $M1$-transition matrix elements in ^{175}Lu. The table is based on the data reported by Hansen *et al., loc.cit.*, Fig. 4-28, and the halflife $\tau_{1/2}=3.2\times10^{-10}$ sec for the 343 keV level (B. I. Deutch, *Nuclear Phys.* **30**, 191, 1962). The quantities in columns three and five are experimentally determined intrinsic matrix elements defined by the relations

$$m(E2)\equiv\left(\frac{16\pi}{5}\right)^{1/2}(2I_i+1)^{-1/2}\langle I_iK_i2\Delta K|I_fK_f\rangle^{-1}\langle K_fI_f\|\ \mathscr{M}(E2)\|K_iI_i\rangle$$

$$m(M1)\equiv\left(\frac{4\pi}{3}\right)^{1/2}(2I_i+1)^{-1/2}\langle I_iK_i1\Delta K|I_fK_f\rangle^{-1}\langle K_fI_f\|\ \mathscr{M}(M1)\|K_iI_i\rangle$$

The signs of the $M1$-matrix elements in column six are obtained from the measured relative phases of the $M1$ and $E2$ amplitudes, adopting a positive sign for the $E2$-matrix elements in column three. For the transitions within the bands, the quantities in columns six and eight equal Q_0 and $(g_K-g_R)K$, respectively. In the analysis, it is assumed that the intrinsic quadrupole moment Q_0 in the $K=5/2$ band has the same value as that measured for the $K=7/2$ band.

▼ The rather large $E2$-matrix elements connecting the bands can be interpreted in terms of a band mixing produced by the Coriolis force. Even a small admixture gives rise to large $E2$-matrix elements on account of the collective moments that are involved. (In fact, one expects quite generally that the Coriolis coupling term will contribute the main part of the $E2$-matrix element for low-frequency $\Delta K=1$ transitions; see p. 282.) The induced ▲ $E2$-transition moment may be obtained from Eq. (4-203) (see also Eqs. (4-67) and (1A-91))

$$\delta\mathcal{M}(E2,\mu)=\left(\frac{5}{64\pi}\right)^{1/2} e\left(\sqrt{6}\,\{\varepsilon_{+1},Q_0\}\mathcal{D}^2_{\mu 1}+[\varepsilon_{+1},Q_0]\{I_-,\mathcal{D}^2_{\mu 0}\}\right)+\mathcal{R}\text{-conj.}$$

$$\approx\left(\frac{15}{8\pi}\right)^{1/2} eQ_0\varepsilon_{+1}\mathcal{D}^2_{\mu 1}+\mathcal{R}\text{-conj.} \qquad (4\text{-}221)$$

In the last expression for $\delta\mathcal{M}(E2,\mu)$, we have assumed the two bands to have the same value of Q_0; the commutator involving Q_0 then vanishes. The induced moment (4-221) gives an I-independent renormalization of the intrinsic moment $\mathcal{M}(E2,\nu=1)$, in agreement with the data in Table 4-19. The observed absolute transition rates determine the quantity $\varepsilon_{+1}Q_0$; for $Q_0=750$ fm^2, as obtained from Coulomb excitation of the $K=7/2$ band (see Fig. 4-25, p. 133), one finds

$$\langle K=7/2|\varepsilon_{+1}|K=5/2\rangle\approx 0.0112\pm 0.0005 \qquad (4\text{-}222)$$

We have adopted a relative phase between the two intrinsic states such that the effective intrinsic $E2$-transition moment (with $\nu=1$) is positive (see Eq. (4-221)); this phase convention is the same as that employed in Table 4-19.

A difference between the Q_0 values for the two coupled bands leads to a modification of the $E2$-intensity relations, which can be derived from the term proportional to $[\varepsilon_{+1},Q_0]$ in Eq. (4-221). The observed agreement with the leading-order intensity relations implies $|Q_0(5/2)-Q_0(7/2)|\lesssim 0.1\,Q_0$.

M1 transitions

The $M1$-matrix elements for transitions between the $K=5/2$ and $K=7/2$ bands exhibit significant deviations from the leading-order intensity relation (see Table 4-19), but the data can be fitted by the generalized relation obtained by including terms linear in $I_\pm$ in the intrinsic moments (see Eq. (4-98))

$$\langle K_f=7/2,I_f\|\mathcal{M}(M1)\|K_i=5/2,I_i\rangle$$

$$=(2I_i+1)^{1/2}\langle I_iK_i11|I_fK_f\rangle\left(M_1+M_2(I_f(I_f+1)-I_i(I_i+1))\right) \qquad (4\text{-}223)$$

with

$$\left.\begin{matrix} M_1=-0.26\pm 0.02 \\ M_2=0.021\pm 0.002\end{matrix}\right\}\left(\frac{3}{4\pi}\right)^{1/2}\frac{e\hbar}{2Mc} \qquad (4\text{-}224)$$

The deviations from the leading-order intensity relations can be ascribed to the Coriolis coupling. The induced $M1$ moment is given by Eq. (4A-31), and the matrix elements M_1 and M_2 occurring in Eq. (4-223) are therefore

▼ given by

$$M_1 = \langle K=7/2|\ \mathscr{M}(M1,\nu=1)|K=5/2\rangle$$

$$+\frac{1}{\sqrt{2}}\langle K=7/2|\{\varepsilon_{+1},\ \mathscr{M}(M1,\nu=0)\}-6[\varepsilon_{+1},\ \mathscr{M}(M1,\nu=0)]|K=5/2\rangle \tag{4-225}$$

$$M_2 = \frac{1}{\sqrt{2}}\langle K=7/2|[\varepsilon_{+1},\ \mathscr{M}(M1,\nu=0)]|K=5/2\rangle$$

The matrix elements involving the Coriolis coupling may receive contributions from the admixing of the two bands as well as from the coupling to other $K=5/2$ and $K=7/2$ bands contributing through nondiagonal matrix elements of $\mathscr{M}(M1,\nu=0)$. The latter contributions are expected to be relatively small in the present example, partly because of the larger energy denominators for the states coupled with large Coriolis matrix elements, and partly because the nondiagonal matrix elements of $\mathscr{M}(M1,\nu=0)$ vanish in the limit of the asymptotic quantum numbers. (See Table 5-3, p. 234.)

The value of M_2 resulting from the mixing of the $K=5/2$ and $K=7/2$ bands is given by

$$M_2 = \frac{1}{\sqrt{2}}\langle K=7/2|\varepsilon_{+1}|K=5/2\rangle$$

$$\times(\langle K=5/2|\ \mathscr{M}(M1,\nu=0)|K=5/2\rangle - \langle K=7/2|\ \mathscr{M}(M1,\nu=0)|K=7/2\rangle) \tag{4-226}$$

The intrinsic $M1$-matrix elements can be expressed in terms of the g factors (see Eqs. (4-84) and (4-86)),

$$\langle K|\ \mathscr{M}(M1,\nu=0)|K\rangle = \left(\frac{3}{4\pi}\right)^{1/2}\frac{e\hbar}{2Mc}(g_K-g_R)K \tag{4-227}$$

The empirical values of (g_K-g_R), determined from $M1$ transitions within the two bands, are listed in Table 4-19. From the observed value of M_2 (see Eq. (4-224)), one thus obtains

$$\langle K=7/2|\varepsilon_{+1}|K=5/2\rangle \approx 0.013\pm0.003 \tag{4-228}$$

which is consistent with the band-mixing amplitude (4-222) determined from the $E2$ transitions.

Coriolis matrix element

Attributing the band mixing to the Coriolis interaction, we obtain from the observed value of ε_{+1}, together with the relations (4-197) and (4-199),

$$\langle K=7/2|h_{+1}|K=5/2\rangle = -\frac{\hbar^2}{2\mathscr{I}_0}\langle K=7/2|j_+|K=5/2\rangle$$

▲ $$\approx -5.0\quad \text{keV} \tag{4-229}$$

▼ An estimate of this matrix element, based on the proton wave functions in Table 5-2, p. 228, for the orbits [404 7/2] and [402 5/2] and the rotational parameter $\hbar^2/2\mathscr{I}_0 = 13.7$ keV (representing the average for ^{174}Yb and ^{176}Hf), yields $\langle K=7/2|h_{+1}|K=5/2\rangle \approx -5.8$ keV. This estimate includes the factor $u(5/2)u(7/2)+v(5/2)v(7/2) \approx 0.9$ representing the effect of the pair correlations (see Eq. (5-43)).

The band mixing also implies corrections to the $M1$ and $E2$ transitions within the bands, as well as to the rotational energies, but because of the small value of ε_{+1}, the effects are too small to have been observed in the present example.

Analysis of E2 transitions between the ground-state band and the $K\pi = 2+$ band (γ vibration) in ^{166}Er (Figs. 4-29 and 4-30)

Extensive studies have been made of the intensity relations for the $E2$ transitions between the ground-state band and the systematically occurring $K\pi = 2+$ bands that have been found as low-lying excitations in most deformed even-even nuclei. As an example, we consider the detailed information available on the $E2$ decay of the $K=2$ band in ^{166}Er. This band is populated in both the decay of the ^{166}Ho ground state ($I\pi = 0-$) and the decay of the ^{166}Ho high-spin isomer ($I\pi = 7-$), which has made possible studies of transitions covering a wide range of I values. The relevant level scheme for ^{166}Er is shown in Fig. 4-29.

I dependence of E2-matrix elements

The γ transitions between the $K=2$ and $K=0$ bands are assumed to be predominantly $E2$ radiation, as suggested by the K-selection rule. For a number of the transitions in Fig. 4-29, angular correlations have verified that the $M1$ admixtures are at most a few percent in amplitude (Reich and Cline, 1965).

The experiments directly measure the relative strength of the transitions originating from each initial state I_i (in the $K=2$ band), and thus provide a number of tests of the $E2$-intensity relations that are independent of any further assumptions. Additional evidence is provided by the observed competition with $E2$ transitions within the $K=2$ band. By assuming the leading-order intensity relations for these intraband transitions, it is possible to compare the $E2$ rates for those transitions between bands that originate from the levels with $I_i = 4, 5, 6$, and 7. The available evidence on the $E2$ amplitudes for the transitions between the $K=2$ and $K=0$ bands in ^{166}Er is brought together in Fig. 4-30.

The $E2$ amplitudes in Fig. 4-30 show systematic deviations from the leading-order intensity relation, which would correspond to a constant ordinate. These deviations can be well accounted for by the generalized

▲ intensity relation (4-98) obtained by including the first-order I-dependent

▼ terms in the intrinsic transition moment

$$B(E2;K=2,I_i \rightarrow K=0,I_f)$$
$$=2M_1^2\langle I_i 22-2|I_f 0\rangle^2(1+a_2(I_f(I_f+1)-I_i(I_i+1)))^2 \qquad (4\text{-}230)$$
$$a_2=-\frac{M_2}{M_1}$$

(Early studies of the $E2$-intensity relations for the decay of γ-vibrational bands were reported by Hansen *et al.*, 1959, who employed a somewhat different, though equivalent, form of the relation (4-230), involving the parameter

$$z_2=\frac{-2M_2}{M_1+4M_2}=\frac{2a_2}{1-4a_2} \qquad (4\text{-}231)$$

▲ The form (4-230) was given by Mikhailov, 1966.)

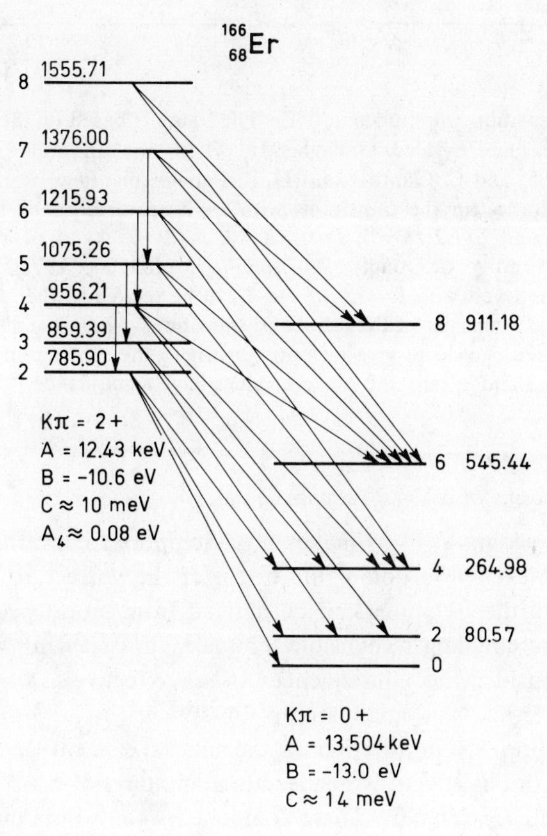

Figure 4-29 Level scheme for ^{166}Er. The data are taken from C. W. Reich and J. E. Cline, *Nuclear Phys.* **A159**, 181 (1970).

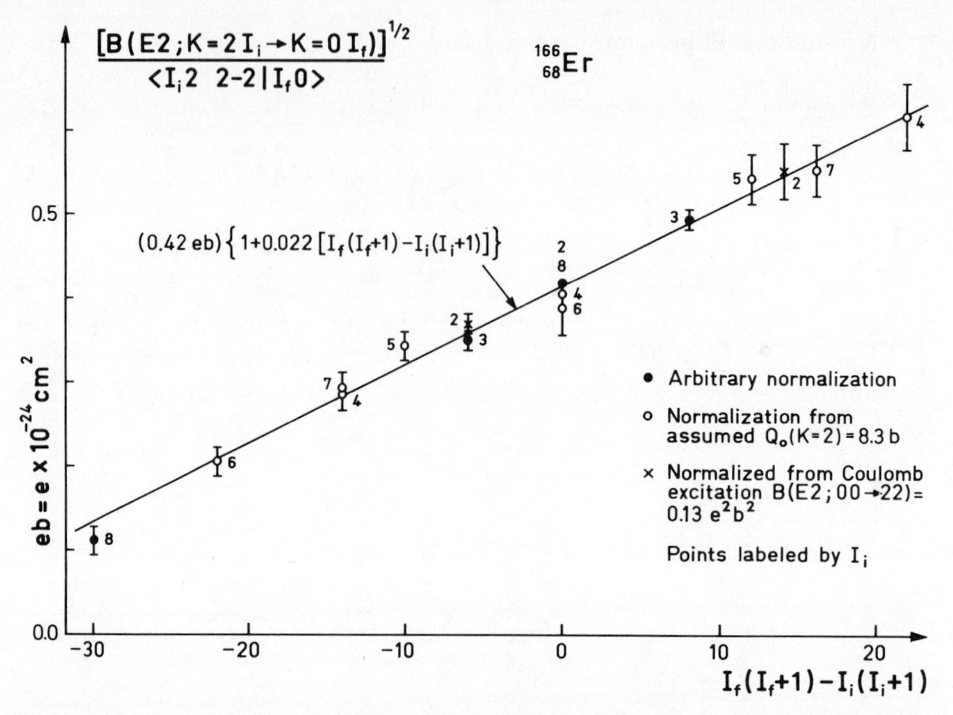

Figure 4-30 $E2$-transition amplitudes in ^{166}Er. The figure is based on the relative intensities of γ transitions measured by C. J. Gallagher Jr., O. B. Nielsen, and A. W. Sunyar, *Phys. Letters* **16**, 298 (1965) and C. Günther and D. R. Parsignault, *Phys. Rev.* **153**, 1297 (1967). The absolute amplitudes for the transitions with $I_i=2$ are determined from the Coulomb excitation measurement $B(E2;\ K=0,\ I=0\rightarrow K=2,\ I=2)=(0.13\pm0.01)e^2b^2$ (J. M. Domingos, G. D. Symons, and A. C. Douglas, *Nuclear Phys.* **A180**, 600, 1972). The absolute rates for the decays of the levels with $I_i=4,5,6$, and 7 can be related to the parameter $Q_0(K=2)$ through the observed intensities of the intraband transitions. The value of $Q_0(K=2)$ adopted in the figure has been chosen to give an optimum fit to the straight-line relation. For the transitions with $I_i=3$ and 8, only the relative values have significance.

▼ *Analysis in terms of $\Delta K=2$ coupling*

The preceding analysis makes no assumptions concerning the structure of the $K=2$ excitation nor of the origin of the correction terms in the $E2$ moments. Further relations can be derived from more specific assumptions regarding the coupling responsible for the matrix element M_2. In particular, we shall consider the consequences of an effective $\Delta K=2$ coupling that directly mixes the two bands (see pp. 149ff.).

An additional contribution to the matrix element M_2 may arise from first-order Coriolis effects, which lead to admixtures of $K=1$ components into the $K=0$ and $K=2$ bands. These components contribute through $E2$-matrix elements with $\Delta K=1$, but the apparent absence of low-lying collective $E2$ modes with $K\pi=1+$ may suggest that these contributions to M_2 are small ▲

▼ compared to those arising from the $\Delta K=2$ coupling. It must be emphasized, however, that the available evidence leaves considerable uncertainty regarding the interesting question of collective effects in the $K=1$ excitations. (The possible significance of the $\Delta K=1$ contributions to M_2 has been discussed by Günther and Parsignault, *loc cit*., Fig. 4-30.)

The contribution to M_2 from the $\Delta K=2$ coupling can be expressed in the form (see Eq. (4-211))

$$M_2=\varepsilon_2\left(\frac{15}{8\pi}\right)^{1/2} eQ_0 \qquad (4\text{-}232)$$

where ε_2 is the reduced amplitude describing the admixture of the two bands

$$\begin{aligned} |\hat{K}=0,I\rangle &\approx |K=0,I\rangle - c(I)|K=2,I\rangle \\ |\hat{K}=2,I\rangle &\approx |K=2,I\rangle + \tfrac{1}{2}c(I)\big(1+(-1)^I\big)|K=0,I\rangle \\ c(I) &= \varepsilon_2(2(I-1)I(I+1)(I+2))^{1/2} \\ \varepsilon_2 &\equiv \langle K=2|\varepsilon_{+2}|K=0\rangle \approx -1.1\times 10^{-3} \end{aligned} \qquad (4\text{-}233)$$

The estimate of ε_2 has been obtained from the value $Q_0=756\ \text{fm}^2$ determined for the ground-state band (Elbek *et al*., 1960) and the value of M_2 given in Fig. 4-30. (We have adopted a phase convention such that the matrix element $\langle K=2|h_{+2}|K=0\rangle$ is negative, corresponding to a labeling of the axes giving $\mathcal{I}_1>\mathcal{I}_2$, in accordance with the conventional notation for triaxial rotors (see Eq. (4-282)). This convention implies a negative sign for ε_2 and $c(I)$, and the observed negative sign of $M_2:M_1$ thus implies a positive value for M_1.) The value of ε_2 corresponds to a coupling matrix element (see Eq. (4-208))

$$\begin{aligned} h_2 &\equiv \langle K=2|h_{+2}|K=0\rangle = \varepsilon_2(E(K=2)-E(K=0)) \\ &\approx -0.8\ \text{keV} \end{aligned} \qquad (4\text{-}234)$$

In the derivation of the intensity relation (4-230) on the basis of the $\Delta K=2$ coupling, it has been assumed that the intrinsic quadrupole moments of the two bands are equal. A difference in the quadrupole moments leads to the more general relation (still retaining only terms linear in ε_2)

$$\begin{aligned} &B(E2;K=2,I_i\rightarrow K=0,I_f) \\ &\quad = 2\langle I_i 2 2 -2|I_f 0\rangle^2\Big(M_1+M_2(I_i(I_i+1)-I_f(I_f+1)) \\ &\qquad + M_3\big((I_i(I_i+1)-I_f(I_f+1))^2-2(I_i(I_i+1)+I_f(I_f+1))\big)\Big)^2 \end{aligned} \qquad (4\text{-}235)$$

▲

with

$$M_2=\varepsilon_2\left(\frac{15}{8\pi}\right)^{1/2} eQ_0(K=2)$$
$$M_3=\varepsilon_2\frac{1}{12}\left(\frac{15}{8\pi}\right)^{1/2} e(Q_0(K=0)-Q_0(K=2)) \tag{4-236}$$

The data in Fig. 4-30 set a limit on the matrix element M_3, which implies $|Q_0(K=2)-Q_0(K=0)|\lesssim 0.1Q_0$. The value of $Q_0(K=2)$ obtained from the analysis in Fig. 4-30 is consistent with this limit.

The band coupling (4-233) implies contributions to the rotational energy terms B and A_4 (see Eq. (4-213))

$$\delta B(K=0)=-2\varepsilon_2^2(E_2-E_0)=-1.8 \text{ eV}$$
$$\delta B(K=2)=\delta A_4(K=2)=-\tfrac{1}{2}\delta B(K=0) \tag{4-237}$$

The observed B value for the ground-state band (see Fig. 4-29) is about an order of magnitude larger than the contribution (4-237) and must therefore be attributed to additional (as yet unidentified) rotational coupling effects; the same conclusion applies to the term $B(K=2)$. However, the contribution to $A_4(K=2)$ resulting from the coupling to the ground-state band is more than an order of magnitude larger than the observed value (see Fig. 4-29). Thus, the present interpretation implies that the effect of couplings to higher-lying $K\pi=0+$ bands cancels the contribution (4-237) to A_4, to a rather high accuracy (see the discussion on p. 166).

The band mixing (4-233) implies small corrections to the intensity relations for $E2$ transitions within the bands. For the $K=0$ band, the modified relation is of the form (4-81) with coefficients given by (4-215)

$$\frac{M_3}{M_1}=-\frac{\varepsilon_2}{\sqrt{12}}\left(\frac{B(E2;0\to K=2,I=2)}{B(E2;0\to K=0,I=2)}\right)^{1/2}$$
$$\approx\frac{1}{12}a_2\frac{B(E2;0\to K=2,I=2)}{B(E2;0\to K=0,I=2)} \tag{4-238}$$
$$\approx -5\times 10^{-5}$$
$$M_2=-6M_3$$

For the transitions $I+2\to I$, the correction to the leading-order intensity relation is sometimes expressed in terms of the parameter α, introduced in Eq. (4-181). The coupling to the $K=2$ band implies a contribution to α given by

▼ (see Eqs. (4-230), (4-232), and (4-238))

$$\delta\alpha = \frac{M_2 + 8M_3}{M_1} \approx -1 \times 10^{-4} \tag{4-239}$$

Such a value of α implies a correction to the ratio $B(E2; I=4 \to I=2)$: $B(E2; I=2 \to I=0)$ of about 0.5%. (Multiple Coulomb excitation studies have yielded an estimate for α of about -10^{-3} for the ground-state band of ^{166}Er (Sayer *et al.*, 1970), which would imply rather large contributions from couplings that have not yet been identified.)

The coupling between the $K=0$ and $K=2$ bands also gives rise to $M1$ transitions between the bands with amplitudes proportional to the amplitude of the $M1$-matrix elements in the bands. If the two bands are assumed to have the same value of g_R, only the part of the $M1$ operator proportional to $(g_K - g_R)I_3$ contributes, and we obtain

$$\begin{aligned} \langle K=2, I_2 \| \mathscr{M}(M1) \| K=0, I_1 \rangle &= ((2I_1+1)(I_1-1)I_1(I_1+1)(I_1+2))^{1/2} \langle I_1 2 1 0 | I_2 2 \rangle M \\ &= (2(2I_1+1)I_1(I_1+1))^{1/2} \langle I_1 1 1 1 | I_2 2 \rangle (M_1 + M_2(I_2(I_2+1) - I_1(I_1+1))) \end{aligned} \tag{4-240}$$

with

$$\begin{aligned} M &= -2\sqrt{2}\left(\frac{3}{4\pi}\right)^{1/2} \frac{e\hbar}{2Mc} (g_K(K=2) - g_R)\varepsilon_2 \\ M_1 &= -4M_2 \\ M_2 &= \tfrac{1}{2}M \end{aligned} \tag{4-241}$$

The last expression in Eq. (4-240) gives the matrix element in the standard form for first-order K-forbidden transitions (see Eq. (4-98)). The analysis of γ-transition intensities within the $K=2$ band in ^{166}Er has yielded the value $|g_K - g_R| \approx 0.11$ (Reich and Cline, 1970), corresponding to $M1:E2$ intensity ratios of the order of 10^{-4} for the interband transitions in Fig. 4-29.

Since the direct band mixing is rather ineffective in producing $M1$ transitions between the bands, due to the small value of $g_K - g_R$, it appears likely that larger contributions to the $M1$ transitions between the bands may arise from first-order Coriolis effects acting through intermediate states with $K=1$. These effects give a matrix element of the form (4-240) with

$$\begin{aligned} M_1 &= \langle K=2 \| [\varepsilon_{+1}, \mathscr{M}(M1, \nu=1)] \| K=0 \rangle \\ M_2 &= 0 \end{aligned} \tag{4-242}$$

▲

▼ This matrix element is of the same form as the leading-order intensity relation for once K-forbidden $M1$ transitions (see Eq. (4-95)). For the $M1$ transitions, therefore, the $\Delta K=2$ and $\Delta K=1$ couplings give different intensity relations and can be distinguished by a measurement of relative intensities. (Evidence supporting an interpretation of the observed $M1$ admixtures in the decay of the $K=2$ band in ^{152}Sm and ^{154}Gd in terms of $\Delta K=1$ couplings has been given by Rud and Nielsen, 1970.)

Interpretation of the $K\pi=2+$ excitation

The low-energy and large $E2$-matrix element for exciting the $K=2$ band suggests that we are dealing with a collective mode involving deviations of the nuclear shape from axial symmetry. (The $B(E2)$ value for exciting the $K=2, I=2$ state is about $28 B_W(E2)$, which is 14 times the appropriate single-particle unit (see p. 549).) Such a collective mode could have the character of a vibration around an axially symmetric equilibrium or might be associated with an equilibrium shape deviating from axial symmetry.

If the equilibrium shape deviates from axial symmetry by more than the zero-point vibrational amplitude, the system may be treated as an asymmetric rotor (see Sec. 4-5). Since the deviations from axial symmetry are small, we may use the approximate expressions of Sec. 4-5c to obtain the three different rotational energy constants A_κ

$$\begin{aligned} E(2_1) &= 3(A_1+A_2) \\ E(2_2) &= A_1+A_2+4A_3 \\ h_2 &= \tfrac{1}{4}(A_1-A_2) \\ &\left(A_\kappa = \frac{\hbar^2}{2\mathcal{I}_\kappa}\right) \end{aligned} \tag{4-243}$$

Employing the energies in Fig. 4-29 and the value (4-234) for h_2, we obtain

$$\begin{aligned} A_1 &= 11.8 \quad \text{keV} \\ A_2 &= 15.0 \quad \text{keV} \\ A_3 &= 190 \quad \text{keV} \end{aligned} \tag{4-244}$$

For a system with ellipsoidal shape, the $E2$ operator involved in the rotational transitions depends on two intrinsic quadrupole moments,

$$\begin{aligned} \mathcal{M}(E2,\mu) &= \left(\frac{5}{16\pi}\right)^{1/2} e\left(Q_0 \mathcal{D}^2_{\mu 0} + Q_2\left(\mathcal{D}^2_{\mu 2}+\mathcal{D}^2_{\mu -2}\right)\right) \\ Q_0 &\equiv \langle\alpha|\sum_p (2x_3^2 - x_1^2 - x_2^2)_p|\alpha\rangle \\ Q_2 &\equiv \sqrt{\tfrac{3}{2}}\,\langle\alpha|\sum_p (x_1^2 - x_2^2)_p|\alpha\rangle \end{aligned} \tag{4-245}$$

▲

▼ where α is the intrinsic state (assumed to be an eigenstate of the operations $\mathscr{R}_\kappa(\pi)$ (see p. 177)). The ratio of Q_2 and Q_0 is a measure of the deviation from symmetry about the 3 axis, which is often expressed in terms of the parameter γ,

$$\tan\gamma = \sqrt{2}\,\frac{Q_2}{Q_0} \tag{4-246}$$

The $E2$-matrix elements obtained from the value of M_1 deduced from the data in Fig. 4-30 and the value of Q_0 quoted on p. 161 yield

$$\begin{aligned} Q_2 &= \left(\frac{16\pi}{5}\right)^{1/2} e^{-1} M_1 = 94\ \text{fm}^2 \\ Q_0 &= 756\ \text{fm}^2 \\ \tan\gamma &= 0.18 \end{aligned} \tag{4-247}$$

For an ellipsoidal shape with a uniform distribution of charge and a sharp surface, the values (4-247) correspond to principal axes with increments $\delta R_1 = -0.07 R_0$, $\delta R_2 = -0.14 R_0$, and $\delta R_3 = 0.21 R_0$, assuming a mean radius $R_0 = 1.2 A^{1/3}$ fm.

In the vibrational interpretation of the $K\pi = 2+$ mode, the excitation energy represents the vibrational frequency ($E(2_2) = \hbar\omega_\gamma$), and the $B(E2)$ value for excitating the $n_\gamma = 1$ band gives a measure of the amplitude of oscillation. Assuming approximately harmonic motion, one obtains

$$\gamma_0 \equiv \langle \gamma^2 \rangle^{1/2} \approx \sqrt{2}\,\frac{\langle K=2|\,\mathscr{M}(E2,\nu=2)|K=0\rangle}{\langle K=0|\,\mathscr{M}(E2,\nu=0)|K=0\rangle} = 0.18 \tag{4-248}$$

for the zero-point amplitude in the nuclear ground state.

The coupling matrix element h_2 measures the dependence of the moment of inertia on the vibrational amplitude γ (see Eqs. (4-234) and (6-296)),

$$\frac{1}{\mathscr{I}}\left(\frac{\partial \mathscr{I}_1}{\partial \gamma}\right)_{\gamma=0} = -\frac{h_2}{\gamma_0}\frac{4\mathscr{I}}{\hbar^2} = 0.7 \tag{4-249}$$

This value is comparable to, though somewhat smaller than, the value obtained by assuming the moments of inertia to be proportional to the square of the eccentricities (see Eq. (6B-17)),

$$\begin{aligned} \left(\frac{1}{\mathscr{I}_1}\frac{\partial \mathscr{I}_1}{\partial \gamma}\right)_{\gamma=0} &= \left(\sin^{-2}\!\left(\gamma - \frac{2\pi}{3}\right)\frac{\partial}{\partial\gamma}\sin^2\!\left(\gamma - \frac{2\pi}{3}\right)\right)_{\gamma=0} \\ &= \frac{2}{\sqrt{3}} = 1.15 \end{aligned} \tag{4-250}$$

▲

▼ The vibrational and rotational interpretations of the $K\pi=2+$ excitation have very different consequences for the higher states corresponding to multiple excitations of the $K\pi=2+$ mode. In the rotational model, the next band has $K\pi=4+$ and an energy four times that of the $K\pi=2+$ band (see Eq. (4-284)), while the superposition of two vibrational quanta leads to bands with $K\pi=0+$ and $4+$ and energies approximately twice the energy of the $n_\gamma=1$ band. In the spectrum of ^{166}Er, there is at present no evidence on the bands representing double excitations. Indirect evidence for the occurrence of a $K=0$, $n_\gamma=2$ band is provided by the observed very small value of the A_4 coefficient in the $K\pi=2+$ band referred to earlier (p. 162). For harmonic vibrations, the coupling to the $K\pi=0+$, $n_\gamma=2$ band exactly cancels the contribution from the $n_\gamma=0$ (ground-state) band; in the asymmetric rotor interpretation, the smallness of the A_4 term as compared with the contribution (4-237) must be attributed to a chance cancellation produced by the coupling to other degrees of freedom.

The available evidence, though inconclusive, thus appears to favor an interpretation of the $K\pi=2+$ excitation in terms of a γ vibration rather than a rotation associated with a static γ deformation. In this connection, it must be emphasized that the application of the asymmetric rotor model assumes the zero-point oscillations in the asymmetry coordinate γ to be small compared with the average value of this coordinate. The present information on the $K=2$ mode of excitation does not provide direct evidence concerning the ratio of these two quantities, but it might be expected that the zero-point oscillations in the γ direction would be of similar magnitude as those in the β direction. The experimentally observed $E2$-matrix elements for exciting the β vibrations are comparable to those exciting the $K\pi=2+$ bands (see the example on p. 551), which may suggest that, even if the equilibrium occurs for $\gamma\neq 0$, the zero-point fluctuations in γ are of similar magnitude as the equilibrium value. (While the models of harmonic vibrations and rotations of a static nonaxial shape represent limiting coupling schemes, the degrees of freedom involved may be treated more generally in terms of a two-dimensional (anharmonic) oscillator with radial variable γ and azimuthal angle associated with rotations about the 3 axis.)

Effects of coupling for large angular momentum

The analysis of the $\Delta K=2$ coupling given above is based on a perturbation treatment that requires the coefficients $c(I)$ in Eq. (4-233) to be small compared with unity. For the largest values of I in Fig. 4-29, this condition is rather well fulfilled ($c(I)\lesssim 0.1$). However, the coupling increases strongly with I, and for $I\approx 20$ the perturbation treatment is no longer valid. For such large values of I, a transition takes place to a new coupling scheme in which the system rotates about the axis with the largest moment of inertia.

The nature of the transition depends on the interpretation of the $\Delta K=2$ ▲ excitations. If this mode corresponds to a rotation about the 3 axis of a fixed

▼ asymmetric shape, the coupling scheme for large angular momenta is that described in Sec. 4-5e, in which the system rotates about the 1 axis. It can be verified that the condition (4-307) for the validity of this coupling scheme is ensured, when the perturbation solution (4-233) becomes inadequate.

If the $\Delta K=2$ excitation corresponds to vibration about an axially symmetric equilibrium shape, the coupling (4-206) implies a tendency for the system to deviate from axial symmetry because of the reduction in the rotational energy that can be obtained by lifting the degeneracy in the moments of inertia perpendicular to the 3 axis. The equilibrium value of γ, as a function of I, can be obtained by considering the sum of the potential energy in γ and the rotational energy. For small γ, the leading-order coefficients in the potential and rotational energy can be obtained from the above analysis,

$$
\begin{aligned}
V(\gamma) &= \frac{1}{2}\left(\frac{\gamma}{\gamma_0}\right)^2 \hbar\omega_\gamma \\
A_1(\gamma) &= A + 2\left(\frac{\gamma}{\gamma_0}\right) h_2 \\
A_2(\gamma) &= A - 2\left(\frac{\gamma}{\gamma_0}\right) h_2 \\
A_3(\gamma) &= \frac{1}{8}\left(\frac{\gamma_0}{\gamma}\right)^2 \hbar\omega_\gamma
\end{aligned}
\qquad (4\text{-}251)
$$

where γ_0 is the zero-point amplitude in γ, given by Eq. (4-248). The terms $V(\gamma)$ and $A_3(\gamma)$ describe the potential and centrifugal energies in the two-dimensional oscillator with radial variable γ and angular momentum $\frac{1}{2}I_3$. The difference between the terms $A_1(\gamma)$ and $A_2(\gamma)$ provides the coupling (4-206), which drives the oscillator away from the $\gamma=0$ equilibrium. For sufficiently small values of I, this coupling can be treated as a perturbation and gives an energy reduction proportional to γ^2, corresponding to a decrease in the effective restoring force in the γ motion. For $I \sim |\hbar\omega_\gamma/h_2|^{1/2}$, the $\gamma=0$ equilibrium becomes unstable, and the perturbation treatment of the coupling breaks down. For larger values of I, there develops an equilibrium with $\gamma > \gamma_0$, and the rotational motion can be approximately treated in terms of the coupling scheme in which the angular momentum is aligned in the direction of the largest moment of inertia ($I \approx I_1$; see Sec. 4-5e). It must be emphasized that, for the very large values of I associated with the transition to a triaxial shape, the rotational motion may be strongly affected by couplings that are not included in the present analysis (see the discussion of the spectrum in the ▲ yrast region on pp. 71 ff.).

▼ *Rotational coupling effects involving the ground-state band and the excited $K\pi = 0+$ band (β vibration) in ^{174}Hf (Figs. 4-31 and 4-32; Tables 4-20 and 4-21)*

An extensive body of data obtained from Coulomb excitation and (α, xn) reactions leading to ^{174}Hf provides evidence on the rotational coupling effects in the $\Delta K = 0$ transitions between the ground-state rotational band and the excited $K\pi = 0+$ band at 828 keV (see Fig. 4-31). The energies of the ground-state band are analyzed in Table 4-20 in terms of an expansion in powers of $I(I+1)$, and the results of this analysis are discussed on p. 69, together with other evidence concerning the behavior of the rotational energies in the ground-state bands of even-even nuclei.

Transitions between the two $K = 0$ bands

The amplitudes for the $E2$ transitions between the excited $0+$ band ($K = 0_2$) and the ground-state band ($K = 0_1$) are shown in Fig. 4-32. It is seen that the data deviate significantly from the leading-order intensity relation, which would correspond to a constant ordinate in Fig. 4-32, but are consistent with the generalized intensity rule, which includes a linear I dependence ▲

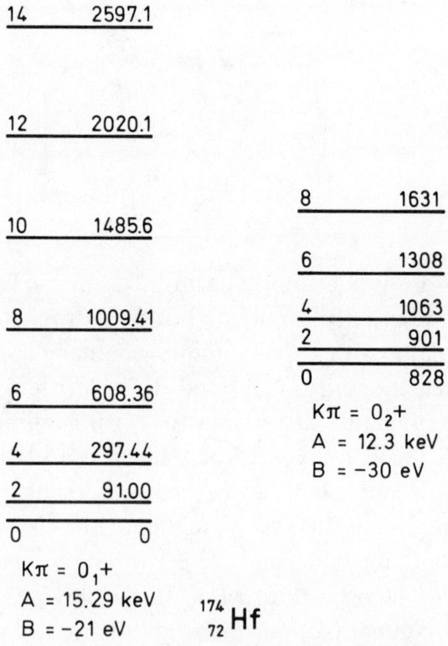

14 2597.1
12 2020.1
10 1485.6
8 1009.41
6 608.36
4 297.44
2 91.00
0 0
$K\pi = 0_1+$
A = 15.29 keV
B = −21 eV

8 1631
6 1308
4 1063
2 901
0 828
$K\pi = 0_2+$
A = 12.3 keV
B = −30 eV

$^{174}_{72}$Hf

Figure 4-31 Low-lying $K\pi = 0+$ bands in ^{174}Hf. The data are taken from J. H. Jett, D. A. Lind, G. D. Jones, and R. A. Ristinen, *Bull. Am. Phys. Soc.* II, **13**, 671 (1968); S. A. Hjorth, S. Jägare, H. Ryde, and A. Johnson (private communication, 1970); H. Ejiri and G. B. Hagemann, *Nuclear Phys.* **A161**, 449 (1971). The A and B coefficients given in the figure have been determined from the energies of the two lowest excited states in each band.

I_{max}	A	$-B(10^{-2})$	$C(10^{-4})$	$-D(10^{-6})$	$E(10^{-8})$	$-F(10^{-10})$	$G(10^{-12})$
2	15.167						
4	15.293	2.106					
6	15.305	2.356	0.96				
8	15.309	2.459	1.54	0.85			
10	15.312	2.523	1.96	1.81	0.69		
12	15.314	2.570	2.30	2.79	1.82	0.45	
14	15.315	2.609	2.60	3.73	3.21	1.38	0.230

Table 4-20 $I(I+1)$ expansion of the ground-state rotational band of ^{174}Hf. The energies of the ground-state rotational band of ^{174}Hf are taken from Fig. 4-31. The expansion coefficients are in keV times the power of 10 indicated at the top of each column. These coefficients have been obtained by fitting the energies up to a given value of I_{max} by a polynomial in $I(I+1)$ with $\frac{1}{2}I_{max}$ terms. We wish to thank I. Hamamoto for help in the preparation of the table.

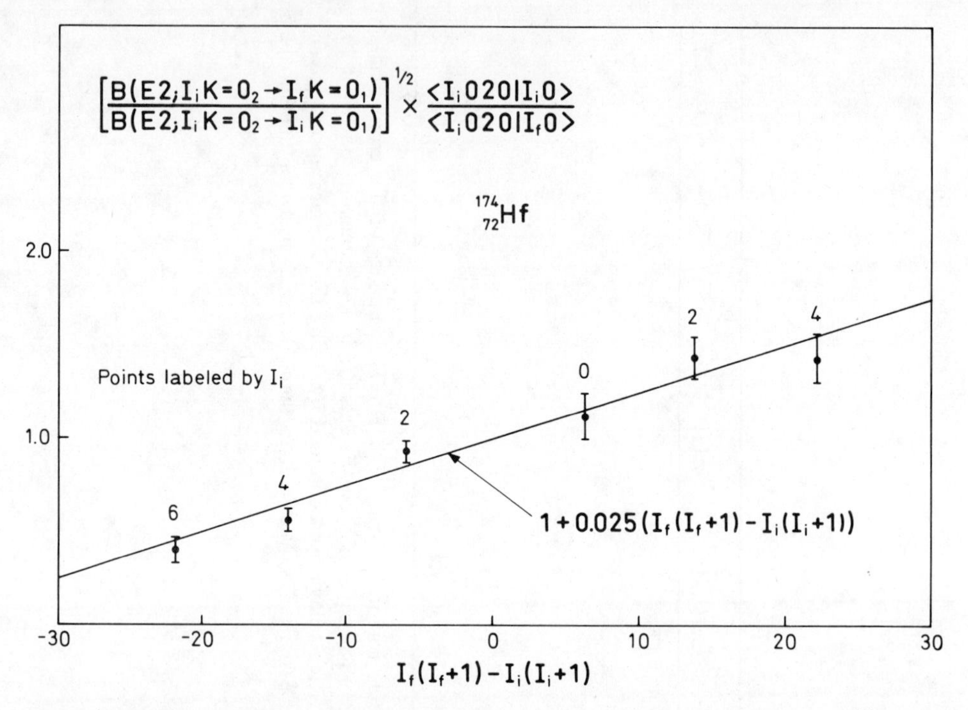

Figure 4-32 $E2$ intensities for transitions between the 0_2+ and 0_1+ bands of ^{174}Hf. The data are taken from Ejiri and Hagemann, *loc. cit.*, Fig. 4-31. The relative phase of the different transitions has not been experimentally determined, and in the construction of the figure it has been assumed that the ratio of the reduced matrix elements, represented by the square root of the $B(E2)$ values, has the same sign as the ratio of the Clebsch-Gordan coefficient.

▼ in the intrinsic moment (see Eq. (4-219))

$$B(E2; K=0_2 I_i \rightarrow K=0_1, I_f)$$

$$= \langle I_i 020 | I_f 0 \rangle^2 (M_1 + M_2(I_i(I_i+1) - I_f(I_f+1)))^2$$

$$= \langle I_i 020 | I_f 0 \rangle^2 M_1^2 (1 + a_0(I_i(I_i+1) - I_f(I_f+1)))^2 \qquad (4\text{-}252)$$

$$a_0 = \frac{M_2}{M_1}$$

where the last expression defines the parameter a_0 often employed in the literature. The empirical data in Fig. 4-32 provide evidence for a value of $a_0 \approx -0.025$. The interpretation of the matrix elements M_1 and M_2 will be considered below. ▲

I	$m(M1)$ $10^{-2}e\hbar/2Mc$	$\|\langle 0_2\|r^2\|0_1\rangle\|$ fm^2
0		12.0 ± 1.5
2	⩽2	12.0 ± 1.0
4	1.6 ± 0.6	10.5 ± 1.5
6	3.2 ± 0.5	11.5 ± 1.5

Table 4-21 $M1$- and $E0$-matrix elements for transitions between the 0_2+ and 0_1+bands of ^{174}Hf. The data are taken from Ejiri and Hagemann, *loc.cit.*, Fig. 4-31. The quantity in the second column is the experimentally determined intrinsic matrix element

$$m(M1) \equiv \left(\frac{4\pi}{3}\right)^{1/2}((2I+1)I(I+1))^{-1/2}$$
$$\times \langle K=0_2, I \| \mathscr{M}(M1) \| K=0_1, I \rangle$$

The sign is determined relative to the $E2$-matrix element, which is taken to be negative (positive M_1, see Eq. (4-219).)

▼ The $M1$-and $E0$-transition amplitudes are listed in Table 4-21. The $E0$-matrix elements are approximately independent of I, which agrees with the leading-order intensity rule for a scalar operator. For the $M1$ transitions, the I-independent moment gives a matrix element proportional to $\langle I010|I0\rangle$, which vanishes. The leading-order contribution to the transition moment is, therefore, linear in the angular momentum; since there is only a single such term (see Eq. (4-89)), it must be proportional to **I**, which has the appropriate symmetry for the $M1$ operator,

$$\mathscr{M}(M1,\mu) = \left(\frac{3}{4\pi}\right)^{1/2} \frac{e\hbar}{2Mc} g_R I_\mu \tag{4-253}$$

where g_R is an operator in the intrinsic variables. The moment (4-253) gives the transition matrix element

$$\langle K=0_2, I_2 \| \mathscr{M}(M1) \| K=0_1, I_1 \rangle$$

$$= \left(\frac{3}{4\pi}\right)^{1/2} \frac{e\hbar}{2Mc} \langle 0_2 | g_R | 0_1 \rangle ((2I_1+1)I_1(I_1+1))^{1/2} \delta(I_1, I_2) \tag{4-254}$$

The $M1$-transition probabilities in Table 4-21 show a strong increase with I, but are barely accurate enough to test the relation (4-254). ▲

▼ *Analysis in terms of coupling between the bands*

The deviations from the leading-order $E2$-intensity relations exhibited in Fig. 4-32 may arise from a number of effects associated with the coupling between intrinsic and rotational motion, but the low-frequency and the rather large matrix element for the excitation of the $K=0_2$ band (see below) suggests that important contributions to the matrix element M_2 arise from the mixing of the two bands described by an effective $\Delta K=0$ coupling. In the following, we shall pursue the consequences of assuming that the entire matrix element M_2 can be interpreted in this manner.

The coupling of the two bands has the form

$$\begin{aligned}|\hat{K}=0_1,I\rangle &\approx |K=0_1,I\rangle - \varepsilon_0 I(I+1)|K=0_2,I\rangle \\ |\hat{K}=0_2,I\rangle &\approx |K=0_2,I\rangle + \varepsilon_0 I(I+1)|K=0_1,I\rangle\end{aligned} \tag{4-255}$$

where the amplitude ε_0 can be obtained from Eq. (4-220)

$$\begin{aligned}\varepsilon_0 \equiv \langle 0_2|\varepsilon_0|0_1\rangle = \left(\frac{16\pi}{5}\right)^{1/2}\frac{M_2}{eQ_0} &\approx a_0\left(\frac{B(E2;0\rightarrow K=0_2,I=2)}{B(E2;0\rightarrow K=0_1,I=2)}\right)^{1/2} \\ &\approx -3\times 10^{-3}\end{aligned} \tag{4-256}$$

The estimate of ε_0 is based on the value of a_0 given above and the transition probabilities $B(E2;K=0_1,I=0\rightarrow K=0_2,I=2)=(0.06\pm 0.01)\times e^2 10^{-48}\text{cm}^4$ and $B(E2;K=0_1,I=0\rightarrow K=0_1,I=2)=5.3\times e^2 10^{-48}$ cm^4, corresponding to $Q_0=730$ fm^2 (Ejiri and Hagemann, *loc.cit.* Fig. 4-31). The sign of ε_0 depends on the relative phase for the intrinsic states 0_1 and 0_2, which has been chosen so as to make the $E2$-matrix element M_1 in Eq. (4-252) positive. The amplitude ε_0 is related to the matrix element of the operator h_0 occurring in the effective $K=0$ coupling Hamiltonian (4-216) and, by means of Eq. (4-218), we obtain

$$\langle 0_2|h_0|0_1\rangle \approx -2.6\,\text{keV} \tag{4-257}$$

The coupling of the two bands gives rise to a correction to the energies of the ground-state band proportional to $I^2(I+1)^2$ and with a coefficient

$$\begin{aligned}\delta B(K=0_1) &= -\varepsilon_0^2(E(K=0_2)-E(K=0_1)) \\ &\approx -8\,\text{eV}\end{aligned} \tag{4-258}$$

which is a little less than half of the experimentally observed value for the B coefficient of the ground-state band (see Table 4-20). The coupling also contributes to the B coefficient of the $K=0_2$ band with an amount of equal magnitude and opposite sign, but comparable contributions of negative sign ▲ may arise from the coupling of the $K=0_2$ band to the higher-lying $K=0$ band

▼ that is expected to result from the double excitation of the mode responsible for the $K=0_2$ band.

The band mixing (4-255) implies corrections to the $E2$-matrix elements within the ground-state rotational band, which can be written in the form

$$\langle K=0_1, I_2 \| \mathscr{M}(E2) \| K=0_1, I_1 \rangle$$

$$= (2I_1+1)^{1/2} \langle I_1 0 2 0 | I_2 0 \rangle \left(\frac{5}{16\pi} \right)^{1/2} eQ_0 (1 + \alpha (I_1(I_1+1) + I_2(I_2+1))) \tag{4-259}$$

where

$$\alpha = -\left(\frac{16\pi}{5} \right)^{1/2} \frac{\varepsilon_0 M_1}{eQ_0} \approx -a_0 \frac{B(E2; 0 \to K=0_2, I=2)}{B(E2; 0 \to K=0_1, I=2)}$$

$$\approx 4 \times 10^{-4} \tag{4-260}$$

where the numerical estimate is based on the experimental data employed in Eq. (4-256). For transition matrix elements ($I_2 = I_1 \pm 2$), the parameter α in Eq. (4-259) is equivalent to that introduced in Eq. (4-181), and the value (4-260) implies corrections of about 1.6% in the ratio $B(E2; 4 \to 2)$: $B(E2; 2 \to 0)$. The available experimental evidence is not sufficiently accurate to identify effects of this magnitude.

From the observed intensity of the $E0$ transitions between the two bands, one can estimate the I dependence of the mean square charge radius of the ground-state band

$$\delta \langle r^2 \rangle = \frac{2\varepsilon_0}{Z} \langle K=0_2 | \sum_{p=1}^{Z} r_p^2 | K=0_1 \rangle I(I+1)$$

$$\approx (\pm) 4 \times 10^{-5} I(I+1) \langle r^2 \rangle \tag{4-261}$$

$$(\langle r^2 \rangle \approx 27 \, \text{fm}^2)$$

where we have used the value of ε_0 from Eq. (4-256) and the average monopole matrix element in Table 4-21. (The sign of this matrix element has not yet been determined.) The magnitude of the I-dependent term (4-261) has not been measured for ^{174}Hf, but somewhat smaller values have been reported for the heavier Hf isotopes on the basis of Mössbauer resonance studies (see the compilation by Shenoy and Kalvius, 1971).

The coupling between the bands also contributes to the $M1$ transition

$$\langle K=0_2, I_2 \| \mathscr{M}(M1) \| K=0_1, I_1 \rangle$$

$$= \left(\frac{3}{4\pi} \right)^{1/2} \frac{e\hbar}{2Mc} \varepsilon_0 (g_R(K=0_1) - g_R(K=0_2))$$

▲

$$\times ((2I_1+1) I_1 (I_1+1))^{1/2} I_1 (I_1+1) \delta(I_1, I_2) \tag{4-262}$$

▼ The matrix element (4-262) is of higher order in I than the leading term (4-254), and with improved accuracy of the experimental data in Table 4-21, it should be possible to determine separately the two contributions to $B(M1)$. An indication of the possible magnitude of the term (4-262) may be obtained by assuming $|g_R(K=0_2)-g_R(K=0_1)|\sim 0.1$, which together with the value (4-256) for ε_0 leads to an $M1$ amplitude that is about a factor of 2 smaller than the values in Table 4-21.

Interpretation of the $K=0_2$ excitation

The $K\pi=0_2+$ excitation in ^{174}Hf is characterized by a rather strong $E2$-excitation probability. (The value of $B(E2;K=0_1,I=0\rightarrow K=0_2,I=2)$ quoted above is about $10B_W(E2)$, where $B_W(E2)$ is the appropriate single-particle unit (see p. 551).) This enhancement suggests that the excitation may be described as a collective vibrational mode associated with variations in the nuclear quadrupole deformation (see Chapter 6, pp. 361 ff.). The amplitude of oscillation may be obtained from the $E2$-transition matrix element; since the transition operator is linear in the deformation parameter β, we have

$$\frac{1}{\beta_0}\langle 0_2|\,\beta-\beta_0|0_1\rangle=\left(\frac{16\pi}{5}\right)^{1/2}\frac{M_1}{eQ_0}$$

$$\approx\left(\frac{B(E2;0\rightarrow K=0_2,I=2)}{B(E2;0\rightarrow K=0_1,I=2)}\right)^{1/2}$$

$$\approx 0.12 \tag{4-263}$$

where β_0 is the equilibrium value of β. (The relation between the quadrupole deformation parameters β and δ is given by Eq. (4-191).)

The $E0$-matrix element in Table 4-21 is a few single-particle units (as defined by Eq. (6-434)) and may be compared with the value associated with a volume-conserving quadrupole vibration (Reiner, 1961; see Eqs. (6-82) and (6-84)),

$$\langle 0_2|\sum_{p=1}^{Z} r_p^2|0_1\rangle=2\beta_0\left(\frac{3}{4\pi}ZR_0^2\right)\left(1+\frac{4}{3}\pi^2\left(\frac{a}{R_0}\right)^2\right)\langle 0_2|\,\beta-\beta_0|0_1\rangle$$

$$\approx 13\ \text{fm}^2 \tag{4-264}$$

using $R_0=6.1$ fm and $a=0.54$ fm (see Eqs. (2-69) and (2-70)).

In the vibrational description, the intrinsic matrix element for the $M1$ transition occurring in Eq. (4-254) can be expressed in the form

$$\langle 0_2|\,g_R|0_1\rangle\approx\left(\frac{\partial g_R}{\partial\beta}\right)_{\beta=\beta_0}\langle 0_2|\,\beta-\beta_0|0_1\rangle \tag{4-265}$$

▲

▼ The values of $B(M1)$ in Table 4-21, if interpreted in terms of the first-order effect (4-254), thus yield $(\partial g_R/\partial\beta)\approx 0.5$.

Similarly, the matrix element of the effective coupling operator h_0 can be expressed in terms of the dependence of the moment of inertia on the eccentricity (see Eq. (6-296)),

$$\langle 0_2|h_0|0_1\rangle = \frac{\partial}{\partial\beta}\left(\frac{\hbar^2}{2\mathscr{I}(\beta)}\right)_{\beta=\beta_0}\langle 0_2|\beta-\beta_0|0_1\rangle \tag{4-266}$$

and the empirical value of h_0 (see Eq. (4-257)) yields

$$\left(\frac{\beta}{\mathscr{I}}\frac{\partial\mathscr{I}}{\partial\beta}\right)_{\beta=\beta_0}\approx 1.5 \tag{4-267}$$

This value is much greater than the eccentricity dependence of the rigid moment of inertia (see Eq. (4-104)), but the effect of the pairing on the moment of inertia is rather strongly dependent on the vibrational amplitude. Such a dependence arises partly from the fact that the excitation energies of the single-particle transitions induced by the Coriolis interaction depend on the nuclear eccentricity. In the simple estimate (4-128) for the moment of inertia, the single-particle transition energy is proportional to β, and the dependence of $\mathscr{I}$ on this energy implies

$$\begin{aligned}\frac{\beta}{\mathscr{I}}\left(\frac{\partial\mathscr{I}}{\partial\beta}\right)_{\Delta\,\mathrm{const}} &= -\frac{x}{(1-g(x))}\frac{dg}{dx}\\ &\approx 0.9\end{aligned} \tag{4-268}$$

$$\left(x=\frac{\hbar\omega_0\delta}{2\Delta}\approx 1.3\right)$$

assuming the values of $\hbar\omega_0\delta$ and Δ given on p. 83. An additional contribution to $\partial\mathscr{I}/\partial\beta$ may arise if the pairing parameter Δ depends on the vibrational amplitude β. ▲

4-5 ROTATIONAL SPECTRA FOR SYSTEMS WITHOUT AXIAL SYMMETRY

The considerations in the preceding sections apply to systems with axial symmetry, for which the collective rotational motion is restricted to take place about axes perpendicular to the symmetry axis. In any quantal system, there will be fluctuations in the shape, which involve momentary excursions away from axial symmetry. These fluctuations are treated above as intrinsic excitations, and their effect on the rotational motion is included in the general expressions derived in Sec. 4-3 (see also the $\Delta K=2$ coupling discussed in Sec. 4-4).

If the system possesses a stable equilibrium shape that deviates from axial symmetry by an amount exceeding the zero-point shape fluctuations, it becomes possible to go a step further in the separation between rotational and intrinsic motion and to consider collective rotations about all axes of the intrinsic structure. In such a situation, the rotational families contain an added dimension, and the rotational relationships are correspondingly more extensive.

The study of rotational motion in nuclei with asymmetric shapes is potentially a field of broad scope. Although, at present, there are no well-established examples of nuclear spectra corresponding to asymmetric equilibrium shapes, it appears likely that such spectra will be encountered in the exploration of nuclei under new conditions (large deformations, angular momentum, isospin, etc.). The discussion in the present section is directed partly toward these potential applications to nuclear spectra, but is also motivated by the fact that the analysis of the more general rotational motion of a nonaxial system gives a broader view of many of the problems treated in the previous sections. (The quantal theory of asymmetric rotors has been extensively developed in connection with the analysis of the spectra of polyatomic molecules; see, for example, the texts by Herzberg, 1945, and by Townes and Schawlow, 1955. The possibility of exploiting the asymmetric rotor in the interpretation of nuclear spectra has been especially emphasized by Davydov and Filippov, 1958; see also the review by Davidson, 1965.)

4-5a Symmetry Classification for Even *A*

In the analysis of rotational motion for a system with an even number of fermions, we shall assume a nondegenerate intrinsic state, as may be expected if the intrinsic Hamiltonian possesses no higher symmetry than rotations of π about one or more axes, in addition to space and time reversal invariance. In fact, in such a situation, as will be discussed below, the intrinsic invariance group has only one-dimensional irreducible representations. (In contrast, for an odd-*A* system, time-reversal invariance in itself implies a twofold degeneracy. For the even-*A* system, multidimensional irreducible representations may occur if the intrinsic invariance group includes an axis of order $n \geqslant 3$ (an axis of order n signifies invariance with respect to rotations of $2\pi/n$). In fact, an axis of order $n \geqslant 3$ together with time reversal (which has nondiagonal matrix elements between the states with complex conjugate eigenvalues for the rotations) or together with an additional rotational axis implies that the intrinsic invariance group contains noncommuting elements.)

Symmetry of Hamiltonian

If we assume the deformation to preserve time-reversal invariance, the nondegenerate intrinsic states will be eigenstates of $\mathscr{T}$ and will have vanishing expectation values for intrinsic operators that are odd under $\mathscr{T}$. Hence, the rotational Hamiltonian, expressed as a function of the components I_κ of the total angular momentum with respect to intrinsic axes, is invariant with respect to the inversion $I_\kappa \rightarrow -I_\kappa$. To leading order, H_{rot} is thus a bilinear expression in the I_κ, which can be brought to a diagonal form by a transformation to principal axes,

$$H_{\text{rot}} = \sum_{\kappa=1}^{3} A_\kappa I_\kappa^2 \tag{4-269}$$

The coefficients A_κ, which represent expectation values for the intrinsic state, may be expressed in terms of the moments of inertia,

$$A_\kappa = \frac{\hbar^2}{2\mathscr{I}_\kappa} \tag{4-270}$$

The Hamiltonian (4-269) is invariant with respect to rotations through the angle π about each of the three principal axes,

$$\mathscr{R}_\kappa(\pi) = \exp\{-i\pi I_\kappa\} \tag{4-271}$$

The commutation relations for the intrinsic components I_κ (see Eq. (1A-91)) imply the identities

$$\begin{gathered}\mathscr{R}_1(\pi) = \mathscr{R}_3(\pi)\,\mathscr{R}_2(\pi) \\ \mathscr{R}_2(\pi)\,\mathscr{R}_3(\pi) = \mathscr{R}_3(\pi)\,\mathscr{R}_2(\pi)\,\mathscr{R}(2\pi) \\ \text{and cyclic permutations}\end{gathered} \tag{4-272}$$

where

$$\mathscr{R}(2\pi) = (\mathscr{R}_1(\pi))^2 = (\mathscr{R}_2(\pi))^2 = (\mathscr{R}_3(\pi))^2 = (-1)^A \tag{4-273}$$

is a rotation of 2π. For even values of A, such a rotation equals the identity, and the operations $\mathscr{R}_\kappa(\pi)$ commute; the more general relations (4-272) also apply to odd-A systems to be considered below.

For even A, the symmetry group of the Hamiltonian consisting of the commuting elements 1, $\mathscr{R}_1(\pi)$, $\mathscr{R}_2(\pi)$, and $\mathscr{R}_3(\pi)$ is the point group conventionally denoted by D_2. The eigenvalues of $\mathscr{R}_\kappa(\pi)$ equal $r_\kappa = \pm 1$, and since $r_1 = r_2 r_3$, the symmetry group has the four (one-dimensional) representations

$$(r_1 r_2 r_3) = \begin{matrix} (+++) \\ (+--) \\ (-+-) \\ (--+) \end{matrix} \tag{4-274}$$

The symmetry quantum numbers r_κ thus label the eigenstates of the asymmetric rotor Hamiltonian (4-269).

Eigenstates

For a system with three different moments of inertia, none of the components I_κ are constants of the motion, but eigenstates with symmetry quantum number r_κ involve only components with even values of I_κ (for $r_\kappa = +1$) or odd I_κ (for $r_\kappa = -1$). In the representation in which I_3 is diagonal (with eigenvalues K), the expansion of the rotational wave function with specified symmetry $r_2 r_3$ takes the form

$$\varphi_{r_2 r_3;\tau IM}(\omega) = \sum_{\substack{K=0,2,\ldots,(r_3=+1) \\ K=1,3,\ldots,(r_3=-1)}} c(r_2 r_3;\tau IK)\chi_{r_2 r_3; IKM}(\omega) \tag{4-275}$$

where τ labels the different states in the band with the same value of I and where

$$\chi_{r_2 r_3; IKM}(\omega) = \left(\frac{2I+1}{16\pi^2}\right)^{1/2} \left(\mathscr{D}^I_{MK}(\omega) + r_2(-1)^{I+K}\mathscr{D}^I_{M-K}(\omega)\right)\left(1+\delta(K,0)\right)^{-1/2} \tag{4-276}$$

The relative phase of the components with opposite K is given by the quantum number r_2 (see Eq. (4-18), where the operation $\mathscr{R}_2(\pi)$ is denoted by $\mathscr{R}_e$). Components with $K=0$ occur only for $r_3 = +1$ and $(-1)^I = r_2$; from this selection rule, it follows that the eigenstates of the rotational Hamiltonian with specified point symmetry $(r_1 r_2 r_3)$ comprise the sets of I values

$$I = \begin{cases} 0, 2^2, 3, 4^3, 5^2, \cdots \\ 1, 2, 3^2, 4^2, 5^3, \cdots \end{cases} \qquad (r_1 r_2 r_3) = \begin{cases} (+++) \\ (+--), (-+-), (--+) \end{cases} \tag{4-277}$$

The four symmetry classes together give $(2I+1)$ states for each value of I.

Constraints associated with intrinsic D_2 symmetry

The form (4-269) for the leading-order rotational Hamiltonian, which leads to the classification of the states in terms of D_2 symmetry, was based only on the assumed time-reversal invariance of the deformation. If, in addition, the deformation, and thus the intrinsic Hamiltonian, possesses D_2 invariance, as for shapes with ellipsoidal symmetry, the rotational degrees of freedom are restricted by constraints similar to those discussed in Sec. 4-2 for axially symmetric systems.

For a deformation with D_2 symmetry, orientations that differ by a rotation $\mathscr{R}_\kappa(\pi)$ are indistinguishable, and these rotations become part of the intrinsic degrees of freedom. The corresponding reduction in the rotational degrees of freedom can be expressed by the constraint $(\mathscr{R}_\kappa(\pi))_e = (\mathscr{R}_\kappa(\pi))_i$, which is similar to the relation (4-11). Thus, for an intrinsic state with quantum numbers r_κ, the rotational spectrum contains only the states belonging to the symmetry class with the same values of r_κ. In contrast, for a deformation that violates D_2 symmetry, the rotations $\mathscr{R}_\kappa(\pi)$ are no longer part of the intrinsic degrees of freedom, and the rotational spectrum contains all the states (4-277) belonging to the four different symmetry classes.

Additional intrinsic symmetries

If the deformation is invariant with respect to finite rotations that are not contained in the D_2 group, the rotational Hamiltonian possesses the invariance of the extended symmetry group that includes the D_2 group, as well as the additional elements of invariance of the deformation. In such a situation, the tensor of inertia will have axial symmetry; in fact, the extended group must contain at least one axis of order $n>2$, and such an axis is seen to be a symmetry axis for the inertial tensor. The existence of a symmetry axis for the inertial tensor implies that the moments of inertia in directions perpendicular to the symmetry axis are equal, and that the component of angular momentum along this axis is a constant of the motion. A system with these properties is referred to as a symmetric top. If the intrinsic structure possesses two or more axes of order $n>2$ (as for tetrahedral or cubic symmetry), the tensor of inertia acquires spherical symmetry ($A_1 = A_2 = A_3$), and the system is referred to as a spherical top. (The occurrence of a symmetry axis of order $n \geqslant 3$ may lead to degeneracy in the intrinsic motion (see p. 176) and thus to the occurrence of terms in the rotational Hamiltonian that are linear in I_κ.)

The eigenstates of the rotational Hamiltonian can be classified in terms of the representations of the extended symmetry group. For a given intrinsic state, the rotational spectrum is restricted to the representations that have the same symmetry quantum numbers as the intrinsic state with respect to the rotations that leave the intrinsic structure invariant. For example, if the intrinsic state has the eigenvalue $\exp\{-2\pi i\nu/n\}$ for a rotation of $2\pi/n$ about an axis of order n, the rotational spectrum contains only states with $K=\nu, \nu \pm n, \nu \pm 2n, \ldots$, with respect to this axis. If the intrinsic state belongs to a multidimensional irreducible representation of the intrinsic invariance group (as for axial symmetry combined with $\mathscr{R}_2(\pi)$ invariance), the wave function involves a linear combination of the different intrinsic states belonging to the given representation, obtained by a symmetrization procedure analogous to that employed for the axially symmetric nuclei (see Eq. (4-19)).

One may also consider the possibility of deformations that are invariant with respect to a combination of a finite rotation and a reflection in space or time, without being invariant under these operations separately. In such a situation, the rotational degrees of freedom are coupled to those associated with the inversion in the manner discussed in Sec. 4-2f. For example, if a deformation of symmetry $Y_{32}+Y_{3-2}$ is superposed on an ellipsoidal deformation, the intrinsic structure is invariant under $\mathscr{R}_3(\pi)$ and the combination $\mathscr{S}_2=\mathscr{R}_2(\pi)\mathscr{P}$, and the intrinsic states can be labeled by the eigenvalues r_3 and s_2 of these two operations. The rotational spectrum contains the representations $(r_3, r_2=\pm 1)$ that have the same value of r_3 as the intrinsic state, and the two representations have opposite parity $(\pi=r_2 s_2)$.

Molecules with identical nuclei

In molecular spectra, restrictions on the rotational degrees of freedom arise when a permutation of identical nuclei can be accomplished by a rotation of the molecule. (For a discussion of the symmetry classification of the states of poly-atomic molecules, see for example, Landau and Lifschitz, 1958, pp. 383ff.) The rotations that either leave the equilibrium positions of the nuclei invariant, or result in permutations of identical nuclei, constitute a subgroup G of the rotations in three dimensions. The group G is contained in the "molecular symmetry group" that characterizes the invariance of the Hamiltonian describing the electronic and vibrational degrees of freedom with respect to the intrinsic coordinate system; in turn, the molecular symmetry group is contained in the invariance group of the rotational Hamiltonian. The elements of G can be expressed either in terms of overall (external) rotations or as internal operations consisting of rotations of electronic and vibrational variables and permutations of identical nuclei. In analogy to the constraints discussed above for nuclear systems, the molecular spectrum is restricted by the condition

$$D(G;\text{rotation})=D(G;\text{electronic})\otimes D(G;\text{vibration})\otimes D(G;\text{permutation}) \quad (4\text{-}278)$$

where the representations $D(G)$ characterize the symmetry with respect to the group G of the different components in the molecular wave function. The factor $D(G;$ permutation) refers to the spatial permutation symmetry of the nuclei, which in turn is determined by the statistics and the wave function for the nuclear spins.

As an example, we consider the molecule C_2H_4 (ethylene, $>C=C<$), in which the four protons lie at the corners of a plane rectangle with the two carbon nuclei (assumed to be ^{12}C) placed symmetrically on an axis bisecting two opposite sides of the rectangle. For this molecule, the symmetry group G consists of the identity and the three rotations $\mathscr{R}_\kappa(\pi)$ about axes through the center of the molecule; each of these rotations involves a pair of tıanspositions of the four protons, $(P_{12}P_{34}), (P_{13}P_{24}), (P_{14}P_{23})$. Two of the rotations also involve an interchange of the ^{12}C nuclei, but the Bose statistics and the spin 0 for these nuclei imply that this interchange always gives a factor $+1$. The invariance group is seen to be $G = D_2$, whose representations are one-dimensional, and the condition (4-278) therefore reduces to the corresponding relation for the eigenvalues, r_κ(rotation) $= r_\kappa$(electronic)$\times r_\kappa$(vibration)$\times r_\kappa$(permutation). Permutations of the H atoms constitute a subgroup of the symmetric group S_4 and, for the protons with spin 1/2 and Fermi statistics, the possible spatial permutation symmetries are $[f]=[1111]$, $[211]$, and $[22]$, corresponding to the conjugate spin permutation symmetries $[\tilde{f}]=[4], [31]$, and $[22]$ associated with the total spin quantum number $S = 2, 1$, and 0, respectively (see Appendix 1C, Vol. I, pp. 115 and 120). The determination of the quantum numbers r_κ(permutation) involves the decomposition of the representations of S_4 into representations of the subgroup that consists of the identity and the three pairs of transpositions (and which is isomorphic with D_2). In the present case, this decomposition can be immediately obtained by noting that the three representations of D_2 involving negative values of r_κ (see Eq. (4-274)) must occur together and that these representations do occur in $[f]=[211]$, for which one of the basis functions may be taken to be symmetric in particles 1 and 2 and antisymmetric in 3 and 4 (see Table 1C-1, Vol. I, p. 128) and therefore an eigenfunction of $P_{12}P_{34}$ with eigenvalue -1. Since the dimensions of $[f]$ are 1, 3, and 2 for [1111], [211], and [22], respectively, as can be seen from the rules given in Vol. I, p. 111, the following decomposition results: A proton spin state with $S=2$, corresponding to $[f]=[1111]$, implies r_κ(permutation)$=(+++)$, while $S=1$ ($[f]=[211]$) yields the representations $(+--)$, $(-+-)$, $(--+)$, and $S=0$ ($[f]=[22]$) the symmetric representation $(+++)$ occurring twice. Thus, from a knowledge of the quantum numbers r_κ for the electronic and vibrational motion, one can determine the possible symmetries r_κ (rotation), for given value of S. The set of states contained in the rotational spectrum then follows from Eq. (4-277).

4-5b Energy Spectra

The determination of the eigenvalues of the rotational Hamiltonian (4-269) with three different moments of inertia requires diagonalization of matrices with dimensions increasing with I. The energy spectrum can be expressed in the form

$$E_{\text{rot}} = \tfrac{1}{2}(A_1 + A_3)I(I+1) + \tfrac{1}{2}(A_1 - A_3)E_{\tau I}(\kappa) \tag{4-279}$$

where $E_{\tau I}(\kappa)$ are the eigenvalues of the matrix

$$H(\kappa) = I_1^2 + \kappa I_2^2 - I_3^2 \tag{4-280}$$

depending on a single asymmetry parameter, conventionally denoted by κ,

$$\kappa = \frac{2A_2 - A_1 - A_3}{A_1 - A_3} \tag{4-281}$$

Since we can label the principal axes such that

$$A_1 \leqslant A_2 \leqslant A_3 \qquad (\mathscr{I}_1 \geqslant \mathscr{I}_2 \geqslant \mathscr{I}_3) \tag{4-282}$$

it is sufficient to determine the eigenvalues of $H(\kappa)$ in the interval

$$-1 \leqslant \kappa \leqslant 1 \tag{4-283}$$

For $\mathscr{I}_2 = \mathscr{I}_1$, corresponding to $\kappa = 1$, the ellipsoid of inertia is an oblate spheroid, while for $\mathscr{I}_2 = \mathscr{I}_3$ and $\kappa = -1$, the ellipsoid of inertia is prolate. (If the moments of inertia have the classical rigid-body values, a prolate (oblate) shape implies an oblate (prolate) ellipsoid of inertia, but this relationship is not in general valid for quantal systems; for example, if the system possesses axial symmetry, the moment of inertia with respect to this axis vanishes, and the ellipsoid of inertia is therefore prolate, irrespective of whether the shape is prolate or oblate.)

The spectrum of eigenvalues of $H(\kappa)$ as a function of κ is shown in Fig. 4-33 for a few values of I. The states are labeled by the symmetry quantum numbers $(r_1 r_2 r_3)$. The set of eigenvalues $E_{\tau I}$ for negative κ is the same, apart from opposite sign, as for positive κ; the inversion involves an interchange of the 1 and 3 axes. (Tables of eigenvalues $E_{\tau I}$ have been given by King *et al.*, 1943 and 1949 ($I \leqslant 12$) and by Erlandsson, 1956 ($13 \leqslant I \leqslant 40$); the properties of the spectrum for large I will be discussed further below.)

If two of the moments of inertia are equal (axially symmetric rotor, $\kappa = \pm 1$), the spectrum consists of a succession of bands with specified K representing the component of the angular momentum along the symmetry axis of the inertial tensor, and the energy within each band is linear in $I(I+1)$. For $\kappa = 1$, the symmetry axis is the 3 axis and $K = I_3$, while for $\kappa = -1$, the 1 axis becomes the symmetry axis and $K = I_1$, as indicated in Fig. 4-33a. A small asymmetry can be taken into account by means of a coupling between bands with different values of K, and the spectrum can be expressed as a power series in $I(I+1)$ (see Sec. 4-5c). For intermediate values of κ, there is no longer a conserved quantum number K, and the expansion in powers of $I(I+1)$ will not in general be a useful tool. Nevertheless, for any value of κ, one can define analytic trajectories that link states with successive values of I into rotational sequences (Regge trajectories). However, the spectrum of the asymmetric rotor with its multiplicity of states for given I reveals ambiguities in the definition of the Regge trajectories. This feature is illustrated by the example considered on pp. 194ff.

In a nucleus with a shape that is invariant with respect to $\mathscr{R}_\kappa(\pi)$, the one-particle orbits are twofold degenerate and the ground-state configuration of an even-even nucleus is obtained by a pairwise filling of conjugate orbits. Such an intrinsic state has the quantum numbers $r_1 = r_2 = r_3 = \pi = +1$ (see the similar arguments for axially symmetric shapes on p. 29). For an asymmetric rotor with this intrinsic symmetry, the low-energy spectrum contains the sequence $I\pi = 0+, 2+, 4+, \ldots$ (as for a $K=0$ band), while the next group of excitations has $I\pi = 2+, 3+, 4+, \ldots$ (as for a $K\pi = 2+$ band). Such a pattern is widely observed (see, for example, the spectrum of ^{166}Er discussed on p. 159 and the spectra of the Os isotopes in Fig. 6-32, p. 536); with the available evidence, however, it has not been possible to establish quantitative relationships in these spectra that are specific to the asymmetric rotor model. The excitation energy of the second 2+ state is never observed to be less than 500 keV, and is therefore considerably larger than the rotational excitation energies in the strongly deformed nuclei. On this basis, the excitations have usually been described in terms of vibrational degrees of freedom, but it must be emphasized that many features of the spectra are not well understood at present (see the discussion of the quadrupole vibrations in Chapter 6, pp. 549 ff. and pp. 554 ff.). The possibility of an interpretation of some of these spectra in terms of shapes with a small asymmetry parameter is further discussed on pp. 164 ff. and in Sec. 4-5c. The crucial evidence for deciding between a rotational and a vibrational interpretation would be provided by the identification of the levels corresponding to a repeated excitation of the mode responsible for the second $I\pi = 2+$ state. In the asymmetric rotor model, these states correspond to a $K\pi = 4+$ band beginning at an energy of between 10/3 and 4 times that of the second $I\pi = 2+$ state. (In the oblate limit ($\kappa = -1$), the second 2+ and third 4+ states have $I_1 = 0$ and energies $A_3 I(I+1)$; in the prolate limit ($\kappa = 1$), the states have $|I_3| = I$ and energies $A_1 I + A_3 I^2$.) For harmonic vibrations about an axially symmetric equilibrium shape, the double excitation gives rise to two bands with $K\pi = 0+$ and 4+, with excitation energies approximately twice that of the second $I\pi = 2+$ state.

Additional tests of the coupling scheme of the asymmetric rotor could be provided by evidence on the intensity rules for transitions involving different states with the same intrinsic configuration. For wave functions of the form (4-275), the matrix elements of tensor operators can be evaluated by a transformation to the intrinsic coordinate frame, as in the derivation of the intensity rules for the axially symmetric rotor (see Sec. 4-3d). The resulting intensity relations depend on the coefficients $c(K)$, which in turn depend on the asymmetry parameter κ, and in general involve a number of

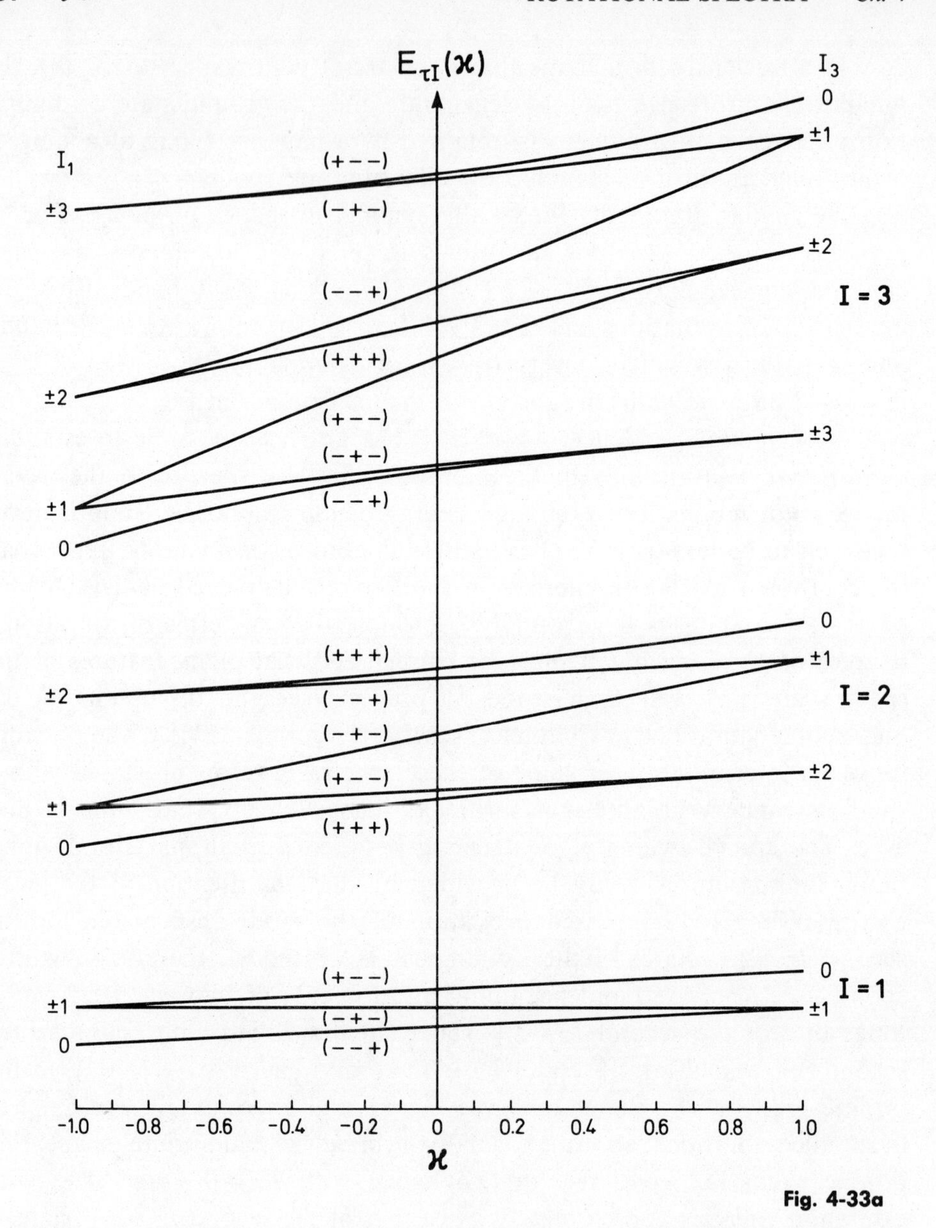

Fig. 4-33a

Figure 4-33 Eigenvalues of the characteristic matrix $H(\kappa)$ for the asymmetric rotor. The eigenvalues are taken from the tables by G. W. King, R. M. Hainer, and P. C. Cross, *J. Chem. Phys.* **11**, 27 (1943) and **17**, 826 (1949) and by G. Erlandsson, *Arkiv Fysik* **10**, 65 (1956) and are shown, for a few values of I, as a function of the asymmetry parameter κ; the states are labeled by the quantum numbers $(r_1 r_2 r_3)$ of the D_2 group.

intrinsic matrix elements, corresponding to different components ν of the tensor operator. These intensity relations are more comprehensive than

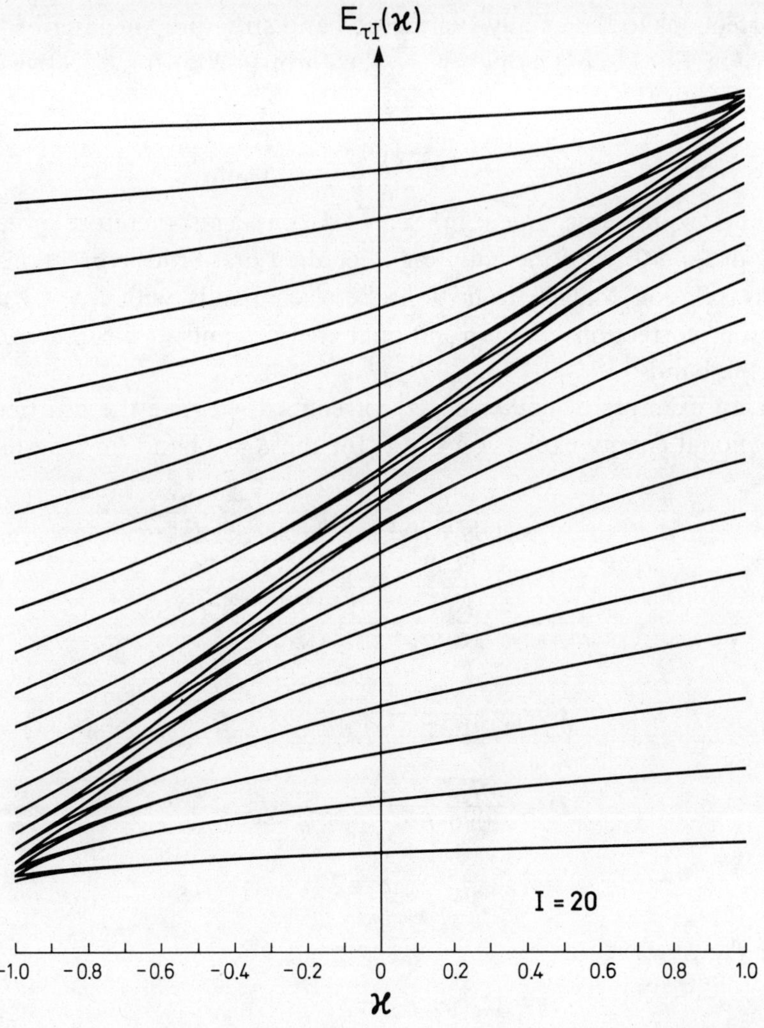

Fig. 4-33b

those for axially symmetric nuclei, on account of the greater number of rotational levels associated with a given intrinsic state of an asymmetric system.

4-5c Systems with Small Asymmetry

If two of the moments of inertia are approximately equal ($\mathscr{I}_1 \approx \mathscr{I}_2$), the spectrum can be separated into bands characterized by the quantum number K $(=I_3)$. In the treatment of such a nearly symmetric rotor, it is convenient to write the Hamiltonian (4-269) in the form

$$\begin{aligned} H_{\text{rot}} &= \tfrac{1}{2}(A_1 + A_2)\mathbf{I}^2 + \tfrac{1}{2}(2A_3 - A_1 - A_2)I_3^2 + H_c \\ H_c &= \tfrac{1}{4}(A_1 - A_2)(I_-^2 + I_+^2) \end{aligned} \tag{4-284}$$

The corrections to the axially symmetric band structure are described by the term H_c in Eq. (4-284), which is of the form (4-206) for effective $\Delta K=2$ couplings with

$$h_{+2}=\tfrac{1}{4}(A_1-A_2) \tag{4-285}$$

From the expressions given in Sec. 4-4, one can therefore obtain the leading-order effects of the coupling, including first-order corrections to the intensity relation for $E2$ transitions between bands with $\Delta K=2$ and second-order corrections to the rotational energies and to the $E2$ transitions within the bands.

As an example of higher-order corrections, we give the coefficients in the rotational energy expansion (4-62) for the $K=0$ band (in a system with $r_2=r_3=+1$),

$$\begin{aligned}
A &= \tfrac{1}{2}(A_1+A_2)+\frac{1}{8}\frac{a^2}{b}+\cdots \\
B &= -\frac{1}{16}\frac{a^2}{b}-\frac{19}{1024}\frac{a^4}{b^3}+\cdots \\
C &= -\frac{3}{1024}\frac{a^4}{b^3}+\cdots \\
D &= \frac{7}{4096}\frac{a^4}{b^3}+\cdots \\
a &\equiv A_1-A_2 \qquad b\equiv 2A_3-A_1-A_2
\end{aligned} \tag{4-286}$$

and for the $K=2$ band,

$$\begin{aligned}
A &= \tfrac{1}{2}(A_1+A_2)+\frac{1}{8}\frac{a^2}{b}+\cdots \\
A_4 &= \frac{1}{32}\frac{a^2}{b}+\cdots \\
B &= \frac{1}{48}\frac{a^2}{b}+\cdots
\end{aligned} \tag{4-287}$$

It must be emphasized that the above expressions for the rotational energy coefficients are based on the leading-order rotational Hamiltonian (4-269), which is quadratic in the I_κ. Additional contributions may arise from higher-order terms in the rotational Hamiltonian involving fourth and higher powers of the angular momentum.

In view of the systematic occurrence of excited $K\pi=2+$ bands in the spectra of strongly deformed even-even nuclei, one may consider the possibility of describing these spectra in terms of a rotor deviating slightly from axial symmetry. The energy of the second 2+ state is typically an order of magnitude larger than the energy of the first excited 2+ state, which implies $A_3 \sim 10A_1$, while the observed magnitude of the $\Delta K=2$ coupling implies $|A_1 - A_2| \sim 10^{-1}A_1$ (see the example discussed on p. 164). The assumption of a slightly asymmetric rotor does not lead to any new relation between the $E2$-matrix elements connecting the $K=0$ and $K=2$ bands, beyond those obtained from the general analysis of $\Delta K=2$ couplings (as treated in Sec. 4-4), but the observed rotational energies may be compared with the relations (4-286) and (4-287). Examples are provided by the spectra of ^{168}Er (Table 4-1, p. 65), ^{166}Er (Fig. 4-29, p. 159), and ^{174}Hf (Table 4-20, p. 169). It is found that the observed higher-order terms in the energies exhibit quite a different pattern from that implied by the relations (4-286) and (4-287). Thus, the B terms in the $K=0$ and $K=2$ bands are observed to be rather similar, in contrast to Eqs. (4-286) and (4-287), which imply $B(K=2) \approx -\frac{1}{3}B(K=0)$. For the C and D terms in the observed $K=0$ bands, one finds $C>0$, $D<0$, and $|C| \gtrsim 10^2|D|$, while Eq. (4-286) implies $C<0$, $D>0$, and $|C| \approx \frac{12}{7}|D|$. The observed A_4 terms in the $K=2$ bands are either very small (compared with $B(K=0)$) or negative, in contrast to the result (4-287), which implies $A_4 \approx \frac{1}{2}|B(K=0)|$. Thus, the observed deviations from the leading-order rotational energies in the low-lying bands of even-even nuclei cannot be ascribed to deviations from axial symmetry in the nuclear shape.

4-5d Symmetry Classification for Odd *A*

The lack of definite evidence for the occurrence of asymmetric rotor patterns in the spectra of even-even nuclei makes it difficult to test the corresponding relationships in the more complex spectra of the odd-A nuclei (see, for example, the comments on the spectrum of $A=25$ on pp. 287 ff.). In the present section, we consider briefly the general features of the odd-A spectra that can be obtained from symmetry considerations. (The analysis of the spectrum of a particle coupled to a nonaxial rotor was considered by Hecht and Satchler, 1962.)

For a system with an odd number of fermions, the intrinsic states are doubly degenerate, if the deformation is invariant under time reversal (Kramers' degeneracy, see Vol. I, p. 19). The rotational Hamiltonian is therefore a 2×2 matrix in the space of the conjugate intrinsic states, which will be denoted by $|\alpha, \rho=1\rangle$ and $|\alpha, \rho=-1\rangle = -\mathscr{T}|\alpha, \rho=1\rangle$ (see Eq. (1B-20)). Such a matrix can be written as a linear combination of the unit matrix ρ_0 and the three Pauli matrices $\boldsymbol{\rho}(=\rho_1, \rho_2, \rho_3)$. Since ρ_0 is even under time reversal, while the components of $\boldsymbol{\rho}$ are odd (see Eq. (1B-23)), the rotational Hamiltonian is of the form

$$H_{\mathrm{rot}} = \rho_0 H_0(I_\kappa) + \boldsymbol{\rho} \cdot \mathbf{H}(I_\kappa) \tag{4-288}$$

where H_0 is even and $\mathbf{H}$ odd with respect to the inversion $I_\kappa \rightarrow -I_\kappa$. If we include only the leading-order terms, linear and bilinear in I_κ, the rotational Hamiltonian can be written

$$H_{\mathrm{rot}} = \sum_{\kappa=1}^{3} A_\kappa \left(I_\kappa^2 - \sum_{k=1}^{3} a_{k\kappa} \rho_k I_\kappa \right) \tag{4-289}$$

where the intrinsic axes have been chosen to coincide with the principal axes of the inertial tensor in H_0. The coefficients $a_{k\kappa}$ in Eq. (4-289) are real (since H_{rot} is Hermitian) and are analogous to the decoupling parameter in $K = 1/2$ bands of axially symmetric nuclei.

If the intrinsic Hamiltonian possesses D_2 symmetry, one can choose basis states $\alpha\rho$ that are eigenstates of $\mathscr{R}_3(\pi)$; since $(\mathscr{R}_3(\pi))^2 = -1$ for an odd-A system (see Eq. (4-273)), the eigenvalues of $\mathscr{R}_3(\pi)$ are $\pm i$. The states $\rho = \pm 1$ have complex conjugate eigenvalues r_3 and, selecting the state $\rho = 1$ as the one with eigenvalue $-i$, we have $\mathscr{R}_3(\pi) = -i\rho_3$. The phases of the states may be determined in accord with the standard convention $\mathscr{R}_2(\pi)\mathscr{T} = +1$. Since $\mathscr{T} = i\rho_2$ (see Eq. (1B-22)), this condition implies $\mathscr{R}_2(\pi) = -i\rho_2$, and we therefore have

$$\mathscr{R}_\kappa(\pi) = -i\rho_\kappa \tag{4-290}$$

as a consequence of the relation $\mathscr{R}_1(\pi) = \mathscr{R}_2(\pi)\mathscr{R}_3(\pi)$ and cyclic permutations. (These relations for the operators $(\mathscr{R}_\kappa(\pi))_i$ acting on the intrinsic variables are the inverse of the relations (4-272) for the corresponding operators $(\mathscr{R}_\kappa(\pi))_e$ acting on the orientation angles.)

The invariance with respect to the operations $\mathscr{R}_\kappa(\pi)$ implies that, in the basis (4-290), the rotational Hamiltonian (4-289) takes the form

$$H_{\mathrm{rot}} = \sum_{\kappa=1}^{3} A_\kappa (I_\kappa^2 - a_\kappa \rho_\kappa I_\kappa) \tag{4-291}$$

corresponding to a diagonal decoupling matrix, $a_{k\kappa} = a_\kappa \delta_{k\kappa}$.

For an odd-A system, the total wave function may involve rotational components with all possible K values $(-I \leqslant K \leqslant I)$ each occurring together with both the intrinsic states $\rho = \pm 1$. The $\mathscr{R}_\kappa(\pi)$ symmetry of the intrinsic motion imposes the selection rules

$$\mathscr{R}_3(\pi) = \exp\{-i\pi K\} = -i\rho_3 = \begin{cases} -i & \rho = 1 \\ i & \rho = -1 \end{cases} \tag{4-292}$$

and the further relation $\mathscr{R}_2(\pi)\mathscr{T} = 1$ leads to wave functions of the form

$$\begin{aligned} \Psi_{\alpha\tau IM} &= \sum_{K = \ldots, -3/2, 1/2, 5/2, \ldots} c(\alpha\tau IK)\psi_{\alpha IKM} \\ \psi_{\alpha IKM} &= \left(\frac{2I+1}{16\pi^2}\right)^{1/2} \left(\Phi_{\alpha,\rho=1}\mathscr{D}^I_{MK}(\omega) - (-1)^{I+K}\Phi_{\alpha,\rho=-1}\mathscr{D}^I_{M-K}(\omega)\right) \end{aligned} \tag{4-293}$$

(In the limit of axial symmetry, the intrinsic state denoted by Φ_K equals $\Phi_{\alpha,\rho=1}$ for $K = 1/2, 5/2, \ldots$ and $\Phi_{\alpha,\rho=-1}$ for $K = 3/2, 7/2, \ldots$. Since $\Phi_{\alpha,\rho=-1} = -\mathscr{T}\Phi_{\alpha,\rho=1}$, the minus sign between the two components of $\psi_{\alpha IKM}$ in Eq. (4-293) corresponds to the plus sign in Eq. (4-19).) For a given value of I, the states (4-293) involve $(I+1/2)$ terms, and the rotational spectrum has the components

$$I = 1/2, (3/2)^2, (5/2)^3, \ldots \tag{4-294}$$

The multiplicity of states equals $(2I+1)$ multiplied by two (as a result of the twofold degeneracy of the intrinsic states) and divided by four (as a result of the invariance with respect to the D_2 group with four elements).

For a system consisting of a particle coupled to an asymmetric rotor, the rotational energy (4-289) is obtained from expression (4-269) for the rotor by replacing I_κ by $R_\kappa = I_\kappa - j_\kappa$. The operator j_κ, considered as a 2×2 matrix in the space $|\alpha, \rho = \pm 1\rangle$, is linear in ρ_k, and hence the coefficient $a_{k\kappa}$ in Eq. (4-289) equals the trace of the product $j_\kappa \rho_k$. If the potential produced by the rotor possesses D_2 symmetry, the rotational Hamiltonian is given by Eq. (4-291) with

$$a_\kappa = \mathrm{Tr}\,(j_\kappa \rho_\kappa) \tag{4-295}$$

The decoupling parameters a_κ depend on the symmetry r_κ of the rotor as well as on the single-particle wave functions. Thus, for example, for a particle state with $j = 1/2$, we have $\mathscr{R}_\kappa(\pi) = r_\kappa \exp\{-i\pi j_\kappa\} = -2ij_\kappa r_\kappa$ (see Eq. (1A-48)), and the relation (4-290) therefore implies $a_\kappa = r_\kappa$. For these values of a_κ, the rotational Hamiltonian (4-291) yields a spectrum of doublets corresponding to vanishing coupling between the $j = 1/2$ particle and the rotor with symmetry r_κ.

4-5e States with Large I

Simple and illuminating solutions for the asymmetric rotor can be obtained for the high angular momentum states in the yrast region. In the classical theory of the asymmetric rotor, the motion reduces to a simple rotation without precession of the axes, if the angular momentum is along the axis corresponding to the largest or smallest moment of inertia. Correspondingly, in the quantal theory, the states of smallest (or largest) energy for given I acquire a simple structure in the limit of large I (Golden and Bragg, 1949).

With the axes labeled such that $\mathscr{I}_1 > \mathscr{I}_2 > \mathscr{I}_3$ $(A_1 < A_2 < A_3)$, the states of lowest energy for given I (the yrast states) have $|I_1| \approx I$. The coupling between the components of the rotational wave function with positive and negative values of I_1 becomes negligible for the states with $|I_1| \approx I \gg 1$, and I_1 may thus be treated as a positive quantity. The angular momentum components perpendicular to the 1 axes obey the commutation relation

$$[I_-, I_+] = 2I_1 \approx 2I \quad (I_\pm \equiv I_2 \pm iI_3) \tag{4-296}$$

and can therefore be treated approximately in terms of boson creation and annihilation operators

$$c^\dagger = \frac{1}{\sqrt{2I}} I_+ \quad c = \frac{1}{\sqrt{2I}} I_- \\ [c, c^\dagger] \approx 1 \tag{4-297}$$

In terms of these variables, the rotational Hamiltonian (4-269) can be written

$$\begin{aligned} H &= A_1 \mathbf{I}^2 + \tfrac{1}{2}(A_2 + A_3 - 2A_1)(I_2^2 + I_3^2) + \tfrac{1}{2}(A_2 - A_3)(I_2^2 - I_3^2) \\ &= A_1 \mathbf{I}^2 + H' \end{aligned} \tag{4-298}$$

with

$$\begin{aligned} H' &= \tfrac{1}{2}\alpha(c^\dagger c + cc^\dagger) + \tfrac{1}{2}\beta(c^\dagger c^\dagger + cc) \\ &= \alpha(n + \tfrac{1}{2}) + \tfrac{1}{2}\beta(c^\dagger c^\dagger + cc) \end{aligned} \tag{4-299}$$

$$n = c^\dagger c \approx I - I_1$$

$$\alpha \equiv (A_2 + A_3 - 2A_1) I$$

$$\beta \equiv (A_2 - A_3) I$$

The number of boson excitations is denoted by n, and the boson vacuum ($n=0$) is the state with $I_1=I$; each quantum carries an angular momentum of -1 unit with respect to the 1 axis. (For an odd-A system, the leading-order Hamiltonian contains additional terms linear in I_κ (see Eq. (4-289)), which may give rise to terms linear in $c^\dagger$ and c; for a system with D_2 symmetry, however, the terms $a_2 I_2 \rho_2$ and $a_3 I_3 \rho_3$ can be neglected in the approximation considered, while $a_1 I_1 \rho_1$ is approximately a constant ($\approx a_1 I$).)

The Hamiltonian (4-298) can be diagonalized by a canonical transformation to the new boson variables $\hat{c}^\dagger$ and $\hat{c}$,

$$\begin{aligned} c^\dagger &= x\hat{c}^\dagger + y\hat{c} \\ \hat{c}^\dagger &= xc^\dagger - yc \end{aligned} \qquad (x^2-y^2=1) \tag{4-300}$$

and one obtains

$$\begin{aligned} H' &= \hbar\omega(\hat{n}+\tfrac{1}{2}) \\ \hat{n} &= \hat{c}^\dagger\hat{c} \end{aligned} \tag{4-301}$$

with the excitation quanta

$$\hbar\omega = (\alpha^2-\beta^2)^{1/2} = 2I((A_2-A_1)(A_3-A_1))^{1/2} \tag{4-302}$$

and the transformation coefficients

$$\left.\begin{matrix} x \\ y \end{matrix}\right\} = \left(\frac{1}{2}\left(\frac{\alpha}{(\alpha^2-\beta^2)^{1/2}} \pm 1\right)\right)^{1/2} \tag{4-303}$$

The stationary states can thus be characterized by the quantum number $\hat{n}$ together with I (and M), and the energy values are

$$E(\hat{n}, I) = A_1 I(I+1) + (\hat{n}+\tfrac{1}{2})\hbar\omega \tag{4-304}$$

The quantum number $\hat{n}$ describes the precessional motion of the axes with respect to the direction of I; for small amplitudes, this motion has the character of a harmonic vibration with frequency ω. If the intrinsic state possesses D_2 symmetry, the spectrum of an even-A rotor is restricted to the states with

$$(-1)^{\hat{n}} = r_1(-1)^I \tag{4-305}$$

while for odd A, the spectrum contains states with all values of $\hat{n}$. The asymptotic spectrum (4-304) for an even-A system can be recognized in the example illustrated in Fig. 4-34, p. 197.

The validity of the approximate solution (4-304) requires (see Eqs. (4-296) and (4-297))

$$I^2 \gg \langle I_2^2 + I_3^2 \rangle = I\langle c^\dagger c + cc^\dagger \rangle \tag{4-306}$$

By using the explicit solution given above, the condition (4-306) can be expressed in the form

$$I \gg (2\hat{n}+1)\frac{A_2+A_3-2A_1}{2(A_3-A_1)^{1/2}(A_2-A_1)^{1/2}} \tag{4-307}$$

If the three moments of inertia are of comparable magnitude (and $\hat{n}$ is small), the condition (4-307) corresponds to $I \gg 1$. If one of the inertial parameters is much larger than the other two ($A_3 \gg A_2, A_1$), as for a nucleus that deviates only slightly from axial symmetry, the transition to the coupling scheme described above, with $K = I_1 \approx I$, occurs for angular momenta of order $(A_3/(A_2-A_1))^{1/2}$, while for lower values of I, the yrast states have $K = I_3 \approx 0$.

The $E2$ moment of the nucleus can be expressed in terms of two intrinsic quadrupole moments Q_0 and Q_2 (see Eq. (4-245)). For the states in the yrast region with $|I_1| \approx I$, we shall define an intrinsic nuclear coordinate system $\mathcal{K}'$ with axes $x' = x_2, y' = x_3, z' = x_1$, so that ϑ, ϕ is the orientation of the 1 axis. Thus, Q_0 is the quadrupole moment with respect to the 1 axis, while Q_2 is a measure of the asymmetry in the shape with respect to this axis. In the representation in which $K = I$ is diagonal, the reduced $E2$-matrix elements are given by

$$\begin{aligned}\langle I'K' \| \mathcal{M}(E2) \| IK \rangle = (2I+1)^{1/2}\left(\frac{5}{16\pi}\right)^{1/2} e(Q_0\langle IK20|I'K'\rangle \\ + Q_2(\langle IK22|I'K'\rangle + \langle IK2-2|I'K'\rangle))\end{aligned} \tag{4-308}$$

For $K \approx I$, the vector addition coefficients in Eq. (4-308) have the approximate values

$$\begin{aligned}&\langle IK20|IK\rangle \approx \langle IK2\pm2|I\pm2\,K\pm2\rangle \approx 1 \\ &\langle IK20|I+1\,K\rangle \approx -\langle I+1\,K20|IK\rangle \approx \left(\frac{3}{I}\right)^{1/2}(I-K+1)^{1/2} \\ &\langle IK22|I+1\,K+2\rangle \approx -\langle I+1\,K+2\,2-2|IK\rangle \approx -\left(\frac{2}{I}\right)^{1/2}(I-K)^{1/2}\end{aligned} \tag{4-309}$$

ignoring terms of order I^{-1} or smaller. In this approximation, the matrix element (4-308) is conveniently expressed in terms of the variables I and

$n = I - K$. Introducing the operator $m(I_1, I_2)$ defined by

$$\langle I', K' = I' - n' \| \mathscr{M}(E2) \| I, K = I - n \rangle$$

$$= (2I+1)^{1/2} \left(\frac{5}{16\pi} \right)^{1/2} e \langle n' | m(I, I') | n \rangle \tag{4-310}$$

we obtain

$$m(I, I') \approx Q_0 \delta(I, I') + Q_2 \delta(I \pm 2, I')$$

$$+ \left(\left(\frac{3}{I} \right)^{1/2} Q_0 c^\dagger - \left(\frac{2}{I} \right)^{1/2} Q_2 c \right) \delta(I+1, I')$$

$$+ \left(- \left(\frac{3}{I} \right)^{1/2} Q_0 c + \left(\frac{2}{I} \right)^{1/2} Q_2 c^\dagger \right) \delta(I-1, I') \tag{4-311}$$

where the operators $c^\dagger$ and c are the creation and annihilation operators (4-297). By means of the transformation (4-300), the operator m can be expressed in terms of the variable $\hat{n}$, in which the Hamiltonian is diagonal to the approximation considered.

The leading-order terms in the operator (4-311) involve no change of $\hat{n}$ and give the static moment

$$Q \approx Q_0 \tag{4-312}$$

as well as the transitions

$$B(E2; \hat{n}I \rightarrow \hat{n}, I \pm 2) \approx \frac{5}{16\pi} e^2 Q_2^2 \tag{4-313}$$

induced by the rotation of the intrinsic Q_2 moment about the 1 axis.

The terms in Eq. (4-311) with $I' = I \pm 1$ are proportional to the vibrational amplitude and involve transitions with $\Delta\hat{n} = \pm 1$,

$$\begin{aligned} B(E2; \hat{n}I \rightarrow \hat{n}-1, I-1) &= \frac{5}{16\pi} e^2 \frac{\hat{n}}{I} \left(\sqrt{3}\, Q_0 x - \sqrt{2}\, Q_2 y \right)^2 \\ B(E2; \hat{n}I \rightarrow \hat{n}+1, I-1) &= \frac{5}{16\pi} e^2 \frac{\hat{n}+1}{I} \left(\sqrt{3}\, Q_0 y - \sqrt{2}\, Q_2 x \right)^2 \end{aligned} \tag{4-314}$$

The strength of these transitions is smaller than for the transitions (4-313) without change in the vibrational quantum number, by a factor of order $\hat{n}/I$, which represents the square of the amplitude of the precessional motion.

The qualitative features of the spectrum of the asymmetric rotor in the yrast region are seen to correspond to sequences of one-dimensional rotational trajectories with strong $E2$ transitions along the trajectories and much weaker transitions connecting the different trajectories. In contrast to the bands of symmetric rotors, which have band heads with $I = K$ (that may give rise to K isomerism), the different trajectories of the asymmetric rotor merge into the single common band with $I = 0$ (or $I = 1/2$), for any intrinsic state. The very tentative evidence on nuclear spectra in the yrast region, for $A \sim 160$ and $I \sim 30$, may suggest the occurrence of the characteristic asymmetric rotor patterns (see pp. 72ff.).

▼

ILLUSTRATIVE EXAMPLES TO SECTION 4-5

Regge trajectories for an asymmetric rotor (Fig. 4-34)

The concept of a Regge trajectory (see Vol. I, p. 13) is intimately connected with that of rotational band structure. One-dimensional trajectories that are analytic functions of the total angular momentum I can be trivially defined for the rigid, axially symmetric rotor, for which the energy is a linear function of $I(I+1)$. Such trajectories can also be quite generally defined in the two-body problem, since the separation between the angular and radial motion leads to a radial wave equation in which the angular momentum only appears as a parameter and can therefore be varied continuously. The characteristic feature of these systems is the existence of an internal axis of rotational symmetry with respect to which one may define the quantum number K; for given values of the three quantum numbers KIM, the rotational motion is completely specified. For the axially symmetric rotor, K is a constant of the motion and labels the trajectories. For a system of two spinless particles, K is constrained to have the value 0, while for two particles with spin, the trajectories are obtained by a diagonalization of a matrix with a number of dimensions corresponding to the possible values of the helicities.

In more general many-body systems, the quantum number K is not a constant of the motion; indeed, there may be no unique way of introducing an intrinsic coordinate system. In such a situation, the definition of the Regge trajectories involves new features, some of which may be illustrated by a study of the asymmetric rotor.

The energy of an asymmetric rotor can be defined as an analytic function of the total angular momentum I by considering the eigenvalue problem as a matrix equation in the representation in which I_3 is diagonal.

▲ The invariance of the rotational Hamiltonian with respect to the D_2 group

▼ (see Sec. 4-5a) implies that the matrix (for an even-A system) separates into four submatrices specified by the symmetry quantum numbers $r_3 = (-1)^{I_3}$ and c_3, of which the latter represents the relative phase of the components with $I_3 = \pm K$, where $K = |I_3|$. For the physical states, we have (see Eqs. (4-273) and (4-276))

$$c_3 = r_2(-1)^{I+K} = r_2 r_3(-1)^I = r_1(-1)^I \tag{4-315}$$

In the representation IKc_3, the matrix elements of the Hamiltonian (4-269) are given by

$$\begin{aligned}
\langle IKc_3|H_{\text{rot}}|IKc_3\rangle &= AI(I+1) + (A_3 - A)K^2 + \tfrac{1}{4}c_3(A_1 - A_2)I(I+1)\delta(K,1) \\
\langle I, K+2, c_3|H_{\text{rot}}|IKc_3\rangle &= \langle IKc_3|H_{\text{rot}}|I, K+2, c_3\rangle \\
&= \tfrac{1}{4}(A_1 - A_2)((I-K-1)(I-K)(I+K+1)(I+K+2))^{1/2}\begin{cases}\sqrt{2} & K=0 \\ 1 & K\neq 0\end{cases} \\
A_\kappa &= \frac{\hbar^2}{2\mathscr{I}_\kappa} \qquad A = \tfrac{1}{2}(A_1 + A_2)
\end{aligned} \tag{4-316}$$

In these matrices, I may be treated as a parameter that can be continuously varied, while the spectrum of K values (K even for $r_3 = +1$ and K odd for $r_3 = -1$) as well as the phase $c_3(=\pm 1)$ are held fixed. (The state $K=0$ occurs only for the trajectory with $r_3 = +1$ and $c_3 = +1$.) For noninteger values of I, the matrix elements ($K \leftrightarrow K+2$) no longer vanish for certain K values ($K = I$ or $I-1$), and we therefore obtain infinite matrices extending to arbitrarily large positive K values. For $I-1 < K < I$, the nondiagonal matrix elements become imaginary; the analytic continuation is obtained from the symmetric (non-Hermitian) matrix, as indicated in Eq. (4-316).

The condition for the existence of discrete eigenvalues for the infinite matrix (4-316) can be determined by considering the asymptotic behavior, for large K, of the coefficients $c(K)$ in the eigenvectors (4-275). Thus, to leading order in K,

$$(A_3 - A)c(K) \underset{K \gg I}{\approx} -\tfrac{1}{4}(A_1 - A_2)(c(K-2) + c(K+2)) \tag{4-317}$$

Hence, the ratio

$$\alpha = \frac{c(K+2)}{c(K)} \tag{4-318}$$

approaches a constant value given by

$$\alpha + \frac{1}{\alpha} = -2\frac{2A_3 - A_1 - A_2}{A_1 - A_2} \tag{4-319}$$

▲

▼ If $\mathscr{I}_3$ is either the largest or the smallest of the three moments of inertia, the right-hand side of Eq. (4-319) is positive and larger than 2 (assuming $\mathscr{I}_2$ to be the intermediate moment). The two roots α_1, α_2 are then real and positive with $\alpha_1 < 1$ and $\alpha_2 > 1$, and $c(K)$ is a superposition of an exponentially increasing and an exponentially decreasing function of K. Thus, in the usual manner, the condition that no exponentially increasing component may be present defines a set of discrete eigenvalues. For the representation based on the intermediate axis, the roots of Eq. (4-319) are complex with $|\alpha_1| = |\alpha_2| = 1$. The inclusion of terms of order K^{-1} in Eq. (4-317) shows that, in this case, $|c(K)|$ is always proportional to K^{-1}; the infinite matrices, therefore, do not have discrete eigenvalues, but a continuous spectrum. (The discussion of the convergence of the solutions is due to R. Cutkosky, private communication. For a more detailed treatment of the analytic structure of the asymmetric rotor spectrum as a function of I, see Talman, 1971.)

For given moments of inertia labeled such that $\mathscr{I}_1 > \mathscr{I}_2 > \mathscr{I}_3$, one can thus define Regge trajectories in two different ways, employing either the I_3 or the I_1 representation. These two sets of trajectories are illustrated in Fig. 4-34 for the moment ratios $\mathscr{I}_1 : \mathscr{I}_2 : \mathscr{I}_3 = 4:2:1$. The trajectories with positive r_3 (based on even I_3 values) and those with positive r_1 (even I_1) are drawn with solid lines, while those with negative r_3 and negative r_1 are dashed. Within each set, the trajectories are labeled by a serial number K, which represents the ordering of the solutions for given I; with increasing values of K, the energy of the I_3 trajectories increases, while the I_1 trajectories decrease in energy. (If the moments of inertia are continuously varied, so that the rotor becomes symmetric with respect to the 3 (or 1) axis, the label K on the I_3 (or I_1) trajectories becomes equal to the eigenvalue of $|I_3|$ (or $|I_1|$).) For $K \neq 0$, there are two trajectories for each value of K, corresponding to the phase c_3 (or c_1) between the components $\pm K$. As can be seen from the c_3-dependent terms in Eq. (4-316), the lowest I_3 trajectory of given K has $r_3 c_3 = -1$; for the I_1 trajectories, the lowest member has $r_1 c_1 = +1$. The splitting of the c doublets increases with increasing values of I, but decreases with increasing K, since the c-dependent terms only affect the matrix elements in Eq. (4-316) with the lowest K values.

The physical states correspond to the intersections of two trajectories with the same set of symmetry quantum numbers $(r_1 r_2 r_3)$; see Eq. (4-315). In the figure, the physical states with symmetry $(r_1 r_2 r_3) = (+++)$ are indicated by solid circles.

The spectrum of the asymmetric rotor acquires a simple structure in the regions corresponding to the highest and lowest states of given I (and $I \gg 1$); in these asymptotic regions, the motion can be approximately described in terms of rotation about a single axis and a vibration-like precessional motion (see Sec. 4-5e). Thus, in the yrast region (lowest energies for given I), the
▲ physical states have the approximately conserved quantum number $K = I_1 \approx I$,

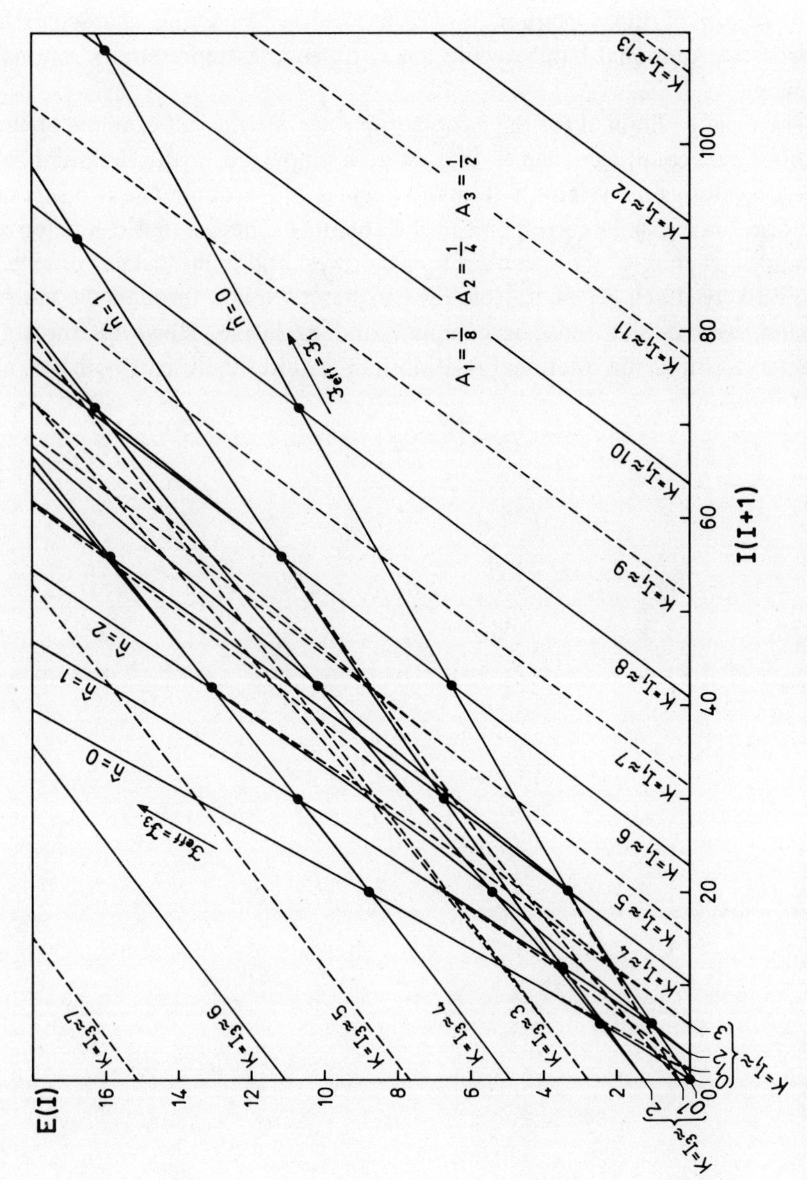

Figure 4-34 Regge trajectories for an asymmetric rotor. We wish to thank H. Lütken and J. D. Talman for help in the preparation of the figure.

▼ corresponding to the serial number of the I_1 trajectories. The states with different I, but with the same intrinsic vibrational motion, lie on the I_3 trajectories, which can be labeled by the vibrational quantum number $\hat{n}$ and have a slope determined by the moment of inertia $\mathscr{I}_1$ (see Fig. 4-34). Hence, in this region of the spectrum, the I_3 trajectories have the characteristic properties of rotational bands, while the I_1 trajectories represent vibrational sequences.

The results obtained for the asymmetric rotor provide an example of the multiplicity of coupling schemes that may be employed in the definition of Regge trajectories in systems with many degrees of freedom. The two sets of trajectories in Fig. 4-34 correspond to the coupling schemes that characterize the simple structure of the spectrum for the lowest and highest states of given I, respectively. Each set of trajectories can be continued through the entire spectrum, and in this sense is complete and exclusive. However, the full characterization of the rotational relationships requires the use of both sets of trajectories. ▲

APPENDIX

4A

Particle-Rotor Model

The simple model consisting of a particle coupled to a rotor provides an approximate description of many of the properties of the low-lying bands in odd-A nuclei and, at the same time, illustrates various general features of rotating systems. It has therefore played an important role in the development of the theory of nuclear rotational spectra. The application of the model to scattering problems is considered in Appendix 5A.

4A-1 The Coupled System

The rotor is assumed to possess axial symmetry and $\mathscr{R}$ invariance (see Sec. 4-2c) and to have the quantum numbers $R_3=0$, $r=+1$, as for the ground states of even-even nuclei (see p. 27). The angular momentum of the rotor is denoted by $\mathbf{R}$, with component R_3 with respect to the symmetry axis, and the spectrum consists of the sequence $R=0,2,4,\ldots$ (see Eq. (4-15)).

The energy of the rotor is taken to be proportional to the square of the angular momentum, corresponding to the Hamiltonian

$$H_{\text{rot}}=\frac{\hbar^2}{2\mathscr{I}_0}(R_1^2+R_2^2)=\frac{\hbar^2}{2\mathscr{I}_0}\mathbf{R}^2 \tag{4A-1}$$

as in the classical expression for a symmetric rigid body constrained to rotate about axes perpendicular to the symmetry axis (the dumbbell). The moment of inertia of the rotor is denoted by $\mathscr{I}_0$. The expression (4A-1) may be regarded as the first term in an expansion of the rotational energy in powers of the angular momentum (see p. 23); higher terms would be associated with the distortion of the rotor produced by the centrifugal and Coriolis forces.

The interaction between the rotor and the particle is described by a potential V, depending on the variables of the particle in the body-fixed system. The potential may be velocity and spin dependent and in particular may include a spin-orbit coupling, but V is assumed to be invariant under space reflection and time reversal, as well as axially symmetric and invariant under the rotation $\mathscr{R}=\mathscr{R}_2(\pi)$.

The present model ignores the possibility of terms in the potential depending on the angular rotational frequency, such as may arise from rotational perturbations of the rotor (centrifugal distortions proportional to ω_{rot}^2, as well as terms linear in ω_{rot} associated with the velocity-dependent interactions; see the discussion on p. 80 and pp. 278 ff.). The Hamiltonian for the coupled system is

$$H = H_{\text{rot}} + T + V \tag{4A-2}$$

where T is the kinetic energy of the particle. The anisotropy of the potential provides the coupling between the motion of the particle and that of the rotor.

4A-2 Adiabatic Approximation

If the rotational frequency is small compared with the excitation frequencies that characterize orbits with different orientations relative to the potential V, the motion of the particle is strongly coupled to the rotor and follows the precessional motion of the axis of the rotor in an approximately adiabatic manner. The coupled system can then be described in terms of a superposition of intrinsic motion, for fixed orientation of the rotor, and a rotational motion of the system as a whole.

4A-2a Wave functions

The intrinsic motion is determined by the wave equation

$$(T+V)\Phi_K(q) = E_K\Phi_K(q) \tag{4A-3}$$

where q represents intrinsic coordinates (including the spin variable). The eigenfunctions Φ_K and eigenvalues E_K of the intrinsic motion are labeled by the quantum number K representing the components I_3 of the total angular momentum; since the rotor has $R_3 = 0$, the quantum number K is equal to the eigenvalue of the component j_3 of the particle angular momentum (which is often denoted by Ω). The intrinsic states are degenerate with respect to the sign of K, as a consequence of the $\mathscr{R}$ invariance, or $\mathscr{T}$ invariance, of the intrinsic Hamiltonian. The states with negative K are labeled by $\bar{K}$, and we employ the phase choice (see Eqs. (4-16) and (4-27))

$$\Phi_{\bar{K}}(q) = \mathscr{T}\Phi_K(q) = \mathscr{R}_{\mathrm{i}}^{-1}\Phi_K(q) = -\mathscr{R}_{\mathrm{i}}\Phi_K(q) \tag{4A-4}$$

The particle is assumed to be a fermion; hence, $\mathscr{R}_{\mathrm{i}}^2 = \mathscr{R}_{\mathrm{i}}(2\pi) = -1$, and K takes the values $1/2, 3/2, \ldots$.

The rotational motion is specified by the quantum numbers KIM, and the total wave function has the form appropriate to an $\mathscr{R}$-invariant system (see Eq. (4-19))

$$\Psi_{KIM} = \left(\frac{2I+1}{16\pi^2}\right)^{1/2}\left(\Phi_K(q)\mathscr{D}^I_{MK}(\omega) + (-1)^{I+K}\Phi_{\bar{K}}(q)\mathscr{D}^I_{M-K}(\omega)\right) \tag{4A-5}$$

where ω represents the orientation of the rotor. The wave functions (4A-5) form a complete orthonormal set, which provides a convenient basis for describing the coupled system, when the adiabatic condition is fulfilled.

The coupling between the angular momenta of the particle and the rotor that is implicit in the wave function (4A-5) can be exhibited by transforming to the representation $(jR)IM$ appropriate to a weakly coupled system. Employing the transformation (1A-34) from the intrinsic to the fixed coordinate system and the coupling relation (1A-43) for the $\mathscr{D}$ functions, one obtains

$$\langle ((ljR)IM|KIM\rangle = \langle ljK|K\rangle \left(\frac{2R+1}{2I+1}\right)^{1/2} \langle jKR0|IK\rangle \frac{1}{\sqrt{2}}\left(1+(-1)^R\right) \tag{4A-6}$$

where the first factor represents the expansion of the intrinsic state $\Phi_K(q)$ onto a spherical basis ljK characterized by the orbital angular momentum l and the total angular momentum j of the particle. The symmetrized form of the wave function (4A-5) ensures that the amplitude (4A-6) is nonvanishing only for $R=0,2,\ldots$. The second and third factors in Eq. (4A-6) give the weight of the different states R of the rotor in the adiabatic coupling scheme.

4A-2b Energy

In order to express the Hamiltonian in the representation (4A-5), we employ the relation $\mathbf{R}=\mathbf{I}-\mathbf{j}$. The angular momentum of the rotor is thereby decomposed into a part $\mathbf{I}$, which rotates the system as a whole and thus acts only on the rotational wave function, and a part $\mathbf{j}$ acting only on the intrinsic variables.

For the energy of the rotor (4A-1), we obtain

$$\begin{aligned} H_{\text{rot}} &= \frac{\hbar^2}{2\mathscr{I}_0}\left((I_1-j_1)^2+(I_2-j_2)^2\right) \\ &= \frac{\hbar^2}{2\mathscr{I}_0}\mathbf{I}^2 + \frac{\hbar^2}{2\mathscr{I}_0}(j_1^2+j_2^2-j_3^2) - \frac{\hbar^2}{2\mathscr{I}_0}(j_+I_- + j_-I_+) \end{aligned} \tag{4A-7}$$

with the notation $I_\pm = I_1 \pm iI_2$ and $j_\pm = j_1 \pm ij_2$. The first term in the Hamiltonian (4A-7) depends only on the total angular momentum and is a constant of the motion. The second term in Eq. (4A-7) represents a recoil energy of the rotor and depends only on the intrinsic variables; this term might therefore be included in equation (4A-3) determining the intrinsic wave function Φ_K. However, since the recoil energy is comparable in magnitude with rotational energies and, therefore, small compared with intrinsic energies, we may neglect the effect on Φ_K in first approximation. The third term in Eq. (4A-7) represents the Coriolis and centrifugal forces acting on the particle in the rotating coordinate system (see Eqs. (4-107) and (4-119)). This term is nondiagonal in K (has $\Delta K = \pm 1$) and provides a coupling between intrinsic and rotational motion. In the special case of $K=1/2$ bands, the Coriolis interaction connects the components with opposite K in the states (4A-5) and hence contributes to the expectation value of the energy.

The expression (4A-7) for H_{rot} leads to a total Hamiltonian, which may be written in the form (see Eq. (4A-2))

$$H = H_0 + H_c \tag{4A-8a}$$

$$H_0 = T + V + \frac{\hbar^2}{2\mathscr{I}_0}\mathbf{I}^2 \tag{4A-8b}$$

$$H_c = -\frac{\hbar^2}{2\mathscr{I}_0}(j_+ I_- + j_- I_+) + \frac{\hbar^2}{2\mathscr{I}_0}(j_1^2 + j_2^2 - j_3^2) \tag{4A-8c}$$

The wave functions (4A-5) are the eigenstates of H_0, and in this basis, the expectation value of the energy is given by (see Eqs. (4-20) and (1A-93))

$$\begin{aligned} E_{KI} &= \langle KIM | H | KIM \rangle = E_K + E_{\mathrm{rot}} \\ E_{\mathrm{rot}} &= \langle KIM | H_{\mathrm{rot}} | KIM \rangle \\ &= \frac{\hbar^2}{2\mathscr{I}_0}\left(I(I+1) + a(-1)^{I+1/2}(I+1/2)\delta(K,1/2)\right) + \frac{\hbar^2}{2\mathscr{I}_0}\langle K | j_1^2 + j_2^2 - j_3^2 | K \rangle \end{aligned} \tag{4A-9}$$

The parameter a in the rotational energy for $K=1/2$ bands has the value

$$a = -\langle K=1/2 | j_+ | \overline{K=1/2} \rangle \tag{4A-10}$$

and is referred to as the decoupling parameter.

4A-2c Matrix elements

The matrix elements of the various operators, such as electric and magnetic multipole moments, can be evaluated, in the basis (4A-5), by means of the general procedure described in Sec. 4-3. For a tensor operator, the transformation to the body-fixed coordinate system is given by Eq. (4-90) and leads to matrix elements involving products of intrinsic and rotational factors.

An example is provided by the $M1$ moment, which can be expressed in terms of the g factors for the various angular momenta in the system (see Eqs. (3-36) and (3-39))

$$\begin{aligned} \mathscr{M}(M1,\mu) &= \left(\frac{3}{4\pi}\right)^{1/2} \frac{e\hbar}{2Mc}(g_R R_\mu + g_l l_\mu + g_s s_\mu) \\ &= \left(\frac{3}{4\pi}\right)^{1/2} \frac{e\hbar}{2Mc}(g_R I_\mu + (g_l - g_R) l_\mu + (g_s - g_R) s_\mu) \end{aligned} \tag{4A-11}$$

For transitions between two states in the same band, the reduced $M1$-matrix element is of the form (4-85) derived in Sec. 4-3c on the basis of symmetry arguments. The present model provides explicit expressions for the intrinsic matrix elements occurring in this general expression,

$$g_K K = \langle K | g_l l_3 + g_s s_3 | K \rangle$$
$$b(g_K - g_R) = \langle K = 1/2 | (g_l - g_R) l_+ + (g_s - g_R) s_+ | \overline{K = 1/2} \rangle \tag{4A-12}$$

The quantity g_K is the average g factor for the intrinsic motion. The occurrence of the parameter b, referred to as the magnetic decoupling parameter, is a special feature of $M1$ transitions within $K = 1/2$ bands.

4A-3 Nonadiabatic Effects

The coupling H_c (see Eq. (4A-8)) gives rise to an interweaving of rotational and intrinsic motion that can be analyzed in terms of a mixing of different rotational bands. (The effects of the band mixing produced by the Coriolis interaction were discussed by Kerman, 1956.)

4A-3a Contributions to energy

For small rotational frquencies, the coupling H_c can be treated by a perturbation expansion (except when special degeneracies occur in the intrinsic spectrum). Acting in second order, H_c gives the energy contribution

$$\delta E_{KI}^{(2)} = -\sum_{\nu} \frac{\langle \nu IM | H_c | KIM \rangle^2}{E_\nu - E_K} \tag{4A-13}$$

where ν labels the bands that are coupled to the state KIM by the interaction H_c. The perturbation (4A-13) gives rise to a term proportional to $I(I+1)$ arising from the second-order effect of the Coriolis coupling. In addition, the expression (4A-13) contains a term independent of I and, for $K = 1/2$ bands, a signature-dependent term linear in I arising from the combined effect of the Coriolis and recoil energies in the coupling H_c (see Eq. (4A-8c)). The coefficient of $I(I+1)$ is

$$\delta A = -\left(\frac{\hbar^2}{2\mathscr{I}_0}\right)^2 \sum_{\nu, K_\nu = K \pm 1} \frac{\langle \nu | j_\pm | K \rangle^2}{E_\nu - E_K} \tag{4A-14}$$

where the summation extends over all bands ν with $K_\nu = K \pm 1$. For $K = 1/2$ bands, the terms in Eq. (4A-14) with $K_\nu = K - 1$ are to be interpreted as excited bands with $K_\nu = 1/2$.

The contribution (4A-14) to the rotational energy can be expressed as a renormalization of the effective moment of inertia

$$\mathscr{I} = \mathscr{I}_0 + \delta\mathscr{I}$$
$$\delta\mathscr{I} = -\frac{2\mathscr{I}_0^2}{\hbar^2}\delta A = 2\hbar^2 \sum_{\nu} \frac{\langle \nu | j_1 | K \rangle^2}{E_\nu - E_K} \tag{4A-15}$$

assuming $\delta\mathscr{I} \ll \mathscr{I}_0$. The increase in the inertia of the system represents the contribution of the particle as it is dragged around by the rotor, and may also be obtained by considering a particle moving in an external field rotated with a constant frequency $\omega_{\text{rot}} = (\mathscr{I}_0)^{-1}\hbar I$ (the cranking model, see pp. 75ff.).

To third order in H_c, we obtain a contribution to the energy of $K=3/2$ bands that is cubic in I,

$$\delta E_{KI}^{(3)} = (-1)^{I+3/2}(I-1/2)(I+1/2)(I+3/2)A_3$$

$$A_3 = -\left(\frac{\hbar^2}{2\mathscr{I}_0}\right)^3 \sum_{\substack{\nu,\nu' \\ K_\nu = K_{\nu'} = 1/2}} \frac{\langle K=3/2|j_+|\nu\rangle\langle\nu|j_+|\bar{\nu}'\rangle\langle\bar{\nu}'|j_+|\overline{K=3/2}\rangle}{(E_\nu - E_K)(E_{\nu'} - E_K)} \tag{4A-16}$$

In addition, the third-order effects of H_c give a contribution to the energy of $K=1/2$ bands proportional to $(-1)^{I+1/2}(I+1/2)I(I+1)$, as well as terms of lower order in I. The fourth-order perturbation contains terms proportional to $I^2(I+1)^2$, and by proceeding to higher orders, one obtains a power series expansion for the energy of the general form (4-62).

The leading-order contribution to the term in the energy expansion involving I^n arises from the Coriolis coupling acting in nth order, and the magnitude of this contribution is given by

$$\delta E^{(n)} \sim \frac{\hbar^2 j}{\mathscr{I}_0}\left(\frac{\hbar j}{\mathscr{I}_0\omega_{\text{int}}}\right)^{n-1} I^n \approx \hbar\omega_{\text{int}}\left(\frac{\omega_{\text{rot}} j}{\omega_{\text{int}}}\right)^n \tag{4A-17}$$

where ω_{int} is a measure of the intrinsic frequencies ($\hbar\omega_{\text{int}} \sim E_\nu - E_K$), and where j represents the magnitude of the matrix elements of j_+. The rotational frequency ω_{rot} is the ratio between the angular momentum $\hbar I$ and the moment of inertia.

If the intrinsic spectrum is of vibrational character, the validity of the rotational coupling scheme requires the adiabatic condition $\omega_{\text{rot}} \ll \omega_{\text{int}} = \omega_{\text{vib}}$ (see the discussion on p. 3). The estimate (4A-17) implies a more restrictive condition for the convergence of the perturbation expansion in the case of particle motion with large values of j. However, for particle orbits with large angular momentum, the elementary effects of the Coriolis interaction produce only relatively small changes in the orientation of the orbit; hence, the smallness of the perturbation of the wave function is not a necessary condition for the convergence of the expansion for the energy and matrix elements in powers of the rotational angular momentum. In fact, the estimate (4A-17) leaves open the possibility of systematic cancellations in the higher-order terms. An example is provided by the motion of a particle in a rotating harmonic oscillator potential, for which the expansion of the energy is determined by the ratio of ω_{rot} to the difference in the oscillator frequencies along different axes, irrespective of the angular momentum of the particle orbit (Valatin, 1956; Hemmer, 1962).

The recoil energy in H_c (see Eq. (4A-8c)) is independent of I and therefore contributes to the coefficient of the I^n-dependent terms only through rotational perturbations of order $n+1$, or higher. The resulting corrections to the coefficient, as estimated in nth order, are therefore of relative magnitude $(\hbar j)^2/\mathscr{I}_0(E_\nu - E_K)$, which is comparable to $\delta\mathscr{I}/\mathscr{I}_0$ (see Eq. (4A-15)). The recoil term has especially simple consequences if the intrinsic excitations involve approximately independent degrees of freedom (spin and orbit in the absence of spin-orbit coupling, two or more independent particles, etc.). In such a situation, the part of the recoil term that couples the different components gives rise to a renormalization of the moment of inertia occurring in the Coriolis coupling.

An example illustrating this effect of the recoil term is provided by the decoupling term in the absence of spin-orbit coupling. For a $K=1/2$ band, with $l_3=\Lambda=0$, the second-order energy arising from the Coriolis and recoil terms, each acting in first order, gives

$$\delta E^{(2)} = (-1)^{I+1/2}(I+1/2)\delta A_1$$

$$\delta A_1 = 2\left(\frac{\hbar^2}{2\mathscr{I}_0}\right)^2 \sum_{\nu, K_\nu=1/2} \frac{\langle K=1/2|j_1^2+j_2^2|\nu\rangle\langle\nu|j_+|\overline{K=1/2}\rangle}{E_\nu - E_K}$$

$$= -2\left(\frac{\hbar^2}{2\mathscr{I}_0}\right)^2 \sum_{\kappa, \Lambda_\kappa=1} \frac{\langle\Lambda=0|l_-|\kappa\rangle\langle\kappa|l_+|\Lambda=0\rangle}{E_\kappa - E_\Lambda}a \qquad \text{(4A-18)}$$

The intermediate states have $K_\nu = 1/2$, $\Sigma(=s_3) = -1/2$ and the orbital part, denoted by κ, has $\Lambda_\kappa = 1$. In the Coriolis matrix element, only the orbital part of j_+ contributes, and the factor a is the decoupling parameter in the unperturbed band, which equals the value of $r(=\pm 1)$ for the $\Lambda=0$ orbital state (see p. 33). The recoil matrix element receives its contribution from the part of $j_1^2+j_2^2$ involving $l_- s_+$. The sum in Eq. (4A-18) is the same as that occurring in the expression for δA for the $\Lambda=0$ band (see Eq. (4A-14) and note the factor $\sqrt{2}$ in the matrix elements (4-92) connecting a $\Lambda=0$ state with a state having $\Lambda \neq 0$). We therefore obtain

$$\delta A_1 = a\delta A \qquad \text{(4A-19)}$$

Hence, the decoupling parameter, defined as the ratio between the (renormalized) A_1 and A coefficients, remains equal to the unperturbed value r, and the spectrum, with the inclusion of the nonadiabatic effects, retains the doublet structure, as required by the absence of spin-orbit coupling.

A further illustration of the role of the recoil term is provided by the contribution to the moment of inertia, for a system consisting of a rotor with two independent particles n and p (a neutron and a proton). In this case, the third-order contribution to the coefficient of the $I(I+1)$-dependent energy that results from the recoil term acting once and the Coriolis term twice (once on n and once on p) is

$$\delta E^{(3)} = 2\left(\frac{\hbar^2}{2\mathscr{I}_0}\right)^3 I(I+1)$$

$$\times\left(\sum_{\substack{\nu_n,\nu_p \\ K(\nu_n)=K_n\pm1 \\ K(\nu_p)=K_p\pm1}} \frac{\langle K_nK_p|(j_p)_\mp|K_n\nu_p\rangle\langle K_n\nu_p|(j_n)_\mp(j_p)_\pm|\nu_nK_p\rangle\langle \nu_nK_p|(j_n)_\pm|K_nK_p\rangle}{(E(\nu_n)-E(K_n))(E(\nu_p)-E(K_p))}\right.$$

$$\left.+\sum_{\substack{\nu_n,\nu_p \\ K(\nu_n)=K_n\pm1 \\ K(\nu_p)=K_p\pm1}}\left(\frac{\langle K_nK_p|(j_n)_\mp(j_p)_\pm|\nu_n\nu_p\rangle\langle \nu_n\nu_p|(j_p)_\mp|\nu_nK_p\rangle\langle \nu_nK_p|(j_n)_\pm|K_nK_p\rangle}{(E(\nu_n)+E(\nu_p)-E(K_n)-E(K_p))(E(\nu_n)-E(K_n))}+(\mathrm{n}\leftrightarrow\mathrm{p})\right)\right)$$

$$=\hbar^2\frac{(\delta\mathscr{I}_n)(\delta\mathscr{I}_p)}{(\mathscr{I}_0)^3}I(I+1) \tag{4A-20}$$

The same term occurs if one adds the separate contributions of n and p to the moment of inertia rather than to the energy itself.

The result (4A-20) corresponds to the fact that, in estimating the rotational energy contribution of the proton, we may consider the system in terms of the proton coupled to an effective rotor that consists of the neutron plus the core. The moment of inertia of this rotor equals $\mathscr{I}_0+\delta\mathscr{I}_n$ and, hence, the Coriolis coupling acting on the proton involves this renormalized moment.

It must be emphasized that corresponding third-order terms involving the Coriolis coupling acting twice on the neutron or on the proton will not, in general, give results equivalent to those arising from the expansion of $(\mathscr{I}_0+\delta\mathscr{I}_n+\delta\mathscr{I}_p)^{-1}$. For a single degree of freedom, the effect of the recoil term can be expressed in the above manner only if the degree of freedom can be repeated, as in the case of vibrational modes.

4A-3b Renormalization of operators

A systematic analysis of the effect of H_c on the various matrix elements for the particle-rotor system may be developed by viewing the perturbation expansion in terms of a canonical transformation

$$\begin{aligned} H' &= \exp\{iS\}H\exp\{-iS\} \\ &= H+i[S,H]-\tfrac{1}{2}[S,[S,H]]+\cdots \end{aligned} \tag{4A-21}$$

that diagonalizes the Hamiltonian in the representation (4A-5).

To first order in $I_{\pm}$, the transformation S is determined by the condition

$$i[S,H_0]=\frac{\hbar^2}{2\mathscr{I}_0}(j_+I_-+j_-I_+) \tag{4A-22}$$

We have ignored the recoil term in H_c, which only produces I-independent effects, to the order considered. From Eq. (4A-22), we obtain

$$S=-i(\varepsilon_+I_-+\varepsilon_-I_+) \tag{4A-23}$$

where the intrinsic operator $\varepsilon_{\pm}$ (denoted in Sec. 4-4 by $\varepsilon_{\pm 1}$) is given by

$$\begin{aligned}[H_0,\varepsilon_{\pm}]&=h_{\pm}\equiv-\frac{\hbar^2}{2\mathscr{I}_0}j_{\pm}\\ \langle K'|\varepsilon_{\pm}|K\rangle&=\frac{\langle K'|h_{\pm}|K\rangle}{E_{K'}-E_K}\end{aligned} \tag{4A-24}$$

(In Eq. (4A-22), only the nondiagonal part of the Coriolis interaction is included.)

With S given by Eq. (4A-23), the transformed Hamiltonian (4A-21) takes the form, to second order in $I_{\pm}$,

$$\begin{aligned}H'&=H_0+(H_c)_{\text{diag}}+\tfrac{1}{2}[(\varepsilon_+I_-+\varepsilon_-I_+),(h_+I_-+h_-I_+)]\\ &=H_0+(H_c)_{\text{diag}}+\tfrac{1}{2}(\{\varepsilon_+,h_-\}-\{\varepsilon_-,h_+\})I_3\\ &\quad+\tfrac{1}{2}([\varepsilon_+,h_-]+[\varepsilon_-,h_+])(I_1^2+I_2^2)+\tfrac{1}{2}[\varepsilon_+,h_+]I_-^2+\tfrac{1}{2}[\varepsilon_-,h_-]I_+^2\end{aligned} \tag{4A-25}$$

as follows from the relation

$$[aA,bB]=\tfrac{1}{2}\{a,b\}[A,B]+\tfrac{1}{2}[a,b]\{A,B\} \tag{4A-26}$$

for operator products in which a and b commute with A and B.

In the Hamiltonian (4A-25), the term proportional to $I_3(=j_3)$ gives a correction to the intrinsic motion. The term proportional to $I_1^2+I_2^2$ gives the renormalization (4A-15) of the moment of inertia and produces a coupling between different bands with $\Delta K=0$. Finally, the terms proportional to I_-^2 and I_+^2 provide a coupling between bands with $\Delta K=\pm 2$.

The effect of the Coriolis coupling on the matrix elements of a given operator F is obtained from the transformed operator

$$F'=\exp\{iS\}F\exp\{-iS\}=F+\delta F \tag{4A-27}$$

with

$$\begin{aligned}\delta F&=i[S,F]+\cdots\\ &=[(\varepsilon_+I_-+\varepsilon_-I_+),F]+\cdots\end{aligned} \tag{4A-28}$$

The transformed operator is to be evaluated in the space of the unperturbed wave functions (4A-5), and thus the consequences of the band mixing caused by the Coriolis coupling are described by the additional term δF in the effective transition operator.

As an illustration, we consider the renormalization of a dipole moment

$$\mathscr{M}(\lambda=1,\mu)=\sum_{\nu}\mathscr{M}_{\nu}\mathscr{D}^{1}_{\mu\nu}(\omega) \tag{4A-29}$$

where the components $\mathscr{M}_{\nu}\equiv\mathscr{M}(\lambda=1,\nu)$ refer to the intrinsic coordinate system (see Eq. (4-90)). For the induced moment $\delta\mathscr{M}$, we obtain from Eq. (4A-28), to leading order,

$$\begin{aligned}\delta\mathscr{M}(\lambda=1,\mu)&=\Big[(\varepsilon_{+}I_{-}+\varepsilon_{-}I_{+}),\sum_{\nu}\mathscr{M}_{\nu}\mathscr{D}^{1}_{\mu\nu}\Big]\\&=\sum_{\Delta K=-2}^{2}\delta\mathscr{M}(\lambda=1,\mu)_{\Delta K}\end{aligned} \tag{4A-30}$$

with

$$\begin{aligned}\delta\mathscr{M}(\lambda=1,\mu)_{\Delta K=0}&=\big[\varepsilon_{+}I_{-},\ \mathscr{M}_{-1}\mathscr{D}^{1}_{\mu-1}\big]+\big[\varepsilon_{-}I_{+},\ \mathscr{M}_{+1}\mathscr{D}^{1}_{\mu1}\big]\\&=2^{-1/2}(\{\varepsilon_{+},\ \mathscr{M}_{-1}\}+\{\varepsilon_{-},\ \mathscr{M}_{+1}\})\mathscr{D}^{1}_{\mu0}\\&\quad+2^{-1/2}([\varepsilon_{+},\ \mathscr{M}_{-1}]-[\varepsilon_{-},\ \mathscr{M}_{+1}])(I_{\mu}-I_{3}\mathscr{D}^{1}_{\mu0})\\&\quad+2^{-3/2}([\varepsilon_{+},\ \mathscr{M}_{-1}]+[\varepsilon_{-},\ \mathscr{M}_{+1}])[\mathbf{I}^{2},\mathscr{D}^{1}_{\mu0}]\\[1em]\delta\mathscr{M}(\lambda=1,\mu)_{\Delta K=\pm1}&=\big[\varepsilon_{\pm}I_{\mp},\ \mathscr{M}_{0}\mathscr{D}^{1}_{\mu0}\big]\\&=2^{-1/2}\{\varepsilon_{\pm},\ \mathscr{M}_{0}\}\mathscr{D}^{1}_{\mu\pm1}\\&\quad+2^{-1/2}[\varepsilon_{\pm},\ \mathscr{M}_{0}]\big([\mathbf{I}^{2},\mathscr{D}^{1}_{\mu\pm1}]\mp\{I_{3},\mathscr{D}^{1}_{\mu\pm1}\}\big)\\[1em]\delta\mathscr{M}(\lambda=1,\mu)_{\Delta K=\pm2}&=\big[\varepsilon_{\pm}I_{\mp},\ \mathscr{M}_{\pm1}\mathscr{D}^{1}_{\mu\pm1}\big]\\&=[\varepsilon_{\pm},\ \mathscr{M}_{\pm1}]I_{\mp}\mathscr{D}^{1}_{\mu\pm1}\end{aligned} \tag{4A-31}$$

We have assumed that the intrinsic moments $\mathscr{M}_{\nu}$ commute with $I_{\pm}$, as is the case for operators (such as the electric dipole moment) that do not explicitly depend on the rotational angular momentum. The $M1$ moment contains a term proportional to $\mathbf{I}$ (see Eq. (4A-11)), but since this term commutes with S, the induced moment may be obtained from the foregoing equations with

$$\mathscr{M}_{\nu}=\left(\frac{3}{4\pi}\right)^{1/2}\frac{e\hbar}{2Mc}((g_{l}-g_{R})l_{\nu}+(g_{s}-g_{R})s_{\nu}) \tag{4A-32}$$

In the evaluation of the terms in Eq. (4A-31), we have employed the relation (1A-91) for the commutator of $I_\pm$ and $\mathscr{D}^\lambda_{\mu\nu}$, and the anticommutator relation (see Eqs. (1A-88) and (1A-89)),

$$\{I_-,\mathscr{D}^1_{\mu-1}\}-\{I_+,\mathscr{D}^1_{\mu 1}\}=2^{1/2}\sum_{\nu=\pm 1}\{I_\nu,\mathscr{D}^1_{\mu\nu}\}$$
$$=2^{3/2}(I_\mu - I_3\mathscr{D}^1_{\mu 0}) \tag{4A-33}$$

as well as the further identity

$$[\mathbf{I}^2,\mathscr{D}^\lambda_{\mu\nu}]=\nu\{I_3,\mathscr{D}^\lambda_{\mu\nu}\}+\tfrac{1}{2}(\lambda(\lambda+1)-\nu(\nu+1))^{1/2}\{I_+,\mathscr{D}^\lambda_{\mu,\nu+1}\}$$
$$+\tfrac{1}{2}(\lambda(\lambda+1)-\nu(\nu-1))^{1/2}\{I_-,\mathscr{D}^\lambda_{\mu,\nu-1}\} \tag{4A-34}$$

The induced moment $\delta\mathscr{M}$ given by Eq. (4A-31) partly contains I-independent terms, which are equivalent to a renormalization of the unperturbed intrinsic moments $\mathscr{M}_\nu$. In addition, $\delta\mathscr{M}$ involves terms linear in $I_\pm$, which give rise to matrix elements with an I dependence different from that of the unperturbed moments. Moreover, these terms contribute to the K-forbidden transitions with $\Delta K=2$, for which the matrix element vanishes in the absence of the Coriolis coupling.

For $M1$ transitions within a band, the two first terms with $\Delta K=0$ in Eq. (4A-31) represent renormalizations of the intrinsic and rotational g factors by the amounts (see Eqs. (4A-11) and (4A-12))

$$K\delta g_K=\langle K|\{\varepsilon_+,((g_l-g_R)l_-+(g_s-g_R)s_-)\}|K\rangle$$
$$\delta g_R=\langle K|[\varepsilon_+,((g_l-g_R)l_-+(g_s-g_R)s_-)]|K\rangle \tag{4A-35}$$

while the first term with $\Delta K=1$ in Eq. (4A-31) gives a renormalization of the b parameter in $K=1/2$ bands (see Eq. (4A-12)),

$$\delta((g_K-g_R)b)=-\langle K=1/2|\{\varepsilon_+,((g_l-g_R)l_3+(g_s-g_R)s_3)\}|\overline{K=1/2}\rangle \tag{4A-36}$$

The terms in Eq. (4A-31) involving $[\mathbf{I}^2,\mathscr{D}^1_{\mu\nu}]$ do not contribute within a band, since the intrinsic matrix elements vanish as a consequence of the behavior of the operators under Hermitian conjugation and time reversal. It is seen that the $M1$ operators in a band, up to terms linear in the Coriolis coupling, have the general form represented by Eq. (4-89).

4A-3c Transformation of coordinates

The canonical transformation that leads to a diagonalization of the Hamiltonian may be viewed as a coordinate transformation to a new set of variables, in terms of which the Hamiltonian is diagonal in the representation (4A-5). If we denote the original variables by $x=(qp,\omega I_\kappa)$, where q and p represent the particle variables relative to the orientation of the rotor, the new variables are

$$x'=\exp\{-iS\}x\exp\{iS\} \tag{4A-37}$$

When expressed in these variables, the Hamiltonian has the form (see Eq. (4A-21))

$$\begin{aligned} H(qp,I_\kappa) &= \exp\{-iS\}H'(qp,I_\kappa)\exp\{iS\} \\ &= H'(q'p',I'_\kappa) \\ &= H'_{\text{intr}}(q'p')+H'_{\text{rot},\nu}(I'_\kappa) \end{aligned} \tag{4A-38}$$

The eigenstates of the effective intrinsic Hamiltonian H'_{intr} are labeled by a set of quantum numbers ν that includes K. The effective rotational Hamiltonian depends on ν and can be expressed as a function of the components I'_κ of the angular momentum with respect to the intrinsic coordinate system with orientation ω'. (The Hamiltonian, being a scalar, cannot depend explicitly on orientation angles.)

The stationary states written in terms of the coordinates (q,ω) are superpositions of wave functions of the form (4A-5) and describe sets of coupled rotational bands. In the new variables (q',ω'), the stationary states retain the form (4A-5), and the couplings are now expressed in terms of the structure of the operators

$$F(x)=F'(x') \tag{4A-39}$$

where F' is the transformed operator (4A-27). For example, the electromagnetic moments, which are rather simple functions of x, have a more complicated form when expressed in the x' variables, as illustrated by the renormalization of the dipole moments given by Eq. (4A-31).

For the first-order transformation (4A-23), the renormalization of the coordinates is given by

$$\begin{aligned} \delta q &= q'-q=-[\varepsilon_+,q]I_- - [\varepsilon_-,q]I_+ \\ \delta\omega &= \omega'-\omega=-\varepsilon_+[I_-,\omega]-\varepsilon_-[I_+,\omega] \end{aligned} \tag{4A-40}$$

For the Euler angles, the evaluation of the commutators yields

$$\begin{aligned} \delta\theta &= \varepsilon_+e^{i\psi}-\varepsilon_-e^{-i\psi} \\ \delta\phi &= \frac{-i}{\sin\theta}(\varepsilon_+e^{i\psi}+\varepsilon_-e^{-i\psi}) \\ \delta\psi &= i\cot\theta(\varepsilon_+e^{i\psi}+\varepsilon_-e^{-i\psi}) \end{aligned} \tag{4A-41}$$

This renormalization implies that the collective orientation angles depend to some extent on the particle variables. In turn, the new intrinsic coordinates q' depend on the rotational frequency. (For a further discussion of collective coordinates for rotation, see Bohr and Mottelson, 1958.)

CHAPTER 5

One-Particle Motion in Nonspherical Nuclei

As discussed in Chapter 4, a considerable class of nuclei exhibit spectra with the characteristic relations in energies and transition intensities that imply an approximate separation between rotational motion of the system as a whole and internal motion with respect to a body-fixed coordinate frame. This separation is a consequence of the departure of the nuclear equilibrium shape from spherical symmetry. The starting point for the description of the intrinsic degrees of freedom in such nuclei is the analysis of one-particle motion in nonspherical potentials with the symmetry and shape indicated by the rotational spectra.[1]

The present chapter begins with a discussion of the deformed potential and the associated one-particle orbits (Sec. 5-1). The main results are contained in the figures and tables illustrating the one-particle spectra as a function of the eccentricity. These spectra provide an immediate classification of the observed low-lying states in the broad class of deformed odd-A nuclei (Sec. 5-2). This classification opens the possibility of exploring many different features of the single-particle motion, including the generalizations of these degrees of freedom that result from pairing and polarization effects and from the coupling to the rotational motion (Sec. 5-3). Thus, the phenomena that form the topic of the present chapter represent a major enlargement of the available data on one-particle motion in the nucleus.

5-1 STATIONARY STATES OF PARTICLE MOTION IN SPHEROIDAL POTENTIAL

5-1a Symmetry and Shape of Nuclear Equilibrium Deformation

The observed quantum numbers of the rotational spectra imply that, in the cases studied so far, the nuclear equilibrium shape possesses axial symmetry and $\mathscr{R}$ invariance, and that the deformation conserves $\mathscr{P}$ and $\mathscr{T}$ symmetry (see the discussion in Sec. 4-3a, pp. 27ff.). The measured moments show that the main deformation is of quadrupole symmetry, corresponding to a spheroidal shape with an eccentricity that is typically of order 0.2 to 0.3; see Fig. 4-25, p. 133. (The significance of the spheroidal

[1]Early studies of nucleonic motion in spheroidal potentials were made by Hill and Wheeler (1953), Moszkowski (1955), Nilsson (1955), and Gottfried (1956). The calculations by Nilsson (*loc. cit.*) played a special role in providing the basis for the classification of the extensive body of data on the spectra of odd-A deformed nuclei (Mottelson and Nilsson (1955, 1959); see also Stephens, Asaro, and Perlman (1959) for the classification of the heavy-element region; Paul (1957), Rakavy (1957), and Litherland *et al.* (1958) for the classification of the (sd) shell nuclei; Kurath and Pičman (1959) for the p shell).

symmetry can be understood in terms of simple features of the nuclear shell structure, as discussed in Chapter 6, pp. 591 ff.)

The considerations in the present chapter are restricted to deformed potentials with shapes corresponding to those encountered in the low-energy nuclear spectra that at present constitute the only source of detailed evidence on one-particle motion in deformed nuclei. The phenomenon of shape isomerism (see p. 26) opens the prospect of extending the studies to potentials with much larger deformations. (Single-particle spectra in very strongly deformed potentials such as are encountered in the fission isomers are considered in Chapter 6, pp. 591 ff.) Moreover, the possibility of instability with respect to odd-parity deformations is suggested by the spectra of heavy nuclei as well as by the mass asymmetry observed for the fission fragments. (See, for example, the discussion of the octupole vibrational mode on pp. 559 ff. For an analysis of single-particle motion as a function of odd-parity deformations in the region of the fission barrier, see Gustafson *et al.* 1971.) The possibility of shapes without axial symmetry represents another potentially important extension of the present studies. (See the discussion in Sec. 4-5b; for an early discussion of nucleonic motion in ellipsoidal potentials, see, for example, Newton 1960.) The interest in such an extension is heightened by the fact that the centrifugal distortion provides a systematic tendency away from axial symmetry for states with large angular momentum (see p. 42). The rotational motion also produces a component in the nuclear potential (Coriolis interaction) that violates time-reversal invariance; for large rotational frequencies, major changes in the structure of the one-particle motion may result (see, for example, the analysis of motion in a rotating harmonic oscillator potential, discussed on pp. 84 ff.). In electronic many-body systems, the occurrence of spin-dependent $\mathcal{T}$-violating potentials results from exchange effects of the Coulomb forces and in particular gives rise to the phenomenon of ferromagnetism (see, for example, the review by Herring, 1966). In atoms and molecules, the valence electrons produce a polarization of the spins of the electrons in the core, which can be described by a spin dependence of the Hartree–Fock potential (see, for example, the review by Watson and Freeman, 1967). In a similar manner, the nuclear spin polarizability can be viewed as an effect produced by a static deformed field violating $\mathcal{T}$ symmetry (see the discussion by Tewari and Banerjee, 1966).

5-1b Deformed Potential

An axially symmetric quadrupole deformation of the spin-independent part of the nuclear potential can be expressed in the form

$$V(r,\vartheta) = V_0(r) + V_2(r)P_2(\cos\vartheta) \tag{5-1}$$

where ϑ is the polar angle of the particle with respect to the nuclear symmetry axis. A deformation of the nucleus that does not affect the diffuseness of the surface to first order can be described in terms of an angular dependence of the radius parameter R in expressions such as (2-169); see the discussion of the leptodermous model on pp. 137ff. For a deformation that transforms the spherical surface $R=R_0$ into a spheroid with a difference of radii equal to $R_0\delta$, we obtain to first order in the deformation parameter δ

$$\begin{aligned} V(r,\vartheta) &\approx V_0 f(r-R(\vartheta)) \\ &\approx V_0 f(r-R_0) - \tfrac{2}{3}\delta R_0 V_0 \frac{\partial f}{\partial r} P_2(\cos\vartheta) \\ R(\vartheta) &= R_0(1+\tfrac{2}{3}\delta P_2(\cos\vartheta)) \end{aligned} \tag{5-2}$$

where the spherical potential is written in the form $V_0 f(r-R_0)$, in terms of the constant V_0 and the radial form factor f. Since the average nuclear field is generated by nucleonic interactions with a range small compared to R, one expects the parameter δ in Eq. (5-2) to be approximately equal to the corresponding parameter (4-72) describing the quadrupole deformation of the nuclear density. (See further comments on this point on p. 354.)

The leading-order effect described by Eq. (5-2) provides the basis for the discussion in the present chapter. Finer details in the potential involve the angular dependence of the surface diffuseness and the possibility of higher multipole components (evidence on the magnitude of the Y_{40} deformation is discussed on p. 140). The neutron excess and the polarization produced by the Coulomb interaction also imply the possibility of a difference in the deformations of the isoscalar and isovector components of the total potential.

The deformation of the spin-orbit potential follows from that of the central potential, if one assumes the spin-orbit coupling to be proportional to the local gradient of a potential, as in the discussion in Chapter 2, Vol. I, p. 218 (see also the derivation based on an average of a two-body spin-orbit interaction in Vol. I, p. 259),

$$\begin{aligned} V_{ls}(r,\vartheta) &= V_{ls} r_0^2 \nabla\left(f_{ls}(r) - \tfrac{2}{3}\delta R_0 \frac{\partial f_{ls}}{\partial r} P_2(\cos\vartheta)\right)\cdot(\mathbf{p}\times\mathbf{s}) \\ &= V_{ls} r_0^2 (\mathbf{l}\cdot\mathbf{s}) \frac{1}{r}\frac{\partial f_{ls}}{\partial r} - \tfrac{2}{3}\delta V_{ls} r_0^2 \frac{R_0}{r}\left\{\left(\frac{\partial^2 f_{ls}}{\partial r^2} - \frac{2}{r}\frac{\partial f_{ls}}{\partial r}\right) P_2(\cos\vartheta)(\mathbf{l}\cdot\mathbf{s}) \right. \\ &\quad \left. + \frac{1}{r}\frac{\partial f_{ls}}{\partial r}(3x_3(\mathbf{p}\times\mathbf{s})_3 - (\mathbf{l}\cdot\mathbf{s}))\right\} \end{aligned} \tag{5-3}$$

In this expression, we have used the form (2-144) for the spin-orbit potential in the spherical nucleus, with f_{ls} representing the radial form factor and the constant V_{ls} giving the strength. Moreover, we have expanded to first order in the deformation, as in the relation (5-2) (see also Eq. (3A-26) for the gradient of a spherical harmonic).

The estimate (5-3) of the effect of the deformation on the spin-orbit potential depends rather sensitively on the structure of the surface region. At present, there is only little empirical evidence that tests this component in the deformed potential, though the polarization measurements in inelastic proton scattering appear to be consistent with Eq. (5-3); see, for example, Sherif (1969).

Since the spin-orbit potential in spherical nuclei is an order of magnitude smaller than the spin-independent potential, the deformed part of V_{ls} is correspondingly smaller than the deformation of the spin-independent potential and may often be neglected in the analysis of the one-particle motion.

5-1c Structure of One-Particle Wave Functions

The $\mathscr{P}$ invariance and axial symmetry of the nuclear potential imply that the parity π and the projection Ω of the total angular momentum along the symmetry axis are constants of the motion for the one-particle states. These states are twofold degenerate, since two orbits that differ only in the sign of Ω represent the same motion, apart from the direction of revolution about the symmetry axis (time-reversal symmetry or $\mathscr{R}$ symmetry). A qualitative survey of the structure of the one-particle states can be obtained by considering limiting situations corresponding to small and large deformations.

For small deformations, the states may be approximately characterized by the quantum numbers nlj appropriate to a spherical potential. To first order in δ, the energy is

$$\begin{gathered}\varepsilon(nlj\Omega) = \varepsilon_0(nlj) - \frac{3\Omega^2 - j(j+1)}{4j(j+1)}\langle V_2(r)\rangle_{nlj} \\ \Omega = \pm\tfrac{1}{2}, \pm\tfrac{3}{2}, \ldots, \pm j\end{gathered} \tag{5-4}$$

where $\varepsilon_0(nlj)$ is the eigenvalue for the spherical potential. The $2j+1$ states of given nlj are seen to be split, with only the $\pm\Omega$ degeneracy remaining. For positive V_2 (negative δ), the potential (5-1) represents an oblate distortion and the perturbation favors the orbits with large $|\Omega|$, corresponding to the fact that these orbits represent states of motion in the equatorial plane.

In addition to the diagonal terms (5-4), the deformed potential (5-1) gives rise to nondiagonal matrix elements with $\Delta\Omega = 0$, $\Delta l = 0, \pm 2$. Since the one-particle spectrum in the spherical potential contains near-lying states that fulfill these selection rules, even rather small deformations may lead to appreciable mixing of states with different values of nlj. In fact, for the deformations typically encountered, the approximation (5-4) may be grossly inadequate.

For large deformations, an approximate description of the one-particle motion can be based on the similarity of the nuclear potential to that of a harmonic oscillator (see, for example, Fig. 2-22, Vol. I, p. 223). A spheroidal deformation can be represented by the anisotropic oscillator potential

$$\begin{aligned} V &= \tfrac{1}{2}M\left(\omega_3^2 x_3^2 + \omega_\perp^2 (x_1^2 + x_2^2)\right) \\ &= \tfrac{1}{2}M\omega_0^2 r^2\left(1 - \tfrac{4}{3}\delta P_2(\cos\vartheta)\right) \end{aligned} \tag{5-5}$$

$$\omega_3 \approx \omega_0\left(1 - \tfrac{2}{3}\delta\right) \qquad \omega_\perp \approx \omega_0\left(1 + \tfrac{1}{3}\delta\right)$$

where ω_0 is the oscillator frequency for the spherical potential, while ω_3 and $\omega_\perp$ represent the oscillator frequencies of the deformed potential, in directions parallel and perpendicular to the symmetry axis, respectively. The deformation transforms the spherical equipotential surfaces into ellipsoids, which have been assumed to have a volume that is independent of deformation ($\omega_3\omega_\perp^2 = \omega_0^3$). To first order in the deformation, the parameter δ defined by Eq. (5-5) equals the difference between the major and minor axes, divided by the mean radius (compare Eq. (4-73) for the deformed density).

For the spheroidal potential (5-5), the motion separates into independent oscillations along the 3 axis and in the (12) plane, and the energy is

$$\varepsilon(n_3 n_\perp) = \left(n_3 + \tfrac{1}{2}\right)\hbar\omega_3 + (n_\perp + 1)\hbar\omega_\perp \tag{5-6}$$

where $n_\perp = n_1 + n_2$ is the number of quanta in the oscillations perpendicular to the symmetry axis. The degenerate states with the same value of $n_\perp$ can be specified by the component Λ of the orbital angular momentum along the symmetry axis, which takes the values

$$\Lambda = \pm n_\perp, \pm(n_\perp - 2), \ldots, \pm 1 \text{ or } 0 \tag{5-7}$$

The nuclear potential deviates from that of a harmonic oscillator partly in the radial dependence associated with the existence of a rather well-defined surface and partly in the occurrence of the spin-orbit interaction. The pattern of the energy shifts associated with the surface effect is considered in the example on pp. 218ff. and is further discussed in the example in Chapter 6, pp. 591ff.; in particular, these shifts remove the degeneracy of the states with different Λ but the same values of $(n_3 n_\perp)$; the energy $\varepsilon(n_3 n_\perp \Lambda)$ decreases with increasing Λ, corresponding to the favoring of the states with large l in the spherical potential (see Vol. I, p. 222).

In the absence of spin-orbit coupling, the single-particle states in the axially symmetric (and time-reversal invariant) potential have a fourfold degeneracy (for $\Lambda \neq 0$), corresponding to the sign of Λ and of the component $\Sigma(=\pm 1/2)$ of the spin along the symmetry axis. The spin-orbit coupling produces a splitting of the states with numerically different values of the component of total angular momentum

$$\Omega = \Lambda + \Sigma \tag{5-8}$$

To leading order, the coupling gives an energy shift of the form

$$\Delta\varepsilon_{ls} = V_{ls} r_0^2 \langle n_3 n_\perp \Lambda | \frac{1}{r}\frac{\partial f_{ls}}{\partial r} | n_3 n_\perp \Lambda \rangle \Lambda\Sigma \tag{5-9}$$

which favors the states with parallel spin and orbit (since V_{ls} is positive). In the estimate (5-9), we have ignored the effect of the deformation on the spin-orbit potential.

Numerical solutions of the one-particle spectrum valid for small as well as large deformation parameters are illustrated in Figs. 5-1 to 5-5 (pp. 221 ff.). The main features of these spectra can be immediately understood in terms of the transition between the coupling schemes $(nlj\Omega)$ and $(n_3 n_\perp \Lambda\Omega)$ considered above. Thus, for small δ, the figures show the linear dependence on δ given by Eq. (5-4), while for large δ, the variation is again approximately linear, but with a slope given by Eq. (5-6). As can be seen from the figures, the transition takes place gradually over a range of values of δ, but for most orbits is nearing completion for the values of δ corresponding to the equilibrium shapes of the strongly deformed nuclei.

In view of the approximate validity of the quantum numbers $n_3 n_\perp \Lambda\Sigma$, it is convenient to label the orbits by these "asymptotic" quantum numbers; the conventional notation is $[Nn_3\Lambda\Omega]$, where $N = n_3 + n_\perp$ is the total oscillator quantum number. The validity of the asymptotic quantum numbers can also be inferred from the wave functions given in Tables 5-2, pp. 228 ff., and 5-9, p. 290. The quantum numbers $n_3, n_\perp$, and Λ imply selection rules in the one-particle matrix elements (see, for example, Table 5-3, p. 234); these asymptotic selection rules are found to play a major role in the distribution of the transition strength for the various multipole operators (see the discussion in Sec. 5-3b).

The validity of the quantum numbers Λ and Σ is a rather general feature of strongly deformed axially symmetric potentials. In fact, the states coupled by the spin-orbit interaction, which are degenerate in the spherical limit, become widely separated for large deformations. (A deformation that violates axial symmetry leads to a loss of the quantum numbers Λ and Σ and to the disappearance of the first-order spin-orbit coupling. In fact, in the potential of lower symmetry, the orbital motion, in the absence of coupling to the spin, becomes nondegenerate, in which case all components of the orbital angular momentum have vanishing expectation value, as a consequence of time-reversal invariance. (Degeneracy in the orbital motion may occur, if the potential retains a symmetry axis of order $n \geqslant 3$, together with time-reversal invariance; see the discussion on p. 176.) This result corresponds to the quenching of the orbital moment known in polyatomic molecules and crystal fields (see, for example, Van Vleck, 1932); for a discussion of orbital angular momentum in molecules and crystals, see Herzberg, 1966, Chapter 1, and Ballhausen, 1962.)

The separation of the orbital motion represented by the quantum numbers n_3 and $n_\perp$ appears as a more specific feature associated with the resemblance of the nuclear potential to that of a harmonic oscillator. The most direct empirical evidence for this separation comes from the near crossing of levels with different $n_3 n_\perp$, but with the same values of Ω and π (see the discussion on p. 230).

The analysis of the one-particle spectra in deformed nuclei can be extended to include the unbound states that manifest themselves in terms of single-particle resonances. The strength associated with a particular single-particle resonance nlj in a spherical potential becomes distributed over many different resonances in the spheroidal potential, each characterized by the quantum numbers $\Omega\pi$, with $\Omega \leqslant j$ and $\pi = (-1)^l$. Thus, one may speak of a splitting of the one-particle strength function produced by the deformation. This effect is illustrated by the neutron s-wave strength function, shown in Fig. 5-6, p. 235. (The general formalism that may be employed in the treatment of scattering problems for deformed potentials is considered in Appendix 5A.)

▼

ILLUSTRATIVE EXAMPLES TO SECTION 5-1

One-particle spectra as function of spheroidal eccentricity (Figs. 5-1 to 5-5; Table 5-1)

An approximate description of the deformed nuclear potential, which has ▲ been extensively employed in the interpretation of single-particle motion in

▼ deformed nuclei, is obtained by a simple modification of the harmonic oscillator (Nilsson, 1955; Gustafson *et al.*, 1967),

$$H = \frac{\mathbf{p}^2}{2M} + \tfrac{1}{2}M\left(\omega_3^2 x_3^2 + \omega_\perp^2 (x_1^2 + x_2^2)\right) + v_{ll}\hbar\omega_0(\mathbf{l}^2 - \langle \mathbf{l}^2 \rangle_N) + v_{ls}\hbar\omega_0(\mathbf{l}\cdot\mathbf{s})$$

$$\langle \mathbf{l}^2 \rangle_N = \tfrac{1}{2}N(N+3) \tag{5-10}$$

For $\omega_3 = \omega_\perp = \omega_0$, the Hamiltonian (5-10) corresponds to a spherical harmonic oscillator with the addition of a spin-orbit coupling of the form (5-3) and a term proportional to $\mathbf{l}^2$. The latter term lifts the degeneracy within each major oscillator shell in such a manner as to favor the states with large l, corresponding to the level ordering in a potential with a more sharply defined surface (see Vol. I, p. 222). The term $\langle \mathbf{l}^2 \rangle_N$ is a constant for each oscillator shell chosen so that the average energy difference between shells is not affected by the $\mathbf{l}^2$ term. Since the $\mathbf{l}^2$ and $\mathbf{l}\cdot\mathbf{s}$ terms represent relatively small perturbations of the oscillator potential, the effect of the deformation on these terms has been neglected. (For large deformations, see the discussion on pp. 593 ff.)

The values of the constants v_{ll} and v_{ls} that are given in Table 5-1 have been obtained from an adjustment of the Hamiltonian (5-10) to the available data on the intrinsic spectra of the deformed nuclei. The approximate values of v_{ll} and v_{ls} can also be derived from the parameters appropriate to spherical Woods-Saxon potentials; see Vol. I, p. 238 and the discussion on pp. 593 ff. (In the light nuclei, the simple form of the spin-orbit coupling in Eq. (5-10) only qualitatively describes the observed doublets; the value of v_{ls} in Table 5-1 for N and Z less than 20 has been chosen to reproduce the splitting of the d doublet in the nuclei after ^{16}O.)

A convenient definition of the deformation parameter for the anisotropic oscillator is

$$\delta_{\text{osc}} = 3\frac{\omega_\perp - \omega_3}{2\omega_\perp + \omega_3} \tag{5-11}$$

which leads to a linear relation for the eigenvalues (5-6),

$$\varepsilon(n_3 n_\perp) = \hbar\bar{\omega}\left(N + \tfrac{3}{2} - \tfrac{1}{3}\delta_{\text{osc}}(2n_3 - n_\perp)\right) \tag{5-12}$$

where $N = n_3 + n_\perp$ is the total oscillator quantum number and $\bar{\omega}$ the mean frequency,

$$\bar{\omega} = \tfrac{1}{3}(\omega_1 + \omega_2 + \omega_3) = \tfrac{1}{3}(2\omega_\perp + \omega_3) \tag{5-13}$$

To leading order, the definition (5-11) corresponds to the parameter employed in Eq. (5-5) and, assuming the same eccentricity of density and the potential, also corresponds to the parameter δ defined in terms of the quadrupole moment (see Eq. (4-72)). However, the various deformation ▲

Figure	Region	$-v_{ls}$	$-v_{ll}$
5-1	N and $Z<20$	0.16	0
5-2	$50<Z<82$	0.127	0.0382
5-3	$82<N<126$	0.127	0.0268
5-4	$82<Z<126$	0.115	0.0375
5-5	$126<N$	0.127	0.0206

Table 5-1 Parameters used in the single-particle potentials of Figs. 5-1 to 5-5.

▼ parameters differ when higher-order terms are included. We have therefore explicitly labeled the parameter in Eq. (5-11) to indicate its definition in terms of the oscillator frequencies. (The relation between δ and the parameter β_2 appearing in the multipole expansion of the radius parameter is given by Eq. (4-191).)

In the diagonalization of the one-particle Hamiltonian, one may employ the oscillator representation $n_3 n_\perp$, for which the l-independent part of H has the diagonal form (5-12). In this representation, the operator $\mathbf{l}$ is given by Eq. (5-26) and is seen to contain terms that conserve the total number of oscillator quanta ($\Delta N=0$) as well as terms with $\Delta N=2$. The latter terms are of order δ relative to the $\Delta N=0$ terms and will be neglected. (See in this connection the comment on p. 230 concerning the crossing of levels with $\Delta N=2$.)

The spectra obtained from a numerical diagonalization of the Hamiltonian (5-10) are illustrated in Figs. 5-1 to 5-5. The eigenvalues are plotted in units of $\hbar\overline{\omega}$ as a function of the deformation parameter δ_{osc}. Apart from terms of order δ_{osc}^2, the mean frequency $\overline{\omega}$ can be identified with the frequency ω_0 estimated for spherical nuclei (see Vol. I, p. 209),

$$\hbar\overline{\omega} \approx \hbar\omega_0 = 41 A^{-1/3} \quad \text{MeV} \tag{5-14}$$

This estimate of $\hbar\omega_0$ is based on an adjustment of the oscillator potential so as to yield the correct mean square radius of the nuclear density distribution. Such an adjustment of the oscillator potential also leads to an energy separation between major shells in the neighborhood of the Fermi level in agreement with that obtained from a Woods-Saxon potential (see, for example, Fig. 2-30, Vol. I, p. 239). This agreement is preserved when the modification of the oscillator potential is taken in the form (5-10).

The empirical information on the nuclear eccentricities comes mainly from the measured $E2$-matrix elements and thus determines the deformation parameter δ defined in Eq. (4-72). As mentioned above, this parameter differs ▲ from δ_{osc} in higher-order terms, but in order to carry the analysis to this

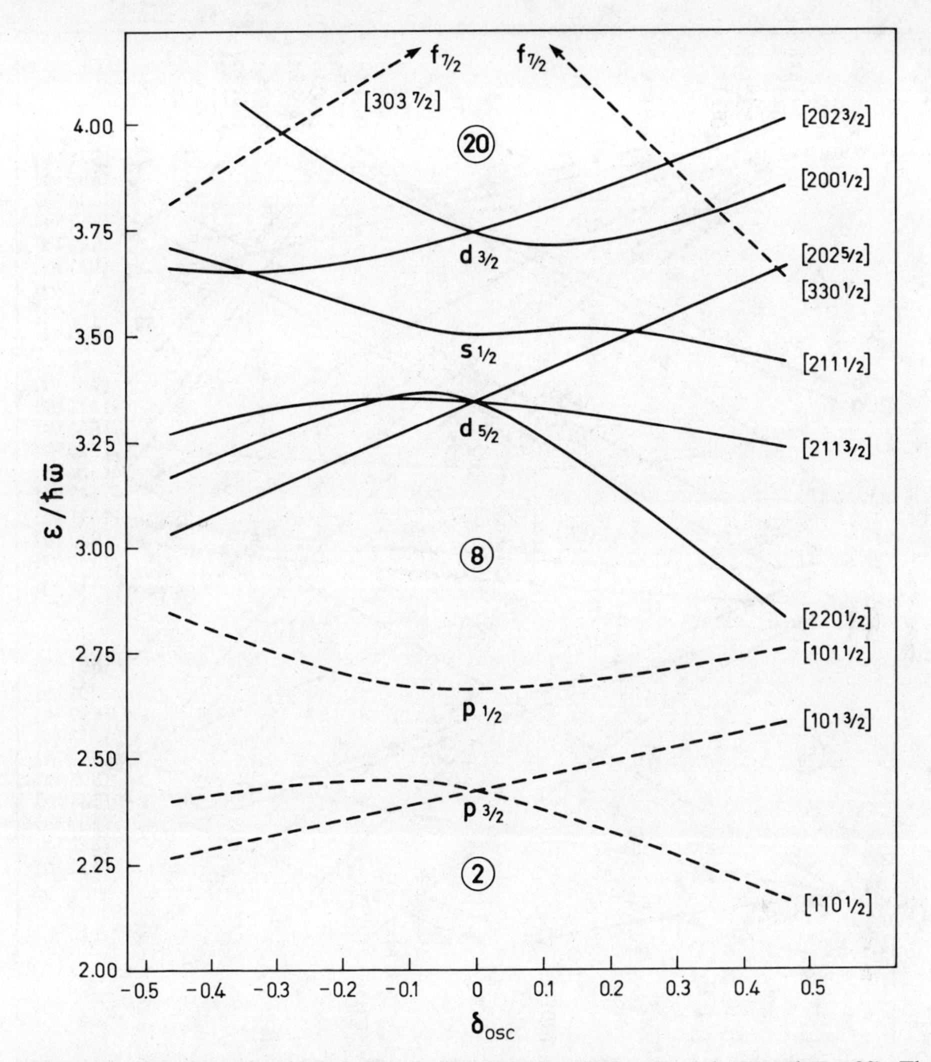

Figure 5-1 Spectrum of single-particle orbits in spheroidal potential (N and $Z < 20$). The spectrum is taken from B. R. Mottelson and S. G. Nilsson, *Mat. Fys. Skr. Dan. Vid. Selsk.* **1**, no. 8 (1959). The orbits are labeled by the asymptotic quantum numbers $[Nn_3\Lambda\Omega]$ referring to large prolate deformations. Levels with even and odd parity are drawn with solid and dashed lines, respectively.

▼ accuracy, it becomes necessary to consider a number of other effects, such as the angular dependence of the surface diffuseness and differences in the eccentricities of the density and potential.

In regard to the latter point, it may be more appropriate to equate the quantities $R_0\delta$ corresponding to the displacement of the surface (see, for example, Table 6-2, p. 354); this relationship would imply values of δ for the potential that are 10–20% smaller than for the density. Since the spectra in ▲

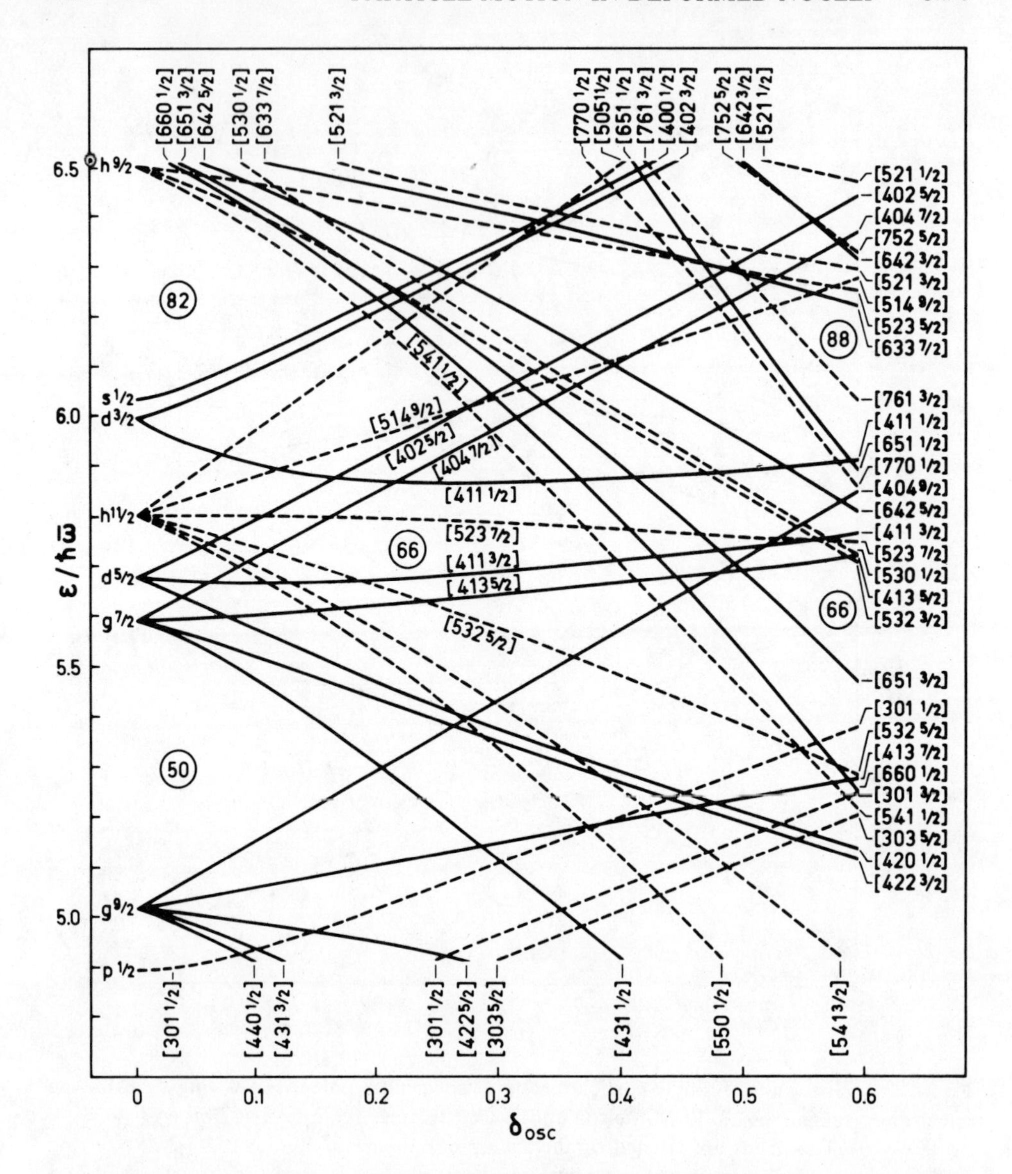

Figure 5-2 Proton orbits in prolate potential ($50 < Z < 82$). The spectra in this and the following figures (Figs. 5-2 to 5-5) are taken from C. Gustafson, I. L. Lamm, B. Nilsson, and S. G. Nilsson, *Arkiv Fysik* **36**, 613 (1967). The orbits are labeled by the asymptotic quantum numbers $[Nn_3\Lambda\Omega]$. Levels with even and odd parity are drawn with solid and dashed lines, respectively. (Erratum: The orbit [301 3/2] is incorrectly labeled [301 1/2] at bottom of figure.)

▼ Figs. 5-1 to 5-5 are not sensitive to such relatively small changes in δ, the applications in the present chapter are based on the assumption that δ_{osc} can be approximately equated to the deformation parameters obtained from Q_0 as ▲ given in Fig. 4-25 and Table 4-15.

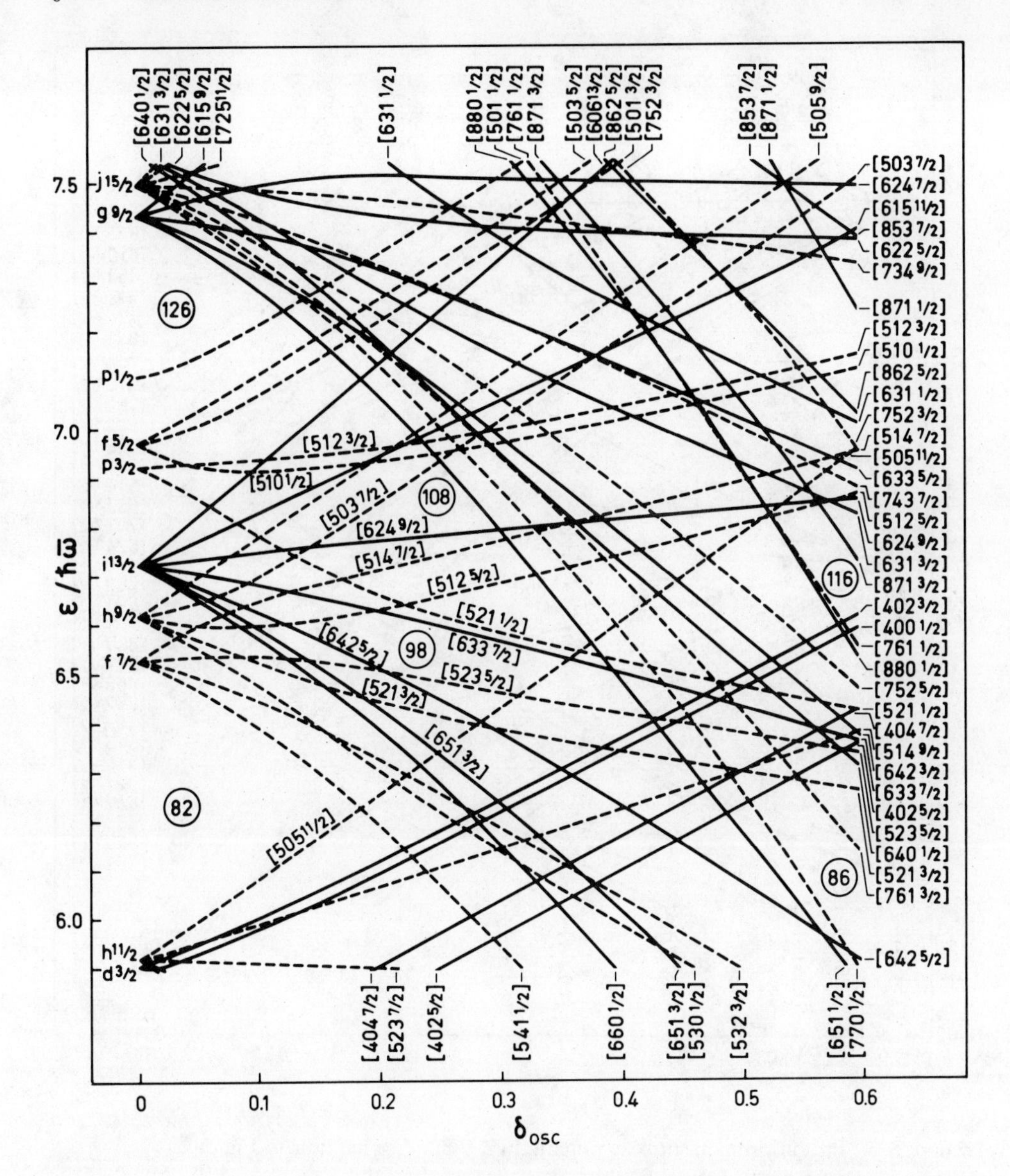

Figure 5-3 Neutron orbits in prolate potential ($82 < N < 126$). See caption to Fig. 5-2.

▼ The main trends of the energy levels in Figs. 5-1 to 5-5 can be understood in terms of the approximate solutions for small and large deformations as discussed in the text, p. 217. Thus, the expectation value of H in the representation $Nlj\Omega$ appropriate to small δ is given by Eq. (5-4), while the ▲ expectation value in the cylindrical oscillator representation $n_3 n_\perp \Lambda\Omega$ is given

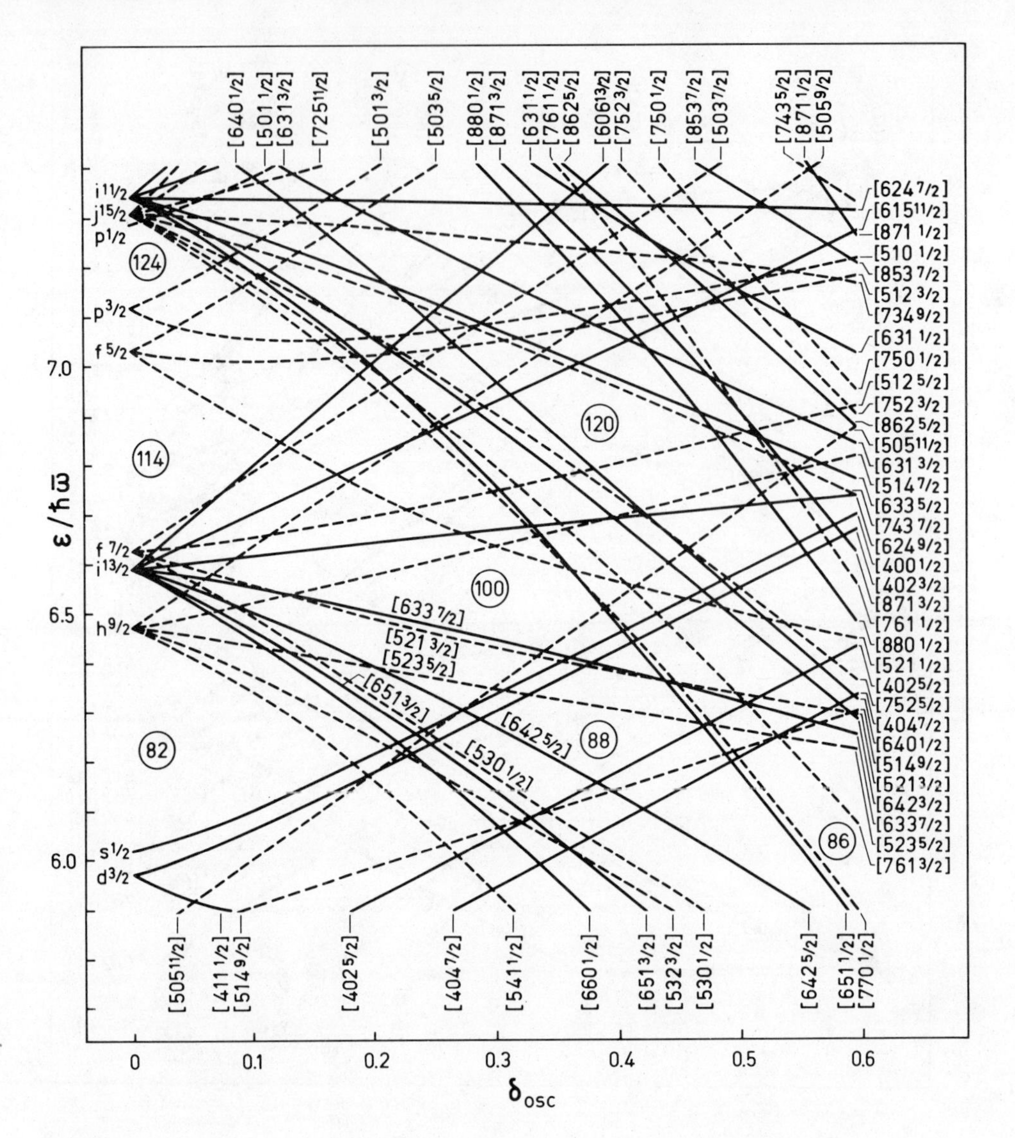

Figure 5-4 Proton orbits in prolate potential ($Z > 82$). See caption to Fig. 5-2.

▼ by

$$\langle n_3 n_\perp \Lambda\Omega | H | n_3 n_\perp \Lambda\Omega \rangle$$

$$= \hbar\bar{\omega}\left(N + \tfrac{3}{2} - \tfrac{1}{3}\delta_{\text{osc}}(3n_3 - N) - v_{ll}\left(\tfrac{1}{2}(2n_3 - N - \tfrac{1}{2})^2 - \Lambda^2 - \tfrac{1}{8}\right) + v_{ls}\Lambda\Sigma\right) \quad (5\text{-}15)$$

$$(N = n_3 + n_\perp \quad \Omega = \Lambda + \Sigma)$$

The deformation characterizing the transition to the asymptotic approximation (5-15) can be estimated by comparing the energy shifts produced by the deformation with those associated with the terms v_{ll} and v_{ls}. The ▲

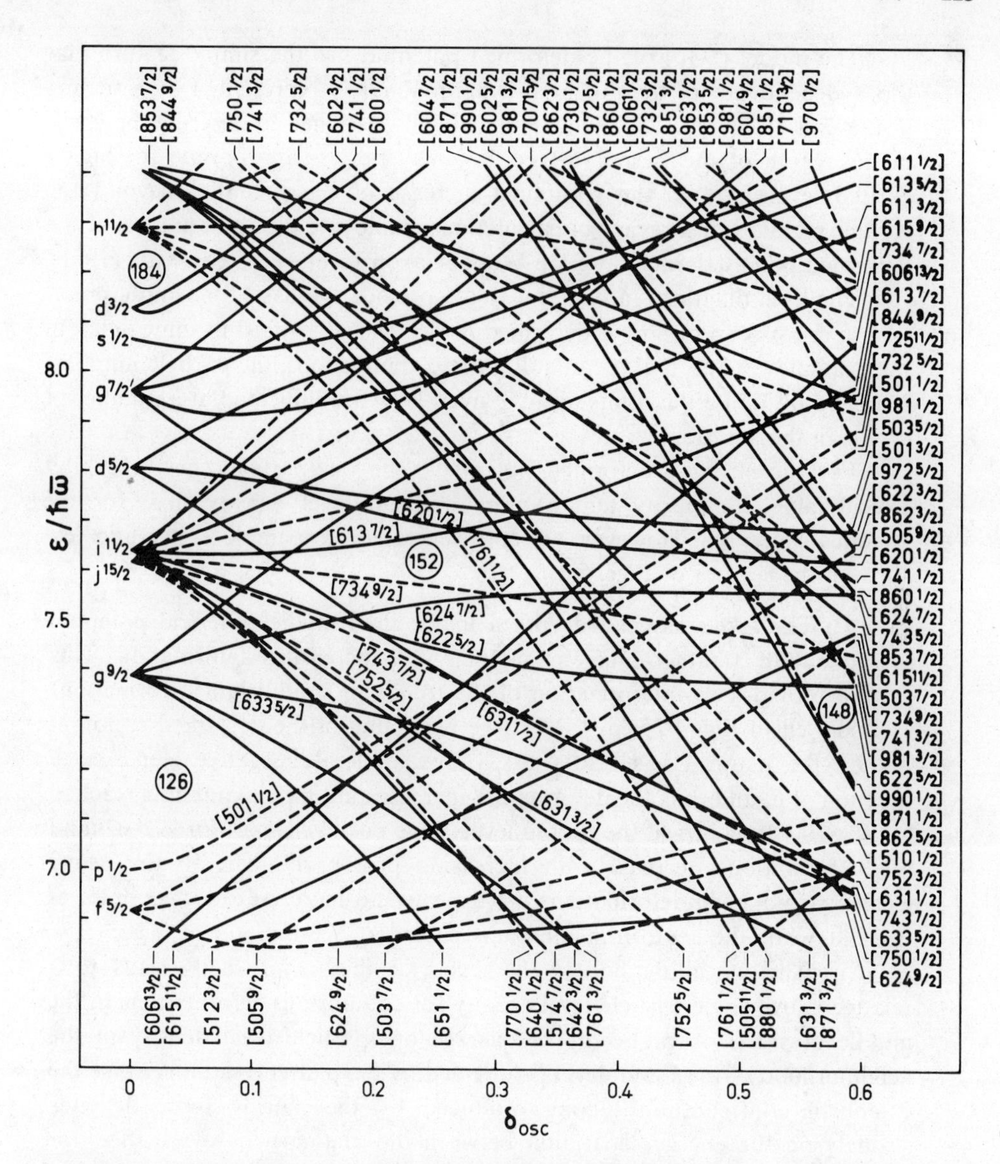

Figure 5-5 Neutron orbits in prolate potential ($N > 126$). See caption to Fig. 5-2.

▼ former are of order $N\hbar\omega_0\delta$, while the latter increase with l and become comparable to the spacings of the major shells for the largest values of the angular momentum. Thus, for typical equilibrium deformations ($\delta \sim A^{-1/3} \sim N^{-1}$), the asymptotic description is applicable to most orbits; however, this description may be inadequate for orbits with especially large values of j. (For example, the sequence of proton orbits [550 1/2], [541 3/2], [532 5/2], etc. in Fig. 5-2 is rather well described as $h_{11/2}$, even for $\delta \approx 0.3$ (as can be seen from Table 5-2), and similarly, the neutron orbits [660 1/2], [651 3/2], etc. in Fig. ▲ 5-3 are rather close to $i_{13/2}$.)

▼ The model (5-10) of the deformed potential has the simple feature that the A dependence of the potential is approximately represented by a scaling of the unit of energy. The model is specifically designed to describe the levels in the region of the Fermi surface, but may be less appropriate for highly excited levels due to the continued increase of the potential beyond the surface region. For a review of calculations employing different models of the deformed potential, including the Woods-Saxon form, see Ogle *et al.* (1971).

The level diagrams in Figs. 5-1 to 5-5 provide the basis for an interpretation of the spectra of odd-A deformed nuclei, as is discussed in some detail in the examples to Sec. 5-3. As an aid in these applications, the particle numbers obtained by filling the orbits up to a given level are indicated at a number of points in the spectra.

For the strongly deformed potentials, the one-particle spectra exhibit considerable complexity and a more uniform level density than for the spherical potential. However, we note several rather simple systematic features in the spectra of Figs. 5-1 to 5-5.

(i) The levels associated with a major shell in the spherical potential spread out in a rather uniform manner with increasing deformation. This pattern provides an interpretation of the trends in equilibrium deformations as exhibited in Fig. 4-25, p. 133. At the beginning of the deformed region at $N \approx 90$, the neutron levels at $\delta \approx 0.3$ have moderate negative slopes as a function of δ, and thus the deformation increases until a maximum is reached for $N \approx 100$. In contrast, the proton levels have positive slopes for $Z > 62$, and the deformations decrease with increasing proton number. In the heavy element region, the deformations are expected to increase to a maximum at $Z \approx 100$, while the neutron maximum is passed at $N \approx 150$.

(ii) The spreading out of the levels of the major shells leads to a relatively low single-particle level density for equilibrium deformations in the middle of shells. A probe of the average one-particle level density in the neighborhood of the Fermi level is provided by the pair correlations, since the extent of configuration mixing produced by the pairing force depends strongly on the energy separation between the one-particle levels. (See, for example, the expression (6-616) for the magnitude of the pair-correlation parameter Δ.) The decreased level density in the deformed nuclei is reflected in a decrease in the magnitude of the odd-even mass differences (see Fig. 2-5, Vol., I, p. 170).

(iii) For the anisotropic harmonic oscillator potential, major degeneracies in the single-particle spectra (shell structure) occur when the oscillator frequencies are in the ratio of (small) integers (see Fig. 6-48, p. 592). The ratio $\omega_\perp : \omega_3 = 2:1$ implies $\delta_{osc} = 0.6$ (see Eq. (5-11)), and the closed-shell numbers in a pure harmonic oscillator potential are $\ldots, 60, 80, 110, 140, \ldots$. The $\mathbf{l}^2$ and $\mathbf{l}\cdot\mathbf{s}$ terms remove the degeneracies of these shells and, as for the spherical ▲ shell structure, shift a few of the orbits with the largest Λ and $\Omega = \Lambda + 1/2$

▼ into the shell below (see the discussion in Chapter 6, p. 597). Indications of the resulting minima in the single-particle level density at nucleon numbers ..., 66, 86, 116, 148,... can be seen in Figs. 5-2 to 5-5. (For such large deformations, it becomes necessary to consider the effect of the deformation on the $\mathbf{l}^2$ and $\mathbf{l}\cdot\mathbf{s}$ terms; see the discussion on pp. 594ff.)

One-particle wave functions for spheroidal potential (Table 5-2)

The diagonalization of the Hamiltonian (5-10), as discussed in the previous example, yields eigenfunctions in the cylindrical oscillator basis $n_3 n_\perp \Lambda\Omega$. Table 5-2a gives a selection of these wave functions for the one-particle spectra drawn in Figs. 5-2 and 5-3. The table is intended to illustrate the structure of the eigenstates and provides a basis for the analysis of the empirical data on the low-energy spectra of the odd-A nuclei in the region $150 < A < 190$. The eigenstates are labeled by the asymptotic quantum numbers $[Nn_3\Lambda\Omega]$ and are given for a deformation $\delta = 0.3$. The evaluation of matrix elements of one-particle operators in the $n_3 n_\perp \Lambda\Omega$ representation is discussed in the following example, pp. 231ff.

Other features of the deformed wave function may be exhibited by transforming the wave functions to a spherical basis $Nl\Lambda\Omega$. This transformation can be carried out in two steps; the first, involving a transition to an isotropic oscillator, is obtained by scaling the wave functions by the factors $(\omega_3/\omega_0)^{1/2}$ and $(\omega_\perp/\omega_0)^{1/2}$ in directions parallel and perpendicular to the symmetry axis, respectively. The second step involves a transformation from the isotropic cylindrical basis to the spherical quantum numbers. (For a discussion of the transformations between the various oscillator representations, see Talman, 1970.)

The scaling of the wave function leads to a considerable increase in the number of significant components of the states in the spherical basis. It is therefore convenient to absorb this part of the transformation into the operators and to exhibit the wave functions resulting from the effect of the second transformation alone. The states given in Table 5-2b are obtained in this manner from those listed in Table 5-2a. In the evaluation of matrix elements on the basis of the wave functions in Table 5-2b, the scaling of the position and momentum operators implies

$$
\begin{aligned}
x_1 &\rightarrow x_1\left(\frac{\omega_0}{\omega_\perp}\right)^{1/2} \approx x_1(1-\tfrac{1}{6}\delta) & p_1 &\rightarrow p_1\left(\frac{\omega_\perp}{\omega_0}\right)^{1/2} \approx p_1(1+\tfrac{1}{6}\delta) \\
x_2 &\rightarrow x_2\left(\frac{\omega_0}{\omega_\perp}\right)^{1/2} \approx x_2(1-\tfrac{1}{6}\delta) & p_2 &\rightarrow p_2\left(\frac{\omega_\perp}{\omega_0}\right)^{1/2} \approx p_2(1+\tfrac{1}{6}\delta) \qquad (5\text{-}16) \\
x_3 &\rightarrow x_3\left(\frac{\omega_0}{\omega_3}\right)^{1/2} \approx x_3(1+\tfrac{1}{3}\delta) & p_3 &\rightarrow p_3\left(\frac{\omega_3}{\omega_0}\right)^{1/2} \approx p_3(1-\tfrac{1}{3}\delta)
\end{aligned}
$$

▲

$N=4$ $[Nn_3\Lambda\Omega]$	$n_3=4$ $\Lambda=0$	$n_3=3$ $\Lambda=1$	$n_3=2$ $\Lambda=0,2$	$n_3=1$ $\Lambda=1,3$	$n_3=0$ $\Lambda=0,2,4$
protons					
413 5/2			−0.342	0.938	0.054
411 3/2		−0.250	−0.223	0.926	0.174
411 1/2	0.130	−0.255	−0.310	0.900	0.108
404 7/2				−0.219	0.976
402 5/2			−0.151	−0.111	0.982
402 3/2		0.061	−0.156	−0.203	0.965
400 1/2	0.026	0.051	−0.179	−0.168	0.968
neutrons					
404 7/2				−0.219	0.976
400 1/2	0.010	0.047	−0.123	−0.213	0.968
402 3/2		0.048	−0.115	−0.218	0.968

$N=5$ $[Nn_3\Lambda\Omega]$	$n_3=5$ $\Lambda=0$	$n_3=4$ $\Lambda=1$	$n_3=3$ $\Lambda=0,2$	$n_3=2$ $\Lambda=1,3$	$n_3=1$ $\Lambda=0,2,4$	$n_3=0$ $\Lambda=1,3,5$
protons						
532 5/2			0.861	0.397	0.310	0.075
523 7/2				0.934	0.312	0.177
514 9/2					0.978	0.211
541 1/2	−0.669	0.605	−0.158	0.396	0.022	0.063
neutrons						
530 1/2	−0.107	−0.618	0.702	0.160	0.291	0.055
505 11/2						1.000
521 3/2		−0.152	−0.475	0.826	0.189	0.184
523 5/2			−0.452	0.850	0.204	0.178
521 1/2	0.181	−0.234	−0.453	0.797	0.184	0.197
512 5/2			−0.141	−0.330	0.919	0.166
514 7/2				−0.331	0.932	0.151
510 1/2	0.022	0.110	−0.207	−0.337	0.898	0.157
512 3/2		0.110	−0.199	−0.335	0.901	0.156

$N=6$ $[Nn_3\Lambda\Omega]$	$n_3=5$ $\Lambda=1$	$n_3=4$ $\Lambda=0,2$	$n_3=3$ $\Lambda=1,3$	$n_3=2$ $\Lambda=0,2,4$	$n_3=1$ $\Lambda=1,3,5$	$n_3=0$ $\Lambda=0,2,4,6$
neutrons						
651 3/2	0.698	0.489	0.463	0.218	0.107	0.026
642 5/2		0.811	0.435	0.368	0.126	0.045
633 7/2			0.889	0.371	0.263	0.058
624 9/2				0.943	0.296	0.151

Table 5-2a Cylindrical basis

Table 5-2 Wave functions for nucleonic motion in prolate potential. The table gives eigenstates for the Hamiltonian (5-10) with parameters as given in Table 5-1 ($50<Z<82$; $82<N<126$) and a deformation $\delta=0.3$. In Table 5-2a, the wave functions are expanded in the basis $Nn_3\Lambda\Omega$, while in Table 5-2b the same states are given in the basis $Nl\Lambda\Omega$. The wave functions are similar to those contained in the more extensive tabulation by S. G. Nilsson, *Mat. Fys. Medd. Dan. Vid. Selsk.* **29**, no. 16 (1955), except for the minor modifications in the parameters of the potential. In Table 5-2a, the phases of the components are chosen in accordance with the convention described on pp. 231 ff. (see especially Eq. (5-27)), and the

$N=4$	$\Sigma=+1/2 \quad \Lambda=\Omega-1/2$			$\Sigma=-1/2 \quad \Lambda=\Omega+1/2$	
$[Nn_3\Lambda\Omega]$	$l=4$	$l=2$	$l=0$	$l=4$	$l=2$
protons					
413 5/2	−0.296	0.179		0.938	
411 3/2	0.418	0.864		−0.140	0.246
411 1/2	−0.163	−0.099	0.297	0.396	0.848
404 7/2	−0.218			0.976	
402 5/2	0.232	0.966		−0.111	
402 3/2	−0.086	−0.193		0.221	0.952
400 1/2	0.147	0.539	0.811	−0.072	−0.160
neutrons					
404 7/2	−0.219			0.976	
400 1/2	0.186	0.563	0.775	−0.104	−0.192
402 3/2	−0.106	−0.196		0.259	0.940

$N=5$	$\Sigma=+1/2 \quad \Lambda=\Omega-1/2$			$\Sigma=-1/2 \quad \Lambda=\Omega+1/2$		
$[Nn_3\Lambda\Omega]$	$l=5$	$l=3$	$l=1$	$l=5$	$l=3$	$l=1$
protons						
532 5/2	0.882	−0.244		0.399	−0.062	
523 7/2	0.939	−0.144		0.312		
514 9/2	0.978			0.211		
541 1/2	−0.354	0.485	−0.335	0.686	−0.232	0.040
neutrons						
530 1/2	0.663	0.179	−0.342	−0.249	0.552	−0.212
505 11/2	1.000					
521 3/2	0.571	0.575	−0.287	−0.278	0.429	
523 5/2	−0.251	0.427		0.861	−0.116	
521 1/2	−0.207	0.081	0.471	0.501	0.629	−0.287
512 5/2	0.415	0.832		−0.255	0.267	
514 7/2	−0.262	0.252		0.932		
510 1/2	0.281	0.672	0.565	−0.153	−0.151	0.323
512 3/2	−0.151	−0.150	0.321	0.358	0.850	

$N=6$	$\Sigma=+1/2 \quad \Lambda=\Omega-1/2$			$\Sigma=-1/2 \quad \Lambda=\Omega+1/2$		
$[Nn_3\Lambda\Omega]$	$l=6$	$l=4$	$l=2$	$l=6$	$l=4$	$l=2$
neutrons						
651 3/2	0.746	−0.377	0.120	0.497	−0.199	0.034
642 5/2	0.829	−0.324	0.055	0.437	−0.120	
633 7/2	0.895	−0.240		0.371	−0.056	
624 9/2	0.945	−0.141		0.296		

Table 5-2b Spherical basis

overall phase is chosen so as to give a positive value for the leading coefficient, $\langle Nn_3\Lambda\Omega|[Nn_3\Lambda\Omega]\rangle$. The phases of the spherical basis, employed in Table 5-2b, correspond to angular wave functions $i^l Y_{l\Lambda}$ and radial wave functions that become positive for $r \rightarrow \infty$ (see Vol. I, p. 360; this phase convention differs by the factor i^l from that employed by Nilsson, *loc.cit.*); the overall phase is the same as in Table 5-2a, which leads to a positive sign for the largest component, for all the states in the table. We wish to thank J. Damgaard, I. L. Lamm, and S. G. Nilsson for help in the preparation of the table.

▼ While the scaling in general implies corrections to the matrix elements of order δ, the matrix elements of the angular momentum within an oscillator shell ($\Delta N = 0$) are only affected to order δ^2.

An expansion in terms of basis states labeled by the total angular momentum j of the particle can be obtained from the wave functions in Table 5-2b by means of the transformation

$$\begin{aligned}\langle Nl, j=l\pm\tfrac{1}{2}, \Omega|\nu\rangle &= \sum_{\Lambda\Sigma} \langle l\Lambda\tfrac{1}{2}\Sigma|j\Omega\rangle\langle Nl\Lambda\Sigma|\nu\rangle \\ &= \pm\left(\frac{l\pm\Omega+\frac{1}{2}}{2l+1}\right)^{1/2}\langle Nl,\Omega-\tfrac{1}{2},\tfrac{1}{2}|\nu\rangle + \left(\frac{l\mp\Omega+\frac{1}{2}}{2l+1}\right)^{1/2}\langle Nl,\Omega+\tfrac{1}{2},-\tfrac{1}{2}|\nu\rangle\end{aligned} \quad (5\text{-}17)$$

where ν $(=[Nn_3\Lambda\Sigma])$ labels the eigenstates. The wave functions of the (sd) shell are given in the $Nlj\Omega$ representation in Table 5-9, p. 290.

The eigenstates in Table 5-2b exhibit the very strong mixing of components with different l values resulting from the deformation. The approximate validity of the asymptotic quantum numbers can be seen from Table 5-2a. For most of the states, the leading component represents 80% or more of the total wave function; however, the states that in the spherical limit have large values of j approach the asymptotic limit more slowly and, for $\delta = 0.3$, have large admixtures. This conclusion is the same as that deduced from the variation of the energies as a function of δ (see p. 222).

As discussed on p. 220, the $\Delta N = 2$ matrix elements of the $\mathbf{l}^2$ and $\mathbf{l}\cdot\mathbf{s}$ terms have been neglected in the diagonalization of the Hamiltonian (5-10). As a result of this approximation, levels with different N, but with the same quantum numbers $\Omega\pi$, exhibit crossings. (See, for example, the $\Omega\pi = 3/2+$ levels [402 3/2] and [651 3/2] in Fig. 5-3 that cross at $\delta \approx 0.3$.) The interaction between these approximately degenerate levels with $\Delta N = 2$ is expected to be small, on account of the large differences in the quantum numbers n_3 and $n_\perp$. Estimates have yielded interaction energies that are of the order of or less than 100 keV, but the results are sensitive to the occurrence of higher multipoles in the deformation as well as to the radial shape of the potential (Nemirovsky and Chapurnov, 1966; Andersen, 1968 and 1972). In the narrow region of δ in which the distance between the unperturbed levels is comparable with their interaction, the orbits become strongly mixed; such a sharing of the properties of the orbits has been observed in the one-particle transfer intensities and has yielded experimental values of the interaction matrix elements of 50–100 keV (Sheline *et al.*, 1967; Tjøm and Elbek, 1967; Grotdal *et al.*, 1970). The smallness of these coupling matrix elements is a rather striking manifestation of the approximate validity of the separation of the motion into oscillations parallel and perpendicular to the symmetry axis. ▲

▼ *Matrix elements for oscillator wave functions in cylindrical basis (Table 5-3)*

In the evaluation of matrix elements for particles moving in a harmonic oscillator potential, it is often convenient to describe the motion in terms of the quanta of oscillation. For a one-dimensional oscillator, the operators $c^\dagger$ and c that create and destroy quanta are linear combinations of the coordinate and momentum of the particle,

$$x = \left(\frac{\hbar}{2M\omega}\right)^{1/2}(c^\dagger + c)$$

$$p = i\left(\frac{\hbar M\omega}{2}\right)^{1/2}(c^\dagger - c) \tag{5-18}$$

$$c^\dagger = \left(\frac{M\omega}{2\hbar}\right)^{1/2}\left(x - \frac{i}{M\omega}p\right)$$

where ω is the oscillator frequency. The nonvanishing matrix elements of $c^\dagger$ are

$$\langle n+1|c^\dagger|n\rangle = (n+1)^{1/2} \tag{5-19}$$

and the normalized wave functions can be written

$$|n\rangle = (n!)^{-1/2}(c^\dagger)^n|n=0\rangle \tag{5-20}$$

For the three-dimensional oscillator, we shall employ the standard phasing of the states, such that $\mathscr{R}\mathscr{T} = +1$, where $\mathscr{R}$ represents a rotation of 180° about the 2 axis. The matrix elements of p_1, x_2, and p_3 are then real and will be taken to be positive in the $(n_1n_2n_3)$ basis, while the matrix elements of x_1, p_2, and x_3 are imaginary. (The phase of the matrix element of x in Eq. (2-154) is that of the one-dimensional oscillator, as in Eq. (5-18).) Taking the operators $c_\kappa^\dagger$ to have real and positive matrix elements, as in Eq. (5-19), one obtains

$$c_1^\dagger = \left(\frac{M\omega_1}{2\hbar}\right)^{1/2}\left(ix_1 + \frac{p_1}{M\omega_1}\right)$$

$$c_2^\dagger = \left(\frac{M\omega_2}{2\hbar}\right)^{1/2}\left(x_2 - \frac{ip_2}{M\omega_2}\right) \tag{5-21}$$

$$c_3^\dagger = \left(\frac{M\omega_3}{2\hbar}\right)^{1/2}\left(ix_3 + \frac{p_3}{M\omega_3}\right)$$

▲

▼ In the limit of spherical symmetry ($\omega_1 = \omega_2 = \omega_3$), the quantities $c_1^\dagger, ic_2^\dagger, c_3^\dagger$ form the Cartesian components of a vector.

For the spheroidal oscillator potential, one may employ a cylindrical representation where the particle states are specified by the number of quanta n_3 in the direction of the symmetry axis and the number of quanta n_+ (and n_-) moving in the (12) plane and carrying $+1$ (and -1) units of angular momentum relative to the 3 axis. The total number of quanta and the 3 component of the total angular momentum are given by

$$N = n_3 + n_+ + n_-$$
$$\Lambda = n_+ - n_- \tag{5-22}$$

In the cylindrical basis ($n_+ n_- n_3$), we shall adopt a phase choice such that the operators $c_+^\dagger, c_3^\dagger, c_-^\dagger$ have positive matrix elements and, in the limit of spherical symmetry, form the spherical components of a vector operator (see Eq. (1A-56)),

$$c_\pm^\dagger = \mp 2^{-1/2}(c_1^\dagger \mp c_2^\dagger)$$
$$= \mp \left(\frac{M\omega_\perp}{4\hbar}\right)^{1/2} \left(i(x_1 \pm ix_2) + \frac{p_1 \pm ip_2}{M\omega_\perp}\right) \tag{5-23}$$
$$c_3^\dagger = \left(\frac{M\omega_3}{2\hbar}\right)^{1/2} \left(ix_3 + \frac{p_3}{M\omega_3}\right)$$

with the inverse relations

$$x_1 \pm ix_2 = \pm i\left(\frac{\hbar}{M\omega_\perp}\right)^{1/2} (c_\pm^\dagger + c_\mp)$$
$$p_1 \pm ip_2 = \mp(\hbar M\omega_\perp)^{1/2}(c_\pm^\dagger - c_\mp)$$
$$x_3 = -i\left(\frac{\hbar}{2M\omega_3}\right)^{1/2} (c_3^\dagger - c_3) \tag{5-24}$$
$$p_3 = \left(\frac{\hbar M\omega_3}{2}\right)^{1/2} (c_3^\dagger + c_3)$$

▲

▼ The wave functions in Table 5-2a are expressed in the basis $Nn_3\Lambda$, in which the matrix elements of the coordinates are (see Eq. (5-24))

$$\langle N'n_3, \Lambda\pm 1|x_1 \pm ix_2|Nn_3\Lambda\rangle = \pm i\left(\frac{\hbar}{2M\omega_\perp}\right)^{1/2}\Big((N-n_3\pm\Lambda+2)^{1/2}\delta(N', N+1)$$

$$+(N-n_3\mp\Lambda)^{1/2}\delta(N', N-1)\Big) \qquad (5\text{-}25)$$

$$\langle N\pm 1, n_3\pm 1, \Lambda|x_3|Nn_3\Lambda\rangle = \mp i\left(\frac{\hbar}{2M\omega_3}\right)^{1/2}(n_3+\tfrac{1}{2}\pm\tfrac{1}{2})^{1/2}$$

The orbital angular momentum is a bilinear expression in the $c^\dagger, c$ operators,

$$l_1 \pm il_2 = (2\omega_3\omega_\perp)^{-1/2}\Big((\omega_3+\omega_\perp)(c_3^\dagger c_\mp + c_\pm^\dagger c_3) + (\omega_3-\omega_\perp)(c_3^\dagger c_\pm^\dagger + c_3 c_\mp)\Big)$$

$$l_3 = c_+^\dagger c_+ - c_-^\dagger c_- \qquad (5\text{-}26)$$

and we obtain the matrix elements

$$\langle Nn_3', \Lambda\pm 1|l_1\pm il_2|Nn_3\Lambda\rangle = \frac{\omega_3+\omega_\perp}{2(\omega_3\omega_\perp)^{1/2}}\Big((n_3+1)^{1/2}(N-n_3\mp\Lambda)^{1/2}\delta(n_3', n_3+1)$$

$$+(n_3)^{1/2}(N-n_3\pm\Lambda+2)^{1/2}\delta(n_3', n_3-1)\Big) \qquad (5\text{-}27)$$

$$\langle Nn_3\Lambda|l_3|Nn_3\Lambda\rangle = \Lambda$$

The components $l_1 \pm il_2$ have additional matrix elements with $\Delta N = \pm 2$, associated with the second term in Eq. (5-26). The $\Delta N = 2$ matrix elements, however, are smaller by a factor δ than the $\Delta N = 0$ matrix elements, corresponding to the fact that the orbital angular momentum becomes a constant of the motion for a spherical potential.

The above relations are convenient for the evaluation of matrix elements of the various multipole operators. For $E1$ and $M1$ moments, the selection rules are summarized in Table 5-3, and the matrix elements follow from Eqs. (5-25) to (5-27). For multipole operators of higher order, the selection rules and matrix elements can be obtained by expressing the operators as polynomials of the coordinates and angular momentum components. (Tables of selection rules for multipole operators for γ and β transitions have been given by Mottelson and Nilsson, 1959.) ▲

Multipole	$\nu=\Delta\Omega$	Operator	$\|\Delta N\|$	$\|\Delta n_3\|$	$\Delta\Lambda$	$\Delta\Sigma$
$E1$	± 1	$x_1 \pm i x_2$	1	0	± 1	0
	0	x_3	1	1	0	0
$M1$	± 1	$l_1 \pm i l_2$	0, 2	1	± 1	0
	± 1	$s_1 \pm i s_2$	0	0	0	± 1
	0	s_3, l_3	0	0	0	0

Table 5-3 Selection rules for $E1$ and $M1$ moments in spheroidal oscillator potential.

▼ *Strength function for slow neutrons interacting with a nonspherical nucleus (Fig. 5-6; Tables 5-4 and 5-5)*

As an illustration of the effect of the nuclear deformation on the one-particle strength function in the resonance region, we consider the model of slow neutrons interacting with a complex square-well potential of spheroidal shape without spin-orbit coupling (Margolis and Troubetzkoy, 1957),

$$\begin{aligned} V(r) &= V + iW \qquad r < R(\vartheta) = R_0(1 + \tfrac{2}{3}\delta P_2(\cos\vartheta)) \\ &= 0 \qquad\qquad\;\; r > R(\vartheta) \end{aligned} \tag{5-28}$$

The angle ϑ is measured from the symmetry axis of the nucleus, which is assumed to have a fixed orientation during the collision (adiabatic approximation; see p. 238). For zero incident energy, the expansion of the wave function in spherical harmonics takes the form

$$\psi(\mathbf{r}) = \sum_l \mathscr{R}_l(r) i^l Y_{l0}(\vartheta)$$

$$\mathscr{R}_l(r) = \begin{cases} A_l j_l(Kr) & r < R(\vartheta) \\ B_l r^{-(l+1)} \quad l > 0 & r > R(\vartheta) \\ 1 - \dfrac{a}{r} \quad l = 0 & r > R(\vartheta) \end{cases} \tag{5-29}$$

$$K = \left(-\frac{2M}{\hbar^2}(V + iW)\right)^{1/2}$$

where j_l is a spherical Bessel function, while the coefficients A_l, B_l, and the scattering length a are complex constants obtained by requiring the continuity of the wave function and its derivative across the nonspherical boundary $R(\vartheta)$. The boundary conditions in Eq. (5-29) are appropriate to the limit of ▲ very small incident energies, for which one can neglect the outgoing waves

▼ with $l \neq 0$; in this limit, the scattering amplitude reduces to the scattering length ($f = -a$; see Eqs. (2-183) and (3F-33)). The cross section for absorption (compound nucleus formation) is given by (see Eq. (3F-72))

$$\sigma_{\text{comp}} = -4\pi\lambda\!\!\!^{-}\,\text{Im}\,a \tag{5-30}$$

which may also be expressed in terms of the s-wave strength function (see Eq. (2-159))

$$\frac{\Gamma_n^{(0)}}{D} = \left(\frac{\sigma_{\text{comp}}}{2\pi^2\lambda\!\!\!^{-2}}\right)_{1\,\text{eV}} = -\frac{\text{Im}\,a}{7.2\times 10^3\ \text{fm}} \tag{5-31}$$

where $\Gamma_n^{(0)}$ is the average neutron width of the s-wave resonances, reduced to a neutron energy of 1 eV, while D is the average spacing of these resonances.

Strength functions, obtained from the potential (5-28), are given in Fig. 5-6 as a function of the mass number. The parameters given in the figure imply

$$|KR_0| = 2.05A^{1/3} \tag{5-32}$$

where K is the inside wave number (see Eq. (5-29)). The strength function peak, shown in Fig. 5-6 for spherical nuclei with $A \approx 150$, corresponds to the 4s resonance ($|KR_0| = \frac{7}{2}\pi$). ▲

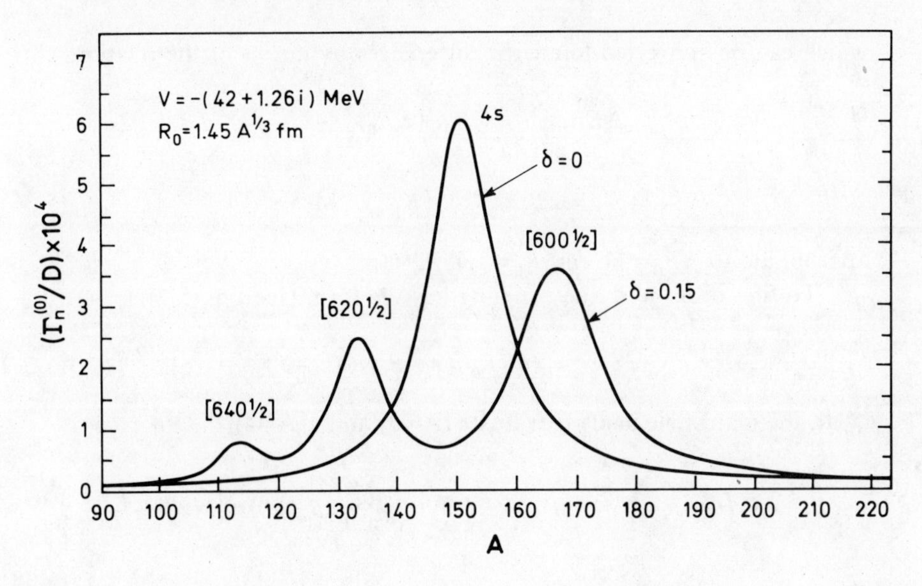

Figure 5-6 Strength function for s-wave neutrons of zero energy incident on a deformed square-well potential. The theoretical analysis is due to B. Margolis and E. S. Troubetzkoy, *Phys. Rev.* **106**, 105 (1957).

▼ As can be seen from Fig. 5-6, the deformation causes a splitting of the 4s resonance into several components. This effect can be attributed to the coupling to the near-lying resonances in the channels with $l=2,4$, etc. and is similar to the effect of the deformation on the bound orbits, which leads to a division of the s-wave strength between the different orbits with $\Lambda\pi=0+$. A simple estimate of the strength function in the neighborhood of the 4s resonance may be obtained by considering the coupling between the states in the $N=6$ shell (4s, 3d, 2g, and 1i). The position of these states, in the absence of coupling, may be characterized by the value of R_0 for which they occur at zero energy. The imaginary part of the potential does not significantly affect the location of the resonances, and for a real spherical potential, the condition for zero energy resonance implies (see Eq. (5-29))

$$R_0\left(\frac{d}{dr}j_l(Kr)\right)_{R_0} = -(l+1)j_l(KR_0) \tag{5-33}$$

which is equivalent to

$$j_{l-1}(KR_0)=0 \tag{5-34}$$

The resulting eigenvalues for the four resonances in the $N=6$ shell are

$$KR_0 = \begin{cases} 11.00 & 4s \\ 10.90 & 3d \\ 10.42 & 2g \\ 9.36 & 1i \end{cases} \tag{5-35}$$

which can be converted to energy differences by means of the relation

$$\begin{aligned} \Delta E &\approx \frac{\hbar^2}{MR_0^2} KR_0 \Delta(KR_0) \\ &\approx 220A^{-2/3}\Delta(KR_0) \quad \text{MeV} \end{aligned} \tag{5-36}$$

appropriate to $KR_0=11$ and $R_0=1.45A^{1/3}$ fm.

To first order in δ, the coupling has the form (see Eq. (5-2))

$$H_c = \tfrac{2}{3}\delta V R_0 \delta(r-R_0) P_2(\cos\vartheta) \tag{5-37}$$

with the matrix elements (see Eqs. (1A-42) and (1A-44))

$$\langle l',\Lambda=0|H_c|l,\Lambda=0\rangle = \tfrac{2}{3}\delta i^{l-l'} V R_0^3 \mathscr{R}_l(R_0)\,\mathscr{R}_{l'}(R_0)\langle l020|l'0\rangle^2\left(\frac{2l+1}{2l'+1}\right)^{1/2} \tag{5-38}$$

In evaluating the coupling matrix elements (5-38), we shall employ radial ▲ wave functions corresponding to resonances at zero energy; this approxima-

▼ tion is justified provided the shifts in the one-particle strength satisfy the condition $\Delta(KR_0) \ll \pi$. For resonance states at zero energy, the radial wave functions at the surface can be obtained from the integral of the square of the wave function over the nuclear volume,

$$\int_0^{R_0} (\mathscr{R}_l(r))^2 r^2 dr = \tfrac{1}{2} R_0^3 (\mathscr{R}_l(R_0))^2 \tag{5-39a}$$

$$= \begin{cases} 1 & l=0 \\ \dfrac{2l-1}{2l+1} & l \neq 0 \end{cases} \tag{5-39b}$$

The relation (5-39a) follows from standard expressions for the integral of the Bessel functions, using the boundary condition (5-33). The relation (5-39b) expresses the normalization condition; for the states with $l \neq 0$, the normalization integral receives an additional contribution from the region outside the nuclear surface, as given by Eq. (3F-56). This region does not contribute for $l=0$ waves, due to the strong reflection at the surface $r = R_0$; in fact, once the neutron has penetrated into the nucleus, it stays there indefinitely (for zero bombarding energy); in contrast, the $l \neq 0$ waves are held within the neighborhood of the nucleus by the centrifugal barrier, but penetrate rather easily back and forth through the surface $r = R_0$. One may also recognize that the square of the wave function at $r = R_0$ can be directly related to the single-particle width for resonance scattering, since the outgoing flux is equal to the density obtained from Eq. (5-39) multiplied by the centrifugal barrier factor v_l and the outside velocity (see Eq. (3F-51) and the discussion in Sec. 3F-2b).

Diagonalizing the Hamiltonian consisting of the diagonal elements given by Eqs. (5-35) and (5-36) and the nondiagonal terms given by Eqs. (5-38) and (5-39), one obtains the eigenvalues and s-wave strengths listed in Table 5-4. The eigenstates have been labeled by the asymptotic quantum numbers $Nn_3\Lambda$. The differences in KR_0 can be related to differences in A by means of

$$\frac{\Delta A}{A} = 3\frac{\Delta(KR_0)}{KR_0} = 1.46 A^{-1/3} \Delta(KR_0) \tag{5-40}$$

and the values in Table 5-4 are seen to account rather well for the distribution of s-wave strength in Fig. 5-6. The widths of the separate resonances are determined by the absorption potential W and are therefore not much affected by the deformation. As mentioned above, the treatment of the coupling has assumed $\Delta(KR_0) \ll \pi$. While this condition is rather well fulfilled for the two major components in the s-wave strength function, it is badly violated for the very weakly excited state with $KR_0 \approx 9$.

The empirical data on the s-wave strength function in the region $140 < A < 180$ show large deviations from the values calculated for a spherical potential (see Fig. 2-26, Vol. I, p. 230), and it is apparent from the results for ▲

the simple model that the effect of the nuclear deformations must be included in the interpretation of these data. However, in a quantitative analysis, the effects of the diffuseness of the potential and of the spin-orbit coupling are expected to be significant.

$[Nn_3\Lambda]$	KR_0	$\langle l=0 \| [Nn_3\Lambda]\rangle^2$
[600]	11.29	0.62
[620]	10.60	0.34
[640]	10.11	0.04
[660]	(9.01)	(6×10^{-5})

Table 5-4 Eigenstates for zero-energy neutrons in deformed potential. The parameters of the potential are the same as those given in Fig. 5-6.

The diffuseness reduces the reflection of the neutrons at the nuclear surface and thus considerably increases the single-particle width and hence the total area under the strength function curve. (Compare the results for a spherical potential in Fig. 5-6 with those in Fig. 2-26, Vol. I, p. 230; see also the treatment of the effect of diffuseness on the single-particle width in Vol. I, p. 445; estimates of neutron strength functions for diffuse spheroidal potentials have been given by Chase *et al.*, 1958.)

The spin-orbit coupling gives rise to a mixing of the $\Lambda=0$, $\Sigma=1/2$ and $\Lambda=1$, $\Sigma=-1/2$ states and thus to an additional splitting of the s-wave strength function. As an illustration of the significance of this additional coupling, the s-wave strength for the $N=6$, $\Omega=1/2$ states, as calculated from the Hamiltonian (5-10), are listed in Table 5-5. It is seen that, for $\delta=0.15$, as in Fig. 5-6, the spin-orbit coupling has a major effect on the distribution of the $l=0$ strength, and even for $\delta=0.30$, as for the largest deformations in the region considered, the effect is considerable. The last column of Table 5-5 gives the distribution of the $l=0$ strength for a spheroidal oscillator potential, which represents an asymptotic situation, in which the effects of the deformation are large compared with the spin-orbit coupling as well as with the separation between the states with the same value of N in the spherical potential.

In the above discussion, the rotational energy of the target nucleus has been neglected. As in the treatment of the bound-state problem, such an approximation can be justified when the rotational frequency is small compared with the frequencies involved in the structure of the intrinsic state. In the treatment of low-energy scattering, it is in general necessary to include the rotational energy in the region outside the nucleus, where it may decisively

$[Nn_3\Lambda\Omega]$	$\langle l=0 \vert [Nn_3\Lambda\,\Omega=1/2]\rangle^2$ $\delta=0.15$	$\delta=0.3$	Asymptotic
[600 1/2]	0.59	0.55	16/35
[611 1/2]	0.17	0.08	0
[620 1/2]	0.08	0.15	8/35
[631 1/2]	0.14	0.11	0
[640 1/2]	0.002	0.03	6/35
[651 1/2]	0.02	0.07	0
[660 1/2]	0.0003	0.01	5/35

Table 5-5 Strength of s waves for neutron orbits with $N=6$. The expansion amplitudes for $\delta=0.15$ and 0.30 are taken from S. G. Nilsson, *Mat. Fys. Medd. Dan. Vid. Selsk.* **29**, no. 16 (1955). The values in the last column refer to a spheroidal harmonic oscillator potential (in the limit of small deformation).

▼ affect the propagation of the particles. In the present case, this correction is of minor significance, since it does not affect the $l=0$ channel and since, in the other channels, the rotational energy is small compared with the centrifugal barrier. (The use of the adiabatic approximation in the more general treatment of scattering by nonspherical systems is discussed in Appendix 5A.) ▲

5-2 CLASSIFICATION OF ODD-A SPECTRA

Aligned coupling scheme

The occurrence of a nonspherical component in the one-particle potential removes the degeneracies that are characteristic of a spherical nucleus and leaves only the twofold degeneracy associated with time reversal invariance. Thus, in an even-even nucleus, the lowest state of independent-particle motion is obtained by a pairwise filling of conjugate orbits and is nondegenerate. This ground-state configuration is characterized by an alignment of the orbital motion of the nucleons with respect to the orientation of the deformed fieid (aligned coupling scheme). In an axially symmetric potential, the pairwise filling of the orbits implies

$$K=\sum_{k=1}^{A}\Omega_k=0 \tag{5-41}$$

The last filled orbit in the ground-state configuration is referred to as the Fermi level. Excited configurations are obtained by transferring one or more particles from the occupied orbits into the unoccupied orbits above the Fermi level. The unoccupied orbits below the Fermi level are referred to as holes, while the occupied orbits above the Fermi level are called particles (see the discussion of the particle-hole representation in Appendix 3B).

In an odd-A nucleus, low-lying configurations are obtained by occupying one of the orbits singly (a single particle or a single hole), while the rest of the nucleons occupy the lowest remaining orbits pairwise. In such a configuration, the values of $K\pi$ are the same as the quantum numbers $\Omega\pi$ of the single particle or single hole. More complicated configurations may be created by exciting one or more of the particles from the paired "core."

Pair correlations. Quasiparticles

A significant modification of the independent-particle motion is to be expected from the interactions that give rise to the nuclear pairing effect. In fact, in heavy nuclei ($A > 100$), these interaction energies, as measured by the odd-even mass difference Δ, are of the order of 1 MeV (see Fig. 2-5, Vol. I, p. 170) and thus several times larger than the spacing between single-particle orbits in the deformed field (see Figs. 5-2 to 5-5). In spite of the rather large configuration mixing that must be expected in such a situation, essential features of the one-particle motion may be preserved, because the pairing effect can be generated by interactions that only give rise to transitions $(\nu_1\bar{\nu}_1)\rightarrow(\nu_2\bar{\nu}_2)$ of two particles in conjugate orbits, as in the $J\pi=0+$ paired two-particle state in a spherical nucleus. (For an analysis of the correlation effect in the paired two-particle state, see the discussion of the ground state of ^{206}Pb on pp. 641 ff.)

In the ground state of an even-even deformed nucleus, the configuration mixing generated by transition matrix elements $(\nu_1\bar{\nu}_1)\rightarrow(\nu_2\bar{\nu}_2)$ produces a coherent superposition of components involving only pairs of particles in conjugate orbits. In an odd-A nucleus, the lowest states involve a single unpaired particle in one of the orbits near the Fermi level; this particle is not affected by the pair interactions of the type considered, and the rest of the particles form a pair correlation similar to that of the ground state of even-even nuclei. Such states in odd-A nuciei can be classified by the quantum numbers of the unpaired particle.

Thus, the pair-correlated system can be described in terms of generalized one-particle excitations referred to as quasiparticles, and the number of quasiparticles is denoted by v. The ground state of an even-even nucleus is the quasiparticle vacuum ($\mathsf{v}=0$), and the low-lying states of odd-A nuclei are single-quasiparticle states ($\mathsf{v}=1$).

The pair correlations are associated with strong configuration mixings involving all the levels ($\nu\bar{\nu}$) within a distance of order Δ from the Fermi level. When the number of these levels is large, a rather simple, collective description of the pair correlations can be obtained in terms of a generalized one-particle potential that acts to create and annihilate pairs of particles in conjugate orbits. (The collective degrees of freedom of such a "pair field" are discussed in Sec. 6-3f, and the quasiparticles that result from one-particle motion in the presence of the pair field are considered in the example on pp. 647 ff.) Though the quasiparticles retain the quantum numbers of the one-particle orbits, their properties are significantly modified by the presence of the pair field, and these modifications will be included in the analysis of the present chapter.

Identification of single-quasiparticle states

The observed spectra of the odd-A deformed nuclei provide extensive tests of the aligned coupling scheme. The low-lying states form rotational bands with quantum numbers $K\pi$ corresponding to the calculated sequence of $\Omega\pi$ values for one-particle motion in a spheroidal potential. Examples are illustrated in Figs. 5-7, p. 254 (^{159}Tb), 5-9, p. 259 (^{175}Yb), 5-12, p. 266 (^{237}Np), 5-14, p. 274 (^{235}U), and 5-15, p. 284 (^{25}Mg and ^{25}Al), as well as in the examples in Chapter 4; see Figs. 4-15, p. 103 (^{169}Tm), 4-16, p. 106 (^{177}Lu and ^{177}Hf), 4-19, p. 112 (^{239}Pu), and 4-28, p. 154 (^{175}Lu). The analyses of the various properties of these bands confirm the identification and provide detailed comparisons with the calculated one-particle orbits (see Sec. 5-3). The systematic classification of the low-lying bands in deformed odd-A nuclei with $150 < A < 190$ is summarized in Tables 5-12, p. 296, and 5-13, p. 299. For nuclei with $19 \leqslant A \leqslant 25$, the classification of the odd-A spectra is given in Table 5-8, p. 286

The sequence of single-particle orbits confirms the approximate equality between the deformation of the potential and that determined from the $E2$-transition probabilities. The evidence for the occurrence of β_4 deformations (see Table 4-16, p. 140) is also supported by more detailed features in the single-particle level separations (see the discussion on p. 298).

It is a striking feature of the observed odd-A spectra in heavy deformed nuclei that the only states occurring in the low-energy spectrum below about 0.5 MeV are those that can be associated with the one-quasiparticle configurations. In the region from 0.5 MeV to about 1 MeV, there occur additional states that appear to represent vibrational excitations superposed on the $\mathsf{v}=1$ configurations (see footnotes to Tables 5-12 and 5-13). At higher energies, a much greater density of levels is observed, among which a number have been identified as configurations with three unpaired particles (see, for example, the high-spin isomers in Fig. 4-16, p. 106). The absence of the three-quasiparticle states in the lower part of the odd-A spectra can be attributed to the pair correlations, which imply that an energy of approximately 2Δ is involved in breaking a pair. (In a similar manner, the two-quasiparticle configurations in the even-even nuclei are not observed below excitation energies of about 1 MeV; see, for example, the spectrum of ^{168}Er shown in Fig. 4-7, p. 63, and ^{234}U shown in Fig. 6-39, p. 554.)

The systematic identification of the one-particle intrinsic states in the odd-A deformed nuclei makes it possible to establish the absolute energy scale for the one-particle spectrum and thus to test the possible occurrence of a velocity dependence in the one-particle potential (effective mass; see Vol. I, pp. 147 ff.). In this analysis, one must take into account the above-mentioned effect of the pair correlations, which implies a systematic reduction of the spacings between the low-lying quasiparticle states. When corrected for this effect, the absolute scale of the observed single-particle energy differences appears to be in agreement with the calculations for a static potential (see the discussion of empirical data on p. 265).

The description of the low-lying states in odd-A nuclei in terms of single quasiparticles is the counterpart, for the deformed nuclei, of the generalized one-particle model for spherical nuclei discussed in Sec. 2-4 (see Vol. I, pp. 224 ff.). The major interaction effect that tends to destroy the quasiparticle coupling scheme is the interaction with the quadrupole shape oscillations. Due to the large amplitude and low frequency of these oscillations in the spherical nuclei, the one-particle description has only qualitative significance. (A measure of the strength of particle-vibration coupling is provided by the parameter f_2 defined by Eq. (6-212); the quadrupole parameters in Figs. 6-28 and 6-29 imply values of f_2 that are typically of order unity.) However, in the spheroidal nuclei, the main part of this coupling is included in the average deformed field, and the amplitudes of the oscillation about the spheroidal shape are much smaller than for most spherical nuclei. (The zero-point amplitudes for the β and γ vibrations are typically of order $\Delta\beta \sim 0.05$ and $\beta_0 \Delta\gamma \sim 0.05$ (see the examples given by Eqs. (4-263) and (4-248)), while the corresponding quantity for quadrupole vibrations of spherical nuclei is typically a factor of two to five times greater). Thus, the one-quasiparticle description has a significantly greater quantitative validity in the deformed than in the spherical nuclei.

In the enumeration of the excited configurations, one must take into account that certain combinations of intrinsic excitations are spurious, in the sense that they are included in the rotational degrees of freedom. The rotational motion is mainly built out of intrinsic excitations with energies of order $\hbar\omega_0\delta$, corresponding to the states that are strongly excited by the Coriolis interaction (see the discussions in Chapter 4, pp. 82 ff. and Chapter 5, pp. 281 ff.); in odd-A nuclei, the spurious excitations are built predominantly out of configurations with three quasiparticles. In heavy nuclei, the spurious excitations involve configurations with excitation energies that are rather large compared with the spacing of the single-particle levels and do not play any role in the classification of the states in Tables 5-12 and 5-13. In the light nuclei, the removal of the spurious intrinsic excitations may be important even in the analysis of the lowest excited intrinsic states (see the example of the $A=19$ spectrum discussed on pp. 288 ff.).

5-3 MOMENTS AND TRANSITIONS

5-3a One-Particle Transfer

The reactions involving the transfer of a single nucleon provide a specific tool for determining the one-particle structure in nuclear configurations. The amplitudes are proportional to the one-particle parentage coefficients (see Appendix 3E) and, for deformed nuclei are obtained by a transformation to the intrinsic coordinate system, exploiting the tensorial properties of the operator $a^\dagger(j)$ (Satchler, 1955; see also Eq. (3E-8)). Thus, for the transitions from the ground-state band of an even-even nucleus ($\mathrm{v}=0$) to a one-quasiparticle state ($\mathrm{v}=1,\nu$), the parentage coefficient is given by (see Eqs. (4-92) and (6-599b))

$$\langle \mathrm{v}=1,\nu; K=\Omega, I_2 \| a^\dagger(j) \| \mathrm{v}=0; K=0, I_1\rangle$$

$$= \sqrt{2}\,(2I_1+1)^{1/2}\langle I_1 0 j\Omega | I_2\Omega\rangle\langle lj\Omega|\nu\rangle u(\nu) \qquad (5\text{-}42)$$

The factor $\langle lj\Omega|\nu\rangle$ gives the expansion of the one-particle intrinsic state ν in spherical components; in Eq. (5-42), the factor $u(\nu)$ is the probability amplitude for the orbit ν to be unoccupied in the ground state of the even-even nucleus. In the absence of pair correlations, the factor u would be unity for all orbits above the Fermi level (particle orbits) and zero for the hole orbits; the pair correlations imply that the amplitude u increases gradually from zero to unity with increasing energy of the orbit relative to the Fermi level. The transition extends over an energy range of order Δ; for a constant static pair field, the dependence of u on the single-particle energy is given by Eq. (6-601); see also Fig. 6-63a. For a pickup reaction leading from an even-even nucleus to a one-quasiparticle state in an odd-A nucleus, the cross section involves the matrix element of the annihilation operator $a(j)$, which is the same as that of $a^{\dagger}(j)$ except for the replacement of $u(\nu)$ by the amplitude $v(\nu)$ for finding the orbit ν occupied in the even-even ground state ($u^2+v^2=1$).

The observed magnitudes of the transfer cross sections populating the low-lying bands in odd-A nuclei provide a direct confirmation of the qualitative validity of the one-particle interpretation (see the example discussed on pp. 258 ff.). The relative intensities of the transitions to the different members of the rotational bands are governed by the coefficients $\langle lj\Omega|\nu\rangle$ and are found to provide a characteristic "fingerprint" for identifying the different intrinsic states and testing the calculated one-particle wave functions (Vergnes and Sheline, 1963; see also the review by Elbek and Tjøm, 1969; examples are shown in Fig. 5-10, p. 260, and in Table 5-11, p. 294).

The effect of the pair correlations in producing a gradual transition between particle-like and hole-like excitations is clearly demonstrated by the one-particle transfer studies (see Fig. 5-10). The analysis of the cross sections for transfer into the same orbit, for a sequence of nuclei with increasing nucleon number, yields values of Δ consistent with those obtained from the odd-even mass difference (see the examples illustrated in Fig. 5-11, p. 264).

For transitions involving unbound states, the one-particle parentage coefficients can also be determined from the partial widths in resonance scattering cross sections (see Appendix 3F). Opportunities for studies of this type are provided by the proton-induced reactions proceeding through the states with isospin $T=M_T+1$ (isobaric analog resonances; see the example referred to on p. 258). Such studies open the possibility of determining transfer amplitudes connecting excited states in both initial and final nuclei, which are not directly accessible in the one-particle transfer reactions.

5-3b Single-Particle Moments and Transitions

Matrix elements for one-quasiparticle states

From the measured moments and transition probabilities for γ and β transitions, one can determine the intrinsic matrix elements $\langle K_2|\mathscr{M}(\lambda\nu)|K_1\rangle$ that occur in the expressions given in Sec. 4-3. For one-quasiparticle states, the intrinsic matrix element can be expressed as a product of the corresponding one-particle matrix element and a factor depending on the pair correlations. (In the case of diagonal matrix elements (transitions within a band), there may be an additional collective contribution representing the expectation value of $\mathscr{M}(\lambda, \nu=0)$ in the $\mathrm{v}=0$ state.) For electromagnetic moments, one obtains (see Eq. (6-610a))

$$\langle \mathrm{v}=1, \Omega_2 | \mathscr{M}(\lambda\nu) | \mathrm{v}=1, \Omega_1 \rangle = \langle \Omega_2 | \mathscr{M}(\lambda\nu) | \Omega_1 \rangle (u_1 u_2 + c v_1 v_2) \quad (5\text{-}43)$$

The last factor in Eq. (5-43) may be derived by noting that a quasiparticle is a linear combination of a particle, with amplitude u, and a hole, with amplitude v (see Eq. (6-599a)). The phase factor $c(=\pm 1)$ describes the transformation of the moment $\mathscr{M}(\lambda)$ under particle-hole conjugation (see Eq. (3-14)); thus, $c=-1$ for $E\lambda$ moments, for which particles and holes have opposite sign, and $c=+1$ for $M\lambda$ moments, for which particles and holes have the same sign.

The reduction in the one-particle transition strength caused by the pair correlations is connected with the occurrence of transitions that, in the uncorrelated system, are forbidden by particle conservation. Thus, in the absence of pair correlations, the transition $\Omega_1 \rightarrow \Omega_2$ occurs as a one-particle or one-hole transition if both orbits are on the same side of the Fermi level; for orbits on opposite sides of the Fermi level, the transition occurs as a particle-hole excitation $0 \rightarrow (\bar{\Omega}_1)^{-1}\Omega_2$. In the pair-correlated system, both types of processes ($\Delta\mathrm{v}=0$ and $\Delta\mathrm{v}=2$) occur for any pair of single-particle orbits, with a summed transition strength equal to the one-particle value (see Eq. (6-610)).

If the moment involves a charge exchange, as for the β transitions, only the particle or hole component contributes, and one obtains (see Eq. (6-599b))

$$\langle \mathrm{v_n}=1, \Omega_\mathrm{n} | \mathscr{M}(\mu_\tau=+1, \lambda\nu) | \mathrm{v_p}=1, \Omega_\mathrm{p} \rangle = \langle \Omega_\mathrm{n} | \mathscr{M}(\mu_\tau=+1, \lambda\nu) | \Omega_\mathrm{p} \rangle u_\mathrm{n} u_\mathrm{p}$$

$$\langle \mathrm{v_n}=1, \Omega_\mathrm{n} | \mathscr{M}(\mu_\tau=-1, \lambda\nu) | \mathrm{v_p}=1, \Omega_\mathrm{p} \rangle = \langle \bar{\Omega}_\mathrm{p} | \mathscr{M}(\mu_\tau=-1, \lambda\nu) | \bar{\Omega}_\mathrm{n} \rangle v_\mathrm{n} v_\mathrm{p}$$

$$(5\text{-}44)$$

(In the above expressions, we have assumed that the pair correlations can be applied to the neutrons and protons separately; the neglect of the np pairing is justified when the correlation parameter Δ is small compared to the difference in the Fermi energies of neutrons and protons. This condition is well fulfilled in the heavy deformed nuclei, $A > 150$.)

For the $M\lambda$ operators, the pair correlations have no effect on the static moments and only weakly affect the matrix elements for low-energy transitions. For $E\lambda$ moments, the pair correlations are expected to strongly reduce the one-particle transition strength in the low-energy spectrum, but the available data do not provide quantitative tests of this feature. The $E1$ transitions are strongly hindered by other selection rules; see the text below and the example (^{159}Tb) discussed on p. 256. The $E2$ transitions are very sensitive to the coupling to the collective degrees of freedom; see, for example, the discussion of ^{235}U on pp. 275 ff. For the β transitions between states close to the Fermi level, the pair correlations imply a reduction in the transition rates by about a factor 4 (see the analysis of these moments on pp. 306 ff.).

Asymptotic selection rules and distribution of one-particle strength

The approximate validity of the asymptotic quantum numbers $Nn_3\Lambda\Sigma$ implies rather far-reaching selection rules governing the matrix elements for transitions between one-particle states (Alaga, 1955). The interpretation of the observed transition rates for β and γ transitions on the basis of these selection rules has played a major role in establishing the classification of the intrinsic states; examples are discussed on p. 256 ($E1$, ^{159}Tb), p. 111 ($E1$, ^{177}Hf), p. 154 ($E2$, ^{175}Lu), p. 262 ($M3$, ^{175}Yb), p. 257 (GT and first forbidden β decay, ^{159}Gd), p. 262, (GT, ^{175}Yb). It is found that the transitions violating the selection rules are systematically hindered by factors of at least 10 to 100, as compared with the unhindered transitions.

The validity of the asymptotic quantum numbers also implies a rather simple pattern for the spectrum of one-particle strength (response function) for the different multipole operators. Thus, the unhindered $E1$ transitions with $\Delta K(=\Delta\Omega)=0$ have $\Delta N=1$, $\Delta n_3=1$, $\Delta\Lambda=0$, $\Delta\Sigma=0$, and transition energies $\Delta\varepsilon$ that cluster around $\hbar\omega_3$, while the $\Delta K=1$ transitions have $\Delta N=1$, $\Delta n_3=0$, $\Delta\Lambda=\pm 1$, $\Delta\Sigma=0$, and $\Delta\varepsilon\approx\hbar\omega_\perp$ (see Table 5-3, p. 234); since these energies are of order 5 to 10 MeV in heavy nuclei, all the low-energy $E1$ transitions are expected to be hindered, in accordance with the empirical data.

In the case of the one-particle moment proportional to the spin operator $\mathbf{s}$ (which is responsible for the β transitions of Gamow-Teller (GT) type and usually gives the main term in $M1$ moments), the asymptotic selection rules for $\Delta K=0$ allow only diagonal matrix elements describing static moments and transitions within a band (as well as β transitions between isobaric analog bands). For these matrix elements, the validity of the quantum number Σ is especially tested by the pattern of g_K values as analyzed in Table 5-14, p. 303. The $\Delta K=1$ matrix elements of the spin operator have $\Delta N=\Delta n_3=\Delta\Lambda=0$, $\Delta\Sigma=\pm 1$, and connect the spin-orbit partners that are separated by energies ranging from 1 to 5 MeV (see Eq. (5-9)); there is not at present any quantitative evidence on the expected strong $M1$ transitions, but the corresponding unhindered GT transitions have been systematically studied (see Table 5-15, p. 307).

The spectrum of the one-particle transition strength for the different multipole operators is of significance not only for the classification of the quasiparticle intrinsic states; it also provides the starting point for the analysis of the collective vibrational modes in terms of the motion of the individual particles. The qualitative pattern derived on the basis of the asymptotic selection rules gives a simple picture of the manner in which the shell structure is expected to affect the collective modes in deformed nuclei (see, for example, the discussion of the systematics of the β- and γ-vibrational modes on pp. 551 ff. and pp. 549 ff., respectively).

Polarization effects

The quantitative analysis of the observed intrinsic matrix elements provides information on the effective single-particle moments. As for the configurations involving single particles outside closed shells, discussed in Chapter 3, the effective moments may be significantly influenced by the polarization of the core produced by the extra particle.

For the hindered transitions, the matrix elements may be sensitive to small uncertainties in the one-particle wave functions as well as to the effects of the Coriolis coupling; the most clear-cut evidence on polarization effects has therefore come from the analysis of unhindered matrix elements. The main body of data is provided by the $M1$-matrix elements within a rotational band and the GT-matrix elements between intrinsic states that are spin-orbit partners. The analysis of the data provides evidence for a renormalized spin g factor in $M1$-matrix elements with $\Delta K=0$ of about $0.7(g_s)_{\text{free}}$, for nuclei with $A>150$ (see the discussion in connection with Table 5-14, p. 303) and of an effective axial vector coupling constant for GT transitions with $\Delta K=1$ of $\approx 0.5(g_A)_{\text{free}}$ (see the discussion in connection with Table 5-15, p. 307). These reductions of the spin moments are of the same magnitude as those observed in the spherical nuclei (see Sec. 3-3b); the larger effect for the charge-exchange moment measured in GT transitions may be understood from the effect of the neutron excess, which implies a larger amplitude of fluctuations of the corresponding spin fields in the even-even core (see p. 307). Since, for large deformations, Σ becomes an approximate constant of the motion, one also expects a stronger renormalization effect for the $\Delta K=1$ spin magnetic moment than for the $\Delta K=0$ moment occurring in g_K (see p. 304; tentative evidence on the $\Delta K=1$

moments is provided by the parameter $(g_K - g_R)b$ in $K = 1/2$ bands). In the analysis of the $A = 25$ spectra, the isovector spin g factor and the GT moment are related by isobaric invariance, and the data are consistent with a common renormalization factor for both of these moments of about 0.8 (see pp. 290ff.).

5-3c Pair Transfer and α Decay

In the heavy element region ($A > 220$), the α-transition intensities provide a valuable tool for establishing the assignments and testing the detailed structure of the intrinsic configurations. The formation of an α particle involves a pair of neutrons and a pair of protons closely correlated in space, and is therefore sensitive to the neutron and proton pair correlations. These correlations imply the occurrence of collective α transitions, by which the α particle is formed from nucleons in paired orbits $(\nu\bar{\nu})$ without affecting the configurations of the quasiparticles. This special class of transitions comprises the ground-state transitions of even-even nuclei and the "favored" transitions in odd-A nuclei, in which the configuration of the quasiparticle does not change. These transitions are found to have reduced intensities two to three orders of magnitude greater than the "unfavored" transitions involving a change of configuration of the quasiparticle. Examples illustrating the dominance of the favored transitions in the α decay of odd-A nuclei are discussed in connection with Fig. 4-19, p. 112, and Fig. 5-13, p. 268. The comparison with the even-even ground-state transitions shows that the favored odd-A transitions are impeded by about a factor of 2 by the presence of the quasiparticle. The major part of this effect appears to arise from the reduction in the pair correlation implied by the presence of an unpaired particle (see the estimate of this effect on p. 272).

The intensities of the unfavored α transitions associated with a change in the quasiparticle configuration mainly reflect the spatial overlap of the two orbits involved (the structure of this matrix element is given by Eq. (5-58); see also Eqs. (5-63) to (5-66)). It is found that quite large values of the angular momentum may be involved ($L \sim 5$–8), partly because the centrifugal barrier is rather ineffective for the α particles, and partly due to selection rules in the α-formation amplitude, which often strongly reduce the intensities of the low L components. As a result, many members of a given rotational band may be populated with comparable reduced intensities (see Fig. 5-13). The calculations based on the one-particle orbits in a deformed potential have been able to reproduce the qualitative features in the observed intensity patterns and have provided significant evidence regarding the one-particle configuration assignments (Poggenburg *et al.*, 1969; see also Fig. 5-13).

The two-nucleon transfer reactions provide a very promising tool for the study of matrix elements similar to those involved in the α-decay process. The available data confirm the expected collective enhancement of the transitions without change in quasiparticle configuration, and potentially may provide detailed information on intrinsic configurations of the type discussed above (see, for example, the review by Broglia *et al.*, 1973).

5-3d Coupling of Particles to Rotational Motion

The rotational motion in odd-A nuclei is influenced by the presence of the last odd particle. This coupling effect provides further evidence on the one-particle motion in deformed nuclei; moreover, it constitutes a link between the motion of the individual particles and the collective rotation of the nucleus as a whole.

In the rotating frame, the individual particles are subject to the Coriolis force, which gives a particle-rotation coupling of the form (see Eq. (4-107) as well as the discussion of the particle-rotor model in Sec. 4A-2b)

$$H_c = -A_0(j_+I_- + j_-I_+) \tag{5-45}$$

$$A_0 \equiv \frac{\hbar^2}{2\mathscr{I}_0}$$

where $\mathscr{I}_0$ is the moment of inertia of the even-even core. The value of $\mathscr{I}_0$ is expected to be similar to that observed in the ground-state bands of neighboring even-even nuclei, but the presence of the last odd particle may to some extent modify the rotational motion of the even-even core (see the discussion on p. 311).

Additional couplings between particle and rotational motion arise if the potential acting on the individual particles is modified by the rotational motion. Terms linear in the rotational frequency occur in the presence of velocity-dependent potentials (effective mass terms and pair fields; see the discussion on pp. 79ff. and pp. 278ff.). The role of these velocity-dependent contributions to the nuclear potential remains to be clarified, but the analysis of the many different phenomena associated with the particle-rotation coupling provides a primary source of evidence on these effects.

First-order Coriolis effects. Decoupling parameter and $\Delta K = 1$ transitions

The expectation value of the Coriolis coupling gives rise to the signature-dependent term in the rotational energies of $K=1/2$ bands (see Eq. 4-61)). Neglecting the difference between $\mathscr{I}_0$ and the moment of inertia $\mathscr{I}$ of the odd-A nucleus, and evaluating the coupling (5-45) in a one-quasiparticle state, we obtain the decoupling parameter

$$
\begin{aligned}
a &= -\langle \mathrm{v}=1,\nu | j_+ | \mathrm{v}=1,\bar{\nu}\rangle \\
&= -\langle \nu | j_+ | \bar{\nu}\rangle \\
&= \sum_{Nlj} (-1)^{j-1/2}(j+\tfrac{1}{2})\langle Nlj,\Omega=\tfrac{1}{2}|\nu\rangle^2 \\
&= (-1)^N \sum_{Nl} \Big(\langle Nl,\Lambda=0,\Sigma=\tfrac{1}{2}|\nu\rangle^2 \\
&\qquad + 2(l(l+1))^{1/2}\langle Nl,\Lambda=0,\Sigma=\tfrac{1}{2}|\nu\rangle\langle Nl,\Lambda=1,\Sigma=-\tfrac{1}{2}|\nu\rangle\Big)
\end{aligned}
\tag{5-46}
$$

where ν labels the one-particle state with $\Omega=1/2$. The diagonal matrix element of j_+, as for the $M1$ moment, is not affected by the pair correlations (see Eq. (5-43)).

Theoretical estimates of the decoupling parameter based on Eq. (5-46) are compared with empirical data in connection with the analysis of the spectra of ^{169}Tm (Fig. 4-15, p. 103), and ^{159}Tb (Fig. 5-7, p. 254). A survey of decoupling parameters is given in Tables 5-16, p. 309 ($150 < A < 180$) and Table 5-8, p. 286 ($19 \leqslant A \leqslant 25$). The decoupling parameters vary considerably from one orbit to another, and provide a valuable identification mark for the different $K=1/2$ bands. The theoretical estimates describe the observed values rather well.

The nondiagonal effects of the Coriolis coupling acting between bands with $\Delta K = 1$ imply corrections to the matrix elements evaluated in the static limit. A favorable opportunity for studying these effects is provided by the $E2$-transition probabilities between bands with $\Delta K = 1$, which receive their main contribution from the mixing of the bands produced by the particle-rotation coupling. Examples are provided by the analysis of the spectra of ^{175}Lu, pp. 154ff., and of ^{235}U, pp. 275ff. The latter case represents an especially significant test, since the orbits involved are connected with large, unhindered Coriolis matrix elements; it is found that expression (5-45) evaluated for the one-quasiparticle states implies an overestimate of the coupling matrix element by about a factor of 2. (Evidence for similar overestimates of the Coriolis matrix elements has been obtained from the analysis of other transition matrix elements involving states coupled by large Coriolis matrix elements; see, for example, the references quoted on p. 318.)

Second-order contribution to rotational energy

Acting in second order, the Coriolis coupling gives a contribution to the energy proportional to $I(I+1)$, which can be viewed as a renormalization of the rotational energy parameter A,

$$\delta A(\nu) = -(A_0)^2 \sum_{\nu'} \langle \nu' | j_{\pm} | \nu \rangle^2 \left(\frac{(uu' + vv')^2}{E(\nu') - E(\nu)} - \frac{(uv' - vu')^2}{E(\nu) + E(\nu')} \right) \tag{5-47}$$

The first term represents the effect of the Coriolis coupling between the one-quasiparticle states $\mathrm{v}=1,\nu$ and $\mathrm{v}=1,\nu'$; the factor involving the pair correlation amplitudes u and v is that given by Eq. (5-43). The second term represents the effect of the odd particle in preventing some of the excitations that contribute to the rotational energy of the even-even system; for these transitions with $\Delta\mathrm{v}=2$, the pair correlation factor can be obtained from Eq. (6-610b). (The renormalization of the moment of inertia can be viewed as a self-energy of the quasiparticle resulting from the particle-rotation coupling. The two terms in Eq. (5-47) correspond to the two diagrams in Fig. 6-11b for the self-energy effects arising from the particle-vibration coupling.) For an orbit with energy close to the Fermi level ($u(\nu)=v(\nu)=(2)^{-1/2}$; $E(\nu)\approx\Delta$; see Eqs. (6-601) and (6-602)), the expression (5-47) gives

$$\delta A(\nu) = -(A_0)^2 2\Delta \sum_{\nu'} \frac{\langle \nu' | j_{\pm} | \nu \rangle^2}{(\epsilon(\nu) - \epsilon(\nu'))^2} \tag{5-48}$$

In the limit $\Delta \to 0$, the renormalization (5-48) vanishes, corresponding to the fact that the moment of inertia of the odd-A nucleus is the average of the moment for the neighboring even-even nuclei.

The observed moments of inertia of odd-A nuclei are systematically larger than those of the ground-state bands of neighboring even-even nuclei (see Fig. 4-12, p. 74, and Table 5-17, p. 312). In the interpretation of this difference, one must take into account that the presence of the odd particle modifies the properties of the even-even core, in particular through the blocking effect on the pair correlations; the discussion of this effect on p. 311 indicates that it may increase the moment of inertia of low-lying one-quasiparticle states by about 15%. While this effect is expected to be approximately the same for all low-lying single-quasiparticle configurations, the contribution (5-48) depends strongly on the orbit involved and becomes especially large for the orbits with large j and small Ω (such as the $N=5$ proton orbits and the $N=6$ neutron orbits for $150<A<180$). Indeed, these orbits are found to have systematically larger moments of inertia than neighboring configurations (see Table 5-17); however, the quantitative estimate encounters the same difficulties as the analysis of the first-order effects discussed above, since the increment in the moment of inertia is overestimated by about a factor of 2 (see the example, ^{235}U, discussed on p. 279). Even larger discrepancies are encountered for excited particle-like and hole-like configurations that are strongly coupled by Coriolis matrix elements (see the discussion on p. 314).

Higher-order terms

The many high-order terms observed in the rotational energies provide further evidence on the Coriolis coupling effects. However, at the present stage, the interpretation of these higher-order effects is of a rather qualitative nature (see, for example, the discussion of the parameter A_7 for the ground-state rotational band and the large increment in the B term of the excited configuration [633 5/2], in the spectrum of ^{235}U, pp. 280ff.).

Summary of evidence on particle-rotation coupling

The empirical determinations of the Coriolis matrix elements, as discussed above, reveal significant deviations from the theoretical estimates of these matrix elements obtained from quasiparticles in a constant static pair field. (This effect was first recognized by Stephens, 1960.) The origin of these discrepancies is not at present understood, but the following points may be relevant in assessing their implications.

1. The effect is well established in a number of cases involving large Coriolis matrix elements that are rather insensitive to the deformation parameters and other detailed features of the one-particle potential.

2. The spin contribution to the Coriolis matrix elements may be expected to be reduced by polarization effects of a type similar to those encountered in the $M1$- and GT-matrix elements. (Tentative evidence for the occurrence of such an effect is obtained from the analysis of the decoupling parameters, p. 310.) However, the main part of the Coriolis matrix elements is associated with the orbital contribution.

3. The appearance of velocity-dependent terms in the potential (effective mass) might suggest a corresponding modification in the orbital part

of the Coriolis coupling; however, this effect is expected to be approximately canceled by the interactions that are required by Galilean invariance (see the discussion on p. 80). Moreover, the empirical Coriolis matrix elements are not compatible with a uniform reduction in all low-energy matrix elements (compare the rather small renormalization of the decoupling parameters with the larger corrections demanded by some of the nondiagonal matrix elements).

4. Evidence concerning Coriolis matrix elements with $\Delta v = 2$ comes from the moments of inertia of the ground-state bands in even-even nuclei. The theoretical estimates based on the quasiparticle description appear to account approximately for the empirical data (see the discussion on pp. 82 ff.). A valuable extension of this evidence would be provided by the exploration of the expected pattern of $K\pi = 1+$ excitations in even-even nuclei (see the discussion on p. 282).

5. The largest overestimates of Coriolis matrix elements in the odd-A nuclei are encountered in the coupling between particle-like and hole-like quasiparticle states, which would vanish in the absence of pair correlations (see p. 314). Little or no discrepancy is observed in diagonal matrix elements (decoupling parameters) and between excited states on the same side of the Fermi surface (see the discussion of the A_7 term in ^{235}U on p. 281). These features are suggestive of the additional coupling effects that may be expected from the dependence of the pair field on the rotational motion (see the discussion on p. 278).

▼

ILLUSTRATIVE EXAMPLES TO SECTION 5-3

The Spectrum of ^{159}Tb (Figs. 5-7 and 5-8)

Intrinsic configurations

The ground-state deformation in ^{159}Tb is prolate with $\delta \approx 0.3$ (see Fig. 4-25) and thus the level spectrum for protons in Fig. 5-2 suggests that the ground state (for $Z=65$) should have the configuration [411 3/2] and that the neighboring orbits [413 5/2], [532 5/2], and [523 7/2] should provide low-lying excited configurations. The lowest observed bands in ^{159}Tb have just the $K\pi$ values corresponding to the first three of these orbits (see Fig. 5-7). The

▲ [523 7/2] configuration has not yet been observed, but would not have been expected in the reactions studied so far. (In the similar spectrum of ^{161}Tb, the [523 7/2] configuration has been identified at an excitation energy of 417 keV; see Table 5-12, p. 296.)

Above 500 keV, the Coulomb excitation reaction has shown the existence of three even-parity bands that are excited from the ground-state band with $B(E2)$ values of order $B_W(E2)$. The band at 972 keV has been tentatively classified as the single-particle configuration [411 1/2], which is expected in this region of the spectrum (see Table 5-12); this assignment is supported by the observed value of the decoupling parameter, which is very similar to that found for the [411 1/2] orbit in neighboring nuclei and which agrees rather well with the theoretical estimate (5-46) (see Table 5-16).

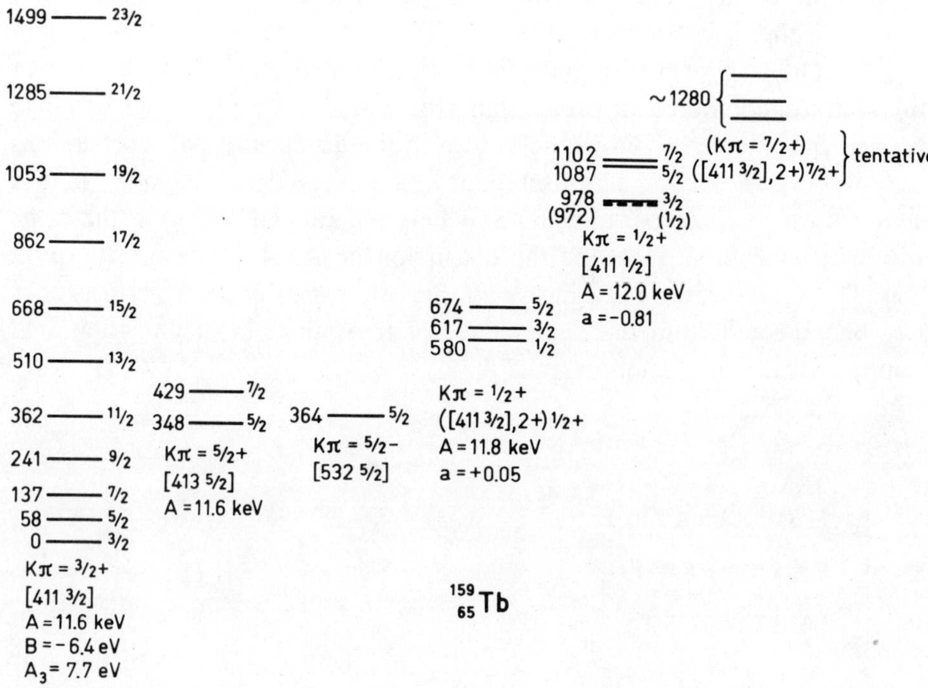

1499 — 23/2
1285 — 21/2
1053 — 19/2
862 — 17/2
668 — 15/2
510 — 13/2
362 — 11/2
241 — 9/2
137 — 7/2
58 — 5/2
0 — 3/2
$K\pi = 3/2+$
[411 3/2]
A = 11.6 keV
B = −6.4 eV
A_3 = 7.7 eV

429 — 7/2
348 — 5/2
$K\pi = 5/2+$
[413 5/2]
A = 11.6 keV

364 — 5/2
$K\pi = 5/2-$
[532 5/2]

674 — 5/2
617 — 3/2
580 — 1/2
$K\pi = 1/2+$
([411 3/2], 2+) 1/2+
A = 11.8 keV
a = +0.05

978 — 3/2
(972) (1/2)
$K\pi = 1/2+$
[411 1/2]
A = 12.0 keV
a = −0.81

1102 — 7/2
1087 — 5/2
($K\pi = 7/2+$)
([411 3/2], 2+) 7/2+
tentative

~1280

$^{159}_{65}$Tb

Figure 5-7 Spectrum of ^{159}Tb. The observed levels of ^{159}Tb are grouped into rotational bands; for each band, the classification of the intrinsic configuration is given together with the inertial parameters that can be deduced from the observed rotational energies. The spectrum is based on the experimental data summarized in *Table of Isotopes* by Lederer *et al.* (1967) and the additional evidence on the high-lying states obtained from the Coulomb excitation studies by R. M. Diamond, B. Elbek, and F. S. Stephens, *Nuclear Phys.* **43**, 560, 1963, and J. S. Greenberg, D. A. Bromley, G. C. Seaman, and E. V. Bishop, in *Proceedings of the Third Conference on Reactions Between Complex Nuclei*. p. 295, eds. A. Ghiorso *et al.*, University of California Press, Berkeley 1963.

▼ The two other strongly populated bands ($K\pi = 1/2+$ at 580 keV and $K\pi = 7/2+$ at 1280 keV) have been tentatively interpreted in terms of $\nu\pi = 2+$ collective excitations (γ vibration) superposed on the one-particle configuration [411 3/2] of the ground state. This interpretation accounts for the rather large $B(E2)$ values, which are comparable to those observed for γ-vibrational excitations in neighboring even-even nuclei, and for the small decoupling parameter of the $K\pi = 1/2+$ band (see Sec. 6-3b). The expected coupling between the single-particle and vibrational motion (see Sec. 6-5) suggests that there may be significant mixing between the two $K\pi = 1/2+$ bands, but the available evidence is inconclusive on this point (Bès and Cho, 1966; Soloviev and Vogel, 1967).

M1 moments in ground-state band

The $M1$-matrix elements measured in the ground-state rotational band (Boehm *et al.*, 1966), together with the ground-state magnetic moment (see reference in caption to Table 5-14, p. 303), yield

$$g_K = 1.83 \pm 0.04$$
$$g_R = 0.42 \pm 0.06 \qquad (5\text{-}49)$$

The single-particle wave function for the orbit [411 3/2] implies (see Eq. (5-86) and Table 5-2)

$$g_K = 0.73 + 0.27(g_s)_{\text{eff}} \qquad (5\text{-}50)$$

As discussed in connection with Table 5-14, the systematics in this region of elements implies $(g_s)_{\text{eff}} \approx 0.7(g_s)_{\text{free}} \approx 4.0$, which yields a value of g_K in good agreement with the empirical value (5-49).

The value of g_R is appreciably larger than the rotational g factor observed in the ground-state bands of neighboring even-even nuclei (see Fig. 4-6, p. 55). A renormalization of g_R results from the first-order Coriolis coupling to $K\pi = 1/2+$ and $5/2+$ bands (see Eq. (4A-35)). The coupling to the low-lying one-particle excitations (at 348 keV and 972 keV) is not expected to give appreciable contributions, since the Coriolis matrix elements are estimated to be rather small. The contribution of the more distant orbits may be estimated by relating the increment in g_K to that in the moment of inertia, $\mathscr{I}$. If we assume that the moment of inertia is primarily associated with orbital motion, the g_R factor may be expressed in terms of the ratio of the moment $\mathscr{I}_{\text{p}}$ associated with the motion of the protons, to the total moment $\mathscr{I}$,

▲
$$g_R \approx \frac{\mathscr{I}_{\text{p}}}{\mathscr{I}} \qquad (5\text{-}51)$$

▼ The observed moments of inertia in odd-A nuclei are systematically larger than in neighboring even-even nuclei, and if we ascribe this increment in an odd-proton nucleus entirely to the orbital motion of the protons (resulting either from the Coriolis coupling or from the decrease in the proton pair correlations associated with the presence of the odd proton, see p. 310), and similarly for an odd-neutron nucleus, we obtain from Eq. (5-51)

$$\delta g_R \approx \begin{cases} \dfrac{\delta \mathscr{I}}{\mathscr{I}}(1-g_R) & \text{odd } Z \\ -\dfrac{\delta \mathscr{I}}{\mathscr{I}} g_R & \text{odd } N \end{cases} \tag{5-52}$$

For the ground-state band in ^{159}Tb, one finds $\delta\mathscr{I}/\mathscr{I} \approx 0.2$ (see Table 5-17) and, since $g_R \approx 0.3$, we obtain $\delta g_R \approx 0.15$, which is of the order of the observed effect.

Rotational energies of the ground-state band

Multiple Coulomb excitation with heavy ions has made it possible to populate high-spin members of the ground-state rotational band in ^{159}Tb, and the energies of these states show a small signature-dependent component, which can be described by the cubic term in Eq. (4-62). In Fig. 5-8, the energies are plotted in such a manner that the expression (4-62), with the inclusion of the three leading-order terms with the coefficients A, B, and A_3, yields two straight lines with slopes $2B \pm A_3$, corresponding to the two values of the signature, $\sigma(I)=(-1)^{I+3/2}=\pm 1$. (The intersection of the lines has the ordinate $A+\frac{1}{2}B$.)

The A_3 term in the energy expansion can be expressed as a third-order perturbation effect of the Coriolis coupling (see Eq. (4A-16) for the particle-rotor model). The evaluation of A_3 is similar to the estimate of the increment to the moment of inertia in odd-A nuclei (see p. 251), but may involve a considerable degree of cancellation between the contributions of different $K=1/2$ intermediate states. It has not yet been possible to give a quantitative interpretation of the observed A_3 terms.

Hindered E1 transitions

The half life of the 364 keV state in ^{159}Tb is $\tau_{1/2}=1.6\times 10^{-10}$ sec, which implies a hindrance by about 10^5 as compared with the single-particle estimate for $E1$-decay rates. A strong hindrance is expected, since the single-particle assignments imply $\Delta n_3=2$, in violation of the asymptotic selection rules given in Table 5-3. An estimate of the $E1$-transition probability on the basis of the wave functions in Table 5-2 yields a value close to the observed one; however, the quantitative agreement must be regarded as fortuitous, since one expects a reduction by about an order of magnitude as a ▲ result of the pair correlations (see Eq. (5-43)) and a further reduction of

▼ similar magnitude due to the $E1$-polarization effect (see Eq. (6-329)). Moreover, in a quantitative estimate, one must take into account the effects of the Coriolis coupling, which may be especially important, since the I-independent matrix element is strongly hindered while, in first order, the Coriolis coupling admixes orbits that contribute with large $E1$-matrix elements ([521 3/2] into the initial state and [422 5/2] into the final state). In fact, the relative intensities of the $E1$ transitions to the 3/2, 5/2, and 7/2 members of the ground-state band deviate in a major way from the I-independent intensity rules, revealing the importance of the Coriolis coupling effect (see the analysis of the similar situation in the decay scheme of ^{177}Hf, Fig. 4-18, p. 110). ▲

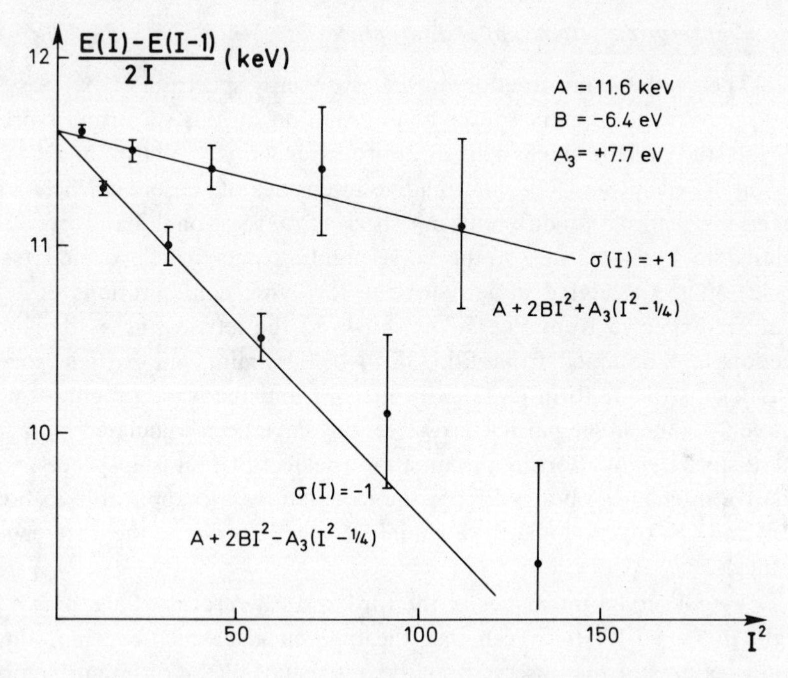

Figure 5-8 Rotational energies in the ground-state band of ^{159}Tb. The energies given in Fig. 5-7 are analyzed to exhibit the contributions of the three leading-order terms (A, A_3, B) in the rotational expansion appropriate to a $K=3/2$ band.

▼ *Selection rules in β decay*

The ground state of $^{159}_{64}$Gd has the configuration [521 3/2] (see Table 5-13), and the β decay strongly favors the [411 3/2] configuration in $^{159}_{65}$Tb ($\log ft = 6.7$) as compared with the [413 5/2] state ($\log ft = 8.2$). This difference is in agreement with the asymptotic selection rules for first forbidden β transitions, for which the transition moment is linear in the coordinates or ▲

▼ momenta (see Eq. (3D-43)). Thus, the transition to the [413 5/2] state, which involves $\Delta\Lambda=2$, is hindered, while the transition to the [411 3/2] state ($\Delta N=\Delta n_3=1$, $\Delta\Lambda=\Delta\Sigma=0$) is unhindered.

The decay of ^{159}Gd to the [532 5/2] band is an allowed GT transition ($\Delta I=1$, no change of parity), and the observed $\log ft=6.6$ implies a hindrance by a factor of about 10^2, as compared with the allowed unhindered transitions (see Table 5-15, p. 307). This hindrance can be understood in terms of the intrinsic configurations, which involve $\Delta n_3=1$, $\Delta\Lambda=1$, in violation of the asymptotic selection rules for GT transitions (see Table 5-3).

Spectrum of ^{175}Yb and evidence obtained from one-particle transfer reactions (Figs. 5-9 to 5-11)

"Fingerprints" from (dp) and (dt) processes

The available information on the low-energy spectrum of ^{175}Yb is shown in Fig. 5-9. The primary source of information on this spectrum is provided by the study of the single-neutron transfer reactions ^{174}Yb(dp) and ^{176}Yb(dt). As discussed in Sec. 5-3a, the relative intensities of the one-particle transfer reactions leading to different members of a rotational band are directly related to the structure of the corresponding one-particle orbital (see Eq. (5-42)). The calculated pattern for the low-lying configurations of $^{175}_{70}\mathrm{Yb}_{105}$ that are expected from Fig. 5-3 is shown to the left in Fig. 5-10. The cross sections are obtained from Eq. (3E-9) by assuming one-particle parentage coefficients of the form (5-42) with $u(\nu)=1$ and the wave functions given in Table 5-2. The single-particle cross sections have been calculated by means of the distorted-wave Born approximation (neglect of multistep processes). The relative intensities obtained from the calculations are applicable to both the (dp) and (dt) reaction (except for minor corrections due to the difference in Q values).

The observed intensities in the (dp) and (dt) processes are drawn to the right in Fig. 5-10. It is seen that the theoretical calculations reproduce the main features of the observed relative intensities of the different members in each band in a striking manner (note the logarithmic scale). This agreement supports the assignment of the intrinsic quantum numbers and provides a qualitative test of the wave functions given in Table 5-2. In a quantitative analysis of the one-particle transfer intensities, one must take into account the effect of the Coriolis coupling (see, for example, Erskine, 1965, and Casten *et al.*, 1972) as well as the possibility of multistep processes in which the transfer is accompanied by inelastic scattering within the rotational bands of both the target and final nucleus (see, for example, Ascuitto *et al.*, 1974).

The one-particle parentage coefficients studied in the (dp) process have also been measured in resonance proton scattering on $^{174}_{70}$Yb proceeding ▲ through the $M_T=33/2$, $T=35/2$ states in $^{175}_{71}$Lu that are isobaric analogs of

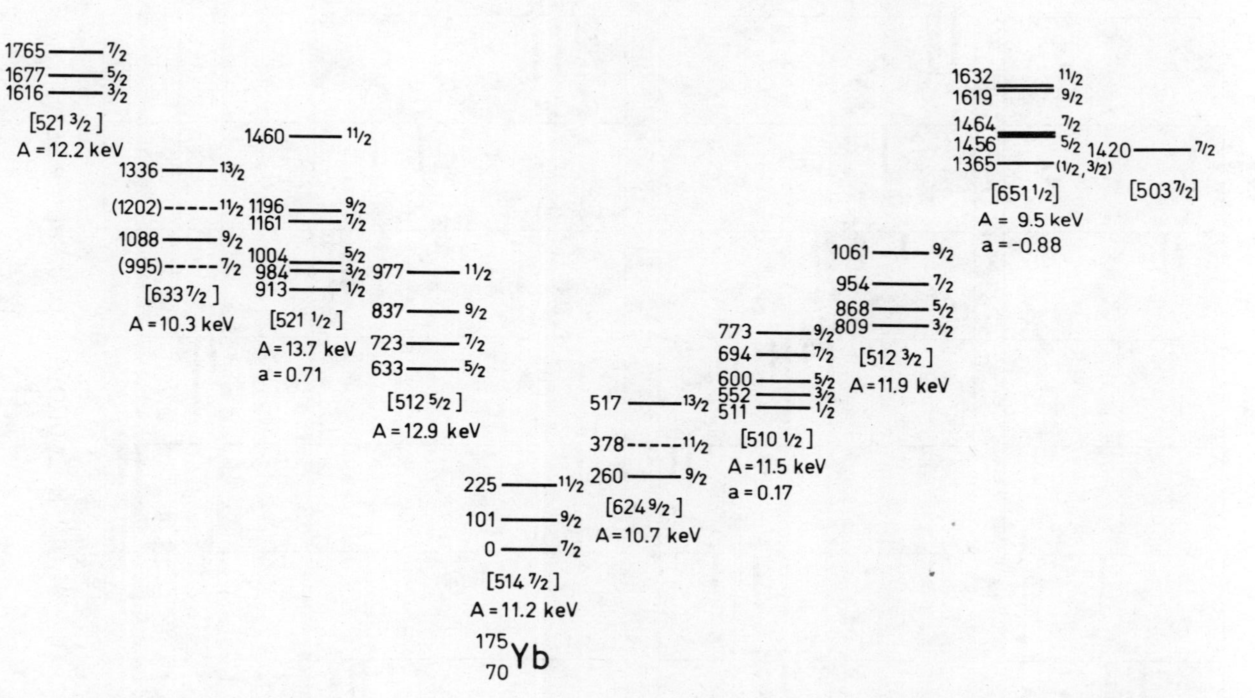

Figure 5-9 Spectrum of ^{175}Yb. The figure gives the known low-energy levels of ^{175}Yb. The levels classified as particle states are drawn to the right of the ground-state band, while hole states are drawn to the left. The experimental data are taken from (dp) and (dt) reaction studies by D. G. Burke, B. Zeidman, B. Elbek, B. Herskind, and M. Olesen, *Mat. Fys. Medd. Dan. Vid. Selsk.* **35**, no. 2 (1966) except for the evidence on the [651 1/2] and [503 7/2] bands, which is taken from the work of S. Whineray, F. S. Dietrich, and R. G. Stokstad, *Nuclear Phys.* **A157**, 529 (1970). In a few rotational bands, some of the low-lying members were not observed, probably because of the low expected intensity of the corresponding particle group (see Fig. 5-10), and in these cases the expected position of the level is drawn as a dotted line in the diagram.

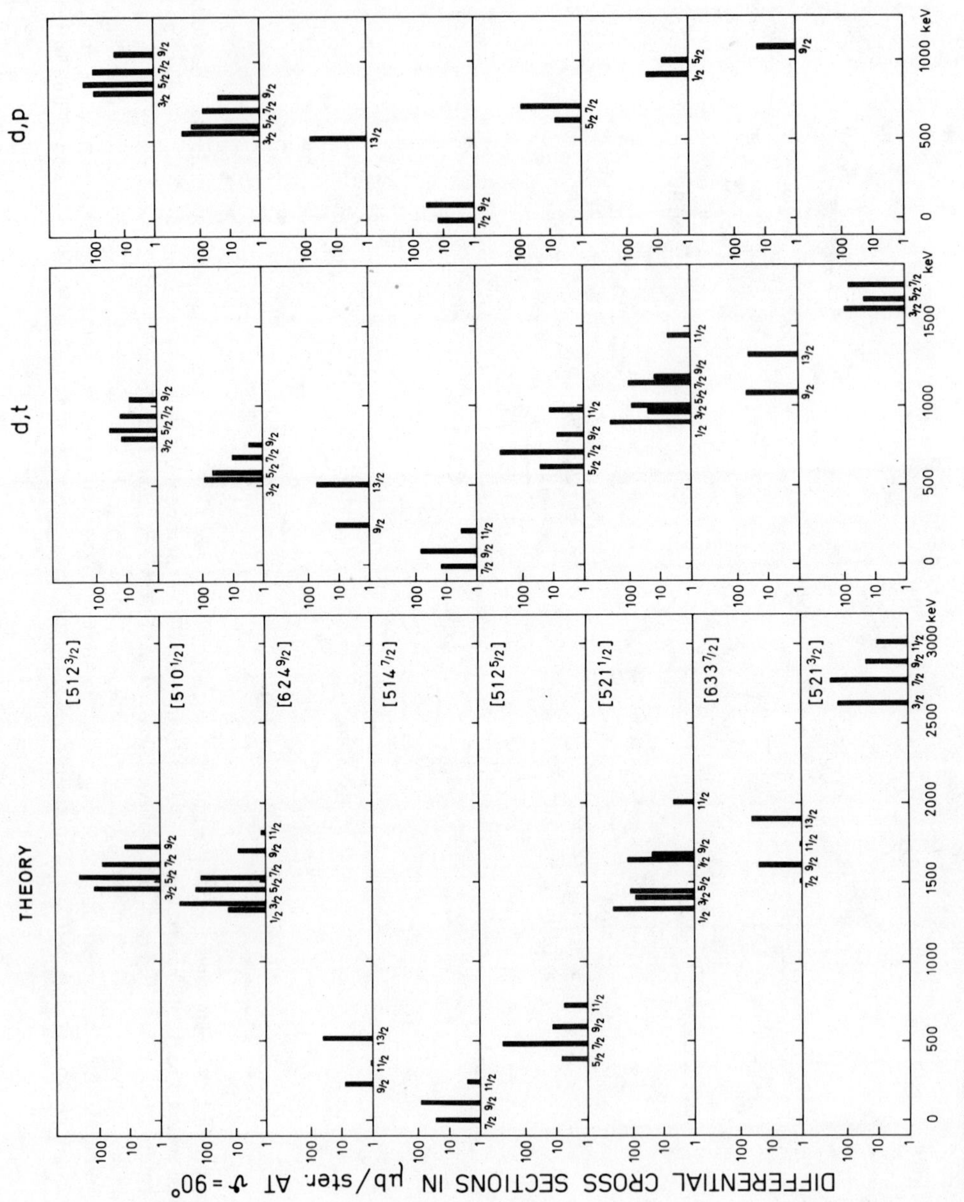
THEORY
d,t
d,p
[512 3/2]
[510 1/2]
[624 9/2]
[514 7/2]
[512 5/2]
[521 1/2]
[633 7/2]
[521 3/2]
DIFFERENTIAL CROSS SECTIONS IN μb/ster. AT ϑ = 90°
keV

Figure 5-10

▼ the low-energy states in ^{175}Yb (Whineray *et al.*, 1970; Foissel *et al.*, 1972; for a similar analysis in a spherical nucleus, see Fig. 1-9 and Table 1-2 in Vol. I, pp. 47 and 48). The evidence from the proton resonances lends added support to the conclusions drawn from the one-particle transfer data. The analog resonances have a rather large width ($\Gamma \sim 100$ keV), which makes it difficult to separate the contributions from the individual levels in a spectrum as complicated as ^{175}Yb. However, the intrinsic simplicity of the resonance scattering process offers advantages. Thus, for example, the first identification of the positive parity [651 1/2] band in ^{175}Yb was based on the observation of strong interference with Coulomb scattering at 90°. The precise determination of the energies of the isobaric analog states also provides evidence concerning the small shifts in the Coulomb energies that are specific to the orbits of the odd particle. (See, in this connection, the discussion of Coulomb energies in the $A=25$ isobaric doublet, p. 294.)

Additional evidence on classification of intrinsic states

Further support for the classification of the intrinsic states in ^{175}Yb can be obtained from the observed rotational energies and transition probabilities. The decoupling parameters a, deduced from the energy levels of the $K=1/2$ bands, have values that are in approximate agreement with those obtained from the wave functions in Table 5-2,

$$a_{\text{calc}} = \begin{cases} 0.9 & [521\ 1/2] \\ -0.2 & [510\ 1/2] \end{cases} \tag{5-53}$$

The observed positive value ($a = +0.17$) for the [510 1/2] band may indicate a reduction of the spin contribution, as a result of polarization effects (see p. 310). For the [651 1/2] orbit, the value of a is rather sensitive to detailed features of the potential. Thus, for the parameters in Table 5-1, one obtains $a = +0.9$ for $\delta = 0.3$ and $a = -0.5$ for $\delta = 0.4$. This rapid variation with δ reflects the strong interaction of the two $\Omega = 1/2$ orbits that in the spherical ▲ limit become $g_{9/2}$ and $i_{11/2}$, respectively. (A negative value of a could also be

---

Figure 5-10 Fingerprints from intensities of (dp) and (dt) reactions leading to $^{175}_{70}$Yb. The observed cross sections of the ^{174}Yb (dp) and ^{176}Yb(dt) reactions ($E_d = 12$ MeV) are taken from Burke *et al.* (*loc.cit.*, Fig. 5-9). The figure gives relative cross sections normalized to the observed intensity of the ground-state group ([514 7/2], $I=7/2$) which, for the (dp) as well as the (dt) reactions, is taken to have half the theoretical value shown in the left-hand part of the figure. The theoretical cross sections are taken from the same reference and have been obtained from a distorted-wave calculation for a (dt) reaction with $Q=0$ at a scattering angle of 90° and assuming the parentage coefficient given in Sec. 5-3a, with the pairing amplitude $v(\nu)=1$. (For comparison, the ground-state transition in ^{176}Yb(dt)^{175}Yb has $Q=-0.62$ MeV.) The patterns of relative intensities obtained from the distorted-wave calculations are very similar for (dp) and (dt) reactions. We wish to thank K. Nybö for help in the preparation of the figure.

▼ obtained for $\delta \approx 0.3$, by a minor modification of the parameters v_{ll} and v_{ls}.) The dominance of the $i_{11/2}$ component over the $g_{9/2}$ component in the observed [651 1/2] orbit, which is primarily responsible for the negative a value, is directly confirmed by the one-particle transfer intensities, which yield $\langle i_{11/2}, \Omega=1/2|[651\ 1/2]\rangle^2 \approx 0.3$ and $\langle g_{9/2}, \Omega=1/2|[651\ 1/2]\rangle^2 \approx 0.1$ (Whineray *et al.*, 1970).

Another conspicuous feature of the rotational energies is the fact that the parameter A is somewhat smaller for the three bands with $N=6$ than for the other bands observed in ^{175}Yb. The increased moment of inertia of these states can be attributed to the large orbital angular momentum carried by the last neutron, and is observed as a systematic feature of such states (see the evidence collected in Table 5-17, p. 312).

The ground state of $^{175}_{70}$Yb is found to decay with an allowed unhindered β transition to an $I\pi=9/2-$state at 396 keV in $^{175}_{71}$Lu ($\log ft=4.7$; see Table 5-15, p. 307); the $9/2-$ state in ^{175}Lu is assigned [514 9/2] (see Table 5-12, p. 296), and thus the observed rate supports the [514 7/2] assignment for the ^{175}Yb ground state. In fact, the rarity of allowed unhindered β processes, which is connected with the very restrictive asymptotic selection rules ($\Delta N = \Delta n_3 = \Delta\Lambda = 0$, $\Delta\Sigma = 0, 1$), makes this assignment very compelling. (The absolute values of the observed matrix elements for the allowed unhindered β transitions are discussed in connection with Table 5-15, p. 307.)

The observed half life of the 511 keV level ($\tau_{1/2}=67$ msec; Lederer *et al.*, 1967) implies $B(M3; 1/2- \rightarrow 7/2-) = 1.8\times 10^2 (e\hbar/2Mc)^2 \text{fm}^4 = 0.11\ B_W(M3)$, in terms of the Weisskopf unit given by Eq. (3C-38). The hindrance can be attributed to the asymptotic selection rules, since the assigned configurations ([510 1/2]→[514 7/2]) imply $\Delta\Lambda=4$. The wave functions given in Table 5-2 lead to a value of $B(M3)$, approximately equal to that observed. However, one expects the matrix element to be significantly reduced by polarization effects similar to those observed in $M1$ and Gamow-Teller transitions, although there is at present almost no empirical evidence on these effects in $M3$ moments.

Sequence of intrinsic states

In Fig. 5-10, the sequence of orbits is that obtained from Fig. 5-3 (for $\delta \approx 0.3$). Thus, the configurations that are expected to appear as particle states are placed above the [514 7/2] ground-state configuration, while the levels below this configuration are expected to appear as holes.

The observed sequence of intrinsic states is seen to be consistent with the single-particle spectrum in Fig. 5-3 for a deformation of $\delta \approx 0.3$, except for the position of the [651 1/2] state. This state is observed at an energy somewhat higher than would be expected on the basis of Fig. 5-3, but the predicted position of this level is sensitive to the separation between the major shells,
▲ which appears to be somewhat underestimated in the potential employed in

▼ Fig. 5-3. (See, for comparison, the spectra in Fig. 3-3, Vol. I, p. 325, showing the single-neutron levels in the region of $N=126$.) In the energy region covered by the experiments, the only missing single-particle configuration is the [523 5/2] hole state, which has been seen for neutron configurations with $N=93$–99 (see Table 5-13, p. 299). This state is expected somewhat above 1 MeV in ^{175}Yb, but would have been difficult to identify in the experiments reported so far.

Effect of pair correlations

The fact that the same one-quasiparticle state can be populated in both stripping and pickup reactions implies that, in the ground states of the even-even nuclei, the orbits near the Fermi level are partially occupied. The normalization employed in Fig. 5-10 assigns equal strength to the (dp) and (dt) reactions to the ground-state transition; it is seen that, with increasing excitation energy, the (dp) reaction dominates for particle-like orbits, while the (dt) reaction dominates for hole-like orbits.

These features receive an immediate interpretation in terms of the pair correlations. A quantitative test of the effect of pair correlations on the one-particle transfer cross sections can be obtained by comparing the cross sections to the same configuration in a sequence of isotopes. The ratios of the cross sections yield the relative values of the coefficients u^2 (or v^2) (see Eq. (5-42)), and Fig. 5-11 gives the results obtained from the measurements of the intensities for two of the prominent transitions observed in the sequence of odd-A Yb isotopes. The main variation in u^2 (and v^2) can be described in terms of a variation of the Fermi level, λ, with respect to a fixed single-particle spectrum. Assuming a linear variation of λ with N, we obtain from Eq. (6-601)

$$v^2(N)=1-u^2(N)$$
$$=\frac{1}{2}\left(1+\frac{N-N_0}{\left((N-N_0)^2+\gamma^2\right)^{1/2}}\right) \tag{5-54}$$

with

$$\gamma=\Delta\left(\frac{\partial\lambda}{\partial N}\right)^{-1}=\frac{2\Delta}{d} \tag{5-55}$$

where Δ is the pairing parameter, while d is the average spacing of the twofold degenerate one-particle levels and N_0 the value of the neutron number for which λ coincides with the level in question. The solid lines in Fig. 5-11, which are seen to describe the empirical data rather well, correspond to Eq. (5-54) with $\gamma=3.8$. The average level spacing d in the region of the Fermi level can be obtained from Fig. 5-3, which yields $d\approx 0.4$ MeV (assuming ▲ $\hbar\omega_0\approx 41A^{-1/3}$ MeV). The observed value of γ thus yields $\Delta\approx 0.8$ MeV, which

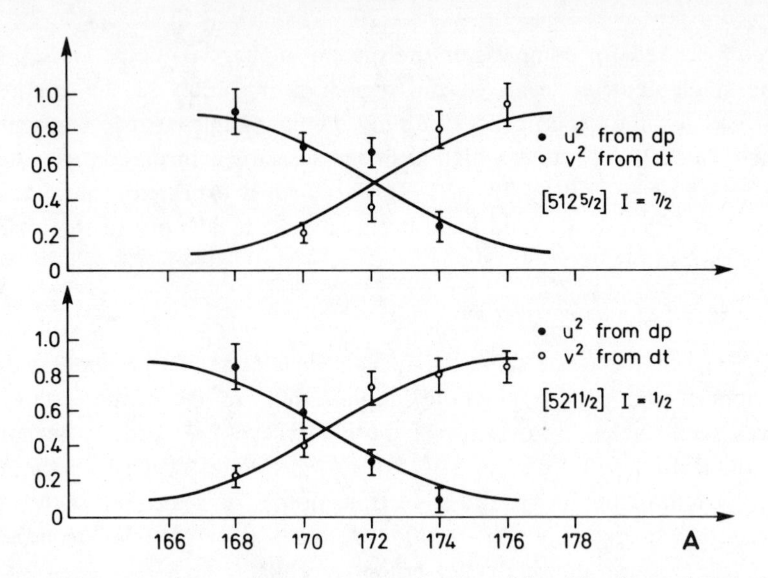

Figure 5-11 Effect of pair correlations on transfer intensities. The figure is based on the empirical data and analysis of Burke *et al.*, *loc.cit.*, Figs. 5-9 and 5-10. The observed cross sections are corrected to constant Q value by employing the theoretical Q dependence obtained on the basis of the distorted-wave Born approximation. The values of u^2 (and v^2) represent the ratios of these corrected cross sections to the assumed one-particle values

$$d\sigma(\text{dp}; [521\ 1/2], I=1/2) = 144\ \mu\text{b/ster.}$$

$$d\sigma(\text{dp}; [512\ 5/2], I=7/2) = 247\ \mu\text{b/ster.}$$

$$d\sigma(\text{dt}; [521\ 1/2], I=1/2) = 575\ \mu\text{b/ster.}$$

$$d\sigma(\text{dt}; [512\ 5/2], I=7/2) = 495\ \mu\text{b/ster.}$$

which refer to $Q=4.0$ MeV for the (dp) reactions and $Q=-1.5$ MeV for the (dt) reactions.

▼ may be compared with the value $\Delta=0.7$ MeV obtained from the odd-even mass differences of the Yb isotopes (see Eq. (2-92) and the neutron separation energies given by Burke *et al.*, *loc. cit.* Fig. 5-9).

Energies of quasiparticle states

The observation in ^{175}Yb of single-particle levels extending over a considerable energy range and the agreement between the calculated and observed sequence of levels makes possible a determination of the absolute energy scale for the single-particle spectrum. Comparison with Fig. 5-3 (for $\delta=0.3$, as determined from the $E2$ moment; see Fig. 4-25) shows that the observed levels in ^{175}Yb cover a range that is calculated to be about $0.6\hbar\omega_0$, which corresponds to an energy interval of about 4.4 MeV. In comparison, the observed energy interval, obtained by adding the particle to the hole ▲ energies, is about 3.0 MeV. This compression of the absolute energy scale is a

▼ systematic feature of the spectra of the odd-A deformed nuclei (Bakke, 1958; see also the survey by Bunker and Reich, 1971).

A compression of the energy scale for single-particle states is a simple consequence of the pair correlations, since the presence of an unpaired nucleon in an orbit ν_1 prevents the pairs of particles from exploiting the configuration $(\nu_1\bar{\nu}_1)$ and thus leads to a loss of the pair correlation energy. For an orbit close to the Fermi level, the resulting energy shift corresponds to the odd-even mass difference Δ, but the energy shift decreases with increasing distance of the orbit from the Fermi level.

For a constant pair field, the quasiparticle energy is (see Eq. (6-602))

$$E(\nu)=\left((\varepsilon(\nu)-\lambda)^2+\Delta^2\right)^{1/2} \tag{5-57}$$

where $\varepsilon(\nu)$ is the one-particle energy in the absence of pair correlations, while λ is the Fermi energy. From Eq. (5-57), it follows that, for a sufficiently large range of excitation energies including both particle and hole excitations, the observed interval of excitations should be reduced by about 2Δ as compared with the range of $\varepsilon(\nu)$ values (see also Fig. 6-63b). The value of Δ obtained from the odd-even mass differences is $\Delta\approx 0.7$ MeV and, when this correction is included, the observed energy scale is seen to be consistent with that of Fig. 5-3. It is significant that the calculated single-particle spectra are based on a velocity-independent potential; thus, the present data do not provide evidence for an expansion of the energy scale, as would be implied by the observed energy dependence of the optical model potential (see, for example, Fig. 2-29, Vol. I, p. 237).

The expression (5-57) for the quasiparticle energies implies a rapid variation in the level density in the neighborhood of the ground state of an odd-A nucleus, but the data in Fig. 5-9 do not show such an effect. The failure to observe this effect in any particular spectrum might be a special feature of the one-particle spectrum $\varepsilon(\nu)$, but the absence of the effect is confirmed by the systematics of the intrinsic states. Thus, the average excitation energy of the first excited state in odd-N nuclei in Table 5-13 is found to be about 140 keV, which is comparable to the average spacing of the intrinsic states within the first MeV in the spectrum of ^{175}Yb, and several times larger than would be expected from the expression (5-57). The singularity in the quasiparticle level density for $\varepsilon(\nu)\approx\lambda$ results from the assumption that the pairing potential Δ is the same for all orbits, as in Eq. (6-597). This assumption is expected to be appropriate in a situation where the pairing potential is produced by a large number of particle configurations uniformly distributed over the nuclear volume. However, in the deformed nuclei, the value of Δ is only a few times larger than the single-particle spacing, and Δ may therefore depend somewhat on the orbit considered. The observed spacings of the low-lying intrinsic orbits appear to give evidence for variations in Δ of the order of 200 keV. ▲

▼ *Spectrum of ^{237}Np and interpretation of α-decay fine-structure intensities (Figs. 5-12 and 5-13)*

The development of α spectroscopy has made it possible to observe very weak components in the fine-structure pattern (of relative intensities below 10^{-6}), and has provided a rich body of data concerning the energy spectra and the nuclear coupling scheme of the heavy elements. As an example, Fig. 5-12 shows the level spectrum of ^{237}Np, which is populated in the α decay of ^{241}Am. The classification of the states is based on the analysis of the intensities of the α fine-structure components (Lederer *et al.*, 1966) and is further supported by the evidence on β- and γ-transition rates, rotational energy parameters, and especially by the studies of the reactions ^{236}U(^{3}He d) and ^{236}U(αt), which provide information similar to that obtained from the one-particle transfer studies of ^{175}Yb illustrated in Fig. 5-10. It is seen that the ▲ intrinsic states in ^{237}Np below 700 keV are just those expected from Fig. 5-4

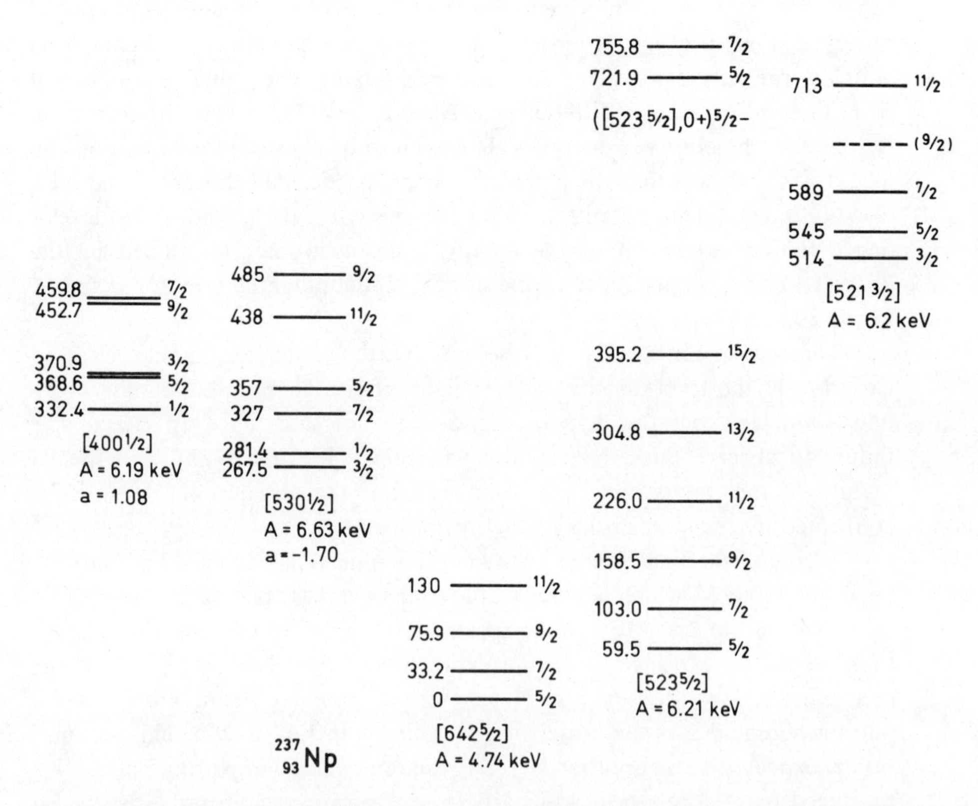

Figure 5-12 Level scheme of ^{237}Np. The level scheme of ^{237}Np is based partly on the decay of ^{241}Am (S. A. Baranov, V. M. Kulakov, and V. M. Shatinsky, *Nuclear Phys.* **56**, 252, 1964, and C. M. Lederer, J. K. Poggenburg, F. Asaro, J. O. Rasmussen, and I. Perlman, *Nuclear Phys.* **84**, 481, 1966), and partly on one-particle transfer studies (Th. W. Elze and J. R. Huizenga, *Phys. Rev.* **C1**, 328, 1970).

▼ in the region $Z \approx 93$, $\delta \approx 0.25$ (see Fig. 4-25, p. 133). The $K\pi = 5/2-$ state at 722 keV cannot be accounted for in terms of a one-quasiparticle state obtained from the spectrum in Fig. 5-4, but can be understood in terms of a β-vibrational excitation superposed on the [523 5/2] configuration (see p. 273).

Matrix element for α decay

The intensities of the observed α fine-structure components are shown in Fig. 5-13, which gives the inverse hindrance factor F^{-1}, defined as the transition probability in units of the quantity $T_0(E_\alpha, Z)$ that describes the ground-state transitions of even-even nuclei (see Eq. (4-164)). As discussed on pp. 115ff., the transition amplitude for α decay can be expressed as a product of two factors; one represents the formation of the α particle at the nuclear surface and the other describes the penetration of the α particle through the Coulomb barrier.

The formation factor involves a generalization to a four-particle cluster of the operator defined in Eq. (3E-17) for the two-nucleon transfer,

$$A_\alpha^\dagger(\mathbf{r}) = \frac{1}{4} \sum_{\substack{\nu_{n1}\nu_{n2} \\ \nu_{p1}\nu_{p2}}} \langle \nu_{n1}\nu_{n2}\nu_{p1}\nu_{p2} | \alpha, \mathbf{r} \rangle a^\dagger(\nu_{p2}) a^\dagger(\nu_{p1}) a^\dagger(\nu_{n2}) a^\dagger(\nu_{n1}) \qquad (5\text{-}58)$$

where the factor $\langle \nu_{n1}\nu_{n2}\nu_{p1}\nu_{p2} | \alpha, \mathbf{r} \rangle$ is the transformation coefficient describing the overlap of the α particle at the position $\mathbf{r}$ with the product state of four nucleons, antisymmetrized in the quantum numbers of the two neutrons and of the two protons. (The sum in Eq. (5-58) may be restricted to $\nu_{n1} < \nu_{n2}$ and $\nu_{p1} < \nu_{p2}$, in which case the factor 1/4 is to be omitted.) The amplitude for the formation of the α particle is proportional to the matrix element of A_α between initial and final intrinsic states, evaluated at the nuclear surface. (The treatment of the α-formation process in terms of the configurations of the individual nucleons has been especially developed by Mang (1957) and by Mang and Rasmussen (1962).) The estimate of the absolute rate of α decay involves the uncertainties associated with the treatment of the surface region as well as of the many-body aspects of the cluster formation problem; we shall therefore especially consider the reduced amplitudes describing the decay probabilities relative to those of the ground-state transitions in even-even nuclei.

Because of the nonspherical nuclear surface, the penetration of the α particle through the classically forbidden region gives rise to anisotropy in the motion of the α particle relative to the nuclear orientation, associated with an exchange of angular momentum between the α particle and the daughter nucleus. Additional exchange of angular momentum results from the nonspherical components of the Coulomb field. A simplified treatment of the coupling between the α particle and the nuclear orientation can be based on ▲ the fact that exchange of angular momentum is mainly confined to a small

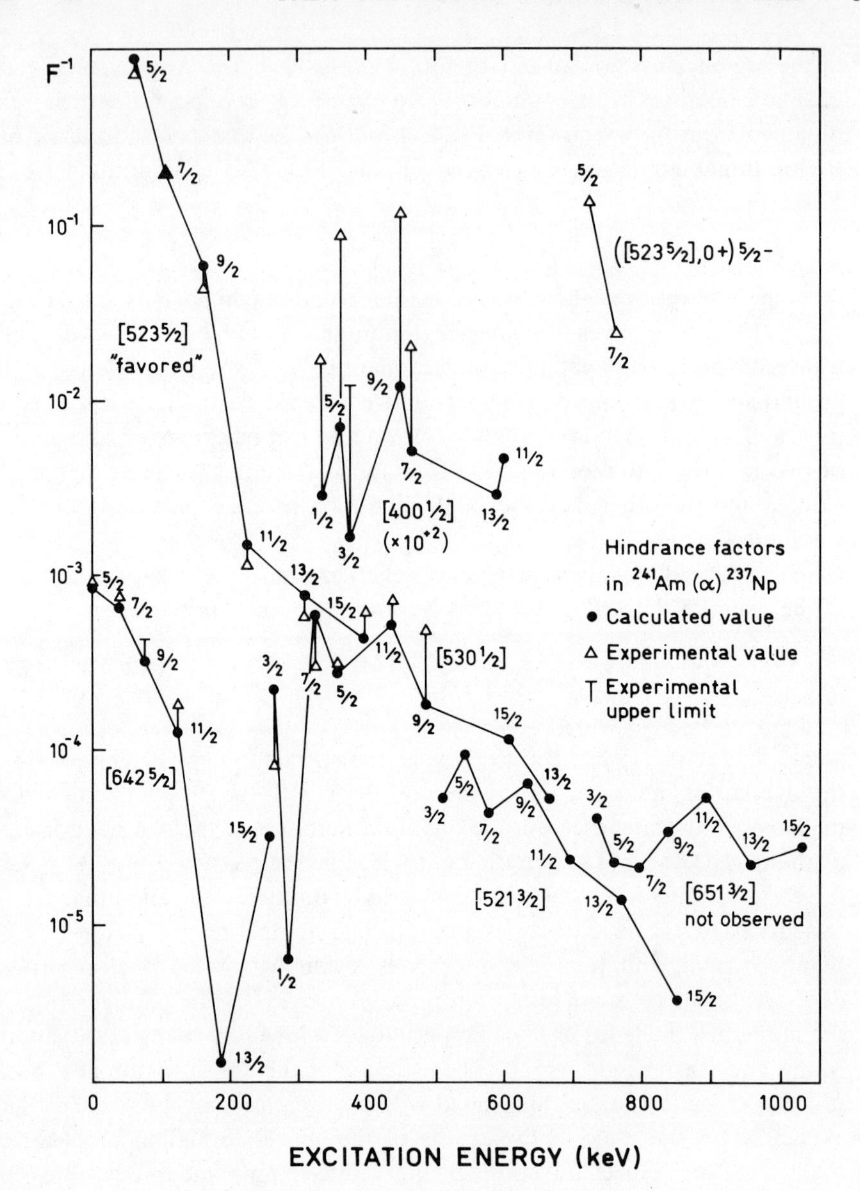

Figure 5-13 Hindrance factors in ^{241}Am(α)^{237}Np. The experimental data are from the references quoted in Fig. 5-12. The hindrance factors are contained in the compilation of Michel (1966); see also Lederer *et al.* (1967). The theoretical calculations of the α-decay probabilities are from J. K. Poggenburg, Ph.D. thesis, University of California, *UCRL* 16187 (1965); see also J. K. Poggenburg, H. J. Mang, and J. O. Rasmussen, *Phys. Rev.* **181**, 1697 (1969). The expected [651 3/2] band has not been observed so far and no significance should be attached to the excitation energies in the figure for this band.

▼ region just outside the nuclear surface during the penetration of which the rotation of the nucleus can be neglected (see the discussion in Chapter 4, p. 116). The amplitude for finding α particles with angular momentum L outside the nonspherical region of the barrier can therefore be described by the coefficients

$$f(L, K_i \to K_f) = \frac{\int \langle K_f | A_\alpha(R; \vartheta, \varphi) | K_i \rangle B(\vartheta) i^L Y_{L, M' = K_i - K_f}(\vartheta, \varphi)\, d\Omega}{\int \langle \nu = 0 | A_\alpha(R; \vartheta, \varphi) | \nu = 0 \rangle B(\vartheta) Y_{00}\, d\Omega} \tag{5-59}$$

where the integrations extend over the different directions ϑ, φ with respect to the nuclear symmetry axis, and where the factor $B(\vartheta)$ describes the ϑ dependence of the wave function imposed by the penetration of the non-spherical nuclear barrier. The coefficients (5-59) are reduced amplitudes, measured in units of the amplitude for the ground-state decay of even-even nuclei.

The penetration through the three-dimensional barrier region can be approximately described by a one-dimensional radial penetration integral, provided the centrifugal potential is small compared to the magnitude of the (negative) kinetic energy ($L \ll \kappa R_0$, where κ is the absolute value of the wave number of the α particle in the region outside the nuclear surface; for ^{237}Np, the value of κR_0 is about 20); thus, in the WKB approximation, one obtains (Fröman, 1957)

$$B(\vartheta) = \exp\left\{ \left(\frac{2M_\alpha}{\hbar^2} \left(\frac{Z_\alpha Z e^2}{R_0} - E_\alpha \right) \right)^{1/2} \frac{2}{3} R_0 \delta P_2(\cos\vartheta) \right.$$
$$- \int_{R_0} \left(\left(\frac{2M_\alpha}{\hbar^2} \left(\frac{Z_\alpha Z e^2}{r} - E_\alpha + \frac{Z_\alpha Q_0 e^2}{2r^3} P_2(\cos\vartheta) \right) \right)^{1/2} \right.$$
$$\left. \left. - \left(\frac{2M_\alpha}{\hbar^2} \left(\frac{Z_\alpha Z e^2}{r} - E_\alpha \right) \right)^{1/2} \right) dr \right\} \tag{5-60a}$$

$$\approx \exp\left\{ \left(\frac{2M_\alpha}{\hbar^2} \frac{Z_\alpha Z e^2}{R_0} R_0^2 \right)^{1/2} \frac{8}{15} \delta P_2(\cos\vartheta) \right\} \tag{5-60b}$$

where M_α and Z_α are, respectively, the mass and charge number of the α particle. In the penetration integral (5-60a), the first term represents the transmission of the α particle from the nuclear surface to a spherical surface of mean radius R_0, while the second term gives the effect of the nonspherical part of the Coulomb field; only the quadrupole component of the nuclear deformation has been included. In Eq. (5-60b), the energy E_α of the α particle ▲

has been neglected compared with the barrier height ($\approx$30 MeV), and the relation $Q_0 = (4/5)R_0^2 Z\delta$ (see Eq. (4-72)) has been employed. For typical values of the deformation in heavy nuclei, the coefficient of $P_2(\cos\vartheta)$ in the exponent in Eq. (5-60b) is of order unity; hence, during the penetration of the nonspherical part of the barrier there is an appreciable probability for the exchange of a few units of angular momentum between the orbital motion of the α particle and the nuclear rotation.

In terms of the reduced amplitudes (5-59), the transition probabilities to the different members of the rotational band in an odd-A nucleus are given by (see Eq. (4-167))

$$T(L, I_i K_i \rightarrow I_f K_f) = T_0(E_\alpha, Z) p_L |\langle I_i K_i L\, K_f - K_i | I_f K_f \rangle f(L, K_i \rightarrow K_f) + (-1)^{I_f + K_f} \langle I_i K_i L - K_f - K_i | I_f - K_f \rangle f(L, K_i \rightarrow \overline{K}_f)|^2 \tag{5-61}$$

where p_L gives the effect of the centrifugal potential on the penetration of the α particle. In the WKB approximation, the factor p_L is given by

$$p_L = \exp\left\{ -2 \int_{R_0} \left(\left(\frac{2M_\alpha}{\hbar^2} \left(\frac{Z_\alpha Z e^2}{r} - E_\alpha \right) + \frac{L(L+1)}{r^2} \right)^{1/2} - \left(\frac{2M_\alpha}{\hbar^2} \left(\frac{Z_\alpha Z e^2}{r} - E_\alpha \right) \right)^{1/2} \right) dr \right\}$$

$$\approx \exp\left\{ -2 \frac{L(L+1)}{\kappa R_0} \right\} \qquad \left(\kappa \approx \left(\frac{2M_\alpha}{\hbar^2} \frac{Z_\alpha Z e^2}{R_0} \right)^{1/2} \right) \tag{5-62}$$

where the last expression has been obtained by setting $E_\alpha = 0$ and retaining only the leading-order term in $L(L+1)$ in the integrand.

Effect of pair correlations

The pair correlations have a major effect on the α-formation process (Mang and Rasmussen, 1962; Soloviev, 1962). For the ground-state transition in even-even nuclei, the parentage coefficient takes the form (see Eqs. (5-58) and (6-599))

$$\langle \mathrm{v}=0 | A_\alpha^\dagger(\mathbf{r}) | \mathrm{v}=0 \rangle = \sum_{\substack{\nu_n > 0 \\ \nu_p > 0}} \langle \nu_n \bar{\nu}_n \nu_p \bar{\nu}_p | A_\alpha^\dagger(\mathbf{r}) | 0 \rangle u(\nu_n) v(\nu_n) u(\nu_p) v(\nu_p) \tag{5-63}$$

The coherence effect in the matrix element (5-63) is similar to that expressing the contribution of the different orbits to the pair fields (see Eq. (6-612)); in fact, if we neglect the finite size of the α particle, the quantity $A_\alpha^\dagger(\mathbf{r})$ involves the product of the local pair fields of neutrons and protons, defined by Eq.

▼ (6-141). In this limit, the matrix elements of $A_\alpha^\dagger(\mathbf{r})$ for creating pairs of nucleons in conjugate orbits all have the same sign, since the wave functions ψ_ν and $\psi_{\bar{\nu}}$, taken at the same space point, are complex conjugate. Thus, the overlap matrix element for the antisymmetrized two-particle state $\nu\bar{\nu}$ has the value

$$\langle \mathbf{r}_1 = \mathbf{r}_2 = \mathbf{r}; S=0 | \nu\bar{\nu}\rangle_a = - \sum_{\Sigma=\pm\frac{1}{2}} |\langle \mathbf{r}, \Sigma | \nu\rangle|^2 \tag{5-64}$$

It is expected that this coherence of the matrix elements will not be significantly affected by the finite size of the α particle, which is not much greater than the wavelength of the nucleons at the Fermi surface, and the available numerical studies confirm this expectation (Mang and Rasmussen, 1962).

The sum in Eq. (5-63) receives contributions from all the orbits in the neighborhood of the Fermi surface that are neither completely filled nor completely empty and which therefore contribute to the pair correlation. A measure of the enhancement is provided by the quantity

$$q \equiv \sum_{\nu>0} u(\nu)v(\nu) \tag{5-65}$$

which can be estimated from the observed strength of the pair correlations. In the approximation of a constant static pair field, the parameter Δ is equal to q multiplied by the average pairing matrix element G (see Eq. (6-613)), which is found to have a magnitude of about $G \approx 100$ keV in heavy nuclei (see, for example, the discussion on p. 653). For Np, the value of Δ is about 0.7 MeV (see Fig. 2-5, Vol. I, p. 170), and we thus obtain a total enhancement of the α-decay rate by a factor $q_n^2 q_p^2 \sim 10^3$.

Favored transitions

The transition matrix element between one-quasiparticle states in odd-Z, even-N nuclei may be obtained from the quasiparticle transformation (6-599)

$$\begin{aligned} \langle \mathrm{v}=1, \nu_p' | A_\alpha^\dagger(\mathbf{r}) | \mathrm{v}=1, \nu_p\rangle &= \langle \mathrm{v}=0 | A_\alpha^\dagger(\mathbf{r}) | \mathrm{v}=0\rangle \delta(\nu_p, \nu_p') \\ &\quad - u(\nu_p')v(\nu_p) \sum_{\nu_n>0} \langle \nu_n \bar{\nu}_n \nu_p' \bar{\nu}_p | A_\alpha^\dagger(\mathbf{r}) | 0\rangle u(\nu_n) v(\nu_n) \end{aligned} \tag{5-66}$$

For the favored decay ($\nu_p = \nu_p'$), the matrix element (5-66) is dominated by the first term, which expresses the coherent contribution from the paired core as in the ground-state transitions of the even-even nuclei. For the unfavored decays, only the second term in Eq. (5-66) contributes; this term does not contain a summation over proton orbits, and the matrix element is therefore reduced, as compared with the favored decay, by a factor q_p (in addition to the effects of selection rules in the one-particle matrix elements considered ▲ below).

▼ In the comparison between the decay rates for favored decays in odd-A and neighboring even-even nuclei, one must take into account partly the fact that the orbit occupied by the last odd particle cannot contribute to the α decay (second term in Eq. (5-66)) and partly the reduction in the pair correlations of the even-even core that results from the presence of the unpaired particle (blocking effect). Both of these effects are contained in the reduction of the sum (5-65) that measures the strength of the pair field; in the $v=1$ states, the sum q extends over the orbits ν that are available for the pair correlations, and thus excludes the orbit of the unpaired particle.

The favored transitions in the α decay of ^{241}Am are observed to be a factor of about 2 slower than the even-even ground-state transitions in neighboring nuclei. Thus, the transition rates for ^{241}Am, analyzed by means of expression (4-167), yield $C_0=0.6$, $C_2=0.3$, and $C_4=0.003$, which may be compared with the values C_L observed in the decay of ^{240}Pu ($C_0=1$ (norm.), $C_2=0.6$, and $C_4=0.01$). The study of the (pt) reaction has provided evidence for a similar reduction of the dineutron formation probabilities in the favored odd-A transitions, as compared with those in even-even nuclei (Oothoudt and Hintz, 1973).

A retardation by a factor of 2 in the intensities of the favored transitions can be interpreted in terms of a reduction by about 30% in the strength Δ of the pair field in odd-A nuclei as compared with that of the ground states of even-even nuclei. A decrease of approximately this magnitude can be understood on the basis of the self-consistency equation (6-614) for Δ (see also Eq. (6-602)); by treating the change in Δ as a small perturbation, one finds that the blocking of an orbit near the Fermi level leads to a reduction of Δ by the amount $d/2$, where d is the average distance between the doubly degenerate orbits in the single-particle potential. From Fig. 5-4, p. 224, it is seen that $d\approx0.3$ MeV, which implies a reduction in Δ by about 20%.

Unfavored transitions

While the relative intensities in the favored band reflect the collective α-formation process for the even-even core (see Eq. (4-167)), the pattern of intensities for the unfavored transitions depends sensitively on the quasiparticle configurations in the parent and daughter nuclei. These patterns, as shown in Fig. 5-13, therefore provide characteristic "fingerprints" that can be used to establish the configuration assignments and to test the intrinsic wave functions. The theoretical values in Fig. 5-13 represent relative intensities calculated by means of Eq. (5-66) from one-particle wave functions obtained from the Hamiltonian (5-10).

The observed hindrance factors for the unfavored transitions in Fig. 5-13 are seen to be of the order of 10^3. The interpretation of these intensities implies that the amplitudes $f(L)$ for large values of L reach values of about 10^{-1}, corresponding approximately to the factor $(q_p)^{-1}$; for small values of L,

▲ the hindrances are in most cases considerably greater, as a result of selection

▼ rules associated with the one-particle quantum numbers. For example, the transition [523 5/2]→[642 5/2] involves a spin flip ($\Delta\Sigma = 1$), and hence the main contribution arises from the amplitude $f(L, K_i \to \bar{K}_f)$ in Eq. (5-61); this amplitude has $\Delta K = 5$ and therefore only contributes for $L \geqslant 5$. In fact, it is found that the $L = 5$ amplitude is also inhibited and that the largest matrix elements are associated with $L = 7$ and 9. (This additional hindrance may be connected with the fact that the transition involves $\Delta n_3 = 2$, in addition to $\Delta\Lambda = 5$.)

The intensities in Fig. 5-13 show a rather marked dependence on the signature $\sigma = (-1)^{I_f + K_f}$ reflecting the interference between the two terms in Eq. (5-61). This signature dependence provides a valuable tool for identification of the different intrinsic configurations; it is seen that the theoretical calculations are able to account for the qualitative features of this effect. No simple rule has yet been given for predicting the sign of the interference effect in terms of the quantum numbers of the one-particle orbits.

The largest discrepancy between the calculated and observed intensities in the α fine-structure occurs in the band [400 1/2], where the calculation, though reproducing the relative intensities moderately well, underestimates the absolute intensities by about an order of magnitude. This discrepancy may be connected with the failure of the harmonic oscillator wave functions to describe correctly the dependence of the radial wave functions on the total oscillator quantum number N.

In the fine-structure of the α decay of ^{241}Am, the $K\pi = 5/2-$ band starting at 722 keV stands out in a striking manner as the only band with reduced intensities within three orders of magnitude of the favored band. The intensity of these transitions suggests an interpretation as a β-vibrational excitation ($\nu\pi = 0+$) built on the [523 5/2] configuration; the α intensity may be compared with that observed in $^{242}_{96}$Cm$(\alpha)^{238}_{94}$Pu, where the transition to the state $n_\beta = 1$, $I\pi = 0+$ at 943 keV has $F^{-1} = 0.14$ (Lederer *et al.*, 1967).

Spectrum of ^{235}U and analysis of Coriolis coupling effects (Fig. 5-14; Tables 5-6 and 5-7)

The spectrum of ^{235}U has been extensively studied by a variety of reactions, including α decay, Coulomb excitation, one-particle transfer, and the (nγ) process. The data summarized in Fig. 5-14 establish more than 50 levels in the region up to about 1 MeV. Despite the great complexity of the spectrum, the high specificity of the intensity patterns observed in the different reactions has made possible a classification into rotational bands and the assignment of intrinsic configurations. The observed one-particle configurations correspond well to those expected for $N \approx 143$ on the basis of Fig. 5-5, p. 225, assuming $\delta \approx 0.25$, as determined from the rotational $B(E2)$ ▲ values (see Fig. 4-25, p. 133).

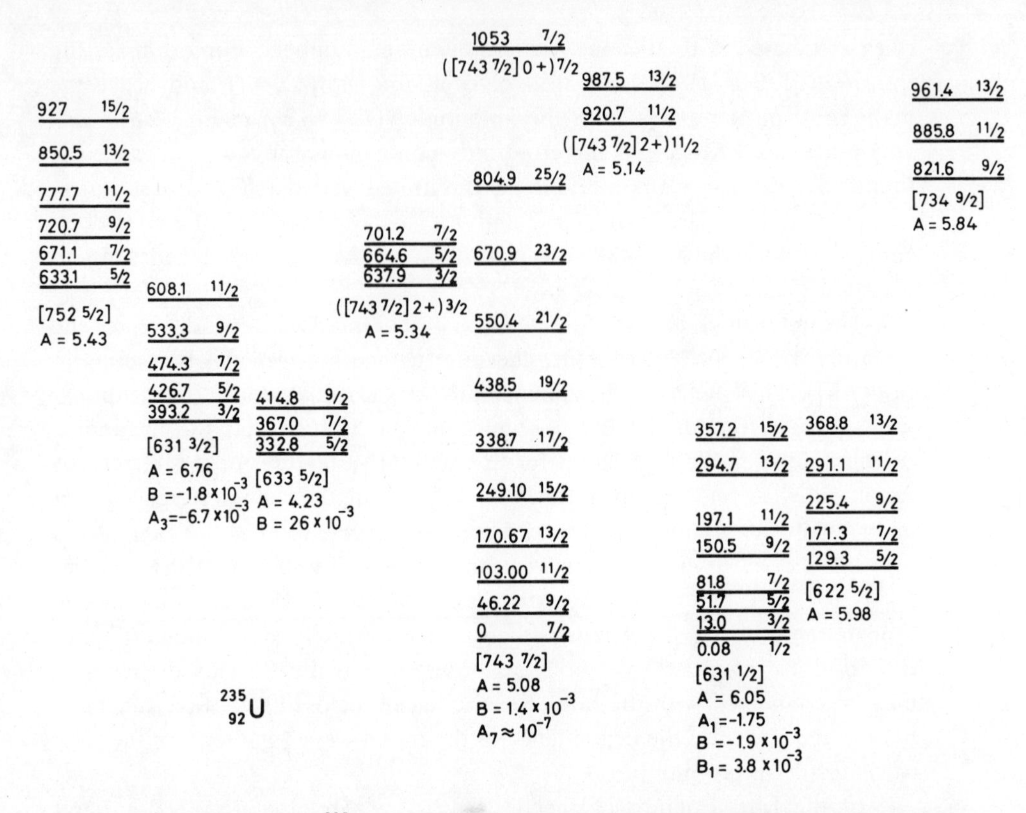

Figure 5-14 Spectrum of ^{235}U. The figure is based on the experimental data obtained from Coulomb excitation of ^{235}U (F. S. Stephens, M. D. Holtz, R. M. Diamond, and J. O. Newton, *Nuclear Phys.* **A115**, 129, 1968), α decay of ^{239}Pu (J. E. Cline, *Nuclear Phys.* **A106**, 481, 1968), one-particle transfer (Th. W. Elze and J. R. Huizenga, *Nuclear Phys.* **A133**, 10, 1969; T. H. Braid, R. R. Chasman, J. R. Erskine, and A. M. Friedman, *Phys. Rev.* **C1**, 275, 1970), and the reaction ^{234}U(nγ) (E. T. Jurney, *Neutron Capture Gamma-Ray Spectroscopy*, Proceedings of the International Symposium held in Studsvik, p. 431, International Atomic Energy Agency, Vienna, 1969). All energies are in keV. The particle-like states are drawn to the right of the ground-state band, while the hole-like states are shown to the left.

▼ Of the single-particle levels expected from Fig. 5-5, only the configuration [624 7/2] is missing in the low-energy spectrum in Fig. 5-14; tentative evidence for the occurrence of this band starting at about 500 keV in ^{235}U has been reported by Braid *et al.* (1970).

In addition to the one-quasiparticle states, the observed spectrum above 600 keV contains a number of states that have been classified in terms of configurations involving vibrational excitations superposed on the lowest single-quasiparticle states (see Fig. 5-14). For the β and γ vibrations based on the ground-state configuration, the interpretation receives support from the ▲ observed rather large $E2$-excitation probabilities (Stephens *et al.*, 1968).

▼ In the present example, we shall especially discuss the evidence on Coriolis coupling effects as revealed by the spectrum of ^{235}U. The $N=7$ orbits are predominantly $j=15/2$ and thus have very large Coriolis matrix elements; both the rotational energies and the $E2$-transition probabilities provide information on the magnitude of these matrix elements (Stephens *et al.*, 1968).

Estimate of Coriolis coupling from E2-matrix elements

A rather direct determination of the first-order particle-rotation coupling can be obtained from the $E2$-transition matrix elements between rotational bands with $\Delta K=1$. For such transitions, the first-order Coriolis coupling gives rise to an I-independent renormalization of the intrinsic $E2$-transition operator (see Eqs. (4-203) and (4-221))

$$\hat{\mathscr{M}}(E2,\nu=\pm 1)=\mathscr{M}(E2,\nu=\pm 1)+\left(\frac{15}{8\pi}\right)^{1/2} eQ_0\varepsilon_{\pm 1} \tag{5-67}$$

where the operator $\varepsilon_{\pm 1}$ gives the admixing of the coupled bands (see Eqs. (4-198) and (4-199)). For one-quasiparticle states, the matrix elements of $\varepsilon_{\pm 1}$ can be expressed in terms of the angular momentum operator $j_{\pm}$ associated with the Coriolis coupling (see Eqs. (4-199) and (4-197))

$$\begin{aligned}\langle \mathrm{v}=1,\Omega\pm 1|\varepsilon_{\pm 1}|\mathrm{v}=1,\Omega\rangle &= -A_0\frac{\langle \mathrm{v}=1,\Omega\pm 1|j_{\pm}|\mathrm{v}=1,\Omega\rangle}{E(\Omega\pm 1)-E(\Omega)}\\ A_0 &= \frac{\hbar^2}{2\mathscr{I}_0}\end{aligned} \tag{5-68}$$

where $\mathscr{I}_0$ is the moment of inertia of the rotating core.

An estimate of the two terms in the effective moment (5-67) shows that the rotational coupling gives a contribution that is about an order of magnitude larger than the intrinsic moment. The relative magnitude of the two contributions is further discussed in the last section of the present example, where it is shown that the Coriolis effect can be interpreted in terms of a polarization charge resulting from the coupling to the rotational motion.[2]

The empirical $B(E2)$ values for the excitations of the $K\pi=5/2-$ and $K\pi=9/2-$ bands in ^{235}U are given in Table 5-7, which also lists the values of $\varepsilon_{\pm 1}$ obtained from these $E2$-matrix elements by means of Eq. (5-67). These determinations of $\varepsilon_{\pm 1}$ include a correction of $+10\%$, corresponding to the ▲ effect of including the intrinsic moment $\mathscr{M}(E2,\nu=\pm 1)$, as estimated on pp.

---

[2]The evidence for a rather strong enhancement of the $E2$ transitions with $\Delta K=1$ was collected by Löbner (1965), and estimates of the matrix element resulting from the Coriolis coupling were given by Faessler (1966).

▼ 281 ff. The determination of $\varepsilon_{\pm 1}$ is based on the value $Q_0=980$ fm^2, as obtained from the static quadrupole moment measured in the μ-mesic atom (Dey *et al.,* 1973).

Theoretical estimates of the matrix elements of $\varepsilon_{\pm 1}$ based on Eq. (5-68) are listed in column four of Table 5-7; the estimates have employed the one-particle matrix elements of $j_{\pm}$ listed in Table 5-6 and the inertial parameter $A_0=7.4$ keV, representing the mean value for the ground-state bands of ^{234}U and ^{236}U. The calculated matrix elements in Table 5-7 include a small correction resulting from the pair correlations (see Eq. (5-43)); if we assume the ground state to be close to the Fermi level, we have $u_1 \approx v_1 \approx 2^{-1/2}$ and (see Eq. (6-601))

$$u_1u_2+v_1v_2 \approx 2^{-1/2}\left(1+\frac{\Delta}{E_2}\right)^{1/2}$$

$$\approx\left(\frac{1}{2}\,\frac{E_2-E_1+2\Delta}{E_2-E_1+\Delta}\right)^{1/2} \tag{5-69}$$

which yields a reduction of the matrix elements by about 0.85 (for $\Delta \approx 0.7$ MeV). The theoretical estimates of $\varepsilon_{\pm 1}$ in Table 5-7 are seen to exceed the values deduced from the observed $B(E2)$-matrix elements by about a factor of 2. ▲

$[Nn_3\Lambda\Omega]$	$\langle\Omega|j_+|\Omega-1\rangle$
[770 1/2]	7.37
[761 3/2]	7.36
[752 5/2]	7.30
[743 7/2]	7.10
[734 9/2]	6.72
[725 11/2]	6.13
[716 13/2]	5.24
[707 15/2]	3.86

Table 5-6 Coriolis matrix elements for odd-parity states in ^{235}U. The single-particle matrix elements of $j_+=j_1+ij_2$ are obtained from the wave functions given by B. R. Mottelson and S. G. Nilsson, *Mat. Fys. Skr. Dan. Vid. Selsk.* **1**, no. 8 (1959), for a deformation of $\delta=0.25$ and a potential with parameters $v_{ll}=-0.02$ and $v_{ls}=-0.1$.

K_i	K_f	$B(E2)$ $e^2\text{fm}^4$	$\langle K_f \vert \varepsilon_{\pm 1} \vert K_i \rangle$ Obs.	Calc.
7/2	5/2	430	3.2×10^{-2}	6.8×10^{-2}
7/2	9/2	320	2.7×10^{-2}	5.5×10^{-2}

Table 5-7 Coriolis mixing amplitudes for the [743 7/2] band in ^{235}U. The $B(E2)$ values listed in column two represent the sum over transitions to different members of the rotational band of the final configuration

$$B(E2) \equiv \sum_{I_f} B(E2; K_i I_i \rightarrow K_f I_f)$$

The data are taken from Stephens *et al.*, *loc.cit.*, Fig. 5-14.

▼ Within the framework of the model involving a quasiparticle coupled to a rotor, the Coriolis coupling effect involves the two quantities, the moment of inertia $\mathscr{I}_0$ of the rotor and the matrix element of $j_{\pm}$ connecting the quasiparticle states. In the above estimate of $\varepsilon_{\pm 1}$, the value of $\mathscr{I}_0$ has been taken from the ground-state bands of neighboring even-even nuclei. There are, however, effects associated with the presence of the odd particle that lead to a modification of the moment of inertia of the even-even core. (See the effect of "blocking" on the pair correlations and the role of the fourth-order term in the rotational energy of the even-even core, as discussed on pp. 310ff.) The quantitative estimates of these effects are somewhat uncertain, but the observed total moment of inertia in the ground-state band of ^{235}U, as analyzed below, appears to imply that in the present case the increase in $\mathscr{I}_0$ cannot exceed 20%.

The estimate of the one-particle matrix element of $j_{\pm}$ is relatively insensitive to the details of the one-particle potential, since the states involved are, to a rather good approximation, described in terms of $j_{15/2}$ orbits. Moreover, the estimated correction due to the pair correlations is small and therefore insensitive to the parameters of the pair field.

The rather large magnitude of the Coriolis coupling for the particular orbits considered raises the question of the possible significance of higher-order effects of these couplings. Such effects might be especially large in the $K=5/2$ and $K=9/2$ bands, if the remaining, as yet unobserved, $j_{15/2}$ orbits (with $\Omega=1/2, 3/2, 11/2$, etc.) are rather close in energy. However, the magnitude of the Coriolis matrix elements that couple these orbits is only a few hundred keV (for $I \leqslant 11/2$), and most of the Coriolis strength removed from the observed transitions would therefore be expected to reappear in transitions to other excited states within half an MeV of the observed excitations. ▲

▼ Such transitions might have been expected to appear in the Coulomb excitation studies and, moreover, would make contributions to the effective moment of inertia in the ground-state band that seem to be excluded by the analysis given below.

The analysis of the particle-rotation coupling, as discussed above, thus appears to provide evidence for significant terms in addition to the Coriolis coupling that can lead to a reduction of the $\Delta K=1$ rotational coupling. Such additional coupling terms, of first order in the rotational frequency and angular momentum, can be expressed in the form

$$H_c = A_0 F_+ I_- + \mathscr{R}\text{-conj.} \tag{5-70}$$

where the single-particle field F_+ and its Hermitian conjugate F_- have the same symmetry as $j_\pm$, with respect to time reversal and $\mathscr{R}$ conjugation. Couplings of the form (5-70) arise from velocity-dependent interactions. Thus, for example, if the single-particle potential of the nonrotating system has a velocity-dependent component $V(p)$, the transformation to the rotating coordinate system gives rise to a term of the form (5-70), with F_+ proportional to l_+, as in the Coriolis coupling. However, the contribution from this term is largely spurious, since the velocity dependence in the potential is to be considered relative to the collective flow (see the discussion on p. 80).

Another coupling of the form (5-70) results from the pair correlations. The pair correlations in the nonrotating system involve particles in states $(\nu,\bar{\nu})$ that are conjugate under time reversal, but in the rotating nucleus, the paired orbits have opposite motion with respect to the collective flow. This modification can be described in terms of a pair field with the symmetry Y_{21} in the intrinsic frame (assuming the deformation of the nuclear shape to have Y_{20} symmetry),

$$F_+ = \int (\rho_2(\mathbf{r}) - \rho_{-2}(\mathbf{r})) f(r) Y_{21}(\vartheta,\varphi) d\tau \tag{5-71}$$

where $\rho_{\pm 2}(\mathbf{r})$ are the pair densities defined by Eqs. (6-141) and (6-142) and where $f(r)$ is real. The quadrupole pair field (5-71) has the matrix elements

$$\begin{aligned}
\langle \mathrm{v}=1,\nu_2|F_+|\mathrm{v}=1,\nu_1\rangle &= \langle \nu_2|f(r)Y_{21}|\nu_1\rangle(u_1 v_2 - v_1 u_2) \\
\langle \mathrm{v}=2,\nu_1\nu_2|F_+|\mathrm{v}=0\rangle &= \langle \nu_2|f(r)Y_{21}|\bar{\nu}_1\rangle(u_1 u_2 + v_1 v_2)
\end{aligned} \tag{5-72}$$

For one-quasiparticle states, the matrix elements (5-72) vanish when the single-particle orbits are degenerate ($u_2 = u_1$); in particular, there is no contribution to the decoupling parameter in $K=1/2$ bands. The $\Delta\mathrm{v}=0$ matrix elements are largest when the states ν_1 and ν_2 are on opposite sides of the Fermi surface. These features are suggestive of the additional coupling effects indicated by the empirical evidence (see the discussion on p. 252). However, it ▲ remains an open question whether a pair field of the form (5-71) can account

▼ quantitatively for the discrepancy between the Coriolis coupling and the observed rotational $\Delta K = 1$ coupling; in this connection, one must also consider the effects of the quadrupole pair field on the moment of inertia of the even-even system (resulting from the $\Delta v = 2$ matrix elements) as well as on the $K\pi = 1+$ intrinsic excitations.

Contribution of Coriolis coupling to moment of inertia

The large difference between the rotational constant of the ground-state band of ^{235}U ($A = 5.1$ keV) and that of the neighboring even-even nuclei ($A_0 = 7.4$ keV) is a characteristic feature of the bands in odd-A nuclei based on orbits with large j (see Table 5-17, p. 312). A contribution to δA arises from the Coriolis couplings to the $K\pi = 5/2-$ and $K\pi = 9/2-$ bands considered above,

$$\delta A = - \sum_{\substack{\nu' \\ (\Omega(\nu') = \Omega(\nu) \pm 1)}} \langle \nu' | \varepsilon_{\pm 1} | \nu \rangle^2 (E(\nu') - E(\nu)) \tag{5-73}$$

If the calculated values of $\varepsilon_{\pm 1}$ in the last column of Table 5-7 are employed in Eq. (5-73), one obtains $\delta A = -5.4$ keV, which is more than twice the observed effect. The contribution (5-73) is to some extent compensated by the fact that the presence of the odd particle prevents some of the excitations that contribute to the moment of inertia of the even core. This effect is given by the second term of Eq. (5-47), and an estimate of this term, using the same approximation as in Eq. (5-69), leads to a reduction of δA by about 10% in the present case. Thus, the evidence from the inertial parameter provides independent evidence for the overestimate of the Coriolis coupling in ^{235}U.

An evaluation of the contribution (5-73) based on the empirical values of $\varepsilon_{\pm 1}$ from the third column of Table 5-7 yields $\delta A \approx -1.1$ keV, which is less than half of the observed effect. Other $\Omega\pi = 5/2-$ and $9/2-$ single-particle states make additional negative contributions to δA, but estimates of these terms, based on the spectrum of Fig. 5-5, indicate that they are much smaller than the ones considered.

Contributions to δA also result from the effects referred to above in the discussion of the Coriolis matrix elements (shift of one-particle transition strength to other levels and reduction of A_0), and the observed magnitude of δA may be used to set limits on these effects. If we first assume A_0 to have the value given by the ground-state bands of neighboring even-even nuclei, the observed couplings to the $K = 5/2$ and $K = 9/2$ bands, as analyzed in Table 5-7, only account for about one quarter of the expected magnitude of the transition intensities of the $j_{\pm}$ operators for the quasiparticle states considered. If the remaining strength is shifted to higher states, the excitation energies of these states must be in excess of about 2 MeV in order that the total contribution to δA not exceed the observed value. A reduction of A_0 due
▲ to the presence of the odd particle would require even higher energies for the

missing Coriolis strength. A lower limit for A_0 is obtained by attributing the entire residual part of δA, after subtraction of the terms (5-73) based on the empirical values of $\varepsilon_{\pm 1}$, to the difference between A_0 in the even-even and odd-A nuclei; the value of A_0 in ^{235}U is then about 20% smaller than in the neighboring even-even nuclei. (The same limit is suggested by the magnitude of A in the low-lying even parity bands of ^{235}U; see Fig. 5-14.)

The inertial parameter to be used in the estimate of the Coriolis matrix elements for the odd particle includes the modification in the rotational properties of the even-even core produced by the presence of the particle, but not the inertial effect (5-73) of the particle itself. As illustrated by the particle-rotor model (see Appendix 4A), the reaction of the particle on the rotor is described by the combined effect of the Coriolis and recoil terms, both involving the moment of inertia of the rotor (see Eq. (4A-8c) and the discussion on p. 205). The recoil term partly implies corrections to the one-particle energies, but in the example discussed above, these energies are taken from the experimental spectrum and therefore include such corrections. In addition, the recoil term modifies the wave functions of the particle so as to reduce the amplitude of the $j=15/2$ component; however, since the states that can be admixed with large matrix elements of $j_1^2+j_2^2$ have excitation energies of more than 5 MeV (see Fig. 5-5), the resulting effect on the Coriolis matrix elements is estimated to be less than a few percent.

A_7 term in ground-state band

Another characteristic feature of the bands based on the single-particle orbits with large j is the relatively large signature-dependent term in the rotational energy. The analysis of the empirical energies yields the value $A_7 \approx (1.0 \pm 0.4) \times 10^{-4}$ eV for the coefficient in the leading-order signature-dependent term in the expansion (4-62). The error is rather large, since the determination must be made in the states of low angular momentum, where the expansion (4-62) is rapidly convergent; higher-order terms become of significance already for $I \approx 17/2$.

The A_7 term is expected to receive its main contribution from the seventh-order Coriolis effect involving the sequence of one-quasiparticle states [743 7/2]→[752 5/2]→[761 3/2]→[770 1/2]→ $\overline{[770\ 1/2]}$ → $\overline{[761\ 3/2]}$ → $\overline{[752\ 5/2]}$ → $\overline{[743\ 7/2]}$,

$$A_7 = (A_0)^7 \frac{\langle 7/2|j_+|5/2\rangle^2 \langle 5/2|j_+|3/2\rangle^2 \langle 3/2|j_+|1/2\rangle^2 \langle 1/2|j_+|\overline{1/2}\rangle}{(E_{7/2}-E_{5/2})^2 (E_{7/2}-E_{3/2})^2 (E_{7/2}-E_{1/2})^2} \tag{5-74}$$

The energies of the $K=3/2$ and $K=1/2$ intermediate states are not known, but can be estimated from the single-particle spectrum in Fig. 5-5 and the expression (5-57) for the effect of the pair correlations. Assuming $\delta = 0.25$, $\Delta = 0.7$ MeV, and $\lambda \approx \varepsilon([743\ 7/2])$, one obtains $E_{3/2} - E_{7/2} \approx 1.05$ MeV and

▼ $E_{1/2}-E_{7/2}\approx 1.4$ MeV. The one-particle matrix elements of j_+ are listed in Table 5-6; the pairing effect is significant only for the $7/2\rightarrow 5/2$ matrix element (for which the correction factor is given by Eq. (5-69)). With these energies and coupling matrix elements and with $A_0=7.4$ keV, one obtains $A_7\approx 10^{-3}$ eV, which is about a factor of 10 greater than the observed value.

The empirical value of A_7 thus further supports the conclusion that the magnitude of the particle-rotation coupling is reduced with respect to the estimate based on the Coriolis interaction. If one employs the empirical value for the amplitudes $\varepsilon_{\pm 1}$ associated with the coupling between the $K=7/2$ and $K=5/2$ bands, the estimate of A_7 exceeds the observed value by about a factor of 2. Thus, it appears that the coupling matrix elements among the excited bands are more nearly in accord with the Coriolis estimates than is the case for the $7/2\leftrightarrow 5/2$ coupling, though the conclusion is tentative, in view of the uncertainty in the location of the expected $K=3/2$ and $K=1/2$ bands.

Inertial parameters in [633 5/2] band

For the low-lying positive parity bands in ^{235}U, the observed moments of inertia exceed those of the neighboring even-even nuclei by about 20%, which is typical for the configurations that do not have exceptionally large single-particle angular momenta (see Table 5-17). The very large moment of inertia for the [633 5/2] band may be associated with the rather strong coupling to the [624 7/2] band, which has been tentatively identified at about 500 keV (see the reference on p. 274). This interpretation receives additional support from the observed large positive fourth-order term in the rotational energy ($B\approx +25$ eV); the coupling of two bands yields $\delta B_1=(\delta A_1)^2(E_2-E_1)^{-1}$, which is approximately fulfilled for $(E_2-E_1)\approx 130$ keV and $\delta A_1\approx -1.8$ keV, corresponding to the difference between the A coefficients for the [633 5/2] band and the lower-lying positive parity bands. Such an interpretation implies a corresponding positive contribution to A and a large negative contribution to B in the [624 7/2] band.

Interpretation of Coriolis effect in E2 moments in terms of polarization charge

The expression (5-67) exhibits the $E2$ moment for $\Delta K=1$ transitions as a sum of an intrinsic moment $\mathscr{M}(E2,\nu=\pm 1)$ and a contribution resulting from the Coriolis coupling. The intrinsic moment would vanish for a pure neutron excitation, but receives polarization contributions from the coupling to intrinsic excitations with $\Delta K=1$. The contribution of the high-frequency quadrupole excitations ($\Delta N=2$) can be estimated on the basis of the model discussed in Chapter 6, which yields $\delta e_{\text{pol}}\approx 0.6e$ for a neutron. (See Eq. (6-386); neglecting corrections of the order of the deformation parameter, the high-frequency polarization effect is independent of ΔK and the same as in a spherical nucleus.) Additional contributions arise from the low-frequency intrinsic excitations ($\Delta N=0$). The main quadrupole strength of the $\Delta N=0$ excitations

▲ with $\Delta K=1$ is associated with the $E2$ moment of the rotational mode and therefore

▼ does not appear as intrinsic excitations (see the analysis of the "spurious" mode in Chapter 6, pp. 446 ff.). If all the $\Delta K=1$ excitations had the same frequency, the rotational mode would collect the entire $E2$ strength associated with these excitations, but in the absence of this degeneracy, the intrinsic spectrum retains some $E2$ strength; estimates of the residual $E2$ strength associated with $\Delta N=0$ yield a contribution to the polarization charge of $\delta e_{\text{pol}} \approx 0.2e$ (Hamamoto, 1971). Thus, we expect a total polarization charge from the intrinsic excitations of about $e_{\text{pol,int}} \approx 0.8e$. (In view of the overestimate of the total $E2$ moment exhibited in Table 5-7, one may consider the possibility that the intrinsic polarization charge, which enters with a sign opposite to that of the Coriolis contribution, has been seriously underestimated. Larger values of $e_{\text{pol,int}}$ could arise if the $K\pi=1+$ excitation spectrum in the even-even nuclei contains relatively low-lying bands that are connected to the ground state with enhanced $E2$-matrix elements.)

The second term in the moment (5-67) can be viewed as a contribution to the total effective charge resulting from the coupling to the rotational motion. The structure of this term can be exhibited by employing the relation between the matrix elements of $j_\pm$ and $Y_{2\pm1}$ implied by the Y_{20} symmetry of the deformed field (see Eq. (5-1))

$$\frac{d}{dt}j_\pm = i[H,j_\pm] = i[V_2(r)P_2(\cos\vartheta), l_\pm]$$

$$= -i\left(\frac{24\pi}{5}\right)^{1/2} V_2(r)Y_{2\pm1}(\vartheta,\varphi) \tag{5-75}$$

where the intrinsic Hamiltonian H may include the interactions responsible for the monopole pair correlations. (In Eq. (5-75), the effect of the deformation on the spin-orbit coupling has been neglected.) For the total effective $E2$ moment, we thus obtain, from Eqs. (5-67), (5-68), and (5-75),

$$\langle \mathrm{v}=1,\Omega\pm1|\hat{\mathscr{M}}(E2,\nu=\pm1)|\mathrm{v}=1,\Omega\rangle = \langle \mathrm{v}=1,\Omega\pm1|r^2Y_{2\pm1}|\mathrm{v}=1,\Omega\rangle(e_{\text{eff,int}}+e_{\text{pol,rot}}) \tag{5-76}$$

with

$$e_{\text{pol,rot}} = \frac{3\hbar^2 Q_0 e}{2\mathscr{I}_0(E(\Omega\pm1)-E(\Omega))^2}\,\frac{\langle\Omega\pm1|V_2(r)\,Y_{2\pm1}|\Omega\rangle}{\langle\Omega\pm1|r^2Y_{2\pm1}|\Omega\rangle} \tag{5-77}$$

The expression (5-77) exhibits the sign and frequency dependence characteristic of a polarization charge resulting from coupling to a collective mode with zero frequency (see Eq. (6-216)). A qualitative estimate of $e_{\text{pol,rot}}$ may be obtained by assuming a deformed oscillator potential (see Eq. (5-5)) and employing the relations (4-72) and (4-104) for Q_0 and $\mathscr{I}_{\text{rig}}$

$$e_{\text{pol,rot}} \approx -2\frac{Ze}{A}\left(\frac{\mathscr{I}_{\text{rig}}}{\mathscr{I}_0}\right)\left(\frac{\hbar\omega_0\delta}{E(\Omega\pm1)-E(\Omega)}\right)^2 \tag{5-78}$$

Since $\hbar\omega_0\delta \approx 1.6$ MeV, while $E(\Omega\pm1)-E(\Omega)\approx 0.7$ MeV and $\mathscr{I}_{\text{rig}} \approx 2\mathscr{I}_0$ for ^{235}U, we obtain $e_{\text{pol,rot}} \approx -8e$. Thus, it is expected that the rotational coupling gives the dominant contribution to the $E2$ moment for all low-frequency $E2$ transitions with ▲ $\Delta K=1$.

▼ The structure of the rotational polarization effect can be further elucidated by noting that the energy $\hbar\omega_0\delta$ appearing in Eq. (5-78) corresponds to the excitation energy $\Delta\varepsilon$ of the main $\Delta K=1$ single-particle transitions out of which the rotational motion is generated (see the discussion of the moment of inertia on the basis of the cranking model, p. 82). If all the single-particle excitations had this single frequency and if there were no pair correlations, one would expect, as mentioned above, that the entire $\Delta K=1$ $E2$ strength would be collected in the rotational mode, and there would be no transition strength remaining in the one-particle excitations. Such a total screening of the single-particle transitions is a consequence of the above analysis, as can be verified most easily by considering the relation corresponding to Eq. (5-78) for the isoscalar quadrupole moment in a nucleus with $N=Z$; in this case, the polarizability coefficient representing the ratio between the induced moment and that of the bare particle is $\chi_{rot}=-2$ (since $\mathscr{I}_0=\mathscr{I}_{rig}$ in the absence of pairing) and thus exactly cancels the moment of the particle as augmented by the high-frequency polarizability ($\chi_{int}=1$, see Eq. (6-370)). The large value of $e_{pol,rot}$ obtained for the $\Delta K=1$ transitions in ^{235}U represents an "overscreening" of the intrinsic moment resulting partly from the fact that the single-particle excitation energy $\Delta\varepsilon$ is rather small compared with $\hbar\omega_0\delta$, and partly from the pair correlations, which reduce the moment of inertia below the rigid-body value and at the same time decrease the energies of the $\Delta v=0$ excitations appearing in the low-energy odd-A spectrum.

One-particle states in deformed (sd)-shell nuclei. Spectra of ^{25}Mg and ^{25}Al (Fig. 5-15; Tables 5-8 to 5-11)

Classification of odd-A spectra with $19 \leqslant A \leqslant 25$

For a large class of configurations in the light nuclei, the qualitative features of the observed spectra can be immediately understood in terms of the aligned coupling scheme based on one-particle motion in a spheroidal potential.[3] The especially well studied spectra of ^{25}Mg and ^{25}Al are shown in Fig. 5-15; it is seen that the energies and $I\pi$ quantum numbers of the observed states are consistent with a classification in terms of rotational bands characterized by quantum numbers $K\pi$. This interpretation is supported by the measured large $E2$-matrix elements within the bands and by the observed approximate validity of the leading-order $E2$- and $M1$-intensity relations (see Table 5-10).

The observed $K\pi$ quantum numbers of the intrinsic states in Fig. 5-15 correspond with those expected from Fig. 5-1 for a prolate deformation with $\delta \approx 0.4$ (see Table 4-15, p. 134). As shown in Table 5-8, such a classification can be extended to the other $T=1/2$ spectra in the beginning of the (sd) ▲ shell, with the exception of $A=17$, for which the low-lying states correspond

[3]See footnote 1, p. 212, for references to the papers establishing the aligned coupling scheme for the light nuclei. The comparison between these wave functions and those obtained from shell-model calculations played an important role in establishing the connection between independent-particle motion and collective rotations (Elliott, 1958; Redlich, 1958). The configuration mixing involved in the aligned coupling scheme has also been derived from a Hartree-Fock treatment of the deformed field (Kelson and Levinson, 1964; see also the review by Ripka, 1968).

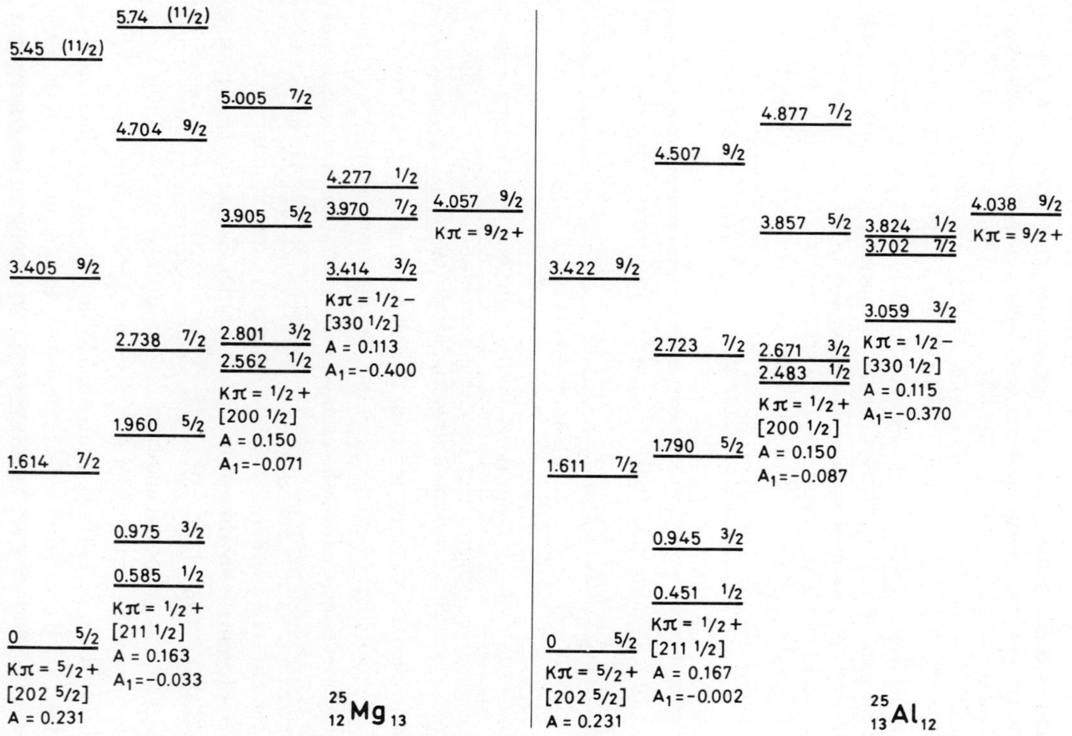

5.45 (11/2)
5.74 (11/2)
5.005 7/2
4.704 9/2
4.277 1/2
3.905 5/2
3.970 7/2
4.057 9/2
$K\pi$ = 9/2+
3.405 9/2
3.414 3/2
$K\pi$ = 1/2−
[330 1/2]
A = 0.113
A_1=−0.400
2.738 7/2
2.801 3/2
2.562 1/2
$K\pi$ = 1/2+
[200 1/2]
A = 0.150
A_1=−0.071
1.960 5/2
1.614 7/2
0.975 3/2
0.585 1/2
$K\pi$ = 1/2+
[211 1/2]
A = 0.163
A_1=−0.033
0 5/2
$K\pi$ = 5/2+
[202 5/2]
A = 0.231
$^{25}_{12}Mg_{13}$

4.877 7/2
4.507 9/2
3.857 5/2
3.824 1/2
3.702 7/2
4.038 9/2
$K\pi$ = 9/2+
3.422 9/2
3.059 3/2
$K\pi$ = 1/2−
[330 1/2]
A = 0.115
A_1=−0.370
2.723 7/2
2.671 3/2
2.483 1/2
$K\pi$ = 1/2+
[200 1/2]
A = 0.150
A_1=−0.087
1.790 5/2
1.611 7/2
0.945 3/2
0.451 1/2
$K\pi$ = 1/2+
[211 1/2]
A = 0.167
A_1=−0.002
0 5/2
$K\pi$ = 5/2+
[202 5/2]
A = 0.231
$^{25}_{13}Al_{12}$

Figure 5-15 Spectra of ^{25}Mg and ^{25}Al. The recognition of rotational band structure in the $A=25$ system followed from the extensive series of experiments at Chalk River (see the survey by A. E. Litherland, H. McManus, E. B. Paul, D. A. Bromley, and H. E. Gove, *Can. J. Phys.* **36**, 378, 1958). The figure is based on the experimental data summarized by Endt and van der Leun, 1967, and by Litherland, 1968. The tentative assignment of $I=11/2$ for the 5.45 and 5.74 MeV states in ^{25}Mg is based on the data of S. Hinds, R. Middleton, and A. E. Litherland, *Proceedings of the Rutherford Jubilee Intern. Conf.*, p. 305, ed. J. B. Birks, Heywood and Co., London 1961. Energies are in units of MeV.

▼ to a single particle outside the closed shells of ^{16}O (see Fig. 3-2b, Vol. I, p. 321).

As in heavier nuclei, the deformations are so great that orbits from adjacent major shells can appear in the low-energy spectra of the same nucleus. Examples in Table 5-8 are provided by the [101 1/2] band occurring in the beginning of the (sd) shell and by the [330 1/2] band in nuclei around $A=25$. (A striking effect of this type in the lighter nuclei is the occurrence of an even-parity ground state for the "p-shell" nucleus ^{11}Be (Alburger *et al.*, 1964), which appears to be a consequence of the strong favoring of the orbit [220 1/2] for prolate deformations.)

Besides the one-quasiparticle states in Table 5-8, there is tentative evidence for collective vibrational excitations at energies of about 3 or 4 MeV in the spectra of the odd-A nuclei in the (sd) shell. An example is provided by the $K\pi=9/2+$ band at 4.05 MeV in ^{25}Mg (see Fig. 5-15). The energy of this band is similar to that of the $K\pi=2+$ band in ^{24}Mg, occurring at 4.23 MeV, and the interpretation as a collective γ vibration is further supported by the relatively strong $E2$ transition probability $B(E2;\ 5/2+\rightarrow 9/2+)=34e^2\ \text{fm}^4 =(1.0\pm 0.2)B_W$, which is comparable with the transition probability $B(E2;\ 0\rightarrow K=I=2)=1.4B_W$ observed in ^{24}Mg. (See the references in the caption to Table 5-10; in the odd-A nucleus, the γ-vibrational excitation leads to two bands with $K=\Omega\pm 2$, each of which carries half the transition strength.)

Table 5-8 includes only states with $T=1/2$. For $T=3/2$ states, the evidence is much less detailed, but it appears that the tendency to establish an aligned coupling scheme is not as pronounced as for the $T=1/2$ states. For example, the spectrum of ^{19}O, in contrast to ^{19}F, does not exhibit simple rotational relationships (see the review by Ajzenberg-Selove, 1972). The weakened significance of the aligned coupling scheme for the $T=3/2$ states may be attributed partly to the effect of the exclusion principle and partly to the role of the isovector interactions. As a result of the exclusion principle, an increase in the neutron excess requires the shifting of particles from orbits that are favored by the deformation to orbits that are less favored. The repulsive character of the isovector field in the nucleus (see the discussion in Vol. I, p. 148) implies that the strength of the deformed field is weakened if the particles outside closed shells have a large neutron excess (see also the discussion in Chapter 6, p. 522).

Shapes of nuclei with (sd) configurations

The strong preference for prolate deformation in the beginning of the (sd) shell can be understood from the fact that the energy of the [220 1/2] orbit decreases more strongly as a function of deformation than do any of the low-lying orbits in a potential of oblate shape (see Fig. 5-1). One can obtain a
▲ simple estimate of the quadrupole moment of the equilibrium shape by

Nucleus	[101 1/2] ($a=0.5$)	[220 1/2] ($a=2.0$)	[211 3/2]	[202 5/2]	[211 1/2] ($a=-0.1$)	[200 1/2] ($a=0.0$)	[202 3/2]	[330 1/2] ($a=-2.5$)
^{19}F	0.11 ($a=1.1$)	0 ($a=3.2$)						
^{21}Ne	2.79 ($a=0.7$)		0		2.80			(4.73)
^{23}Na	2.64 ($a=0.8$)	(4.43)	0		2.39 ($a=0.0$)			
^{25}Mg				0	0.58 ($a=-0.2$)	2.56 ($a=-0.5$)		3.41 ($a=-3.5$)

Table 5-8 Intrinsic states in deformed odd-A nuclei with $19 \leqslant A \leqslant 25$. The table gives the excitation energies (in MeV) of the lowest member of the bands associated with the various single-particle orbits, labeled $[Nn_3\Lambda\Omega]$. The classification is based on the data and analyses given by: ^{19}F (Ajzenberg-Selove and Lauritsen, 1959); ^{21}Ne (A. J. Howard, J. G. Pronko, and C. A. Whitten, Jr., *Phys. Rev.* **184**, 1094, 1969; A. A. Pilt, R. H. Spear, R. V. Elliott, D. T. Kelly, J. A. Kuehner, G. T. Ewan, and C. Rolfs, *Can. J. Phys.* **50**, 1286, 1972); ^{23}Na (J. Dubois, *Nuclear Phys.* **A104**, 657, 1967); ^{25}Mg (see Fig. 5-15).

employing the relation (4-186) for the harmonic oscillator potential. Thus, for ^{20}Ne, with the configuration $[220\ 1/2]^4$ in addition to the closed shells, one finds, from Eq. (4-72), the moment

$$Q_0 = 42\ \text{fm}^2 \tag{5-79}$$

assuming the mean square radius $\langle r^2\rangle = \frac{3}{5}(1.2\times A^{1/3}\ \text{fm})^2$. The value (5-79) is comparable to the intrinsic moment deduced from the rotational $B(E2)$ value (see Table 4-15, p. 134; quantal corrections associated with the zero-point fluctuations in the orientation of the intrinsic coordinate system lead to an increase of Q_0 by about 20% (see Eq. (4-145). There is also experimental evidence for a rather large $E4$ moment for ^{20}Ne, which can be understood in terms of the shape of the [220 1/2] orbit (see the discussion in Chapter 4, p. 141).

For a harmonic oscillator potential, one would expect a preference for oblate shapes at the end of the $N=2$ shell, associated with the hole states with the same asymptotic quantum number [220 1/2], now referring to the oblate potential. However, the spin-orbit force, which implies a tendency for a subshell structure at N or $Z=16$, considerably reduces the tendency to deformation in the latter part of the (sd) shell, and the spectra in this region do not exhibit well-developed rotational patterns like those for $19 \leqslant A \leqslant 25$. Tentative evidence for occurrence of oblate deformations in the region of $A=28$ is provided by the quadrupole moment of the $2+$ state of ^{28}Si ($Q=+17$ fm^2; Häusser *et al.,* 1969), corresponding to $Q_0=-60$ fm^2 and $\delta \approx -0.4$. A striking confirmation of the oblate shape is the observation of a well-developed rotational band with $K\pi=7/2-$ in ^{29}Si (Viggars *et al.,* 1973). As seen from Fig. 5-1, such a band corresponds to the configuration [303 7/2], expected for oblate deformations, while for a prolate deformation, the lowest odd-parity band has $K\pi=1/2-$, as in the spectrum of $A=25$ (see Fig. 5-15, p. 284).

In the region around $A=24$, the single-particle level structure in Fig. 5-1 suggests that there may be a tendency toward shapes of ellipsoidal symmetry, since the two orbits [211 3/2] and [211 1/2] are strongly coupled by a one-particle field of Y_{22} symmetry. For a harmonic oscillator potential, the lowest configuration for $A=24$ is $[n_1=0, n_2=0, n_3=2]^4[n_1=1, n_2=0, n_3=1]^4$, in addition to the closed shells, and the equilibrium shape obtained from Eqs. (4-113) and (4-115) implies

$$\langle x_1^2\rangle : \langle x_2^2\rangle : \langle x_3^2\rangle = 20^2 : 16^2 : 28^2 \tag{5-80}$$

corresponding to the intrinsic moments

$$Q_0 = \frac{Z}{A}\Big\langle \sum_{k=1}^{A} (2x_3^2 - x_1^2 - x_2^2)_k \Big\rangle = 55\ \text{fm}^2$$

$$Q_2 = \left(\frac{3}{2}\right)^{1/2} \frac{Z}{A}\Big\langle \sum_{k=1}^{A} (x_1^2 - x_2^2)_k \Big\rangle = 11\ \text{fm}^2 \tag{5-81}$$

▼ and the asymmetry parameter

$$\tan\gamma = \sqrt{2}\,\frac{Q_2}{Q_0} = 0.27 \qquad (5\text{-}82)$$

The value (5-81) for Q_0 is in agreement with the experimental data (see Table 4-15, p. 134), and the value of Q_2 is comparable with that obtained from the $E2$-transition probability for exciting the second $2+$ state at 4.23 MeV (see Eq. (4-245) and the $B(E2)$ value quoted on p. 285). However, the available evidence on the spectrum of ^{24}Mg leaves open the question whether the $K=2$ band represents the rotation of a triaxial shape or a vibrational excitation (see the discussion of the similar issue in the spectrum of ^{166}Er, pp. 164ff.).

One may also attempt to bring to bear the evidence from odd-A nuclei in the region of $A=24$ to the question of axial symmetry (see, for example, Chi and Davidson, 1963). For a nucleus with ellipsoidal symmetry, each band in an odd-A nucleus has the states $I=1/2,(3/2)^2,(5/2)^3,\ldots$ (see Eq. (4-294)). The occurrence of three $I=1/2$ states in the low-energy spectrum of $A=25$ therefore implies that at least three different intrinsic configurations are involved; such an interpretation introduces a rather large number of parameters, and it has not yet been possible to establish relationships that are specific to the asymmetric rotor coupling scheme.

Separation of spurious states associated with rotational motion

In light nuclei, the rotational degrees of freedom are associated with the motion of rather few particles and the problem of the redundance in the total degrees of freedom is therefore encountered in an especially acute form. We shall illustrate aspects of the problem by considering the lowest states for configurations with three particles in the (sd) shell. The significance of finite particle number for nuclear rotations is further discussed in the example in Chapter 4, pp. 84ff.

The intrinsic excitations that are involved in the rotational motion are those generated from lower configurations by the action of the operators $J_{\pm}$ associated with infinitesimal rotations. Thus, if we consider the ground state of ^{19}F with the configuration $[220\ 1/2]^3$, the rotation operators produce linear combinations of configurations consisting mainly of ($[220\ 1/2]^2[211\ 3/2]$) and ($[220\ 1/2]^2[211\ 1/2]$). These configurations would be expected to provide the lowest-lying even-parity excitations in the independent-particle approximation. Taking into account the degeneracy of each orbit with respect to the sign of Ω and the different ways of distributing the neutrons and protons among the different orbits, one obtains, from these configurations, five states with $K=1/2$, four states with $K=3/2$, and one state with $K=5/2$. Of these, one $K=1/2$ and one $K=3/2$ have $T=3/2$, while the rest have $T=1/2$. The rotational degree of freedom corresponds to an isoscalar mode of excitation with $\Delta K=1$ and therefore uses one linear combination of the $K=1/2$ states with $T=1/2$, and one with $K=3/2$, $T=1/2$. Hence, we are left with the ▲

▼ following nonspurious intrinsic excitations: $(K=1/2,\ T=1/2)^3$, $(K=3/2,\ T=1/2)^2$, $(K=5/2,\ T=1/2)$, $(K=1/2,\ T=3/2)$, and $(K=3/2,\ T=3/2)$.

In the case considered, all the nonspurious intrinsic excitations are expected to have rather high excitation energies as a result of the residual interactions between the nucleons. These interactions favor states with low total isospin (as manifested, for example, by the symmetry potential in the nucleus), and this feature can be seen as a special case of the tendency to favor states that are symmetric under exchange of the space coordinates of the interacting particles. (This tendency is expressed in terms of effective nucleonic interactions depending on the space-exchange operator; see, for example, the Serber force considered in Vol. I, p. 258.) The ground state of ^{19}F, in the limit of the asymptotic quantum numbers, has a completely symmetric orbital wave function $[220]^3$ coupled to the completely antisymmetric spin-isospin wave function ($[f]=[3]$ in orbital space and $[f]=[111]$, $(S,T)=(1/2,1/2)$ in spin-isospin space; see Table 1C-5, Vol. I, p. 134.) The excited orbital configuration ($[220]^2[211]$) gives rise to a single state that is fully symmetric ($[f]=[3]$), and this state is the one obtained from the ground state by the action of the rotation operator J_+. It therefore follows that all the nonspurious states enumerated above belong to the lower orbital symmetry $[f]=[21]$. (The configuration $[220]^2[211]$ gives rise to a single band with $\Lambda=1$; coupling this band to the spin-isospin states $(S,T)=(1/2,1/2)$, $(1/2,3/2)$, and $(3/2,1/2)$ (see Table 1C-5), we obtain the multiplicity of bands enumerated above.) A systematic procedure for separating between the rotational and intrinsic degrees of freedom for particle motion in a harmonic oscillator potential can be based on the SU_3 classification (Elliott, 1958; for the p-shell configurations, this classification is given in Table 1C-4, Vol. I, p. 132); methods appropriate to particle motion in potentials of a more general type are discussed in Chapter 6, pp. 446 ff.

In the spectrum of ^{19}F, the only states, besides those of the ground-state band, with well-established $(sd)^3$ configurations are the $T=3/2$ states that are isobaric analogs of the low-lying states of ^{19}O. These lowest $T=3/2$ states in ^{19}F occur at about 7.5 MeV, which is considerably larger than the one-particle energy difference between the orbits [220 1/2] and [211 3/2] in Fig. 5-1. These $T=3/2$ states do not exhibit the rotational patterns characteristic of the aligned coupling scheme (see p. 285). The lowest $T=1/2$ intrinsic excitations in ^{19}F comprise a $K\pi=1/2-$ band associated with the $p^{-1}(sd)^4$ configuration (see Table 5-8) and a tentatively assigned $K\pi=3/2+$ band starting at 3.91 MeV (Dixon *et al.*, 1971); the low energy and regular rotational band structure of the positive parity excitations suggest that the shell-model configuration is predominantly $p^{-2}(sd)^5$ or $p^{-4}(sd)^7$.

Another spurious mode of excitation is that associated with the center-of-mass motion. Since the generator of translations is the total momentum **P**,
▲ the spurious excitations have the quantum numbers $\Delta K=0$ or 1 and $\pi=-1$.

▼ The odd-parity excitations in Table 5-8 have this symmetry, but very small matrix elements of **P**, since this operator obeys the selection rules $\Delta n_3 = 1$, $\Delta\Lambda = 0$ or $\Delta n_3 = 0$, $\Delta\Lambda = 1$. Hence, the properties of these excitations are not expected to be significantly affected by the removal of the spurious degree of freedom associated with the center-of-mass motion.

Decoupling parameters

The classification in Table 5-8 is further supported by measurements of various properties of the intrinsic states. Thus, the experimentally determined decoupling parameters for the $K=1/2$ bands are given in Table 5-8 and can be compared with the theoretical values listed together with the configuration. The theoretical estimates are obtained from Eq. (5-46) together with the wave functions in Table 5-9, and are seen to provide an interpretation of the characteristic differences observed for the decoupling parameters of the different bands. For the orbits [220 1/2] and [330 1/2], which have especially large moments of inertia, higher-order rotational coupling effects can be important. Such effects may be responsible for the theoretical underestimate of a for these bands; see the discussion on p. 308. ▲

$[Nn_3\Lambda\Omega]$	$j=1/2$	$j=3/2$	$j=5/2$	$j=7/2$
[101 1/2]	0.920	0.392		
[220 1/2]	−0.523	−0.285	0.803	
[211 3/2]		−0.236	0.972	
[202 5/2]			1.000	
[211 1/2]	0.419	0.735	0.533	
[200 1/2]	0.743	−0.615	0.265	
[202 3/2]		0.972	0.236	
[330 1/2]	0.279	−0.646	−0.188	0.685

Table 5-9 Single-particle wave functions for nuclei with $19 \leqslant A \leqslant 25$. The table gives the expansion coefficients $\langle Nlj\Omega|\nu\rangle$ of the one-particle orbits ν labeled by the asymptotic quantum numbers $[Nn_3\Lambda\Omega]$. The wave functions are obtained from the Hamiltonian (5-10) employing the parameters of Table 5-1 and the deformation $\delta=0.4$. The phases are as in Table 5-2b and Eq. (5-17).

▼ *M1- and GT-matrix elements for $A=25$*

The empirical evidence on $M1$ moments in the [202 5/2] and [211 1/2] bands in ^{25}Mg and ^{25}Al is collected in Table 5-10 and is analyzed on the basis of expressions (4-87) and (4-88). For the $K=5/2$ bands, the constancy of the quantity $(g_K - g_R)$ derived from the different matrix elements provides tests of the rotational coupling scheme. ▲

	^{25}Mg	^{25}Al
Measured quantity	$K=5/2$ [202 5/2]	
$Q(5/2)$	$Q_0 \approx 60$	
$B(E2;7/2\rightarrow 5/2)$	$Q_0 = 55 \pm 6$	$Q_0 \approx 50$
$B(E2;9/2\rightarrow 7/2)$	$Q_0 = 30 \pm 15$	$Q_0 = 45 \pm 15$
$B(E2;9/2\rightarrow 5/2)$		$Q_0 = 40 \pm 15$
$\mu(5/2)$	$g_K = -0.68 - 0.4(g_R - 0.5)$	
$B(M1;7/2\rightarrow 5/2)$	$g_K - g_R = -1.3 \pm 0.1$	$g_K - g_R = 1.4 \pm 0.2$
$B(M1;9/2\rightarrow 7/2)$	$g_K - g_R = -1.5 \pm 0.3$	$g_K - g_R = 1.4 \pm 0.4$
	$(g_K)_{\text{calc}} = -0.76x$	$(g_K)_{\text{calc}} = 0.8 + 1.11x$
	$K=1/2$ [211 1/2]	
$B(E2;3/2\rightarrow 1/2)$	$Q_0 \approx 60$	$Q_0 \approx 55$
$B(E2;5/2\rightarrow 3/2)$	$Q_0 \approx 75$	$Q_0 \approx 75$
$B(E2;5/2\rightarrow 1/2)$	$Q_0 \approx 65$	$Q_0 \approx 110$
$B(E2;7/2\rightarrow 3/2)$	$Q_0 \approx 60$	$Q_0 \approx 60$
$B(M1;3/2\rightarrow 1/2)$	$\begin{cases} g_K - g_R = 1.2 \pm 0.3 \\ b(g_K - g_R) = 0.2 \pm 0.3 \end{cases}$	
$B(M1;5/2\rightarrow 3/2)$		$\begin{cases} g_K - g_R = -1.4 \pm 0.3 \\ b(g_K - g_R) = -0.1 \pm 0.3 \end{cases}$
$B(M1;7/2\rightarrow 5/2)$		
	$(g_K)_{\text{calc}} = 2.5x$	$(g_K)_{\text{calc}} = 1.6 - 3.6x$
	$(b(g_K - g_R))_{\text{calc}} = -0.03 + 0.7x$	$(b(g_K - g_R))_{\text{calc}} = 0.2 - 1.0x$

Table 5-10 $E2$- and $M1$-matrix elements for $A=25$. The values of Q_0 (in fm^2) and of the magnetic g factors are derived from the moments and intraband transition probabilities in column one. The calculated values are based on the wave functions in Table 5-9, and the quantity x represents the ratio between the effective value of g_s and that for a free nucleon. The empirical data are taken from the review by Litherland (1968) and from J. F. Sharpey-Schafer, R. W. Ollerhead, A. J. Ferguson, and A. E. Litherland, *Can. J. Phys.* **46**, 2039 (1968), N. Anyas-Weiss and A. E. Litherland, *Can. J. Phys.* **47**, 2609 (1969), and T. K. Alexander, O. Häusser, A. B. McDonald, and G. T. Ewan, *Can. J. Phys.* **50**, 2198 (1972), as well as from references quoted in the latter article.

▼ The empirical values of g_K and b may be compared with the values calculated from expressions (5-86) and (5-87a) by means of the wave functions in Table 5-9. The calculated values in the table assume $g_l = (g_l)_{\text{free}}$ and $g_R = 0.5$, and are given in terms of the parameter x representing the ratio of $(g_s)_{\text{eff}}$ and $(g_s)_{\text{free}}$. For the $K=5/2$ orbit, the observed value of g_K can be ▲ accounted for by taking $x \approx 0.9$.

▼ Such a small value of the spin polarization effect is, however, in conflict with the information obtained from the GT transition probability in the β decay of ^{25}Al. From the observed reduced lifetime (ft=3900) for the ground-state transition, one obtains from Eqs. (3-46), (3-50), and (4-91)

$$\langle K=5/2, M_T=1/2|\sum \sigma_3 t_+|K=5/2, M_T=-1/2\rangle = 0.75 \qquad (5\text{-}83)$$

to be compared with the value 1.0 obtained from the wave function in Table 5-9 for the [202 5/2] configuration. The matrix element (5-83) is related by rotational invariance in isospace to the isovector spin contribution to g_K, and the rate of the β decay thus implies $x \approx 0.75$. The discrepancy in the value of the spin renormalization deduced from magnetic moments and β decay is similar to the situation encountered in the one-particle $d_{5/2}$ configuration in $A=17$ (see Vol. I, p. 346). In both cases, the discrepancy can be resolved by introducing a renormalization of the orbital g factor of about $\delta g_l \approx -0.1\tau_z$. (Evidence for such an effect in the magnetic moments in the region of ^{208}Pb is referred to on p. 486.)

For the $K=1/2$ orbit in Table 5-10, the observed value of g_K is consistent with $x \approx 0.7$, but is not of sufficient accuracy to test for the renormalization of g_l. For the b parameter, the evidence may suggest a larger spin polarization effect. Such a larger reduction in the effective moment for the transverse spin component is expected as a general consequence of the approximate validity of the asymptotic quantum numbers, which tends to reduce the magnitude of the spin fluctuations in the longitudinal direction in comparison with the fluctuations in the transverse direction (see the discussion on p. 304). In the case of the ^{24}Mg core, the transverse spin fluctuations are especially strong on account of the large matrix elements connecting the filled [211 3/2] and the empty [211 1/2] orbits.

The β decay of ^{25}Al has a weak branch to the $I=7/2$ member of the ground-state band of ^{25}Mg. The reported intensity of about 0.1% (see Endt and van der Leun, 1967) implies $ft \approx 2\times 10^5$ and leads to an intrinsic matrix element that is about a third of the value (5-83). Such a large violation of the rotational intensity relation appears surprising, especially in view of the agreement exhibited by the $M1$-matrix elements.

The conflicting character of the evidence obtained from the GT and $M1$ matrix elements for the $I=5/2$ to the $I=7/2$ transition can be more precisely exhibited by employing the expression (1-65) for the magnetic moment operator together with the rotational invariance in isospace, which yield the relation

$$\begin{aligned}
&\langle I=7/2, M_T=1/2\|\sum t_+\sigma\|I=5/2, M_T=-1/2\rangle \\
&\quad = \frac{1}{4.71}\Big(-\langle I=7/2, M_T=1/2\|\mu\|I=5/2, M_T=1/2\rangle \\
&\qquad +\langle I=7/2, M_T=-1/2\|\mu\|I=5/2, M_T=-1/2\rangle \\
&\qquad -\langle I=7/2, M_T=1/2\|\sum \tau_z l\|I=5/2, M_T=1/2\rangle\Big) \\
&\quad = 1.8-0.21\langle I=7/2, M_T=1/2\|\sum \tau_z l\|I=5/2, M_T=1/2\rangle \qquad (5\text{-}84)
\end{aligned}$$ ▲

where we have inserted the observed $M1$-matrix elements for ^{25}Mg and ^{25}Al, given in Table 5-10. The empirical value for the reduced GT matrix element (5-84) is about 0.35 and, hence, the above relation requires a value of about 7 for the isovector orbital matrix element. Such a large value seems improbable; in fact, the isovector orbital matrix element receives no contributions from the even-even core of ^{24}Mg (which has $T=0$), and the odd neutron gives the contribution 2.6 in the rotational coupling scheme.

One-particle transfer for $A=25$

The strong population of the low-lying states of ^{25}Mg in the ^{24}Mg(dp) reaction confirms the interpretation of the states in terms of one-particle motion with respect to the ^{24}Mg core. The reaction amplitudes yield the parentage factors, from which one can determine the expansion coefficients $\langle lj\Omega|\nu\rangle$ for the one-particle states (see Eq. (5-42)). The probabilities $\langle lj\Omega|\nu\rangle^2$ obtained in this way are compared in Table 5-11 with the theoretical values corresponding to the wave functions given in Table 5-9. In the theoretical estimates, no effects of pair correlations have been included, since in the present case the occurrence of a collective pair field seems unlikely. In fact, the separation between the filled and empty single-particle levels in Fig. 5-1 is about twice the parameter Δ determined from the odd-even mass differences (see Fig. 2-5, Vol. I, p. 170); this feature is confirmed by the absence of hole states in the observed spectrum up to about 4 MeV of excitation.

The theoretical values in Table 5-11 provide a qualitative interpretation of the observed intensity patterns; in a quantitative comparison, one must consider multistep processes as well as effects of the Coriolis coupling. The importance of multistep processes is suggested by the fact that the inelastic scattering of deuterons and protons, leading to rotational excitations of the ground-state band, have cross sections similar in magnitude to the elastic scattering, except for the forward angles (see, for example, Schulz *et al.*, 1970). Studies of the effects of multistep processes in ^{24}Mg(dp) have been made by Schulz *et al.*, 1970; Mackintosh, 1970; Braunschweig *et al.*, 1971. Rather direct evidence on these processes is provided by the excitation of the $I=7/2$ member of the ground-state band of ^{25}Mg (Middleton and Hinds, 1962; Hosono, 1968). A small amplitude for this transition results from the admixture of $N=4$ components in the intrinsic one-particle wave function. This component can be estimated from the matrix element of the deformed potential (5-5) (see Eqs. (3A-14) and (2-154))

$$\langle N=4, g_{7/2}, \Omega=5/2|\tfrac{2}{3}\delta M\omega_0^2 r^2 P_2|N=2, d_{5/2}, \Omega=5/2\rangle = \left(\frac{2}{21}\right)^{1/2}\hbar\omega_0\delta \quad (5\text{-}85)$$

With $\delta\approx 0.4$ and $\Delta\varepsilon\approx 2\hbar\omega_0$, we obtain a $g_{7/2}$ amplitude of about 0.06, which contributes a direct (dp) cross section of less than 10^{-3} of the ground-state transition. The observed intensity is of the order of 10^{-1} times the ground-state transition; thus, it appears that multistep processes provide the main mechanism for populating the $I=7/2$ level.

ν		$\langle lj\Omega\|\nu\rangle^2$			
		$j=1/2$	$j=3/2$	$j=5/2$	$j=7/2$
[202 5/2]	obs			(1.0)	*a*
	calc			1.0	
[211 1/2]	obs	0.23	0.44	0.19	*a*
	calc	0.18	0.54	0.28	
[200 1/2]	obs	0.08	0.40	0.02	
	calc	0.55	0.38	0.07	
[330 1/2]	obs	0.2	0.6		1.4
	calc	0.08	0.42	0.04	0.47

(*a*) Weak groups with intensities less than 10% of the ground-state transition and with approximately isotropic angular distribution.

Table 5-11 Evidence on one-particle states in ^{25}Mg obtained from the ^{24}Mg(dp) reaction. The table lists the square of the expansion amplitude, $\langle lj\Omega|\nu\rangle$, obtained from the observed ^{24}Mg(dp) differential cross sections at 25°, for deuterons of 15 MeV. The experimental data and the DWBA analysis are taken from B. Čujec, *Phys. Rev.* **136**, B1305 (1964); the single-particle cross sections have been normalized to the ground-state transition, which is assumed to have $\langle d_{5/2}, \Omega=5/2|\nu\rangle = 1$.

▼ The Coriolis coupling is especially large for the [330 1/2] band, which contains a large component of $f_{7/2}$. This coupling appears to be responsible for the observed large intensity to the $I=7/2$ member of this band. In the limit where the Coriolis coupling is large compared with the energy spacings of all the $f_{7/2}$ orbits, the whole $f_{7/2}$ strength (corresponding to four in the units employed in Table 5-11) is concentrated on the lowest $I=7/2$ state (spherical coupling scheme). An estimate of the Coriolis matrix element between the $I=7/2$ states of the [330 1/2] and [321 3/2] bands yields values of several MeV, which is comparable to the energy difference between the bands. A matrix element of this magnitude implies an intermediate coupling scheme, consistent with the observed intensity of the $f_{7/2}$ one-particle transfer process.

Coulomb energies

The detailed correspondence between the energy levels in ^{25}Mg and ^{25}Al, as exhibited in Fig. 5-15, provides striking evidence for the charge symmetry of nuclear forces (see Vol. I, p. 35). The quantitative comparison reveals ▲ smaller differences that may be attributed to the dependence of the nuclear

▼ charge (and magnetic moment) distribution on the quantum numbers of the different states. Because of the richness of the data and the simplicity of the coupling scheme for the deformed nuclei, the analysis of these effects is potentially a source of valuable information on the structure of the states. A detailed analysis is not yet available, and we shall confine ourselves to a few comments of a qualitative nature.

The intrinsic excitation energies in ^{25}Al differ from those in ^{25}Mg by amounts of the order of 100 keV, with a sign that implies that the Coulomb energy for an orbit along the symmetry axis ($n_3 \approx N$) is less than for an orbit in the equatorial plane ($n_\perp \approx N$). Thus, the sign is the same as the difference between the Coulomb field at the pole and the equator of a uniformly charged prolate spheroid; however, the quantitative estimate involves a delicate balance between the expectation value of the quadrupole component in the Coulomb field, which favors equatorial orbits, and the opposite effect of the isotropic part of the Coulomb field, which favors orbits along the symmetry axis, since these have somewhat larger radial excursions, as a consequence of the deformation of the nuclear field.

The quantitative interpretation of Coulomb energy shifts is expected to be influenced in an important manner by polarization effects similar to those affecting other one-particle matrix elements (Auerbach *et al.*, 1969; Damgaard *et al.*, 1970). Thus, the coupling of the individual particles to the collective monopole isovector excitations of the core may be expected to reduce significantly the isovector density at the position of the particle, and to redistribute it more uniformly over the nucleus as a whole. (The effect is similar to the polarization charge discussed in connection with the dipole and quadrupole isovector modes in Chapter 6, pp. 486 and pp. 514ff.)

Survey of intrinsic states in deformed odd-A nuclei with $150 < A < 188$ (Tables 5-12 and 5-13)

The spectra discussed in the previous examples illustrate the classification of deformed odd-A spectra on the basis of the intrinsic states of Figs. 5-1 to 5-5. For the odd-A nuclei in the region $150 < A < 188$, the systematic features of this classification are shown in Tables 5-12 and 5-13, which collect the available data on the location of intrinsic states. The one-particle orbitals are labeled by the asymptotic quantum numbers $Nn_3\Lambda\Omega$, and the numbers appearing in the tables are the excitation energies (in keV) of the band head ($I = K$) of the corresponding configuration. The assignments of the observed intrinsic states are based partly on the correspondence with the expected values of $\Omega\pi$, and partly on the more detailed features of the bands. Among these, the one-particle transfer intensities play an especially important role (see the example discussed on pp. 258ff.). Further evidence is provided by the rotational energies, moments, and transition probabilities, as discussed in the preceding examples. (See also the systematics of the g_K values in Table 5-14,

	532 5/2	413 5/2	411 3/2	523 7/2	411 1/2	404 7/2	514 9/2	402 5/2	541 1/2	
$^{153}_{63}Eu$	97	0	103							
^{155}Eu	104	0	246							
$^{155}_{65}Tb$	227	271	0							
^{157}Tb	326		0							*a*
^{159}Tb	364	348	0		(970)					*a*
^{161}Tb	480	315	0	417						
$^{161}_{67}Ho$	827			0	211					*a*
^{163}Ho				0	298	440				
^{165}Ho		995	362	0	429	716				*a*
$^{167}_{69}Tm$				293	0	179			(175)	
^{169}Tm				379	0	316				*a*
^{171}Tm				425	0	636				*a*
$^{171}_{71}Lu$				662	208	0	470	296	71	
^{173}Lu					425	0		357	128	
^{175}Lu						0	396	343	358	
^{177}Lu					570	0	150	458		

	532 5/2	413 5/2	411 3/2	523 7/2	411 1/2	404 7/2	514 9/2	402 5/2	541 1/2	
$^{177}_{73}$Ta						0	74	70	(217)	
^{179}Ta					520	0	31	239	750	
^{181}Ta					615	0	6	482		
^{183}Ta						0	73	459		
$^{181}_{75}$Re					826		262	0	432	
^{183}Re						851	496	0	702	
^{185}Re					880	872	387	0	1045	*a*
^{187}Re					626		206	0		*a*

(*a*) Evidence for occurrence of low-lying ($E<700$ keV) γ-vibrational excitations (^{157}Tb, ([411 3/2], 2+)1/2+, 598 keV; ^{159}Tb, ([411 3/2], 2+)1/2+, 581 keV; ^{161}Ho ([523 7/2], 2+)3/2−, 593 keV; ^{165}Ho ([523 7/2], 2+)3/2−, 515 keV; ([523 7/2], 2+)11/2−, 689 keV; ^{169}Tm ([411 1/2], 2+)3/2+, 571 keV; ^{171}Tm ([411 1/2], 2+)3/2+, 676 keV; ^{185}Re ([402 5/2], 2+)1/2+, 646 keV; ^{187}Re ([402 5/2], 2+)1/2+, 512 keV; ([514 9/2], 2+) 5/2−, 686 keV.)

Table 5-12 Classification of single-particle states in odd-proton nuclei with $63 \leqslant Z \leqslant 75$. The table lists the intrinsic states that have been identified as predominantly one-quasiparticle configurations. The entry gives the excitation energy (in keV) of the band head ($I=K=\Omega$) associated with the specified orbits. For a few bands, the $I=K$ member has not yet been identified, and for these, the energies in the table (given in parentheses) represent extrapolated values based on the observed higher members of the band. The table is based on the review by M. E. Bunker and C. W. Reich, *Rev. Mod. Phys.* **43**, 348 (1971); evidence for the [541 1/2] band in ^{167}Tm has been given by A. Johnson, K.-G. Rensfelt, and S. A. Hjorth, AFI Annual Report 1969, p. 23 (Research Institute for Physics, Stockholm).

▼ p. 303, the β-decay *ft* values in Table 5-15, p. 307, the decoupling parameters in Table 5-16, p. 309, and the moments of inertia in Table 5-17, p. 312.)

The sequence of the observed orbits in Tables 5-12 and 5-13 corresponds approximately to the theoretical spectra of Figs. 5-2 and 5-3 for $\delta \approx 0.3$, which represents a typical value of the equilibrium deformation for this region (see Fig. 4-25, p. 133). At these deformations ($\delta \sim A^{-1/3}$), one encounters crossings of orbits that originate in different major shells in the spherical potential. (Examples are provided by the presence of the [400 1/2], [402 3/2], and [505 11/2] neutron orbits at the beginning and the [541 1/2] proton orbit and [651 1/2] neutron orbit at the end of the region considered.)

The variations in δ, as well as in the higher multipoles in the nuclear shape, give rise to systematic variations of the orbits as observed in different nuclei. Thus, for example, the decrease in δ in the region beyond $A \sim 170$ (see Fig. 4-25) implies that the orbits from the higher shell ([541 1/2] for protons and [651 1/2] for neutrons) increase in energy relative to the other orbits. In the middle of the region considered, the main variation in the nuclear shape is associated with a decrease in β_4 (with increasing A), while δ remains approximately constant (see the evidence in Table 4-16, p. 140). This trend in β_4 favors the orbits that lie somewhat out of the equatorial plane as compared with those ($n_3 = 0$) lying in the plane (see Eq. (4-192) and the following discussion); such an effect is reflected in the observed energies of the proton orbits [411 1/2] and [514 9/2] relative to the [404 7/2] and [402 5/2] orbits. (The importance of the β_4 deformations for the level sequence in the isotopes of Tm and Lu has been discussed by Ekström *et al.*, 1971, 1972.)

The quantitative description of the odd-A spectra in terms of one-particle orbits in the deformed static potential is limited by the coupling to the rotational motion as well as to other intrinsic degrees of freedom. The Coriolis coupling is in most cases only a small perturbation (for not too large values of I), but for some of the orbits with large values of j (and small values of Ω), the coupling may become so large that the states cannot be described in terms of an intrinsic configuration with a well-defined value of K. This situation occurs for the neutron orbits referred to in footnotes *d* and *e* to Table 5-13. The associated bands exhibit major deviations from normal rotational spectra, but it appears that the observed patterns can result from the effect of the Coriolis force acting within the system of one-quasiparticle states (see, for example, Elbek, 1969, and Hjorth *et al*, 1970).

In the extensive investigations that underlie the classification in Tables 5-12 and 5-13, there is no evidence for states below 500 keV that cannot be associated with the one-quasiparticle configurations. In the energy region somewhat above 500 keV, a few additional intrinsic states have been found; these can be interpreted in terms of quadrupole vibrational excitations ▲ superposed on a lower-lying one-particle state (see footnotes to Tables 5-12

	530 1/2	532 3/2	400 1/2	660 1/2	402 3/2	651 3/2	505 11/2	521 3/2	642 5/2
$^{153}Sm_{91}$		127	415		321	0	98	36	
^{155}Gd	422		368		269	105	121	0	
^{157}Dy	(500)	399	388		307	235	199	0	
$^{157}Gd_{93}$	(809)	700	684		475		425	0	64
^{159}Dy	(735)	627	564		418	549	354	0	178
^{161}Er			481		369	463	396	0	(230)
$^{159}Gd_{95}$	(1120)	1109	973	780	744		681	0	68
^{161}Dy	(825)		608	774	550	679	486	75	0
^{163}Er	(815)		541		463		444	104	69
$^{161}Gd_{97}$								313	
^{163}Dy	(1250)		1057	736	857	1084	495	422	251
^{165}Er	(990)		746	507	534		591	243	47
$^{165}Dy_{99}$								574	
^{167}Er			1135		1086		1052	753	812
^{169}Yb								660	591
$^{169}Er_{101}$							1394	714	
^{171}Yb								902	
$^{171}Er_{103}$									
^{173}Yb								1224	
^{175}Hf									
$^{175}Yb_{105}$								1620	
^{177}Hf									
^{179}W									

Table 5-13

	523 5/2	633 7/2	521 1/2	512 5/2	514 7/2	624 9/2	510 1/2	512 3/2	503 7/2	
$^{153}Sm_{91}$			696							*d*
^{155}Gd	321		560							*a,d,f*
^{157}Dy	344		463	(900)						*d*
$^{157}Gd_{93}$	435		704							*a*
^{159}Dy	310		538	(1000)			(1440)			*d,e*
^{161}Er	172									*d,e*
$^{159}Gd_{95}$	146		507	873			1602			*a,b,d*
^{161}Dy	26		367	799			(1276)			*a,d*
^{163}Er	0		346	609			1074			*e*
$^{161}Gd_{97}$	0	446	356	809		800	1309			*b*
^{163}Dy	0	(417)	351	719			1159			*a,d*
^{165}Er	0		297	478			920			*d*
$^{165}Dy_{99}$	534	0	108	184			570	1258		*f*
^{167}Er	668	0	208	346			763	1384		*e,f*
^{169}Yb	570	0	24	191	960					*b*
$^{169}Er_{101}$	850	244	0	92	823		562	1082		
^{171}Yb		95	0	122	835	935	945			*b*
$^{171}Er_{103}$			195	0	531	378	706	906		
^{173}Yb		351	399	0	637		1031	1340		*b*
^{175}Hf		207	126	0	348					
$^{175}Yb_{105}$		(995)	920	639	0	265	514	811		*b*
^{177}Hf		746	560	509	0	321	590	804	1058	
^{179}W		478	222	430	0	309				

$^{177}Yb_{107}$				104	0	332	709	1226	*b*
^{179}Hf		614	518	214	0	375	720	872	
^{181}W	(954)	385	366	409	0	458	726	662	
$^{183}W_{109}$		936	905	1072	623	0	209	453	*c*
$^{185}W_{111}$		1013	888	1058	716	24	0	244	*c*
^{187}Os						0	10	100	*c*

Table 5-13 continued

(*a*) Evidence for the occurrence of the [404 7/2] orbit (^{155}Gd, 1295 keV; ^{157}Gd, 1825 keV; ^{159}Gd, 1960 keV; ^{161}Dy, 1416 keV; ^{163}Dy, 1840 keV).

(*b*) Evidence for the occurrence of the [651 1/2] orbit (^{159}Gd, 1977 keV; ^{161}Gd, 1489 keV; ^{169}Yb, 1590 keV; ^{171}Yb, 1610 keV; ^{173}Yb, 1630 keV; ^{175}Yb, 1366 keV; ^{177}Yb, 1380 keV).

(*c*) Evidence for the occurrence of the [615 11/2] orbit (^{183}W, 310 keV; ^{185}W, 198 keV; ^{187}Os, 257 keV).

(*d*) The Coriolis coupling between the [660 1/2] and [651 3/2] orbits implies strong band mixing. The $I\pi = 1/2+$ state has been identified in the cases listed by virtue of its $l=0$ one-particle transfer strength; the main part of the strength results from the [400 1/2] component admixed by $\Delta N = 2$ couplings, which are found to be of the order of 100 keV (see p. 230). In a similar manner, the $I\pi = 3/2+$ states are identified by $l=2$ transfer strength associated with [402 3/2] admixtures.

(*e*) The observed rotational energies of the [642 5/2] band indicate strong Coriolis mixing with the [651 3/2] and [660 1/2] configurations.

(*f*) Evidence for rather low-lying excitations ($E < 700$ keV) of β- or γ-vibrational character (^{155}Gd, ([521 3/2], 0+)3/2−, 593 keV; ^{165}Dy, ([633 7/2], 2+)3/2+, 539 keV; ^{167}Er, ([633 7/2], 2+)3/2+, 531 keV).

Table 5-13 Classification of single-particle states in odd-neutron nuclei with $91 \leqslant N \leqslant 111$. The table is similar to Table 5-12 and is based on the review by Bunker and Reich (*loc.cit.* Table 5-12); additional evidence derived from one-particle transfer reactions is taken from: Gd isotopes (P. O. Tjøm and B. Elbek, *Mat. Fys. Medd. Dan. Vid. Selsk.* **36**, no. 8, 1967); Dy isotopes (T. Grotdal, K. Nybø, and B. Elbek *ibid.* **37**, no. 12, 1970); Er isotopes (P. O. Tjøm and B. Elbek *ibid.* **37**, no. 7, 1969); W isotopes (R. F. Casten, P. Kleinheinz, P. J. Daly, and B. Elbek *ibid.* **38**, no. 13, 1972); [651 1/2] orbit in Yb isotopes (S. Whineray, F. S. Dietrich, and R. G. Stokstad, *Nuclear Phys.* **A157**, 529, 1970).

▼ and 5-13). At higher excitation energies, a greater density of states is observed, including three-quasiparticle states as well as vibrational excitations and single-particle states. The systematic interpretation of the spectra above about 1 MeV is still somewhat tentative due to the great variety of possible coupling effects between the elementary modes of excitation; however, on the basis of the one-particle transfer studies, it has been possible to identify a considerable number of individual states in the region of 1–2 MeV that appear to carry the main part of the strength associated with specific single-particle configurations. (See, for example, references in Table 5-13; for preliminary evidence on the one-particle strength function at still higher energies, see, for example, Back *et al.* 1974.)

Magnetic g factors for odd-A nuclei with $150<A<190$ (Table 5-14)

The gyromagnetic factors g_K and g_R, for a rotational band with $K \neq 1/2$, can be obtained by combining a measured magnetic moment with an $M1$-transition matrix element (see Eq. (4-87)). For $K=1/2$ bands, the $M1$-matrix elements involve the additional parameter b (see Eq. (4-88)), and the analysis requires the determination of an additional moment or transition probability (see, for example, the analysis of $M1$ moments in the $K=1/2$ band of ^{169}Tm in Table 4-7, p. 104). Table 5-14 contains the available data on the $M1$ parameters of rotational bands for odd-A nuclei with $150<A<190$.

As can be seen from Table 5-14, the g_K factor provides a characteristic property that can be exploited in establishing the classification of the one-particle configurations. The large variations in $g_K - g_R$ directly manifest themselves in the $E2:M1$ ratios for the rotational transitions with $\Delta I = 1$. Thus, the small values of $g_K - g_R$ for the neutron configurations [523 5/2] and [514 7/2] imply that the $\Delta I = 1$ transitions within a rotational band built on these states have large $E2$ admixtures (of the order of 50% or more); this feature provides a clear identification of these orbits. (See, for example, the striking difference in the $E2: M1$ ratio for the transitions within the [514 7/2] and [624 9/2] bands in ^{177}Hf, as exhibited in Table 4-8, p. 108.)

For the one-quasiparticle states, the g_K factor can be estimated from the expression (see, for example, Eq. (4A-12))

$$g_K = \frac{1}{\Omega}\langle \Omega | \, g_l l_3 + g_s s_3 | \Omega \rangle$$

$$= \frac{1}{\Omega}\left(g_l \Omega + (g_s - g_l)\langle \Omega | s_3 | \Omega \rangle \right) \tag{5-86}$$

where g_s and g_l are the spin and orbital g factors for the last odd nucleon. The factor resulting from the pair correlations is unity for these diagonal matrix elements (see Eq. (5-43)). For $\Omega = 1/2$, the magnetic decoupling parameter is ▲

Nucleus	Orbit	g_R	$(g_K)_{\text{obs.}}$	$(g_K)_{\text{calc.}}$	$(g_s)_{\text{eff}}/(g_s)_{\text{free}}$
		Odd-proton configurations			
^{153}Eu	413 5/2	0.47	0.67	0.30	0.57
^{159}Tb	411 3/2	0.42	1.83	2.28	0.71
^{165}Ho	523 7/2	0.43	1.35	1.53	0.72
^{169}Tm	411 1/2	0.41	−1.57	−2.44	0.79
			0.32*	−0.05*	0.47*
^{175}Lu	404 7/2	0.31	0.73	0.41	0.55
^{181}Ta	404 7/2	0.29	0.78	0.41	0.48
^{185}Re	402 5/2	0.42	1.61	1.90	0.74
^{187}Re	402 5/2	0.41	1.63	1.90	0.76
		Odd-neutron configurations			
^{155}Gd	521 3/2	0.32	−0.48	−0.61	0.79
^{157}Gd	521 3/2	0.26	−0.53	−0.61	0.87
^{161}Dy	642 5/2	0.21	−0.34	−0.45	0.76
^{161}Dy	523 5/2	0.32	0.17	0.39	0.44
^{163}Dy	523 5/2	0.27	0.25	0.39	0.64
^{167}Er	633 7/2	0.18	−0.26	−0.39	0.67
^{171}Yb	521 1/2	0.28	1.43	1.75	0.82
			−0.48*	−0.79*	0.71*
^{173}Yb	512 5/2	0.28	−0.49	−0.56	0.87
^{177}Hf	514 7/2	0.26	0.21	0.40	0.52
^{179}Hf	624 9/2	0.22	−0.22	−0.35	0.63

Table 5-14 Magnetic g factors for odd-A nuclei ($150 < A < 190$). The experimental data are taken from F. Boehm, G. Goldring, G. B. Hagemann, G. D. Symons, and A. Tveter, *Phys. Letters* **22B**, 627 (1966); additional values for Eu, Dy, Hf, and Re are obtained from the branching ratios and magnetic moments summarized in the reviews by Rogers (1965) and Lederer *et al.* (1967). The static moments for ^{177}Hf and ^{179}Hf are from S. Büttgenbach, M. Herschel, G. Meisel, E. Schrödl, and W. Witte, *Phys. Letters* **43B**, 479 (1973); the parameters for ^{169}Tm are those quoted on p. 102. For the $K=1/2$ bands, the quantities marked with an asterisk are the observed and calculated values of $(g_K - g_R)b$, and in the last column, the effective g_s factor derived from this transverse matrix element in units of $(g_s)_{\text{free}}$. We wish to thank I. L. Lamm for help in the preparation of this table.

▼ given by (see Eq. (4A-12))

$$(g_K - g_R)b = \langle \Omega = 1/2 | (g_l - g_R) l_+ + (g_s - g_R) s_+ | \overline{\Omega = 1/2} \rangle \tag{5-87a}$$

$$= -(g_l - g_R)a + (g_s - g_l)\langle \Omega = 1/2 | s_+ | \overline{\Omega = 1/2} \rangle \tag{5-87b}$$

▲

$$= -(g_l - g_R)a - \tfrac{1}{2}(-1)^l (g_s + g_K - 2g_l) \tag{5-87c}$$

▼ The relations (5-87b) and (5-87c) between the magnetic decoupling parameter b and the rotational decoupling parameter a given by Eq. (5-46) are specific to one-particle intrinsic states. In obtaining the relation (5-87c), we have exploited the identity

$$\begin{aligned}\langle\Omega=1/2|s_+|\overline{\Omega=1/2}\rangle \\ &=\langle\Lambda=0,\Sigma=1/2|s_+|\overline{\Lambda=0,\Sigma=1/2}\rangle\langle\Lambda=0,\Sigma=1/2|\Omega=1/2\rangle^2 \\ &=-(-1)^l\left(\tfrac{1}{2}+\langle\Omega=1/2|s_3|\Omega=1/2\rangle\right)\end{aligned} \qquad (5\text{-}88)$$

where $\langle\Lambda=0,\Sigma=1/2|\Omega=1/2\rangle^2$ is the probability for $s_3=\Sigma=1/2$, in the state $|\Omega=1/2\rangle$, and where $(-1)^l$ is the quantum number r for the $|\Lambda=0\rangle$ state, which equals the parity of the one-particle orbit.

The empirical data on g_K and b are compared in Table 5-14 with theoretical estimates based on Eqs. (5-86) and (5-87a), the wave functions in Table 5-2, and the g factors g_s and g_l corresponding to free nucleons. The empirical g_K values deviate systematically from these calculations, in a manner that can be approximately accounted for in terms of a spin-polarization effect of the type discussed in connection with the $M1$ moments of configurations involving a single particle outside of closed shells (see Vol. I, pp. 337ff.). The last column in Table 5-14 gives the value of $(g_s)_{\text{eff}}/(g_s)_{\text{free}}$, for which the one-particle wave functions of Table 5-2 yield the observed values of g_K. It is seen that the observed values of g_K indicate a polarization effect of a magnitude

$$(g_s)_{\text{eff}}\approx 0.7(g_s)_{\text{free}} \qquad (5\text{-}89)$$

For the bands with $K=1/2$, the information on the parameter $(g_K-g_R)b$ in Table 5-14 is also seen to provide evidence for a significant reduction in the effective g_s factor. The analysis in Table 5-14 assumes $g_l=(g_l)_{\text{free}}$; a renormalization of g_l by the amount $\delta g_l\approx -0.1\tau_z$, as suggested by other evidence (see pp. 292 and 486) would lead to a modification in the renormalization factors for g_s by amounts of the order of 0.1, for the orbits with largest Λ.

The spin-polarization effect can be described in terms of the coupling to excitations of the even-even core produced by spin-dependent fields (see the discussion in Sec. 6-3e). The renormalization of the longitudinal g factor g_K is associated with $\Delta K=0$ excitations, while the transverse matrix element $(g_K-g_R)b$ is affected by the $\Delta K=1$ excitations. For large deformations, the quantum numbers Λ and Σ become constants of the motion, and the spin fluctuations of $\Delta K=0$ type vanish; one may therefore expect that the longitudinal value of $(g_s)_{\text{eff}}$ is less affected by the renormalization than is the ▲ transverse value (Bochnacki and Ogaza, 1966; evidence for a difference in the

▼ transverse and longitudinal renormalization was discussed by Bodenstedt and Rogers, 1964). The fact that the values of $(g_s)_{\text{eff}}$, for the longitudinal moment, are comparable with, though perhaps somewhat larger than, the values observed in spherical nuclei (see Vol. I, p. 344) indicates that the quenching of the spin fluctuations, characteristic of the large deformation limit, is only partially achieved for the nuclear equilibrium deformations. (A further reduction in the spin polarization of the deformed nuclei, as compared with the closed-shell configurations, arises from the pair correlations. Since the pairing enhances the probability for finding nucleons in spin singlets, these correlations tend to reduce the amplitude of spin fluctuations.)

It should be emphasized that the above description of the renormalization effects in the $M1$ operator must be considered as a rather qualitative approximation, since the effective moment may contain terms of a new structure, which are not present in the moment of a single free particle. Thus, with the inclusion of spin-polarization effects, the operator μ_{eff} involved in the calculation of g_K contains the terms

$$(\mu_{\text{eff}})_{\Delta K=0} = g_{l0} l_3 + g_{s0} s_3 + g'_{s0}(s_3 Y_{20}) + g''_{s0} \frac{1}{\sqrt{2}} (s_{+1} Y_{2-1} + s_{-1} Y_{21}) \tag{5-90}$$

where the extra index 0 on the g factors refers to the value of ΔK. In a spherical nucleus, the effective spin-dependent moment contains only a single tensor involving Y_2 (see Eq. (3-44)). In the expression (5-90), we have neglected possible smaller terms involving higher spherical harmonics. The transverse component of μ_{eff}, which occurs in the $M1$ transitions between bands with $\Delta K = 1$ and in the calculation of the magnetic decoupling parameter for $K = 1/2$ bands, contains the terms

$$(\mu_{\text{eff}})_{\Delta K=1} = g_{l1} l_{+1} + g_{s1} s_{+1} + g'_{s1}(s_{+1} Y_{20}) + g''_{s1}(s_3 Y_{21}) + \mathscr{R}\text{-conj.} \tag{5-91}$$

In a phenomenological analysis, all the g factors appearing in Eqs. (5-90) and (5-91) appear as independent parameters describing the effective $M1$ operator. The determination of the separate contributions to the effective $M1$ moment would provide valuable information on the structure of the spin-dependent fields in the nucleus.

The most striking feature of the g_R values in Table 5-14 is the systematic tendency for the odd-proton values to exceed those of odd-neutron nuclei. The values of g_R for the ground-state bands of the even-even nuclei are given in Fig. 4-6, p. 55, and are seen to be intermediate between the values for the odd-Z and odd-N nuclei. The increments of the rotational g factors for the odd-A nuclei can be related to the fact that the moments of inertia of these nuclei are systematically greater than those of the even-even nuclei, which implies that the last odd particle contributes a significant fraction of the total ▲ current associated with the collective rotation in an odd-A nucleus. (An

▼ approximate relationship between the increments in the g_R factor and the moment of inertia is discussed in connection with the example, ^{159}Tb, considered on p. 255.)

Matrix elements for unhindered Gamow-Teller β transitions (Table 5-15)

The only allowed β transitions with nonvanishing matrix elements for the asymptotic wave functions are the transitions between members of an isobaric multiplet (unchanged orbit for the odd particle) and the transitions between a spin-orbit doublet $\Omega=\Lambda-1/2 \leftrightarrow \Omega=\Lambda+1/2$ (with other quantum numbers unchanged). An example of the first type ($^{25}\text{Al}\rightarrow^{25}\text{Mg}$) is discussed on p. 292; such transitions are energetically forbidden for nuclei with a neutron excess. In heavy nuclei, the Fermi levels for neutrons and protons belong to different major shells, and transitions between spin-orbit partners only occur for the orbits with very large j, for which the spin-orbit energy is of the same order as the energy difference between major shells. These transitions are therefore rather rare, and the observation of an allowed unhindered β decay provides a rather unique identification of the one-particle orbits involved (see the example, ^{175}Yb, discussed on p. 262).

The ft values for the transitions between spin-orbit partners, occurring in nuclei with $150<A<185$, are collected in Table 5-15 and are seen to cluster around the value $\log ft \approx 4.7$. Other allowed β transitions, which violate the asymptotic selection rules, are found to have $\log ft$ values in the range 6–8 (see the example ($^{159}\text{Gd}\rightarrow^{159}\text{Tb}$) discussed on p. 258 and the compilations by Mottelson and Nilsson, 1959, and Lederer *et al.*, 1967).

The ft values in Table 5-15 refer to transitions between band heads $(K, I=K \rightleftarrows K+1, I=K+1)$ and yield the transition probabilities (see Eqs. (3-46), (3-50), and (4-91))

$$\langle K+1|t_{\pm}\sigma_{+}|K\rangle^2\left(\frac{2I_{\mathrm{f}}+1}{2I_{>}+1}\right)=\frac{8.3\times 10^3}{ft\,(\mathrm{sec})} \tag{5-92}$$

where $I_>$ is the larger of I_{i} and I_{f}. The intrinsic matrix elements derived from the observed ft values thus fall in the range

$$|\langle K+1|t_{\pm}\sigma_{+}|K\rangle| \approx 0.4 \tag{5-93}$$

In the asymptotic limit, the one-particle matrix element of the operator $t_{\pm}\sigma_{\pm}$ equals 2; for the orbits in Table 5-15, the wave functions in Table 5-2 imply matrix elements that are about 10% smaller than this limiting value.

The correction for pair correlation is given by Eq. (5-44), and implies a reduction of the intrinsic matrix element by a factor of 2, if both the initial and final orbits are close to the Fermi energy ($u \approx v \approx 2^{-1/2}$ for $\varepsilon(\nu)\approx\varepsilon_F$). (In some cases in Table 5-15, the transitions go to excited states, and in these

▲ cases, the pair correlation factor is somewhat closer to unity; see, for

▼ example, the estimates given by Soloviev, 1961, and Żylicz *et al.*, 1967.) With the correction for the pair correlations, the calculated Gamow-Teller (GT) matrix element for transitions between one-quasiparticle states is seen to be about a factor of two larger than the measured value.

Polarization effects analogous to the spin polarization discussed for the $M1$-matrix elements are expected to lead to a significant reduction in the GT-matrix elements (see the discussion of polarization effects in $M1$ moments, p. 304, and in GT transitions in spherical nuclei, Vol. I, p. 347). The above evidence appears to suggest that the spin-polarization effect is about twice as large for the GT transitions as for the g_K factors. This difference may be qualitatively understood from the effects of neutron excess and deformation on the renormalization factors. If the even-even core has $N=Z$, the isovector part of the spin magnetic moment is related to the GT moment by a rotation in isospace (see Vol. I, p. 346 and the example ($A=25$) discussed on p. 292). With increasing neutron excess, the strength of the charge-exchange spin fluctuations increases, and one must expect a larger renormalization effect for the GT moment than for the $M1$ operator. Moreover, the nuclear deformation implies a reduction of the renormalization ▲ effect for the longitudinal ($\Delta K=0$) $M1$ moment (see the discussion on p. 304).

Transition	Parent	Daughter	log*ft*
$[523\ 7/2]_p \leftrightarrow [523\ 5/2]_n$	$^{159}_{67}$Ho	$^{159}_{66}$Dy (310)	~4.8
	$^{161}_{64}$Gd	$^{161}_{65}$Tb (418)	4.85 ± 0.04
	$^{161}_{67}$Ho	$^{161}_{66}$Dy (26)	4.8 ± 0.2
	$^{163}_{68}$Er	$^{163}_{67}$Ho (0)	4.83 ± 0.01
	$^{165}_{68}$Er	$^{165}_{67}$Ho (0)	4.64 ± 0.02
	$^{165}_{70}$Yb	$^{165}_{69}$Tm	4.8 ± 0.1
	$^{167}_{67}$Ho	$^{167}_{68}$Er (668)	4.8 ± 0.2
	$^{167}_{70}$Yb	$^{167}_{69}$Tm (293)	4.55 ± 0.05
	$^{169}_{67}$Ho	$^{169}_{68}$Er (850)	4.7
$[514\ 9/2]_p \leftrightarrow [514\ 7/2]_n$	$^{175}_{70}$Yb	$^{175}_{71}$Lu (396)	4.7
	$^{179}_{74}$W	$^{179}_{73}$Ta (31)	4.6
	$^{181}_{76}$Os	$^{181}_{75}$Re (262)	4.4

Table 5-15 Allowed unhindered β transitions in deformed odd-A nuclei with $150<A<185$. Table 5-15 contains the β transitions that are classified as "allowed unhindered" on the basis of the configuration assignments in Tables 5-12 and 5-13. The numbers in parentheses in column three give the excitation energy (in keV) of the state in the daughter nucleus. The log*ft* values are based on the critical review by J. Żylicz, P. G. Hansen, H. L. Nielsen, and K. Wilsky, *Arkiv Fysik* **36**, 643 (1967).

▼ *Decoupling parameters in rotational bands with $K=1/2$ (Table 5-16)*

Table 5-16 shows the systematics of the observed decoupling parameters for the low-lying single-quasiparticle configurations ($E<500$ keV) in the region $150<A<190$. For the nuclei with $19 \leqslant A \leqslant 25$, the corresponding data are contained in Table 5-8, p. 286. (Decoupling parameters for bands in actinide nuclei are given in Figs. 4-19 (^{239}Pu), 5-12 (^{237}Np), and 5-14 (^{235}U)). It is seen that the decoupling parameter has approximately the same value for a given quasiparticle state in different nuclei and provides an identification mark for the $K=1/2$ configurations.

The calculated values of a given in Table 5-16 have been obtained from Eq. (5-46), employing the wave functions in Table 5-2, which correspond to $\delta=0.3$. In the asymptotic limit, the decoupling parameter equals $a=0$ (for $\Lambda=1$) and $a=(-1)^N=\pi$ (for $\Lambda=0$). The observed and calculated values of a differ appreciably from these limiting values, reflecting the sensitivity of this parameter to the finer details of the wave functions. The orbits with large values of a are composed predominantly of a single j value ($f_{7/2}$ for [330 1/2] and $h_{9/2}$ for [541 1/2]), which gives a contribution $a(j)=(-1)^{j-1/2}(j+1/2)$ (see Eq. (5-46)).

The linear term in I in the rotational energy is an especially simple manifestation of the coupling between intrinsic and rotational motion; the approximate agreement between the theoretical estimates and observed values of the decoupling parameters in Table 5-16 is significant as a test of the assumed structure of the particle-rotation coupling (see the discussion on p. 253). In a more detailed analysis of the decoupling parameters, one must consider both the possibility of polarization effects and the consequences of the difference of the moment of inertia between the even-even and odd-A nuclei.

The expression (5-46) for a assumes that the inertial parameter A in the $K=1/2$ band is the same as the parameter A_0 describing the rotational energy of the even-even core; in cases where A and A_0 differ, the theoretical estimate of a involves the additional factor A_0/A (see Eqs. (4-61) and (5-45)). For most of the orbits in Table 5-16, this correction is rather small; in fact, the values of A are only 10–20% smaller than the values of A_0 observed in the ground-state bands of even-even nuclei (see Table 5-17, p. 312), and a significant part of this difference may be attributed to the modification of A_0 resulting from the presence of the odd particle (see the discussion on p. 311).

For the [541 1/2] proton orbit, however, the values of A are especially small (see Table 5-17), and this increment in the moment of inertia appears to result from the strong Coriolis coupling to the [532 3/2] and [530 1/2] orbits, which are expected to be rather close in energy (see Fig. 5-2); such an interpretation implies an increase in the theoretical estimate of a of about ▲ 30%, which would represent a significant improvement in comparison with

Orbit	Nucleus	E(keV)	$a_{obs.}$
$[411\ 1/2]_p$	^{165}Ho	429	−0.44
$a_{calc} = -0.9$	^{167}Tm	0	−0.72
$(a_{orb} = -1.0)$	^{169}Tm	0	−0.77
	^{171}Tm	0	−0.86
	^{171}Lu	208	−0.71
	^{173}Lu	425	−0.75
$[541\ 1/2]_p$	^{167}Tm	(175)	3.6*
$a_{calc} = 3.0$	^{171}Lu	71	4.0*
$(a_{orb} = 3.5)$	^{173}Lu	128	4.2
	^{175}Lu	358	4.2
	^{175}Ta	(80)	5.1*
	^{177}Ta	(230)	5.7*
$[521\ 1/2]_n$	^{161}Gd	356	0.31
$a_{calc} = 0.9$	^{163}Dy	351	0.26
$(a_{orb} = 1.2)$	^{165}Er	297	0.56
	^{165}Dy	108	0.58
	^{167}Er	208	0.70
	^{169}Yb	24	0.80
	^{169}Er	0	0.83
	^{171}Yb	0	0.85
	^{173}Hf	0	0.82
	^{171}Er	195	0.62
	^{173}Yb	399	0.73
	^{175}Hf	126	0.75
$[510\ 1/2]_n$	^{177}Yb	332	0.24
$a_{calc} = -0.2$	^{179}Hf	375	0.16
$(a_{orb} = +0.7)$	^{181}W	458	0.59
	^{183}W	0	0.19
	^{185}Os	0	0.02
	^{185}W	24	0.10
	^{187}W	146	−0.00

Table 5-16 Decoupling parameters for single-particle configurations ($150 < A < 190$). The table lists the excitation energies E of the band heads ($I = K = 1/2$) and the decoupling parameters obtained from the energies of the $I = 1/2$, 3/2, and 5/2 members of the bands. For some of the [541 1/2] bands, this information is not available; for these, the decoupling parameters a (marked with an asterisk) and the extrapolated band-head energies (in parentheses) are based on the lowest observed members of the band. The experimental data are taken from the references given in Tables 5-12 and 5-13.

▼ the observed values (Hjorth and Ryde, 1970). In this comparison, it should also be borne in mind that the value of a for the [541 1/2] band is rather sensitive to the nuclear deformation. The theoretical value in Table 5-16 refers to $\delta = 0.3$, but the derivative has the large value $da/d\delta \approx -10$; the variation of δ as seen in Fig. 4-25 therefore contributes significantly to the variation of the observed a values for this orbit.

The contribution of the nucleonic spin to the decoupling parameter is influenced by polarization effects similar to those discussed for the magnetic moments (see pp. 304ff.). Since the operator involved is s_+, the relevant polarizability is that of the transverse isoscalar spin field ($\Delta K = 1, \pi = +1, \tau = 0$). In contrast, the magnetic moment is especially sensitive to the isovector spin fields. There is at present almost no empirical evidence on the effective interaction for the isoscalar spin fields.

The relative contribution of spin and orbit to the calculated decoupling parameters can be obtained from the values of $a_{\text{orb}} = -\langle \Omega = 1/2 | l_+ | \overline{\Omega = 1/2} \rangle$ listed in Table 5-16. For most of the configurations, the main contribution to a results from the orbital matrix elements, but for the [510 1/2] configuration, the spin makes an appreciable contribution, and the empirical values appear to give evidence for a polarization effect that may reduce the spin contribution by as much as a factor of 2.

A renormalization of the matrix elements of the orbital angular momentum can arise from velocity-dependent interactions as considered on p. 304 in connection with the interpretation of the orbital g factors. However, similar renormalization effects of the isoscalar orbital moment are expected to be smaller, on account of the approximate validity of local Galilean invariance (see p. 80). Evidence on the matrix elements of l_+ is especially provided by the decoupling parameters for the orbits [411 1/2] and [521 1/2]; the data in Table 5-16 may suggest a slight reduction of the matrix elements of l_+ (bearing in mind that the calculated values of a for these orbits may be somewhat increased in magnitude by spin-polarization effects and by the difference between A and A_0).

Moments of inertia of rotational bands in odd-A nuclei with $150 < A < 190$ (Table 5-17)

Evidence on the inertial parameters $A = \hbar^2/2\mathcal{I}$ for one-quasiparticle states of odd-A nuclei is given in Table 5-17. The table also lists the average inertial parameter for the neighboring even-even nuclei, and it is seen that the moments of inertia of the odd-A bands exceed those of the even-even nuclei by amounts that are typically of order 20%, but show large variations (see also Fig. 4-12).

Systematic effects that increase the moments of inertia for low-lying bands in odd-A nuclei arise from the pair correlations, which are responsible ▲ for reducing the moments of inertia of the ground-state bands in even-even

▼ nuclei by a factor of about 2 below the rigid-body value (see the discussion in Chapter 4, pp. 82 ff.). The presence of the odd particle leads to a reduction of the pair correlation parameter Δ and hence of the rotational parameter A; an additional increase in the moment of inertia results from the Coriolis coupling between the one-quasiparticle states.

Evidence for a reduction in Δ^2 by about a factor of 2 in the low-lying bands of odd-A nuclei, as compared with the $\mathrm{v}=0$ bands of even-even nuclei, has been obtained from the analysis of α-decay intensities and two-particle transfer cross sections (see p. 248); such a reduction in Δ can be approximately accounted for in terms of the "blocking effect" of the last odd particle (see p. 272). A reduction of Δ^2 by a factor of 2 (for the neutrons or the protons) implies, according to the estimate (4-128), an increase in the moment of inertia of about 15%.

The increase in the moment of inertia that arises from second-order effects of the Coriolis coupling acting on the last odd particle is given by Eq. (5-47), which also includes the effect of the particle in preventing some of the excitations that contribute to the moment of inertia of the even-even core. For orbits near the Fermi level, one can employ the expression (5-48), and a qualitative estimate is obtained by using the asymptotic form of the intrinsic wave functions. The Coriolis coupling induces transitions with $\Delta n_\perp = \pm 1$, $\Delta n_3 = \mp 1$, which have one-particle excitation energies $\Delta\varepsilon = \pm\hbar(\omega_\perp - \omega_3) = \pm\hbar\omega_0\delta$, and one thus obtains (see Eq. (5-27) for the matrix elements of $l_\pm$)

$$\delta A = -\frac{4A_0^2\Delta}{(\hbar\omega_0\delta)^2}\left(2n_3(N-n_3)+N+n_3\right) \tag{5-94}$$

In the estimate (5-94), we have neglected the much smaller contributions from the nucleon spin as well as from the orbital transitions with $\Delta N=2$. The expression (5-94) gives values for $-\delta A/A_0$ ranging from about 5% to 20%.

For the proton orbits with $N=5$ and neutron orbits with $N=6$ included in Table 5-17, the value (5-94) based on the asymptotic representation gives an underestimate for δA. These orbits can be approximately described in terms of $\mathrm{h}_{11/2}$ and $\mathrm{i}_{13/2}$ one-particle motion with Coriolis matrix elements $(\langle j,\Omega+1|j_+|j\Omega\rangle = (j(j+1)-\Omega(\Omega+1))^{1/2})$ that, for $\Omega \ll j$, are appreciably larger than the estimate (5-27); moreover, the energy denominators, as given by Eq. (5-4), are seen to be smaller than in the asymptotic limit by a factor of approximately Ω/j (the smallness of these energy differences for $\Omega \ll j$ can be seen from Figs. 5-2 and 5-3). Thus, the evaluation of Eq. (5-48) for the $N=5$ proton orbits and the $N=6$ neutron orbits in Table 5-17 gives values for δA that in some cases become comparable to A_0.

The observed moments of inertia for low-lying bands in odd-A nuclei clearly exhibit an especially large increment in $\mathscr{I}$ for the $N=5$ proton and $N=6$ neutron configurations, which can be attributed to the importance of

▲ the Coriolis effect in these orbits. For the other low-lying orbits in Table 5-17,

Nucleus	*e-e*	532 5/2	413 5/2	411 3/2	523 7/2	411 1/2	404 7/2	514 9/2	402 5/2	541 1/2
$^{153}_{63}Eu$	20.4	7.7	**11.9**	13.9						
^{155}Eu	14.3	9.2	**11.2**	12.3						
$^{155}_{65}Tb$	21.8			**13.1**						
^{157}Tb	15.7	4.5		**12.2**						
^{159}Tb	13.8		11.6	**11.6**		12.0				
^{161}Tb	13.0	15.1	11.4	**11.2**						
$^{161}_{67}Ho$	15.7				**11.0**	(12.5)				
^{163}Ho	14.3				**10.2**					
^{165}Ho	12.8		12.1	11.6	**10.5**	11.7	11.7			
$^{167}_{69}Tm$	14.0				10.1	**12.4**	12.9			$\approx$11
^{169}Tm	13.7				10.4	**12.4**	13.0			
^{171}Tm	13.2					**12.0**				
$^{171}_{71}Lu$	14.9				14.0	13.2	**13.6**	11.3	14.2	(10.8)
^{173}Lu	14.1					12.7				8.6
^{175}Lu	13.7						**12.6**		12.8	10.0
^{177}Lu	14.6					14.2	**13.5**	12.6	13.4	

Nucleus	*e-e*	532 5/2	413 5/2	411 3/2	523 7/2	411 1/2	404 7/2	514 9/2	402 5/2	541 1/2
$^{177}_{73}Ta$	16.5						**14.6**	13.5	14.5	(12.5)
^{179}Ta	16.4						**14.9**			
^{181}Ta	16.1						**15.1**	13.9		
^{183}Ta							**15.9**		16.2	
$^{181}_{75}Re$	19.2							14.9	**16.8**	
^{183}Re	18.3						16.8	15.3	**16.3**	
^{185}Re	20.7							15.3	**17.9**	
^{187}Re	23.1							16.6	**19.2**	

Table 5-17a Proton orbits

▼ it is more difficult to discern the effects of the Coriolis coupling, since the correction to the moment of inertia arising from the modification of Δ is expected to be comparable to or greater than the estimate (5-94).

A quantitative analysis of the Coriolis contribution to $\mathscr{I}$ can be given in cases where the energies of the relevant one-particle configurations are known from other evidence. This information is available for a few cases involving the orbits with large values of j, and it is found that, in these cases, the estimated Coriolis coupling leads to values of δA that are about a factor of 2 larger than those observed (see the example, ^{235}U, discussed on pp. 279ff.)

For the orbits with large j, an additional contribution leading to a significant further increase of the effective moment of inertia may arise from the fourth-order term in the rotational energy of the even-even core. This term yields the energy $B(\mathbf{I}-\mathbf{j})^4$ (see the similar treatment of the second-order term (4A-7) in the particle–rotor model), which implies a contribution

$$\delta A = 4B\left(j(j+1) - \tfrac{3}{2}K^2\right) \tag{5-95}$$

to the coefficient in the rotational energy proportional to $I(I+1)$. In the evaluation of the contribution (5-95), one must employ the coefficient B that describes the energy of the even-even core in the presence of the odd particle.

Additional evidence that the Coriolis interaction overestimates the rotational coupling matrix element between one-quasiparticle states is provided by a class of excited configurations in Table 5-17. The one-particle states linked by unhindered matrix elements of $l_{\pm}$ (having $\Delta n_3 = \pm 1, \Delta n_{\perp} = \mp 1$) have energy differences that are typically of order 2 MeV, but in cases where the Fermi energy lies approximately midway between the two, the quasiparticle states may become almost degenerate at an excitation energy of the order of 0.5 MeV. In the absence of pair correlations, the Coriolis matrix element between these particle and hole states would vanish, but for the quasiparticle states, the matrix element remains finite, though the value is reduced by the factor $u_1u_2 + v_1v_2 \approx \Delta/E$, where E is the quasiparticle energy (5-57) approximately equal to the sum of Δ and the excitation energy with respect to the ground state. Table 5-17 contains many examples of such strongly coupled near-lying configurations, for which the expression (5-47) yields large contributions to the moments of inertia. Examples where the estimated values of $\delta A/A_0$ exceed $\pm 50\%$ (negative for the lower state and positive for the higher state) are provided by the orbits [532 5/2] and [523 7/2] in ^{161}Tb, [532 3/2] and [523 5/2] in ^{157}Dy, [633 7/2] and [624 9/2] in ^{179}W, [521 1/2] and [510 1/2] in ^{177}Hf and in ^{181}W, and [521 1/2] and [512 3/2] in ^{179}Hf. The observed moments of inertia of these bands, in Table 5-17, do not show the expected large effects of the Coriolis coupling; in some cases, the data set limits to δA that appear to be at least an order of magnitude ▲ smaller than the theoretical estimates. In a number of cases, the study of

Nucleus	*e-e*	530 1/2	532 3/2	505 11/2	521 3/2	642 5/2	523 5/2	633 7/2	651 1/2
$^{153}Sm_{91}$	16.9		11.1		11.0				
^{155}Gd	17.7	8.3		12.4	**12.0**				
^{157}Dy	19.7	(8.7)	11.0		**12.2**		11.3		
$^{157}Gd_{93}$	14.1	(7.7)	10.2		**10.9**	7.4	11.7		
^{159}Dy	15.5	(7.4)	12.6		**11.3**		12.2		
^{161}Er	19.0			14.0	**11.9**		13.5		
$^{159}Gd_{95}$	12.9	(7.7)	9.8		**10.1**	(7.3)	11.6		7.3
^{161}Dy	13.9	(7.5)			11.4	**6.3**	11.1		
^{163}Er	16.1	(8.9)		13.1	12.0		**12.0**		
$^{161}Gd_{97}$					(10.4)		**10.4**	(7.1)	7.6
^{163}Dy	12.8	7.0			10.7	5.0	**10.5**	(9.2)	
^{165}Er	14.3	(10.2)			10.6		**11.0**		
$^{165}Dy_{99}$								**9.3**	
^{167}Er	13.6				11.5		11.0	**8.8**	
^{169}Yb	14.3				12.4		11.1	**7.9**	
$^{169}Er_{101}$	13.3				11.0		12.4	8.4	
^{171}Yb	13.6				(14.8)			8.0	
$^{171}Er_{103}$									
^{173}Yb	12.9				10.8			6.9	
^{175}Hf	14.9								
$^{175}Yb_{105}$	13.2				12.2			(10.3)	
^{177}Hf	15.1							11.4	
^{179}W	17.8							10.6	

Table 5-17b Neutron orbits

Nucleus	*e-e*	521 1/2	512 5/2	514 7/2	624 9/2	510 1/2	512 3/2	503 7/2	615 11/2
$^{153}Sm_{91}$	16.9	13.6							
^{155}Gd	17.7	13.5							
^{157}Dy	19.7	12.8	(12.9)						
$^{157}Gd_{93}$	14.1	12.2						10.4	
^{159}Dy	15.5	12.6	11.1			(12.4)			
^{161}Er	19.0								
$^{159}Gd_{95}$	12.9	11.5	10.7			11.6			
^{161}Dy	13.9	11.8	11.6			(11.3)			
^{163}Er	16.1	13.3	12.8			11.3			
$^{161}Gd_{97}$		10.5	11.4			11.4			
^{163}Dy	12.8	10.2	11.7			13.0			
^{165}Er	14.3	12.6	13.9			13.1			
$^{165}Dy_{99}$		10.6	11.1			11.1	12.0		
^{167}Er	13.6	11.2	11.9			11.6	11.9		
^{169}Yb	14.3	11.7	12.5			12.4			
$^{169}Er_{101}$	13.3	**11.8**	12.1	12.0		11.7	12.6		
^{171}Yb	13.6	**12.0**	12.2	12.6		14.0			
$^{171}Er_{103}$		12.0	**10.9**	12.7	10.0	11.5	13.2		
^{173}Yb	12.9	12.1	**11.2**	12.5		11.7	12.8		
^{175}Hf	14.9	13.5	**11.6**	14.1					

Nucleus	*e-e*	521 1/2	512 5/2	514 7/2	624 9/2	510 1/2	512 3/2	503 7/2	615 11/2
$^{175}Yb_{105}$	13.2	13.7	12.7	**11.6**	(10.7)	11.6	12.0		
^{177}Hf	15.1		13.8	**12.6**	9.6		14.8		
^{179}W	17.8	15.0	14.3	**13.3**					
$^{177}Yb_{107}$				12.3	**11.2**	12.3	12.8		
^{179}Hf	15.5	13.1	13.8	13.6	**11.2**	13.2	13.5		
^{181}W	16.8	14.6	15.8	13.3	**10.3**	15.1	16.2		
$^{183}W_{109}$	17.6	18.2	13.9	16.2	(14.0)	**13.0**	16.6	15.8	13.7
$^{185}W_{111}$	19.5	16.7	14.0	17.9	12.7	21.1	**13.2**	16.4	14.3
^{187}Os	24.4					**23.7**	13.1		

Table 5-17b continued

Table 5-17 Moments of inertia of rotational bands associated with one-particle configurations in nuclei with $150 < A < 190$. The tables list the rotational energy constant $A = \hbar^2/2\mathscr{I}$ (in keV) obtained from the spacing of the two lowest members of the band (for $K = 1/2$, three lowest members). Where the lowest members have not been observed, the rotational constants have been estimated from higher members of the band and are listed in parentheses. Values for ground-state configurations are printed in bold face. The experimental data are taken from the references given in Tables 5-12 and 5-13 (pp. 296 and 299). The bands referred to in footnotes *a* and *c* to Table 5-13 as strongly Coriolis mixed are not well described by the rotational energy expansion and are not included in the present table. (As a criterion for inclusion in the table, we have required the moments of inertia determined from the two lowest energy intervals to be equal to within 20%.) The rotational constants in column two represent the average of the two adjacent even-even nuclei, obtained from the compilation by Lederer *et al.*, 1967; see also Fig. 4-12, p. 74.

▼ other properties of the levels has provided additional confirmation of the strong reduction in the Coriolis coupling effect for these orbits. (See, for example, the discussion of the allowed unhindered β decay leading to ^{161}Tb by Żylicz *et al.*, 1966, and the analysis of one-particle transfer reactions in the W isotopes by Casten *et al.*, 1972.) The effects responsible for a reduction of the Coriolis interaction have not yet been identified; however, the especially large reduction in the rotational coupling matrix element between states on opposite sides of the Fermi surface is suggestive of the coupling to the rotationally induced pair field (see p. 278).

Besides the inertial contribution associated with the large j orbits, the effect of the Coriolis coupling on the last odd particle can be recognized in a number of additional features of the data in Table 5-17.

1. Large Coriolis matrix elements connecting rather close-lying single-particle states occur for the lowest orbits coming down from the higher spherical shell. These orbits, which occur as excited configurations, are observed to have very large moments of inertia. (Examples are provided by the [541 1/2] proton orbit and the [651 1/2] neutron orbit. In a similar manner, the large moment of inertia observed for the [530 1/2] neutron-hole orbit can be attributed to the coupling to the expected near-lying [541 1/2] configuration.)

2. For excited configurations, the coupling to lower-lying quasiparticle states yields a negative contribution to the moment of inertia. A striking example is the $h_{11/2}$ orbit [532 5/2] in ^{161}Tb, whose anomalously small moment of inertia appears to reflect the coupling to the lower-lying [523 7/2] configuration. (Though the observed effect is rather large, it is appreciably smaller than the theoretical estimate for one-quasiparticle states; see the discussion above.)

3. The single-particle spectrum contains a number of close-lying levels with $\Delta\Omega = 1$, such as the [402 5/2] and [404 7/2] proton orbits and the [510 1/2] and [512 3/2] neutron orbits. The matrix elements of $j_{\pm}$ are forbidden by the asymptotic quantum numbers, and the estimated values are of order unity or less. However, in situations where the quasiparticle states come very close together, the coupling may imply significant contributions to the rotational energy. Examples are provided by the differences in the moments of inertia of the [510 1/2] and [512 3/2] configurations in ^{183}W, ^{185}W, and ^{187}Os. The observed differences interpreted by means of Eq. (5-47) imply Coriolis matrix elements $\langle[512\ 3/2]\|j_+\|[510\ 1/2]\rangle \approx 1$, which agrees rather well with

▲ the theoretical estimate based on the wave functions in Table 5-2.

APPENDIX

5A

Scattering by Nonspherical Systems

In a scattering process involving a nonspherical nucleus, the deformation of the potential implies a coupling between the rotational degrees of freedom of the target and the orbital motion of the projectile. In the present appendix, we consider the general framework for treating such a generalized optical model, which may also be viewed as a prototype for a broad class of scattering problems in which internal degrees of freedom of the target are explicitly included. (For a review of the extension of the optical model to both vibrational and rotational excitations in nuclei, see, for example, Hodgson, 1971).

The Hamiltonian describing potential scattering by a nonspherical object can be written in the form

$$H = T + V + H_{\text{rot}} \tag{5A-1}$$

where T is the kinetic energy of the particle and V the deformed potential, which may involve absorptive terms describing compound nucleus formation. The rotational energy is represented by H_{rot}, whose eigenstates may be specified by the total angular momentum R of the target nucleus (and, for nonaxial rotors, by an additional quantum number, τ; see Eq. (4-9)). The present model is seen to be the analog for scattering problems of the particle-rotor model considered in Appendix 4A.

5A-1 Treatment in Terms of Coupled Channels

A general method for treating the scattering problem defined by the Hamiltonian (5A-1) is obtained by expanding the wave function in the form

$$\Psi_{IM} = \sum_{ljR} \mathscr{R}_{ljRI}(\mathrm{r})\Phi_{ljRIM} \tag{5A-2}$$

where the channels are specified by the orbital and total angular momenta (l and j) of the particle, the rotational angular momentum (R) of the target nucleus, and the total angular momentum ($\mathbf{I} = \mathbf{j} + \mathbf{R}$). For a reflection-symmetric target, the value of l for given j is specified by the parity quantum number, and the sum in Eq. (5A-2) therefore involves only the two variables j and R for given $I\pi$. The wave functions Φ describe the dependence on the spin and angular coordinates of the particle and the orientation angles of the target. The Schrödinger equation for the state (5A-2) leads to a set of coupled differential equations for the radial functions

$$\left(-\frac{\hbar^2}{2M}\left(\frac{d^2}{dr^2}-\frac{l(l+1)}{r^2}\right)+E_{\text{rot}}(R)-E\right)r\mathscr{R}_{ljRI}(r)$$
$$+\sum_{l'j'R'}\langle l'j'R'I|V|ljRI\rangle r\mathscr{R}_{l'j'R'I}(r)=0 \qquad \text{(5A-3)}$$

where $E_{\text{rot}}(R)$ is the eigenvalue of H_{rot} and E the total energy.

The matrix element of V in Eq. (5A-3) is a function of the radial variable and may be obtained by expanding the potential in spherical harmonics. Thus, for an axially symmetric nucleus, the spin-independent part of the potential can be expressed in the form

$$V(r,\vartheta')=\sum_{\lambda}V_{\lambda}(r)P_{\lambda}(\cos\vartheta') \qquad \text{(5A-4)}$$

where ϑ' is the polar angle of the particle with respect to the nuclear symmetry axis. The matrix elements of $P_{\lambda}(\cos\vartheta')$ are obtained by standard recoupling techniques (see Eqs. (1A-72), (1A-43), and (1A-46), as well as the wave function (4-7) for an axially symmetric nucleus),

$$\langle l'j'R'IM|P_{\lambda}(\cos\vartheta')|ljRIM\rangle$$
$$=(-1)^{I+j+R'}(4\pi)^{1/2}(2\lambda+1)^{-1/2}(2R+1)^{1/2}\langle RK_R\lambda 0|R'K_R\rangle\begin{Bmatrix} j & R & I\\ R' & j' & \lambda\end{Bmatrix}\langle l'j'\|Y_{\lambda}\|lj\rangle \qquad \text{(5A-5)}$$

where K_R is the component of R on the nuclear symmetry axis ($K_R=0$ for an even-even target). The reduced matrix elements of Y_{λ} may be obtained from Eq. (3A-14).

The scattering matrix in the $(ljR)I$ representation is obtained by solving the coupled radial equations (5A-3), for given I (and parity), subject to the boundary condition that incident waves occur only in the entrance channel (labeled by 1)

$$\mathscr{R}_{l_1j_1R_1I}\underset{r\to\infty}{=}\frac{1}{r}\left(\exp\{-i(k_1r-\tfrac{1}{2}\pi l_1)\}-\langle l_1j_1R_1I|S|l_1j_1R_1I\rangle\exp\{i(k_1r-\tfrac{1}{2}\pi l_1)\}\right) \qquad \text{(5A-6)}$$

$$\mathscr{R}_{ljRI}\underset{r\to\infty}{=}\begin{cases}-\left(\dfrac{k_1}{k}\right)^{1/2}\langle ljRI|S|l_1j_1R_1I\rangle\dfrac{1}{r}\exp\{i(kr-\tfrac{1}{2}\pi l)\} & E_{\text{p}}>0\\ & (ljR)\neq(l_1j_1R_1)\\ \alpha_{ljRI}\dfrac{1}{r}\exp(-\kappa r) & E_{\text{p}}<0\end{cases}$$

where E_p is the energy of the particle in the channel considered

$$E_p = E - E_{rot}(R)$$

$$= \frac{\hbar^2}{2M}\begin{cases} k^2 & E_p > 0 \\ -\kappa^2 & E_p < 0 \end{cases} \tag{5A-7}$$

The coefficients α_{ljRI} characterize the asymptotic wave functions in the closed channels ($E_p < 0$), while the amplitudes of the outgoing waves in the open channels define the S-matrix elements.

The scattering amplitude for specified directions of the ingoing and outgoing particles can be obtained from the S-matrix elements in Eq. (5A-6) by transforming to a representation specified by the momentum $\mathbf{p} = \hbar\mathbf{k}$, the helicity h, and the angular momentum quantum numbers RM_R of the target (see Eqs. (3F-5) and (1B-31)),

$$f((\mathbf{p}hRM_R)_1 \to (\mathbf{p}hRM_R)_2) = -2\pi i (k_1 k_2)^{-1/2} \sum_{l_1 j_1 l_2 j_2 IM} \langle (\hat{\mathbf{p}}hRM_R)_2 | (ljR)_2 IM \rangle$$

$$\times (\langle (ljR)_2 I | S | (ljR)_1 I \rangle - \delta((ljR)_1, (ljR)_2)) \langle (ljR)_1 IM | (\hat{\mathbf{p}}hRM_R)_1 \rangle \tag{5A-8a}$$

$$\langle \hat{\mathbf{p}}hRM_R | ljRIM \rangle = \sum_m \langle jmRM_R | IM \rangle \langle l0\tfrac{1}{2}h | jh \rangle \left(\frac{2l+1}{8\pi^2} \right)^{1/2} \mathscr{D}^j_{mh}(\hat{\mathbf{p}}) \tag{5A-8b}$$

The corresponding differential cross section is given by (see the expression (1B-34) which, in the helicity representation, acquires an extra factor 2π, as explained in Vol. I, p. 101)

$$d\sigma((\mathbf{p}hRM_R)_1 \to (\mathbf{p}hRM_R)_2) = 2\pi \frac{k_2}{k_1} |f(1 \to 2)|^2 d\Omega \tag{5A-9}$$

with the element of solid angle $d\Omega = 2\pi \sin\vartheta \, d\vartheta \, d\varphi$.

The total cross section can be obtained from the optical theorem (see Eq. (2-90), which also acquires an additional factor 2π in the helicity representation)

$$\sigma_{tot} = \frac{8\pi^2}{k_1} \operatorname{Im} f((\mathbf{p}hRM_R)_1 \to (\mathbf{p}hRM_R)_1)$$

$$= \frac{16\pi^3}{k_1^2} \sum_{ljl_1 j_1 IM} \langle (\hat{\mathbf{p}}hRM_R)_1 | ljR_1 IM \rangle$$

$$\times (\delta((lj),(l_1 j_1)) - \operatorname{Re}\langle ljR_1 I | S | l_1 j_1 R_1 I \rangle) \langle l_1 j_1 R_1 IM | (\hat{\mathbf{p}}hRM_R)_1 \rangle \tag{5A-10a}$$

$$\langle (\hat{\mathbf{p}}hRM_R)_1 | ljR_1 IM \rangle = \left(\frac{2l+1}{8\pi^2} \right)^{1/2} \langle l0\tfrac{1}{2}h_1 | jh_1 \rangle \langle jh_1 R_1 M_{R_1} | IM \rangle \tag{5A-10b}$$

In Eq. (5A-10b), the axis of space quantization has been taken to coincide with the direction of incidence, in which case $\mathscr{D}^{j}_{mh}(\hat{\mathbf{p}}_1)=\delta(m,h)$. The total cross section (5A-10a) includes direct processes as well as compound nucleus formation. The total cross section for direct processes (including elastic scattering) is obtained by integration over the final states in Eq. (5A-9). Thus, for specified incident helicity h_1 and target polarization $(RM_R)_1$, one obtains

$$\sigma_{\text{dir}}=\sum_{(hRM_R)_2} 2\pi\frac{k_2}{k_1}\int d\Omega_2|f((\mathbf{p}hRM_R)_1\rightarrow(\mathbf{p}hRM_R)_2)|^2$$

$$=\frac{8\pi^3}{k_1^2}\sum_{R_2}\sum_{ljl_1j_1l_2j_2IM}\langle(\hat{\mathbf{p}}hRM_R)_1|ljR_1IM\rangle(\langle ljR_1I|S^{\dagger}|l_2j_2R_2I\rangle-\delta((ljR_1),(l_2j_2R_2)))$$

$$\times(\langle l_2j_2R_2I|S|l_1j_1R_1I\rangle-\delta((l_1j_1R_1),(l_2j_2R_2)))\langle l_1j_1R_1IM|(\hat{\mathbf{p}}hRM_R)_1\rangle \quad \text{(5A-11)}$$

The part of the cross section (5A-11) with $R_2\neq R_1$ represents direct rotational excitations. The difference between the cross sections (5A-10) and (5A-11) is the cross section for compound nucleus formation, which can be expressed in the form

$$\sigma_{\text{comp}}=\sigma_{\text{tot}}-\sigma_{\text{dir}}$$

$$=\frac{8\pi^3}{k_1^2}\sum_{ljl_1j_1IM}\langle(\hat{\mathbf{p}}hRM_R)_1|ljR_1IM\rangle$$

$$(\delta((lj),(l_1j_1))-\langle ljR_1I|S^{\dagger}S|l_1j_1R_1I\rangle)\langle l_1j_1R_1IM|(\hat{\mathbf{p}}hRM_R)_1\rangle \quad \text{(5A-12)}$$

in terms of the deviation of the S matrix from unitarity resulting from the absorptive part in the optical potential.

5A-2 Adiabatic Approximation

If the rotational motion can be neglected during the time of the collision, one obtains an approximate solution by considering the scattering for a fixed orientation of the nucleus,[4] as in the treatment of the bound-state problems (see Appendix 4A). Denoting the scattering amplitude for a target with orientation ω by $f((\mathbf{p}h)_1\rightarrow(\mathbf{p}h)_2;\ \omega)$, the amplitude for a scattering process in which the target is initially in the state with quantum numbers $R_1(M_R)_1$ and finally in the state $R_2(M_R)_2$ is given by the matrix element

$$f((\mathbf{p}hRM_R)_1\rightarrow(\mathbf{p}hRM_R)_2)=\langle R_2(M_R)_2|f((\mathbf{p}h)_1\rightarrow(\mathbf{p}h)_2;\ \omega)|R_1(M_R)_1\rangle \quad \text{(5A-13)}$$

The scattering for fixed orientation is a one-body problem (though involving the three-dimensional geometry); hence, the Schrödinger equation reduces to a

[4]The adiabatic treatment was applied to nuclear scattering problems by Drozdov (1955; 1958) and Inopin (1956). For strongly absorbed particles, the treatment leads to simple relations between elastic and inelastic scattering (Blair, 1959; Blair *et al.*, 1962; Austern and Blair, 1965).

classical problem encountered in the scattering of electromagnetic and acoustical waves (see, for example, the text by Van de Hulst, 1957). For an axially symmetric potential, the one-particle orbits have the constant of motion Ω, representing the projection of the angular momentum on the symmetry axis.

The one-body problem resulting from the adiabatic approximation may be treated in terms of coupled channels in a manner similar to the more general analysis given above, but with the simplification resulting from the reduction in the number of coupled channels. (For an axially and reflection-symmetric potential, the number of coupled channels for given $\Omega\pi$ is specified by the single variable j (for particles with $s=0$ or $1/2$), while the nonadiabatic treatment involves a coupling of channels specified by the two variables jR, for fixed $I\pi$. For a nonsymmetric body, the adiabatic treatment involves the coupling of channels specified by three angular momentum quantum numbers, such as ljm, while the general treatment involves coupling of channels specified by four such quantum numbers, lj for the particle and $R\tau$ for the rotor.)

In the $lj\Omega$ representation for scattering by an axially symmetric potential, the one-particle wave function has the form

$$\psi_\Omega = \sum_{lj} \mathscr{R}_{lj\Omega}(r)(i^l Y_l \chi)_{(l\frac{1}{2})j\Omega} \tag{5A-14}$$

where the last factor is the spin-angular wave function (see Eq. (3A-1)). The coupling to the nuclear deformation involves the matrix element of $P_\lambda(\cos\vartheta')$, which can be obtained from Eq. (3A-14). The scattering matrix may be found by solving coupled radial differential equations analogous to those given by Eq. (5A-3) with the boundary condition

$$\mathscr{R}_{lj\Omega} \underset{r\to\infty}{=} \frac{1}{r}\left(\delta(ll_1)\delta(jj_1)\exp\{-i(kr-\tfrac{1}{2}\pi l)\} - \langle lj\Omega|S|l_1 j_1\Omega\rangle \exp\{i(kr-\tfrac{1}{2}\pi l)\}\right) \tag{5A-15}$$

where k is the wave number, which is the same in all channels, in the approximation considered.

For fixed orientation of the target, the scattering amplitude can be expressed in the form (see Eqs. (1B-31) and (3F-5))

$$f((\mathbf{p}h)_1\to(\mathbf{p}h)_2;\omega) = -2\pi i(k_1k_2)^{-1/2} \sum_{(ljm)_1(ljm)_2} \langle(\hat{\mathbf{p}}h)_2|(ljm)_2\rangle \times\left(\langle(ljm)_2|S(\omega)|(ljm)_1\rangle - \delta((ljm)_1,(ljm)_2)\right)\langle(ljm)_1|(\hat{\mathbf{p}}h)_1\rangle \tag{5A-16a}$$

$$\langle\hat{\mathbf{p}}h|ljm\rangle = \langle l0\tfrac{1}{2}h|jh\rangle\left(\frac{2l+1}{8\pi^2}\right)^{1/2} \mathscr{D}^j_{mh}(\hat{\mathbf{p}}) \tag{5A-16b}$$

where the S-matrix elements in the ljm representations may be obtained by a transformation from the intrinsic system,

$$\langle l_2 j_2 m_2 | S(\omega) | l_1 j_1 m_1 \rangle = \sum_{\Omega} \left(\mathscr{D}^{j_2}_{m_2\Omega}(\omega) \right)^* \langle l_2 j_2 \Omega | S | l_1 j_1 \Omega \rangle \mathscr{D}^{j_1}_{m_1\Omega}(\omega) \quad \text{(5A-17)}$$

Carrying out the integration over ω in Eq. (5A-13) with $f(1\rightarrow 2;\omega)$ given by Eqs. (5A-16) and (5A-17), one obtains a scattering amplitude of the form (5A-8) with the S-matrix elements in the factored form

$$\langle (ljR)_2 I | S | (ljR)_1 I \rangle = \sum_{\Omega} \frac{\left((2R_1+1)(2R_2+1)\right)^{1/2}}{2I+1} \langle j_2 \Omega R_2 K_R | I K_R + \Omega \rangle$$

$$\times \langle l_2 j_2 \Omega | S | l_1 j_1 \Omega \rangle \langle j_1 \Omega R_1 K_R | I K_R + \Omega \rangle \quad \text{(5A-18)}$$

The result (5A-18), which expresses the consequences of the adiabatic approximation, can also be directly obtained by noting that the total wave function, in this approximation, is given by the product of the intrinsic wave function (5A-14) and the rotational wave function $\mathscr{D}^I_{MK}$, with $K = K_R + \Omega$; the transformation of this wave function to the $(ljR)IM$ representation is given by Eq. (4A-6).

The adiabatic approximation as described above assumes the rotational energies to be small compared to the other energies in the problem. If the energy of the ingoing or outgoing particles is small, or if the penetration through the centrifugal and Coulomb barriers is sensitive to energy differences of the order of E_{rot}, the treatment as given above requires modifications. However, the essential results can be retained if the adiabatic treatment is appropriate for describing the motion of the particle inside the nucleus and in the nonspherical part of the barrier; one can then join the adiabatic wave function in the internal region where the angular momentum exchanges take place onto the wave function in the external region, which can be described in terms of uncoupled motion in the different channels $(ljR)I$, and for which the rotational energy is therefore readily taken into account (see, for example, the treatment of slow neutron scattering, pp. 234ff., and of α decay, pp. 115ff.).

CHAPTER 6

Vibrational Spectra

6-1 INTRODUCTION

Occurrence of collective vibrations

For a great variety of many-body systems, it is possible to describe the excitation spectra in terms of elementary modes of excitation representing the different, approximately independent, fluctuations about equilibrium.[1] The nature of these fluctuations depends on the internal structure of the system. Thus, the elementary modes may be associated with excitations of individual particles or they may represent collective vibrations of the density, shape, or some other parameter that characterizes the equilibrium configuration.

In the description in terms of normal modes, the excitation spectrum is obtained by superposing individual quanta associated with the elementary modes of excitation. To a first approximation, the quanta are considered as noninteracting entities, and the deviations from this idealization can be taken into account in terms of interactions between the normal modes. These interactions provide the natural limitation to such a description.

Examples of collective vibrations in quantal systems are well known from the study of molecules. The atoms of the molecule form an approximately rigid structure, and the low-energy internal excitations correspond to the normal modes of vibration of this structure. In macroscopic solid bodies, the vibrations of the equilibrium lattice are the elastic waves.

In systems that are described, to a first approximation, by independent-particle motion in an average field (nuclei, electrons in metals, etc.), collective vibrations may also occur as a result of interactions between the particles giving rise to correlations in the particle motion and associated oscillations in the average density and field. An example is provided by the density oscillations (plasmons) in an electron gas. The extent to which the fluctuations separate into collective and individual-particle modes may depend on rather detailed features of the single-particle spectrum, as well as on the strength and character of the interactions. The elementary modes of excitation of such systems may therefore partly comprise well-defined collective vibrations involving the correlated motion of a large number of particles and partly excitations primarily involving only a single particle or a few degrees of freedom of the particle motion.

In the nucleus, the possibility of collective shape oscillations is strongly suggested by the fact that some nuclei are found to have nonspherical

---

[1] The concept of elementary excitations was introduced by Landau (1941) in connection with the analysis of the excitation spectrum of superfluid liquid helium.

equilibrium shapes (see Chapter 4), whereas others, such as the closed-shell nuclei, have an equilibrium with spherical shape. One might thus expect to find intermediate situations in which the shape undergoes rather large fluctuations away from the equilibrium shape.

Indeed, a striking feature of the energy spectra of almost all nuclei is the occurrence of low-lying states that are strongly excited by electric quadrupole processes (see, for example, Fig. 4-5, p. 48). In the regions of nonspherical nuclei, these states are members of the ground-state rotational band, but for the remaining nuclei, it appears that we are dealing with collective vibrations in the nuclear shape.

At higher excitation energies, a number of other vibrational modes have been identified. These correspond partly to shape oscillations of different multipole order and partly to fluctuations in which the neutrons move collectively with respect to the protons. In addition to the modes with classical analogs, there occur in the nuclear spectra vibrational modes involving charge exchange or excitation of the nucleonic spins as well as oscillations in the pair field involving the creation or annihilation of two nucleons.

The rich possibilities for vibrational modes in nuclei and the problems associated with the interactions of the excitation quanta present a field of great scope that involves a diversity of issues concerning the structure of quantal many-body systems. In order to help the reader to find his way through this diversity, we give in the following a preview of the organization of the material in the present chapter and of the main themes of the discussion.

Preview of topics associated with nuclear vibrational motion

A starting point in the analysis of vibrational motion is the establishment of general properties that follow from the symmetry of the equilibrium configuration and the nature of the interactions. This problem is analogous to the phenomenological analysis of rotational motion in Chapter 4. However, the vibrational motion is so profoundly affected by the shell structure of the single-particle motion that we have found it necessary to include from the outset the concepts involved in relating the collective motion to the degrees of freedom of the individual particles. Therefore, the discussion in the present chapter involves a microscopic as well as a macroscopic description of the vibrational motion. The discussion of the microscopic treatment will focus attention on the insight it provides into the qualitative properties of the vibrational modes and their couplings. The extent to which this framework may give a quantitative account of the nuclear modes is an issue that remains to be clarified, and that goes beyond the scope of the present chapter.

As a first step in the discussion of the nuclear vibrational modes, we consider, in Sec. 6-2, simple basic features of vibrations in a quantal system. Sections 6-2a and 6-2b summarize the complementary relationships between the phenomenological descriptions in terms of amplitudes and the operators that create vibrational quanta. Section 6-2c deals with the mechanism that is involved in the generation of collective modes out of excitations of individual particles. The establishment of a coherent motion of many nucleons can be understood in terms of the oscillating potential generated by a collective vibration of the nucleonic density. The phenomenon is illustrated by a highly simplified model of the particle degrees of freedom, which makes it possible to isolate the essential physical effects.

The classification of the nuclear vibrations in terms of symmetry properties is the topic of Sec. 6-3. Because of the many different dimensions in which a nucleus may vibrate, the classification involves a variety of different symmetry quantum numbers, including multipolarity, intrinsic spin and orbital angular momentum, isospin, and nucleon number. Some of these modes can be compared with vibrations in a classical system and, indeed, the liquid-drop model has continued to provide important inspiration in the development of the subject. (The normal modes of a liquid drop are analyzed in Appendix 6A.) The scope and limitations of these classical pictures of the nuclear vibrations are summarized in Sec. 6-3; a more detailed analysis of the effects of shell structure is given in the examples at the end of the chapter.

Section 6-3 includes a discussion of the fission process as an extension of the phenomena associated with small-amplitude shape oscillations. The treatment of the large deformations encountered in the fission process involves the challenging problem of combining bulk properties of nuclear matter with the microscopic effects resulting from the shell structure.

The vibrational modes whose quanta consist of correlated pairs of nucleons (Sec. 6-3f) are intimately connected with the nuclear pairing effect, which has far-reaching consequences for the low-energy nuclear properties, as discussed in previous chapters. The nuclear pair correlations can be viewed in terms of a condensate of pair quanta (deformation of the pair field) similar to that of superfluid macroscopic Fermi systems. The single-particle degrees of freedom in the presence of such a condensate are the quasiparticles, whose properties are derived in the example on pp. 647 ff.

The existence of algebraic identities involving the multipole moments leads to sum rules for the corresponding transition probabilities, which are the topic of Sec. 6-4. Such sum rules provide rather general tools for the analysis of complex systems in many different fields of quantal physics. Their inclusion in the present chapter is motivated by the fact that the strongly enhanced excitations of the nuclear vibrational modes often exhaust an appreciable fraction of the total sum rule unit for simple multipole fields.

While the discussion in Secs. 6-2 and 6-3 refers to the approximation of independent harmonic vibrations, the interactions of the normal modes are considered in Secs. 6-5 and 6-6. A basic element in these interactions is the effect of the vibration on the motion of the individual particles. Such a coupling, which arises from the average potential generated by the vibrational motion, is invoked already in Sec. 6-2c as the mechanism responsible for producing the collective motion itself. The more systematic study of the particle-vibration coupling is taken up in Sec. 6-5, as a basis for interpreting a wide variety of effects associated with the interplay of vibrational motion and individual-particle degrees of freedom, such as the effective charges and moments of single-particle states and the interaction energy between a particle and a vibrational quantum. Carried to higher order, the particle-vibration coupling leads to interactions between the vibrational quanta. The phenomenological analysis of anharmonicities in the vibrational modes and the related problem of the coupling between different collective modes is the topic of Sec. 6-6.

Examples illustrating the properties of the vibrational modes are collected at the end of the chapter, rather than at the end of each section. This arrangement makes it possible to bring together the discussion of all the different properties of a given mode. The discussion of the fission process builds on a broader view of shell structure in the single-particle spectrum than that given in Chapter 2; the tools that may be employed in such a generalized treatment of shell structure are discussed in a set of examples on pp. 578ff.

6-2 QUANTAL THEORY OF HARMONIC VIBRATIONS

The vibrations in a quantal system can be described in terms of collective coordinates representing the amplitudes of density fluctuations about equilibrium. The equations of motion have the same form as in classical theory, and the quantal features can be obtained by the canonical procedure of quantization. The quanta of excitation may also be taken as the starting point for the description; the basic dynamic variables are then the operators associated with the creation and annihilation of quanta.

6-2a Creation Operators for Excitation Quanta

A vibrational mode of excitation is characterized by the property that it can be repeated a large number of times. The nth excited state of a specified mode can thus be viewed as consisting of n individual quanta. The quanta obey Bose statistics, since for a given number of quanta, there is only a single state, as for identical particles with a totally symmetric wave function. Such a system of bosons can be treated in terms of the operators $c^\dagger$ and c that create and annihilate a quantum of excitation.

If the excitations can be superposed without a modification of the vibrational mode, the quanta can be regarded as noninteracting entities. In this approximation, the boson operators are defined by the relation

$$c^\dagger|n\rangle = (n+1)^{1/2}|n+1\rangle \tag{6-1}$$

where $|n\rangle$ is the state with n quanta of excitation. The factor $(n+1)^{1/2}$ in Eq. (6-1) implies that the transition probability $\langle n+1|c^\dagger|n\rangle^2$ is equal to $n+1$. The physical significance of this "boson factor" may be illustrated by considering the decay $|n+1\rangle \to |n\rangle$; if the process involves the quanta acting independently, the total decay rate is proportional to the number of quanta, $n+1$.

The defining equation (6-1) leads to the commutation relation

$$[c, c^\dagger] = 1 \tag{6-2}$$

and the number of quanta is represented by the operator

$$n_{\text{op}} = c^\dagger c \tag{6-3}$$

The excited states $|n\rangle$ can be constructed from the ground state $|n=0\rangle$ by acting n times with the creation operator

$$|n\rangle = (n!)^{-1/2}(c^\dagger)^n|n=0\rangle \tag{6-4}$$

For noninteracting quanta, each with an energy $\hbar\omega$, the Hamiltonian has the form

$$H = \hbar\omega c^\dagger c + E(n=0) \tag{6-5}$$

and the equation of motion for $c^\dagger$ is

$$\dot{c}^\dagger = \frac{i}{\hbar}[H.c^\dagger] = i\omega c^\dagger \tag{6-6}$$

Thus, the boson operators are harmonic functions of time,

$$c^\dagger(t) = \exp\{i\omega t\} c^\dagger(t=0) \tag{6-7}$$

Anharmonic effects in the vibrational motion can be treated in terms of interactions between the quanta (see Sec. 6-6). For some of the nuclear vibrational modes, these effects are of importance even for the lowest states in the spectrum.

6-2b Vibrational Amplitudes

The variation of the density associated with a vibrational mode can be characterized by an amplitude α, representing the displacement from equilibrium. In the present section, we consider vibrations with real (Hermitian) amplitudes, such as standing waves in a deformable system. Modes with quanta that carry nonvanishing values of a conserved quantum number, like charge, nucleon number, or a component of angular momentum, are described by non-Hermitian amplitudes. The extension of the present treatment to cover the non-Hermitian modes will be considered in Sec. 6-3.

For small values of α, the vibrational energy may be expressed in powers of α and its time derivative $\dot{\alpha}$. To leading order, we have

$$E(\alpha, \dot{\alpha}) = \tfrac{1}{2}C\alpha^2 + \tfrac{1}{2}D\dot{\alpha}^2 \tag{6-8}$$

(For oscillations about equilibrium, there are no linear terms in α; moreover, terms proportional to $\dot{\alpha}$ or $\alpha\dot{\alpha}$ violate time reversal symmetry and cannot occur if we assume the equilibrium to be invariant under time reversal.)

The expression (6-8) represents a harmonic oscillator. The first term is the potential energy of deformation V, and the coefficient C is referred to as the restoring force parameter. The second term in Eq. (6-8) is the kinetic energy T, and the quantity D is referred to as the mass parameter.[2]

---

[2]The conventional notation for the vibrational mass parameter is B. We have felt it desirable, however, to adopt a different notation in order to avoid confusion with the reduced transition probability.

Introducing the momentum variable

$$\pi = \frac{\partial}{\partial \dot{\alpha}}(T - V) = D\dot{\alpha} \tag{6-9}$$

we obtain the Hamiltonian

$$H = \tfrac{1}{2}D^{-1}\pi^2 + \tfrac{1}{2}C\alpha^2 \tag{6-10}$$

The relations (6-8) to (6-10) are the same as for a classical vibrator. For a quantal system, the Hermitian operators α and π satisfy the canonical commutation relation

$$[\pi, \alpha] = -i\hbar \tag{6-11}$$

and the energy spectrum is given by the familiar relation for a harmonic oscillator

$$E(n) = (n + \tfrac{1}{2})\hbar\omega \tag{6-12}$$

with the classical frequency

$$\omega = \left(\frac{C}{D}\right)^{1/2} \tag{6-13}$$

The vibrational wave functions $\varphi_n(\alpha)$ have the form

$$\varphi_n(\alpha) = (2\pi)^{-1/4}(2^n n!\,\alpha_0)^{-1/2} H_n\left(2^{-1/2}\frac{\alpha}{\alpha_0}\right)\exp\left\{-\frac{1}{4}\frac{\alpha^2}{\alpha_0^2}\right\} \tag{6-14}$$

where H_n is the nth Hermite polynomial ($H_0(x) = 1$, $H_1(x) = 2x$, $H_2(x) = 4x^2 - 2, \cdots$), while α_0 is the zero-point amplitude

$$\begin{aligned} \alpha_0 &\equiv \langle n=0|\alpha^2|n=0\rangle^{1/2} = \langle n=1|\alpha|n=0\rangle \\ &= \left(\frac{\hbar}{2D\omega}\right)^{1/2} = \left(\frac{\hbar\omega}{2C}\right)^{1/2} = \left(\frac{\hbar^2}{4CD}\right)^{1/4} \end{aligned} \tag{6-15}$$

(In Eqs. (6-14) and (6-15), the matrix elements of α are assumed to be real; for states with the standard phase convention (1-39), the reality of the matrix elements requires α to be invariant under the transformation $\mathscr{R}\mathscr{T}$. If α changes sign under $\mathscr{R}\mathscr{T}$, the standard phasing is obtained by inserting a factor i^n in the wave function (6-14), and the matrix elements of α become imaginary; the transformation (6-16) then also involves an extra factor i; see, for example, Eqs. (5-21) and (6-51).)

In the harmonic approximation, the transformation from the variables α, π to the variables $c^\dagger$, c, introduced in Sec. 6-2a, is given by

$$\begin{aligned} \alpha &= \alpha_0(c^\dagger + c) \\ \pi &= \frac{i\hbar}{2\alpha_0}(c^\dagger - c) \qquad (6\text{-}16) \\ c^\dagger &= \frac{1}{2\alpha_0}\alpha - \frac{i\alpha_0}{\hbar}\pi = \frac{1}{2\alpha_0}\left(\alpha - \frac{i}{\omega}\dot{\alpha}\right) \end{aligned}$$

The transformation corresponds to the decomposition of the amplitudes α and π into positive and negative frequency parts, with time dependence $\exp\{-i\omega t\}$ and $\exp\{i\omega t\}$, respectively; see Eq. (6-7). (The relation (6-16) is the same as Eq. (5-18) for particle motion in a harmonic oscillator potential.)

For a system with many degrees of freedom, the total wave function can be expressed in terms of the vibrational coordinate α and additional coordinates q describing the remaining degrees of freedom (other vibrational modes, one-particle excitations, etc.). In the approximation of independent normal modes of excitation, the total wave function can be written in the form

$$\Psi_{\sigma,n}(\alpha,q) = \Phi_\sigma(q)\varphi_n(\alpha) \qquad (6\text{-}17)$$

where $\Phi_\sigma(q)$, specified by a set of quantum numbers σ, is referred to as the intrinsic state. The vibrational wave function $\varphi_n(\alpha)$ will, in general, depend on the intrinsic quantum numbers σ. For example, in a molecule, the vibrational potential energy is strongly dependent on the electronic state, since the molecular binding is determined by the few valence electrons. In nuclei, the vibrational properties may involve a large number of nucleons, of the order of the number of particles in a major shell, in which case they are expected to be less sensitive to the quantum numbers of the last few nucleons.

The functional relationship between the collective coordinates α and the particle variables (coordinates, momenta, and spins) depends on the detailed dynamics of the system and is determined by the criterion that the wave function can be approximately represented by the separated form (6-17), when expressed in the variables q, α. The vibrational coordinates α may approximately coincide with simple multipole moments of the system (or, in the case of molecules, with nuclear separations); however, the fluctuations of the actual coordinate α about these macroscopic values are in general important for ensuring the separation of the motion. (See the discussion of the collective orientation angles for rotating nuclei in Sec. 4A-3c and the analysis of collective vibrational variables in terms of one-particle excitations in Sec. 6-5h. The special conditions under which the collective coordinate can be expressed in terms of the position coordinates of the particles (point transformation) are discussed in the example on p. 510).

6-2c Collective Motion Generated by Vibrational One-Body Potential

The occurrence of collective vibrational modes in a system governed by independent-particle motion can be understood in terms of the variations in the average one-body potential produced by an oscillation in the nucleonic density. Such variations in the one-particle potential give rise to excitations in the nucleonic motion, and a self-sustained collective motion is obtained if the induced density variations are equal to those needed to generate the oscillating potential.

For small amplitudes of oscillation, the variations in the one-body potential are proportional to the amplitude α and can be written in the form

$$\delta V = \kappa\alpha F(x) \tag{6-18}$$

where $F(x)$ is a one-particle operator describing the dependence of the potential on the nucleonic variables x (space, spin, isospin; F may also be velocity dependent, as in the deformation of the spin-orbit potential and for nonlocal single-particle potentials). The operator F, which will often be referred to as the vibrational field, is partly characterized by the symmetry quantum numbers of the vibrational modes, such as the multipolarity, isospin, etc. (see Sec. 6-3). The full structure of F is ultimately determined by the requirement of self-consistency, as in the case of the static nuclear potential. The parameter κ in Eq. (6-18) is a coupling constant that characterizes the relationship between potential and density for the mode considered.

The response of the particle motion to an oscillating potential of the form (6-18) can be expressed in terms of the spectrum of one-particle excitations produced by the field F. Examples of such one-particle response functions for fields of different multipolarity are illustrated in Figs. 6-16 and 6-17, pp. 465 ff. A prominent feature of these spectra is the concentration of transition strength in rather narrow regions of excitation energy. This concentration is connected with the fact that the motion in the nuclear potential (in contrast to the electron motion in atoms or metals) is characterized by a rather well-defined period common to the particles in the last filled shells.

In the present section, we consider the collective motion that arises from the action of the field coupling in a system with degenerate one-particle excitation. This simple model contains the essential features involved in establishing the microscopic description of collective modes in terms of particle degrees of freedom and can be readily generalized.

In the following discussion, it may be simplest to think in terms of excitation with respect to a ground state consisting of closed shells. The excitations produced by the field F then involve a particle and a hole. The treatment, however, is immediately applicable to configurations with partly filled shells, since the pair correlations lead to a nondegenerate ground state ($\mathrm{v}=0$), the excitations of which can be described in terms of states involving two-quasiparticle configurations ($\mathrm{v}=2$). (The properties of quasiparticle excitations are considered in the example on pp. 647ff.)

The action of the field F on the ground state, $|\mathrm{v}=0\rangle$, of the one-particle motion produces the excitation

$$F|\mathrm{v}=0\rangle=\sum_i |i\rangle\langle i|F|\mathrm{v}=0\rangle$$
$$F\equiv\sum_k F(x_k) \tag{6-19}$$

where i labels the individual particle-hole (or two-quasiparticle) configurations. In a situation where the states $|i\rangle$ are degenerate, the excitation (6-19) is an eigenstate of the one-particle Hamiltonian. If this excitation involves many different single-particle modes i, it can be repeated a large number of times, and the resulting set of states can be associated with harmonic vibrational motion, as described in Sec. 6-2a. The quanta of excitation have the properties of approximately noninteracting bosons and can be described by the creation and annihilation operators $c^{(0)\dagger}$ and $c^{(0)}$, with the Hamiltonian

$$\begin{aligned} H^{(0)} &= \hbar\omega^{(0)}c^{(0)\dagger}c^{(0)} \\ &= n^{(0)}\hbar\omega^{(0)} \end{aligned} \tag{6-20}$$

where $\hbar\omega^{(0)}$ is the common excitation energy of the degenerate particle-hole states $|i\rangle$, while $n^{(0)}$ is the number of quanta of this mode. The superscript (0) refers to the fact that the quanta so far considered represent coherent independent-particle motion without inclusion of the field interaction. As emphasized above, the crucial condition for the description in terms of boson-like quanta is the occurrence of many individual components $|i\rangle$ in the coherent state (6-19). The enumeration of these components involves states such as $((n_1l_1j_1m_1)^{-1}(n_2l_2j_2m_2))$ fully specified in the quantum numbers of the particle and hole; thus, in general, a coupled state $(j_1^{-1}j_2)\lambda$ contributes a number of different components i.

The collective mode associated with the boson operators $c^{(0)}$ can also be described in terms of a deformation coordinate α, as in Sec. 6-2b. It is convenient to choose the normalization of α in such a manner that the zero-point amplitude $\alpha_0^{(0)}$ is equal to that of the field F

$$\alpha_0^{(0)} \equiv \langle n^{(0)}=1|\alpha|n^{(0)}=0\rangle$$

$$= \langle n^{(0)}=1|F|n^{(0)}=0\rangle = \left(\sum_i \langle i|F|\mathrm{v}=0\rangle^2\right)^{1/2} \tag{6-21}$$

The state $|n^{(0)}=0\rangle$ is the ground state $|\mathrm{v}=0\rangle$ for one-particle motion, while the state $|n^{(0)}=1\rangle$ is the normalized state obtained by multiplying the excitation (6-19) by the normalization factor $(\alpha_0^{(0)})^{-1}$. (The expression (6-21) assumes that the matrix elements of F (and of α) are real; see the comment on p. 332.)

In terms of the amplitude α, the Hamiltonian (6-20) takes the form

$$\begin{aligned} H^{(0)} &= \tfrac{1}{2}\left(D^{(0)}\right)^{-1}\pi^2 + \tfrac{1}{2}C^{(0)}\alpha^2 - \tfrac{1}{2}\hbar\omega^{(0)} \\ \pi &= D^{(0)}\dot{\alpha} \end{aligned} \tag{6-22}$$

with the restoring force and mass parameters (see Eqs. (6-13) and (6-15))

$$\begin{aligned} C^{(0)} &= \frac{\hbar\omega^{(0)}}{2\left(\alpha_0^{(0)}\right)^2} \\ D^{(0)} &= \frac{\hbar}{2\omega^{(0)}\left(\alpha_0^{(0)}\right)^2} \end{aligned} \tag{6-23}$$

In the space of states described by the quantum number $n^{(0)}$, the field F has matrix elements connecting states with $\Delta n^{(0)}=1$, all of which are equal to those of α, in the approximation considered; these matrix elements are of collective character, giving rise to transition probabilities that are enhanced with respect to single-particle magnitude by a factor that is of the order of the number of components i contributing to the excitation (6-19) (see Eq. (6-21)). The field F may have additional matrix elements in the space of $n^{(0)}$ as well as matrix elements that connect this space with other modes of the one-particle motion; these additional matrix elements lead to anharmonic effects in the vibrational motion as well as to couplings to other degrees of freedom and will be ignored in the present section.

In the approximation in which the matrix elements of F are the same as those of α, the field coupling (6-18) can be treated in terms of a contribution to the vibrational Hamiltonian of the form

$$H' = \frac{1}{2}\sum_k \kappa\alpha F(x_k)$$
$$= \frac{1}{2}\kappa F^2 = \frac{1}{2}\kappa\alpha^2 \qquad (6\text{-}24)$$

where the factor 1/2 results from the fact that the potential (6-18) is produced by two-body interactions, which are counted twice when the potential is summed over all particles.

The total Hamiltonian is obtained by adding the contributions (6-22) and (6-24),

$$H = H^{(0)} + H' = \frac{1}{2}D^{-1}\pi^2 + \frac{1}{2}C\alpha^2 - \frac{1}{2}\hbar\omega^{(0)} \qquad (6\text{-}25a)$$
$$= n\hbar\omega + \frac{1}{2}(\hbar\omega - \hbar\omega^{(0)}) \qquad (6\text{-}25b)$$

with

$$C = C^{(0)} + \kappa$$
$$\qquad (6\text{-}26)$$
$$D = D^{(0)}$$

and

$$\hbar\omega = \hbar\left(\frac{C}{D}\right)^{1/2} = \hbar\omega^{(0)}\left(1 + \frac{\kappa}{C^{(0)}}\right)^{1/2} \qquad (6\text{-}27a)$$

$$\underset{|\kappa| \ll C}{\approx} \hbar\omega^{(0)} + \kappa\langle n^{(0)} = 1|F|n^{(0)} = 0\rangle^2 \qquad (6\text{-}27b)$$

Thus, the effect of the field coupling (6-18) amounts to a modification of the restoring force for the vibrational mode, while the mass parameter is unaffected.

The quantum number n in Eq. (6-25b) represents the number of quanta in the interacting system. The energy of the ground state ($n=0$) is given by the last term in Eq. (6-25b). If expanded in powers of the interaction strength κ, the leading term in the ground-state energy gives the expectation value of the interaction energy H' in the unperturbed ground state, $n^{(0)}=0$ (see Eq. (6-27b)); the higher-order terms result from the modification of the ground state arising from the interaction. This modification finds expression in the altered zero-point amplitude

$$\alpha_0 \equiv \langle n=1|\alpha|n=0\rangle = \langle n=1|F|n=0\rangle$$

$$= \left(\frac{\hbar\omega}{2C}\right)^{1/2} = \alpha_0^{(0)}\left(\frac{\omega^{(0)}}{\omega}\right)^{1/2} = \alpha_0^{(0)}\left(1 + \frac{\kappa}{C^{(0)}}\right)^{-1/4} \qquad (6\text{-}28)$$

It is seen from Eq. (6-28) that the interactions do not affect the oscillator strength $\hbar\omega|\langle n=1|F|n=0\rangle|^2$ for the field F. (The conservation of oscillator strength follows directly from the fact that the interaction commutes with the field F; see the discussion of oscillator sum rules in Sec. 6-4a.)

A field coupling with $\kappa>0$ corresponds to interaction effects opposing the density variations and leads to an increase in the collective frequency as compared with the single-particle frequency ($\omega>\omega^{(0)}$). Interactions favoring the density variations have $\kappa<0$ and lower the collective frequency. For large negative values of κ approaching the value $-C^{(0)}$, the vibrational frequency goes to zero, corresponding to instability of the assumed equilibrium state with respect to deformations of the type considered. For still larger negative κ, the interactions establish an equilibrium with $\alpha\neq 0$. A schematic representation of potential energy functions $V(\alpha)$, referring to different values of κ relative to $C^{(0)}$, is given in Fig. 6-1. For a coupling approaching the critical value for instability ($\kappa\approx -C^{(0)}$), one must in general expect the occurrence of vibrational motion with large anharmonicity.

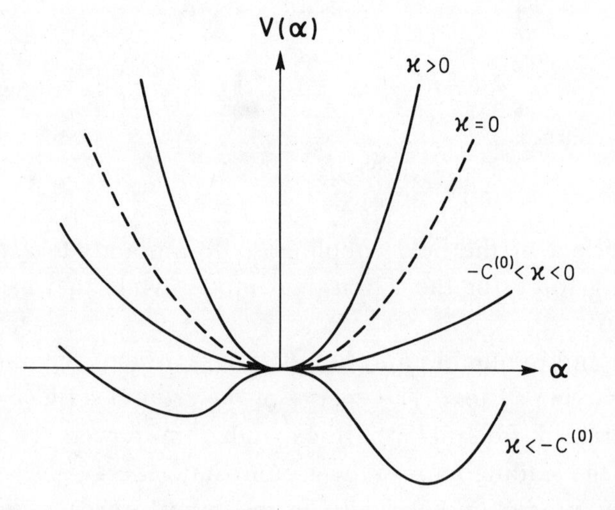

Figure 6-1 Potential energy functions. The figure gives a schematic illustration of the nuclear potential energy as a function of the deformation parameter α. The various curves correspond to different values of the field coupling constant κ.

In the nuclear vibrations, interaction effects of both positive and negative sign are encountered. Examples of modes with attractive interactions are provided by the nuclear shape oscillations (see pp. 520ff.); the low-frequency quadrupole mode is observed to exhibit instability corresponding to the occurrence of nuclei with static deformations. Examples of modes with repulsive interactions are provided by the neutron-proton polarization oscillations (see p. 481) and the isovector spin excitations (see p. 638).

The diagonalization of the Hamiltonian with inclusion of the field coupling can also be viewed in terms of a transformation from the boson variables $c^{(0)\dagger}$, $c^{(0)}$ to new variables $c^\dagger$, c associated with the vibrational quanta in the interacting system. The Hamiltonian (6-25) is a quadratic form in the variables $c^{(0)\dagger}$, $c^{(0)}$,

$$H = \hbar\omega^{(0)} c^{(0)\dagger} c^{(0)} + \tfrac{1}{2}\kappa\big(\alpha_0^{(0)}\big)^2 \big(c^{(0)\dagger} + c^{(0)}\big)^2 \tag{6-29}$$

and can be diagonalized by means of a linear transformation to new boson variables $c^\dagger$, c,

$$\begin{aligned} c^\dagger &= X c^{(0)\dagger} - Y c^{(0)} \\ c^{(0)\dagger} &= X c^\dagger + Y c \end{aligned} \tag{6-30}$$

The requirement that the transformation (6-30) preserves the boson commutation relations (6-2) leads to the normalization condition

$$X^2 - Y^2 = 1 \tag{6-31}$$

for the (real) amplitudes X and Y. The condition that the Hamiltonian in the new variables is diagonal in the number of quanta implies

$$X = \frac{\kappa\alpha_0\alpha_0^{(0)}}{\hbar(\omega - \omega^{(0)})} \qquad Y = -\frac{\kappa\alpha_0\alpha_0^{(0)}}{\hbar(\omega + \omega^{(0)})} \tag{6-32}$$

and

$$H = \tfrac{1}{2}(\hbar\omega - \hbar\omega^{(0)}) + \hbar\omega c^\dagger c \tag{6-33}$$

The presence of the amplitude Y in the transformation (6-30) reflects the modification in the ground state that was discussed above in terms of the modified zero-point oscillation. Expressed in the boson variables, the relation between the ground states takes the form

$$\begin{aligned} |n=0\rangle &= (1-K^2)^{1/4} \exp\{\tfrac{1}{2} K c^{(0)\dagger} c^{(0)\dagger}\} |n^{(0)}=0\rangle \\ &\approx |n^{(0)}=0\rangle + 2^{-1/2} K |n^{(0)}=2\rangle + \cdots \\ K &= \frac{Y}{X} = \frac{\omega^{(0)} - \omega}{\omega^{(0)} + \omega} \end{aligned} \tag{6-34}$$

as can be verified by acting with the Hamiltonian (6-29).

Though the discussion in the present section has employed collective variables such as the vibrational amplitude and the boson operators, the treatment is fully microscopic, since the collective variables are expressed in terms of the degrees of freedom of the individual particles. Thus, the operator $c^{(0)\dagger}$, producing a quantum of the basic excitation (6-19), can be written

$$c^{(0)\dagger} = \left(\alpha_0^{(0)}\right)^{-1} \sum_i \langle i|F|\mathsf{v}=0\rangle A_i^\dagger \tag{6-35}$$

where the operator $A_i^\dagger$ creates the particle-hole state $|i\rangle$ when acting on the ground state $|\mathsf{v}=0\rangle$ of the single-particle motion. For the operators of the normal mode, we obtain from Eqs. (6-30), (6-32), and (6-35)

$$\begin{gathered} c^\dagger = \sum_i \left(X_i A_i^\dagger - Y_i A_i\right) \\ X_i = \frac{\kappa\alpha_0\langle i|F|\mathsf{v}=0\rangle}{\hbar(\omega-\omega^{(0)})} \qquad Y_i = -\frac{\kappa\alpha_0\langle i|F|\mathsf{v}=0\rangle}{\hbar(\omega+\omega^{(0)})} \end{gathered} \tag{6-36}$$

From these relations, the various properties of the collective states can be expressed in terms of matrix elements for the one-particle states.

The interaction (6-24) resulting from the field coupling can also be interpreted as an effective two-body interaction

$$V(1,2) = \kappa F(1)F(2) \tag{6-37}$$

The analysis in the present section is thus equivalent to the study of correlation effects in the many-body system produced by such a "separable" interaction. The interaction (6-37) represents only a single component of the total effective interaction between nucleons in the nucleus. Thus, a short-range nucleonic interaction, $V(|\mathbf{r}_1-\mathbf{r}_2|)$, can generate a variety of different collective modes, each of which can be treated in terms of an interaction of the form (6-37).

In the discussion in the present volume, we shall attempt to establish properties of the field couplings on the basis of empirical evidence concerning the static nuclear potential and the modes of excitation. The problem of relating the effective interactions responsible for the collective modes to the interactions between free nucleons involves the many subtle correlations that must be taken into account in the analysis of the connection between the two-body forces and the static potentials in the nucleus (see Sec. 2-5). However, considerable scope attaches to the information that may be obtained from the analysis of this relationship, due to the need for guidance concerning the structure and strength of the many different fields involved in the nuclear collective modes.

The above discussion of the microscopic structure of collective modes has been simplified through the assumption of degeneracy in the one-particle response function and the neglect of anharmonic effects. Such a simplified treatment provides a first approximation to the analysis of many of the nuclear modes. The generalization to a nondegenerate single-particle response function is straightforward and is given in Sec. 6-5h. The treatment of anharmonicities involves a variety of different effects and is the topic of Sec. 6-6.

6-3 NORMAL MODES OF NUCLEAR VIBRATION

The various modes of vibration in the nucleus can be characterized by the symmetry quantum numbers of the vibrational quanta and the related symmetry properties of the density and field variations accompanying the vibrational motion. Some of the symmetries considered are only partially satisfied by the nucleonic interactions; moreover, the equilibrium configuration, on which the vibrations are superposed, does not in general possess the full invariance of the Hamiltonian. Thus, an essential element in the classification of the nuclear vibrations is the analysis of symmetry breaking occurring in a variety of different forms.

6-3a Shape Oscillations. Spherical Equilibrium

Vibrational amplitudes. Density variations

For a spherical equilibrium, the vibrational modes can be characterized by the multipole quantum number, which represents the angular momentum of a vibrational quantum. Each mode λ is $(2\lambda+1)$-fold degenerate, corresponding to the different values of the component μ of the vibrational angular momentum. The set of creation operators $c^{\dagger}(\lambda\mu)$ with $\mu=-\lambda, -\lambda+1,\ldots,+\lambda$ forms a spherical tensor (see the definition of such tensors in Vol. I, p. 80). The vibrational amplitudes $\alpha_{\lambda\mu}$ with corresponding tensorial properties describe the expansion of the density fluctuations in spherical harmonics. The theory of oscillation of a liquid drop provides the classical example of such a description (see Appendix 6A).[3]

[3]The quantization of the vibrations in the liquid-drop model was considered by Flügge (1941) and Fierz (1943).

In the present section, we consider vibrational modes associated with variations in the total particle density (summed over spin and isospin variables)

$$\rho(\mathbf{r}) = \sum_{k=1}^{A} \delta(\mathbf{r} - \mathbf{r}_k) \tag{6-38}$$

To leading order in the vibrational amplitudes, the density variations are linear in $\alpha_{\lambda\mu}$, and rotational invariance leads to the form

$$\begin{aligned} \delta\rho(\mathbf{r}) &= f_\lambda(r) \sum_\mu Y^*_{\lambda\mu}(\vartheta, \varphi)\alpha_{\lambda\mu} \\ &= (-1)^\lambda (2\lambda+1)^{1/2} f_\lambda(r)(Y_\lambda \alpha_\lambda)_0 \end{aligned} \tag{6-39}$$

for a mode with symmetry λ. The radial form factor $f_\lambda(r)$ depends on the internal structure of the vibrating system and will be discussed below (p. 343). The density operator is a function of the intrinsic variables as well as of the collective amplitudes $\alpha_{\lambda\mu}$, and Eq. (6-39) is to be regarded as an average of the density variations with respect to the intrinsic motion (specified by the wave function $\Phi_\sigma(q)$ in Eq. (6-17)).

The density operator (6-38) is Hermitian, and the spherical tensor $\alpha_{\lambda\mu}$ is therefore self-adjoint. With a suitable choice of overall phase, we have

$$\alpha^\dagger_{\lambda\mu} = (-1)^\mu \alpha_{\lambda -\mu} \tag{6-40}$$

The radial function $f_\lambda(r)$ in Eq. (6-39) is then real.

If the equilibrium is invariant under space reflection and time reversal, the transformation of $\alpha_{\lambda\mu}$ under $\mathscr{P}$ and $\mathscr{T}$ is given by

$$\begin{aligned} \mathscr{P}\alpha_{\lambda\mu}\mathscr{P}^{-1} &= (-1)^\lambda \alpha_{\lambda\mu} \\ \mathscr{T}\alpha_{\lambda\mu}\mathscr{T}^{-1} &= (-1)^\mu \alpha_{\lambda-\mu} = \alpha^\dagger_{\lambda\mu} \end{aligned} \tag{6-41}$$

These relations follow from the fact that $\rho(\mathbf{r})$ transforms into $\rho(-\mathbf{r})$ under $\mathscr{P}$ and is invariant with respect to $\mathscr{T}$. (A departure from $\mathscr{P}$ or $\mathscr{T}$ symmetry of the equilibrium on which the vibrations are superposed would imply a doublet structure of the energy levels, which is not observed in the nuclear spectra; see the discussion in Sec. 4-2e.)

The lowest vibrational modes are expected to have density variations with no radial nodes and may be referred to as shape oscillations. A simple model of such a mode is obtained by considering deformations that distort the radius parameter R, while leaving the surface diffuseness independent of angle (see Eq. (4-189)),

$$\rho(\mathbf{r},\alpha_{\lambda\mu}) \approx \rho_0(r) - R_0 \frac{\partial \rho_0}{\partial r} \sum_{\mu} Y^*_{\lambda\mu}\alpha_{\lambda\mu} \tag{6-42}$$

where $\rho_0(r)$ is the equilibrium density. The density variation given by Eq. (6-42) is of the form (6-39) with the radial form factor

$$f_\lambda(r) = -R_0 \frac{\partial \rho_0}{\partial r} \tag{6-43}$$

For $\lambda = 1$, the deformation (6-42) gives a translation without change in shape, and the $\lambda = 1$ shape oscillation is therefore represented by the center-of-mass motion. For a system with constant density and a sharp surface, the density distortion (6-42) is equivalent to the surface deformation described by Eq. (6A-1). (It should be emphasized that at present there exists no direct experimental evidence testing the assumption that the surface diffuseness is independent of deformation.)

Hamiltonian

In the harmonic approximation, the Hamiltonian is a quadratic function of the vibrational amplitudes and the canonically conjugate momenta, as in Eq. (6-10). For a vibrational mode λ, the requirement of rotational invariance leads to the form

$$\begin{aligned} H &= \tfrac{1}{2} D_\lambda^{-1} \sum_\mu \pi^\dagger_{\lambda\mu}\pi_{\lambda\mu} + \tfrac{1}{2} C_\lambda \sum_\mu \alpha^\dagger_{\lambda\mu}\alpha_{\lambda\mu} \\ &= \tfrac{1}{2}(-1)^\lambda (2\lambda+1)^{1/2} \left(D_\lambda^{-1} (\pi_\lambda \pi_\lambda)_0 + C_\lambda (\alpha_\lambda \alpha_\lambda)_0 \right) \end{aligned} \tag{6-44}$$

The equations of motion give (see the hermiticity relation (6-40))

$$\pi_{\lambda\mu} = D_\lambda \dot{\alpha}^\dagger_{\lambda\mu} = D_\lambda (-1)^\mu \dot{\alpha}_{\lambda -\mu} \tag{6-45}$$

and the oscillation frequency

$$\omega_\lambda = \left(\frac{C_\lambda}{D_\lambda} \right)^{1/2} \tag{6-46}$$

The Hamiltonian (6-44) represents the independent oscillations of the $(2\lambda + 1)$ degenerate modes associated with a shape deformation of order λ.

The vibrational motion can be described in terms of traveling or standing waves, as in the classical analysis of the oscillations of a liquid drop (see Sec. 6A-1c).

The quantal features of the vibrations can be expressed in terms of the creation and annihilation operators $c^\dagger(\lambda\mu)$ and $c(\lambda\mu)$ that satisfy the boson commutation relations (see Eq. (6-2))

$$\begin{aligned} [c(\lambda\mu),c(\lambda\mu')]=[c^\dagger(\lambda\mu),c^\dagger(\lambda\mu')]=0 \\ [c(\lambda\mu),c^\dagger(\lambda\mu')]=\delta(\mu,\mu') \end{aligned} \tag{6-47}$$

The number of quanta with specified $\lambda\mu$ is represented by the operators (see Eq. (6-3))

$$(n_{\lambda\mu})_{\text{op}}=c^\dagger(\lambda\mu)c(\lambda\mu) \tag{6-48}$$

and the Hamiltonian is a sum of terms of type (6-5) (see also Eq. (6-12)),

$$H=\hbar\omega_\lambda\sum_\mu\left(c^\dagger(\lambda\mu)c(\lambda\mu)+\tfrac{1}{2}\right) \tag{6-49}$$

The amplitudes $\alpha_{\lambda\mu}$ and their conjugates $\pi_{\lambda\mu}$ obey the canonical commutation relations

$$\begin{aligned} [\alpha_{\lambda\mu},\alpha_{\lambda\mu'}]=[\pi_{\lambda\mu},\pi_{\lambda\mu'}]=0 \\ [\pi_{\lambda\mu},\alpha_{\lambda\mu'}]=-i\hbar\delta(\mu,\mu') \end{aligned} \tag{6-50}$$

and are linearly related to the $c^\dagger(\lambda\mu), c(\lambda\mu)$ operators, as in Eq. (6-16),

$$\begin{aligned} \alpha_{\lambda\mu}&=i^{-\lambda}(\alpha_\lambda)_0\left(c^\dagger(\lambda\mu)+c(\overline{\lambda\mu})\right) \\ \pi_{\lambda\mu}&=i^{\lambda-1}\frac{\hbar}{2(\alpha_\lambda)_0}\left(c(\lambda\mu)-c^\dagger(\overline{\lambda\mu})\right) \\ c^\dagger(\lambda\mu)&=\frac{i^\lambda}{2(\alpha_\lambda)_0}\alpha_{\lambda\mu}-i^{\lambda+1}\frac{(\alpha_\lambda)_0}{\hbar}(-1)^\mu\pi_{\lambda-\mu} \\ &=\frac{i^\lambda}{2(\alpha_\lambda)_0}\left(\alpha_{\lambda\mu}-\frac{i}{\omega_\lambda}\dot{\alpha}_{\lambda\mu}\right) \end{aligned} \tag{6-51}$$

While $(\alpha_\lambda)_0$ represents the zero-point amplitude for the individual modes $\lambda\mu$, the total zero-point amplitude β_λ of multipole order λ is given by

$$\begin{aligned}\beta_\lambda^2 &= (2\lambda+1)(\alpha_\lambda)_0^2 = \langle n_\lambda = 0|\sum_\mu \alpha_{\lambda\mu}^\dagger \alpha_{\lambda\mu}|n_\lambda = 0\rangle \\ &= |\langle n_\lambda = 1\|\alpha_\lambda\|n_\lambda = 0\rangle|^2 \\ &= (2\lambda+1)\frac{\hbar}{2D_\lambda\omega_\lambda} = (2\lambda+1)\frac{\hbar\omega_\lambda}{2C_\lambda} = (2\lambda+1)\frac{\hbar}{2(C_\lambda D_\lambda)^{1/2}}\end{aligned} \tag{6-52}$$

where $|n_\lambda = 0\rangle$ is the ground state. In Eq. (6-51), we have assumed the standard phasing for creation operators (see Eq. (1A-87))

$$c^\dagger(\overline{\lambda\mu}) \equiv \mathscr{T} c^\dagger(\lambda\mu)\mathscr{T}^{-1} = (-1)^{\lambda+\mu} c^\dagger(\lambda - \mu) \tag{6-53}$$

With this phase convention, the hermiticity and time reversal transformation for $\alpha_{\lambda\mu}$ (see Eqs. (6-40) and (6-41)) determine the phases in Eq. (6-51) to within a factor of ± 1.

It is seen from Eq. (6-51) that the amplitude $\alpha_{\lambda\mu}$ is associated with the creation of a quantum with angular momentum component μ and the annihilation of a quantum with component $-\mu$. More generally, if the quanta have a nonvanishing value for any set of additive quantum numbers γ, the amplitude α_γ is a combination of $c^\dagger(\gamma)$ and $c(-\gamma)$. For the oscillations associated with the density (6-38), the quanta have vanishing values for all additive quantum numbers except μ, corresponding to the fact that the amplitudes $\alpha_{\lambda\mu}$ are self-adjoint spherical tensors.

The angular momentum of the vibrational motion can be expressed as a bilinear form in the amplitudes $\alpha_{\lambda\mu}$ and $\dot{\alpha}_{\lambda\mu}$, or in the operators $c^\dagger(\lambda\mu)$ and $c(\overline{\lambda\mu})$. This form is uniquely determined by the vector character of the angular momentum operator (see the analogous expression (1A-86) for tensor operators for fermions and Eq. (1A-63) for the reduced matrix element of the angular momentum)

$$\begin{aligned}I_\mu &= \frac{1}{\sqrt{3}}\langle\lambda\|I\|\lambda\rangle\left(c^\dagger(\lambda)c(\bar{\lambda})\right)_{(\lambda\bar{\lambda})1\mu} \\ &= \left(\frac{1}{3}\lambda(\lambda+1)(2\lambda+1)\right)^{1/2}\sum_{\mu'\mu''}\langle\lambda\mu'\lambda\mu''|1\mu\rangle c^\dagger(\lambda\mu')c(\overline{\lambda\mu''})\end{aligned} \tag{6-54}$$

If the amplitudes (α, π) are used as variables, one obtains the canonical expression (6A-12) for the vibrational angular momentum.

The quanta of a density oscillation of symmetry λ have the parity

$$\pi = (-1)^{\lambda} \tag{6-55}$$

The parity quantum number follows from the $\mathscr{P}$ transformation (6-41) for $\alpha_{\lambda\mu}$, which also applies to the $c^{\dagger}(\lambda\mu)$ operators, on account of the linear relationship (6-51).

Spectrum

In the harmonic approximation, the excitation energy equals

$$E(n_\lambda) - E(n_\lambda = 0) = n_\lambda \hbar\omega_\lambda \tag{6-56}$$

where n_λ is the number of quanta

$$n_\lambda = \sum_\mu n_{\lambda\mu} \tag{6-57}$$

For given n_λ, the values of the total vibrational angular momentum I can be obtained by coupling the angular momenta λ of the individual quanta, taking into account the Bose statistics. Thus, for $n_\lambda = 2$, the symmetric states have

$$I = 0, 2, \ldots, 2\lambda \tag{6-58}$$

The enumeration of I values, for given n_λ, can be obtained by a simple counting of the number of states with different values of the component of total angular momentum,

$$M = \sum_\mu \mu n_{\lambda\mu} \tag{6-59}$$

For quadrupole oscillations ($\lambda = 2$), the states with $n_2 \leq 6$ are listed in Table 6-1. For $n_2 > 3$, there may be several states with the same quantum numbers n_2, I (and M); these can be further specified by the seniority quantum number v (see pp. 691 ff.). The explicit construction of many-phonon states in terms of parentage coefficients is discussed in Sec. 6B-4.

Description of vibrations in terms of rotational and intrinsic shape degrees of freedom

A deformation of multipole order λ can be described in terms of $(2\lambda - 2)$ parameters that characterize the intrinsic shape together with the three angular variables (Euler angles) that specify the orientation of the deformed shape with respect to a fixed coordinate system. Thus, the $2\lambda + 1$ vibrational degrees of freedom can be viewed in terms of $2\lambda - 2$ intrinsic

	I												
n_2	0	1	2	3	4	5	6	7	8	9	10	11	12
0	1												
1			1										
2	1		1		1								
3	1		1	1	1		1						
4	1		2		2	1	1		1				
5	1		2	1	2	1	2	1	1		1		
6	2		2	1	3	1	3	1	2	1	1		1

Table 6-1 Enumeration of states with many quadrupole phonons. The table lists the number of states with $n_2(\leqslant 6)$ quadrupole quanta and total angular momentum I.

modes representing oscillations in the shape and the three rotational modes associated with a change of orientation, for fixed intrinsic shape. In this way, the rotational degrees of freedom appear as special superpositions of vibrational modes. For quadrupole vibrations, the transformation of the amplitudes and the Hamiltonian into shape and angle variables is given in Appendix 6B. In this case, the shape has ellipsoidal symmetry and can be characterized by a parameter β giving the magnitude of the total deformation and a parameter γ that measures the deviation from axial symmetry.

The representation in terms of the shape and angle variables has especially simple consequences in situations where the intrinsic shape performs small oscillations about a nonspherical equilibrium; the motion then separates into rotation and intrinsic shape vibrations, and the spectrum exhibits rotational band structure, as discussed in Chapter 4. This situation occurs if the potential energy contains anharmonic terms that give rise to a pronounced minimum for deformed shape. The analysis of the anharmonicities in terms of such a picture provides the connection between the vibrational spectra of spherical nuclei and the rotation-vibration spectra of deformed nuclei. (See pp. 448 ff. and p. 681.)

Even for a potential energy that favors the spherical shape, as for harmonic vibrations, the equilibrium shape becomes nonspherical for sufficiently large values of the angular momentum, as a result of the centrifugal forces. This effect is especially pronounced in the yrast region of the vibrational spectrum. (See Sec. 6B-3; the rotational band structure and the classification of the intrinsic excitations in the yrast region are illustrated, for $\lambda = 2$, in Fig. 6B-2, p. 686.) An alternative way of viewing the simplicity of the yrast region can be based on the fact that the alignment of the angular momenta of the individual quanta implies the existence of a condensate with many identical quanta. (See the analogous situation for pair quanta described on pp. 392ff.)

$E\lambda$ moments

A shape vibration of order λ, with density variation (6-39), is characterized by large values of the multipole moment of the same order

$$\begin{aligned}\mathscr{M}(\lambda\mu) &= \int \rho(\mathbf{r}) r^\lambda Y_{\lambda\mu}(\vartheta,\varphi)\, d\tau \\ &= \int r^{\lambda+2} f_\lambda(r)\, dr\, \alpha_{\lambda\mu}\end{aligned} \tag{6-60}$$

The moment (6-60) refers to the total particle density; the electric moment $\mathscr{M}(E\lambda,\mu)$ is obtained by replacing $\rho(\mathbf{r})$ and $f_\lambda(r)$ by the corresponding charge density functions.

For the shape vibrations, it is expected that the ratio of proton to neutron density remain approximately constant (see the discussion on p. 382), in which case

$$\mathscr{M}(E\lambda,\mu) \approx \frac{Ze}{A}\,\mathscr{M}(\lambda\mu) \tag{6-61}$$

If we further assume the radial form factor (6-43), the $E\lambda$ moment can be expressed in the form

$$\begin{aligned}\mathscr{M}(E\lambda,\mu) &\approx -\frac{Ze}{A} R_0 \int r^{\lambda+2} \frac{\partial \rho_0}{\partial r}\, dr\, \alpha_{\lambda\mu} \\ &= \frac{\lambda+2}{4\pi} ZeR_0 \langle r^{\lambda-1}\rangle \alpha_{\lambda\mu}\end{aligned} \tag{6-62}$$

where the radial moment refers to the spherical distribution ρ_0. One may also express the electric multipole moment in the form

$$\mathscr{M}(E\lambda,\mu) = \frac{3}{4\pi} ZeR^\lambda \alpha_{\lambda\mu} \tag{6-63}$$

with the effective radius R defined by

$$3R^\lambda = (\lambda+2) R_0 \langle r^{\lambda-1}\rangle \tag{6-64}$$

The relation (6-63) is the same as for a surface deformation in a system with constant density and a sharply defined radius R (see Eq. (6A-2)). The diffuseness of the nuclear surface implies that the parameter R defined by

Eq. (6-64) depends somewhat on the value of λ. The evaluation of the moments $\langle r^n \rangle$ has been considered in Vol. I, p. 160, for a density distribution of the Woods-Saxon shape.

Even though the density variations associated with the nuclear vibrations may differ from the form (6-42), it is often convenient to use relation (6-63) as a normalization of the vibrational amplitude. (For an axially symmetric quadrupole deformation, the amplitude α_{20} is equal to the total deformation parameter β, which in turn is related by Eq. (4-191) to the eccentricity parameter δ, defined by Eq. (4-72).)

With the normalization (6-63), the transition probability for exciting a vibrational quantum is given by (see Eqs. (6-51) and (6-52))

$$\begin{aligned} B(E\lambda; n_\lambda=0\rightarrow n_\lambda=1) &= \left(\frac{3}{4\pi} ZeR^\lambda\right)^2 \beta_\lambda^2 \\ &= (2\lambda+1)\left(\frac{3}{4\pi} ZeR^\lambda\right)^2 \frac{\hbar}{2D_\lambda\omega_\lambda} \qquad (6\text{-}65) \\ &= (2\lambda+1)\left(\frac{3}{4\pi} ZeR^\lambda\right)^2 \frac{\hbar\omega_\lambda}{2C_\lambda} \end{aligned}$$

In the harmonic approximation, the matrix elements of $\mathscr{M}(E\lambda,\mu)$ obey the selection rule $\Delta n_\lambda = \pm 1$, and all ratios between transition matrix elements are specified. The reduced matrix element is given by

$$\begin{aligned} &\langle n_\lambda+1, \zeta_{n+1} I_{n+1} \| \mathscr{M}(E\lambda) \| n_\lambda \zeta_n I_n \rangle \\ &\quad = i^{-\lambda} \frac{3}{4\pi} ZeR^\lambda \left(\frac{\hbar}{2D_\lambda\omega_\lambda}\right)^{1/2} \langle n_\lambda+1, \zeta_{n+1} I_{n+1} \| c^\dagger(\lambda) \| n_\lambda \zeta_n I_n \rangle \qquad (6\text{-}66) \end{aligned}$$

where the vibrational states are specified by the number n_λ of phonons, the total angular momentum I_n, and additional quantum numbers ζ_n needed to distinguish between states with the same n_λ and I_n. The reduced matrix element of the creation operator is the parentage coefficient discussed in Sec. 6B-4.

For independent quanta, the total decay rate is the sum of the decay rates for the individual quanta,

$$\sum_{\zeta_{n-1}, I_{n-1}} B(E\lambda; n_\lambda \zeta_n I_n \rightarrow n_\lambda - 1, \zeta_{n-1} I_{n-1}) = n_\lambda B(E\lambda; n_\lambda = 1 \rightarrow n_\lambda = 0) \qquad (6\text{-}67)$$

This relation can also be viewed as an expression of the sum rule for the parentage coefficients that follows from the relation (6-48).

Occurrence of shape oscillations

The study of the low-energy spectra of the even-even nuclei has revealed the systematic occurrence of $I\pi = 2+$ and $3-$ states with properties suggesting a vibrational interpretation. The collective character of the excitations is implied by the large transition probabilities and by the fact that the properties vary rather smoothly with N and Z. The systematics of the excitation energies of the $2+$ and $3-$ modes are shown in Fig. 2-17a, b, Vol. I, pp. 196–197, and Fig. 6-40, p. 560, respectively. The transition probabilities are an order of magnitude larger than the single-particle unit, and if interpreted as shape oscillations, as in Eq. (6-65), correspond to zero-point amplitudes β_2 and β_3 typically of order 0.2 (see Fig. 4-5, p. 48, for values of $B(E2)$ and Table 6-14, p. 561, for values of $B(E3)$). The available evidence on the radial form factor for the transition moment suggests a strong peaking at the nuclear surface, as for shape oscillations (see, for example, Heisenberg and Sick, 1970).

There is also some evidence for $I\pi = 4+$ collective density oscillations. (See, for example, the $4+$ excitations in the Cd isotopes ($\hbar\omega_4 = 2.3$ MeV) with cross sections about 10 times the single-particle magnitude (Koike *et al.*, 1969) and in ^{208}Pb ($\hbar\omega_4 = 4.3$ MeV, $B(E4; 4 \rightarrow 0) \approx 25 B_W$; Ziegler and Peterson, 1968).)

Evidence on multiple excitations is at present confined to the quadrupole mode. It is found that the spectra contain states with $I\pi = 0+, 2+,$ and $4+$ (see Eq. (6-58)), with qualitative features corresponding to double excitations ($n_2 = 2$). However, the quantitative values of the energies and $E2$-matrix elements show quite large deviations from the relations (6-56) and (6-66) for harmonic vibrations. (The interpretation of these anharmonic effects is discussed on pp. 541 ff.; see also Sec. 6-6a.) So far, the evidence concerning states with large numbers of quanta is confined to the leading trajectory with $I = 2n_2$, and thus it has not been possible to test the expected pattern of the spectrum in the yrast region.

The occurrence of $\lambda\pi = 2+$ and $3-$ excitations as the lowest collective modes is characteristic of a system with spherical equilibrium and with a stiffness against compression that is large compared with the stiffness for shape deformations. The liquid-drop model provides a prototype for such a system (see Appendix 6A). However, the quantitative properties of the observed nuclear excitations differ significantly from those deduced from such a macroscopic description. In particular, the properties of the quadrupole vibrations are profoundly affected by the shell structure. In nuclei with many particles outside of closed shells, the quadrupole mode has very low frequency and large amplitude, which is related to the instability phenomena manifested in the occurrence of nuclei with non-spherical equilibrium shapes. With the approach to closed shells, this mode becomes progressively weaker, and in the nuclei with closed shells of neutrons and protons, no low-frequency quadrupole mode occurs. The lowest-frequency shape oscillations in such nuclei are of octupole type.

The quantitative comparison with the liquid-drop model can be exhibited in terms of the restoring force and mass parameters C_λ and D_λ, which can be determined from the observed values of $\hbar\omega_\lambda$ and $B(E\lambda)$ by means of Eqs. (6-46) and (6-65). For the quadrupole mode, these parameters are shown in Figs. 6-28, p. 529, and 6-29, p. 531. It is seen that the restoring force parameter C_2 is an order of magnitude larger than the liquid-drop value near closed shells, but becomes much smaller than this value with the approach to the regions where the nuclei possess stable quadrupole deformations. The mass parameter D_2 is an order of magnitude larger than the liquid-drop value based on the assumption of irrotational flow.

The inadequacy of the liquid-drop analogy can be attributed to the long mean free path for the motion of the individual nucleons in the nucleus. For such a system, the collective modes cannot be described in terms of energy densities determined by local density variations and collective flow.

The structure of the collective modes in a system with independent-particle motion can be analyzed in terms of the coupling of the nucleons to the oscillating potential generated by the collective deformation, as considered in Sec. 6-2c. On this basis, one obtains a simple understanding of the main qualitative features of the observed quadrupole modes (see pp. 520 ff.) and octupole modes (see pp. 559 ff.). The occurrence of collective modes of very low frequency and the related instability of the nuclear shape are associated with the approximate degeneracies in the spectrum of single-particle motion in the spherical potential. The general conditions for the occurrence of such degeneracies are considered on pp. 579 ff; the dominance of the quadrupole instability, as observed in nuclei, can be seen as a consequence of the approximate validity of the harmonic oscillator potential for the single-particle motion (see p. 591).

The low-frequency quadrupole oscillations are mainly associated with the transitions of the particles within partially filled shells, but one expects additional quadrupole excitations associated with transitions between major shells. The occurrence of such additional high-frequency quadrupole strength can be directly inferred from the oscillator sum rule (see Sec. 6-4a), since the observed low-frequency mode only accounts for about 10% of the oscillator sum (see Fig. 6-29). The expected properties of these high-frequency quadrupole shape oscillations and the available experimental evidence are discussed on pp. 508 ff. (The expected high-frequency octupole strength is considered on pp. 556 ff.)

The modes so far discussed have nodeless radial form factors, corresponding to shape oscillations. Density variations with radial nodes correspond to compressive oscillations, but there is at present no experimental evidence concerning such collective excitations in the nucleus. Compressive modes may occur with $\lambda\pi = 0+,\ 1-,\ 2+, \ldots$, but because of the radial nodes, these excitations give rise to rather weak multipole moments (see Sec. 6A-3); however, they may be excited in inelastic scattering processes involving large momentum transfers or strongly absorbed particles. (The experimental study of these modes would provide information on the nuclear compressibility.)

In the description of collective motion in macroscopic quantum fluids, one distinguishes between the hydrodynamic and collision-free regimes, depending on whether the relaxation time for the quasiparticles is small or large compared with the period of the collective oscillation (see, for example, Pines and Nozières, 1966). The relaxation is due to the interactions with the thermal excitations and vanishes in the limit of low temperatures. Thus, the nuclear modes considered in the present chapter belong to the collision-free regime and are to be compared with "zero sound" rather than with "first sound," which is characteristic of the hydrodynamic regime with local thermodynamic equilibrium. The nucleus provides the opportunity for studying collective modes in a quantum fluid with wavelengths comparable to the dimensions of the system and with properties strongly influenced by the quantization of the particle motion in the system as a whole. It is possible that similar phenomena may be encountered in plasma oscillations in atoms (see, for example, Fano and Cooper, 1968, Amusia *et al.*, 1971, and Wendin, 1973).

Vibrational field

A density variation of multipole order λ produces a variation of the nuclear potential which, to first order in the deformation, is of the same multipolarity and hence of the form

$$\delta V = -k_\lambda(r) \sum_\mu Y^*_{\lambda\mu}(\vartheta, \varphi)\alpha_{\lambda\mu} \tag{6-68}$$

For a shape oscillation, one may expect the total potential $V + \delta V$ to be approximately obtained by a deformation of the average static potential. As discussed in Chapter 5, such an assumption accounts rather well for the potential observed in the strongly deformed nuclei. A deformation of the central potential of the same form as the density distortion (6-42) leads to the value

$$k_\lambda(r) = R_0 \frac{\partial V}{\partial r} \tag{6-69}$$

for the form factor in Eq. (6-68).

The deformed field (6-68) implies that the shape vibrations are strongly excited in the inelastic scattering of nucleons as well as of other nuclear projectiles (d, α, π, K, etc.). Indeed, the energy spectra of inelastically scattered particles are dominated by a few prominent groups that can be identified with the excitation of the lowest modes of shape oscillation (see the example in Fig. 6-2).[4]

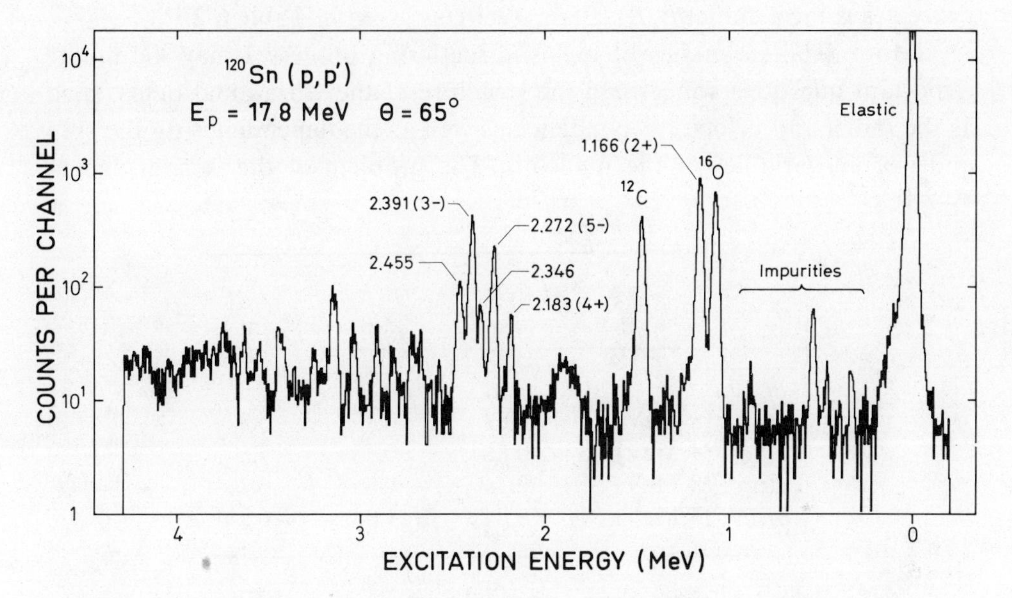

Figure 6-2 Inelastic scattering of protons on ^{120}Sn. The figure shows the energy spectrum of protons scattered through 65° by a target of ^{120}Sn and observed in a Si detector. The weak group below 1 MeV, as well as the strong groups labeled ^{12}C and ^{16}O, correspond to elastic scattering on various impurities. The broad group at about 1.9 MeV results from the escape of the gamma ray following the excitations of the 2+ state in ^{28}Si in the detector. The groups corresponding to the lowest states in ^{120}Sn are labeled by the excitation energy in MeV and by the spin and parity, where these are known. Most of the pronounced peaks above 2.5 MeV have also been identified as excited states in ^{120}Sn. The data are from the experiment reported by O. N. Jarvis, B. G. Harvey, D. L. Hendrie, and J. Mahoney, *Nuclear Phys.* **A102**, 625 (1967). We wish to thank B. G. Harvey for additional information concerning this experiment.

---

[4]The selective excitation of these levels was first observed in proton scattering (Cohen, 1957; Cohen and Rubin, 1958) and soon after in deuteron and α scattering (Yntema and Zeidman, 1959; Fulbright *et al.*, 1959). The amplitudes of vibrational motion deduced from the measured inelastic cross sections were found to agree rather well with those derived from measurements of the $E\lambda$-transition rates (McDaniels *et al.*, 1960; Buck, 1963; Satchler *et al.*, 1963), thus providing significant confirmation of the interpretation of these states in terms of shape oscillations.

The values of the transition matrix elements β_2 and β_3, derived from the inelastic cross sections in Fig. 6-2, are given in Table 6-2 and compared with values obtained from other inelastic processes as well as from the $E\lambda$-transition moments. It is seen that the deformation parameters agree rather well with each other. The comparison is not unique because the average radius parameters for the different excitation processes are somewhat different. These differences are connected with the nonlinear relationship between potential and density and the strong absorption of the composite projectiles. Thus, if the physically significant quantity is the displacement of the surface, $\delta R = R_0 \alpha_{\lambda\mu} Y^*_{\lambda\mu}$, the quantity to be compared for the different processes is the product $\beta_\lambda R_0$ (Blair, 1960), as given in Table 6-2.

More detailed studies of inelastic scattering processes may illuminate important questions concerning the structure of the vibrational fields, such as the radial and velocity dependence as well as the dependence on the spin and isospin variables of the nucleons. The problem of the isospin depen-

	$\beta_2 R_0$ fm	$\beta_3 R_0$ fm	R_0 fm
$B(E\lambda)$	0.64	0.87	5.9
σ(dd′; 15 MeV)	0.69	0.80	5.8
$\sigma(\alpha\alpha'$; 34 MeV)	0.82	0.90	7.5
σ(pp′; 18 MeV)	0.74	0.86	6.2
σ(nn′; 14 MeV)	0.74	1.05	6.2

Table 6-2 Comparison of vibrational amplitudes for ^{120}Sn as obtained from different excitation processes. The table compares the amplitudes obtained from a number of different processes that excite the low-frequency quadrupole and octupole modes in ^{120}Sn ($\hbar\omega_2 = 1.17$ MeV and $\hbar\omega_3 = 2.39$ MeV). The inelastic neutron scattering was studied on a natural mixture of tin isotopes. The radius parameter R_0 in the last column is that assumed in the analysis of the different processes. The experimental data and analyses are taken from: Coulomb excitation (P. H. Stelson, F. K. McGowan, R. L. Robinson, and W. T. Milner, *Phys. Rev.* **C2**, 2015, 1970; D. G. Alkazov, Y. P. Gangrskii, I. K. Lemberg, and Y. I. Undralov, *Izv. Akad. Nauk. SSSR, Ser. Fiz.* **28**, 232, 1964); dd′ (R. K. Jolly, *Phys. Rev.* **139**, B318, 1965); $\alpha\alpha'$ (I. Kumabe, H. Ogata, T. H. Kim, M. Inoue, Y. Okuma, and M. Matoba, *J. Phys. Soc. Japan* **25**, 14, 1968); pp′ (Jarvis *et al., loc. cit.*, Fig. 6-2; nn′ (P. H. Stelson, R. L. Robinson, H. J. Kim, J. Rapaport, and G. R. Satchler, *Nuclear Phys.* **68**, 97, 1965).

dence is discussed on pp. 382 ff. As regards the spin dependence, one expects a term associated with the distortion of the spin-orbit potential. If this potential is taken to be of the form (2-144), a deformation of the type assumed in Eq. (6-42) yields (see Eq. (3A-26) and compare with the corresponding term (5-3) for axially deformed nuclei)

$$\begin{aligned}\delta V_{ls} &= -(2\lambda+1)^{1/2} V_{ls} r_0^2 \nabla\left(R_0 \frac{\partial f}{\partial r}(Y_\lambda \alpha_\lambda)_0\right)\cdot(\mathbf{p}\times\mathbf{s}) \\ &= -(2\lambda+1)^{1/2} V_{ls} r_0^2 \frac{R_0}{r}\left(\left(\frac{\partial^2 f}{\partial r^2} - \frac{\lambda}{r}\frac{\partial f}{\partial r}\right)(Y_\lambda \alpha_\lambda)_0 (\mathbf{l}\cdot\mathbf{s})\right. \\ &\quad \left. + \lambda^{1/2}(2\lambda+1)^{1/2}\frac{\partial f}{\partial r}((Y_{\lambda-1}, (\mathbf{p}\times\mathbf{s}))_\lambda \alpha_\lambda)_0\right)\end{aligned} \tag{6-70}$$

Evidence for such a component in the vibrational field has been obtained from the analysis of polarization effects in inelastic scattering (see, for example, Sherif, 1969).

The potential generated by the vibration may in addition contain terms proportional to $\dot{\alpha}_{\lambda\mu}$, with a one-particle field that is odd under time reversal (combined with Hermitian conjugation). Terms of this type arise from the velocity dependence of the average one-particle potential and can be obtained by expressing the nucleonic velocities relative to the local collective flow (local Galilean invariance, Belyaev (1963)). Examples of such coupling terms are discussed in connection with the analysis of the high-frequency quadrupole mode (pp. 511 ff.) and of the center-of-mass mode (p. 445).

The vibrational field can also be studied through the resulting coupling between the vibration and the motion of individual particles in the nuclear bound states. The various effects of this particle-vibration coupling will be considered in Sec. 6-5. The experimental data provide evidence for a coupling of approximately the strength (6-69); see, for example, the discussion of effective charges for $E2$ moments in odd-A nuclei, pp. 517 and 533, and the particle-octupole coupling in ^{209}Bi and ^{209}Pb, pp. 565 ff.

In the discussion in Sec. 6-2c of the interaction effects in the vibrational motion, the vibrational coupling is expressed in the form (6-18), which, in the case of a mode of multipole order λ, corresponds to

$$\delta V = \kappa_\lambda \sum_\mu \alpha_{\lambda\mu} F^\dagger_{\lambda\mu}(x) \tag{6-71}$$

where κ_λ is a coupling constant and where $F_{\lambda\mu}$ is so normalized that

$$\langle \sum_k F_{\lambda\mu}(x_k) \rangle = \int \delta\rho F_{\lambda\mu} d\tau = \alpha_{\lambda\mu} \tag{6-72}$$

when averaged for the deformed density distribution. For the coupling (6-68) with the form factor (6-69), one may employ the dimensionless field

$$F_{\lambda\mu} = -\kappa_\lambda^{-1} R_0 \frac{\partial V}{\partial r} Y_{\lambda\mu} \tag{6-73}$$

The corresponding coupling constant is obtained from an evaluation of the average (6-72) with $\delta\rho$ given by Eq. (6-42)

$$\begin{aligned} \kappa_\lambda &= \int R_0 \frac{\partial V}{\partial r} R_0 \frac{\partial \rho_0}{\partial r} r^2 dr \\ &= -\frac{R_0^2}{4\pi} \int \frac{1}{r^2} \frac{\partial}{\partial r}\left(r^2 \frac{\partial V}{\partial r} \right) \rho_0 d\tau \end{aligned} \tag{6-74}$$

(Note that κ and V are negative for attractive fields.)

Qualitative orientation regarding the effect of the shell structure on the shape oscillations may be obtained by considering vibrational fields with $F_{\lambda\mu}$ equal to the multipole operator $r^\lambda Y_{\lambda\mu}$, in conjunction with an average potential of harmonic oscillator form. For this model, the spectrum of single-particle excitations and the resulting collective modes have an especially simple and transparent character. For the oscillator potential

$$V = \tfrac{1}{2} M \omega_0^2 r^2 \tag{6-75}$$

a deformation of the standard form (6-71) with

$$F_{\lambda\mu} = r^\lambda Y_{\lambda\mu} \tag{6-76}$$

can be written

$$\delta V = \frac{\kappa_\lambda}{M\omega_0^2} r^{\lambda-1} \frac{\partial V}{\partial r} \sum_\mu Y_{\lambda\mu}^* \alpha_{\lambda\mu} \tag{6-77}$$

For $\lambda = 2$, the deformed potential represents an anisotropic oscillator (see Eq. (5-5)). If the equidensity and equipotential surfaces suffer identical deformations, the density variation $\delta\rho$ is given by Eq. (6-77), with V replaced by ρ_0. The normalization condition (6-72) then yields

$$\kappa_\lambda = -\frac{4\pi}{2\lambda+1} \frac{M\omega_0^2}{A\langle r^{2\lambda-2} \rangle} \tag{6-78}$$

The quadrupole and octupole modes for the oscillator model with the coupling (6-78) are discussed on pp. 508 ff. and pp. 556 ff., respectively.

M1 moments

The magnetic dipole moment associated with a shape-vibrational mode λ is, to leading order, proportional to α_λ and $\dot{\alpha}_\lambda$, as follows from time reversal and rotational symmetry. The unique vector formed by coupling α_λ and $\dot{\alpha}_\lambda$ is the same as that involved in the vibrational angular momentum (given by Eq. (6-54) or (6A-12)), and we therefore have

$$\boldsymbol{\mu} = g_\lambda \mathbf{I} \tag{6-79}$$

with g_λ a constant, for a given vibrational mode. Hence, to leading order, all the states in the vibrational spectrum have the same static g factor. Moreover, since the total angular momentum I is a constant of the motion, there are no $M1$ transitions.

For the quadrupole vibrations, there is extensive evidence testing the forbiddenness of the $M1$-matrix element for the transition $(n_2=2, I=2) \to (n_2=1, I=2)$. The $B(M1)$ values are in most cases found to be less than $10^{-2}(e\hbar/2Mc)^2$; see the examples in Figs. 6-30, p. 532, and 6-32, p. 536. (A compilation of $M1{:}E2$ mixing ratios for these transitions has been given by Krane, 1974.[5])

The value of g_λ is a property of the collective flow associated with the vibrational motion. For a shape vibration, one may expect the main contribution to arise from orbital motion of the particles and the ratio of charge to mass in the flow to be approximately the same as for the static density. We then obtain

$$g_\lambda \approx \frac{Z}{A} \tag{6-80}$$

as for a uniformly charged liquid drop. The main empirical evidence concerns the $n_2=1$, $I\pi=2+$ quadrupole excitation, for which the g values are found to be approximately of the magnitude (6-80), although significant variations are observed (see Fig. 4-6, p. 55). In a few cases, measurements have been made of the g factor for the $n_3=1$, $I\pi=3-$ excitations (^{16}O, $\hbar\omega_3=6.1$ MeV, $g=0.55\pm0.03$, Randolph *et al.*, 1973; ^{208}Pb, $\hbar\omega_3=2.6$ MeV, $g=0.58\pm0.14$, Bowman *et al.*, 1969).

---

[5]The strong hindrance of the $M1$-matrix element for the transition between the first and second 2+ states was among the earliest systematic features to be recognized in the spectra of the even-even nuclei (Kraushaar and Goldhaber, 1953).

E0 moments

Electric moments of multipole order different from that of the vibrational motion must involve second or higher powers of the vibrational amplitudes. Thus, the $E0$ moment

$$m(E0)=\int r^2\rho_{el}(\mathbf{r})d\tau \tag{6-81}$$

is, to leading order, quadratic in the vibrational amplitude and may involve two terms, proportional to $(\alpha_\lambda\alpha_\lambda)_0$ and $(\dot{\alpha}_\lambda\dot{\alpha}_\lambda)_0$, respectively. If the frequency of the vibrational motion is small compared with single-particle excitations (adiabatic situation), the nucleonic density distribution for given α_λ is approximately the same as for a static deformation, and one therefore expects $m(E0)$ to depend only weakly on $\dot{\alpha}_\lambda$. (For a discussion of the adiabatic approximation, see Sec. 6-6a.) Under these conditions, there is a unique second-order term in $m(E0)$,

$$m(E0)=m(E0,\alpha=0)+k\sum_{\mu}|\alpha_{\lambda\mu}|^2 \tag{6-82}$$

and all matrix elements can be expressed in terms of the single parameter k. Thus, for example,

$$\langle n_\lambda IM|m(E0)|n_\lambda IM\rangle-\langle n_\lambda=0|m(E0)|n_\lambda=0\rangle=kn_\lambda\frac{\hbar}{D_\lambda\omega_\lambda} \tag{6-83a}$$

$$\langle n_\lambda=2,I=0|m(E0)|n_\lambda=0\rangle=k(\lambda+\tfrac{1}{2})^{1/2}\frac{\hbar}{D_\lambda\omega_\lambda} \tag{6-83b}$$

One may attempt to estimate the coefficient k by considering a deformation represented by a bodily displacement of the surface, without change of the interior density or of the diffuseness normal to the surface (see Eq. (4-188)). The expansion (4-189) of this density distribution, to second order in the amplitudes $\alpha_{\lambda\mu}$, yields an $E0$ moment with

$$k=\frac{3}{4\pi}ZeR_0^2\left(2-\frac{2}{3}\frac{\langle r^{-2}\rangle\langle r\rangle}{\langle r^{-1}\rangle}+\frac{1}{2}\lambda(\lambda+1)\left(1-\frac{2}{3}\langle r^{-1}\rangle R_0\right)\right) \tag{6-84a}$$

$$\approx\frac{3}{4\pi}ZeR_0^2\left(1+\frac{\pi^2}{3}(\lambda-1)(\lambda+2)\left(\frac{a_0}{R_0}\right)^2\right) \tag{6-84b}$$

where, in Eq. (6-84b), we have used the radial moments (2-65) for the Woods-Saxon form factor. It must be emphasized, however, that the monopole moment associated with a shape oscillation is sensitive to possible variations in the radial density distribution, and the result (6-84) is therefore only expected to give the order of magnitude of the effect. Evidence on the value of k can be derived from the measured $E0$-matrix element for the $n=2$, $I=0 \rightarrow n=0$ transition in ^{114}Cd (see Fig. 6-30, p. 532); the value of k determined by means of Eq. (6-83b) is found to be about a factor of 2 smaller than the estimate (6-84). For the deformed nucleus ^{174}Hf, the observed value of k for the $K\pi=0+$ monopole vibration agrees rather well with the estimate (6-84); see p. 174. (Additional, more indirect estimates of monopole moments associated with a quadrupole deformation have been obtained from isotope shifts in atomic spectra and are found to be of the order of the estimate (6-84) although somewhat smaller; see, for example, the evidence on ^{152}Sm discussed in Vol. I, p. 164.)

The terms in the density of second order in the deformation $\alpha_{\lambda\mu}$ also give rise to electric moments of multipolarity $2,4,\ldots,2\lambda$. Thus, for a quadrupole vibration, one obtains a second-order contribution to the $E2$ moment, which gives rise to matrix elements with $\Delta n_2=0$ and ± 2. Similar effects, however, can occur as a result of anharmonicity in the vibrational motion, and the analysis of terms in the moment going beyond the leading order must therefore be considered together with higher-order terms in the Hamiltonian (see Sec. 6-6a).

Superposition of vibrational and particle motion[6]

If the intrinsic state possesses a finite angular momentum J (as in odd-A or odd-odd nuclei, or for excited configurations in even-even nuclei), a vibrational excitation with angular momentum R gives rise to a multiplet of states with total angular momentum

$$I=|J-R|,|J-R|+1,\ldots,J+R \qquad (6\text{-}85)$$

The particle-vibration coupling produces energy separations within the multiplet (6-85) and a mixing of states with different quantum numbers JR; these coupling effects will be considered in Sec. 6-5.

In the limit of weak coupling, the various nuclear moments can be expressed as a sum of contributions from the vibrational and intrinsic

---

[6]The spectra and moments resulting from the superposition of approximately independent-particle and vibrational motion were considered at an early stage in the study of collective nuclear motion (Foldy and Milford, 1950; Bohr and Mottelson, 1953; see also the review and references quoted by Alder *et al.*, 1956, pp. 538 and 539). The generality of such a coupling scheme (core-excitation model) was emphasized by de-Shalit (1961).

degrees of freedom, and the matrix elements can be evaluated by the general expression (1A-72a) for multipole operators in weakly coupled systems.

Large collective matrix elements are associated with $E\lambda$ transitions with $\Delta n_\lambda = \pm 1$ and no change of intrinsic state. In the weak coupling situation, only the vibrational moment contributes to such transitions and the $B(E\lambda)$ value is directly related to the vibrational transition rate,

$$B(E\lambda;\, n_\lambda RJI \rightarrow n_\lambda \pm 1, R'JI')$$
$$= (2I'+1)(2R+1)\begin{Bmatrix} R & J & I \\ I' & \lambda & R' \end{Bmatrix}^2 B(E\lambda;\, n_\lambda R \rightarrow n_\lambda \pm 1, R') \quad (6\text{-}86)$$

The sum of the transition probabilities to the different members I' of the final multiplet is equal to $B(E\lambda;\, n_\lambda R \rightarrow n_\lambda \pm 1, R')$, as can be formally verified by employing the completeness relation (1A-17).

The states within a multiplet are connected by $M1$-matrix elements,

$$\langle RJI' \| \mathscr{M}(M1) \| RJI \rangle = \left(\frac{3}{4\pi}\right)^{1/2} \frac{e\hbar}{2Mc} \Bigg(g_J (I(I+1)(2I+1))^{1/2} \delta(I,I')$$
$$+ (-1)^{R+J+I+1}(g_\lambda - g_J)(2I+1)^{1/2}(2I'+1)^{1/2}(R(R+1)(2R+1))^{1/2} \begin{Bmatrix} R & J & I \\ I' & 1 & R \end{Bmatrix} \Bigg)$$
$$(6\text{-}87)$$

The $M1$ transitions are seen to be typically of single-particle magnitude, since the g factors for the particle motion will usually differ significantly from those of the collective motion.

The multiplet of states involving an octupole quantum superposed on the ground state of ^{209}Bi provides an example of the pattern corresponding to weak particle-vibration coupling (see Fig. 6-42, p. 572). The test of the intensity relation for excitation of the multiplet by inelastic scattering, which is equivalent to Eq. (6-86), is shown in Table 6-16, p. 573. In other cases, the coupling is found to be much stronger and to imply appreciable interweaving of intrinsic and vibrational motion (see, for example, the quadrupole excitations in the odd Cd isotopes illustrated in Fig. 6-30, p. 532).

6-3b Vibrations about Spheroidal Equilibrium

The symmetry of the equilibrium state plays a major role in determining the pattern of the vibrational modes. In the present section, we consider the properties of vibrations with respect to an equilibrium shape with axial symmetry and invariance under reflection in a plane perpendicular to the symmetry axis, as for the dominant nuclear deformations.

Symmetry classification

For an axially symmetric equilibrium, the normal modes can be characterized by the quantum number ν, representing the component of vibrational angular momentum along the symmetry axis. The reflection symmetry provides the additional quantum number π. Moreover, a shape with axial and reflection symmetry is invariant with respect to a rotation $\mathscr{R}$ through an angle of 180° about an axis perpendicular to the symmetry axis (see the discussion of $\mathscr{R}$ symmetry in Chapter 4, p. 8). The $\mathscr{R}$ symmetry implies a degeneracy of modes with $\pm\nu$ (as also follows from $\mathscr{T}$ invariance); for $\nu=0$, the modes can be classified by the quantum number

$$r = \pm 1 \tag{6-88}$$

associated with the $\mathscr{R}$ transformation of the vibrational excitation. For a shape vibration with $\nu=0$, we have $r=\pi$, since a shape with axial symmetry is invariant under the transformation $\mathscr{R}\mathscr{P}$ (reflection in a plane containing the symmetry axis).

If the intrinsic angular momentum, in the absence of vibration, has the value $K_0=0$, as for the ground-state configuration of even-even nuclei, the excitation of a quantum with $\nu=0$ gives rise to a $K=0$ band with $(-1)^I=r$ (see Eq. (4-14)). For $\nu\neq0$, the conjugate modes $(\pm\nu)$ together generate a single band with $K=|\nu|$ and $I=K, K+1, \ldots$ (see Eq. (4-19)).

For intrinsic states with $K_0\neq0$, the excitation of a quantum with $\nu\neq0$ gives rise to two bands with $K=|K_0+\nu|$ and $|K_0-\nu|$. The degeneracy between these two bands is removed by the particle-vibration coupling as well as by the coupling to the rotational motion. For $K=1/2$ bands, the rotational decoupling parameter a vanishes for a vibrational excitation $(n_\nu=1)$ with $\nu\neq0$, while for $\nu=0$ (see Eq. (4-60) and note that $\mathscr{R}\mathscr{T}=1$ for the vibrational excitation)

$$a(n_{\nu=0}=1, K=K_0=1/2) = r a(n_{\nu=0}=0, K_0=1/2) \tag{6-89}$$

in terms of the quantum number r for the vibrational mode.

A schematic illustration of the geometry of the simplest even-parity shape oscillations and the associated spectra is given in Fig. 6-3. The deformations with respect to equilibrium are taken to be of quadrupole type ($\lambda=2$). The deformation amplitudes in the intrinsic coordinate system are denoted by $a_{\lambda\nu}$, and for quadrupole deformations are conventionally expressed in terms of the deformation parameters β and γ (see Eq. (6B-2) and Fig. 6B-1). The spheroidal equilibrium shape has $\gamma=0$ (prolate) or $\gamma=\pi$ (oblate), and total deformation parameter β_0. For small oscillations about prolate equilibrium, we have

$$\begin{aligned} a_{20} &= \beta\cos\gamma \approx \beta_0 + (\beta-\beta_0) \\ a_{22} &= \frac{1}{\sqrt{2}}\beta\sin\gamma \approx \frac{1}{\sqrt{2}}\beta_0\gamma \end{aligned} \tag{6-90}$$

The $\nu=0$ vibrations preserve axial symmetry and are referred to as β vibrations. The oscillations with $\nu=\pm 2$ (γ vibrations) break the axial symmetry and lead to nuclear shapes of ellipsoidal type.

A quadrupole deformation with $|\nu|=1$ is equivalent to a rotation about an axis perpendicular to the symmetry axis, without change in shape (see Fig. 6-3). This mode, therefore, is "spurious" in the sense that it does not occur as an intrinsic (nonrotational) excitation. The absence of the $\lambda=2$, $|\nu|=1$ mode is analogous to the absence of shape oscillations with $\lambda=1$, the degrees of freedom of which correspond to center-of-mass motion; see p. 343. The removal of the spurious $|\nu|=1$ mode in the deformed nuclei is considered on pp. 446 ff.

In the spectra of many deformed nuclei, one has observed low-lying excitations with $K\pi r=0++$ and $K\pi=2+$ that may be approximately described as β and γ vibrations. (See the examples discussed on pp. 158 ff. (γ vibration in ^{166}Er) and pp. 168 ff. (β vibration in ^{174}Hf); for the discussion of the systematics of these modes, see pp. 548 ff.) The superposition of β- and γ-vibrational excitations and quasiparticle degrees of freedom is encountered in many of the spectra of the odd-A deformed nuclei (see footnotes to Tables 5-12, p. 296, and 5-13, p. 299).

Odd-parity collective modes of low frequency corresponding to octupole shape oscillations have also been observed as a systematic feature in the spectra of even-even deformed nuclei (see the evidence discussed on pp. 559 ff.). The occurrence of odd-parity shape oscillations in the saddle-point spectrum plays an important role in the interpretation of asymmetry and anisotropy in the fission process (see the discussion in Chapter 4, p. 126).

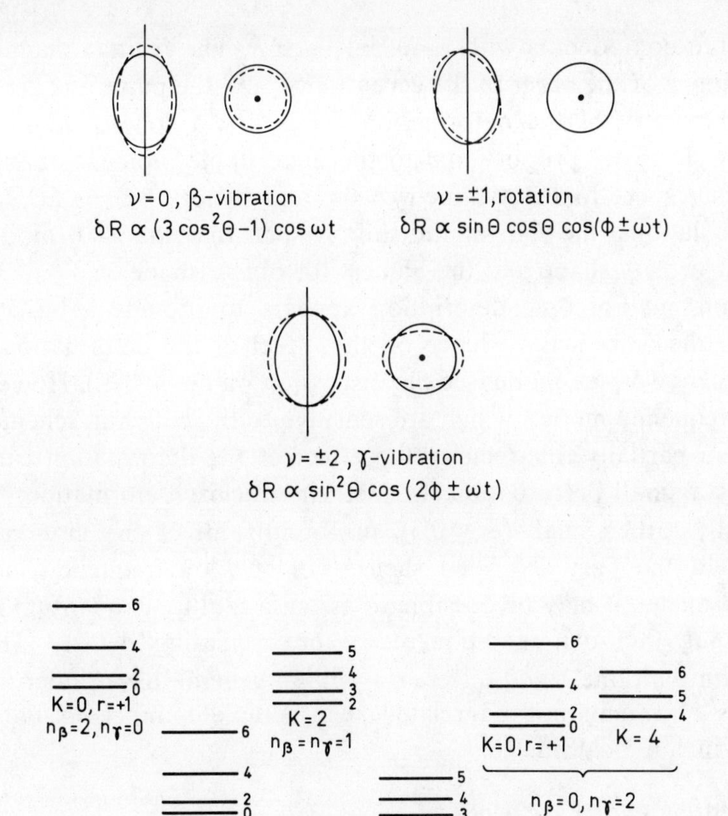

Figure 6-3 Quadrupole shape oscillations in a spheroidal nucleus. The upper part of the figure shows projections of the nuclear shape in directions perpendicular and parallel to the symmetry axis. The lower part of the figure shows the spectrum associated with excitations of one or two quanta. The relative magnitude of $\hbar\omega_\beta$ and $\hbar\omega_\gamma$ depends on the internal structure, and no significance is to be attached to the value used in the figure. For the bands with two quanta, it has been assumed that the vibrational motion is harmonic with no interactions between the β- and γ- vibrational quanta, and that the rotational energy is proportional to $I(I+1)-K^2$.

Frequency splitting due to static deformation

In the classical description of the oscillations of a system with a small equilibrium eccentricity, the effect of the deformation may be treated as a perturbation in the equations determining the eigenmodes (see the discussion of the liquid-drop model in Sec. 6A-4d). Each mode of multipole order

λ splits into components with $\nu=0,\pm 1,\ldots,\pm\lambda$. The relative magnitude of the splitting is of the order of the eccentricity, and the pattern is determined by the symmetry of the deformation. Thus, for a spheroidal shape, the frequency shifts are proportional to the quadrupole moment of the vibrations with respect to the symmetry axis, and hence to $3\nu^2-\lambda(\lambda+1)$. For shape oscillations, the sign of the shift is such that the $\nu=0$ mode is the lowest for prolate shape and the highest for oblate shape.

Such a macroscopic description appears to provide a basis for interpreting the empirical evidence on the effect of the deformation on the high-frequency dipole mode (see the discussion on pp. 490ff.). However, for the low-frequency modes, which are sensitive to the coupling scheme of the particles in partially filled shells, the effect of the deformation cannot be viewed as a small perturbation. In fact, the nuclear deformations, though numerically rather small ($\delta \lesssim 0.3$), profoundly affect the motion of the nucleons in the partially filled shells. For the low-frequency modes in deformed nuclei, it may be possible to assign a multipole quantum number λ describing the main component in the oscillating density. However, modes with the same λ and different ν will, in general, have rather different properties and may not be related in a simple manner to the modes observed in spherical nuclei.

Transition matrix elements

A shape oscillation $\lambda\nu$ is characterized by a large $E\lambda$-transition probability for exciting a vibrational quantum. With the same normalization as in Eq. (6-63), the $E\lambda$ moment and the $B(E\lambda)$ value take the form (see Eqs. (4-91) and (4-92))

$$\mathscr{M}(E\lambda,\mu)=\frac{3}{4\pi}ZeR^{\lambda}\sum_{\nu}a_{\lambda\nu}\mathscr{D}^{\lambda}_{\mu\nu}(\omega) \tag{6-91a}$$

$$B(E\lambda;\, K_0, n_{\lambda\nu}=0, I_1\rightarrow K_0, n_{\lambda\nu}=1, K=|K_0\pm\nu|, I_2)$$

$$=\left(\frac{3}{4\pi}ZeR^{\lambda}\right)^2 |\langle n_{\lambda\nu}=1|a_{\lambda\nu}|n_{\lambda\nu}=0\rangle|^2\langle I_1K_0\lambda\pm\nu|I_2K_0\pm\nu\rangle^2\begin{cases}2 & K_0=0,\nu\neq 0\\ 1 & \text{otherwise}\end{cases} \tag{6-91b}$$

where the zero-point amplitude is related in the usual way to the frequency and mass parameter

$$|\langle n_{\lambda\nu}=1|a_{\lambda\nu}|n_{\lambda\nu}=0\rangle|^2=\frac{\hbar}{2D_{\lambda\nu}\omega_{\lambda\nu}} \tag{6-92}$$

The expression (6-91b) represents the leading-order intensity rule corresponding to I-independent moments; the I-dependent corrections to the moments, arising from the coupling between the rotational and vibrational motion, are discussed in Sec. 6-6c.

The observed $E2$-transition probabilities for the excitation of β and γ vibrations are typically five to ten times the appropriate single-particle units (see pp. 551 ff. and pp. 549 ff.). Though enhanced, these transition probabilities are considerably smaller than for the quadrupole vibrations in spherical nuclei. Thus, the vibrational amplitudes $\beta-\beta_0$ and $\beta_0\gamma$ derived from the $E2$-matrix elements are of the order of 0.05. With such small amplitudes, only relatively few particles participate in the vibrational motion.

6-3c Collective Motion in Fission Process

In the preceding sections, we have considered the elementary modes associated with small-amplitude shape oscillations. In contrast to these modes, the fission process involves collective motion of large amplitude. The occurrence of the fission mode as a spontaneous process, or as a reaction induced by low-energy projectiles, reflects the near instability of heavy nuclei resulting from the long-range Coulomb repulsion between the protons (Meitner and Frisch, 1939; Bohr and Wheeler, 1939; Frenkel, 1939). The fission process has occupied a unique place in the development of nuclear physics, but should be recognized as part of a wider range of phenomena involving large-scale nuclear deformations and collective flow that are now becoming accessible in the study of reactions produced by accelerated heavy ions.

The early studies of the fission process focused attention on the macroscopic features that can be described in terms of the liquid-drop model of the nucleus. More recently, the discovery of the fission isomers has brought into focus the far-reaching consequences of the nuclear shell structure for many features of the fission process.

Macroscopic features of the potential energy surface. Fission barrier

Many of the qualitative features associated with large-scale nuclear deformations are determined by bulk properties of the nuclear matter. These features depend smoothly on the particle numbers N and Z and on the deformation parameters α, as in the liquid-drop model. Thus, for a deformation that approximately conserves volume, the main terms in this "macroscopic" part of the nuclear energy $\tilde{\mathscr{E}}(\alpha)$ can be associated with the surface and Coulomb energies

$$\tilde{\mathscr{E}}(\alpha) \approx \mathscr{E}_{\text{surf}} + \mathscr{E}_{\text{Coul}} \tag{6-93}$$

where $\mathscr{E}_{\text{surf}}$ is proportional to the increase in surface area with a proportionality constant that can be related to the parameter b_{surf} in the semi-empirical mass formula (2-12); see Eq. (6A-19). The term $\mathscr{E}_{\text{Coul}}$ expresses the decrease in Coulomb energy associated with the deformation.

For small deformations with respect to spherical shape, both the surface and the Coulomb energy are quadratic functions of the deformation parameters. The ratio of the coefficients in $\mathscr{E}_{\text{Coul}}$ and $\mathscr{E}_{\text{surf}}$ is proportional to Z^2/A, and for a quadrupole deformation is given by the quantity (see Eq. (6A-24))

$$\begin{aligned} x &= \frac{3}{10}\frac{e^2}{r_0} b_{\text{surf}}^{-1} \frac{Z^2}{A} \\ &\approx 0.0203 \frac{Z^2}{A} \qquad (r_0 = 1.25 \text{ fm}; \quad b_{\text{surf}} = 17 \text{ MeV}) \end{aligned} \tag{6-94}$$

referred to as the fissility parameter. The values of r_0 and b_{surf} used in Eq. (6-94) are those obtained from the semi-empirical mass formula (see Eq. (2-14)). For $x < 1$, the bulk energy $\tilde{\mathscr{E}}(\alpha)$ has a minimum for spherical shape, but for a critical value of Z^2/A, corresponding to $x = 1$, the spherical shape becomes unstable with respect to quadrupole deformations; the parameters of Eq. (6-94) yield

$$\left(\frac{Z^2}{A}\right)_{\text{crit}} \approx 49 \tag{6-95}$$

For $x < 1$, the energy $\tilde{\mathscr{E}}(\alpha)$ increases with deformation for small α, but eventually reaches a saddle point (the fission barrier) beyond which the energy decreases as the system approaches division into two or more fragments. The calculation of the potential energy function and associated fission barriers on the basis of the liquid-drop model is discussed in Sec. 6A-2. The resulting barrier heights and saddle-point shapes are shown in Fig. 6A-1 as a function of the fissility parameter x. For a system with angular momentum, the barrier is lowered as a result of the centrifugal forces, and the saddle point acquires a triaxial shape; for sufficiently large values of the angular momentum, the system has no stable equilibrium (see the discussion on pp. 663 ff.).

In Fig. 6-56, p. 616, the empirical values of the fission barriers are compared with those estimated on the basis of the liquid-drop model. Since the fission barriers of the heavy nuclei represent the small differences between large surface and Coulomb terms, the qualitative agreement in Fig. 6-56 implies that for such large deformations, the restoring force due to the nuclear interactions is given rather well by the surface tension as determined from the nuclear masses. (For example, for ^{238}U, with $x \approx 0.73$, the calculated fission barrier is about 10 MeV, while the increase in the surface energy is about 100 MeV.)

Evidence on the saddle-point shape has been obtained from the analysis of angular distributions of the fission fragments (see the example on pp. 617 ff.). The deformations derived from such analyses decrease with increasing values of Z^2/A, and are also quantitatively in approximate agreement with the predictions of the liquid-drop model, though the data may suggest an increase of the numerical coefficient in Eq. (6-94) by about 10% and a corresponding decrease of $(Z^2/A)_{\text{crit}}$ in Eq. (6-95).

Effect of shell structure on potential energy surface. Shape isomers

Important contributions to the nuclear potential energy function may result from the effects of shell structure.[7] The nonuniformities in the one-particle eigenvalue spectrum associated with the shell structure imply that the nuclear energy will not vary smoothly with particle number, as in the description in terms of bulk properties, but will exhibit specific variations depending on the degree of filling of the shells. Such effects have already been encountered in the occurrence of nonspherical equilibrium shapes (see especially the discussion in Chapter 4, pp. 135 ff.) and in the properties of the low-frequency shape vibrations (see pp. 351 ff.).

The contribution of shell structure to the potential energy function (the shell-structure energy) can be obtained from an analysis of relative energies of nucleonic configurations in a potential of given shape. Such relative energies are given by the one-particle energies of the occupied states, and the shell-structure energy is therefore contained in the sum

$$\mathscr{E}_{\text{ip}} = \sum_{k=1}^{A} \varepsilon_k \tag{6-96}$$

---

[7]The possibility of shell-structure effects in the fission process was considered from the early period of the nuclear shell model, in particular in relation to the asymmetry in the mass division (see, for example, Hill and Wheeler, 1953, and the comments on this early discussion by Mayer and Jensen, 1955, p. 37). The successful classification of one-particle orbits in deformed equilibrium configurations provided a basis for extending the analysis to the highly deformed shapes involved in the fission process (Strutinsky, 1966; Gustafson *et al.*, 1967). A crucial issue in this development was the separation between the smooth behavior associated with the bulk properties of nuclear matter and the more specific effects associated with the nuclear shell structure, which is contained in the sum of the single-particle energies (Myers and Swiatecki, 1966; Strutinsky, 1966).

It should be emphasized that the independent-particle energy $\mathscr{E}_{\mathrm{ip}}$ does not in any sense represent an estimate of the total nuclear energy. However, the deviations of $\mathscr{E}_{\mathrm{ip}}$ from a smooth function of particle numbers represents an approximation to the shell-structure energy, which is expected to be valid to the extent that the occupancy of the orbits corresponds to that of the independent-particle description. A more formal derivation of this result, in the self-consistent field approximation, is given in the fine print below. The effects of the pair correlation can be included by replacing the independent-particle sum (6-96) by the total energy for independent-quasiparticle motion in the presence of a pairing potential (see Eq, (6-603)).

For a system containing many particles, the main part of $\mathscr{E}_{\mathrm{ip}}$ varies smoothly with particle number. This smooth part, $\tilde{\mathscr{E}}_{\mathrm{ip}}$, can be identified by considering the asymptotic behavior of $\mathscr{E}_{\mathrm{ip}}$ in the limit of large particle numbers. The shell-structure energy $\mathscr{E}_{\mathrm{sh}}$ is then obtained as the difference

$$\mathscr{E}_{\mathrm{sh}} = \mathscr{E}_{\mathrm{ip}} - \tilde{\mathscr{E}}_{\mathrm{ip}} \tag{6-97}$$

between $\mathscr{E}_{\mathrm{ip}}$ and the asymptotic function $\tilde{\mathscr{E}}_{\mathrm{ip}}$.

For a saturating system, such as the nucleus, for which the linear scale of the potential increases asymptotically as $A^{1/3}$, the function $\tilde{\mathscr{E}}_{\mathrm{ip}}$ can be expressed as an expansion of the same type as that involved in the semi-empirical mass formula (2-12). The coefficients of the various terms could be obtained, for example, by a numerical evaluation of $\mathscr{E}_{\mathrm{ip}}$, as a function of particle number, for any given form of the potential; such a procedure would be analogous to the determination of the parameters b_{vol}, b_{surf}, etc. in Eq. (2-12) from the empirical masses. The asymptotic part of $\mathscr{E}_{\mathrm{ip}}$ can also be obtained by analytical methods. Thus, the leading terms in the particle number (the volume terms) are given by the semiclassical, or Fermi gas, approximation (see Eq. (6-511)). The next terms, proportional to $A^{2/3}$, correspond to an energy per unit surface area of the potential and can be evaluated by considering the continuum solutions defined by an infinite plane surface (Swiatecki, 1951; Siemens and Sobieczewski, 1972).

An alternative way of characterizing the asymptotic form $\tilde{\mathscr{E}}_{\mathrm{ip}}$ can be based on the asymptotic expression $\tilde{g}(\varepsilon)$ for the one-particle level density in the specified potential, as discussed in the example on pp. 606ff. (The determination of the smooth function $\tilde{\mathscr{E}}_{\mathrm{ip}}$ currently employed in the literature is based on an energy average of the single-particle level density; Strutinsky, 1966. For an infinite potential, the energy averaging and the extraction of the asymptotic form in particle number are equivalent; see the discussion on p. 606. However, for a finite potential, the equivalence requires that the distance between shells be small compared with the separation energy.)

The magnitude of $\mathscr{E}_{\rm sh}$ is of the order of the degeneracy of the shells times the energy separation between successive shells. While the distance between shells is quite generally of the order of the basic single-particle frequency $\varepsilon_F A^{-1/3}$, the degeneracy of the shells reflects the symmetries of the potential. (See the example on pp. 579 ff., which discusses the characterization of shells in the single-particle spectrum and the intimate relation of the shell structure to the occurrence of closed classical orbits.)

The major shell-structure effects in the spherical nuclear potential imply contributions to the nuclear energy that can be observed in the masses of the nuclear ground states ($\mathscr{E}_{\rm sh} \sim -10$ MeV in heavy closed-shell nuclei; see the example on pp. 598 ff.); the large fission barriers observed for nuclei in the region of ^{208}Pb (see Fig. 6-56, p. 616) can also be interpreted in terms of the large negative shell-structure energy in the ground states of these nuclei.

In deformed potentials, expecially large shell-structure effects occur if the potential resembles that of an anisotropic harmonic oscillator with rational ratios of frequencies (see the discussion on pp. 584 ff. and Table 6-17). A potential with the symmetry $\omega_\perp : \omega_3 = 2:1$ corresponds to a deformation in the region of the fission barrier for heavy nuclei ($Z \approx 92$), and the major shell-structure effect associated with this symmetry appears to account for the occurrence of the shape isomers identified in the fission process. Experimental evidence concerning the fission isomers is discussed in the example on pp. 623 ff. The more detailed analysis of the shell structure in the nuclear potential with ratio of axes 2:1 is considered in the example on pp. 591 ff.; the closed-shell configuration with neutron number $N = 148$ in this potential corresponds to the region of greatest stability for the fission isomers.

A striking feature of the fission process is the asymmetry in the masses of fragments observed for low-energy fission of nuclei in the region $90 \lesssim Z \lesssim 100$. (For a survey of the experimental data, see Hyde, 1964.) The liquid-drop model does not provide an explanation of this phenomenon (see the discussion in Sec. 6A-2), and it appears likely that the mass asymmetry must be attributed to shell-structure effects that, for the nucleon numbers involved, favor deformations deviating from reflection symmetry, at an appropriate stage of the fission process (Möller and Nilsson, 1970; Pauli *et al.*, 1971).

The argument for identifying the shell-structure energy as part of the sum of single-particle energies can be given a simple derivation for a system in which the interactions can be treated in terms of the Hartree-Fock approximation (Strutinsky, 1968). In this approximation, the total energy can be written (see Eq. (3B-34))

$$\begin{aligned}
\mathscr{E} &= \mathscr{E}_{\text{kin}} + \mathscr{E}_{\text{pot}} \\
\mathscr{E}_{\text{kin}} &= \int dx\,dx'\langle x'|T|x\rangle\langle x|\rho|x'\rangle_0 = \sum_{i=1}^{A}\langle \nu_i|T|\nu_i\rangle \\
\mathscr{E}_{\text{pot}} &= \tfrac{1}{2}\int dx_1\,dx_1'\,dx_2\,dx_2'\langle x_1'x_2'|V|x_1x_2\rangle_a\langle x_1|\rho|x_1'\rangle_0\langle x_2|\rho|x_2'\rangle_0 \\
&= \tfrac{1}{2}\int dx\,dx'\langle x'|U|x\rangle\langle x|\rho|x'\rangle_0 = \tfrac{1}{2}\sum_{i=1}^{A}\langle \nu_i|U|\nu_i\rangle
\end{aligned} \tag{6-98}$$

with the self-consistent potential

$$\langle x'|U|x\rangle = \int dx_1\,dx_1'\langle x'x_1'|V|xx_1\rangle_a\langle x_1|\rho|x_1'\rangle_0 \tag{6-99}$$

The energies in Eq. (6-98) are expressed in terms of the one-particle density matrix ρ (see Eq. (2A-37)), and the subscript zero indicates the expectation value for the ground-state configuration, in which the particles occupy the A lowest one-particle orbits ν_i in the self-consistent potential U.

The density $\langle x|\rho|x'\rangle_0$ and the associated one-particle potential can be written in the form

$$\begin{aligned}
\langle x|\rho|x'\rangle_0 &= \tilde{\rho}(x,x') + \delta\rho(x,x') \\
\langle x'|U|x\rangle &= \langle x'|\tilde{U}|x\rangle + \langle x'|\delta U|x\rangle
\end{aligned} \tag{6-100}$$

where the quantity $\tilde{\rho}(x,x')$ is a smooth function of particle number, obtained by considering the asymptotic limit of large particle numbers. The potential $\tilde{U}$ is obtained from the relation (6-99) by replacing the Hartree-Fock density $\langle x_1|\rho|x_1'\rangle_0$ by the smooth part $\tilde{\rho}(x_1,x_1')$. Ignoring terms of order $(\delta\rho)^2$, the total energy (6-98) can be written

$$\begin{aligned}
\mathscr{E} &= -\tilde{\mathscr{E}}_{\text{pot}} + \mathscr{E}_{\text{ip}} \\
\tilde{\mathscr{E}}_{\text{pot}} &= \tfrac{1}{2}\int dx\,dx'\langle x'|\tilde{U}|x\rangle\tilde{\rho}(x,x') \\
\mathscr{E}_{\text{ip}} &= \int dx\,dx'\langle x'|T+\tilde{U}|x\rangle\langle x|\rho|x'\rangle_0 \\
&= \sum_{i=1}^{A}\langle \nu_i(U)|T+\tilde{U}|\nu_i(U)\rangle \approx \sum_{i=1}^{A}\langle \nu_i(\tilde{U})|T+\tilde{U}|\nu_i(\tilde{U})\rangle = \sum_{i=1}^{A}\varepsilon_i(\tilde{U})
\end{aligned} \tag{6-101}$$

where the notation $\nu_i(U)$ and $\nu_i(\tilde{U})$ refers to one-particle states in the potentials U and $\tilde{U}$, respectively. The expectation values of the Hamiltonian $T+\tilde{U}$ are stationary with respect to small variations of the one-particle wave functions and, in the last line of Eq. (6-101), the states $\nu_i(U)$ can therefore be replaced by $\nu_i(\tilde{U})$, to the accuracy considered.

Since $\tilde{\mathscr{E}}_{\text{pot}}$ in Eq. (6-101) is a smooth function of particle number, the shell-structure energy is contained in the sum of single-particle eigenvalues evaluated for the potential $\tilde{U}$. The above derivation is similar to the proof that, in the independent-particle approximation, the energy difference between two neighboring configurations is equal to the difference between the sums of one-particle energies evaluated in the same average potential.

Temperature dependence of shell-structure energy

For finite temperatures, the effect of the shell structure on the nuclear energy and level density can be obtained by an evaluation of the thermodynamic functions on the basis of the independent-particle level spectrum. (The tools for such a statistical analysis are described in Appendix 2B.) The thermodynamic functions depend on an average of the single-particle level spacing in an energy interval of order of the temperature, on each side of the Fermi level; the shell-structure effects are therefore expected to decrease strongly when the temperature approaches values comparable with the energy spacing between the shells. Examples illustrating this behavior are considered on pp. 607 ff. Evidence on the shell-structure effect in the nuclear level densities is discussed on p. 610.

For finite temperatures, the generalization of the potential energy function that determines the driving force in the fission process is the free energy

$$F(\alpha, T) = \mathscr{E}(\alpha) - TS \tag{6-102}$$

where S is the entropy and T the temperature. Examples illustrating the shell-structure contribution to the free energy as a function of temperature are shown in Fig. 6-54, p. 610; the shell-structure effects in the fission barrier are expected to become small for temperatures of about 1 MeV, corresponding to an excitation energy of about 30 MeV in a heavy nucleus (see the discussion on p. 621).

In order to exhibit the significance of the free energy, we follow the standard procedure of thermodynamics and consider a reversible process by which the nucleus changes its deformation and thereby performs the work $p\,d\alpha$ on some other system, such as an external electric field. The quantity p is the force with which the nucleus acts on the external system. For such a process, the change in the nuclear energy is given by

$$d\mathscr{E} = dQ - p\,d\alpha \tag{6-103}$$

where dQ is the heat transfer to the nucleus. The change in entropy is

$$dS = \frac{dQ}{T} = \frac{1}{T}(d\mathcal{E} + p\,d\alpha) \tag{6-104}$$

and thus, from Eq. (6-102),

$$dF = -S\,dT - p\,d\alpha \tag{6-105}$$

The force p therefore satisfies the relation

$$p = -\left(\frac{\partial F}{\partial \alpha}\right)_T \tag{6-106}$$

which shows the free energy as the generalized potential energy function. From Eq. (6-104), it is also seen that

$$p = T\left(\frac{\partial S}{\partial \alpha}\right)_{\mathcal{E}} \tag{6-107}$$

and thus the equilibrium toward which the system moves is characterized by a maximum in S for given energy, or equivalently, by a minimum in F for constant temperature.

Motion with respect to saddle point. Fission channels

The reaction rate for the fission process can be described in terms of the transmission through the available fission channels, in analogy to the description of molecular reaction rates (Bohr and Wheeler, 1939). The fission channels represent states with specified quantum numbers for all the degrees of freedom except for the motion across the barrier. The spectrum of channels is expected to exhibit rotational band structure, with quantum numbers determined by the symmetry of the saddle-point deformation (see the discussion in Chapter 4, pp. 36 ff.). The intrinsic quantum numbers labeling the different channels comprise vibrational as well as quasiparticle modes of excitation (see, for example, the discussion on pp. 125 ff. of the low-energy channels contributing to the photofission process).

If the collective motion toward the saddle point can be separated from the intrinsic degrees of freedom, as in an adiabatic process, the transmission coefficient can be obtained from the one-dimensional vibrational equation in the fission variable α. A simple estimate is provided by the "harmonic" approximation, which assumes a parabolic form for the barrier and a mass parameter independent of α,

$$H = \tfrac{1}{2}D^{-1}\pi_\alpha^2 - \tfrac{1}{2}C\alpha^2 \tag{6-108}$$

where α is measured with respect to the unstable saddle-point equilibrium.

For the Hamiltonian (6-108), the transmission factor P is given by (Hill and Wheeler, 1953; see also Wheeler, 1963)

$$P=\left(1+\exp\left\{-2\pi\frac{E}{\hbar\omega}\right\}\right)^{-1}$$
$$\omega=\left(\frac{C}{D}\right)^{1/2} \tag{6-109}$$

where the energy E is measured with respect to the saddle point. The derivation of Eq. (6-109) is given in the fine print below.

The transmission factor describes the gradual opening of a fission channel as the energy approaches the fission barrier. The observed energy variations of the fission cross sections in the threshold region can be approximately expressed in terms of transmission factors of the form (6-109) with values for $\hbar\omega$ of the order of 0.5–1 MeV and thus of the same order of magnitude as the energies of quadrupole shape oscillations in spheroidal nuclei. Examples are provided by the photofission cross section of ^{238}U illustrated in Fig. 4-23, p. 126, and the analysis of the neutron-induced fission of ^{241}Pu discussed on p. 633. It should be emphasized, however, that the degree of validity of the adiabatic approximation, which underlies the one-dimensional treatment, remains an open question (see, in this connection, the evidence for coupling between the shape deformation and pairing degrees of freedom exhibited by the properties of the β vibrations, pp. 552 ff.).

If the nucleus is vibrating with an energy below the fission threshold (so that $P \ll 1$), the fission rate is equal to the product of P and the frequency with which the vibrational motion approaches the saddle point. The fission width is therefore given by

$$\Gamma_f=\frac{\hbar\omega_{vib}}{2\pi}P \tag{6-110}$$

where ω_{vib} is the vibrational frequency (which is 2π times the inverse period).

For excitation energies comparable with the fission barrier, the fission widths resulting from the individual channels become distributed as strength functions in the dense spectrum of states of the compound nucleus. The width of the strength function is determined by the rate of damping of the vibrational motion in the fission mode; in situations where the damping time is short compared to the vibrational period, the average fission widths can be obtained from a statistical argument (Bohr and Wheeler, 1939) analogous to that employed in deriving the average neutron widths for a black nucleus (see Vol. I, p. 229)

$$\Gamma_{\mathrm{f}} = \frac{D}{2\pi} \sum_{c} P_c \tag{6-111}$$

where D is the average spacing of levels of given $I\pi$, and where the sum extends over the fission channels with these quantum numbers. The relation (6-111) provides a basis for interpreting the evidence on fission widths in terms of the channel spectrum. (For evidence obtained from slow neutron resonances, see, for example, Lynn, 1968a; for higher excitations, see, for example, Vandenbosch and Huizenga, 1958.)

The eigenfunctions of the Hamiltonian (6-108) can be expressed in terms of the parabolic cylinder function D_ν (see, for example, Erdelyi, 1953, vol. II, pp. 116ff.), and the eigenstate representing motion incident from the side of negative α is given by

$$\begin{gathered}\varphi(\alpha) = \text{const}\, D_\nu\left(\frac{\alpha}{\alpha_0}\exp\left\{-i\frac{\pi}{4}\right\}\right) \\ \alpha_0 = \left(\frac{\hbar}{2D\omega}\right)^{1/2} \qquad \omega = \left(\frac{C}{D}\right)^{1/2} \qquad \nu = i\frac{E}{\hbar\omega} - \frac{1}{2}\end{gathered} \tag{6-112}$$

with the asymptotic form, for large $|\alpha|$,

$$D_\nu\left(\frac{\alpha}{\alpha_0}\exp\left\{-i\frac{\pi}{4}\right\}\right) \underset{|\alpha|\gg\alpha_0}{\approx} \begin{cases} \left(\frac{\alpha}{\alpha_0}\right)^\nu \exp\left\{\frac{\pi}{4}\frac{E}{\hbar\omega} + i\frac{\pi}{8} + i\frac{1}{4}\left(\frac{\alpha}{\alpha_0}\right)^2\right\} & \alpha > 0 \\ \left|\frac{\alpha}{\alpha_0}\right|^\nu \exp\left\{-\frac{3\pi}{4}\frac{E}{\hbar\omega} - i\frac{3\pi}{8} + i\frac{1}{4}\left(\frac{\alpha}{\alpha_0}\right)^2\right\} & \\ \quad - \frac{(2\pi)^{1/2}}{\Gamma(-\nu)}\left|\frac{\alpha}{\alpha_0}\right|^{-\nu-1} \exp\left\{-\frac{\pi}{4}\frac{E}{\hbar\omega} - i\frac{7\pi}{8} - i\frac{1}{4}\left(\frac{\alpha}{\alpha_0}\right)^2\right\} & \alpha < 0 \end{cases} \tag{6-113}$$

For $\alpha < 0$, the last term in Eq. (6-113) is the incident wave propagating in the direction of the positive axis, while the first term is the reflected wave. The ratio of intensities in the transmitted and incident waves yields the transmission coefficient (6-109).

The expression (6-109) for the barrier penetration can also be obtained by a treatment based on the WKB approximation. For $E < 0$, the penetration involves the phase integral

$$S=\hbar^{-1}\int|\pi_\alpha|\,d\alpha=\frac{2}{\hbar}(2D)^{1/2}\int_0^{|2E/C|^{1/2}}\left(|E|-\tfrac{1}{2}C\alpha^2\right)^{1/2}d\alpha=-\pi\frac{E}{\hbar\omega} \qquad (6\text{-}114)$$

and, to leading order, the penetration probability is given by $\exp\{-2S\}$. When the energy is close to the top of the barrier, an improved treatment, with the inclusion of the next higher terms in the WKB approximation, leads to the penetrability $(1+\exp\{2S\})^{-1}$, and this expression, considered as a function of E, is found to be valid also for positive values of E. (See, for example, Fröman and Fröman, 1965, p. 90.)

6-3d Isospin of Vibrations. Polarization and Charge Exchange Modes

Vibrations in nuclei with $N=Z$ ($T_0=0$)

For the ground-state configuration of an even-even nucleus with $N=Z$, the intrinsic motion is rotationally invariant in isospace, if we ignore the effect of the Coulomb field. The vibrational modes can then be classified by the quantum number τ representing the isospin of the quantum of excitation. Vibrational quanta built out of particle-hole excitations or correlated pairs of particles have $\tau=0$ or 1.

The eigenvalue of τ_z is denoted by μ_τ, and modes with $\mu_\tau=0$ can be characterized by the eigenvalue r_τ for the charge symmetry operation $\mathscr{R}_\tau$, which exchanges neutrons and protons. The operation $\mathscr{R}_\tau$ can be represented by a rotation of 180° about the y axis in isospace (see Eq. (1-59)), and for a quantum with $\mu_\tau=0$, we have

$$r_\tau=(-1)^\tau \qquad (6\text{-}115)$$

(If we restrict ourselves to modes with $\tau=0$ and 1 (and $\mu_\tau=0$), the quantum numbers r_τ and τ are synonymous.) The isoscalar modes are charge symmetric and involve neutrons and protons moving in phase, as for the shape oscillations considered in Sec. 6-3a. The isovector modes with $\mu_\tau=0$ are charge antisymmetric and involve oscillations of neutrons and protons against each other (neutron-proton polarization).

Excitation modes with $\tau=0$ and 1 are illustrated in Fig. 6-4; the modes with $\mu_\tau\neq 0$ are excited by charge exchange processes, which produce states in isobaric nuclei. For a given mode with $\tau=1$, the excitations with different components μ_τ belong to an isobaric triplet (analog states). The

Coulomb energy implies a splitting of the vibrational frequencies

$$\hbar\omega(\tau,\mu_\tau)\approx\hbar\omega(\tau,\mu_\tau=0)-\mu_\tau\,\Delta E_{\mathrm{Coul}} \tag{6-116}$$

where the shift ΔE_{Coul} of the Coulomb energy per unit Z may be approximately estimated from the expression (2-19). For the charge exchange modes, the frequencies in Eq. (6-116) represent the difference between the binding energies of the states with $n=1$ and $n=0$, which also equals the Q values for excitation of the vibrational quantum in an (np) or (pn) reaction. If the excitation frequencies are defined by the mass difference of the $n=1$ and $n=0$ states, the quantity $\mu_\tau(M_{\mathrm{n}}-M_{\mathrm{p}})c^2$ must be added to the expression (6-116).

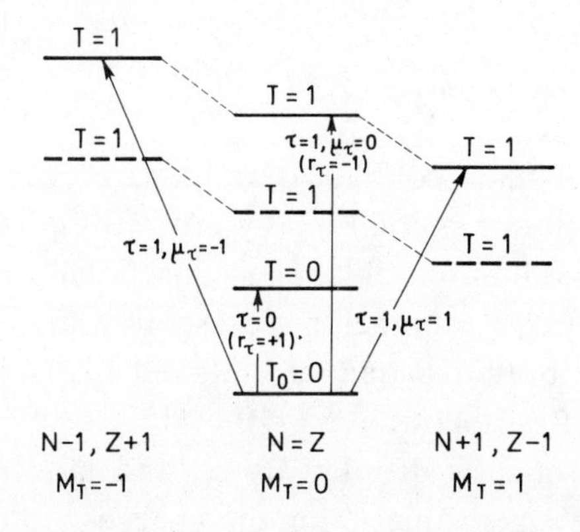

Figure 6-4 Isospin of vibrational excitations in nucleus with $T_0=0$. The figure illustrates the excitation of $\tau=0$ and $\tau=1$ vibrational modes in a nucleus with zero isospin in the ground state. Isobaric analog states are connected with thin broken lines. The ground states of the adjacent nuclei and their analog state in the target are indicated by dashed lines. No quantitative significance is to be attached to the relative position of the levels.

States involving several quanta belonging to the same mode are totally symmetric in the isospin and spin-orbital variables (Bose statistics). Thus, two $\tau=1$ quanta of multipole order λ added to a nucleus with $I_0=0$ and $T_0=0$ form the states

$$\begin{aligned} &I=0,2,\ldots,2\lambda &\qquad T=0,2\\ &I=1,3,\ldots,2\lambda-1 &\qquad T=1 \end{aligned} \tag{6-117}$$

The degeneracy with respect to T, which would hold in the harmonic approximation, is lifted by the systematic tendency of the nucleonic interactions to favor states with low T. (See, for example, the coupling effect given by Eq. (6-129).)

For an arbitrary number of quanta, the totally symmetric states can be classified in terms of representations $[f]$ of the permutation symmetry in the separate isobaric and orbital spaces. The possible representations and the associated quantum numbers T and I can be obtained by the methods discussed in Appendix 1C and correspond to classifications according to U_3 and $U_{2\lambda+1}$, respectively.

Isovector densities and fields

While the shape oscillations are associated with variations in the total (or isoscalar) density (6-38), the neutron-proton polarization modes involve oscillations in the isovector density

$$\rho_1(\mathbf{r}) = \rho_{\mathrm{n}}(\mathbf{r}) - \rho_{\mathrm{p}}(\mathbf{r}) \tag{6-118}$$

representing the difference between neutron and proton densities. The density (6-118) is the $\mu_\tau = 0$ component of the isovector

$$\rho_{\tau=1,\mu_\tau}(\mathbf{r}) = \sum_k 2t_{\mu_\tau}(k)\delta(\mathbf{r}-\mathbf{r}_k) \tag{6-119}$$

expressed in terms of the spherical components of the nucleonic isospin

$$t_{\mu_\tau} = \begin{cases} +2^{-1/2}(t_x - it_y) & \mu_\tau = -1 \\ t_z & \mu_\tau = 0 \\ -2^{-1/2}(t_x + it_y) & \mu_\tau = +1 \end{cases} \tag{6-120}$$

The charge exchange modes are associated with oscillations in the $\mu_\tau = \pm 1$ components of the isovector density (6-119). (One can also combine the three density functions (6-119) and the isoscalar density $\rho_{\tau=0}$ given by Eq. (6-38) into a 2×2 density matrix $\rho(\mathbf{r},\mathbf{t})$; see Sec. 2A-7.)

For a mode with quantum numbers $\lambda\mu\tau\mu_\tau$, the density variations have the form

$$\delta\rho_{\tau\mu_\tau}(\mathbf{r}) = f_{\lambda\tau}(r)Y^*_{\lambda\mu}(\hat{\mathbf{r}})\alpha_{\lambda\mu\tau\mu_\tau} \tag{6-121}$$

where the amplitudes $\alpha_{\lambda\mu\tau\mu_\tau}$ are tensors in isospace as well as in orbital space. (The hermiticity condition for such tensors can be expressed in terms of the operation $\mathcal{F} = \mathcal{R}_\tau^{-1}\mathcal{T}$; see Eq. (3B-14).)

The density variations (6-121) give rise to the multipole moments

$$\mathscr{M}(\tau\mu_\tau,\lambda\mu) \equiv \int \rho_{\tau\mu_\tau}(\mathbf{r}) r^\lambda Y_{\lambda\mu}(\hat{\mathbf{r}})\, d\tau$$
$$= \begin{cases} \sum_k r_k^\lambda Y_{\lambda\mu}(\hat{\mathbf{r}}_k) & (\tau=0) \\ \sum_k 2t_{\mu_\tau}(k) r_k^\lambda Y_{\lambda\mu}(\hat{\mathbf{r}}_k) & (\tau=1) \end{cases} \tag{6-122}$$
$$= \alpha_{\lambda\mu\tau\mu_\tau} \int dr f_{\lambda\tau}(r) r^{\lambda+2}$$

The $\tau=0$ moment is the same as that of the total particle density given by Eq. (6-60).

The electric multipole moment is a combination of isoscalar and isovector moments

$$\mathscr{M}(E\lambda,\mu) = \int \rho_{\text{el}}(\mathbf{r}) r^\lambda Y_{\lambda\mu}(\hat{\mathbf{r}})\, d\tau$$
$$= \frac{e}{2}\left(\mathscr{M}(\tau=0,\lambda\mu) - \mathscr{M}(\tau=1,\mu_\tau=0,\lambda\mu)\right) \tag{6-123}$$

Thus, the polarization modes ($\tau=1, \mu_\tau=0$), as well as the shape oscillations ($\tau=0$), may generate collective oscillations of electric multipole moments.

The isovector density variations give rise to corresponding variations in the isovector potential, which can be estimated from the isovector component in the static nuclear potential. In fact, the symmetry term in the potential, which is linear in $N-Z$, represents the effect of a relatively small isovector density superposed on the total isoscalar density ρ_0. From the expression (2-26) for the nuclear potential, we obtain in this manner (for $\mu_\tau=0$)

$$\delta V = \tfrac{1}{2} V_1 t_z \frac{\delta\rho_1}{\rho_0} \tag{6-124}$$

where the symmetry potential V_1 is repulsive (positive) and of order 100 MeV. The relation (6-124) is based on the assumption that both the static and the vibrational isovector fields are dominantly volume effects; the connection between isovector density and potential in the surface region poses important unsolved problems. (A related element of uncertainty in the description of the isovector potential is connected with the fact that the analysis leading to the average static potential V_1 does not correctly ensure the observed equality of the radii of the neutron and proton density distributions; see the comments on p. 518.) The possible role of velocity-dependent isovector interactions in modifying the inertial parameter of isovector modes is discussed on pp. 483 ff.

For $\lambda \neq 0$, the lowest isovector mode is expected to be nodeless, and a simple model for exploring these modes can be based on a vibrational field of the form

$$F_{\lambda\mu,\tau=1,\mu_\tau=0} = 2t_z r^\lambda Y_{\lambda\mu} \tag{6-125}$$

extending over the nuclear volume. If we normalize the amplitude so that the average value of F equals α (see Eq. (6-72)), the density variation is given by

$$\delta\rho_1 = \frac{4\pi\rho_0}{A\langle r^{2\lambda}\rangle} \sum_\mu r^\lambda Y^*_{\lambda\mu} \alpha_{\lambda\mu,\tau=1,\mu_\tau=0} \tag{6-126}$$

and the potential (6-124) takes the standard form (6-71) with

$$\kappa_{\lambda,\tau=1} = \frac{\pi V_1}{A\langle r^{2\lambda}\rangle} \tag{6-127}$$

The collective modes resulting from the fields (6-125) and couplings (6-127) are treated on pp. 480 ff. ($\lambda = 1$), pp. 512 ff. ($\lambda = 2$), and pp. 558 ff. ($\lambda = 3$).

Occurrence of neutron-proton polarization modes

The prime example of a $\tau = 1$ vibration is the dipole resonance ($\lambda\pi = 1-$) observed in the nuclear photo-effect and discussed in the example on pp. 474 ff. The isovector character of this mode can be inferred from the large electric dipole moment, which implies that the protons and neutrons are oscillating against each other. (For $\lambda = 1$, an isoscalar mode involves no electric dipole moment, since the isoscalar dipole moment is proportional to the center-of-mass coordinate and therefore unaffected by intrinsic excitations.) The resonance frequencies and the $E1$- transition strengths for the isovector dipole mode are shown in Figs. 6-19, p. 476, and 6-20, p. 479, and are seen to vary smoothly with N and Z.

The properties of the dipole mode in nuclei with $A \gtrsim 50$ can be rather well described in terms of an isovector dipole field with a strength given by Eq. (6-124); see pp. 480 ff. It is also possible to describe the mode in terms of the polarization oscillations of a two-fluid liquid-drop model (see Sec. 6A-4). The similarity of the results obtained from these two apparently very different descriptions is a special feature of the dipole mode, which can be attributed to the fact that the main single-particle dipole transition strength is concentrated in a single, rather narrow, region of the excitation spectrum. (The comparison of the liquid-drop and microscopic descriptions of the dipole mode is discussed on pp. 488 ff.)

The rather high frequency of the dipole mode ($\hbar\omega \sim 10$–25 MeV) places it in a region of the spectrum with very great level density, and the collective motion is observed to be rather strongly damped ($\Gamma \sim 4$–10 MeV); the observed resonance line thus represents a strength function. No detailed interpretation of the observed width and line structure has yet been established, but various effects that may contribute are discussed on pp. 503 ff.

There is so far no direct evidence concerning other isovector density modes; the significance of the $\lambda\pi = 0+$ mode for the violation of isobaric symmetry is discussed in Vol. I, pp. 171 ff., and the expected properties of the $\lambda\pi = 2+$ mode resulting from the interaction (6-127) are considered on pp. 512 ff.

Effect of neutron excess

In a nucleus with neutron excess and associated ground-state isospin $T_0 = M_{T_0} = (N-Z)/2$, the excitation of an isovector mode gives rise to a triplet of states $(T_0, n_{\tau=1} = 1)TM_T$ with $T = T_0 - 1, T_0, T_0 + 1$. In the absence of coupling between T_0 and the vibrational motion, the components of the triplet would be degenerate, apart from the Coulomb energy difference between isobaric analog states; moreover, the multipole matrix element for exciting the states would be obtained by a vector coupling in isospace (Fallieros *et al.*, 1965)

$$\langle (T_0, n_\tau = 1)TM_T | \mathscr{M}(\tau\mu_\tau) | T_0 M_{T_0} \rangle = m_0 \langle T_0 M_{T_0} \tau\mu_\tau | TM_T \rangle \qquad (6\text{-}128)$$

with m_0 independent of T, as well as of the components M_{T_0}, μ_τ, and M_T.

However, major coupling effects are associated with the isovector fields in the nucleus. Thus, the neutron excess gives rise to a static isovector potential that acts on the isospin carried by the vibrational motion in such a manner as to favor states with low total isospin. The coupling energy can be estimated from Eq. (2-29),

$$H' = \frac{V_1}{A}(\boldsymbol{\tau} \cdot \mathbf{T}_0) = \frac{V_1}{2A}\left(T(T+1) - T_0(T_0+1) - \tau(\tau+1)\right) \qquad (6\text{-}129)$$

Since the value of V_1 is of the order of 100 MeV, the interaction energies (6-129) become quite large, even for moderate values of T_0.

Moreover, the presence of the neutron excess implies that the configurations available for the charge exchange modes will be different from those in the $\mu_\tau = 0$ modes, as a result of the exclusion principle (see Fig. 6-22, p. 494). The effect can be expressed in terms of the coupling between the excess neutrons and the oscillating isovector field in the vibrational motion (in contrast to the interaction (6-129) associated with the static monopole isovector field); this coupling gives rise to modifications in the intensity relations (6-128), as well as in the vibrational energies.

Quite generally, the leading-order energy shift produced by the neutron excess is linear in T_0 and, therefore, of the form

$$H' = a(\boldsymbol{\tau} \cdot \mathbf{T}_0) \tag{6-130}$$

as in Eq. (6-129). For the dipole mode, the coupling through the isovector dipole field gives a negative contribution to the coefficient a, which is estimated to compensate about half of the positive coefficient in Eq. (6-129); see Eq. (6-349).

For a tensor operator $\mathscr{M}(\tau\mu_\tau)$, the matrix element (6-128) with inclusion of corrections linear in T_0 has the general form

$$\begin{aligned}\langle (T_0, n_\tau = 1) T M_T | \mathscr{M}(\tau\mu_\tau) | T_0 M_{T_0} \rangle \\ = \langle T_0 M_{T_0} \tau\mu_\tau | T M_T \rangle \big(m_0 + m_1 (T(T+1) - T_0(T_0+1) - \tau(\tau+1)) \big)\end{aligned} \tag{6-131}$$

In fact, since $\mathscr{M}(\tau = 1, \mu_\tau)$ is a vector in isospace, the term linear in T_0 and in the creation operator $\mathbf{c}^\dagger$ for the vibrational quantum must be proportional to $\mathbf{T}_0 \times \mathbf{c}^\dagger$, which in turn is proportional to $(\mathbf{T}_0 \cdot \boldsymbol{\tau})\mathbf{c}^\dagger$, when acting on a state with $n_\tau = 0$. (Note that $\boldsymbol{\tau} = -i\mathbf{c}^\dagger \times \mathbf{c}$.) For the dipole mode, an estimate of m_1 is given by Eq. (6-359).

If the neutron excess covers a major part of the single-particle levels involved in the vibrational motion, modes differing in the orientation of the vibrational isospin relative to T_0 have quite different properties (see, for example, the dipole modes illustrated in Fig. 6-23, p. 497). The consequences of isobaric invariance are then restricted to those associated with the existence of the total isospin quantum number T. The pattern of states with a single quantum of excitation is illustrated in Fig. 6-5; the modes are labeled by the quantum number $\Delta T = T - T_0$.

Transitions induced by operators with isobaric tensor order τ leading to different components of an isobaric multiplet are related by the Wigner-Eckart theorem in isospace (see Eq. (1A-132))

$$\langle T M_T | \mathscr{M}(\tau\mu_\tau) | T_0 M_{T_0} \rangle = \langle T_0 M_{T_0} \tau\mu_\tau | T M_T \rangle (2T+1)^{-1/2} \langle T \| \mathscr{M}(\tau) \| T_0 \rangle \tag{6-132}$$

where the last factor is a reduced matrix element. For $T_0 \gg 1$ and $M_{T_0} = T_0$

(as in nuclear ground states), the vector addition coefficient in Eq. (6-132) has the asymptotic value

$$\langle T_0 T_0 1 \mu_\tau | T = T_0 + \Delta T, T_0 + \mu_\tau \rangle \underset{T_0 \gg 1}{\approx} \begin{cases} 1 & \Delta T = \mu_\tau \\ (T_0)^{-1/2} & \Delta T = \mu_\tau + 1 \\ 2^{-1/2} T_0^{-1} & \Delta T = \mu_\tau + 2 \end{cases} \tag{6-133}$$

Thus, the transitions leading to the fully aligned states with $M_T = T$ (indicated by arrows in Fig. 6-5) are strong compared with those leading to the isobaric analog states. (The dominance of the transitions with $\Delta T = \mu_\tau$ has a simple interpretation in terms of a semiclassical picture of the isospin coupling.)

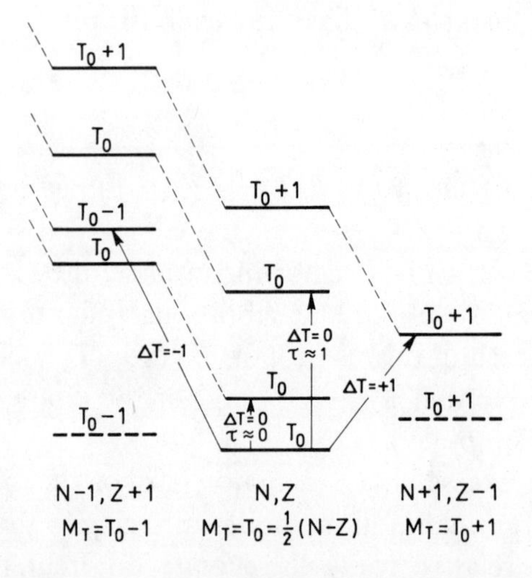

Figure 6-5 Isospin of vibrational excitations in nucleus with neutron excess. The figure gives a schematic illustration of the pattern of states formed by adding a vibrational quantum, with $\Delta T = +1$, 0, or -1, to a nucleus with a neutron excess. No quantitative significance is to be attached to the relative positions of the excitations with different T. Isobaric analog states are connected with thin broken lines. The ground states of the nuclei with $M_T = T_0 \pm 1$ are indicated by dashed lines.

In the presence of the neutron excess, the modes with $\Delta T = 0$ are mixtures of $\tau = 0$ and $\tau = 1$ excitations, but the strong neutron-proton forces imply a tendency for the collective density oscillations to acquire a "macroscopic" τ symmetry. Thus, for a given spatial symmetry, one expects the lowest mode to approximately represent a shape oscillation with preservation of the local neutron-proton ratio ($\tau \approx 0, r_\tau \approx +1$); such a mode generates average fields and moments that are predominantly isoscalar. The orthogonal mode, with higher frequency, corresponds to motion of the neutrons with respect to protons ($\tau \approx 1, r_\tau \approx -1$), and the associated fields and moments are predominantly isovector. (See the discussion of the dipole and quadrupole modes on pp. 487 ff. and pp. 513 ff., respectively.) For the shape oscillations, the predominantly $\tau = 0$ character has been tested by a comparison between $(\alpha\alpha')$ scattering, which determines the amplitude of the isoscalar field, and electromagnetic excitation, which determines the amplitude of proton oscillation (see Table 6-2 and the systematic compilation by Bernstein, 1970).

The isobaric structure of the $\Delta T = 0$ modes presents an example of symmetry breaking with very different effects at the macroscopic and microscopic levels. Though the isospin symmetry is approximately conserved at the macroscopic level (for $N - Z \ll A$), the symmetry may be completely violated at the microscopic level of the individual excitations out of which the mode is constructed. The symmetry breaking is a feature of the neutron excess in the "vacuum state" (equilibrium configuration) rather than of the interactions responsible for the collective mode.

6-3e Collective Modes Involving Spin Degree of Freedom[8]

The spin dependence of the density oscillations associated with a nuclear vibration can be characterized by the quantum number σ, which represents the spin angular momentum of a vibrational quantum. Modes associated with particle-hole excitations or two-particle correlations have $\sigma = 0$ or 1. The total multipole order λ results from the coupling of σ to the orbital angular momentum κ of the vibrational quantum.

The parity of the vibrational quanta is determined by the orbital multipole order

$$\pi = (-1)^\kappa \tag{6-134}$$

and, for specified $\lambda\pi$, the possible quantum numbers are

$$\begin{array}{ll} \lambda \neq 0 & \left\{ \begin{array}{lll} \pi = (-1)^\lambda & \kappa = \lambda & \sigma = 0, 1 \\ \pi = (-1)^{\lambda+1} & \kappa = \lambda \pm 1 & \sigma = 1 \end{array} \right. \\ \lambda = 0 & \left\{ \begin{array}{lll} \pi = +1 & \kappa = 0 & \sigma = 0 \\ \pi = -1 & \kappa = 1 & \sigma = 1 \end{array} \right. \end{array} \tag{6-135}$$

[8]Early discussions of collective oscillations involving nuclear spin densities were given by Wild (1955) and Glassgold *et al.* (1959). The expected occurrence of $\lambda\pi = 1+$ modes, resulting from the strong spin-orbit coupling, was pointed out by Kurath (1963).

Except for $\lambda=0$, a mode of given $\lambda\pi$ can thus be formed by two different $\kappa\sigma$ combinations.

If only central forces acted between the nucleons (giving rise to a ground state with $S=L=0$), the vibrational quanta would have definite spin and orbital quantum numbers $\kappa\sigma$, and the associated vibrational field would have the same values of κ and σ as the density oscillations. However, the presence of the noncentral forces implies a coupling between densities and fields with the two different symmetries associated with the same set of quantum numbers $\lambda\pi$.

Such a coupling is already implied by the rather strong spin-orbit interaction in the average one-body potential (symmetry breaking associated with the vacuum state). In fact, the density oscillations generated by the individual one-particle excitations, $l_1 j_1 \rightarrow l_2 j_2$, out of which the collective mode $\lambda\pi$ is built, involve components with both values of $\kappa\sigma$ (and with relative amplitude determined by the jj to LS recoupling coefficient $\langle (l_1\ 1/2)j_1, (l_2\ 1/2)j_2; \lambda | (l_1 l_2)\kappa, (1/2\ 1/2)\sigma; \lambda \rangle$).

In addition, the interactions associated with the spin-dependent deformations may involve components that are nondiagonal in the quantum numbers $\kappa\sigma$. Interactions that couple the channels $\kappa=\lambda-1$, $\sigma=1$ and $\kappa=\lambda+1$, $\sigma=1$ result from effective noncentral two-body forces that are bilinear in the nucleonic spins (such as the tensor interaction), while forces that are linear in the nucleonic spins (such as the two-body spin-orbit interaction) couple the modes $\kappa=\lambda$, $\sigma=0$ and $\kappa=\lambda$, $\sigma=1$. The deformation of the spin-orbit potential described by Eq. (6-70) is an example of a coupling term of the latter type.

The effective interactions and the resulting pattern of collective motion for the spin-dependent fields are, as yet, rather poorly known. A static density deformation with $\sigma=1$ is odd under time reversal (as exhibited more formally in the fine print below); hence, the average potential in the nucleus does not provide estimates of the potential generated by a $\sigma=1$ deformation like those in Eqs. (6-74) and (6-124) for the spin-independent fields (with $\tau=0$ and 1, respectively).

Information on the spin-dependent interactions comes from the polarization phenomena that modify the effective spin g factors for $M1$ moments (see Vol. I, pp. 337ff., and Chapter 5, p. 304) and the effective coupling constants for β transitions of GT type (see Vol. I, p. 347, and Chapter 5, p. 307). The analysis of these polarization effects indicates an interaction in the $\kappa=0$, $\sigma=1$, $\tau=1$ channel that is repulsive and has a strength comparable to that of the $\sigma=0$ couplings discussed in the previous sections.

The collective modes generated by fields with $\kappa=0$, $\sigma=1$ are considered in the example on pp. 636ff. The available information on collective $M1$ transitions provides evidence for an interaction in the $\tau=1$ channel comparable to that inferred from the effective g factors. The evidence on the interaction in the $\tau=0$ channel is at present very tentative (see, for example, the indication of a repulsive effect provided by the analysis of the decoupling parameters in rotational bands with $K=1/2$, discussed on p. 310). The absence of strongly attractive spin-dependent fields is supported by the fact that no low-lying collective spin excitations are observed in the nucleus. (In liquid ^{3}He, the short-range repulsion, which is much more effective in the ^{1}S than in the ^{3}P channel, can be represented by an interaction favoring the parallel orientation of the nuclear spins (see the analogous nuclear effect discussed in Vol. I, p. 259); this attractive spin-spin interaction leads to a major modification in the response function for spin-dependent fields, referred to as the paramagnon effect; see, for example, the review of the properties of liquid ^{3}He by Wheatley, 1970.)

The coupling between $\kappa=\lambda$, $\sigma=0$ and $\kappa=\lambda$, $\sigma=1$ modes implies the possible occurrence of $\sigma=1$ fields in the low-frequency collective modes with $\pi=(-1)^\lambda$ that were discussed above in terms of collective shape oscillations (with $\sigma=0$). In fact, the single-particle excitations from which these modes are built have energies of the same order as the spin-orbit splittings, and hence the pair of orbits with $j=l\pm 1/2$ enter with quite different amplitudes. Thus, at the microscopic level, the $\kappa\sigma$ symmetry is completely violated, but the interplay of many orbits may lead to cancellations in the long-wavelength $\sigma=1$ density, resulting in a macroscopic $\sigma\approx 0$ symmetry (compare the similar situation for the τ symmetry in nuclei with large neutron excess discussed on p. 382). The dominance of the $\sigma=0$ fields in the low-frequency shape oscillations is strongly suggested by the close relationship between these modes and the rotational excitations of the deformed nuclei; the observed rotational bands establish the time reversal invariance of the static nuclear deformations (see Chapter 4, p. 27), which must therefore have $\sigma=0$, aside from velocity-dependent terms such as the spin-orbit potential. Evidence for the validity of an approximate $\sigma=0$ symmetry in the low-frequency shape oscillations is provided by the analysis of the particle-vibration coupling in odd-A nuclei (see the example discussed on p. 566).

A deformation of symmetry $\sigma=1$ can be described by spin-dependent density functions analogous to those employed for the isospin-dependent deformations; see Eqs. (6-119) and (6-121). The deformed density is conveniently expressed in terms of the density-matrix formalism (see Sec. 2A-7),

$$\delta\rho(\mathbf{r},\mathbf{s}) = f_{\kappa,\sigma=1,\lambda}(r)(Y_\kappa \sigma)^\dagger_{(\kappa 1)\lambda\mu}\alpha_{\kappa,\sigma=1,\lambda\mu} \tag{6-136}$$

for a mode with quantum numbers κ, $\sigma = 1$, $\lambda\mu$. Under time reversal, $\rho(\mathbf{r},\mathbf{s})$ goes into $\rho(\mathbf{r},-\mathbf{s})$; hence, the amplitudes $\alpha_{\kappa,\sigma=1,\lambda\mu}$ are odd under time reversal combined with Hermitian conjugation (see Eq. (6-41) for the corresponding relation for the $\sigma=0$ amplitudes).

6-3f Two-Nucleon Transfer Modes. Pair Vibrations

The nuclear modes of excitation can be characterized by a further quantum number α, the nucleon transfer number, which gives the change in nucleon number associated with the excitation of a quantum.[9] The vibrations considered in the previous sections conserve the number of nucleons (although they may transform neutrons into protons or *vice versa*) and thus have $\alpha=0$; when viewed in terms of the nucleonic degrees of freedom, these collective modes are built out of particle-hole excitations (see Sec. 6-2c). Collective modes with $\alpha = \pm 2$ are associated with correlations of pairs of particles or holes. The tendency of nucleons to form correlated pairs with angular momentum and parity 0+ has already been encountered as an important feature in the nuclear coupling scheme for configurations with particles in unfilled shells (see, for example, Vol. I, p. 210, and Chapter 5, pp. 240ff.).

Correlated nucleon pairs may form units with properties characteristic of vibrational quanta, and in the present section, we consider the general results that follow from treating pair addition (or removal) as an elementary mode of excitation.[10] In this analysis, one must distinguish between the closed-shell nuclei (pp. 387ff.) and the configurations removed from closed shells, where the 0+ pairs form a condensate (as discussed on pp. 392ff.).

---

[9]The notation α is chosen as the Hellenization of the Latin A, with a view to the analogous relationships for the other vibrational quantum numbers σ and τ. The advantages of continuing this tradition seem to outweigh the risk of confusing the quantum number α with the generic vibrational amplitude α.

[10]Collective nuclear modes associated with particle transfer were considered in early versions of the present volumes (see, for example, Bohr, 1964, 1968). The topic was developed by Bès and Broglia, 1966; see also the discussion of the occurrence of the pair addition mode in connection with the phase transition from superfluid to normal Fermi systems (Högaasen-Feldman, 1961). A major step in the experimental study of the pair transfer modes was the observation of strongly enhanced ground-state transitions in (tp) reactions in superfluid nuclei (Middleton and Pullen, 1964); excited states of pair vibrational type were identified in the region of ^{208}Pb by Bjerregaard *et al.* (1966a).

The occurrence of such a condensate in the nucleus implies many similarities with the description of macroscopic superfluid systems, though the small size of the nucleus prevents the occurrence of superflow (see the discussion in fine print on pp. 398 ff.).

Pair vibrations in closed-shell nuclei

The spectra of nuclei with two particles added or removed from a closed-shell configuration exhibit low-lying states with major correlations in the motion of the two particles. In particular, this is the case for the $I\pi = 0+$ ground states involving configurations with pairs of identical particles or holes. The correlation can be understood as a consequence of a short-range attractive force acting between the nucleons, which produces a superposition of single-particle configurations of such a nature as to build up the spatial overlap of the two particles or holes. (See, for example, the analysis of the ground state of ^{206}Pb discussed on pp. 641 ff.) When the correlation involves the superposition of a large number of different configurations of the two particles, the addition or removal of the correlated pair may be treated as an elementary mode of excitation (pair vibration), which can be repeated and combined with other modes in the description of the total excitation spectrum.

In the present discussion, we shall especially consider pairs with total angular momentum and parity $\lambda\pi = 0+$, though the analysis can be readily extended to other channels. In a heavy nucleus, where neutrons and protons fill orbits with different $j\pi$ values, low-energy quanta with $\alpha = \pm 2$, $\lambda\pi = 0+$ can be formed only from pairs of identical particles.

The pattern of $\lambda\pi = 0+$ neutron pair vibrations is schematically illustrated in Fig. 6-6. The spectrum involves two elementary modes with neutron transfer numbers $\alpha = +2$ and $\alpha = -2$, corresponding to the formation of the ground states of the nuclei with neutron numbers $N_0 + 2$ and $N_0 - 2$, where N_0 is the neutron number of the closed-shell configuration. The states are labeled by $(n_{\alpha=-2}, n_{\alpha=+2})$, giving the number of quanta of the two modes. If the interaction between the quanta can be neglected (harmonic approximation), the energies of the vibrational states can be written

$$E = \hbar\omega_{-2} n_{-2} + \hbar\omega_{+2} n_{+2} \tag{6-137}$$

in the energy scale employed in Fig. 6-6.

If the shell closure at N_0 is associated with a large gap between the single-particle orbits above and below N_0, the $\alpha = +2$ quanta, to a first approximation, involve the nucleonic orbits above N_0, while the $\alpha = -2$

quanta involve the orbits below N_0. However, the quantitative properties of the quanta may be modified significantly by even relatively weak zero-point vibrational oscillations in the closed-shell state associated with the virtual excitation of $0+$ pairs of neutrons across the shell gap. (See the discussion on pp. 644ff. of the renormalization of the pair interaction.)

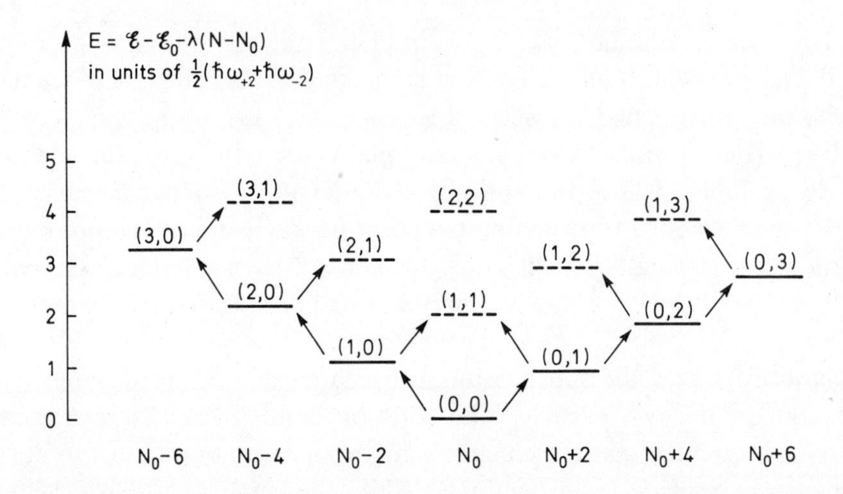

Figure 6-6 Neutron pair vibrations with $\lambda\pi = 0+$. The states are labeled by the quantum numbers $(n_{\alpha=-2}, n_{\alpha=+2})$. Solid lines represent ground states, and the arrows indicate strong two-particle transfer transitions from these states. The spectrum illustrated corresponds to the harmonic approximation, in which the excitation quanta are regarded as noninteracting entities. The energy E plotted in the figure is the total energy relative to the value $\mathscr{E}_0$ for the closed-shell configuration (ground state for $N = N_0$), from which is subtracted a linear function of the neutron number. The structure of the plot is independent of the coefficient λ in the linear function, but often it is convenient to choose for λ the mean value of the first single-particle levels $\varepsilon(j_>)$ and $\varepsilon(j_<)$ above and below the gap associated with the closed-shell configuration. With this choice of λ, the energies of the quanta are $\hbar\omega_{\pm 2} = \varepsilon(j_>) - \varepsilon(j_<) - \Delta\mathscr{B}_{\pm}$, where $\Delta\mathscr{B}_{\pm}$ are the binding energies of the pairs of particles (or holes) relative to the values $-2\varepsilon(j_>)$ (or $2\varepsilon(j_<)$) for independent-particle motion; in the figure, it has been assumed that $\Delta\mathscr{B}_+$ is somewhat larger than $\Delta\mathscr{B}_-$, as in the region of ^{208}Pb illustrated in Fig. 6-62, p. 646).

Since the pair addition quanta result from the spatial correlation of the particles, this mode is strongly excited by the two-particle transfer reactions, such as the (tp) process, that involve the transfer of two nucleons closely correlated in space (see Sec. 3E-2). These processes therefore play a role in the study of pair vibrations similar to that of the inelastic scattering reactions in the study of shape vibrations with $\alpha = 0$. To leading order in the vibrational amplitude (harmonic approximation), the transition operator for a two-nucleon addition process is linear in the operators $c^{\dagger}_{\alpha=2}$ and $c_{\alpha=-2}$

that create a quantum with $\alpha=2$ and annihilate a quantum with $\alpha=-2$, and the matrix elements can be obtained from the general relations (see Eq. (6-1))

$$\begin{aligned}\langle n_{-2}, n_{+2}+1|c^{\dagger}_{\alpha=2}|n_{-2}, n_{+2}\rangle &= (n_{+2}+1)^{1/2} \\ \langle n_{-2}-1, n_{+2}|c_{\alpha=-2}|n_{-2}, n_{+2}\rangle &= (n_{-2})^{1/2}\end{aligned} \tag{6-138}$$

Evidence concerning the neutron pair addition and removal modes in ^{208}Pb is discussed in the example on pp. 645 ff. The quanta are found to approximately retain their identity when combined with other excitation quanta. The observed pair vibrational spectrum, shown in Fig. 6-62, p. 646, is qualitatively as expected for harmonic vibration, but also reveals effects of anharmonicity. (For a review of evidence concerning pair vibrations, see Broglia *et al.*, 1973.)

For a closed-shell configuration with $N_0 = Z_0$, $T_0 = 0$, the pair modes involving nn, pp, and np are related by isobaric invariance. The quanta with $\lambda\pi = 0+$ have unit isospin, $\tau = 1$, and the excitation of n quanta of the same type ($\alpha = +2$ or $\alpha = -2$) gives rise to states with total isospin

$$T = n, n-2, \ldots, 0 \text{ or } 1 \tag{6-139}$$

as for the l values of a three-dimensional harmonic oscillator (see Eq. (2-150)). (The states (6-139) can also be classified in terms of the representation $(\lambda\mu) = (n0)$ of SU_3, as for the symmetric states of the configuration p^n; see Vol. I, pp. 131 ff.)

In the harmonic approximation, the relative transition amplitudes for two-particle transfer reactions can be obtained from the parentage factors

$$\langle n+1, T'M_T'|c^{\dagger}(\tau=1, \mu_\tau)|nTM_T\rangle$$

$$= \langle TM_T 1\mu_\tau|T'M_T'\rangle \begin{cases} \left(\dfrac{(T+1)(n+T+3)}{2T+3}\right)^{1/2} & T' = T+1 \\ \left(\dfrac{T(n-T+2)}{2T-1}\right)^{1/2} & T' = T-1 \end{cases} \tag{6-140}$$

which are Clebsch-Gordan coefficients in SU_3. These parentage factors may also be derived, as for the three-dimensional oscillator, in terms of a product of a radial matrix element (see Eq. (2-154)) and the angular matrix element $\langle l'm'|Y_{1\mu}|lm\rangle$. The phases of the matrix elements in Eq. (6-140) follow the standard choice for the harmonic oscillator, as defined on pp. 231 ff.

If $\alpha=-2$ and $\alpha=+2$ quanta are superposed, one obtains a multiplicity of states, which can be enumerated by employing, for example, the quantum numbers $(n_{-2},T_{-2};\ n_{+2},T_{+2})T$ with $T=T_{+2}+T_{-2}, T_{+2}+T_{-2}-1, \cdots, |T_{+2}-T_{-2}|$. The states with the same number of quanta, but different values of T, are separated in energy by the rather strong interactions favoring states of low isospin (see the coupling (6-129)).

In the pattern of 0+ excitations observed in two-particle transfer reactions in the region around ^{56}Ni, some of the observed features can be understood in terms of pair vibrations; however, the anharmonic effects appear to play a major role in these spectra (Nathan, 1968; Hansen and Nathan, 1971).

Pair densities and potentials

The features of the pair vibrations considered in the preceding section are the immediate consequence of the existence of correlated pairs as elementary modes of excitation. As in the case of the $\alpha=0$ modes, the pair vibrations exhibit both particle and field aspects, corresponding to the description in terms of quanta and amplitudes, respectively. The amplitudes of the pair vibrations provide a measure of densities and potentials that act by creating and annihilating pairs of particles, corresponding to the elementary processes in the pair-vibrational motion. These fields play a role similar to that of the deformed one-particle densities and potentials in the $\alpha=0$ modes. Though the more systematic discussion of the nucleonic correlations involved in the pairing phenomena is a subject for Chapter 8, the pair fields are briefly considered in the present section, since they provide the unifying concept for viewing the many different manifestations of the nuclear pair correlations encountered in this volume.

The density that creates two identical nucleons at the same space point is given by

$$\rho_{\alpha=2}(\mathbf{r})=a^{\dagger}(\overline{\mathbf{r},m_s=\tfrac{1}{2}})a^{\dagger}(\mathbf{r},m_s=\tfrac{1}{2}) \tag{6-141}$$

$$\left(a^{\dagger}(\overline{\mathbf{r},m_s=\tfrac{1}{2}})=-a^{\dagger}(\mathbf{r},m_s=-\tfrac{1}{2})\right)$$

in terms of the one-particle creation operators $a^{\dagger}$ (see Appendix 2A). The density operator (6-141) and its Hermitian conjugate that annihilates two particles

$$\rho_{\alpha=-2}(\mathbf{r})=\rho^{\dagger}_{\alpha=2}(\mathbf{r}) \tag{6-142}$$

are the counterparts of the local one-particle density operator

$$\rho_{\alpha=0}(\mathbf{r})=\sum_{m_s}a^{\dagger}(\mathbf{r}m_s)a(\mathbf{r}m_s)=\sum_k\delta(\mathbf{r}-\mathbf{r}_k) \tag{6-143}$$

that has been employed in the description of deformation effects in shape vibrations. Operators that are bilinear in $a^\dagger$ and a, such as the $\alpha=0$ density (6-143) and the corresponding one-particle potential, are associated with transitions of a single particle (or a hole) and the creation (or annihilation) of a particle-hole pair; the generalized $\alpha=2$ one-particle density (6-141) and corresponding potentials create a pair of particles (or annihilate a pair of holes) and have matrix elements connecting a hole state with a particle state (scattering of a hole into a particle).

A pair-vibrational mode of angular momentum λ involves deformations of the densities $\rho_{\pm 2}$ of corresponding multipole order. Since the local density (6-141) creates two identical particles at the same space point, the pair is in a spin-singlet state ($\sigma=0$), and the associated quanta have total angular momentum equal to the orbital momentum ($\lambda=\kappa$). The parity is $\pi=(-1)^\lambda$, since $\rho_2(\mathbf{r})\rightarrow\rho_2(-\mathbf{r})$, under the transformation $\mathscr{P}$.

For the monopole mode, the pair density is spatially isotropic. There is little evidence concerning the radial form factor; we shall consider a mode associated with a pair density that is approximately constant over the nuclear volume. The amplitude may thus be characterized by the monopole moment

$$M_2\,(=M_{\alpha=2,\lambda=0})=\int\rho_2(\mathbf{r})\,d\tau \tag{6-144}$$

Expanding the creation operators $a^\dagger(\mathbf{r}m_s)$ in terms of operators $a^\dagger(\nu)$ creating a particle in the shell-model orbit ν (see Eq. (2A-34)), the moment (6-144) becomes

$$M_2=\sum_{\nu>0}a^\dagger(\bar{\nu})a^\dagger(\nu) \tag{6-145}$$

where the notation $\nu>0$ indicates that the sum includes only a single term for each degenerate pair of single-particle levels $(\nu\bar{\nu})$.

The amplitude α_2 $(=\alpha_{\alpha=2,\lambda=0})$ of a monopole deformation of the pair density depends linearly on the operators $c_2^\dagger$ and c_{-2} that create and annihilate quanta of the two modes with nucleon transfer number ± 2 (see p. 345),

$$\begin{aligned}\alpha_2&=(\alpha_2)_0c_2^\dagger+(\alpha_{-2})_0c_{-2}\\ \alpha_{-2}=\alpha_2^\dagger&=(\alpha_2)_0c_2+(\alpha_{-2})_0c_{-2}^\dagger\end{aligned} \tag{6-146}$$

where the coefficients $(\alpha_2)_0$ and $(\alpha_{-2})_0$ are the zero-point amplitudes determined by the matrix elements of M_2,

$$\begin{aligned}(\alpha_2)_0&=\langle n_2=1|M_2|n_2=0\rangle\\ (\alpha_{-2})_0&=\langle n_{-2}=1|M_2^\dagger|n_{-2}=0\rangle\end{aligned} \tag{6-147}$$

The matrix elements of M_2 are real, for states with the standard phasing, since M_2 is invariant under time reversal as well as under rotations. The zero-point amplitudes and the frequencies for the two modes with $\alpha = \pm 2$ differ from each other, since there is no symmetry connecting the two modes with transfer number ± 2. Hence, the amplitudes α_2 and α_{-2} are not commutable variables.

The pair density produces a corresponding pair potential, which acts on the nucleonic degrees of freedom by creating (and annihilating) pairs of particles. The assumption that the $\lambda = 0$ pair mode is associated with the simple field M_2 leads to a potential of the form (compare Eqs. (6-71) and (6-72))

$$\begin{aligned} \delta V_{\text{pair}} &= -G\alpha_2^\dagger \sum_{\nu>0} a^\dagger(\bar{\nu})a^\dagger(\nu) + \text{H. conj.} \\ \alpha_2 &= \langle M_2 \rangle \end{aligned} \tag{6-148}$$

where G is a coupling constant. The interaction (6-148) provides a coupling between the vibrational motion and the motion of the individual particles, and its role in organizing the collective pair correlation is similar to that played by the particle-vibrational coupling in the shape vibrations (see the example discussed on pp. 643 ff.).

Static pair deformation. Pair rotations

In nuclei with many particles outside of closed shells, a large number of 0+ pairs occur in the nuclear ground state. Such an assembly of identical quanta, referred to as a condensate, can be described in terms of a static deformation of the field that creates these quanta. The deformation of the pair field M_2 becomes large compared with the zero-point fluctuations when the matrix elements of M_2 for addition of a pair to the condensate become large compared with other matrix elements of M_2; the magnitude of M_2 is then approximately a constant, as can be seen by evaluating the expectation value of $M_2^\dagger M_2$ and higher powers of this quantity. In the harmonic approximation, the matrix elements for adding a pair to the condensate increase as the square root of the number of quanta (see Eq. (6-138)) and thus become the dominant ones for $n_2 \gg 1$ (and $n_2 \gg n_{-2}$); the magnitude of these matrix elements may be significantly modified by anharmonic effects but, provided a large number of single-particle configurations are involved in the correlation of each pair, the condensate is expected to lead to a considerable enhancement of the pair addition mode and to matrix elements that do not change significantly in going from one nucleus to the next. In such a situation, we can express M_2

in the form

$$M_2 = |M_2| \exp\{i\phi\} \tag{6-149}$$

where $|M_2|$ is approximately a constant (a c number), while ϕ represents the phase of the pair moment. From the commutation relation

$$\left[\frac{N}{2}, M_2\right] = M_2 \tag{6-150}$$

where N is the number of particles, it follows that the angle ϕ is conjugate to the number of pairs

$$\left[\frac{N}{2}, \phi\right] = -i \tag{6-151}$$

We can therefore view ϕ as the orientation angle of the pair deformation in a space, where the number of pairs plays the role of the angular momentum. Such a space is referred to as a gauge space.[11]

The occurrence of a static pair deformation that is large compared with the zero-point fluctuations has consequences similar to those that follow from a static deformation of the nuclear shape. In particular, the motion separates approximately into rotational and intrinsic components. The angular variable in the rotational motion is the phase angle ϕ of the pair deformation, which describes the orientation in gauge space. The intrinsic motion comprises the oscillations in the magnitude of the pair deformation and additional degrees of freedom describing motion relative to the rotating pair field. Such a separation corresponds to wave functions of the form

$$\Psi_{N,\sigma} = \Phi_\sigma(q)(2\pi)^{-1/2} \exp\left\{i\frac{N}{2}\phi\right\} \tag{6-152}$$

where $\Phi_\sigma(q)$ represents the intrinsic motion and where the "rotational"

---

[11]The concept of the gauge variable has its origin in classical electrodynamics, where the electric and magnetic fields are invariant with respect to a transformation that adds the gradient of an arbitrary function Λ to the potentials. This freedom can be used to conveniently "gauge" the potentials employed in a particular calculation. In the quantal description, a gauge transformation of the potentials is accompanied by a change $(e/\hbar c)\Lambda$ in the phase of the fields that annihilate particles of charge e and create the corresponding antiparticles. For the special type of gauge transformations with Λ independent of the space-time coordinates, the potentials are unaffected, but the requirement of gauge invariance of the Hamiltonian ensures conservation of charge; the quantization of charge corresponds to the possibility of treating the gauge variable ϕ as an angular coordinate.

quantum number N takes on a set of values differing by an even number

$$N = N_1, N_1 \pm 2, N_1 \pm 4, \ldots \tag{6-153}$$

The separation of the motion expressed by a wave function of the form (6-152) is associated with the occurrence of sequences of states with different values of the particle number N, but with approximately the same intrinsic structure σ. In particular, the ground states of even-even nuclei form such a sequence, when the structure of the pair quanta in the condensate varies only slowly as a function of the quantum number N.

The selection rule $\Delta N = 2$ in the spectrum (6-153) corresponds to the condition $\Delta I = 2$ for $K = 0$ bands in nuclei with $\mathscr{R}$-invariant deformations (see Eq. (4-12)). Thus, a formal basis for the selection rule in N may be seen in the symmetry of the pair deformation with respect to a rotation through the angle 2π in gauge space. (A gauge rotation through the angle ϕ_0 is generated by the operator (compare Eq. (1-10))

$$\mathscr{G}(\phi_0) = \exp\left\{ -i \frac{N}{2} \phi_0 \right\} \tag{6-154}$$

Hence, $\mathscr{G}(2\pi) = (-1)^N$, and $(-1)^{N_1}$ is the eigenvalue of the intrinsic state with respect to this operation.)

The existence of a large static pair deformation manifests itself very directly in the pattern of two-particle transfer intensities. Thus, in nuclei removed from closed shells, by far the strongest transitions are found to go between members of the pair rotational band, with a transition amplitude proportional to the pair deformation. In contrast, in the region of closed shells, the transitions involving the different modes ($\Delta n_2 = \pm 1$ and $\Delta n_{-2} = \mp 1$) are of comparable magnitude (see Fig. 6-6), with transition amplitudes of the order of the zero-point amplitude of the pair field.

The static pair deformation produces a major modification in the nucleonic motion. The resulting one-particle degrees of freedom can be expressed in a simple manner in terms of quasiparticles that are hybrids of particles and holes (see the treatment on pp. 647ff.). The collective intrinsic modes of excitation are also modified in an essential manner and no longer possess a definite value of the nucleon transfer number α (compare the violation of the multipole symmetry λ in the vibrational modes of a nonspherical nucleus); in particular, the low-frequency shape oscillations may involve significant field components with $\alpha = \pm 2$, in addition to the $\alpha = 0$ deformation of the potential (see the discussion of β vibrations on p. 552).

The rotational frequency of the deformed pair field is given by the canonical equation

$$\dot{\phi} = \frac{2}{\hbar}\frac{\partial H}{\partial N} = \frac{2\lambda}{\hbar} \tag{6-155}$$

where λ is the chemical potential representing the average increase in energy per added particle. For rotational motion in ordinary space with relatively small values of the angular momentum I, the rotational frequency is small compared with the intrinsic frequencies (adiabatic condition), and it is possible to expand the properties of the system in powers of the angular momentum. In contrast, the frequency of the rotation in gauge space is never small compared to intrinsic frequencies and always has a major effect on the intrinsic motion. However, the assumption of a static pair deformation, large compared to its fluctuations, implies that the properties of the system do not vary appreciably for changes in N of a few units; thus, one can develop an expansion in the various matrix elements in powers of $N - N_1$. This situation is similar to that encountered for rotational motion with very high values of the angular momentum I (see, for example, the treatment of the quadrupole vibrations in the yrast region in Sec. 6B-3).

The above discussion assumes a scalar pair field. New aspects of the condensate are encountered if the boson belongs to a degenerate multiplet, since then the formation of the condensate involves a choice of direction (symmetry breaking) in the corresponding space. In the nucleus, such a situation is encountered when neutrons and protons form pairs involving equivalent orbits; the pair field is then an isotriplet, and the condensate implies a deformation in isospace with resulting rotational band structure in the variables T and A (Ginocchio and Weneser, 1968; Dussel *et al.*, 1971). In liquid ^{3}He, the occurrence of an anisotropic condensate for temperatures in the millidegree region appears to be associated with pairing in the ^{3}P state (see Leggett, 1972). Pairing in P states may also occur in the high density nuclear matter in the interior of neutron stars (see Hoffberg *et al.*, 1970).

Superfluidity

As already indicated, the pair correlation effects in nuclei are intimately related to the phenomena of superfluidity (including superconductivity) in macroscopic systems. Indeed, the concepts developed for the treatment of superconductivity in terms of correlations in the electronic motion provided the key to the collective significance of the pairing effect that had been known since the early days of nuclear physics (Bohr, Mottelson, and Pines, 1958). In the present section, we include a brief discussion of the properties of superflow in macroscopic systems in relation to the concepts employed above.

The characteristic feature of the superfluid phase is the presence of a large number of identical bosons in a single quantum state, the condensate. In superconductors, the bosons are the pairs of bound electrons that form at the Fermi surface (Cooper, 1956), and are analogous to the pair quanta in the nucleus. In He II, the condensate involves ^{4}He atoms with zero momentum (when the fluid is at rest). As in the nucleus, the consequences of the condensate are described in terms of a deformation in the field that creates or annihilates bosons in the condensate (Bogoliubov, 1947; Bardeen, Cooper, and Schrieffer, 1957; see also the review by Nozières, 1966). For the discussion of superflow, the central concept is the phase angle ϕ of the condensate field, which represents the orientation in gauge space and which is conjugate to the number of bosons in the condensate. (The condensate field (order parameter) was introduced in the phenomenological theory of superconductivity by Ginzburg and Landau, 1950, before the microscopic interpretation of this field had been discovered; see, for example, Anderson, 1969.)

The collective enhancement of the processes involving the addition or removal of bosons from the condensate, which in the nucleus are studied in the two-particle transfer reactions (pair rotational mode), strikingly manifests itself in the phenomena associated with the transmission of a supercurrent between two superconductors separated by a thin barrier (Josephson, 1962; Anderson, 1964). In such a junction, the tunneling of electron pairs through the barrier can be represented by a coupling that destroys a boson on one side of the barrier and adds a boson to the condensate on the other side. The operator that creates a boson is proportional to $\exp\{i\phi\}$ (see, for example, Eq. (6-149)), and the coupling therefore takes the form

$$H' = a\cos(\phi_1 - \phi_2 + \delta) \tag{6-156}$$

where ϕ_1 and ϕ_2 are the gauge angles of the two superconductors, while δ is the phase shift of the electrons that is associated with the tunneling. The constant a is a measure of the intensity of the electron tunneling through the barrier. In the presence of the coupling (6-156), the equations of motion for the phases ϕ_1 and ϕ_2 and the electron number operators N_1 and N_2 are (see Eq. (6-155))

$$\dot{\phi}_1 = \frac{2}{\hbar}\lambda_1 \qquad \dot{\phi}_2 = \frac{2}{\hbar}\lambda_2$$

$$\begin{aligned} \dot{N}_1 = -\dot{N}_2 &= -\frac{2}{\hbar}\frac{\partial H'}{\partial \phi_1} = 2\hbar^{-1}a\sin(\phi_1 - \phi_2 + \delta) \\ &= -2\hbar^{-1}a\sin\left(\frac{2eV}{\hbar}(t - t_0) - \delta\right) \end{aligned} \tag{6-157}$$

where λ_1 and λ_2 are the chemical potentials for the two superconductors, and V is the electrostatic potential across the junction, which determines the difference in the chemical potentials $(-eV = \lambda_1 - \lambda_2)$. The equations (6-157) describe the current between the superconductors, which results from the phase difference of the two condensates and which therefore has a frequency given by the difference in the pair-rotational frequencies on the two sides of the barrier. Precision measurements of the frequency of the alternating current in such a junction have led to the most accurate available determination of the ratio between e and h (Taylor *et al.*, 1969).

The flow of a supercurrent through a superconductor, or the superflow of He II, involves a condensate in which the phase angle ϕ varies across the volume of the superfluid system. Such a variation is associated with a corresponding variation in the chemical potential in accordance with the canonical equation (see Eq. (6-155))

$$\nabla\dot{\phi} = \frac{1}{\hbar}\nabla\lambda_{\mathrm{B}} = -\frac{1}{\hbar}\mathbf{F}_{\mathrm{B}} \tag{6-158}$$

where λ_{B} is the chemical potential of the boson ($\lambda_{\mathrm{B}}=2\lambda$ for superfluid Fermi systems) and $\mathbf{F}_{\mathrm{B}}$ the force acting on the bosons of the condensate. Equation (6-158) exhibits ϕ as a velocity potential corresponding to the flow

$$\mathbf{v} = -\frac{\hbar}{M_{\mathrm{B}}}\nabla\phi(\mathbf{r}) \tag{6-159}$$

where M_{B} is the mass of the boson ($M_{\mathrm{B}}=2M$ for fermion condensates). For systems of charged particles, the current includes an additional term proportional to the magnetic vector potential (see Eq. (6-162)). In a homogeneous system at zero temperature, the flow (6-159) represents translational motion of the entire fluid in the region of the space point $\mathbf{r}$, corresponding to a wave function in which all the single-particle states of the stationary fluid have been multiplied by the factor $\exp\{-i(M/M_{\mathrm{B}})\phi(\mathbf{r})\}$; thus, the bosons of the condensate all acquire the same spatially dependent phase.

In the case of stationary flow, the continuity equation implies

$$\nabla^2\phi(\mathbf{r}) = 0 \tag{6-160}$$

for a system with constant average density. The solutions to Eq. (6-160) involve currents that either flow in and out of the superfluid domains, or that circulate in a multiply connected domain, such as in a superconducting torus or around a region in which the superfluidity has been destroyed (vortex motion). In multiply connected domains, the one-valuedness of the condensate field implies

$$\oint \nabla\phi\cdot d\mathbf{s} = 2\pi\nu \qquad \nu = 0, \pm 1, \pm 2, \ldots \tag{6-161}$$

where the path of integration is any closed loop lying completely within the superfluid region. The condition (6-161) corresponds to a quantization of the circulation $\oint \mathbf{v}\cdot d\mathbf{s}$, in units of h/M_{B} (Onsager, 1954; Feynman, 1955). In the flow with collective circulation number ν, each boson of the condensate moves in a state with average angular momentum $\nu\hbar$; thus, a change in ν involves a simultaneous change of state of a macroscopic number of particles. The quantum number ν therefore labels different thermodynamic phases of the superfluid; the high stability of the supercurrent can be viewed as a macroscopic isomerism (see in this connection Bohr and Mottelson, 1962).

For superconductors, the charge $-2e$ of the bosons implies that the phase ϕ of the condensate transforms into $\phi+(2e/\hbar c)\Lambda$ under a change of gauge for the electromagnetic potentials (see footnote 11, p. 393). The gauge-invariant velocity of the supercurrent is therefore obtained from Eq. (6-159) by adding a term involving the magnetic vector potential A

$$\mathbf{v} = -\frac{\hbar}{2M}\left(\nabla\phi - \frac{2e}{\hbar c}\mathbf{A}\right) \tag{6-162}$$

The form (6-162) may also be recognized as expressing the canonical relation $M_{\mathrm{B}}\mathbf{v} = \mathbf{p} - (e_{\mathrm{B}}/c)\mathbf{A}$ between velocity and momentum (for a particle of mass $M_{\mathrm{B}} = 2M$ and charge $e_{\mathrm{B}} = -2e$). In a stationary situation, $\nabla \cdot \mathbf{A}$ vanishes, and Eq. (6-160) therefore remains valid. In a multiply connected domain, the condition (6-161) now implies

$$\oint \mathbf{v} \cdot d\mathbf{s} - \frac{e}{Mc}\Phi = \frac{h}{2M}\nu \tag{6-163}$$

where $\Phi = \oint \mathbf{A} \cdot d\mathbf{s}$ is the magnetic flux enclosed by the path of integration. In superconductors of dimensions large compared to the penetration distance for magnetic fields, the current vanishes in the interior, and Eq. (6-163) implies the quantization of the flux Φ (London, 1950). The experimental observation of flux quantization in units of $hc/2e$ (Deaver and Fairbank, 1961; Doll and Näbauer, 1961) provided direct confirmation that the superconductivity is associated with a condensate of bosons, each carrying a charge of 2 units (Byers and Yang, 1961; Onsager, 1961).

Superflow with the properties discussed above is possible if $\phi(\mathbf{r})$ is approximately constant over spatial regions characteristic of the correlations in the superfluid phase. The wave functions of all the effectively interacting particles are then multiplied by approximately the same phase factor, and the correlation structure is not affected significantly. In He II, the characteristic dimension, referred to as the coherence length, is of the order of the interatomic distance a_0, as follows from the fact that the critical temperature is comparable with the energy $\hbar^2/Ma_0^2$. In Fermi systems, the coherence length ξ is of the order of the spatial extent of the correlated pairs, and is thus determined by the momentum interval δp of the single-particle states that contribute to the correlation

$$\xi \sim \frac{\hbar}{\delta p} \sim \frac{\hbar v_{\mathrm{F}}}{\Delta} \tag{6-164}$$

where v_{F} is the Fermi velocity and 2Δ the energy gap resulting from the binding of the pairs.

In nuclei, the pairs cannot be localized within dimensions smaller than the nuclear radius R. Any greater localization would require a space of single-particle states comprising several major shells and could be achieved only if the binding energy of a pair were large compared to the spacing between major shells ($\Delta \gg \hbar\omega_0$; compare also the estimate (6-164) of the coherence distance, which yields $\xi \sim (\hbar\omega_0/\Delta)R$.) Thus, the striking phenomena of quantized superflow do not appear to have direct counterparts in the nuclei. However, these phenomena may play an important role in the dynamics of nuclear matter occurring in neutron stars (see Ruderman, 1972).

6-4 SUM RULES FOR MULTIPOLE OSCILLATOR STRENGTH

In the analysis of the spectra of complex systems, it is often useful to exploit general relations that follow from algebraic relations between operators and can be expressed in the form of sum rules. The present section brings together various aspects of these relations that may be employed in the analysis of the pattern of collective excitations. Some readers may wish to proceed with the discussion of the dynamics of collective modes and return to the material in the present section when needed for the specific applications. (The use of oscillator sums as units for vibrational strength is considered in Sec. 6-4b.)

6-4a Classical Oscillator Sums

Moments depending only on spatial coordinates

The oscillator strength associated with a moment F is equal to the transition probability multiplied by the excitation energy, and the sum of the oscillator strengths can be expressed in the form (assuming F to be a real (Hermitian) operator)

$$S(F) \equiv \sum_a (E_a - E_0)|\langle a|F|0\rangle|^2$$

$$= \tfrac{1}{2}\langle 0|[F,[H,F]]|0\rangle \qquad (6\text{-}165)$$

where a labels the complete set of excited states that can be reached by operating with F on the initial state 0.

If F is a one-particle moment depending only on the spatial coordinates

$$F = \sum_k F(\mathbf{r}_k) \qquad (6\text{-}166)$$

and if the interactions do not explicitly depend on the momenta of the particles, the commutator in Eq. (6-165) receives contributions only from the kinetic energy, and we obtain

$$S(F) = \langle 0|\sum_k \frac{\hbar^2}{2M_k}(\nabla_k F(\mathbf{r}_k))^2|0\rangle \qquad (6\text{-}167)$$

which expresses the oscillator sum as the expectation value of a one-body operator.

The classic example of such a sum rule refers to the dipole excitations in atoms, for which the transition moment is linear in the spatial coordinate $\mathbf{r}_k$ of the electron. For $F=\mathbf{r}$, the sum (6-167) depends only on the number of particles and their masses. In atomic physics, it is conventional to define the oscillator strength for a transition $0\rightarrow a$ as

$$f_{0a}=\frac{2m}{\hbar^2}(E_a-E_0)|\langle a|\sum_k z_k|0\rangle|^2 \tag{6-168a}$$

$$\frac{1}{2I_0+1}\sum_{M_0} f_{0a}=\frac{8\pi}{9}\frac{m}{e^2\hbar^2}(E_a-E_0)B(E1;0\rightarrow a) \tag{6-168b}$$

where m is the electron mass; in Eq. (6-168b), the oscillator strength averaged over the different orientations of the initial state is expressed in terms of the reduced transition probability $B(E1)$. The normalization of the oscillator strength in Eq. (6-168) implies that the sum $\Sigma_a f_{0a}$ is equal to the number of electrons in the atom. The response of the atom to an oscillating weak electric field of long wavelength can be described by considering the atom as a system of harmonic oscillators (the virtual oscillators) with mass m, frequencies corresponding to those observed in the absorption spectrum $\omega_{0a}=\hbar^{-1}(E_a-E_0)$, and the values $(f_{0a})^{1/2}e$ of the charge.[12]

For a multipole field

$$F_{\lambda\mu}=f(r)Y_{\lambda\mu}(\vartheta,\varphi) \tag{6-169}$$

the oscillator sum (6-167) can be evaluated by means of the gradient formula (see Eq. (3A-26))

$$\nabla f(r)Y_{\lambda\mu}=\left(\frac{\lambda}{2\lambda+1}\right)^{1/2}\left((\lambda+1)\frac{f(r)}{r}+\frac{df}{dr}\right)(Y_{\lambda-1}\mathbf{e})_{(\lambda-1,1)\lambda\mu}$$
$$+\left(\frac{\lambda+1}{2\lambda+1}\right)^{1/2}\left(\lambda\frac{f(r)}{r}-\frac{df}{dr}\right)(Y_{\lambda+1}\mathbf{e})_{(\lambda+1,1)\lambda\mu} \tag{6-170}$$

---

[12]The f sum rule for the virtual oscillators played a significant role in the development of quantum mechanics, since it had been derived from a classical analysis of the response of an atom to a high-frequency external field (Thomas, 1925; Kuhn, 1925). On the basis of the correspondence principle, it was anticipated that these asymptotic relations would have a range of validity extending beyond their classical derivation. With the introduction of matrix mechanics, it was confirmed that the sum rule follows immediately from the commutation relations of p and q, as in the above derivation (Heisenberg, 1925).

which yields

$$\sum_{\mu} \nabla\big(f(r) Y^{*}_{\lambda\mu}(\vartheta,\varphi)\big)\cdot \nabla\big(f(r) Y_{\lambda\mu}(\vartheta,\varphi)\big) = \frac{2\lambda+1}{4\pi}\left(\left(\frac{df}{dr}\right)^2 + \lambda(\lambda+1)\left(\frac{f}{r}\right)^2\right) \tag{6-171}$$

We therefore obtain

$$\begin{aligned} S(F_\lambda) &\equiv \sum_{\alpha I} (E_{\alpha I} - E_0) B(F_\lambda; 0 \rightarrow \alpha I) \\ &= \sum_{\alpha I M \mu} (E_{\alpha I} - E_0) |\langle \alpha I M | F_{\lambda\mu} | 0\rangle|^2 \\ &= \langle 0 | \frac{\hbar^2}{2M} \sum_{\mu, k} |\nabla_k F_{\lambda\mu}(\mathbf{r}_k)|^2 | 0\rangle \\ &= \frac{2\lambda+1}{4\pi} \frac{\hbar^2}{2M} A \left\langle \left(\frac{df}{dr}\right)^2 + \lambda(\lambda+1)\left(\frac{f}{r}\right)^2 \right\rangle \end{aligned} \tag{6-172}$$

where $B(F_\lambda; 0 \rightarrow \alpha I)$ is the reduced transition probability for the field F_λ, while the last factor in Eq. (6-172) represents the average per particle in the ground state of the system of A particles, each having a mass M. In Eq. (6-171), the summation over μ implies a scalar coupling of the two multipole moments; sum rules involving tensorial couplings of the matrix elements are considered in Sec. 6-4c.

The special significance of the sum rule with linear energy weighting derives from the fact that this sum can be expressed as an expectation value of a one-body operator and is therefore relatively insensitive to the detailed correlations in the initial state. Sums with other energy weightings can also be expressed in terms of expectation values in the initial state, but will in general involve two-body or many-body operators and thus be much more sensitive to correlations. For example, the sum over the transition probabilities can be expressed in the form

$$\sum_{a} |\langle a | F | 0\rangle|^2 = \int F^*(\mathbf{r}) \rho_0(\mathbf{r}, \mathbf{r}') F(\mathbf{r}') d\tau\, d\tau' + \int \rho_0(\mathbf{r}) |F(\mathbf{r})|^2 d\tau \tag{6-173}$$

where $\rho_0(\mathbf{r})$ is the one-particle density in the initial state 0, while $\rho_0(\mathbf{r}, \mathbf{r}')$ is the expectation value of the two-particle density function (see Eq. (2-33)). These more general sums may be difficult to evaluate theoretically, but an experimental determination may be possible in terms of the response of the system to external perturbations (scattering cross sections, polarizabilities, etc.). For example, the static polarizability for a perturbation proportional to F can be expressed in terms of the sum $(E_a - E_0)^{-1}|\langle a|F|0\rangle|^2$; see Eq. (6-238).

The velocity-dependent components in the nucleonic interactions imply corrections to the oscillator sum rules derived above. However, the Galilean invariance implies that the interactions commute with the isoscalar dipole field $F = x$; hence, the corrections to the sum rule depend on the second spatial derivative of F and are of order $(ka)^2$, where a is the range of the interaction and k the wave number of the field F.

Even in the absence of velocity-dependent nucleonic interactions, the average self-consistent one-body potential will in general exhibit velocity dependence (see, for example, the discussion of the effective mass in Chapter 2, p. 257). In such a situation, the one-particle excitations of the shell model give an oscillator sum in disagreement with the identities (6-167) and (6-172). This error will be compensated by the effective interactions that are not included in the average one-body potential. The explicit construction of these additional terms is discussed, for the isoscalar dipole and quadrupole modes, on pp. 445 and 511, respectively.

The occurrence of a static pair potential also violates the Galilean invariance in the generalized one-particle motion. The additional interactions required by the sum rule are considered on p. 446, in connection with the analysis of the pushing model. (See also the discussion of the cranking model for the rotational motion on p. 83.)

It is important to emphasize that the sum rules derived above treat the particles as elementary, and thus neglect possible contributions from internal degrees of freedom of the particles. Though the transition moments for the excitation of such internal degrees of freedom will be small if the size of the particles is small compared with the dimensions of the system, the high frequency of such excitations may lead to large contributions to the oscillator sum. A hypothetical model illustrating this point would be an atom in which the electrons were composed of two strongly bound particles, each with approximately half the electronic mass. Assuming one of these constituents to be neutral and the other to carry the full electronic charge, the f sum for the electric dipole moment would be doubled by the high-frequency internal excitations.

In the nucleus, the internal structure of the nucleons is exhibited by the spectrum of excited baryonic states and the possibility of meson production (see, for example, Figs. 1-11, Vol. I, p. 57, and 1-12, Vol. I, p. 63). At energies corresponding to these excitations, one must expect important contributions to the sum rules, which are not included in the estimate (6-172); it is therefore necessary to specify the energy interval over which the sum is extended. Well-defined applications of the sum rules in the form given above can be made only when the oscillator strength associated with the motion of the nucleons themselves is exhausted for excitation energies that lie well below those associated with internal excitations of the nucleons. The oscillator strength for single-particle motion in the average potential, as well as for the known or expected collective modes, does occur at energies much smaller than those corresponding to the baryonic excitations, but there remain open questions concerning the effect of the very short-range nucleonic correlations on the distribution of oscillator strength (see, in this connection, the discussion in Sec. 6-4d).

$E\lambda$ moments

The electric multipole moment for a system of nucleons

$$\mathscr{M}(E\lambda,\mu)=e\sum_k\left(\left(\tfrac{1}{2}-t_z\right)r^\lambda Y_{\lambda\mu}\right)_k \tag{6-174}$$

depends on the isobaric variable; the oscillator sum may therefore be affected by the occurrence of charge exchange components in the nucleonic interactions. In the present section, we consider the $E\lambda$ sum rules obtained by including only the kinetic energy terms in the Hamiltonian in Eq. (6-165); the resulting values are referred to as the classical oscillator sums $S(E\lambda)_{\rm class}$. The effect of the charge exchange interactions will be discussed in Sec. 6-4d.

The classical sum rules discussed in the present section also ignore the effects of velocity-dependent interactions. While these interactions have rather minor effects on the sum rules for isoscalar moments, as a result of the Galilean invariance (see p. 402), the sum rules for isovector (and $E\lambda$) moments may be significantly modified. (The modification of the $E1$ oscillator strength in the dipole mode resulting from effective velocity-dependent interactions is considered on pp. 483 ff.)

The electric dipole sum rules for nuclear excitations involve the $E1$ operator referred to the center of mass, which can be written in the form (see Eq. (3C-35))

$$\mathscr{M}(E1,\mu)=e\sum_k\left(\left(\frac{N-Z}{2A}-t_z\right)rY_{1\mu}\right)_k \tag{6-175}$$

and from Eq. (6-172) we therefore obtain

$$\begin{aligned} S(E1)_{\rm class} &= \frac{9}{4\pi}\frac{\hbar^2}{2M}\frac{NZ}{A}e^2 \\ &= 14.8\frac{NZ}{A}e^2\,{\rm fm}^2\,{\rm MeV} \end{aligned} \tag{6-176}$$

The removal of the center-of-mass degree of freedom is responsible for the factor N/A in Eq. (6-176). The oscillator strength associated with the nuclear center-of-mass motion appears as a contribution to the atomic oscillator sum, resulting from the nuclear recoil; since the contribution of each particle to the oscillator sum is proportional to the square of its charge and inversely to its mass, the nucleus gives a contribution $Z^2(m/AM)$ to the atomic f sum. This contribution is seen, from Eq. (6-168b), to correspond to the center-of-mass contribution removed from the oscillator sum (6-176) for internal nuclear excitations.

For the higher multipoles, the center-of-mass correction is expected to be of order $ZA^{-\lambda}$ (or less) and will be neglected in the following (see Vol. I, p. 342, for a discussion of recoil effects for the $E2$ moment). From the $E\lambda$ moment (6-174) and the relation (6-172), we thus obtain

$$S(E\lambda)_{\text{class}} = \frac{\lambda(2\lambda+1)^2}{4\pi}\frac{\hbar^2}{2M}Ze^2\langle r^{2\lambda-2}\rangle_{\text{prot}} \qquad \lambda \geqslant 2 \tag{6-177}$$

where the last factor is the radial average for the protons in the initial state.

For $E0$ transitions, the leading-order moment is proportional to r^2 (see Eq. (3C-20)), and one obtains (see Eq. (6-167))

$$\begin{aligned} S(E0) &\equiv \sum_a (E_a - E_0)|\langle a|m(E0)|0\rangle|^2 \\ &= \sum_a (E_a - E_0)|\langle a|e\sum_k ((\tfrac{1}{2} - t_z)r^2)_k|0\rangle|^2 \end{aligned} \tag{6-178a}$$

$$S(E0)_{\text{class}} = \frac{2\hbar^2}{M}Ze^2\langle r^2\rangle_{\text{prot}} \tag{6-178b}$$

for the corresponding sum rule.

The $E\lambda$ moment is the sum of an isoscalar and an isovector part. For the $\tau=0$ and $\tau=1$ moments (6-122), one obtains the corresponding classical oscillator sums

$$\begin{aligned} S(\tau=0,\lambda)_{\text{class}} = S(\tau=1,\mu_\tau=0,\lambda)_{\text{class}} &= \frac{\lambda(2\lambda+1)^2}{4\pi}\frac{\hbar^2}{2M}A\langle r^{2\lambda-2}\rangle \\ &\approx \frac{A}{Ze^2}S(E\lambda)_{\text{class}} \qquad \lambda \geqslant 2 \end{aligned} \tag{6-179a}$$

$$\begin{aligned} S(\tau=0,\lambda) &\equiv \sum_{\alpha I}(E_{\alpha I} - E_0)B(\tau=0,\lambda;\, 0\to\alpha I) \\ S(\tau=1,\mu_\tau=0,\lambda) &\equiv \sum_{\alpha I}(E_{\alpha I} - E_0)B(\tau=1,\mu_\tau=0,\lambda;\, 0\to\alpha I) \end{aligned} \tag{6-179b}$$

In the last line of Eq. (6-179a), we have assumed the protons and the neutrons to have the same value of $\langle r^{2\lambda-2}\rangle$.

6-4b Vibrational Oscillator Strength in Sum Rule Units

Spherical nuclei

The oscillator sums provide natural units for measuring the strength of collective excitations. Thus, for shape vibrations in a spherical nucleus, we obtain (see Eqs. (6-65) and (6-177))

$$\hbar\omega_\lambda B(E\lambda;\, n_\lambda=0\rightarrow n_\lambda=1)=(2\lambda+1)\left(\frac{3}{4\pi}ZeR^{\lambda}\right)^2\frac{\hbar^2}{2D_\lambda}$$

$$=\frac{D_\lambda(\text{irrot})}{D_\lambda}\frac{Z}{A}S(E\lambda)_{\text{class}} \qquad (6\text{-}180)$$

where

$$D_\lambda(\text{irrot})=\frac{3}{4\pi}\frac{1}{\lambda}AMR^2 \qquad (6\text{-}181)$$

is the mass parameter for a surface oscillation in a liquid drop with irrotational velocity field (see Eq. (6A-31)). In Eq. (6-180), we have employed the estimate (6-64) for the radial average in $S(E\lambda)$, neglecting effects of the diffuseness of the nuclear surface.

If we assume an approximately constant ratio of proton and neutron densities in the vibrational motion, the excitations have $\tau\approx 0$ and

$$B(\tau=0,\lambda)\approx\left(\frac{A}{Ze}\right)^2 B(E\lambda) \qquad (6\text{-}182)$$

in which case the relation (6-180) is equivalent to

$$\hbar\omega_\lambda B(\tau=0,\lambda;\, n_\lambda=0\rightarrow n_\lambda=1)\approx\frac{D_\lambda(\text{irrot})}{D_\lambda}S(\tau=0,\lambda)_{\text{class}} \qquad (6\text{-}183)$$

for the $\tau=0$ vibrational oscillator strength (see Eq. (6-179)).

The expressions (6-180) and (6-183) imply that the ratio of the vibrational mass parameter to that associated with irrotational flow is a measure of the extent to which the given mode exhausts the sum rule. The liquid-drop model represents a limiting case, in which the entire $\tau=0$ oscillator sum is concentrated on a single mode of surface oscillation. (The significance of the irrotational flow in a classical system is connected with the fact that this velocity field gives the least kinetic energy for given time dependence of the density; Lamb, 1916, p. 45.)

In the nucleus, the transition strength of given multipole order may be divided among several modes, on account of the shell structure (see p. 507).

Each mode then contains only a fraction of the oscillator sum, and the mass parameter is larger than for irrotational flow. For the low-frequency shape oscillations, the measured oscillator strength represents at most about 10% of $S(E\lambda)_{\text{class}}$; see Fig. 6-29, p. 531, ($\lambda=2$) and Table 6-14, p. 561, ($\lambda=3$). The main part of the oscillator strength is expected to be associated with high-frequency modes (see pp. 508 ff. and pp. 556 ff.). In the case of the isovector dipole transitions, the main oscillator strength is found to be concentrated on a single mode, which approximately exhausts the classical sum rule (see Fig. 6-20, p. 479); a similar situation is expected for the high-frequency quadrupole modes. The collective flow associated with these modes is explicitly exhibited in Eq. (6-373b) and is seen to be of irrotational type.

Nonspherical nuclei

In a deformed nucleus, the oscillator strength of the collective excitations is associated partly with rotational, partly with vibrational, excitations. The $E2$ rotational oscillator strength for an even-even nucleus with axial symmetry is given by (see Eq. (4-68))

$$(E_2-E_0)B(E2;K_0=0,I_0=0\rightarrow K_0=0,I_0=2)$$

$$=\frac{3\hbar^2}{\mathscr{I}}\frac{5}{16\pi}e^2Q_0^2=\frac{2}{5}\frac{\mathscr{I}(\text{irrot})}{\mathscr{I}}\frac{Z}{A}S(E2)_{\text{class}} \qquad (6\text{-}184)$$

where the moment of inertia

$$\mathscr{I}(\text{irrot})=\tfrac{2}{5}AMR^2\delta^2=3D_2(\text{irrot})\beta^2 \qquad (6\text{-}185)$$

is that associated with a rotational motion described by irrotational flow. (See Eqs. (6A-81) and (6-181) as well as relations (4-72), (4-73), and (4-191) for the deformation parameters δ and β; in Eq. (6-185), we have neglected corrections due to the diffuseness of the nuclear surface, as well as higher-order terms in β.)

For an odd-A nucleus (and, more generally, for an arbitrary initial state I_0), one may evaluate the oscillator sum for the rotational transitions in terms of the double commutator between the Hamiltonian and the multipole operator, as in the derivation of Eq. (6-165). Assuming the leading-order intensity relations for the $E2$ transitions in the band, and rotational energies proportional to $I(I+1)$, we obtain the same results as in

Eq. (6-184) (see Eqs. (4A-34), (1A-39), and (1A-91))

$$\sum_I (E(K_0I) - E(K_0I_0))B(E2; K_0I_0 \to K_0I)$$

$$= \tfrac{1}{2}\langle K_0I_0M_0 | \sum_\mu \left[\mathscr{M}^\dagger(E2,\mu), [H_{\text{rot}}, \mathscr{M}(E2,\mu)]\right] | K_0I_0M_0\rangle$$

$$\approx \frac{5}{32\pi} e^2Q_0^2 \frac{\hbar^2}{2\mathscr{I}} \langle K_0I_0M_0 | \sum_\mu \left[\mathscr{D}^{2*}_{\mu 0}, \left[\mathbf{I}^2, \mathscr{D}^2_{\mu 0}\right]\right] | K_0I_0M_0\rangle$$

$$= \frac{15}{8\pi} e^2Q_0^2 \frac{\hbar^2}{2\mathscr{I}} \tag{6-186}$$

For the quadrupole shape vibrations of the β and γ type (see Fig. 6-3, p. 363), we obtain from Eq. (6-91), for an even-even nucleus,

$$\hbar\omega_\beta B(E2; n_\beta=0, I=0 \to n_\beta=1, I=2) = \frac{1}{5}\frac{D_2(\text{irrot})}{D_\beta}\frac{Z}{A}S(E2)_{\text{class}}$$

$$\hbar\omega_\gamma B(E2; n_\gamma=0, I=0 \to n_\gamma=1, I=2) = \frac{2}{5}\frac{D_2(\text{irrot})}{D_\gamma}\frac{Z}{A}S(E2)_{\text{class}} \tag{6-187}$$

$$\hbar\omega_{\text{rot}} B(E2; n=0, I=0 \to n=0, I=2) = \frac{2}{5}\frac{D_2(\text{irrot})}{D_{\text{rot}}}\frac{Z}{A}S(E2)_{\text{class}}$$

For comparison, we have added the rotational oscillator strength (6-184) and have defined a rotational mass parameter D_{rot} by the relation

$$\mathscr{I} = 3D_{\text{rot}}\beta^2 \tag{6-188}$$

The factors 1/5, 2/5, and 2/5 in Eq. (6-187) correspond to the fact that, of the five quadrupole degrees of freedom, one is associated with vibrations of the β type, two with vibrations of the γ type, and two with the rotational motion (see Fig. 6-3).

For initial states with $(K_0I_0) \neq (00)$, the values (6-187) are obtained by a summation over the transitions leading to the different members of the vibrational bands. For the γ excitations in a nucleus with $K_0 \neq 0$, the bands with $K = K_0 \pm 2$ each contribute one half of the value given by Eq. (6-187).

In the deformed nuclei, the main quadrupole oscillator strength in the low-energy spectrum is found to be associated with rotational excitations. The empirical moments of inertia are about five times larger than $\mathscr{I}$(irrot);

see Fig. 4-12, p. 74. Thus, the oscillator strength of the rotational transitions is about 5% of $S(E2)_{\text{class}}$, which is comparable to the value for the low-frequency quadrupole vibrational mode in spherical nuclei (see Fig. 6-29, p. 531). Oscillator strength of a few percent of $S(E2)_{\text{class}}$ is associated with the observed low-energy $\nu\pi = 0+$ and $2+$ quadrupole excitations in deformed nuclei, which thus have mass parameters that are not only large compared with the irrotational values, but also several times larger than the rotational mass parameter (see the discussion on pp. 552 and 550).

6-4c Tensorial Sums

The sum rules discussed in Sec. 6-4a are expressed in terms of scalar moments in the ground state. For a system with an anisotropic density distribution in the initial state, the oscillator strength depends on the direction of the angular momentum of excitation relative to that of the initial state. One may express this dependence in terms of sum rules involving a tensorial coupling of multipole moments. The more general sum rules of this type have so far had only few applications in the analysis of experimental data, but in the present section, we consider some examples illustrating the nature of these relations. The last example in the section concerns the consequences of the asymmetry in isospace resulting from the neutron excess.

A straightforward extension of the scalar sum rule (6-172) is obtained by coupling the two multipole operators involved in the double commutator to a tensor of rank λ_0. For the isoscalar moments, we have

$$\begin{aligned}
&\left[\left[\mathscr{M}(\tau=0,\lambda),H\right],\mathscr{M}(\tau=0,\lambda)\right]_{(\lambda\lambda)\lambda_0\mu_0}\\
&\quad=\frac{\hbar^2}{M}\sum_k\left(\nabla_k(r_k^\lambda Y_\lambda(k))\cdot\nabla_k(r_k^\lambda Y_\lambda(k))\right)_{(\lambda\lambda)\lambda_0\mu_0}\\
&\quad=-(2\lambda+1)\left(\frac{(2\lambda+\lambda_0+1)(2\lambda+\lambda_0)(2\lambda-\lambda_0)(2\lambda-\lambda_0-1)}{4\pi(2\lambda_0+1)}\right)^{1/2}\\
&\qquad\times\langle\lambda-1\,0\;\lambda-1\,0|\lambda_0 0\rangle\frac{\hbar^2}{2M}\sum_{k=1}^{A}\left(r^{2\lambda-2}Y_{\lambda_0\mu_0}\right)_k
\end{aligned}\tag{6-189}$$

As in Eq. (6-172), we have included only the contribution from the kinetic energy. In deriving the last line of Eq. (6-189), we have used the relation

$$\nabla r^{\lambda} Y_{\lambda\mu} = (\lambda(2\lambda+1))^{1/2} r^{\lambda-1} (Y_{\lambda-1}\mathbf{e})_{(\lambda-1,1)\lambda\mu} \tag{6-190}$$

which is a special case of Eq. (6-170).

Matrix elements of the commutator in Eq. (6-189) can be evaluated by means of the general relation for products of tensor operators

$$\langle I_2 \| (F_{\lambda_1} G_{\lambda_2})_{\lambda_0} \| I_1 \rangle$$

$$= \sum_{\alpha I} (-1)^{I_1+I_2+\lambda_0} (2\lambda_0+1)^{1/2} \begin{Bmatrix} \lambda_1 & \lambda_2 & \lambda_0 \\ I_1 & I_2 & I \end{Bmatrix} \langle I_2 \| F_{\lambda_1} \| \alpha I \rangle \langle \alpha I \| G_{\lambda_2} \| I_1 \rangle \tag{6-191}$$

derived by a summation over a complete set of intermediate states specified by the quantum numbers αI and a recoupling $I_1,(\lambda_1\lambda_2)\lambda_0; I_2 \rightarrow (I_1\lambda_2)I,\lambda_1; I_2$ (see Eqs. (1A-9) and (1A-19)). For the expectation value in the state I_0, one obtains from Eqs. (6-189) and (6-191), by using the relation (1A-79) for Hermitian conjugation of the reduced matrix element and the symmetry $c_H = (-1)^{\lambda}$ of the multipole moment $\mathscr{M}(\lambda)$,

$$\sum_{\alpha I} (E_{\alpha I} - E_0) B(\tau=0,\lambda; I_0 \rightarrow \alpha I) \begin{Bmatrix} \lambda & \lambda & \lambda_0 \\ I_0 & I_0 & I \end{Bmatrix} (-1)^{I_0+I} \left(1+(-1)^{\lambda_0}\right)$$

$$= -\frac{2\lambda+1}{(2\lambda_0+1)(2I_0+1)} ((2\lambda+\lambda_0+1)(2\lambda+\lambda_0)(2\lambda-\lambda_0)(2\lambda-\lambda_0-1))^{1/2}$$

$$\times \langle \lambda-1\, 0\, \lambda-1\, 0 | \lambda_0 0 \rangle (4\pi)^{-1/2} \frac{\hbar^2}{2M} \langle I_0 \| \sum_{k=1}^{A} (r^{2\lambda-2} Y_{\lambda_0})_k \| I_0 \rangle \tag{6-192}$$

In the special case of $\lambda = \lambda_0 = 2$, the tensor sum rule (6-192) relates the static quadrupole moment of the state I_0 to the oscillator strength of the quadrupole transitions from this state. For a vibrational spectrum, both sides of Eq. (6-192) vanish in the harmonic approximation. In fact, the static quadrupole moment vanishes in every state (corresponding to the selection rule $\Delta n = \pm 1$ for the quadrupole operator, see p. 349), and it can be verified that the harmonic relations embodied in Eq. (6-66) ensure the vanishing of the left-hand side of Eq. (6-192). (This result can also be directly obtained by noting that in the harmonic approximation, the moment $\mathscr{M}(\lambda\mu)$, as well as its time derivative, are linear functions of the operators $c^{\dagger}(\lambda\mu)$ and $c(\overline{\lambda\mu})$; hence, the double commutator is a c number, which must vanish for $\lambda_0 \neq 0$.) Thus, the sum rule (6-192) relates the static quadrupole moment in a vibrational state I_0 to the anharmonic effects in the vibrational transitions and the additional quadrupole transitions leading to states outside the vibrational spectrum.

For a nonspherical nucleus, the right-hand side of the tensor sum rule (6-192), for $\lambda=\lambda_0=2$, can be related to the intrinsic quadrupole moment. Thus, for $K=0$, $I_0=2$, we obtain the value $(4\pi)^{-1}(\hbar^2/M)\,Q_0\,(A/Z)$ where the last factor results from the fact that the sum rule has been written for the isoscalar moment, while Q_0 represents the electric moment. On the left-hand side, the transitions within the band contribute an amount $(3/28\pi)(Q_0 A/Z)^2(\hbar^2/2\mathscr{I})$, assuming the leading-order rotational intensity relation (4-68). For example, for ^{166}Er, ($Q_0 \approx 756$ fm^2, $\hbar^2/2\mathscr{I} \approx 13.4$ keV (see p. 161)), the rotational transitions account for about 26% of the tensor sum. The transitions to the γ-vibrational band, as given in Fig. 4-30, p. 160, contribute an amount of about -13% of the sum rule. Thus, about 90% of the sum are associated with transitions that have not yet been observed; in fact, the main contribution to the tensor sum rule is expected from the high-frequency quadrupole modes with $K=0, 1$, and 2 (see p. 548), as for the scalar sum rule for the quadrupole oscillator strength.

Another type of tensorial sum rule, which can be evaluated in terms of one-body operators, is obtained by considering the commutator of the time derivatives of two multipole moments (Dothan *et al.*, 1965). In the case of the quadrupole moment, the coupling to $\lambda_0=1$ can be expressed in terms of the orbital angular momentum

$$\left[\dot{\mathscr{M}}(\tau=0,\lambda=2),\,\dot{\mathscr{M}}(\tau=0,\lambda'=2)\right]_{(\lambda\lambda')1\mu} = -\frac{15\sqrt{10}}{4\pi}\frac{\hbar^2}{M^2}L_\mu \qquad (6\text{-}193)$$

Taking the expectation value in the state I_0, one obtains, by steps like those employed in the derivation of the sum rule (6-192),

$$\sum_{\alpha I}(E_{\alpha I}-E_0)^2 B(\tau=0,\lambda=2;\,I_0\rightarrow\alpha I)\big(I(I+1)-I_0(I_0+1)-6\big)$$

$$=\frac{150}{4\pi}\frac{\hbar^4}{M^2}I_0(I_0+1)\frac{\langle I_0\|L\|I_0\rangle}{\big(I_0(I_0+1)(2I_0+1)\big)^{1/2}} \qquad (6\text{-}194)$$

For the low-lying collective excitations in even-even nuclei, the total angular momentum is expected to be mainly associated with orbital motion,

$$\langle I_0 \| L \| I_0 \rangle \approx \langle I_0 \| I \| I_0 \rangle = \left(I_0(I_0+1)(2I_0+1)\right)^{1/2} \qquad (6\text{-}195)$$

in which case the last factor in Eq. (6-194) is close to unity.

For a vibrational spectrum, the sum rule (6-194), as in the case of the relation (6-192), expresses a constraint on the anharmonicities in the spectrum. (For a harmonic spectrum, the left-hand side of Eq. (6-194) vanishes, as follows from the arguments on p. 409.)

For a deformed nucleus, the transitions within the band yield the contribution $(225/\pi)(Q_0A/Z)^2(\hbar^2/2\mathscr{I})^2$ for $K=0$, $I_0=2$, assuming the leading-order intensity relations. In the case of ^{166}Er considered above, this contribution amounts to about 36% of the tensorial sum (6-194). The transitions to the γ-vibrational band yield a small negative contribution of about -2%; thus, as for the other energy-weighted sum rules, the main contribution must be associated with (as yet unobserved) high-frequency transitions. (The contributions to the sum (6-194) from excited bands vanish to the extent that the rotational energies and deviations from leading-order intensity rules can be neglected, but even small deviations become significant because of the strong energy weighting.)

The operators $\dot{\mathscr{M}}(\tau=0,\lambda=2,\mu)$ together with the angular momentum L_μ close under commutation and form the generators of the noncompact Lie group $SL(3,R)$ (Dothan *et al.*, 1965); among the representations of this group are sets of states forming a single rotational band, as for axially symmetric nuclei, as well as sets corresponding to the more extensive degrees of freedom of the asymmetric rotor (Weaver *et al.*, 1973). The assumption that the nuclear states carry representations of this group would imply that sum rules of the type (6-194) would be exhausted by transitions within a single representation; the observation that the main part of the sum rule is associated with transitions to higher-frequency states corresponds to the fact that the collective orientation angles, from which the actual bands can be generated, are more complicated functions of the nucleonic variables than the orientation of the mass tensor and its time derivative.

The energy-weighted sum rules (6-192) and (6-194) receive their main contributions from the high-frequency excitations, but sum rules that weight the low-energy transitions more strongly can be obtained by considering the multipole moments rather than their time derivatives. Thus, the commutability of the different tensor components μ, of a multipole moment $\mathscr{M}(\lambda\mu)$ depending only on the spatial coordinates of the particles, implies the relation (Belyaev and Zelevinsky, 1970; see Eq. (6-191))

$$\left(1-(-1)^{\lambda_0}\right)\sum_{\alpha I}\begin{Bmatrix}\lambda & \lambda & \lambda_0\\ I_1 & I_2 & I\end{Bmatrix}\langle I_2\|\mathscr{M}(\lambda)\|\alpha I\rangle\langle\alpha I\|\mathscr{M}(\lambda)\|I_1\rangle=0 \tag{6-196}$$

For a vibrational spectrum, the sum rule (6-196) implies constraints on the anharmonic effects, in addition to those considered above. In a deformed nucleus, the relations (6-196), for the transitions within a single rotational band, are equivalent to the validity of the leading-order intensity rules; thus, the assumption that strong multipole transitions only occur between states of a single band implies the rotational coupling scheme developed in Chapter 4 (Belyaev and Zelevinsky, 1970; see also Kerman and Klein, 1963).

An element of asymmetry in isobaric space results from the neutron excess. Sum rules with tensorial structure in isospace may thus be employed to characterize the asymmetry of the excitations induced by the different components μ_τ of isovector moments. An example is provided by the commutation relation for the moments (6-122),

$$\begin{aligned}&[\mathscr{M}(\tau=1,\mu_\tau=-1,\lambda),\mathscr{M}(\tau=1,\mu_\tau=+1,\lambda)]_{(\lambda\lambda)0}\\ &\qquad=\frac{(-1)^\lambda}{\pi}(2\lambda+1)^{1/2}\sum_k (t_z r^{2\lambda})_k\end{aligned} \tag{6-197}$$

which leads to the sum rule

$$\begin{aligned}&\sum_{\alpha I} B(\tau=1,\mu_\tau=-1,\lambda;\,0\to\alpha I)-\sum_{\alpha I} B(\tau=1,\mu_\tau=+1,\lambda;\,0\to\alpha I)\\ &\qquad=\frac{1}{2\pi}(2\lambda+1)\left(N\langle r^{2\lambda}\rangle_{\text{neut}}-Z\langle r^{2\lambda}\rangle_{\text{prot}}\right)\end{aligned} \tag{6-198}$$

The difference between the $\mu_\tau=+1$ and $\mu_\tau=-1$ transition strength becomes large in heavy nuclei; thus, for the dipole mode, the $\mu_\tau=+1$ strength almost vanishes for the β stable nuclei beyond ^{208}Pb.

6-4d Charge Exchange Contributions to $E\lambda$ Oscillator Sum

The dependence of the electric moments on the isobaric variables implies contributions to the $E\lambda$ sum rules from the charge exchange components of the nucleonic interactions (Levinger and Bethe, 1950; see also the review by Levinger, 1960). Thus, an isovector interaction

$$V_{\text{exch}}(1,2)=(\boldsymbol{\tau}_1\cdot\boldsymbol{\tau}_2)V(\mathbf{r}_1\boldsymbol{\sigma}_1,\mathbf{r}_2\boldsymbol{\sigma}_2) \tag{6-199}$$

gives a contribution to the oscillator sum $S(E\lambda)$ amounting to

$$\delta S(E\lambda)=e^2\frac{2\lambda+1}{4\pi}\langle 0|\sum_{pn}P_{pn}\left(r_p^{2\lambda}+r_n^{2\lambda}-2r_p^{\lambda}r_n^{\lambda}P_\lambda(\cos\vartheta_{pn})\right)V(\mathbf{r}_p\boldsymbol{\sigma}_p,\mathbf{r}_n\boldsymbol{\sigma}_n)|0\rangle \tag{6-200}$$

The sum in Eq. (6-200) is extended over all pairs involving a neutron and a proton, and P_{pn} is the operator exchanging space and spin variables of the neutron and proton ($P_{pn}=+1$ for a symmetric pair ($T=0$) and -1 for an antisymmetric pair ($T=1$)). The factor in the parentheses in Eq. (6-200) equals $(\mathbf{r}_p-\mathbf{r}_n)^2$ for $\lambda=1$ and always becomes zero for $\mathbf{r}_p=\mathbf{r}_n$, corresponding to the fact that the effect vanishes as the range of exchange of the charged mesons goes to zero. The physical basis for the additional oscillator strength (6-200) may be seen in the fact that the charge exchange interactions imply a transfer of charge without a motion of the nucleons themselves, and hence effectively modify the mass of the charge carriers.

The sum over *np* pairs in Eq. (6-200) involves a large degree of cancellation, as a result of the exchange operator P_{np}. For distances short compared to the Fermi wavelength, the relative orbital motion is symmetric, and the greater statistical weight of the $S=1$ as compared with the $S=0$ states implies a predominance of $T=0$ ($P_{np}=+1$) pairs. However, for distances large compared with the Fermi wavelength, the number of symmetric and antisymmetric *np* pairs becomes approximately equal (see, for example, the discussion of the two-body correlation function for a Fermi gas in Sec. 2-1h).

The isovector component in the average static nuclear potential can be represented by an effective two-body force (see Eq. (2-29))

$$V_{\text{exch}}=(\boldsymbol{\tau}_1\cdot\boldsymbol{\tau}_2)\frac{V_1}{4A} \tag{6-201}$$

For such an interaction of separable type, the exchange contribution (6-200) to the dipole sum can be rather easily evaluated in the independent-particle approximation for the state $|0\rangle$. Thus, for the monopole interaction (6-201), the terms r_p^2 and r_n^2 in Eq. (6-200) only contribute when the neutron and proton occupy the same orbit, while the term $2\mathbf{r}_p\cdot\mathbf{r}_n$ only contributes when the particles occupy different orbits. The cancellation between these two terms implies that δS can be expressed in terms of matrix elements between occupied and unoccupied orbits,

$$\delta S(E1) = \frac{3}{16\pi} e^2 \frac{V_1}{A} \sum_{\substack{\nu_1 \text{ occ} \\ \nu_2 \text{ unocc}}} |\langle \nu_2 | \mathbf{r} | \nu_1 \rangle|^2 \tag{6-202a}$$

$$= \frac{3}{16\pi} e^2 \frac{V_1}{A} \sum_a |\langle a | \sum_k (\mathbf{r}\tau_z)_k | 0 \rangle|^2 \tag{6-202b}$$

$$= \frac{V_1}{A} \sum_{\alpha I} B(E1; 0 \rightarrow \alpha I) \tag{6-202c}$$

where ν labels the single-particle orbits, including the quantum number m_t; it has been assumed that $N = Z$. Since the $E1$-oscillator strength is mainly concentrated in the energy region corresponding to the separation between major shells ($\Delta E \sim \hbar\omega_0 \sim 41 A^{-1/3}$ MeV), it is seen that the exchange contribution (6-202c) is of relative order $A^{-2/3}$ compared to the total dipole sum (6-176). This result (6-202c) could also have been immediately deduced from the fact that the simple interaction (6-201), when summed over all the particles, equals $\frac{1}{2} V_1 A^{-1} T(T+1)$. Such an interaction has no effect on the wave functions, and merely increases the energy of an isovector excitation ($T = 0 \rightarrow T = 1$) by the amount $V_1 A^{-1}$, corresponding to the result (6-202c).

The effect of the isovector dipole field, which plays an essential role in determining the collective features of the dipole mode (see pp. 480ff.), can be treated in a manner similar to the monopole interaction discussed above, and again yields only a small correction to the sum rule, of relative order $A^{-2/3}$. The same result applies to the charge exchange components of the velocity-dependent isovector dipole field considered on pp. 483ff. (These results may appear surprising, in view of the major effect of the isovector dipole field on the frequency of the dipole mode; in fact, however, this collective effect is produced by the $\mu_\tau = 0$ component of the dipole interaction, which commutes with the $E1$ moment.)

Larger exchange contributions to the oscillator sum rules can arise from the effect of a short-range nucleonic interaction of the type (6-199). A qualitative estimate of these contributions can be obtained by using the wave functions of the Fermi gas model (Levinger and Bethe, 1950). Assuming $N = Z$ and equal occupancy of spin up and spin down, the contribution to the dipole sum can be expressed in the form

$$\delta S(E1) = \frac{3e^2}{32\pi} A \rho_0 \int (V_\tau(r) + 3 V_{\sigma\tau}(r)) r^2 C^2(k_F r)\, d\tau \tag{6-203}$$

where $C^2(k_F r)$ is the two-body correlation function (see Sec. 2-1h); for a neutron and a proton, the two-particle density equals $\rho_n\rho_p(1+\pi C^2)$, where π is the parity in the relative motion, while $\rho_n=\rho_p=\frac{1}{2}\rho_0$. The *np* pair has $T=0$ and $T=1$ with equal weight, and the relative weight of $\pi=\pm 1$ is obtained from the statistical weight $(2S+1)$ of the spin channels. The central force exchange components V_τ and $V_{\sigma\tau}$ are related to the (S,T) parametrization of the two-body interaction by the expression (1-88); the noncentral forces do not contribute in the approximation considered. The contribution (6-203) vanishes as the range of the force goes to zero, and also becomes small for interactions of very long range, since $C^2(k_F r)$ goes to zero as $(k_F r)^{-4}$. This strong decrease of $C^2(k_F r)$ can be recognized as the cause of the smallness of the exchange contribution obtained from the field interaction (6-201) considered above.

For the two-body interaction given in Fig. 2-35 (Hamada-Johnston interaction), an evaluation of the expression (6-203), by integration from the hard-core radius c, yields $\delta S(E1)\approx+0.4S(E1)_{\text{class}}$. It need hardly be emphasized that such an estimate has only qualitative significance. Partly, the interactions strongly modify the relative motion of the nucleons at short distances; partly, contributions to $\delta S(E1)$ may arise from the tensor forces, which have no effect in the Fermi gas approximation. In addition, there remain uncertainties as regards the structure of the interactions, as well as of the $E1$ operator, for very short distances between the nucleons. Finally, it must be emphasized that the present analysis is based on a model that assumes a complete separation between nucleonic and mesonic degrees of freedom. It cannot be expected that such a model will be adequate to describe the oscillator strength for excitation energies comparable with those characteristic of the internal degrees of freedom of the nucleons.

The $E1$-sum rule, with the inclusion of the contributions from the charge-exchange and velocity-dependent interactions, refers to the total oscillator sum associated with all the modes of excitation arising from the nucleonic degrees of freedom. A separate question is the frequency distribution of the oscillator strength and, in particular, the problem of the strength associated with the giant $E1$ resonance in photoabsorption. The latter strength can be related to the velocity dependence of the effective interactions between particles in the orbits near the Fermi level (see the discussion on pp. 483ff.). However, the relation of these effective interactions to the basic nucleonic forces is not well understood at present.

Since the charge exchange contribution to the dipole sum rule can be attributed to the charged meson degrees of freedom, the question arises of the relationship between this contribution and the cross sections for photomeson processes. The dipole oscillator sum can be expressed in terms of the integrated photon absorption cross section (see Eq. (6-310)), and one may therefore attempt to establish a connection of the type in question in terms of a dispersion relation that connects the forward scattering amplitude at zero frequency to the total cross section integrated over all energies (Gell-Mann *et al.*, 1954). The total nuclear photoabsorption cross section integrated over all energies diverges, but it has been suggested that a convergent integral may be obtained by considering the difference $\sigma_A - Z\sigma_p - N\sigma_n$ between the nuclear cross section σ_A and that of the individual nucleons (σ_p and σ_n). However, the discovery of important coherence effects in the nuclear photoabsorption extending to very high energies (see the discussion on p. 479) implies contributions to the subtracted integral from the region up to 20 GeV that, in heavy nuclei, are more than an order of magnitude larger than the contribution corresponding to the classical dipole sum. Thus, the integral of the difference between nuclear and nucleonic cross sections is likely to diverge, and in any case fails to establish the relation between the charge-exchange contributions to the nuclear photoeffect and the photomeson processes.

6-5 PARTICLE-VIBRATION COUPLING

The variation in the average nuclear potential associated with the collective vibrations provides a coupling between the vibrational degrees of freedom and those of the individual particles. In the present section, we consider the various effects arising from this coupling, such as renormalization of the properties of the particles and of the vibrational quanta, and effective interactions between these elementary modes of excitation. The analysis also leads naturally to a self-consistent description of the vibrational motion in terms of the particle degrees of freedom and thus provides a generalization of the schematic treatment of this relationship given in Sec. 6-2c.

The particle-vibration coupling in the nucleus is analogous to the displacement potentials employed in the analysis of various condensed atomic systems (electron-phonon coupling in metals (see, for example, Pines, 1963), particle-phonon coupling in ^{3}He-^{4}He mixtures (Bardeen, Baym, and Pines, 1967)). For interacting fermion systems, such as the electron gas in metals, liquid ^{3}He, and nuclear matter, the elementary modes of excitation may be discussed in terms of the theory of Fermi liquids, as developed by Landau (see, for example, the text by Pines and Nozières, 1966). This formulation operates with a phenomenological effective interaction between the quasiparticles from which the coupling between the particles and the collective modes can be derived. The description of nuclear dynamics in terms of the concepts employed in the theory of Fermi liquids has been developed by Migdal (1967).

6-5a Coupling Matrix Elements

The leading-order particle-vibration coupling is linear in the vibrational amplitude α. For a mode with a real (Hermitian) amplitude, the coupling can be expressed in the form (see Eq. (6-18))

$$H' = \kappa\alpha F \tag{6-204}$$

The structure of the one-particle field F and the magnitude of the coupling constant κ are discussed in Sec. 6-3 for the different nuclear modes.

The coupling (6-204) produces a scattering of the particle with the emission or absorption of a quantum, and the matrix element for these processes is given by (in the representation corresponding to independent elementary excitations; see Eqs. (6-16) and (6-1))

$$\langle \nu_2, n+1 | H' | \nu_1, n \rangle = \kappa\alpha_0 (n+1)^{1/2} \langle \nu_2 | F | \nu_1 \rangle \tag{6-205}$$

where the particle states are denoted by ν_1 and ν_2, and where α_0 denotes the zero-point amplitude of the vibrational motion. (In Eq. (6-205), the matrix elements of α (and F) are assumed to be real; see the comment on p. 332.) The interaction (6-204) also gives rise to processes whereby a particle-hole pair is created or annihilated by the field interaction

$$\begin{aligned}\langle \nu_1^{-1}\nu_2, n=0 | H' | n=1 \rangle &= \langle \nu_1^{-1}\nu_2, n=1 | H' | n=0 \rangle \\ &= \kappa\alpha_0 \langle \nu_1^{-1}\nu_2 | F | 0 \rangle \\ &= \kappa\alpha_0 \langle \nu_2 | F | \bar{\nu}_1 \rangle \end{aligned} \tag{6-206}$$

In the last step, we have employed the relation (3B-19); see also the diagrams in Fig. 3B-1, Vol. I, p. 371. (The matrix elements (6-205) and (6-206) refer to fields with vanishing nucleon transfer number ($\alpha = 0$). For the pair fields, the corresponding matrix elements involve the creation or annihilation of two particles; see the coupling (6-148).)

In nuclei with many particles outside of closed shells, the effect of the pair correlations on the matrix elements of the field coupling can be taken into account by expressing the field F in terms of the quasiparticle variables (see Eq. (6-610))

$$\langle \mathrm{v}=1, \nu_2;\ n=1 | H' | \mathrm{v}=1, \nu_1;\ n=0 \rangle = \kappa\alpha_0 (u_1 u_2 + c v_1 v_2) \langle \nu_2 | F | \nu_1 \rangle \tag{6-207a}$$

$$\langle \mathrm{v}=2, \nu_1\nu_2;\ n=0 | H' | \mathrm{v}=0;\ n=1 \rangle = \kappa\alpha_0 (v_1 u_2 - c u_1 v_2) \langle \nu_2 | F | \bar{\nu}_1 \rangle \tag{6-207b}$$

where the phase factor c equals -1 for spin- and velocity-independent fields, while $c=+1$ for fields that depend linearly on the spin or velocity. The additional factor resulting from the pair correlations can be viewed as representing an interpolation between one-particle configurations ($u=1$, $v=0$) and one-hole configurations ($u=0, v=1$).

In a spherical nucleus, the coupling involves the scalar product of the tensors $\alpha_{\lambda\mu}$ and $F_{\lambda\mu}$ (see Eq. (6-68) for shape vibrations)[13]

$$\begin{aligned} H' &= -k_\lambda(r)\sum_\mu Y^*_{\lambda\mu}(\vartheta,\varphi)\alpha_{\lambda\mu} \\ &= (-1)^{\lambda+1}(2\lambda+1)^{1/2}k_\lambda(r)(Y_\lambda\alpha_\lambda)_0 \end{aligned} \tag{6-208}$$

where $r\vartheta\varphi$ are the polar coordinates of the particle. The matrix element of the coupling (6-208), for the scattering of a particle with the excitation of a quantum, is given by (see Eqs. (1A-72), (3A-14), and (6-51))

$$\begin{aligned} h(j_1,j_2\lambda) &\equiv \langle j_2, n_\lambda=1; I=j_1, M=m_1|H'|j_1m_1\rangle \\ &= (-1)^{j_1+j_2}(2j_1+1)^{-1/2}(2\lambda+1)^{-1/2}\langle j_2\|k_\lambda Y_\lambda\|j_1\rangle\langle n_\lambda=1\|\alpha_\lambda\|n_\lambda=0\rangle \\ &= -i^{l_1+\lambda-l_2}\left(\frac{2\lambda+1}{4\pi}\right)^{1/2}\langle j_1\tfrac{1}{2}\lambda 0|j_2\tfrac{1}{2}\rangle\left(\frac{\hbar\omega_\lambda}{2C_\lambda}\right)^{1/2}\langle j_2|k_\lambda(r)|j_1\rangle \end{aligned} \tag{6-209}$$

with the parity selection rule that requires $l_1+\lambda-l_2$ to be even. The matrix element (6-209) obeys the symmetry relation

$$h(j_1,j_2\lambda) = (-1)^{j_1+\lambda-j_2}\left(\frac{2j_2+1}{2j_1+1}\right)^{1/2} h(j_2,j_1\lambda) \tag{6-210}$$

For processes involving the creation of a particle-hole pair, we obtain the matrix elements (see Eq. (3B-25))

$$\begin{aligned} \langle (j_1^{-1}j_2)\lambda\mu|H'|n_{\lambda\mu}=1\rangle &= -\left(\frac{2j_1+1}{2\lambda+1}\right)^{1/2} h(j_1,j_2\lambda) \\ \langle (j_1^{-1}j_2)\lambda, n_\lambda=1;\ I=0|H'|0\rangle &= -(2j_1+1)^{1/2}h(j_1,j_2\lambda) \end{aligned} \tag{6-211}$$

[13] The effect of this coupling in admixing particle and vibrational degrees of freedom was considered by Foldy and Milford (1950) and Bohr (1952); see also the references quoted in footnote 6, p. 359.

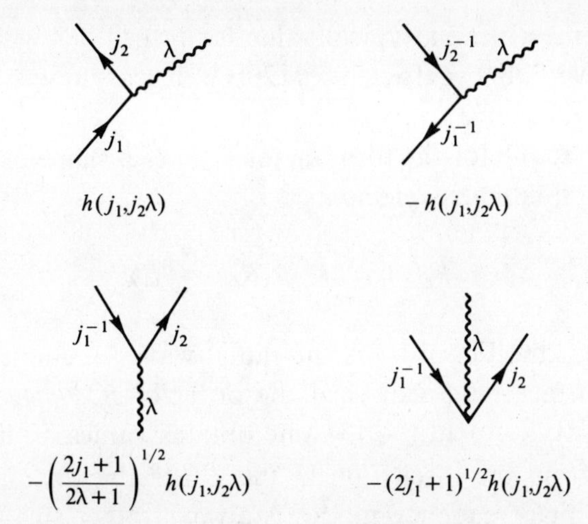

Figure 6-7 Diagrams illustrating first-order coupling between particle and vibration. The phase relations are those appropriate to the coupling to shape vibrations.

The enhancement of these matrix elements by a factor of $(2j_1+1)^{1/2}$, as compared with the matrix element (6-209), expresses the fact that each of the $(2j_1+1)$ particles in the filled shell can be excited by the interaction with the vibrational field. The basic first-order matrix elements (6-209) and (6-211) are illustrated by the diagrams in Fig. 6-7.

When j_1 and j_2 are larger than, or comparable with λ, the vector addition coefficient $\langle j_1 \frac{1}{2} \lambda 0 | j_2 \frac{1}{2} \rangle$ in the coupling matrix element (6-209) is governed by an approximate selection rule, which strongly inhibits the transitions involving spin flip ($j_1 = l_1 \pm 1/2 \rightarrow j_2 = l_2 \mp 1/2$), as compared with transitions that preserve the relative orientation of spin and orbit. This selection rule, which has a simple semiclassical interpretation, results from the assumption that the vibrational field acts only on the spatial coordinate of the particle; thus, the small spin-flip matrix elements may be sensitive to possible spin-dependent components in the vibrational field (see Sec. 6-3e).

A dimensionless parameter that is often useful in characterizing the strength of the particle-vibration coupling is obtained by dividing a standard coupling matrix element by $\hbar\omega_\lambda$. Thus, for a shape vibration, we may employ the parameter (see Eq. (6-209))

$$f_\lambda = \left(\frac{2\lambda+1}{16\pi}\right)^{1/2} \left(\frac{\hbar\omega_\lambda}{2C_\lambda}\right)^{1/2} \frac{\langle k_\lambda \rangle}{\hbar\omega_\lambda} \tag{6-212}$$

In this definition of f_λ, we have taken the vector addition coefficient to have the value 1/2, representing a typical value for transitions without spin flip. (For the case of $\lambda=2$, the coefficient $\langle j\frac{1}{2}20|j\frac{1}{2}\rangle$ has asymptotically, for large j, the numerical value 1/2.)

The estimate (6-69) for the form factor $k_\lambda(r)$ of a shape oscillation leads to the radial coupling matrix elements

$$\langle j_2|k_\lambda(r)|j_1\rangle \approx \langle j_2|R_0\frac{\partial V}{\partial r}|j_1\rangle \tag{6-213}$$

For the states near the Fermi level, the radial wave functions in the region of the nuclear surface have values of the order of $R^{-3/2}$ (see Eq. (3-22)). Thus, for the matrix element (6-213), one obtains values of about 50 MeV.

As follows from the discussion in the illustrative examples, the parameter f_λ for the high-frequency modes is always rather small compared to unity ($f_\lambda \sim A^{-1/3}$, since C_λ is proportional to A and ω_λ to $A^{-1/3}$ (see, for example, Eqs. (6-367) and (6-369) and note that the amplitudes used in Eqs. (6-208) and (6-365) differ by a factor of the order of AR^2, which is proportional to $A^{5/3}$)). For the low-frequency shape oscillations, the values of f_3 range mostly in the interval 0.1–0.5, while f_2 is strongly dependent on the shell structure and may become larger than unity. In situations where $f_\lambda \ll 1$, one can treat the coupling by a perturbation expansion, but the large values encountered for the low-energy quadrupole mode imply a major interweaving of particle and vibrational degrees of freedom. For $f_2 \gg 1$, the particle produces a static shape deformation that is large compared with zero-point fluctuations, and the coupled system can be treated by a separation between rotational and intrinsic degrees of freedom. (For the low-frequency quadrupole mode, the empirical frequencies and restoring force parameters are summarized in Fig. 2-17, Vol. I, p. 196, and Fig. 6-28, p. 529, respectively. Examples of the zero-point amplitudes of octupole vibrations are given in Table 6-14, p. 561.)

6-5b Effective Moments

Because of the large transition moments associated with the vibrational excitations, the particle-vibration coupling gives rise to important modifications in the effective one-particle moments. As a result of the coupling, the single-particle states are clothed in a cloud of quanta; to first order in the coupling, the clothed (or renormalized) state $\hat{\nu}_1$ is given by

$$|\hat{\nu}_1\rangle \approx |\nu_1\rangle - \sum_{\nu_2}\frac{\langle \nu_2, n=1|H'|\nu_1\rangle}{\Delta E_{21}+\hbar\omega}|\nu_2, n=1\rangle \tag{6-214}$$

where ΔE_{21} is the single-particle excitation energy (which is equal to $\varepsilon(\nu_2)-\varepsilon(\nu_1)$ in the absence of pair correlations). If we consider the matrix elements of the field operator F between the clothed single-particle states, the inclusion of the vibrational moment gives a simple renormalization of the transition moment

$$\langle \nu_2|\hat{F}|\nu_1\rangle \equiv \langle \hat{\nu}_2|F|\hat{\nu}_1\rangle = (1+\chi_F)\langle \nu_2|F|\nu_1\rangle \tag{6-215}$$

with the coefficient χ_F given by (see Eq. (6-28))

$$\begin{aligned}\chi_F &= -2\kappa\alpha_0^2 \frac{\hbar\omega}{(\hbar\omega)^2-(\Delta E_{21})^2}\\ &= -\frac{\kappa}{C}\frac{(\hbar\omega)^2}{(\hbar\omega)^2-(\Delta E_{21})^2}\end{aligned} \tag{6-216}$$

The same renormalization factor applies to matrix elements of F for the creation of a particle-hole pair, as well as to matrix elements involving quasiparticles. The renormalization of single-particle moments is illustrated by the diagrams in Fig. 6-8.

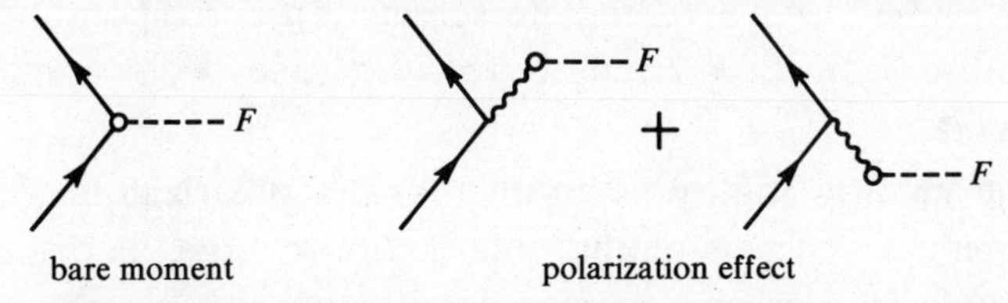

Figure 6-8 Renormalization of single-particle moment resulting from particle-vibration coupling.

The ratio χ_F between the induced moment and that of the single particle is referred to as the polarizability coefficient. The simple form (6-216) applies to moments that are proportional to the field coupling. More generally, the polarizability coefficient involves the ratio between the one-particle matrix elements of the field coupling and of the moment and may thus depend more explicitly on the one-particle states involved (see, for example, Eq. (6-218)).

For low-frequency transitions ($|\Delta E| \ll \hbar\omega$), the polarizability (6-216) approaches the static limit, in which the nuclear deformation α adjusts to the instantaneous value of the moment F_p contributed by the particle p,

$$\alpha = -\frac{\kappa}{C}F_p \tag{6-217}$$

corresponding to the minimum for the potential energy $\frac{1}{2}C\alpha^2 + \kappa\alpha F_p$. The magnitude of χ_F in the static limit is of the order of the dimensionless coupling constant f multiplied by the ratio of collective and single-particle transition moments of F, and thus may be comparable with or greater than unity, even for small values of f. (For the high-frequency shape oscillations, the values of f are of order $A^{-1/3}$, while the collective moment is of order $A^{1/3}$, in single-particle units; hence, χ is of order unity.)

The sign of $\chi_F(\Delta E=0)$ is opposite to that of κ, since the static polarization effect produced by an attractive coupling ($\kappa<0$) is in phase with the single-particle moment, while a repulsive coupling ($\kappa>0$) implies opposite phases for the polarization effect and the one-particle moment (screening of the one-particle moment). As in the familiar classical motion of an externally driven oscillator, the induced moment changes sign with respect to the driving force for frequencies such that $|\Delta E|>\hbar\omega$. (For the model considered in Sec. 6-2c, in which the collective motion is formed by the coherence of degenerate particle excitations with unperturbed frequencies $\omega^{(0)}$, it is seen from Eq. (6-216) together with Eqs. (6-26) and (6-27a) that $\chi_F(\Delta E_{21}=\hbar\omega^{(0)})=-1$ (complete screening), corresponding to the fact that the entire transition strength for the field F is contained in the collective mode.)

For the electric multipole moments, the polarizability coefficient is usually expressed in terms of the polarization charge, as discussed in Sec. 3-3a. The polarization charge associated with the coupling to a shape vibration ($\tau\approx 0$) involves the ratio of the matrix elements of the coupling $-k_\lambda(r)Y_{\lambda\mu}\alpha^\dagger_{\lambda\mu}$ to that of the multipole moment $r^\lambda Y_{\lambda\mu}$, and by the same procedure as employed in the derivation of Eq. (6-216), we obtain (see Eq. (6-63) for the normalization of the amplitude for shape vibrations)

$$(e_{\text{pol}})_{\lambda,\tau\approx 0} = \frac{\langle j_2|k_\lambda(r)|j_1\rangle}{\langle j_2|r^\lambda|j_1\rangle}\frac{3}{4\pi}\frac{ZeR^\lambda}{C_\lambda}\frac{(\hbar\omega_\lambda)^2}{(\hbar\omega_\lambda)^2-(\Delta E_{21})^2}$$

$$= \frac{8\pi}{3(2\lambda+1)}\frac{B(E\lambda;\,0\rightarrow\lambda)}{ZeR^\lambda\hbar\omega_\lambda}\frac{\langle j_2|k_\lambda(r)|j_1\rangle}{\langle j_2|r^\lambda|j_1\rangle}\frac{(\hbar\omega_\lambda)^2}{(\hbar\omega_\lambda)^2-(\Delta E_{21})^2} \tag{6-218}$$

In the second form of Eq. (6-218), the restoring force parameter C_λ has been expressed, by means of Eq. (6-65), in terms of the vibrational frequency and transition probability. (For nuclei with $N > Z$, the isovector component in $k_\lambda(r)$ implies that the predominantly isoscalar polarization charge (6-218) contains a small isovector part; see the discussion on pp. 513 ff.)

The isovector modes contribute a $\tau = 1$ component to the effective charge. For the coupling given by Eqs. (6-124) and (6-121) involving the radial form factor $f_\lambda(r)$, we obtain

$$(e_{\text{pol}})_{\lambda,\tau=1} = t_z \frac{eV_1}{4\rho_0 C_\lambda} \frac{\langle j_2 | f_\lambda(r) | j_1 \rangle}{\langle j_2 | r^\lambda | j_1 \rangle} \frac{(\hbar\omega_\lambda)^2}{(\hbar\omega_\lambda)^2 - (\Delta E_{21})^2} \int f_\lambda(r) r^{\lambda+2} dr \quad (6\text{-}219)$$

using the expressions (6-122) and (6-123) for the electric multipole moment.

For electric dipole transitions, the main polarization effect is associated with the high-frequency isovector mode. The repulsive coupling in this mode leads to an estimated reduction in the low-frequency $E1$ one-particle transition moments by a factor of about 0.3, which appears to be compatible with the tentative empirical evidence. (See the discussion on pp. 486 ff.; corresponding evidence on the charge exchange dipole moment is considered on pp. 501 ff.)

For $E2$ transitions, the isoscalar polarization effect is strongly shell-structure dependent. For nuclei with doubly closed shells, only the high-frequency mode contributes, and the polarization charges are of order unity; see Table 6-9, p. 517. For nuclei with particles in unfilled shells, the large-amplitude low-frequency shape oscillations give the main contribution, and the expected and observed values of the effective charge range from about 5 to 20 (see Table 6-10, p. 533, and Fig. 6-28, p. 529). The very large values of e_{eff} reflect the near instability of the spherical shape (see the discussion on pp. 520 ff.). The isovector $E2$-polarization charges are expected to result mainly from the high-frequency collective mode, which is estimated to reduce the isovector charge by more than a factor of 2 (see Eq. (6-382)). The tentative evidence is shown in Table 6-9.

For $E3$ moments, the main polarization effect is expected to arise from the rather low-frequency shape vibration. Estimates of this contribution in the region of ^{208}Pb give values for e_{pol} of order $5e$, which is in agreement with the available data; see Table 6-15, p. 565. (For an estimate of the polarization charge in $E4$ transitions, see Hamamoto, 1972.)

The general expression (6-216) for the polarizability coefficient can also be applied to spin-dependent fields that are involved in the renormalization of the $M1$ moments (see Chapter 3, Vol. I, pp. 337 ff., and Chapter 5, pp. 304ff.). The relation between $(g_s)_{\text{eff}}$ and the $\lambda\pi=1+$, $\tau=1$, $\mu_\tau=0$ mode is discussed in the example on pp. 636ff.

The polarizability coefficients discussed above characterize the response of the nucleus to the force exerted by the additional particle; the same coefficients can be used to describe the response of the nucleus to an external field proportional to F (see pp. 436ff.). The polarizability coefficients can thus also be determined from scattering by charged particles and from energy shifts in exotic atoms. The interaction potential resulting from the static polarizability is given by Eq. (6-292), which contains a scalar term as well as a tensor component that gives rise to inelastic scattering processes.

6-5c One-Particle Transfer Matrix Elements

The one-particle transfer processes provide a method for directly measuring the small admixed amplitudes produced by the particle-vibration coupling. Thus, the admixing of the particle state j_1 and the one-phonon state $(j_2, n_\lambda=1)$ with $I=j_1$ implies that the latter state can be populated in a one-particle transfer starting from the closed-shell configuration (see Eqs. (6-209) and (6-214)),

$$\langle j_2, n_\lambda=1;\ I=j_1, M=m_1|\hat{a}^\dagger(j_1m_1)|0\rangle = \frac{h(j_1 j_2 \lambda)}{\varepsilon(j_2)-\varepsilon(j_1)+\hbar\omega_\lambda} \tag{6-220}$$

The notation $\hat{a}^\dagger(j_1m_1)$ indicates that we are dealing with the renormalized operator whose matrix elements between unperturbed states include the effects of the particle-vibration coupling (see Eq. (6-215)). An example of a transfer process involving the coupling to the octupole mode in ^{209}Bi is discussed on p. 571.

For the levels j_1 below the Fermi surface $(\varepsilon(j_1)<\varepsilon_F)$, a similar effect occurs as a result of the matrix element (6-211), which admixes the configuration $(j_1^{-1}j_2)\lambda, n_\lambda=1;\ I=0$ into the ground state

$$|\hat{0}\rangle \approx |0\rangle + \sum_{j_1 j_2} \frac{(2j_1+1)^{1/2} h(j_1 j_2 \lambda)}{\varepsilon(j_2)-\varepsilon(j_1)+\hbar\omega_\lambda} |(j_1^{-1}j_2)\lambda, n_\lambda=1;\ I=0\rangle \tag{6-221}$$

where $\varepsilon(j_2)-\varepsilon(j_1)$ is the energy associated with the creation of the particle-hole configuration $(j_1^{-1}j_2)$. The resulting contribution to the transfer matrix element is again given by Eq. (6-220). (In the derivation of this result, one may use the coupling relation (1A-22) together with the particle-hole transformation (3-7).) Thus, we could also have obtained the result by neglecting the presence of the particles in the closed shells. The diagrams in Fig. 6-9 illustrate the intimate relation between the two contributions to the transfer matrix elements for $\varepsilon(j_1) > \varepsilon_F$ and $\varepsilon(j_1) < \varepsilon_F$.

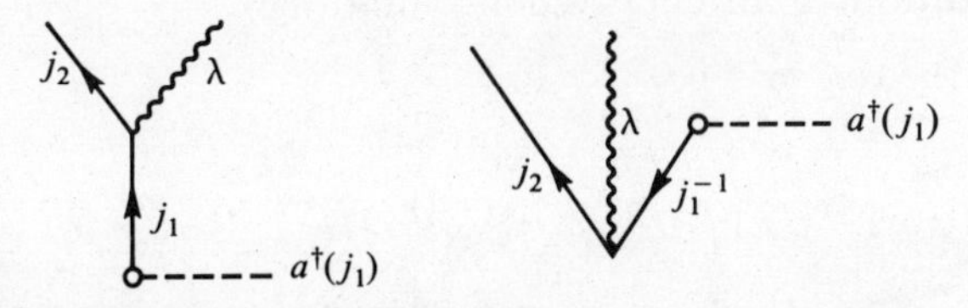

Figure 6-9 Diagrams for particle transfer with excitation of a phonon.

For pair-correlated systems, the transfer matrix element associated with a one-quasiparticle state j_1 receives contributions from both diagrams in Fig. 6-9; the contribution of the first diagram is multiplied by the factor $u(j_1)$ (see Eq. (6-599)), in addition to the factor $u(j_1)u(j_2)+cv(j_1)v(j_2)$ associated with the scattering vertex (see Eq. (6-207a)), while the contribution of the second diagram has the factor $v(j_1)$, in addition to the factor $v(j_1)u(j_2)-cu(j_1)v(j_2)$ for the pair creation vertex (see Eq. (6-207b)). Moreover, the energy denominators are modified by the replacement of single-particle energies by quasiparticle energies.

One-particle transfer starting from an odd-A nucleus ($\mathsf{v}=1, n=0$) and leading to a vibrational excitation ($\mathsf{v}=0, n=1$) is described by diagrams similar to those in Fig. 6-9 and provides a measure of the contribution of the individual-particle configurations in the vibrational motion (see the example discussed on pp. 536 ff.).

6-5d Particle-Phonon Interaction Energy

In second order, the particle-vibration coupling generates an effective interaction between a particle and a phonon, which lifts the degeneracy of the multiplet $(j_1\lambda)I$ formed by the superposition of a particle and a vibrational quantum. There are four terms in the particle-phonon interaction, corresponding to the different types of intermediate states (see the diagrams in Figs. 6-10a–d)

$$V = V^{(a)} + V^{(b)} + V^{(c)} + V^{(d)} \tag{6-222}$$

To exhibit the structure of the different terms in Eq. (6-222), we first ignore the angular momentum coupling and label the particle (and hole) states by 1 $(=j_1 m_1)$, 2, etc., and the vibrational quanta by γ $(=\lambda\mu)$, γ', etc. The four second-order interaction terms leading from an initial state 1γ to a final state $1'\gamma'$ thus involve the following products of the first-order matrix elements (in addition to an energy denominator):

$$\langle 1'\gamma'|H'|2\rangle\langle 2|H'|1\gamma\rangle \tag{6-223a}$$

$$\begin{aligned}\langle 1'\gamma'|H'|2^{-1}1'\gamma'1\gamma\rangle\langle 2^{-1}1'\gamma'1\gamma|H'|1\gamma\rangle &= -\langle 1'\gamma'|H'|2^{-1}1\gamma 1'\gamma'\rangle\langle 2^{-1}1'\gamma'1\gamma|H'|1\gamma\rangle\\ &= -\langle 0|H'|2^{-1}1\gamma\rangle\langle 2^{-1}1'\gamma'|H'|0\rangle\\ &= -\langle 1'\gamma'|H'|\bar{2}\rangle\langle\bar{2}|H'|1\gamma\rangle\end{aligned} \tag{6-223b}$$

$$\begin{aligned}\langle 1'\gamma'|H'|2\gamma'\gamma\rangle\langle 2\gamma'\gamma|H'|1\gamma\rangle &= \langle 1'\gamma'|H'|2\gamma\gamma'\rangle\langle 2\gamma'\gamma|H'|1\gamma\rangle\\ &= \langle 1'|H'|2\gamma\rangle\langle 2\gamma'|H'|1\rangle\end{aligned} \tag{6-223c}$$

$$\begin{aligned}\langle 1'\gamma'|H'|12^{-1}1'\rangle\langle 12^{-1}1'|H'|1\gamma\rangle &= -\langle 1'\gamma'|H'|1'2^{-1}1\rangle\langle 12^{-1}1'|H'|1\gamma\rangle\\ &= -\langle\gamma'|H'|2^{-1}1\rangle\langle 2^{-1}1'|H'|\gamma\rangle\\ &= -\langle 1'|H'|\bar{2}\gamma\rangle\langle\bar{2}\gamma'|H'|1\rangle\end{aligned} \tag{6-223d}$$

In the intermediate states, we have commuted particles (and holes) and quanta, taking into account that the states are symmetric with respect to interchange of bosons and antisymmetric in the fermions. Moreover, the evaluation of the matrix elements in Eqs. (6-223b) and (6-223d) involves the crossing relation, by which a hole, 2^{-1}, in the final (or initial) state can be replaced by a particle state, $\bar{2}$, in the initial (or final) state (see Fig. 3B-1, Vol. I, p. 371). The final form of the matrix elements in Eqs. (6-223b) and (6-223d) exhibits the intimate relation to the matrix elements in Eqs. (6-223a) and (6-223c), respectively, reflecting the topology of the diagrams in Fig. 6-10.

If the particles or the quanta in the final and initial states are the same $(1=1'$ or $\gamma=\gamma')$, some of the intermediate states involve two identical particles or quanta. In such a situation, the use of states with the appropriate permutation symmetry would modify the evaluation of the matrix elements in Eq. (6-223). For example, in Eq. (6-223c), the matrix elements to a symmetrized intermediate state with two identical quanta $(\gamma'=\gamma)$ have an additional factor $\sqrt{2}$ (see Eq. (6-1)), and the second-order energy contribution is therefore multiplied by a factor of 2 (boson factor).

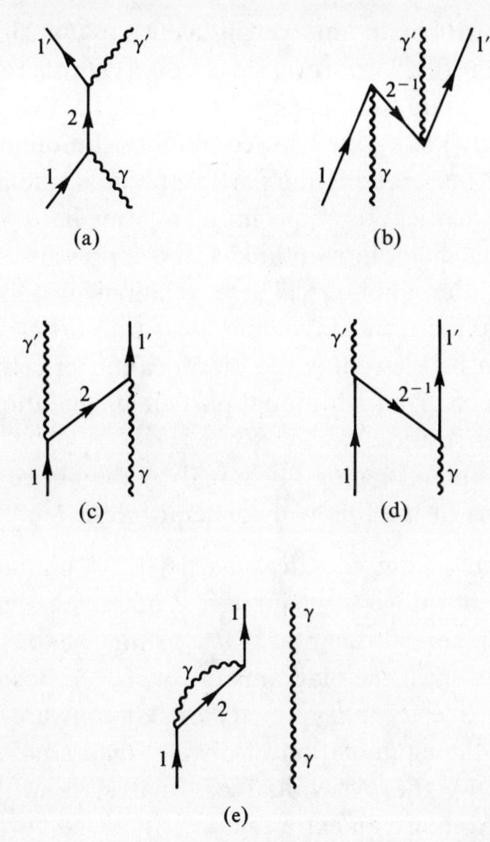

Figure 6-10 Diagrams for particle-phonon interaction. The unlinked diagram (e) represents a self-energy term for the particle, which is not to be included in the particle-phonon interaction.

In terms of the diagrams, this factor of 2 is expressed by the occurrence of two different diagrams 6-10c and 6-10e that involve the same intermediate state and differ only in the way, in which the vertices are attached to the identical phonons in the intermediate state. These two diagrams each contribute equally to the total energy of the particle-phonon state. However, diagram 6-10e, which involves unconnected parts—and is referred to as an unlinked diagram—represents an interaction effect that is already present for a configuration of a single particle (self-energy term; see Sec. 6-5e). This term is not to be included in the particle-phonon interaction energy, which represents the difference between the total energy of the particle-phonon state and the separate energies of the single-particle and single-phonon states. Hence, the contribution to the particle-phonon interaction is correctly obtained by ignoring the unlinked diagram 6-10e and counting the diagram 6-10c with unit weight, as in Eq. (6-223c). In a similar way, the matrix element (6-223d), for $1 = 1'$, would vanish for an antisymmetrized intermediate state (exclusion principle). In the diagrammatic bookkeeping, this result is obtained by combining the opposite contributions of diagram 6-10d representing the particle-phonon interaction and the diagram, in which the identical fermions in the intermediate state have been interchanged. The latter diagram is unlinked and represents a contribution to the self-energy of the phonon. (These relationships imply that the particle-phonon interaction terms involving identical particles or quanta in intermediate states can be viewed as the modification of self-energy terms arising from the presence of an additional particle or quantum through the requirements of symmetrization.)

The diagrammatic bookkeeping illustrated by the above examples can be summarized in terms of the following general rules.

(i) In diagrams involving identical particles or phonons, it is not necessary to explicitly introduce the symmetry requirements in intermediate states, provided the identical quanta play distinguishable roles in the diagram (as is the case for all the diagrams in Fig. 6-10). Boson factors and the exclusion of states due to antisymmetry of fermions are automatically taken into account by the enumeration of distinct diagrams, each of which is counted with unit weight. An initial or final state with n identical phonons does however require an extra factor $(n!)^{1/2}$ (see, for example, Fig. 6-12). Diagrams involving identical quanta that play indistinguishable roles also require additional weight factors that express the symmetry; see the factor $n!$ for the n indistinguishable bosons in Fig. 6-36c and the factor $1/2$ for the indistinguishable fermions in Fig. 6-61a.

(ii) The linked diagrams directly give the contributions to a given matrix element associated with the interaction of the elementary excitations, as distinct from the effects that represent contributions from the constituent parts of the states considered. The latter effects appear as unlinked diagrams.

(iii) In evaluating higher-order terms, one must include diagrams in which the initial or final state reappears as an intermediate state that would give rise to a zero-energy denominator. Such diagrams represent the renormalization of the wave functions of initial or final states and are evaluated by the standard procedure of quantal perturbation theory. (An example is illustrated by the last diagram in Fig. 6-37.)

These rules for a systematic evaluation of the perturbation expansion of field-coupling effects are well known from the field theory of elementary particles and other many-body systems (see, for example, Wick, 1955). As illustrated by the examples considered in the present section, the formalism based on the particle-vibration coupling appears to include in a consistent way the effects resulting from the identity of the nucleons appearing in the particle degrees of freedom and in the collective modes. (For a more systematic discussion of this point based on the analysis of special models, see Bès *et al.*, 1974).

For angular momentum-coupled states $(j_1\lambda)I$, it is seen that the matrix elements illustrated by Figs. 6-10a and 6-10b are proportional to $\delta(j_2,I)$. The matrix elements shown in Figs. 6-10c and 6-10d can be evaluated by performing a recoupling in the intermediate states. Thus, for the energy splitting within the multiplet, we obtain (see Eq. (6-209))

$$\langle j_1,n_\lambda=1;\,IM|V^{(a)}+V^{(b)}|j_1,n_\lambda=1;\,IM\rangle$$

$$=\sum_{j_2}\frac{h^2(j_2,j_1\lambda)}{\varepsilon(j_1)-\varepsilon(j_2)+\hbar\omega_\lambda}\delta(j_2,I) \tag{6-224a}$$

$$\langle j_1,n_\lambda=1;\,IM|V^{(c)}+V^{(d)}|j_1,n_\lambda=1;\,IM\rangle$$

$$=\sum_{j_2}\frac{h^2(j_1,j_2\lambda)}{\varepsilon(j_1)-\varepsilon(j_2)-\hbar\omega_\lambda}\langle (j_2\lambda')j_1,\lambda;\,I|(j_2\lambda)j_1,\lambda';\,I\rangle$$

$$=\sum_{j_2}\frac{h^2(j_1,j_2\lambda)}{\varepsilon(j_2)-\varepsilon(j_1)+\hbar\omega_\lambda}(2j_1+1)\begin{Bmatrix}\lambda & j_1 & j_2\\ \lambda & j_1 & I\end{Bmatrix} \tag{6-224b}$$

In the sum over j_2, the states above the Fermi surface contribute to $V^{(a)}$ and $V^{(c)}$, while the states below the Fermi surface contribute to $V^{(b)}$ and $V^{(d)}$. It is seen that the contribution from the states below the Fermi surface is the same as it would have been in the absence of the particles in the closed shells. The interaction is therefore the same as for a phonon coupled to a single particle moving in the nuclear potential. (This relation does not apply in the presence of pair correlations; for a quasiparticle, the interaction with a phonon receives contributions from all four diagrams, for each single-particle orbit j_2.)

The octupole excitations in nuclei adjacent to ^{208}Pb provide a favorable opportunity for studying the interaction between particles and vibrational quanta (see the discussion on pp. 570ff.). The coupling to the low-frequency quadrupole oscillations is usually rather strong, and only limited progress has been made in the quantitative analysis of the spectra of the odd-*A* nuclei. The particle-vibration interaction, however, provides a qualitative interpretation of some of the features of these spectra (see, in particular, the discussion of the observed low-energy states with $I=j-1$, where j is the angular momentum of the one-particle orbit, pp. 540ff.).

6-5e Self-Energies

The particle-vibration coupling acting in second order also gives rise to energy shifts in states involving a single particle (or hole) or a single quantum, as well as a contribution to the energy of the closed-shell configuration (the "vacuum state"). The corresponding diagrams are shown in Fig. 6-11.

The energy shift of the closed-shell configuration associated with the virtual excitation of a particle-hole pair $(j_1^{-1}j_2)$ and a quantum λ (Fig. 6-11a) can be obtained from the matrix element (6-211),

$$\delta \mathscr{E}_0 = -\frac{(2j_1+1)h^2(j_1,j_2\lambda)}{\varepsilon(j_2)-\varepsilon(j_1)+\hbar\omega_\lambda} \tag{6-225}$$

The self-energy of a single particle receives contributions corresponding to the two diagrams in Fig. 6-11b, associated with the coupling to orbits j_2 above and below the Fermi level,

$$\delta\varepsilon(j_1) = \begin{cases} \dfrac{h^2(j_1,j_2\lambda)}{\varepsilon(j_1)-\varepsilon(j_2)-\hbar\omega_\lambda} & \varepsilon(j_2)>\varepsilon_F \\ -\dfrac{h^2(j_1,j_2\lambda)}{\varepsilon(j_2)-\varepsilon(j_1)-\hbar\omega_\lambda} & \varepsilon(j_2)<\varepsilon_F \end{cases} \tag{6-226}$$

In the second term, the minus sign is associated with the interchange of the particles in the intermediate state, as in Eqs. (6-223b) and (6-223d). The energy shifts (6-226) are of the order of magnitude of $f_\lambda^2\hbar\omega_\lambda$, in terms of the dimensionless coupling parameter f_λ given by Eq. (6-212). Thus, for the coupling to the high-frequency modes, $\delta\varepsilon \sim \varepsilon_F A^{-1}$, but larger shifts may arise from the coupling to the low-frequency quadrupole and octupole modes (see, for example, the discussion on pp. 564ff.). The one-particle self-energy decreases the energy of the lowest particle states as well as the energy of the lowest hole states and thus acts to reduce the gap between occupied and unoccupied orbits (Hamamoto and Siemens, 1974); this effect may play a role in reconciling the observed level spacing near the Fermi surface, which is consistent with a velocity-independent potential (see, for

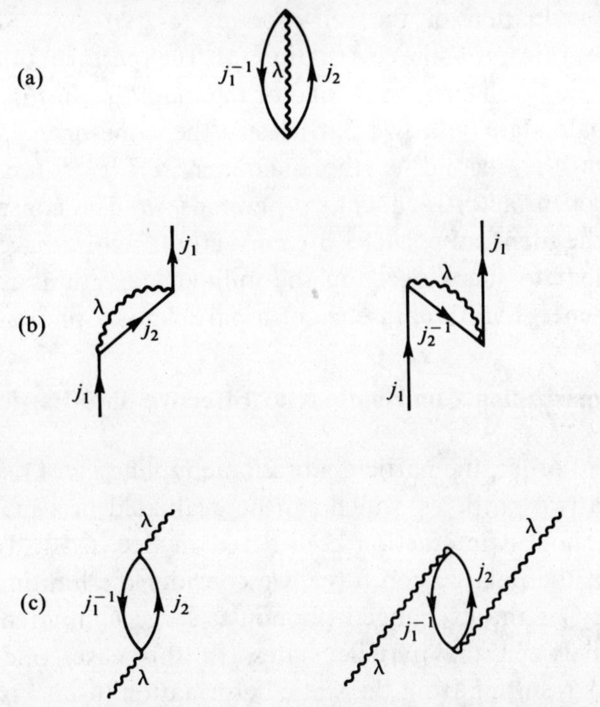

Figure 6-11 Self-energy terms. Diagram (a) represents the vacuum self-energy, while (b) and (c) represent the self-energy term for a particle and a phonon, respectively.

example, Fig. 3-3, Vol. I, p. 325 and Chapter 5, p. 265) with the evidence for an appreciable velocity dependence, as indicated by the optical potential (see, for example, Fig. 2-29, Vol. I, p. 237) and the energies of the deeply bound hole states (see, for example, Fig. 3-5, Vol. I, p. 328).

The phonon self-energy is represented by the diagrams in Fig. 6-11c,

$$\begin{aligned}\delta\hbar\omega_\lambda &= -\frac{2j_1+1}{2\lambda+1}h^2(j_1,j_2\lambda)\left(\frac{1}{\varepsilon(j_2)-\varepsilon(j_1)-\hbar\omega_\lambda}+\frac{1}{\varepsilon(j_2)-\varepsilon(j_1)+\hbar\omega_\lambda}\right)\\ &= -\frac{2j_1+1}{2\lambda+1}h^2(j_1,j_2\lambda)\frac{2(\varepsilon(j_2)-\varepsilon(j_1))}{(\varepsilon(j_2)-\varepsilon(j_1))^2-(\hbar\omega_\lambda)^2}\end{aligned} \tag{6-227}$$

The energy shift (6-227) represents the change in the phonon frequency arising from the coupling to a particular particle-hole configuration $(j_1^{-1}j_2)$. By including the coupling to all particle and hole degrees of freedom, one can obtain a self-consistency condition for the frequency of the vibrational motion (see Sec. 6-5h).

In the evaluation of the self-energies (6-226) and (6-227), we have followed the rules for the evaluation of the perturbation diagrams as discussed on p. 428. Thus, the second of the diagrams in Fig. 6-11b involves an intermediate state with two particles in the same orbit j_1m_1, and, in the evaluation of the second of the diagrams in Fig. 6-11c, we have not included a boson factor for identical phonons $\lambda\mu$. The consequences of the statistics of the identical particles are correctly taken into account, since the same intermediate states occur in the unlinked diagrams representing the vacuum self-energy in the presence of a particle or a phonon.

6-5f Polarization Contributions to Effective Two-Particle Interactions

In second order, the particle-vibration coupling gives rise to an interaction between two particles, which can be evaluated in a manner similar to the particle-phonon interaction considered in Sec. 6-5d. To illustrate the magnitude of the polarization force, we consider the limiting case in which the frequency of the exchanged phonon is large compared to the energy differences between the particle states. In this case, one can view the interaction as resulting from the static deformation (6-217) produced by the first particle acting on the second. Thus, for a mode of multipolarity λ, one obtains (see Eq. (6-68))

$$V_\lambda(1,2) = -\frac{2\lambda+1}{4\pi C_\lambda} k_\lambda(r_1) k_\lambda(r_2) P_\lambda(\cos\vartheta_{12}) \qquad (6\text{-}228)$$

The order of magnitude of the polarization interaction (6-228) is given by $f_\lambda^2 \hbar\omega_\lambda$ (as for the particle self-energy). For the high-frequency modes, we have $f_\lambda^2 \hbar\omega_\lambda \sim \varepsilon_F A^{-1}$, which is comparable to the average interaction between nucleons in the nucleus ($\sim V_0 A^{-1}$). The magnitude of the polarization interaction can be seen directly from the fact that the deformation of the closed shells produced by a single particle implies polarization moments comparable with the bare moments of the particles (see p. 510); hence, the corresponding contribution to the polarization interaction (6-228) is similar in magnitude to the direct force.

The polarization interaction resulting from the coupling to the low-frequency modes may be considerably larger than the bare force; since the frequencies of these modes may be comparable with the particle frequencies, it may be necessary to go beyond the static approximation (6-228), as in the evaluation of the particle-phonon interaction.

6-5g Higher-Order Effects

The first- and second-order terms considered in Secs. 6-5b to 6-5f represent leading-order effects of the particle-vibration coupling. The inclusion of higher-order terms partly gives corrections to these effects and partly leads to a variety of additional interactions, such as anharmonicity of the vibrational motion and coupling between phonons belonging to different modes.[14]

As an example of higher-order effects, we consider the coupling between states with one and two phonons. This interaction is of third order in the particle-vibration coupling and is illustrated by the diagrams in Fig. 6-12 involving a closed triangular loop of particle lines. Additional diagrams contributing to this process are obtained by interchanging particles and holes (reversing the direction of the arrow in the closed loop) and by interchanging the two outgoing phonons.

The contribution of the first diagram in Fig. 6-12 to the amplitude of the two-phonon state admixed into the one-phonon state is given by the standard third-order perturbation result

$$\begin{aligned} c^{(1)} &= \frac{\langle f|H'|a_2\rangle\langle a_2|H'|a_1\rangle\langle a_1|H'|i\rangle}{(E_i-E_f)(E_i-E_{a_2})(E_i-E_{a_1})} \\ &= \frac{\langle\gamma'|H'|1^{-1}3\rangle\langle 3\gamma''|H'|2\rangle\langle 1^{-1}2|H'|\gamma\rangle}{(\hbar\omega_\gamma-\hbar\omega_{\gamma'}-\hbar\omega_{\gamma''})(\hbar\omega_\gamma-\hbar\omega_{\gamma''}-\varepsilon_3+\varepsilon_1)(\hbar\omega_\gamma-\varepsilon_2+\varepsilon_1)} \end{aligned} \tag{6-229}$$

where i and f label the initial and final states and a_1, a_2 the two intermediate states. In a similar manner, the contribution of the additional diagrams can be obtained.

A case of special interest involves the coupling of the one-phonon state to the state of two phonons of the same type, which can occur for quanta of

---

[14]Anharmonic effects resulting from the coupling of particle excitations to the quadrupole vibrations were considered by Scharff-Goldhaber and Weneser (1955) and Raz (1959). The large anharmonicity of the quadrupole mode was exhibited by Coulomb excitation studies of the $n_2=2$ states (see, for example, McGowan, 1959) and was strikingly emphasized by the observation of static quadrupole moments for the $n_2=1$ state with a magnitude comparable to that of the transition moments (de Boer *et al.*, 1965). These discoveries stimulated the development of systematic methods for the treatment of anharmonic effects (Belyaev and Zelevinsky, 1962; Marumori *et al.*, 1964; Tamura and Udagawa, 1964; Alaga and Ialongo, 1966 (see also Alaga, 1969); Dreizler *et al.*, 1967; Ottaviani *et al.*, 1967; Sørensen, 1967; Kumar and Baranger, 1968; Marshalek, 1973. For the development based on the particle-vibration coupling, see also references quoted in footnote 28, p. 571).

even parity and even value of the angular momentum λ,

$$|\hat{n}_\lambda = 1\rangle \approx |n_\lambda = 1\rangle + c|n_\lambda = 2, I = \lambda\rangle \tag{6-230}$$

This interaction represents the leading-order anharmonic effect in the vibrational motion, and its consequences will be considered in Sec. 6-6a. From Eq. (6-229) and the matrix elements (6-209) and (6-211) for shape oscillations, we obtain, by the recoupling $j_1,(j_3\lambda'')j_2;\ \lambda \to (j_1 j_3)\lambda', \lambda'';\ \lambda$,

$$c^{(1)}(j_1^{-1} j_2 j_3) = \frac{K(j_1 j_2 j_3)}{\hbar\omega_\lambda(\varepsilon(j_3) - \varepsilon(j_1))(\varepsilon(j_2) - \varepsilon(j_1) - \hbar\omega_\lambda)} \tag{6-231}$$

with

$$K(j_1 j_2 j_3) = 2^{1/2}(2\lambda+1)^{-1/2}(2j_1+1)^{1/2}(2j_2+1)^{1/2}(2j_3+1)^{1/2}$$
$$\times \begin{Bmatrix} j_1 & j_2 & \lambda \\ \lambda & \lambda & j_3 \end{Bmatrix} h(j_1 j_2 \lambda) h(j_2 j_3 \lambda) h(j_3 j_1 \lambda) \tag{6-232}$$

The factor $2^{1/2}$ is associated with the identity of the two phonons in the final state (see Eq. (6B-36)).

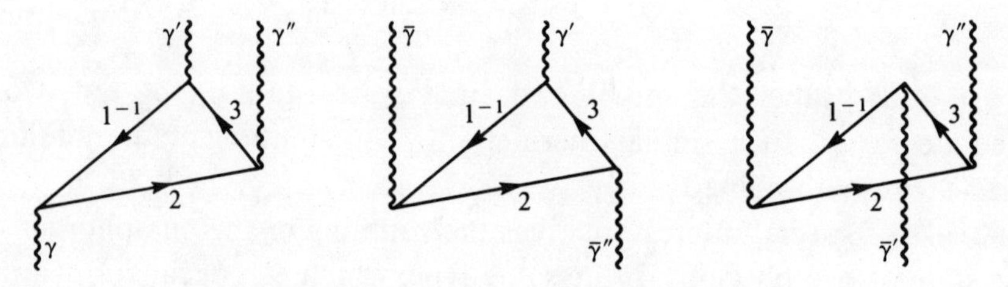

Figure 6-12 Coupling between states with one and two phonons.

The second and third diagrams in Fig. 6-12 are obtained by interchanging $c^\dagger(\lambda\mu)$ and $c(\overline{\lambda\mu})$ at two of the vertices, and thus involve the same matrix elements of H' as in Eq. (6-229); see Eq. (6-51). The angular momentum coupling is the same in view of the symmetries (1A-11) of the vector addition coefficients. These contributions therefore differ from Eq. (6-231) only in the energy denominators, and the resultant amplitude becomes

$$c(j_1^{-1}j_2j_3)=\sum_{p=1}^{3}c^{(p)}(j_1^{-1}j_2j_3)$$

$$=\frac{K(j_1j_2j_3)}{\hbar\omega_\lambda}\left(\frac{1}{\Delta E_{31}(\Delta E_{21}-\hbar\omega_\lambda)}+\frac{1}{\Delta E_{31}(\Delta E_{21}+\hbar\omega_\lambda)}+\frac{1}{(\Delta E_{31}+2\hbar\omega_\lambda)(\Delta E_{21}+\hbar\omega_\lambda)}\right) \tag{6-233}$$

$$(\Delta E_{ik}\equiv\varepsilon(j_i)-\varepsilon(j_k))$$

For $j_2 \neq j_3$, there are two sets of terms with j_2 and j_3 interchanged $(c(j_1^{-1}j_2j_3)+c(j_1^{-1}j_3j_2))$, which again only differ in the energy denominators, since $K(j_1j_2j_3)$ is symmetric with respect to any permutation of the three angular momenta. Finally, there are contributions to the total amplitude c in Eq. (6-230) from intermediate states with one particle and two holes, which are of the same form as Eq. (6-233), except for an overall change of sign resulting from the opposite sign of the coupling of a particle as compared with a hole.

The evaluation of the third-order terms considered above illustrates the general procedure that can be employed in the evaluation of the higher-order terms in the perturbation expansion. The treatment of the particle-vibration coupling H' by a perturbation expansion can also be expressed in terms of a series of canonical transformations leading to the diagonalization of the total Hamiltonian H_0+H', where H_0 represents the independent motion of particles and vibrational quanta. (See the similar analysis of the particle-rotation coupling in Chapter 4, pp. 147 ff.)

6-5h Normal Modes Generated by Particle-Vibration Coupling

In the preceding parts of Sec. 6-5, we have considered some of the consequences of the particle-vibration coupling in renormalizing the properties of the elementary modes of excitation and producing interactions between them. The systematic treatment of the particle-vibration coupling amounts to a nuclear field theory, which incorporates in a consistent manner the consequences arising from the fact that the quanta are built out of the same degrees of freedom as are the particle modes of excitation. Thus, the antisymmetry between the particles that are treated explicitly and those that are involved in the collective modes is expressed in terms of exchange interactions such as that illustrated by Fig. 6-10d (see the comments on p. 428); the inclusion of these exchange interactions at the same time ensures the orthogonality of the states built out of different elementary modes.

The consequences of the particle-vibration coupling considered so far have involved the effects associated with the coupling to individual configurations of the particles. Extending the treatment to include the interplay of all the particle degrees of freedom, one obtains a description of the vibrational modes in terms of self-consistent oscillations of the nuclear density and potential.[15] The present section contains a derivation of the normal modes based on such an approach, which represents a generalization of the schematic treatment in Sec. 6-2c. We also briefly indicate the connection of this formulation to a treatment based on effective two-body interactions.

Response function and random-phase approximation

The analysis of the self-consistent modes of nuclear excitation can be based on a study of the response function, which describes the polarization of the nucleus in the presence of a time-dependent external field.[16] Such a field, acting on the nuclear moment F, produces a coupling of the form

$$H' = \kappa \alpha_{\text{ext}} F \tag{6-234}$$

where α_{ext} is the amplitude of the external field, which may be so normalized that κ is the nuclear field coupling constant employed above (see Eq. (6-204)).

For a weak field α_{ext} with a harmonic time dependence

$$\alpha_{\text{ext}} = (\alpha_{\text{ext}})_0 \cos \omega_{\text{ext}} t \tag{6-235}$$

---

[15]The treatment of nuclear vibrations in terms of the motion of the individual particles was first formulated on the basis of a calculation of potential and kinetic energies for a slowly varying average nuclear field (Inglis, 1955; Araújo, 1956, 1959; Ferrell, 1957; Griffin and Wheeler, 1957; Griffin, 1957; Belyaev, 1959). The general treatment of the normal modes of nuclear excitation based on the random-phase approximation was developed by Glassgold *et al.* (1959), Ferrell and Fallieros (1959), Goldstone and Gottfried (1959), Takagi (1959), Ikeda *et al.* (1959), Arvieu and Veneroni (1960), Baranger (1960), Kobayashi and Marumori (1960), Marumori (1960), and Thouless (1961).

[16]The response of matter to electric fields is described by the dielectric constant considered as a function of frequency and wave number. This response function was employed in a self-consistent treatment of the properties of an electron gas by Lindhard (1954) and Nozières and Pines (1958). For a discussion of the response function as a general tool in the study of quantum liquids, see, for example, the text by Pines and Nozières (1966).

the perturbed nuclear state is given by

$$|\rangle = |0\rangle - \frac{\kappa}{2\hbar}(\alpha_{\text{ext}})_0 \sum_a \left(\frac{\exp\{-i\omega_{\text{ext}}t\}}{\omega_a - \omega_{\text{ext}}} + \frac{\exp\{i\omega_{\text{ext}}t\}}{\omega_a + \omega_{\text{ext}}} \right) F_a |a\rangle \qquad (6\text{-}236)$$

$$F_a \equiv \langle a|F|0\rangle \qquad \hbar\omega_a \equiv E_a - E_0$$

where a labels the nuclear excitations, with energies $\hbar\omega_a$, and where the state $|0\rangle$ represents the ground state of the system. (The solution (6-236) corresponds to an adiabatic switching on of the field; more generally, the solution to the wave equation involves additional terms giving rise to a time-independent induced moment, which, however, is not of significance for the following discussion.)

The induced moment F obtained from Eq. (6-236) is

$$\langle F \rangle = \chi(\omega_{\text{ext}})\alpha_{\text{ext}} = \chi(\omega_{\text{ext}})(\alpha_{\text{ext}})_0 \cos\omega_{\text{ext}} t \qquad (6\text{-}237)$$

with

$$\chi(\omega_{\text{ext}}) \equiv -\frac{2\kappa}{\hbar} \sum_a \frac{|F_a|^2 \omega_a}{(\omega_a)^2 - (\omega_{\text{ext}})^2} \qquad (6\text{-}238)$$

The polarizability coefficient χ is expressed as a sum over the nuclear excitations, in which each term is of the same form as derived above for vibrational modes associated with the field F (see Eq. (6-216), in which ΔE_{21} corresponds to $\hbar\omega_{\text{ext}}$, while $\alpha_0 = |F_a|$; see the relation (6-28) for the vibrational matrix element).

The states a in the expression (6-238) are the normal modes that include the effect of the internal nuclear fields. Thus, for a nuclear mode associated with an oscillating field F, the total field acting on the nucleons involves the sum of the external field with amplitude α_{ext} and the internal field with amplitude given by the induced moment (6-237),

$$\alpha = \alpha_{\text{ext}} + \langle F \rangle = (1+\chi)\alpha_{\text{ext}} \qquad (6\text{-}239)$$

The self-consistency condition is obtained by noting that the nuclear polarization can also be derived by considering the total field, of amplitude α, acting on the individual nucleons

$$\langle F \rangle = \chi^{(0)}\alpha \qquad (6\text{-}240)$$

with

$$\chi^{(0)}(\omega) = -\frac{2\kappa}{\hbar}\sum_i \frac{|F_i|^2\omega_i}{\omega_i^2 - \omega^2} \tag{6-241}$$

$$F_i \equiv \langle \mathrm{v}=2,i|F|\mathrm{v}=0\rangle \qquad \hbar\omega_i \equiv E(\mathrm{v}=2,i) - E(\mathrm{v}=0)$$

The ground state of the independent-particle motion, which may include pair correlations, is the quasiparticle vacuum state $\mathrm{v}=0$, while the states excited by the one-particle operator F are the two-quasiparticle states $\mathrm{v}=2,i$.

From the relations (6-239) and (6-240), we obtain

$$\chi = \chi^{(0)}(1+\chi) \tag{6-242}$$

or

$$\chi(\omega) = \frac{\chi^{(0)}(\omega)}{1-\chi^{(0)}(\omega)} \tag{6-243}$$

which expresses the response of the system of interacting nucleons in terms of the response of the individual nucleons.

The relation (6-243) determines the eigenfrequencies and transition matrix elements appearing in the expression (6-238) for χ. The eigenfrequencies ω_a are the poles of χ and, hence, the roots of the equation

$$\chi^{(0)}(\omega_a) = -\frac{2\kappa}{\hbar}\sum_i \frac{|F_i|^2\omega_i}{\omega_i^2 - \omega_a^2} = 1 \tag{6-244}$$

while the transition matrix elements are given by the residues

$$|F_a|^2 = -\frac{\hbar}{\kappa}\left(\left(\frac{\partial}{\partial\omega}\chi^{(0)}\right)_{\omega=\omega_a}\right)^{-1}$$

$$= \frac{\hbar^2}{4\kappa^2}\frac{1}{\omega_a}\left(\sum_i \frac{|F_i|^2\omega_i}{(\omega_i^2-\omega_a^2)^2}\right)^{-1} \tag{6-245}$$

The relations (6-244) and (6-245) are seen to be generalizations of the expressions (6-27) and (6-28) that apply to the special case, in which all intrinsic states i are degenerate. The above treatment of the normal modes is often referred to as the random-phase approximation.[17]

It is seen that the oscillator sum is not affected by the interactions

$$\sum_a |F_a|^2 \omega_a = \sum_i |F_i|^2 \omega_i \tag{6-246}$$

as follows most simply by taking the limit of the relation (6-243) as $\omega \rightarrow \infty$. (See in this connection footnote 12 on p. 400. The identity (6-246) can also be seen as a consequence of the sum rules derived in Sec. 6-4a, since the field interaction is equivalent to a two-particle interaction $\kappa F(1)F(2)$, and thus commutes with the moment F.)

The eigenmodes obtained from Eq. (6-244) partly represent collective vibrational modes associated with the field F and partly a spectrum of two-quasiparticle excitations as modified by the field coupling. This modification can also be obtained by considering the effects of the particle-vibration coupling, as in Sec. 6-5b. The coupling admixes the particle and collective degrees of freedom in such a manner as to partially transfer the one-particle transition strength onto the collective modes. (The degenerate model considered in Sec. 6-2c corresponds to the limiting situation in which the entire transition strength appears in the collective mode.) Since the present analysis has focused exclusively on the effects arising from the coupling to the field F, it is clear that additional interaction effects may be important in the study of properties of excitations other than the collective modes associated with this particular field.

It must be emphasized that the present treatment of the normal modes employs a first-order perturbation approximation in evaluating the response of the nucleus. The expression (6-238) will always be valid for a sufficiently weak external field. In the self-sustained normal modes, however, the oscillating field acting on a nucleon has a finite amplitude. The use of the linear approximation (6-240) is, therefore, only valid provided the zero-point amplitude of the vibrational motion is sufficiently small. As discussed for the degenerate model (p. 336), the higher-order effects of the field on the particle motion give rise to anharmonic terms in the vibrational Hamiltonian. (See, for example, the discussion of these terms for the low-frequency quadrupole shape oscillations, pp. 523ff.)

---

[17]This name goes back to the early use of such an approximation in the treatment of the collective properties of an electron gas (Bohm and Pines, 1953) and refers to the neglect of correlations (due to the assumption of random phases) in density components with wave numbers different from those of the collective field.

If the collective mode (as is always the case for the high-frequency vibrations) occurs in a region of the spectrum where the levels i form a continuum, or have a high density, the strength of the collective mode becomes distributed over a finite energy interval. The resulting strength function can be viewed in terms of the coupling of the collective mode to the particle degrees of freedom with $\omega_i \approx \omega_a$; if this coupling, per unit energy interval, can be regarded as approximately constant over the width of the collective mode, the strength function has a line shape corresponding to a Breit-Wigner resonance with a width given by

$$\Gamma_a = \frac{2\pi}{D}\langle H_{ai}\rangle^2 = 2\pi(\kappa F_a)^2 \frac{\langle F_i^2\rangle}{D} \tag{6-247}$$

where $H_{ai} = \kappa F_a F_i$ is the coupling matrix element, while D is the average spacing and $\langle F_i^2\rangle$ the average strength of the levels i. The result (6-247) is a consequence of the relations (6-244) and (6-245), but is most directly obtained from the general analysis of strength function phenomena in Appendix 2D (see Eqs. (2D-10) and (2D-11)).

In macroscopic systems, the damping of a collective mode associated with its decay into a particle-hole pair as in Eq. (6-247) is referred to as Landau damping (see, for example, Pines and Nozières, 1966). This type of damping was first considered in connection with the collective modes of a classical plasma, which can transfer energy to individual particles even in the absence of collisions, when these particles have a velocity equal to the wave velocity of the collective mode (Landau, 1946; see also the texts by Ginzburg, 1964, and Clemmow and Dougherty, 1969).

Microscopic description of vibrational quanta

The treatment of the normal modes described in the preceding section provides an analysis of the collective vibrational variables in terms of the degrees of freedom of the individual particles. As for the simple model in Sec. 6-2c, this relationship can be explicitly exhibited in terms of the transformation between the set of operators $A_a^\dagger$, A_a that create and annihilate the normal modes and the set $A_i^\dagger$, A_i that create and annihilate the two-quasiparticle configurations $\text{v}=2, i$. (The creation and annihilation operators for a vibrational mode are usually denoted by $c^\dagger$ and c, but in the present section, we use the notation $A_a^\dagger$, A_a for the total spectrum of normal modes, which includes excitations that are not of collective character.)

The assumption that the response of the particle motion to the field F can be treated in terms of first-order effects of F on the individual modes of excitation i implies a linear relationship between $(A_a^\dagger, A_a)$ and $(A_i^\dagger, A_i)$. The transformation between the two sets of variables can be expressed in the form

$$
\begin{aligned}
A_a^\dagger &= \sum_i \left(X_{ai} A_i^\dagger - Y_{ai} A_i \right) \\
A_a &= \sum_i \left(- Y_{ai} A_i^\dagger + X_{ai} A_i \right)
\end{aligned}
\tag{6-248a}
$$

$$
\begin{aligned}
A_i^\dagger &= \sum_a \left(X_{ai} A_a^\dagger + Y_{ai} A_a \right) \\
A_i &= \sum_a \left(Y_{ai} A_a^\dagger + X_{ai} A_a \right)
\end{aligned}
\tag{6-248b}
$$

where the amplitudes X_{ai} and Y_{ai}, which are real when both sets of states a and i have standard phases, satisfy the orthonormality relations

$$
\begin{aligned}
&\sum_i \left(X_{ai} X_{a'i} - Y_{ai} Y_{a'i} \right) = \delta(a,a') \\
&\sum_i \left(X_{ai} Y_{a'i} - Y_{ai} X_{a'i} \right) = 0
\end{aligned}
\tag{6-249a}
$$

$$
\begin{aligned}
&\sum_a \left(X_{ai} X_{ai'} - Y_{ai} Y_{ai'} \right) = \delta(i,i') \\
&\sum_a \left(X_{ai} Y_{ai'} - Y_{ai} X_{ai'} \right) = 0
\end{aligned}
\tag{6-249b}
$$

If we take Eq. (6-248a) as defining the transformation matrices X and Y, the orthonormality relation (6-249a) can be obtained by expressing the commutators of the $A_a^\dagger, A_a$ variables in terms of the commutators for the $A_i^\dagger, A_i$ variables, and taking the expectation value in the ground state 0. The validity of the perturbation treatment implies that in this state, each quasiparticle mode is only excited with small probability, and hence

$$
\langle 0 | [A_i, A_{i'}^\dagger] | 0 \rangle \approx \delta(i,i')
\tag{6-250}
$$

The commutators $[A_i^\dagger, A_{i'}^\dagger]$ and $[A_i, A_{i'}]$ vanish identically, as a consequence of the anticommutation relations for the operators that create (or annihilate) individual quasiparticles. The relations (6-248a) and (6-249a) imply the inverse relations (6-248b) and (6-249b).

The amplitudes X_{ai} and Y_{ai} can be obtained by considering the wave function for independent-particle motion perturbed by the field F oscillating with frequency ω_a. The perturbation of the wave function produced by an externally applied field coupling of the form (6-234) can be expressed as a unitary transformation ($\exp\{-iS\} \approx 1 - iS$) with (see the analogous derivation of the perturbed wave function (6-236) in the a representation)

$$S = -i\frac{\kappa}{2\hbar}(\alpha_{\text{ext}})_0 \sum_i \left(\frac{\exp\{-i\omega_{\text{ext}}t\}}{\omega_i - \omega_{\text{ext}}} + \frac{\exp\{i\omega_{\text{ext}}t\}}{\omega_i + \omega_{\text{ext}}} \right) F_i A_i^\dagger + \text{H. conj.} \tag{6-251}$$

When ω_{ext} equals the frequency ω_a of a normal mode, the nucleonic motion induced by the external field is the same as in the self-sustained oscillations resulting from the nuclear field coupling. The operator $S(\omega_{\text{ext}} = \omega_a)$ can therefore be expressed in terms of the variables $A_a^\dagger$, A_a for the particular mode considered. The positive frequency part of S creates quanta and is proportional to $A_a^\dagger$, while the negative frequency part is proportional to A_a; the amplitudes are determined by the self-consistency condition $\chi^{(0)}(\omega_a) = 1$, which implies

$$\delta\langle F \rangle = i[S, F] = \alpha_{\text{ext}} \qquad \left(F = \sum_a F_a A_a^\dagger + \text{H. conj.} \right)$$
$$S(\omega_{\text{ext}} = \omega_a) = \frac{i}{2} \frac{(\alpha_{\text{ext}})_0}{F_a^*} A_a^\dagger \exp\{-i\omega_a t\} + \text{H. conj.} \tag{6-252}$$

Introducing the transformation (6-248a) into Eq. (6-252) and comparing with the expression (6-251), one obtains

$$X_{ai} = \frac{\kappa}{\hbar} \frac{F_a F_i^*}{\omega_a - \omega_i} \qquad Y_{ai} = -\frac{\kappa}{\hbar} \frac{F_a F_i}{\omega_a + \omega_i} \tag{6-253}$$

The amplitudes (6-253) are illustrated by the diagrams in Fig. 6-13.

The self-consistency condition for the eigenfrequencies ω_a is illustrated in Fig. 6-14. The relation between the diagrams in this figure expresses the condition that the moment F in the vibrational mode be equal to that generated by the transitions of the individual particles. This condition,

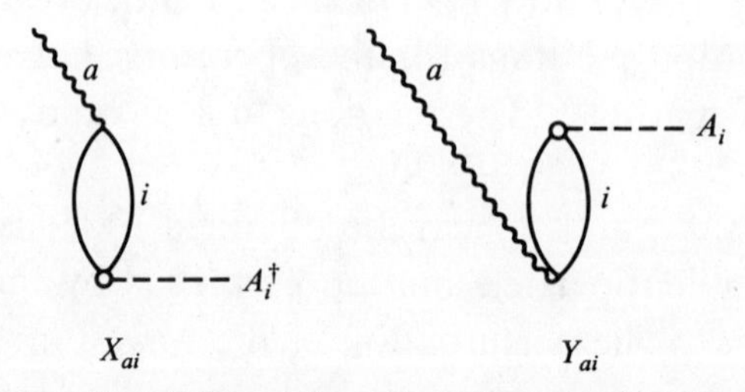

Figure 6-13 Amplitudes of particle configurations in normal modes.

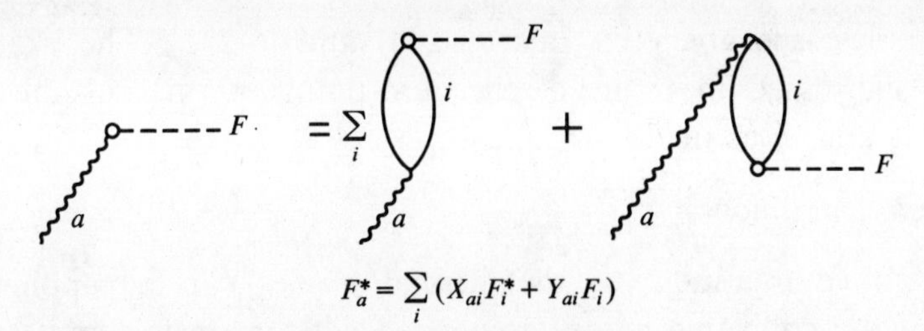

Figure 6-14 Self-consistency condition for normal modes.

together with the values (6-253) for the amplitudes X_{ai} and Y_{ai}, yields the eigenvalue equation (6-244).

The relation (6-248) between the set of variables $A_i^\dagger, A_i$ that describe the independent-particle motion and the set $A_a^\dagger, A_a$ describing the normal modes can be viewed as a canonical transformation between boson variables. A treatment of the noncollective variables, such as the $A_i^\dagger$ and A_i, in terms of boson operators is justified if the states on which these operators act involve only a small amplitude for the occupancy of any given quasiparticle state, as in the relation (6-250). In the collective modes that are built out of many different quasiparticle excitations, this condition will be fulfilled even for states with several quanta; however, the eigenstates a that are approximately described in terms of a single $\nu=2$ configuration represent modes that can only be excited a single time.

The treatment of the collective modes generated by the field coupling can also be formulated as a diagonalization of an effective two-body interaction of the form (6-37). In the approximation that treats the quasiparticle operators $A_i^\dagger, A_i$ as boson variables, the Hamiltonian takes the form

$$\begin{aligned} H &= \sum_i \hbar\omega_i A_i^\dagger A_i + \tfrac{1}{2}\kappa F^2 \\ F &= \sum_i (F_i A_i^\dagger + F_i^* A_i) \end{aligned} \tag{6-254}$$

This Hamiltonian is diagonalized by a canonical transformation of the form (6-248) with the coefficients given by Eq. (6-253),

$$\begin{aligned} H &= \tfrac{1}{2}\sum_a \hbar\omega_a - \tfrac{1}{2}\sum_i \hbar\omega_i + \sum_a \hbar\omega_a A_a^\dagger A_a \\ F &= \sum_a (F_a A_a^\dagger + F_a^* A_a) \end{aligned} \tag{6-255}$$

where the eigenfrequencies ω_a and the field matrix elements F_a, for the normal modes, are given by Eqs. (6-244) and (6-245). The ground-state energy is expressed in terms of the zero-point energies of the effective oscillators i and a, as in Eq. (6-33).

Translational motion

The above treatment of the normal modes can be applied to the analysis of the collective modes associated with intrinsic symmetry breaking. For these modes, the structure of the field coupling can be obtained from the static potential by employing the invariance of the total Hamiltonian.

The shell-model potential violates the translational invariance of the total Hamiltonian and, thus, single-particle excitations can be produced by a field proportional to the total center-of-mass coordinate. The translational invariance can be restored by including the effects of the collective field generated by a small displacement α of the nucleus. Such a displacement, in the x direction, gives rise to the coupling

$$H' = -\alpha \frac{\partial V}{\partial x} \tag{6-256}$$

where V is the average one-particle potential. The coupling (6-256) can also be written in the standard form

$$H' = \kappa \alpha F \tag{6-257}$$

with (compare Eq. (6-74))

$$F = -\frac{1}{\kappa}\frac{\partial V}{\partial x}$$

$$\kappa = \int \frac{\partial V}{\partial x}\frac{\partial \rho_0}{\partial x} d\tau = -A\left\langle \frac{\partial^2 V}{\partial x^2} \right\rangle \tag{6-258}$$

corresponding to a normalization of α such that $\langle F \rangle = \alpha$. The spectrum of normal modes generated by the field coupling (6-257) contains an excitation mode with zero frequency, as follows from Eq. (6-244) and the relation

$$\sum_k \frac{\partial V}{\partial x_k} = \frac{i}{\hbar}[P_x, H_0] \tag{6-259}$$

which implies

$$\begin{aligned} \sum_i \frac{|F_i|^2}{\hbar\omega_i} &= \frac{i}{2\hbar\kappa^2}\langle \mathrm{v}=0| \sum_k \left(\left[p_x, \frac{\partial V}{\partial x} \right] \right)_k |\mathrm{v}=0\rangle \\ &= \frac{1}{2\kappa^2}\langle \mathrm{v}=0| \sum_k \frac{\partial^2 V}{\partial x_k^2} |\mathrm{v}=0\rangle = -\frac{1}{2\kappa} \end{aligned} \tag{6-260}$$

where $\nu=0$ represents the ground state for independent-quasiparticle motion. The appearance of the zero-frequency mode reflects the fact that the inclusion of the coupling (6-256) restores the translational invariance of the total Hamiltonian, to leading order in α.

The additional roots of the normal modes equation (6-244) represent $\lambda\pi=1-$ nuclear excitations. These excitations have been modified by the field coupling, which ensures that they are orthogonal to the "spurious" degree of freedom associated with the center-of-mass motion.

The mass parameter for the translational mode, which is equivalent to the oscillator strength (see, for example, Eq. (6-15)), can be obtained from the relation (6-245),

$$\begin{aligned} D_a &= \frac{\hbar}{2}\left(\omega_a|F_a|^2\right)^{-1} \\ &= \frac{2}{\hbar}\sum_i \frac{|\langle \nu=2,i|\partial V/\partial x|\nu=0\rangle|^2}{\omega_i^3} = \frac{2}{\hbar}\sum_i \frac{|\langle \nu=2,i|P_x|\nu=0\rangle|^2}{\omega_i} \\ &= \frac{i}{\hbar}M\langle \nu=0|\sum_k ([p_x,x])_k|\nu=0\rangle = AM \end{aligned} \tag{6-261}$$

In the derivation of the last line of Eq. (6-261), we have assumed the total momentum P_x of the nucleons to be equal to the nucleonic mass multiplied by the sum of the time derivatives of the individual position coordinates, as implied by Galilean invariance (see Eq. (1-20)).

The presence of a velocity dependence in the one-body potential violates the Galilean invariance assumed in the above derivation. This violation must be corrected for by including an additional field coupling resulting from a uniform collective motion with velocity $\dot{\alpha}$ in the x direction

$$H' = -M\dot{\alpha}\frac{\partial V}{\partial p_x} = \frac{i}{\hbar}M\dot{\alpha}[x,V] \tag{6-262}$$

which restores the Galilean invariance to leading order. (The generator of a Galilei transformation is given by Eq. (1-17).) A treatment of the combined effect of the two couplings (6-256) and (6-262) can be obtained from a straightforward extension of the normal modes analysis for a single field and yields a zero frequency mode with $D_a=AM$ as guaranteed by the Galilean invariance, combined with translation invariance.

For a velocity-dependent single-particle potential described by an effective mass term, the field coupling (6-262) is proportional to the momentum of the particle. The quasiparticle Hamiltonian describing particle motion in the presence of a pair field is also effectively velocity dependent, since the particles are paired in states that are conjugate under time reversal in the stationary coordinate system. The coupling (6-262) connected with this velocity dependence is obtained from the commutator

$$\frac{i}{\hbar}[x, V_{\text{pair}}] = \frac{i}{\hbar}\left[\int \rho_0(\mathbf{r})x\,d\tau, -\Delta\int(\rho_2(\mathbf{r}')+\rho_{-2}(\mathbf{r}'))d\tau'\right]$$

$$= -\frac{2\Delta}{\hbar}\int i(\rho_2(\mathbf{r})-\rho_{-2}(\mathbf{r}))x\,d\tau \tag{6-263}$$

where ρ_0 and $\rho_{\pm 2}$ are the one-particle and pair densities (see Eqs. (6-141) to (6-143)), while the pair potential V_{pair} is given by Eq. (6-597); see also Eqs. (6-144) and (6-145). The field in Eq. (6-263) is seen to be proportional to the part of the dipole pair moment that is odd under time reversal combined with Hermitian conjugation. The excitations produced by this field involve the matrix elements

$$\langle \text{v}=2,\nu_1\nu_2|\int(\rho_2(\mathbf{r})-\rho_{-2}(\mathbf{r}))x\,d\tau|\text{v}=0\rangle = (u_1u_2+v_1v_2)\langle\nu_2|x|\bar{\nu}_1\rangle \tag{6-264}$$

as can be derived from the quasiparticle transformation (6-599).

The creation operator $A_a^\dagger$ for the zero-frequency mode is proportional to the total momentum P_x, as follows from Eqs. (6-248a) and (6-253) together with the relation (6-259). However, the proportionality constant becomes infinite as ω_a goes to zero, reflecting the infinite amplitude of a vibrational mode with zero frequency. Despite the formal problems involved in going to this limit, it appears that the separation of this mode from the remaining degrees of freedom is correctly described by the above treatment, since one may consider the problem for a coupling constant κ slightly smaller than the value (6-258) corresponding to translational invariance. In the absence of the coupling, the zero-point amplitude of the center-of-mass motion is of order $\alpha_0^{(0)} \sim A^{-2/3}R$, and it is therefore possible to choose κ such that the zero-point amplitude is large compared to $\alpha_0^{(0)}$, but not so large as to invalidate the perturbation element in the random-phase approximation.

Rotational motion

For a deformed nucleus, the rotational invariance can be restored by considering the coupling resulting from a small angular rotation α about the intrinsic 1 axis,

$$H' = -\alpha\frac{\partial V}{\partial\varphi_1} = \kappa\alpha F \tag{6-265}$$

with

$$\kappa = -A \langle \frac{\partial^2 V}{\partial \varphi_1^2} \rangle$$
$$F = -\frac{1}{\kappa} \frac{\partial V}{\partial \varphi_1} \tag{6-266}$$

The angle φ_1 represents the azimuthal angle with respect to the axis of rotation. The random-phase equation (6-244) again has a zero-frequency solution as a consequence of the relation

$$\sum_k \left(\frac{\partial V}{\partial \varphi_1} \right)_k = i[J_1, H_0] \tag{6-267}$$

where $\hbar J_1$ is the component of total angular momentum conjugate to the angle φ_1. The inertial parameter for the zero-frequency mode is found to be

$$D_a = 2\hbar^2 \sum_i \frac{|\langle \mathrm{v}=2, i | J_1 | \mathrm{v}=0 \rangle|^2}{\hbar \omega_i} \tag{6-268}$$

in accordance with the result derived by means of the cranking model (see Eq. (4-110)). The additional roots of the random-phase equation (6-244) correspond to the intrinsic $K\pi = 1+$ excitations, as modified by the Coriolis coupling so as to remove the spurious degree of freedom represented by the rotational motion. If the one-particle potential is velocity dependent, one must add additional couplings proportional to the rotational frequency in analogy to the coupling (6-262) in the pushing model; see the discussion of the effective mass term in the cranking model on p. 80 and of the frequency-dependent pair field on pp. 83 and 278.

6-6 ANHARMONICITY IN VIBRATIONAL MOTION. COUPLING OF DIFFERENT MODES

In the present section, we shall consider the phenomenological analysis of the properties of the vibrational modes that go beyond the harmonic approximation. The couplings considered illustrate some of the tools that may be employed in attacking the broad class of problems associated with the coupling of the many different nuclear modes. A basis for a microscopic estimate of the parameters appearing in such an analysis is provided by the treatment of the particle-vibration coupling considered in the previous section.

The anharmonic effects considered in the present section concern the low-frequency quadrupole mode, for which there is a considerable body of evidence regarding multiple excitations. The data imply large deviations from the harmonic approximation, which are connected with the especially large zero-point amplitude for this collective mode. The pair vibrations also exhibit appreciable anharmonic effects (see the discussion on pp. 645 ff.). However, for the modes involving particle excitations between different major shells (such as the high-frequency dipole and quadrupole modes, and the relatively low-frequency octupole mode in closed-shell nuclei), the anharmonicity is expected to be rather small, on account of the large number of different particle configurations that may contribute to these modes (see, for example, the discussion of the coupling between octupole quanta on pp. 567 ff.)

6-6a Anharmonic Effects in Low-Frequency Quadrupole Mode

The anharmonic effects in a vibrational mode can be expressed in many different forms. In the present section, we consider two different approaches. The first involves a phenomenological parameterization of the effective interactions between phonons and of the nonlinear terms in the transition operators. In the second approach, one attempts to characterize the main anharmonic terms in the Hamiltonian expressed in terms of the vibrational amplitude.

Effective interactions between phonons

A phenomenological analysis of anharmonicities can be based on a representation in which the Hamiltonian has been brought to a form that is diagonal in the number of phonons. In this representation, the anharmonic effects can be expressed as power series in the phonon creation and annihilation operators (Brink *et al.*, 1965). Such an approach to the description of anharmonicity effects is similar to the phenomenological treatment of the coupling between rotational and intrinsic motion in Sec. 4-3.

For the quadrupole mode, the leading-order anharmonic terms in the vibrational energy spectrum are of the form $c_2^\dagger c_2^\dagger c_2 c_2$ and represent an interaction between pairs of phonons. The interaction can be expressed in the form

$$H' = \sum_{R=0,2,4} \sum_{M=-R}^{R} \tfrac{1}{2} V_R (c_2^\dagger c_2^\dagger)_{RM} (\bar{c}_2 \bar{c}_2)_{\overline{RM}} \qquad (6\text{-}269)$$

where R is the total angular momentum of the pair of phonons, and where we use the abbreviated notations $c_2^\dagger$ and $\bar{c}_2$ for the spherical tensors $c^\dagger(\lambda=2)$ and $c(\overline{\lambda=2})$; see Eq. (6-53).

The interaction (6-269) involves three parameters V_R, which represent the phonon-phonon interaction in the $n=2$ levels,

$$V_R = E(n=2, I=R) - 2E(n=1) \tag{6-270}$$

For the levels with $n \geqslant 3$, the interaction (6-269) can be evaluated by means of parentage coefficients (see pp. 688 ff.); thus, for $n=3$, we have

$$\langle n=3, IM|H'|n=3, IM\rangle = (2I+1)^{-1} \sum_R V_R \langle n=3, I\|c_2^\dagger\|n=2, R\rangle^2 \tag{6-271}$$

in terms of the one-phonon parentage coefficients given in Table 6B-1, p. 691. We also note the simple result

$$\langle n, I=2n, M|H'|n, I=2n, M\rangle = \tfrac{1}{2}n(n-1)V_{R=4} \tag{6-272}$$

for the states of maximum angular momentum $I_n = 2n$.

The evidence on quadrupole vibrational levels with $n \geqslant 3$ is very incomplete. For a number of nuclei, the sequence of levels with $I=2n$ has been established up to rather large values of n and can be fitted by an interaction of the form (6-272); see Fig. 6-33, p. 537. In a few cases, $I\pi = 3+$ levels have been observed, but the position deviates appreciably from that predicted for the $n=3$, $I=3$ level based on the relations (6-270) and (6-271); see Table 6-13, p. 542. One may attempt to attribute these deviations to higher-order anharmonic terms corresponding to the interaction between three or more phonons. In an expansion in powers of $c_2^\dagger$ and c_2, the next term is the three-phonon interaction which has the form $((c_2^\dagger c_2^\dagger c_2^\dagger)_R(\bar{c}_2\bar{c}_2\bar{c}_2)_R)_0$ with $R=0$, 2, 3, 4, and 6. It is not yet clear, however, whether such an expansion provides a useful description.

E2 moments

The expansion of the $E2$ moment in terms of the tensor operators $c_2^\dagger$ and $\bar{c}_2$ can be written in the form, taking into account the invariance of $\mathscr{M}(E2,\mu)$ under time reversal followed by Hermitian conjugation,

$$\begin{aligned}\mathscr{M}(E2,\mu) = {} & m_{10}(c_{2\mu}^\dagger + \bar{c}_{2\mu}) + m_{11}(c_2^\dagger\bar{c}_2)_{2\mu} + 2^{-1/2}m_{20}\left((c_2^\dagger c_2^\dagger)_{2\mu} + (\bar{c}_2\bar{c}_2)_{2\mu}\right) \\ & + \sum_{R=0,2,4} 2^{-1/2} m_{21,R}\left(((c_2^\dagger c_2^\dagger)_R\bar{c}_2)_{2\mu} + (c_2^\dagger(\bar{c}_2\bar{c}_2)_R)_{2\mu}\right) \\ & + 6^{-1/2}m_{30}\left((c_2^\dagger c_2^\dagger c_2^\dagger)_{2\mu} + (\bar{c}_2\bar{c}_2\bar{c}_2)_{2\mu}\right)\end{aligned} \tag{6-273}$$

The transformation properties of $\mathscr{M}(E2,\mu)$ and $c^\dagger(2\mu)$ under time reversal (see Eq. (6-53)) imply that the coefficients m in Eq. (6-273) are real.

In the harmonic approximation, the $E2$ moment satisfies the selection rule $\Delta n = \pm 1$; the leading-order anharmonic terms give rise to matrix elements with $\Delta n = 0$ and 2. The $\Delta n = 0$ terms comprise static moments, as well as transitions among members of a multiplet with given n, and the matrix elements of $(c_2^\dagger \bar{c}_2)_2$ are given by the relation (6B-42). Thus, for $n=1$ and $n=2$, we obtain the nonvanishing matrix elements given in Table 6-3. Experimental evidence testing the predicted relation between $\Delta n = 0$ transition matrix elements and static moments is discussed on p. 542.

		$n=1$		$n=2$	
		$I=2$	$I=0$	$I=2$	$I=4$
$n=1$	$I=2$	$5^{1/2}$			
$n=2$	$I=0$			2	
	$I=2$		2	$-(3/7)5^{1/2}$	$12/7$
	$I=4$			$12/7$	$(3/7)110^{1/2}$

Table 6-3 Anharmonic terms in $E2$ moments for quadrupole vibrational transitions. The table gives the reduced matrix elements of the quadrupole operator $(c_2^\dagger \bar{c}_2)_{\lambda=2,\mu}$.

For $\Delta n = 2$ transitions induced by the moment (6-273), the matrix elements involving the $n=0$ and 1 states are given by

$$\langle n=2, I=2 \| 2^{-1/2}(c_2^\dagger c_2^\dagger)_2 \| n=0 \rangle = 5^{1/2}$$

$$\langle n=3, I \| 2^{-1/2}(c_2^\dagger c_2^\dagger)_2 \| n=1 \rangle = (-1)^I \langle n=3, I \| c_2^\dagger \| n=2, R=2 \rangle \tag{6-274}$$

where the parentage coefficient can be obtained from Table 6B-1, p. 691.

The terms in the effective moment (6-273) of third order in $(c_2^\dagger, \bar{c}_2)$ partly give rise to transitions with $\Delta n = 3$ and partly give modifications of the leading-order intensity relations for the $\Delta n = 1$ transitions; thus, for the transitions from $n=1$ to $n=2$, we obtain

$$\langle n=2, I=R \| \mathscr{M}(E2) \| n=1 \rangle = \left(2^{1/2}(2R+1)^{1/2} m_{10} + 5^{1/2} m_{21,R} \right) \tag{6-275}$$

From the parameters $m_{21,R}$ determined empirically on the basis of this relation, the $\Delta n = 1$ transition probabilities involving states with $n \geqslant 3$ can be predicted. The available data are insufficient to test these relationships.

The expression (6-273) is based entirely upon the symmetry of the $E2$ operator under rotations and time reversal. If one were to assume that the important matrix elements of the $E2$ moment were all contained within the vibrational spectrum, there would be additional restrictions on the coefficients in expansion (6-273) arising from the operator identities satisfied by the $E2$ moment. In particular, the commutability of the different components of $\mathscr{M}(E2,\mu)$ would imply restrictions of this type, which can also be expressed by the requirement that the sum rule (6-196) be exhausted within the vibrational spectrum.

Potential and kinetic energies. Adiabatic approximation

An alternative approach to the description of anharmonic effects may be based on an analysis of the potential and kinetic vibrational energies $V(\alpha_2)$ and $T(\alpha_2,\dot{\alpha}_2)$, expressed in terms of the amplitudes α_2 and their time derivative $\dot{\alpha}_2$.[18]

In the harmonic approximation, the energy is a quadratic form in the amplitudes α and $\dot{\alpha}$; the anharmonic effects in the spectrum can be described by terms in the Hamiltonian that are cubic, quartic, etc., in α_2 and $\dot{\alpha}_2$ (or π_2). The large number of such terms that are permitted by rotational and time reversal invariance restricts the usefulness of this general expansion.

The number of significant terms in the expansion of the Hamiltonian is further reduced in situations where the vibrational frequency is small compared with the quasiparticle excitation frequencies. Under such adiabatic conditions, the anharmonic terms in the potential energy are expected to dominate, since the vibrational motion is slow, but of large amplitude. (In the microscopic analysis based on the particle-vibration coupling, the adiabatic approximation corresponds to the neglect of the phonon frequency in the excitation energies of the intermediate states occurring in the denominators of expressions such as Eq. (6-233). In this approximation, diagrams that result from each other by interchanging the creation of a phonon with the annihilation of the time-reversed phonon at the same vertex contribute equally to the effective Hamiltonian. The sum of these two terms is proportional to the amplitude α_2, and thus the effective interaction is a function $V(\alpha_2)$ of this amplitude.) The expected dominance of anharmonic terms in the potential energy is consistent with the compar-

---

[18]Early treatments of the effects of the leading-order anharmonic terms in the nuclear quadrupole vibrational Hamiltonian were given by Bès (1961), Kerman and Shakin (1962), Belyaev and Zelevinsky (1962), and Chang (1964).

able magnitude of kinetic and potential energies in an oscillatory motion, since the adiabatic vibrations correspond to a situation where the harmonic term in the potential energy (the restoring force) nearly vanishes and where therefore higher-order terms in the potential energy become relatively large.

In the adiabatic situation, the $E2$ moment is also expected to depend only on the amplitude α_2; one can then exploit the freedom in the definition of α_2 to obtain an $E2$ moment that is linear in α_2. (Under more general conditions, it is not possible to choose a deformation coordinate α_2 that is linear in $\mathscr{M}(E2)$, since the $E2$ operator acting within the vibrational spectrum in general does not satisfy the algebra (6-50) assumed for the amplitudes α_2; see the comments in small print on p. 451.)

The perturbation treatment of cubic and quartic terms in the potential energy is considered on pp. 545ff. The resulting expressions are compared, on pp. 543ff., with experimental evidence on the anharmonicities in the low-frequency quadrupole mode. The available data do not appear to be well represented by the lowest-order effects of these anharmonic terms in the potential energy.

The failure of the analysis of the anharmonicity effects in terms of modifications in the potential energy expressed as a power series in the vibrational amplitude appears to reflect the rapid approach to instability of the spherical shape, as particles are added to closed-shell configurations (see the discussion on pp. 520ff.). Indeed, the amplitudes of the quadrupole oscillations, as measured in the $E2$-transition probabilities, are comparable to the static deformations of the nuclei with rotational spectra. In such circumstances, one must expect the vibrational motion to be associated with major changes in the nucleonic coupling scheme affecting not only the potential energy, but also the vibrational mass parameters. A purely phenomenological analysis of such spectra, intermediate between those of harmonic vibrations and rotation-vibration spectra about a statically deformed shape, is difficult because of the large number of parameters involved (see Appendix 6B). An extensive effort is currently being devoted to obtaining guidance from the microscopic analysis.

In the yrast region ($I \approx I_{max} = 2n \gg 1$), the vibrational motion is expected to separate into rotational and intrinsic components; the qualitative classification of the spectrum is therefore less sensitive to the details of the anharmonic terms in the Hamiltonian (see the discussion on p. 687). Experimental evidence concerning this part of the spectrum might thus be especially valuable in helping to sort out the variety of different anharmonic effects.

6-6b Coupling between Quadrupole and Dipole Modes

Closely related to the anharmonic effects in the spectrum of a single mode are the interactions between the quanta of different modes. As a prototype of such couplings, we consider, in the present section, the interaction between dipole and quadrupole oscillations.

The leading-order coupling between the dipole and quadrupole modes is of third order in the amplitudes. The requirements of rotational invariance and symmetry with respect to space reflection and time reversal restrict the third-order terms to the following combinations

$$H' = k_1(\alpha_1\alpha_1\alpha_2)_0 + k_2(\dot{\alpha}_1\dot{\alpha}_1\alpha_2)_0 + k_3(\alpha_1\dot{\alpha}_1\dot{\alpha}_2)_0 \tag{6-276}$$

(Higher-order coupling effects have been considered by Huber *et al.*, 1967.)

The significance of the coefficients k_1, k_2, k_3 in Eq. (6-276) can be seen by viewing the coupling H' as the effect of the quadrupole deformation and its time derivative on the dipole oscillations. Since the frequency of the quadrupole oscillations is small compared to that of the dipole mode, one expects the time dependence of the quadrupole deformation to be relatively unimportant and, hence, the last term in Eq. (6-276) to be small. For fixed values of $\alpha_{2\mu}$, the first two terms in the coupling (6-276) imply a splitting of the dipole mode into three components associated with the three principal axes in the ellipsoidal nuclear shape, and the parameters k_1 and k_2 characterize the effect of the deformation on the restoring force and mass parameter for the three eigenmodes.

The observed properties of the dipole mode in spherical nuclei suggest that the oscillator strength is rather insensitive to the deformation. In fact, in heavy nuclei, the dipole mode is found to approximately exhaust the classical oscillator sum rule, which is independent of the dimensions of the system (see Eq. (6-176) and the empirical data in Fig. 6-20, p. 479). Conservation of oscillator strength for each component of the dipole mode in a deformed nucleus would imply that the mass parameter is unaffected by the deformation, and hence that the second term in Eq. (6-276) can be neglected.

The magnitude of the coefficient k_1 can be estimated from a scaling argument based on the fact that the dipole frequency is approximately inversely proportional to the nuclear radius ($\omega \propto A^{-1/3}$, see Fig. 6-19, p. 476). Hence, a deformation with increments δR_κ of the principal axes, $\kappa = 1,2,3$, leads to frequency shifts $\delta\omega_{1\kappa}$ and variations $\delta C_{1\kappa}$ in the restoring force parameters given by

$$\frac{\delta C_{1\kappa}}{C_1} \approx 2\frac{\delta\omega_{1\kappa}}{\omega_1} \approx -2\frac{\delta R_\kappa}{R} \tag{6-277}$$

where C_1, ω_1, and R refer to the spherical nucleus. The estimate (6-277) for $\delta C_{1\kappa}$ corresponds to the coupling coefficient (the relation between δR_κ and α_2 is given by Eq. (6B-4))

$$k_1 = -\frac{5}{2}\left(\frac{3}{2\pi}\right)^{1/2} C_1 \approx -1.7C_1 \tag{6-278}$$

The value of the numerical coefficient in Eq. (6-278) is somewhat model dependent. Thus, the liquid-drop description of the dipole oscillation in an ellipsoidal nucleus gives values of $\delta\omega_{1\kappa}$ smaller than the estimate (6-277) by a factor 0.91 (see Eq. (6A-75)), and a corresponding reduction of k_1.

The splitting of the dipole frequencies caused by a static quadrupole deformation can be directly observed in the photoabsorption of deformed nuclei (see the example illustrated in Fig. 6-21, p. 491). The separation between the resonance frequencies is found to be in approximate agreement with the estimate (6-277); see Table 6-7, p. 493.

In a spherical nucleus, the coupling term (6-276) implies an interweaving of the quadrupole and dipole motion with significant consequences for the width and line shape of the dipole resonance (Le Tourneux, 1965) as well as for the dipole polarizability of the nucleus (see below). The interaction energy (6-276) is of the order of magnitude $\alpha_2\hbar\omega_1$, which may be larger than $\hbar\omega_2$, but which is small compared with the dipole excitation energy. In first approximation, it is therefore sufficient to consider the coupling between states with the same number of dipole quanta, which is given by (see Eq. (6-51))

$$H' = \frac{k_1\hbar\omega_1}{C_1}\left(\frac{\hbar\omega_2}{2C_2}\right)^{1/2} (c_1^\dagger \bar{c}_1(c_2^\dagger + \bar{c}_2))_0 \tag{6-279}$$

For $n_1 = 0$, the coupling vanishes, and for $n_1 = 1$, the matrix elements are given by (see Eq. (1A-72))

$$\langle n_2+1, R'; n_1=1; IM|H'|n_2, R; n_1=1; IM\rangle$$

$$= (-1)^{R+I+1}\begin{Bmatrix} 1 & R & I \\ R' & 1 & 2 \end{Bmatrix}\langle n_2+1, R'\|c_2^\dagger\|n_2, R\rangle k_1\left(\frac{\hbar\omega_1}{C_1}\right)\left(\frac{\hbar\omega_2}{2C_2}\right)^{1/2} \tag{6-280}$$

where R is the angular momentum of the quadrupole vibrational motion.

The parentage factors for the quadrupole quanta can be evaluated by methods such as those discussed in Sec. 6B-4.

The strength of the coupling (6-279) may be represented by a dimensionless parameter η, which characterizes the ratio between the matrix elements (6-280) and $\hbar\omega_2$,

$$\eta \equiv -\frac{k_1}{C_1}\frac{\omega_1}{\omega_2}\left(\frac{\hbar\omega_2}{2C_2}\right)^{1/2}$$

$$\approx 1.7\frac{\omega_1}{\omega_2}(\alpha_2)_0 \tag{6-281}$$

where $(\alpha_2)_0$ is the zero-point amplitude for the quadrupole shape oscillations (see Eq. (6-52)). In nuclei with large quadrupole vibrational amplitudes, the parameter η is considerably larger than unity (typically of order 5); in the treatment of the coupling, it is therefore in general necessary to go beyond a perturbation calculation.

For large values of η, one can obtain an approximate description of the line shape for photoabsorption by considering the process for fixed values of $\alpha_{2\mu}$ and subsequently averaging over the zero-point quadrupole motion in the initial state (Semenko, 1964; Kerman and Quang, 1964; Le Tourneux, 1965). In fact, for $\eta \gg 1$, the transition strength is distributed over an energy interval large compared with $\hbar\omega_2$; thus, we may consider incident wave packets with a time spread short compared with the period of the quadrupole motion. During this time interval, one can neglect the time dependence of $\alpha_{2\mu}$, and the probability per unit energy for absorption of a quantum E is proportional to

$$P(E) \equiv \frac{1}{3}\sum_{\kappa=1}^{3}\int\int \varphi_0^2(\beta)\delta(E-\hbar\omega_{1\kappa}(\beta,\gamma))\beta^4 d\beta|\sin 3\gamma|d\gamma \tag{6-282}$$

where $\varphi_0(\beta)$ is the ground-state wave function for quadrupole oscillations (see Eq. (6B-20a); the volume element in β,γ space is given by Eq. (6B-18)). The frequencies

$$\omega_{1\kappa}(\beta,\gamma) = \omega_1\left(1+\left(\frac{2}{15}\right)^{1/2}\frac{k_1}{C_1}\beta\cos\left(\gamma-\frac{2\pi}{3}\kappa\right)\right) \tag{6-283}$$

correspond to the three dipole resonances for an ellipsoidal nucleus with shape parameters β and γ (see Eqs. (6-277) and (6B-4)). Carrying out the integrals in Eq. (6-282), one obtains

$$P(E)=\left(\frac{5}{3\pi}\right)^{1/2}(\eta\hbar\omega_2)^{-1}\left(\left(\frac{45x^2}{4\eta^2}-1\right)\exp\left\{-\frac{15x^2}{4\eta^2}\right\}+2\exp\left\{-\frac{15x^2}{\eta^2}\right\}\right)$$

$$x\equiv\frac{E-\hbar\omega_1}{\hbar\omega_2} \qquad (6\text{-}284)$$

The quantity $P(E)$ is normalized to unity when integrated over E, and $P(E)dE$ thus represents the sum of intensities $\langle n_1=1,n_2=0|i\rangle^2$ for the eigenstates i, with $n_1=1$ (and fixed value of M), in the energy interval considered.

More generally, one can estimate the width of the dipole strength function resulting from the coupling (6-279) by evaluating the second moment

$$\begin{aligned}\langle(E-\hbar\omega_1)^2\rangle &= \sum_i (E_i-\hbar\omega_1)^2\langle n_1=1,n_2=0|i\rangle^2 \\ &= \langle n_1=1,n_2=0|(H')^2|n_1=1,n_2=0\rangle \\ &= \tfrac{1}{3}\eta^2(\hbar\omega_2)^2=0.20\beta_2^2(\hbar\omega_1)^2 \qquad (6\text{-}285)\end{aligned}$$

where β_2^2 is the mean square quadrupole fluctuation in the ground state (see Eq. (6-52)). The mean position of the transition strength is not affected by the coupling, since the expectation value of H' vanishes in the state $n_1=1$, $n_2=0$.

The second moment is a useful measure of the line broadening in the present case, since the sum over i in Eq. (6-285) is rather insensitive to the contributions from the distant tails of the strength function. (This situation contrasts to the case of Lorentzian line shapes, for which the second moment diverges and therefore does not provide a measure of the line width; see the discussion in Vol. I, p. 305.) Indeed, all the higher moments $\langle(E-\hbar\omega_1)^n\rangle$, for even n, are seen to be of order $(\eta\hbar\omega_2)^n$, corresponding to a strength function decreasing more rapidly than any power of the energy. (See the Gaussian form of the strength function (6-284) for large η.)

The detailed structure of the dipole absorption line can be obtained from a numerical diagonalization of the coupled Hamiltonian. An example is illustrated in Fig. 6-15. The strong coupling approximation is seen to account for the gross features of the line broadening in Fig. 6-15, though it does not reproduce the finer details of the calculated spectrum.

The estimate (6-285) of the contribution of the dipole-quadrupole coupling to the width of the dipole resonance is compared with the observed widths in Fig. 6-25, p. 503. The effect of the coupling appears to represent a significant part of the total width, but in most cases other interactions make comparable contributions.

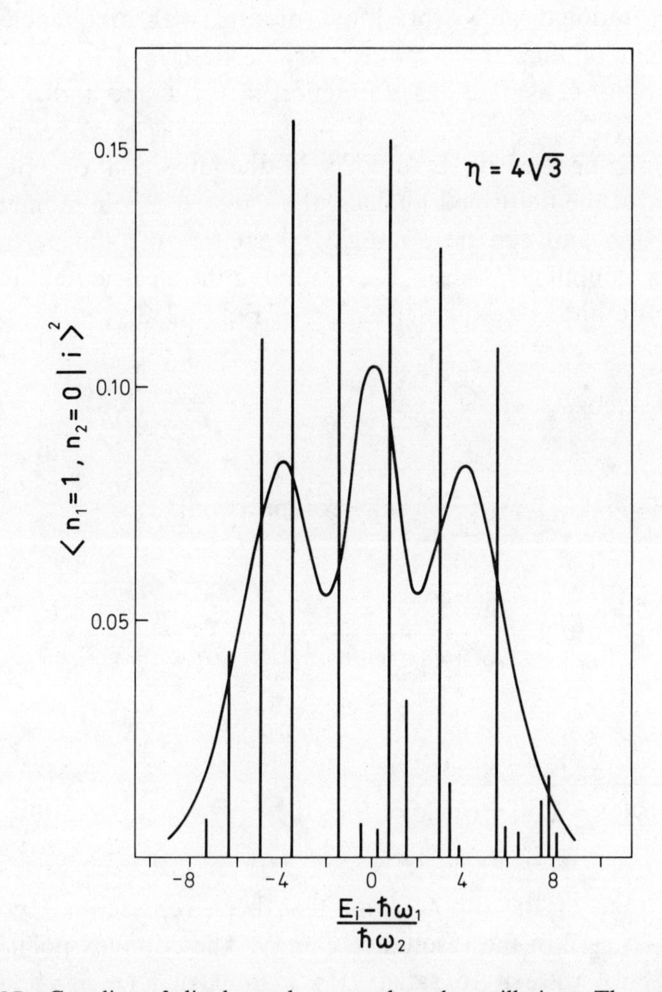

Figure 6-15 Coupling of dipole mode to quadrupole oscillations. The matrix representing coupled dipole and quadrupole vibrations has been diagonalized for states with a single dipole quantum ($n_1=1$) and total angular momentum $I=1$, with the inclusion of all components with up to 14 quadrupole quanta. The length of lines for the various eigenvalues E_i measures the square amplitude of the component with $n_2=0$ and thus gives the relative strength of the transition probability $B(E1; 0 \rightarrow i)$ for dipole absorption from the ground state. The solid curve corresponds to the strong-coupling approximation given by Eq. (6-284). The figure is taken from J. Le Tourneux, *Mat. Fys. Medd. Dan. Vid. Selsk.* **34**, no. 11 (1965).

The dipole-quadrupole coupling can also be studied in second-order processes that involve the virtual excitation of the dipole mode and lead to rotational or vibrational excitations of the nucleus (Raman scattering of γ rays (Baldin, 1959; Marić and Möbius, 1959; see also the review by Fuller and Hayward, 1962a); Coulomb excitation of quadrupole vibrations arising from second-order dipole excitations (Eichler, 1964; MacDonald, 1964; see also the review by de Boer and Eichler, 1968)). In these processes, the excitation of rotational and vibrational modes, with frequencies small compared to that of the dipole mode, may be described in terms of the dipole polarizability considered as a function of the shape and orientation of the nucleus.

If the polarizing electric field **E** is in the direction of one of the principal axes κ of the deformed nucleus, the induced dipole moment **d** has the same direction and can be obtained by considering the perturbation produced by the coupling $H' = -d_\kappa E_\kappa$. Assuming the electric field to have a frequency ω_E, one finds (see the derivation of Eq. (6-238))

$$E_\kappa = (E_\kappa)_0 \cos \omega_E t$$

$$\langle d_\kappa \rangle = p_\kappa(\omega_E) \tfrac{1}{2} (E_\kappa)_0 \exp\{-i\omega_E t\} + \text{complex conj.} \tag{6-286a}$$

$$p_\kappa(\omega_E) = |\langle n_{1\kappa} = 1 | d_\kappa | 0 \rangle|^2 \left(\frac{1}{\hbar(\omega_{1\kappa} - \omega_E) - \frac{1}{2} i \Gamma_{1\kappa}} + \frac{1}{\hbar(\omega_{1\kappa} + \omega_E) + \frac{1}{2} i \Gamma_{1\kappa}} \right)$$

$$= \frac{2|\langle n_{1\kappa} = 1 | d_\kappa | 0 \rangle|^2 \hbar\omega_{1\kappa}}{(\hbar\omega_{1\kappa})^2 + \frac{1}{4}\Gamma_{1\kappa}^2 - (\hbar\omega_E)^2 - i\hbar\omega_E \Gamma_{1\kappa}} \tag{6-286b}$$

where the damping of the dipole states has been represented by an imaginary term $-i\Gamma_{1\kappa}/2$ in the resonance energy. The complex polarizability coefficient (6-286b) is seen to satisfy the symmetry $p_\kappa(-\omega) = p_\kappa^*(\omega)$ connected with the reality of the field E_κ and the moment d_κ. For $\Gamma_{1\kappa} = 0$, the result (6-286b) is equivalent to the expression (6-216), but differs by a factor $-\kappa$, corresponding to the fact that the coupling involved in the definition of the polarizability coefficient χ is $-\kappa\alpha F$.

The $E1$-oscillator strength for each component of the dipole mode is expected to be approximately independent of deformation (see p. 453);

hence, ignoring the small shift in the resonance frequency, of order Γ^2,

$$p_\kappa = p\frac{\omega_1^2-\omega_E^2-i\omega_E\hbar^{-1}\Gamma_1}{\omega_{1\kappa}^2-\omega_E^2-i\omega_E\hbar^{-1}\Gamma_{1\kappa}}$$

$$p=\frac{8\pi}{9}B(E1;n_1=0\to n_1=1)\frac{\hbar\omega_1}{(\hbar\omega_1)^2-(\hbar\omega_E)^2-i\hbar\omega_E\Gamma_1} \tag{6-287}$$

$$\approx\frac{NZe^2}{MA(\omega_1^2-\omega_E^2-i\omega_E\hbar^{-1}\Gamma_1)}$$

in terms of the polarizability p for the undeformed nucleus with resonance energy $\hbar\omega_1 - i\Gamma_1/2$. In the second expression for p, we have assumed the classical sum-rule value (6-176) for the oscillator strength of the dipole mode.

For an electric field with arbitrary direction relative to the nuclear orientation, the polarizability can be described, in Cartesian coordinates, as a symmetric second-rank tensor with eigenvalues p_κ and principal axes coinciding with the intrinsic nuclear coordinate system. In a spherical tensor decomposition, the polarizability involves a scalar $p^{(0)}$ and a quadrupole tensor $p_\mu^{(2)}$,

$$\langle d_\mu\rangle = p^{(0)}E_\mu + (p^{(2)}E)_{(21)1_\mu} \tag{6-288}$$

where d_μ and E_μ are the spherical components of the vectors **d** and **E** (see Eq. (1A-56)). The spherical tensor components of the polarizability are related to the eigenvalues p_κ and orientation angles ϕ,θ,ψ by a transformation of standard type

$$p^{(0)}=\tfrac{1}{3}(p_1+p_2+p_3)$$

$$p_\mu^{(2)}=-\tfrac{1}{3}\left(\tfrac{5}{2}\right)^{1/2}\Big((2p_3-p_1-p_2)\mathscr{D}^2_{\mu0}(\phi,\theta,\psi)$$

$$+\left(\tfrac{3}{2}\right)^{1/2}(p_1-p_2)\left(\mathscr{D}^2_{\mu2}(\phi,\theta,\psi)+\mathscr{D}^2_{\mu-2}(\phi,\theta,\psi)\right)\Big) \tag{6-289}$$

(An analogous relationship expresses the quadrupole deformations $\alpha_{2\mu}$ in terms of the nuclear orientation angles and the increments δR_κ of the nuclear axes; see Eqs. (6B-1), (6B-2), and (6B-4).)

In situations where the deformation has only a small effect on the polarizability, one may employ an expansion in the nuclear deformation, which yields, to first order (see Eqs. (6-287) and (6-277)),

$$p_\kappa = p\left(1 + \frac{2\omega_1^2}{\omega_1^2 - \omega_E^2}\frac{\delta R_\kappa}{R}\right) \tag{6-290}$$

where damping terms have been ignored. The corresponding tensor polarizabilities are

$$\begin{aligned} p^{(0)} &= p \\ p_\mu^{(2)} &= -5(2\pi)^{-1/2} p \frac{\omega_1^2}{\omega_1^2 - \omega_E^2}\alpha_{2\mu} \end{aligned} \tag{6-291}$$

in terms of the quadrupole deformation amplitude $\alpha_{2\mu}$ (see Eqs. (6B-2) and (6B-4)).

The dependence of the polarizability on the orientation and shape of the nucleus implies a coupling to the rotational and vibrational degrees of freedom. Thus, the scattering amplitude for γ rays on nuclei is proportional to the induced dipole moment; the relative probabilities for elastic and inelastic scattering are therefore determined by the absolute squares of the matrix elements of the polarizability coefficients. (See the corresponding adiabatic treatment of the scattering of a particle from a nonspherical nucleus in Appendix 5A, pp. 322 ff.) Evidence on the Raman scattering of γ rays on deformed nuclei is discussed on p. 492.

The dipole polarizability also gives rise to an interaction between the nucleus and a charged particle, which may affect scattering cross sections and energy levels of bound systems. If the periods of the relative motion are large compared with that of the dipole mode, the interaction can be expressed in terms of the static polarizability, and one obtains from Eqs. (6-288) and (6-291)

$$V_{\text{pol}} = -\frac{1}{2}\langle \mathbf{d}\rangle\cdot\mathbf{E} = -\frac{1}{2}\left(\frac{Z_1 e}{r^2}\right)^2 p(\omega_E = 0)\left(1 + 2\sum_\mu \alpha_{2\mu} Y_{2\mu}^*(\vartheta,\varphi)\right) \tag{6-292}$$

where r, ϑ, φ are the polar coordinates of the particle, with charge $Z_1 e$.

6-6c Coupling between Vibration and Rotation

In the deformed nuclei, the vibrational modes have been considered, in first approximation, with respect to a static nonspherical equilibrium (Sec. 6-3b). The finite frequency of the rotational motion implies a coupling between the vibrational and rotational degrees of freedom, which may be analyzed in the general manner discussed in Chapter 4.

The leading-order rotation-vibration coupling is linear in the rotational angular momentum ($\Delta K=1$; see Eq. (4-196)). Such a term arises from the Coriolis force acting on the intrinsic angular momentum (see Eq. (4-197)),

$$H' = -\frac{\hbar^2}{2\mathscr{I}_0}(R_+ I_- + R_- I_+) \tag{6-293}$$

where $R_\pm = R_1 \pm iR_2$ are the components of the vibrational angular momentum. If the nucleus possesses modes with different ν that can be approximately represented as different orientations of a mode of multipole order λ (see the discussion on pp. 363 ff.), the matrix elements of R_+ for shifting a quantum from ν to $\nu+1$ can be expressed in terms of λ and ν, and one obtains (see Eqs. (1A-4) and (1A-93)),

$$\langle n_{\nu+1}=1, K=\nu+1, IM | H' | n_\nu = 1, K=\nu, IM \rangle$$

$$= -\frac{\hbar^2}{2\mathscr{I}_0}((\lambda-\nu)(\lambda+\nu+1)(I-\nu)(I+\nu+1))^{1/2} \begin{cases} 2^{1/2} & \nu=0 \\ 1 & \nu>0 \end{cases} \tag{6-294}$$

The estimate (6-294) is expected to apply to the high-frequency modes, for which the effect of the deformation is a relatively small perturbation. For these modes, however, the Coriolis coupling is very small compared with the energy separations of the modes with different ν. (While these energy separations are of order $\hbar\omega_0\delta \sim \varepsilon_F A^{-2/3}$, the Coriolis coupling is of order $\lambda I \varepsilon_F A^{-5/3}$.) For the dipole mode, the weakness of the rotational coupling is confirmed by the fact that the ratio of the oscillator strengths for exciting the $\nu=0$ and $\nu=1$ modes is found to be in agreement with the value 1:2, as obtained by neglecting the rotational motion (see, for example, Fig. 6-21, p. 491).

For the low-frequency modes, the energy separations $\hbar(\omega_{\nu+1}-\omega_\nu)$ are much smaller (typically of the order of a few hundred keV). Couplings of the order of magnitude (6-294) may therefore lead to a major decoupling of the vibrational motion from the static deformation, tending to concentrate the transition strength onto the lowest vibrational excitation of the given multipole order, as in a spherical nucleus. However, the low-frequency modes are affected in a major way by the deformation (see the discussion on p. 364), and the estimate (6-294) is not expected to have quantitative validity. Effects of the Coriolis coupling in low-frequency octupole bands are discussed on p. 578. For the low-frequency quadrupole mode in spheroidal nuclei, there are no first-order rotational couplings of the form (6-294), since the $\nu=1$ mode corresponds to the rotational motion itself (see p. 362).

The couplings of second order in the rotational angular momentum have $\Delta K=0$ and 2 and are of the form (4-216) and (4-206), respectively. When the vibrational motion is adiabatic with respect to the intrinsic motion ($\hbar\omega_{\text{vib}} \ll \Delta E_{\text{intr}}$), the intrinsic operators h_0 and h_{+2} can be expressed in terms of the dependence of the moment of inertia on the vibrational amplitudes α,

$$h_0 = \frac{\hbar^2}{2}\left(\frac{1}{\mathscr{I}(\alpha)} - \frac{1}{\mathscr{I}(\alpha_{\text{eq}})}\right) \qquad \left(\mathscr{I}^{-1} = \frac{1}{2}(\mathscr{I}_1^{-1} + \mathscr{I}_2^{-1})\right)$$

$$h_{+2} = \frac{\hbar^2}{8}\left(\frac{1}{\mathscr{I}_1(\alpha)} - \frac{1}{\mathscr{I}_2(\alpha)}\right) \tag{6-295}$$

where α_{eq} denotes the axially symmetric equilibrium shape, and where $\mathscr{I}(\alpha)$ is the moment of inertia for a static deformation α.

For the quadrupole vibrations, an expansion of the coupling parameters (6-295) yields, to first order in the vibrational amplitudes, $\beta-\beta_0$ and γ (see Eq. (6-90)),

$$h_0 \approx -\frac{\hbar^2}{2}(\beta-\beta_0)\left(\frac{1}{\mathscr{I}^2}\frac{\partial \mathscr{I}}{\partial \beta}\right)_{\beta=\beta_0}$$

$$h_{+2} \approx -\frac{\hbar^2}{4}\gamma\left(\frac{1}{\mathscr{I}^2}\frac{\partial \mathscr{I}_1}{\partial \gamma}\right)_{\gamma=0} \tag{6-296}$$

We have employed the relation $\mathscr{I}_1(\gamma)=\mathscr{I}_2(-\gamma)$, which follows from the symmetry of the deformation (see Eq. (6B-15)). The matrix elements of h_0 and h_{+2} can be determined from the analysis of the relative $E2$ intensities for the transitions between the vibrational excitations and the ground-state band. For the coefficients a_0 and a_2 in Eqs. (4-252) and (4-230), one obtains the relations (see Eqs. (4-208), (4-211), (4-218), (4-220), and (4-231))

$$a_0 = -\frac{\hbar^2}{2\mathscr{I}}\frac{\beta_0}{\hbar\omega_\beta}\left(\frac{1}{\mathscr{I}}\frac{\partial \mathscr{I}}{\partial \beta}\right)_{\beta=\beta_0} \tag{6-297a}$$

$$a_2 = \frac{z_2}{2+4z_2} \tag{6-297b}$$

$$z_2 = -\sqrt{3}\,\frac{\hbar^2}{\mathscr{I}}\frac{1}{\hbar\omega_\gamma}\left(\frac{1}{\mathscr{I}}\frac{\partial \mathscr{I}_1}{\partial \gamma}\right)_{\gamma=0} \tag{6-297c}$$

We have used here the expressions (see Eqs. (6-90) and (6-91a))

$$\mathscr{M}(E2,\nu)=\left(\frac{5}{16\pi}\right)^{1/2} eQ_0\begin{cases}\beta_0^{-1}(\beta-\beta_0) & \nu=0\\ 2^{-1/2}\gamma & \nu=2\end{cases} \tag{6-298}$$

for the intrinsic $E2$ moment of the β- and γ-vibrational motion, relative to the static quadrupole moment Q_0.

Examples of the analysis of the coupling between the rotational motion and the β and γ vibrations are discussed on pp. 160ff. (γ vibration in ^{166}Er) and on pp. 172ff. (β vibration in ^{174}Hf). The values of $\mathscr{I}^{-1}(\partial\mathscr{I}_1/\partial\gamma)$ and $\beta\mathscr{I}^{-1}(\partial\mathscr{I}/\partial\beta)$ are found to be of order unity. For rigid rotation, these logarithmic derivatives are of the order of the deformation parameter β_0, but the observed larger values are consistent with the expected strong dependence of the moment of inertia on the deformation (see, for example, Eqs. (4-249) and (4-267)).

The expansion of the effective moments of inertia in Eq. (6-295) gives rise to a multitude of coupling terms of higher order. The terms of second order in the vibrational amplitude include couplings that are diagonal in the vibrational quantum numbers and represent the change in moment of inertia resulting from the increase in $\langle\alpha^2\rangle$ associated with a vibrational quantum. The second-order terms also include couplings that shift a quantum between modes with $\Delta\nu=0, \pm 2$, such as the coupling between β- and γ-vibrational bands. Second-order inertial terms of this type are expected to be an order of magnitude smaller than the first-order terms (6-296); however, for the β and γ vibrations, the observed second-order matrix elements are in many cases of the order of 1 keV, as for the first-order terms. (The moments of inertia for bands with $n_\beta=1$ and $n_\gamma=1$ can be obtained from the figures referred to on p. 549 (γ vibrations) and p. 551 (β vibrations). Evidence on the matrix element $\langle n_\gamma=1|h_2|n_\beta=1\rangle$ has been reported by Rud *et al.*, 1971, ^{152}Sm and ^{154}Gd; Baader, 1970, ^{158}Gd; Günther *et al.*, 1971, ^{182}W and ^{184}W; Stephens *et al.*, 1963, ^{238}U and ^{232}Th.) This failure of the description in terms of a classical adiabatic expansion may be connected with the rather few quasiparticle configurations that effectively contribute to these excitations. In fact, for two-quasiparticle excitations, the couplings involved ($\Delta\text{v}=0$) are expected to be rather large (see, for example, the increments in the moment of inertia for quasiparticle excitations discussed on p. 251). It is also possible that the large coupling effects are related to the additional degrees of freedom that are observed in the $K\pi=0+$ and $2+$ channels (see p. 554).

ILLUSTRATIVE EXAMPLES TO CHAPTER 6

Response Function

Single-particle excitation spectra for multipole fields in a nucleus with $Z=46$, $N=60$ (Figs. 6-16 and 6-17; Tables 6-4 and 6-5)

The starting point for the microscopic description of the nuclear vibrational motion is the analysis of the one-particle excitation spectrum produced by a field with the structure characteristic of the mode in question (see Sec. 6-2c). In the present example, we consider the distribution of transition strength associated with multipole fields of the type $F=r^{\lambda}Y_{\lambda\mu}$ evaluated for single-particle motion in a static spherical potential. (For spin-dependent fields and pair fields, see the examples on pp. 636ff. and pp. 641ff., respectively.) The response of the nucleus to a field F can be expressed in terms of the polarizability coefficient χ considered as a function of the frequency of the field. The nuclear excitation frequencies are the poles of this response function, while the transition strength gives the residues of the poles (see Eq. (6-241) for the response function for independent-particle motion).

One-particle energies

The spectra in Figs. 6-16 and 6-17 refer to a nucleus with $A=106$ and $Z=46$ and have been obtained from a modified harmonic oscillator potential (Nilsson, 1955),

$$V=\tfrac{1}{2}M\omega_0^2r^2+v_{ls}\hbar\omega_0(\mathbf{l}\cdot\mathbf{s})+v_{ll}\hbar\omega_0(\mathbf{l}^2-\langle\mathbf{l}^2\rangle_N) \tag{6-299}$$

with the parameters

$$\begin{aligned}
\hbar\omega_0&=41A^{-1/3}\ \text{MeV}=8.7\quad\text{MeV}\\
v_{ls}&=-0.10\\
v_{ll}&=-\begin{cases}0 & N=0,1,2\\ 0.0175 & N=3\\ 0.0225 & N=4,5,6\\ 0.020 & N=7,8\end{cases}
\end{aligned} \tag{6-300}$$

The term in the potential (6-299) proportional to $\mathbf{l}^2$ describes the deviations from the harmonic oscillator degeneracies resulting from the more sharply defined surface of the nuclear potential, and the values of v_{ll} in Eq. (6-300)

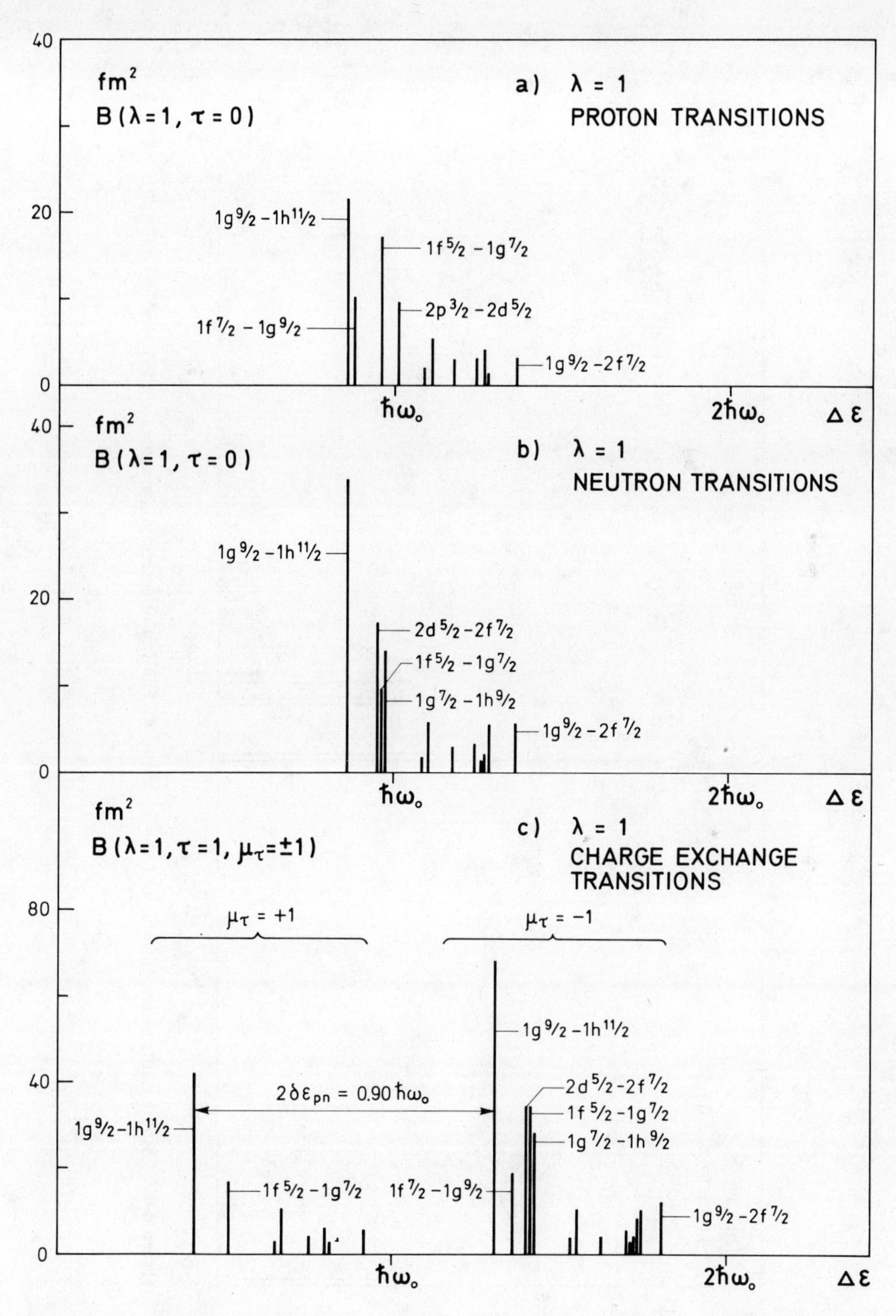

Figure 6-16 Single-particle excitations with $\lambda = 1$. The figure refers to a nucleus with $Z = 46$ and $N = 60$. We wish to thank C. J. Veje and Jens Damgaard for help in the preparation of this and the following figure.

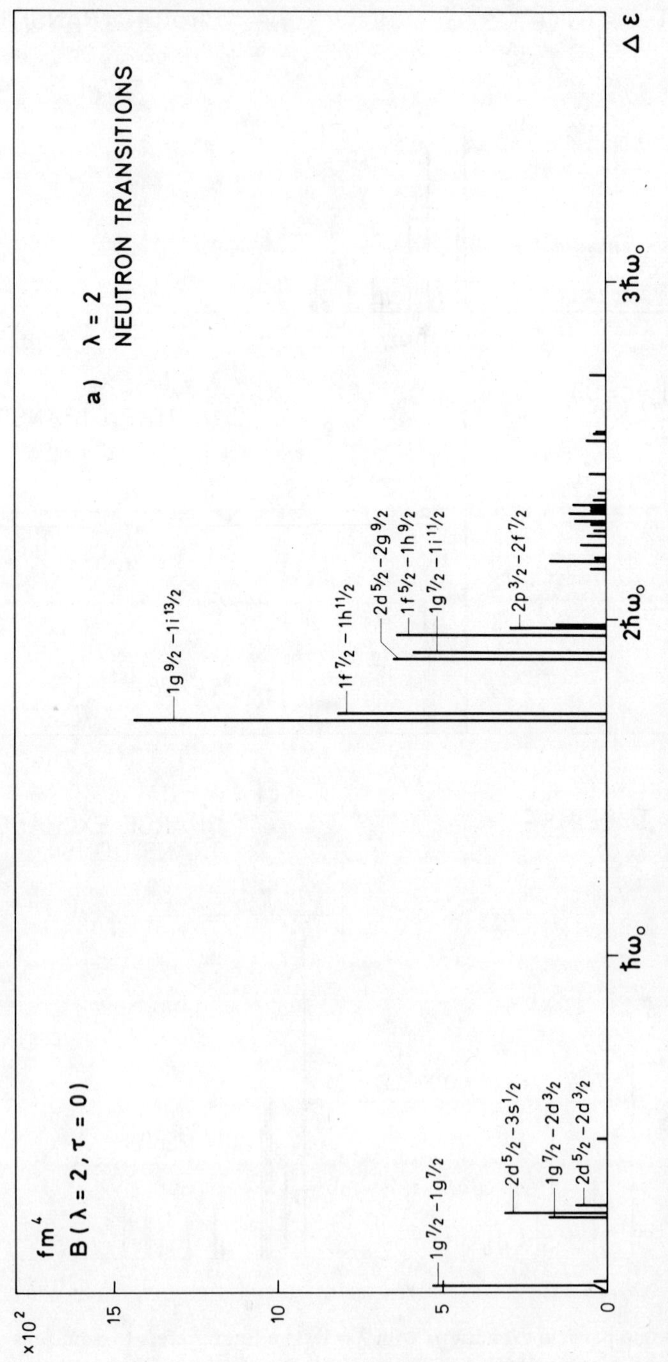

Figure 6-17 Single-neutron excitations with λ = 2, 3, and 4. The figure refers to a nucleus with $N = 60$.

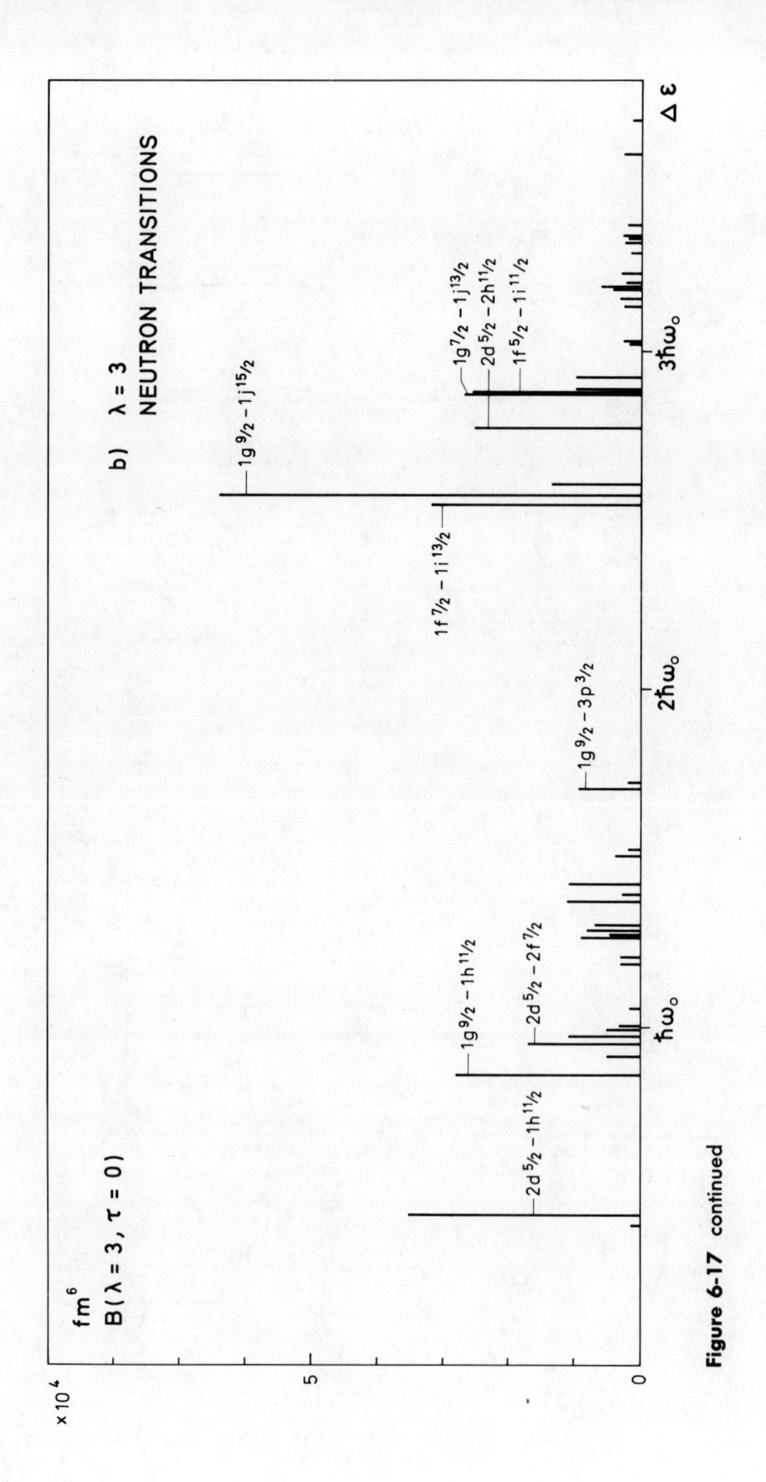

Figure 6-17 continued

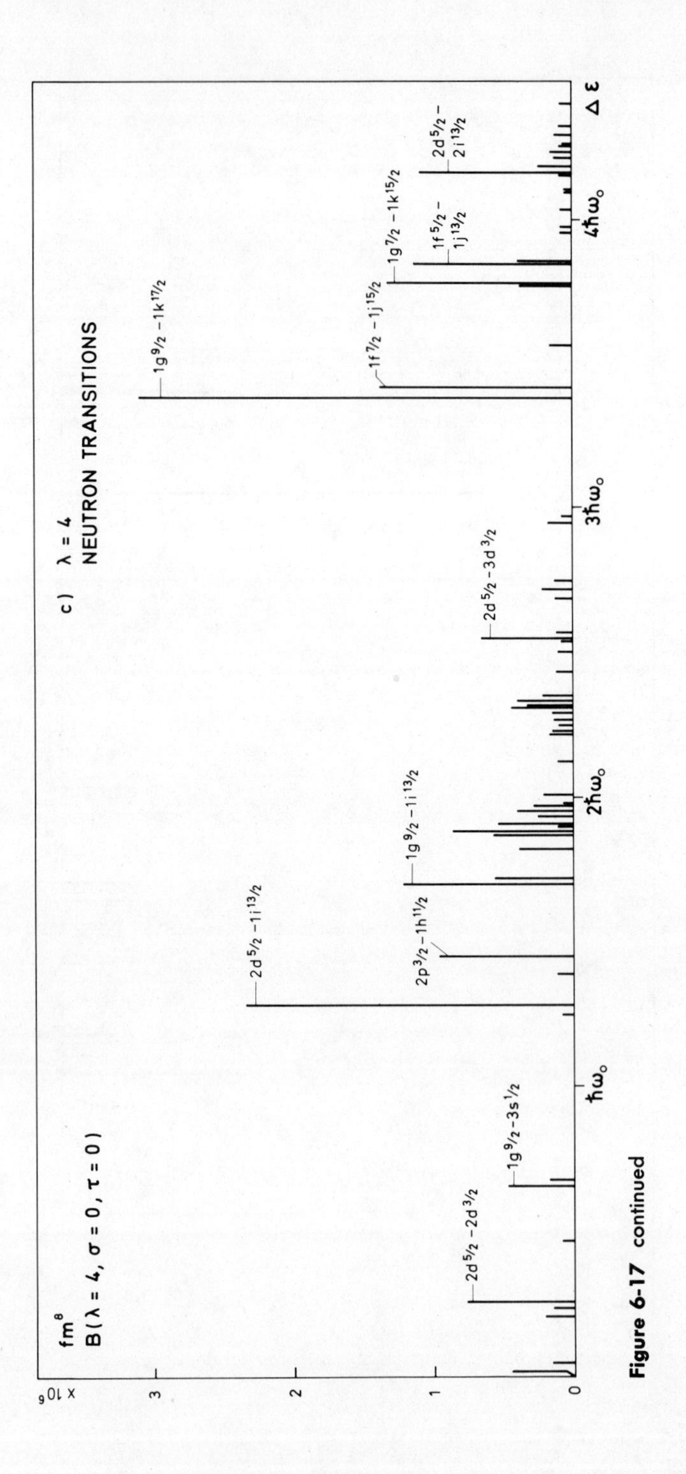

Figure 6-17 continued

▼ are in approximate agreement with those estimated for a Woods-Saxon potential (see Fig. 6-49, p. 593). The principal oscillator quantum number is denoted by N, and $\langle \mathbf{l}^2 \rangle_N$ is the average value for the levels in the major oscillator shell considered,

$$\langle \mathbf{l}^2 \rangle_N = \tfrac{1}{2} N(N+3) \tag{6-301}$$

With this term subtracted, the distance between the centroids of two adjacent major shells remains equal to $\hbar\omega_0$ (Gustafson *et al.*, 1967).

The potential (6-299) approximately reproduces the observed energies of the one-particle states in the neighborhood of the Fermi level. Since the discussion in the present section is only intended to illustrate the qualitative patterns in the one-particle response function, no attempt has been made to adjust the parameters of the potential so as to reproduce the more detailed features of the one-particle spectra.

The ground-state configuration of the nucleus considered is assumed to involve the closed shells $Z=40$ and $N=50$ and, in addition, the proton configuration $(1g_{9/2})^6$ and the neutron configuration $(2d_{5/2})^6(1g_{7/2})^4$ (see Fig. 2-30, Vol. I, p. 239).

The energies of charge exchange transitions ($\mu_\tau = \pm 1$) in Fig. 6-16 are shifted relative to the $\mu_\tau = 0$ transitions by the amount $-0.45\,\hbar\omega_0\,\mu_\tau$. The shift is given by the energy separation $\delta\varepsilon_{\text{pn}}$ between corresponding levels of neutrons and protons,

$$\delta\varepsilon_{\text{pn}} = \varepsilon(j,\text{proton}) - \varepsilon(j,\text{neutron}) \tag{6-302}$$

which is the sum of the contributions from the Coulomb energy and the symmetry potential associated with the neutron excess. Since the nucleus is close to the line of β stability, the Fermi levels for neutrons and protons are approximately equal ($\varepsilon(g_{9/2},\text{ proton}) \approx \varepsilon(g_{7/2},\text{ neutron})$) and, hence,

$$\begin{aligned}\delta\varepsilon_{\text{pn}} &= \varepsilon(g_{7/2},\text{neutron}) - \varepsilon(g_{9/2},\text{neutron}) \\ &= -\tfrac{9}{2} v_{ls}\hbar\omega_0 = 0.45\,\hbar\omega_0\end{aligned} \tag{6-303}$$

Transition probabilities

The ordinates in Figs. 6-16 and 6-17 give the multipole strength of the excitations. For transitions from a completely filled shell, the reduced transition probability is (see Eqs. (3B-25) and (1A-67))

$$B(\lambda; 0 \rightarrow (j_1^{-1} j_2)\lambda) = (2j_1+1) B_{\text{sp}}(\lambda; j_1 \rightarrow j_2) \tag{6-304a}$$

▲
$$B_{\text{sp}}(\lambda; j_1 \rightarrow j_2) = (2j_1+1)^{-1} |\langle j_2 \| r^\lambda Y_\lambda \| j_1 \rangle|^2 \tag{6-304b}$$

The single-particle reduced matrix element can be evaluated by means of Eq. (3A-14). The factor $(2j_1+1)$ in Eq. (6-304a) represents the number of particles in the filled shell contributing to the transition strength.

For the transitions involving the partly filled shells, we have assumed a ground state in which the neutrons and protons are separately coupled to a total angular momentum zero, and the multipole strength given in the figure is based on the expressions

$$B(\lambda;(j_2^n)_0\rightarrow(j_1^{-1}j_2^{n+1})_\lambda)=\frac{2j_1+1}{2j_2+1}(2j_2+1-n)B_{\text{sp}}(\lambda;j_1\rightarrow j_2) \quad \text{(6-305a)}$$

$$B(\lambda;(j_1^n)_0\rightarrow(j_1^{n-1}j_2)_\lambda)=nB_{\text{sp}}(\lambda;j_1\rightarrow j_2) \quad \text{(6-305b)}$$

$$B(\lambda;(j^n)_0\rightarrow(j^n)_\lambda)=\frac{2n}{2j-1}(2j+1-n)B_{\text{sp}}(\lambda;j\rightarrow j) \quad \text{(6-305c)}$$

The expression (6-305a) represents a linear interpolation between the result (6-304a), for $n=0$, and the value zero for $n=2j_2+1$. Similarly, the expression (6-305b) is a linear interpolation between the results for $n=0$ and $n=2j_1+1$. The transition probability (6-305c) vanishes for $n=0$ and $n=2j+1$, and the expression given represents a quadratic interpolation based on the value for $n=2$, which may be obtained from Eq. (1A-72a); the factor 4 reflects the coherence of the two particles, each of which would contribute $B_{\text{sp}}(\lambda;j\rightarrow j)$. (The expressions (6-305) are also equal, apart from terms of relative order j^{-1}, to those obtained from the quasiparticle coupling scheme (see Eq. (6-307)) by noting that for a configuration j^n the amplitudes of the pair-correlated wave function are given by $v^2=n(2j+1)^{-1}$ and $u^2=(2j+1-n)(2j+1)^{-1}$.)

The figures give the response function for the neutrons and protons separately. (For a $\tau=0$ or a $\tau=1$ field, the response function is the sum of the contributions of the neutrons and the protons.) For charge exchange transitions, Fig. 6-16 gives the transition probabilities $B(\lambda=1,\tau=1,\mu_\tau=\pm1)$, which are obtained by multiplying the expressions (6-304) and (6-305) for corresponding orbits by a factor 2, representing the square of the matrix element $\langle \text{n}|\tau_{+1}|\text{p}\rangle$ (see Eq. (6-122)).

The qualitative pattern of the spectra in Figs. 6-16 and 6-17 can be understood in terms of the selection rules involving the harmonic oscillator quantum number N

$$\begin{aligned} &\Delta N=1 && \lambda=1\\ &\Delta N=0,2 && \lambda=2\\ &\Delta N=1,3 && \lambda=3\\ &\Delta N=0,2,4 && \lambda=4 \end{aligned} \quad \text{(6-306)}$$

Thus, for the dipole transitions with $\mu_\tau=0$, the excitation frequencies are roughly equal to the oscillator frequency $\hbar\omega_0$. For the quadrupole transitions,

▼ the excitations cluster in two groups with energies of about 0 and $2\hbar\omega_0$. For $\lambda=3$, the selection rule (6-306) suggests a clustering around excitation energies of $\hbar\omega_0$ and $3\hbar\omega_0$; such a tendency is also exhibited by the spectrum in Fig. 6-17b. The spin-orbit coupling and the $\mathbf{l}^2$ term in the potential (6-299) give rise to a rather low-energy octupole transition ($d_{5/2}\rightarrow h_{11/2}$) involving orbits entirely within the $50<N<82$ major shell. (The occurrence of this low-frequency component in the octupole response function can be seen as part of the systematic tendency toward new types of shell structure in heavy nuclei; see pp. 591 ff.)

Another significant selection rule in the spectra in Figs. 6-16 and 6-17 results from the fact that the fields considered are spin independent. Thus, when the multipole order is small compared with the j values of the single-particle states, the spin-flip transitions ($j_1=l_1\pm 1/2\rightarrow j_2=l_2\mp 1/2$) are weak compared with the transitions that preserve the relative orientation of spin and orbit ($j_1=l_1\pm 1/2\rightarrow j_2=l_2\pm 1/2$). The selection rule is contained in the vector addition coefficient $\langle j_1\frac{1}{2}\lambda 0|j_2\frac{1}{2}\rangle$ that appears in the evaluation of the reduced matrix element in Eq. (6-304); see Eq. (3C-34).

Wave function for collective mode

From the single-particle response function, one can obtain the structure of the collective mode generated by the field F in the manner described in Secs. 6-2c and 6-5h. In the following examples, we employ such a procedure to estimate the frequencies and transition probabilities for the collective modes associated with $\lambda=1$, 2, and 3.

The analysis in terms of the response function and the field coupling also yields the microscopic structure of the collective mode expressed in terms of the amplitudes X and Y that characterize the excitation of a quantum in terms of the variables of the one-particle motion (see Eq. (6-36) and the more general expression (6-253) that applies to multifrequency single-particle response functions). As an example of such an analysis, Table 6-4 gives the amplitude for the dipole mode obtained from the response function in Fig. 6-16. The field associated with the dipole mode is mainly the isovector $F=x\tau_z$, but also includes a small isoscalar component proportional to the neutron excess, which is required to make the mode orthogonal to the center-of-mass motion (see Eq. (6-328)). For the single-particle transitions $j_1\rightarrow j_2$ listed in columns 1 and 2, the excitation energies in units of the oscillator quantum are listed in column 3. The next two columns give the transition strengths (for neutrons and protons, respectively) in units of the corresponding single-particle transition probabilities. The excitations involving the promotion of a particle from j_1 to j_2 are denoted by (12), and we assume a coupling scheme for the particles outside closed shells that leads to the expressions (6-305) for $B(0\rightarrow(12))$. The single-particle matrix elements of

▲ F may be obtained from Eqs. (2-154) and (3A-14); see also Eq. (3B-25). The amplitudes $X(12)$ and $Y(12)$ in the last columns of Table 6-4 are given by Eq. (6-253); the collective frequency is taken from the estimate (6-315), which gives the value $\omega_a = 1.92\omega_0$ (in rather good agreement with the observed photoresonance frequency in this region of elements (see Fig. 6-19, p. 476)).

			$\frac{B(0\to(12))}{B_{sp}(j_1\to j_2)}$		Neutrons		Protons	
j_1	j_2	$\frac{\varepsilon_2-\varepsilon_1}{\hbar\omega_0}$	n	p	$X(12)$	$Y(12)$	$X(12)$	$Y(12)$
$2p_{1/2}$	$3s_{1/2}$	1.09	2	2	−0.088	−0.024	0.114	0.032
	$2d_{3/2}$	1.11	2	2	0.167	0.045	−0.218	−0.059
$2p_{3/2}$	$3s_{1/2}$	1.24	4	4	0.152	0.033	−0.198	−0.043
	$2d_{3/2}$	1.26	4	4	0.092	0.019	−0.120	−0.025
	$2d_{5/2}$	1.01	—	4	—	—	−0.261	−0.081
$1f_{5/2}$	$2d_{3/2}$	1.18	6	6	0.132	0.031	−0.172	−0.041
	$2d_{5/2}$	0.93	—	6	—	—	0.034	0.012
	$1g_{7/2}$	0.97	3	6	0.184	0.061	−0.339	−0.112
$1f_{7/2}$	$2d_{5/2}$	1.28	—	8	—	—	−0.238	−0.048
	$1g_{7/2}$	1.32	4	8	0.056	0.010	−0.103	−0.019
	$1g_{9/2}$	0.87	—	3.2	—	—	−0.221	−0.083
$2d_{5/2}$	$3p_{3/2}$	1.28	6	—	0.213	0.043	—	—
	$2f_{5/2}$	1.30	6	—	0.089	0.017	—	—
	$2f_{7/2}$	0.95	6	—	0.255	0.086	—	—
$1g_{7/2}$	$2f_{5/2}$	1.26	4	—	0.126	0.026	—	—
	$2f_{7/2}$	0.92	4	—	−0.015	−0.006	—	—
	$1h_{9/2}$	0.96	4	—	0.229	0.077	—	—
$1g_{9/2}$	$2f_{7/2}$	1.36	10	6	0.238	0.040	−0.241	−0.041
	$1h_{9/2}$	1.41	10	6	0.092	0.014	−0.093	−0.014
	$1h_{11/2}$	0.86	10	6	0.326	0.124	−0.329	−0.126

$\omega_a = 1.92\omega_0$ $\kappa = 0.0244 M\omega_0^2$ $\Sigma X^2 = 0.479$ $\Sigma X^2 = 0.617$

$|F_a|^2\hbar\omega_a = 102\hbar^2/2M$ $\Sigma Y^2 = 0.042$ $\Sigma Y^2 = 0.055$

Table 6-4 Single-particle amplitudes in collective dipole motion. The amplitudes $X(12)$ and $Y(12)$ for the components in the dipole mode, associated with single-particle transitions $j_1 \to j_2$, have been obtained from a normal modes treatment. The example refers to a nucleus with $Z = 46$, $N = 60$ and a single-particle spectrum as illustrated in Fig. 6-16.

▼ While the matrix elements $F(12)$ and F_a are imaginary, the amplitudes $X(12)$ and $Y(12)$ are real. The normalization of the amplitudes $X(12)$ and $Y(12)$ involves the quantity κF_a, which may be determined from the condition (6-249a). The quantity κ needed to give the assumed collective frequency may be found from the eigenvalue equation (6-244) and is also listed in the table, together with the oscillator strength $|F_a|^2\hbar\omega_a$ of the collective mode.

Since the spread in the single-particle frequencies in Table 6-4 is rather small compared with the frequency shift that results from the interaction, the wave function for the collective state is similar to that associated with a degenerate single-particle excitation spectrum. In particular, the collective mode is found to collect about 98% of the oscillator strength associated with the field F. (The velocity-dependent terms in the potential (6-299) imply that the total oscillator strength is not exactly equal to the classical oscillator sum (6-167), which equals $4NZ/A$ in units of $\hbar^2/2M$; however, the magnitude of this deviation is only a fraction of a percent, as can be verified by an evaluation of $\Sigma|F(12)|^2(\varepsilon_2-\varepsilon_1)$. See, in this connection, the approximate equivalence between the $\mathbf{l}^2$ term and the perturbation produced by a potential proportional to r^4, as discussed on p. 594.)

Effect of pair correlations

The low-frequency behavior of the response functions is modified in an essential manner by the pair correlations, which have not been included in Figs. 6-16 and 6-17. The strength of the pair correlations is characterized by the parameter Δ, which for the nucleus considered ($Z=46$, $N=60$) has the values $\Delta_{\mathrm{n}}=1.3$ MeV and $\Delta_{\mathrm{p}}=1.4$ MeV for the neutrons and protons, respectively (see Fig. 2-5, Vol. I, p. 170). The one-particle motion in the presence of the pair correlations is described by the quasiparticle variables as discussed on pp. 647 ff.; the properties of the quasiparticle states are given in Table 6-5, for the one-particle orbits in the two major shells closest to the Fermi surface. The table gives the quasiparticle energies E (see Eq. (6-602)) and the occupation probabilities v^2 (see Eq. (6-601)); the chemical potential involved in these expressions has been obtained from Eq. (6-611).

The main consequence of the pair correlations for the response function is the shift of the zero-frequency and low-frequency transitions up to the quasiparticle excitation energies E_1+E_2, which are of the order of 2Δ. The strength of the transitions is given by (see Eq. (6-610b))

$$B(\lambda;\mathrm{v}=0\rightarrow\mathrm{v}=2,(j_1j_2)\lambda)=(u_1v_2+v_1u_2)^2(2j_1+1)B_{\mathrm{sp}}(\lambda;j_1\rightarrow j_2) \quad (6\text{-}307)$$

It is seen that the pair correlations imply the occurrence of additional transitions corresponding to orbits, which in the absence of the correlations would be both occupied or both unoccupied; except when the orbits are close to the Fermi surface, these transitions are relatively weak. ▲

Orbit	ε	$Z=46$ ($\Delta_p=1.4, \lambda_p=-2.7$) E	v^2	$N=60$ ($\Delta_n=1.3, \lambda_n=1.3$) E	v^2
$2p_{3/2}$	-8.1	5.5	0.98	9.4	0.99
$1f_{5/2}$	-7.4	4.9	0.98	8.8	0.99
$2p_{1/2}$	-6.8	4.3	0.97	8.1	0.99
$1g_{9/2}$	-2.9	1.4	0.57	4.4	0.98
$2d_{5/2}$	$+0.7$	3.7	0.04	1.4	0.71
$1g_{7/2}$	$+1.0$	4.0	0.03	1.3	0.61
$3s_{1/2}$	$+2.7$	5.6	0.02	1.9	0.13
$2d_{3/2}$	$+2.9$	5.8	0.02	2.0	0.11
$1h_{11/2}$	$+4.6$	7.4	0.01	3.5	0.03

Table 6-5 Properties of quasiparticle states. The one-particle energies in column two are obtained from the potential (6-299) and are measured relative to the energy of the $N=4$ oscillator level. All energies are in MeV. The table refers to a nucleus with $Z=46$, $N=60$.

▼ Features of Dipole Modes ($\lambda\pi=1-$)

Photoresonance in spherical nuclei (Figs. 6-18 to 6-20)

Systematics of energy

The first vibrational mode to be observed in nuclei was the "giant ▲ resonance" excited by photoabsorption.[19] It is found that, in all nuclei, the

[19]Early measurements with γ rays from proton capture reactions had established the existence of a strong nuclear photoeffect (Bothe and Gentner, 1939). The existence of the dipole resonance, often referred to as the "giant resonance," was recognized after the invention of the betatron had made available γ-ray sources of much greater intensity and flexibility (Baldwin and Klaiber, 1947 and 1948).

Even before the discovery of the photoresonance, it was recognized by Migdal (1944) that a mean excitation frequency for dipole absorption could be derived from the nuclear polarizability, which in turn is related to the symmetry energy in the mass formula. After the discovery of the rather sharp resonance, more detailed models of the collective motion of neutrons with respect to protons were developed by Goldhaber and Teller (1948), Jensen and Jensen (1950), and Steinwedel and Jensen (1950).

Evidence for the role of the single-particle degrees of freedom in the photoeffect was provided by the observation that the ratio of (γp) to (γn) processes in heavy nuclei is many orders of magnitude larger than would follow from a statistical treatment (Hirzel and Wäffler, 1947); this evidence prompted the investigation of photoabsorption processes with the direct emission of protons (Courant, 1951). With the growing evidence for nuclear shell structure, more detailed interpretations of the dipole mode were attempted in terms of single-particle excitations (see especially Wilkinson, 1956). For a time, it was felt that the independent-particle and collective descriptions represented opposite and mutually exclusive interpretations (see the discussion at the Glasgow Conference, Weisskopf, 1955, and Wilkinson, 1955).

A significant step toward a clarification of this problem was the recognition that, in a system with degenerate single-particle excitations, collective motion can occur without the introduction of correla-

absorption cross section exhibits a strong maximum in the energy region from 10 to 25 MeV, depending on the mass number. An example is shown in Fig. 6-18, while the systematics of the resonance frequencies is given in Fig. 6-19.

The Au cross section shown in Fig. 6-18 has been measured with monoenergetic γ rays produced by the annihilation of positrons in flight. The main mode of decay of the photoexcited nucleus is by neutron emission, since

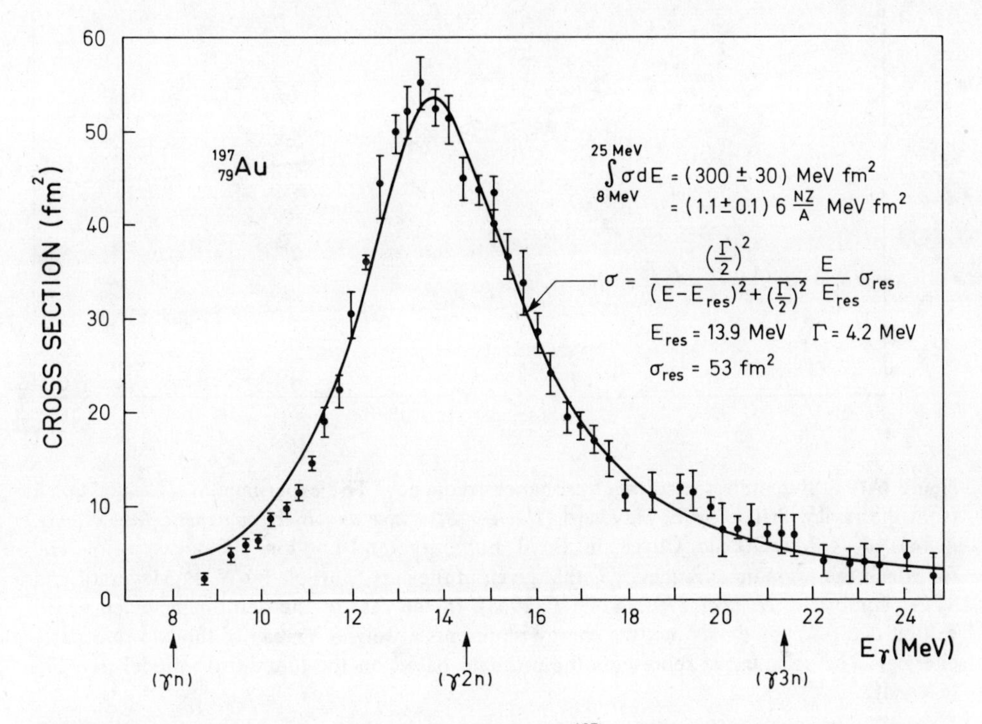

Figure 6-18 Total photoabsorption cross section for ^{197}Au. The experimental data are from S. C. Fultz, R. L. Bramblett, J. T. Caldwell, and N. A. Kerr, *Phys. Rev.* **127**, 1273 (1962). The solid curve is of Breit-Wigner shape with the indicated parameters.

---

tions beyond those implied by the identity of the particles. (This feature is the starting point for the discussion in Sec. 6-2c.) This point was first encountered in connection with the analysis of independent-particle motion in a harmonic oscillator potential. For such a system, the complete degeneracy of all the excitations produced by the center-of-mass coordinate implies that, even in the absence of interactions, the ground-state wave function of the many-body system can be expressed as a product of a function depending only on the relative coordinates and a collective center-of-mass wave function describing zero-point motion in a harmonic oscillator potential (Bethe and Rose, 1937; Elliott and Skyrme, 1955).

A similar argument, applied to the motion of neutrons with respect to protons, shows that the collective dipole mode can be represented by a superposition of degenerate particle-hole excitations (Brink, 1957). The effect of the interactions between the nucleons in collecting the dipole strength of individual nucleonic excitations and shifting it to higher frequencies was exhibited by a shell-model calculation for ^{16}O (Elliott and Flowers, 1957), and the systematic character of this effect was emphasized by Brown and Bolsterli (1959). The discussion in the present example focuses attention on the dipole field and its relation to the phenomenological symmetry potential. A similar relationship is exploited in the treatment based on the theory of Fermi liquids (see Migdal, 1967).

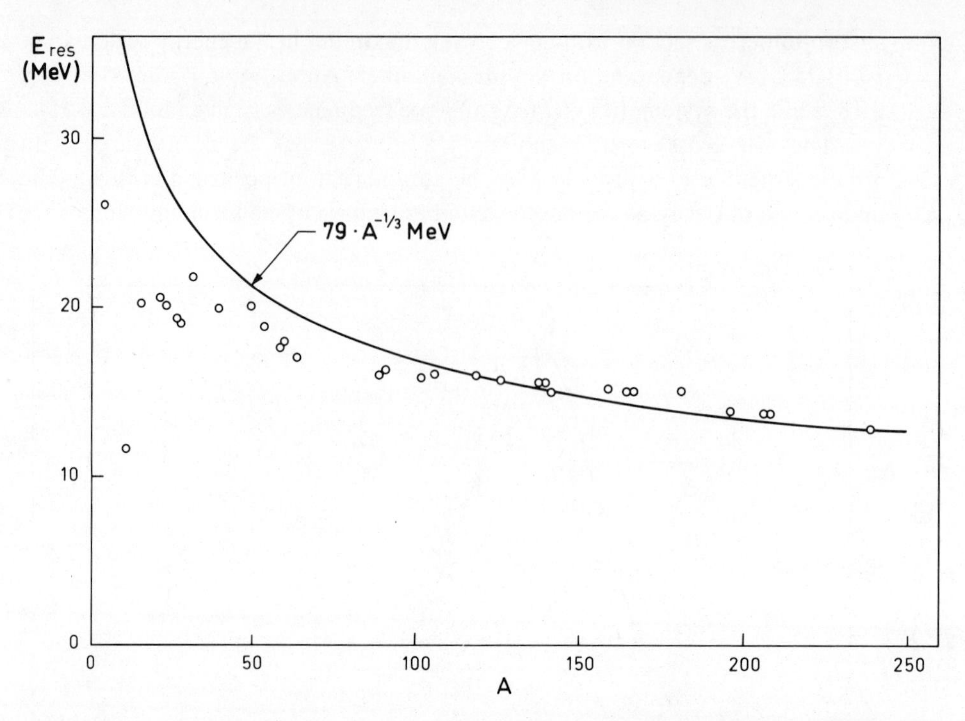

Figure 6-19 Systematics of dipole resonance frequency. The experimental data are taken from the review article by E. Hayward (*Nuclear Structure and Electromagnetic Interactions*, p. 141, ed. N. MacDonald, Oliver and Boyd, Edinburgh and London, 1965), except for ^{4}He, for which the resonance frequency is that given in the survey article by W. E. Meyerhof and T. A. Tombrello, *Nuclear Phys.* **A109**, 1 (1968). In the case of the deformed nuclei, which exhibit two resonance maxima, the energy represents a weighted mean of the two resonance energies. The solid curve represents the estimate based on the liquid-drop model (see Eq. (6A-65)).

▼ the Coulomb barrier strongly inhibits emission of charged particles; thus, the absorption cross section can be deduced from a measurement of the neutron yield. Above about 15 MeV, the $(\gamma,2n)$ process contributes significantly; the magnitude of this cross section has been determined by measuring the yield of two neutrons in coincidence.

For the example shown in Fig. 6-18, the energy variation of the absorption cross section can be rather well described by a Breit-Wigner resonance function with Γ_γ proportional to E^3, as for dipole absorption. (The cross section can also be represented by a resonance line of Lorentz shape; see the small print below.) It should be emphasized, however, that the line shape depends on the couplings that are responsible for the damping (see the discussion on pp. 503 ff.); thus, in the region far away from resonance, there is no reason to expect the Breit-Wigner form to remain valid. For other nuclei, the photoabsorption resonance exhibits additional structure (see Figs. 6-21 ▲ and 6-26).

▼ When the photon wave number ($k \approx (200 \text{ fm})^{-1} E(\text{MeV})$) is small compared with the inverse nuclear radius, the main interaction is expected to be of electric dipole type. The $E1$ character of the observed photoresonance can be directly verified from the magnitude of the measured cross sections, which yields the reduced multipole transition probability for the excitation of the observed mode. Thus, from Eqs. (3C-16) and (3F-13), we have

$$\int_{\text{res}} \sigma(E)\,dE = (2\lambda+1)\frac{\pi^2}{k_{\text{res}}^2}\Gamma_\gamma(\lambda\pi)$$

$$= \begin{cases} 0.40 E_{\text{res}} B(E1;0\to\text{res}) \text{ MeV fm}^2 & \lambda\pi = 1- \\ 3.1\times10^{-7}(E_{\text{res}})^3 B(E2;0\to\text{res}) \text{ MeV fm}^2 & \lambda\pi = 2+ \\ 4.4\times10^{-3} E_{\text{res}} B(M1;0\to\text{res}) \text{ MeV fm}^2 & \lambda\pi = 1+ \end{cases} \tag{6-308}$$

where the energies are measured in MeV and the reduced transition probabilities in units of $e^2\text{fm}^{2\lambda}$ for $E\lambda$ and $(e\hbar/2Mc)^2$ for $M1$. If the nuclear ground state has a spin $I_0 \neq 0$, the cross sections and matrix elements in Eq. (6-308) refer to sums over final states with all angular momenta that can be reached by absorption of a photon of the multipolarity considered. Comparison of Eq. (6-308) with the resonance parameters in Fig. 6-18 shows that if the observed cross section is interpreted in terms of $E1$ absorption, the value of the oscillator strength is approximately equal to the dipole sum rule given by Eq. (6-176); if interpreted in terms of $E2$ absorption, the oscillator strength exceeds the $E2$ sum rule (6-177) by a factor of about 25. An interpretation in terms of $M1$ would imply $B(M1;\ 0\to \text{res}) = 4.9\times10^3\ (e\hbar/2Mc)^2$. The total $M1$-excitation strength is expected to be of the order of $(e\hbar/2Mc)^2$ times the number of occupied orbits $j = l + 1/2$, for which the spin-orbit partners, $j = l - 1/2$, are empty (number of unsaturated spins; see Eqs. (3C-37) and (6-304)). In Au, this number is of order 20; thus, the observed cross section exceeds that expected for $M1$ absorption by more than a factor 100.

The scattering of γ rays on a dipole oscillator with excitation energy E_{res} and damping Γ yields a total cross section of Lorentz shape, as for the corresponding classical scattering problem,

$$\sigma = \frac{3\pi}{k^2}\Gamma_\gamma \text{Im}\left\{\frac{1}{E_{\text{res}} - E - \frac{1}{2}i\Gamma} + \frac{1}{E_{\text{res}} + E + \frac{1}{2}i\Gamma}\right\}$$

$$= \frac{3\pi}{2}\frac{(\Gamma_\gamma)_{\text{res}}}{k_{\text{res}}^2}\frac{E\Gamma}{E_{\text{res}}}\left(\frac{1}{(E-E_{\text{res}})^2 + \frac{1}{4}\Gamma^2} - \frac{1}{(E+E_{\text{res}})^2 + \frac{1}{4}\Gamma^2}\right)$$

$$= \sigma_0 \frac{E^2\Gamma^2}{(E_0^2 - E^2)^2 + E^2\Gamma^2} \tag{6-309}$$

$$\Gamma_\gamma = (\Gamma_\gamma)_{\text{res}}\left(\frac{E}{E_{\text{res}}}\right)^3 \qquad E_0^2 \equiv E_{\text{res}}^2 + \tfrac{1}{4}\Gamma^2 \qquad \sigma_0 \equiv \sigma(E_0) = \frac{6\pi}{\Gamma}\left(\frac{\Gamma_\gamma}{k^2}\right)_{\text{res}}$$ ▲

▼ The Lorentz shape represents the superposition of two terms of Breit-Wigner type corresponding to the fact that the electric field of a photon is real and therefore contains negative as well as positive frequencies. The scattering amplitude for the photon field thus obeys the (crossing) symmetry $f(-E)=f^*(E)$. (The same symmetry applies to the polarizability coefficients for real fields; see, for example, Eq. (6-286).)

In the analysis of nuclear photoresonances (individual resonances as well as the giant resonance), one may often omit the negative energy resonance term, since it is relatively insignificant for $E \approx E_{\text{res}}$ and since, away from resonance, there may be more important additional contributions to the cross section arising from other neighboring and distant resonances. In the example illustrated in Fig. 6-18, the difference between the Breit-Wigner and the Lorentz resonance forms amounts to an approximately constant cross section of only a few millibarns.

Oscillator strength

As noted in the analysis of Fig. 6-18, the observed photoabsorption cross sections imply that the oscillator strength of the dipole resonance is comparable to the classical sum rule value $S(E1)_{\text{class}}$ given by Eq. (6-176). The systematics of the oscillator strength is shown in Fig. 6-20. The oscillator sum $S(E1)$ plotted in the figure has been obtained from the integrated total photo cross section

$$\begin{aligned} \int \sigma_{\text{tot}} dE &= \frac{16\pi^3}{9\hbar c} S(E1) \\ &\approx 6 \frac{NZ}{A} \frac{S(E1)}{S(E1)_{\text{class}}} \text{ MeV fm}^2 \qquad (6\text{-}310) \\ S(E1) &\equiv \sum_a (E_a - E_0) B(E1; 0 \to a) \end{aligned}$$

For a single narrow resonance, the relation (6-310) follows from Eq. (6-308); since the total cross section is linear in the forward scattering amplitude (optical theorem; see Vol. I, p. 167), the relation (6-310) holds for an arbitrary superposition of resonances. It must be emphasized that the relation (6-310) neglects photoabsorption by other multipoles than $E1$; moreover, for photon wavelengths comparable with the nuclear radius, the $E1$-transition moment is given by Eq. (3C-12), which includes retardation effects.

While the oscillator strength in the dipole resonance is approximately equal to $S(E1)_{\text{class}}$, the available evidence implies that the photoabsorption cross section integrated to $E_\gamma \approx 140$ MeV exceeds the classical sum rule value by a factor of the order of 2. (See the review by Hayward, 1965, and the more recent data by Ahrens *et al.*, 1972.) For $E_\gamma \sim 100$ MeV, the photoabsorption process involves momentum components in the nuclear wave functions that are much larger than those associated with the single-particle motion and that must be attributed to the short-range nucleonic correlations. Direct support ▲ for this interpretation comes from the observation that a large fraction of the

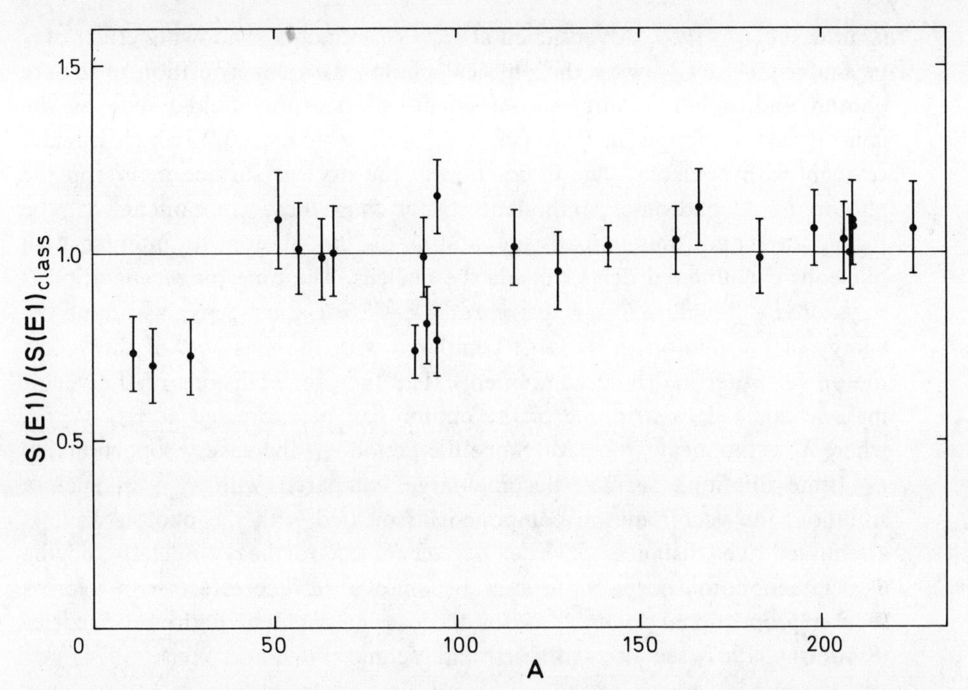

Figure 6-20 Total oscillator strength for dipole resonance. The observed total oscillator strength for energies up to 30 MeV is given in units of the classical sum rule value. For the nuclei with $A > 50$, the integrated oscillator strengths have been obtained from measurements of neutron yields produced by monochromatic γ rays (S. C. Fultz, R. L. Bramblett, B. L. Berman, J. T. Caldwell, and M. A. Kelly, in *Proc. Intern. Nuclear Physics Conference*, p. 397, ed.-in-chief R. L. Becker, Academic Press, New York, 1967). The photoscattering cross sections have been ignored, since they contribute only a very small fraction of the total cross sections. For the lighter nuclei, the yield of (γp) processes must be included and the data are from: ^{12}C and ^{27}Al (S. C. Fultz, J. T. Caldwell, B. L. Berman, R. L. Bramblett, and R. R. Harvey, *Phys. Rev.* **143**, 790, 1966); ^{16}O (Dolbilkin *et al.*, *loc.cit.*, Fig. 6-26). For the heavy nuclei ($A > 50$), other measurements have yielded total oscillator strengths that are about 20% larger than those shown in the figure (see, for example, Veyssière *et al.*, 1970).

▼ absorption processes leads to the emission of high-energy correlated neutron-proton pairs (see Stein *et al.*, 1960, and references given there). The observed excitation functions and correlations are found to be compatible with the assumption that the absorption is a two-nucleon process, as in the photodisintegration of the deuteron (Levinger, 1951).

For photon energies of several hundred MeV, the absorption cross section again rises (Roos and Peterson, 1961), and the main process appears to be photomeson production proceeding through the excitation of individual nucleons to baryonic resonance states. At still higher energies, the absorption cross section for heavy nuclei is found to be significantly smaller than the ▲ value corresponding to independent absorption by the individual nucleons

▼ (Caldwell *et al.*, 1969; Alvensleben *et al.*, 1970). Such a shadowing effect may be understood by viewing the physical photon as a superposition of a bare photon and small amplitude components of hadronic fields, such as the neutral vector mesons ρ, φ, ω (see Fig. 1-12, Vol. I, p. 63), which interact strongly with nucleons. The interaction at the nuclear surface may strip the photon of its hadronic components faster than these components can be regenerated and thus impair the ability of the photon to interact with nucleons encountered deeper inside the nucleus. The time for regeneration is $\tau_{\text{reg}} \sim \hbar(\Delta E)^{-1}$, with $\Delta E \sim (E_\gamma^2 + (m_V c^2)^2)^{1/2} - E_\gamma \approx (m_V c^2)^2 / 2E_\gamma$, assuming the energy of the photon to be large compared with the mass m_V of the vector meson (or other hadronic component). The time for collisions of the vector meson that lead to stripping of the photon can be expressed as $\tau_{\text{col}} \sim \lambda_V / c$, where λ_V is the mean free path. Since the period τ_{reg} increases proportional to E_γ (time dilation), it may become large compared with τ_{col}; in such a situation, the vector meson component associated with the photon wave is attenuated over distances of order $\lambda_V (\approx 2 \text{ fm} \ll R$ for heavy nuclei), and the associated photoabsorption process becomes a surface effect with a cross section proportional to $A^{2/3}$, as for the case of incident hadronic particles (Stodolsky, 1967; see also Gottfried and Yennie, 1969, and Weise, 1974).

Analysis of dipole mode in terms of coupling to individual particles

The main features of the nuclear dipole mode can be understood as a result of the interplay of the nucleonic excitations and the isovector potential that they generate. A qualitative description can be obtained by assuming this collective field to be proportional to the dipole moment

$$F = \sum_{k=1}^{A} (x\tau_z)_k \tag{6-311}$$

as suggested by the fact that the low-energy photoabsorption is dominated by the single giant resonance.

The single-particle excitation spectrum produced by the field (6-311) is illustrated in Fig. 6-16, p. 465 (for $Z = 46, N = 60$), and it is seen that the dipole strength clusters around the energy $\hbar\omega_0$ that would characterize the transitions in a harmonic oscillator potential. This qualitative feature of the independent-particle response to a dipole field is also apparent in the observed one-particle spectra collected in Figs. 3-2b to 3-2f, Vol. I, pp. 321 ff. (The appearance of a rather well-defined characteristic frequency in the single-particle response function is not sensitively dependent on the nuclear potential; in fact, this feature persists even for a potential with a sharp surface (infinite square well); Lushnikov and Zaretsky, 1965.) The concentration of the one-particle dipole strength provides an immediate explanation of the fact that there is only a single collective mode associated with the dipole absorption. ▲

▼ Since the single-particle response to the dipole field corresponds to almost degenerate excitations, we can obtain a simple description of the collective mode by employing the results of Sec. 6-2c together with the estimate of the strength of the isovector field coupling based on the observed static symmetry potential in the nucleus (see Sec. 6-3d). In the absence of the interactions, the mass and restoring force for the coherent dipole motion can be determined from the sum rule relations (6-23) and (6-167)

$$\frac{\hbar^2}{2D^{(0)}}=\hbar\omega_0|\langle n^{(0)}=1|F|n^{(0)}=0\rangle|^2=\frac{\hbar^2}{2M}A \tag{6-312}$$

which yields

$$\begin{aligned} D^{(0)}&=A^{-1}M \\ C^{(0)}&=\omega_0^2D^{(0)}=A^{-1}M\omega_0^2\approx 41A^{-5/3}\text{ MeV fm}^{-2} \end{aligned} \tag{6-313}$$

$$(\hbar\omega_0=41A^{-1/3}\text{ MeV})$$

The coupling constant associated with the field (6-311) can be obtained from Eq. (6-127), which yields (note the difference of a factor $(3/4\pi)^{1/2}$ in the normalization of F employed in Eqs. (6-125) and (6-311))

$$\begin{aligned} \kappa&=\frac{V_1}{4A\langle x^2\rangle} \\ &\approx 113A^{-5/3}\text{ MeV fm}^{-2}\approx 2.8A^{-1}M\omega_0^2 \end{aligned} \tag{6-314}$$

$$\left(\langle x^2\rangle\approx\tfrac{1}{5}R^2\approx\tfrac{1}{5}(1.2A^{1/3}\text{fm})^2\qquad V_1\approx 130\text{ MeV}\right)$$

We have employed the value of V_1 indicated by the differences of the binding energies of protons and neutrons near the Fermi level (see Eq. (2-182) and the evidence discussed in Vol. I, pp. 326–332). With the estimates (6-313) and (6-314), the frequency and transition strength of the normal mode are given by (see Eq. (6-27a))

$$\hbar\omega=\hbar\omega_0\left(\frac{C^{(0)}+\kappa}{C^{(0)}}\right)^{1/2}\approx 80A^{-1/3}\text{ MeV} \tag{6-315a}$$

$$\alpha_0^2=|\langle n=1|F|0\rangle^2|=\frac{\hbar A}{2M\omega}=0.26A^{4/3}\text{ fm}^2 \tag{6-315b}$$

The estimate (6-315a) of the dipole frequency agrees rather well with the empirical values of E_{res} for heavy nuclei (see Fig. 6-19) and thus supports the interpretation of the main interaction effect in the dipole mode in terms of the coupling to an isovector field of the strength implied by the symmetry

▼ potential. The observed oscillator strength of the dipole mode is approximately equal to the classical sum rule value (see Fig. 6-20), as in the above description. (The effect of velocity-dependent interactions in modifying the oscillator strength of the dipole mode is discussed below.)

For the nuclei with $A \lesssim 50$, the resonance frequencies in Fig. 6-19 are systematically below the estimate (6-315). However, for these nuclei, it is difficult to interpret the observed cross sections in terms of a single resonance frequency, due to the existence of a large high-energy shoulder extending beyond the peak in the cross section (Ahrens *et al.*, 1972). The significant photoabsorption in the region 30 MeV $< E_\gamma <$ 50 MeV for these nuclei is also reflected in the deficiency of the oscillator strength in the neighborhood of the peak in the cross section (see Fig. 6-20).

Despite the success of the simple model employed above, one must emphasize the many points that require further study in connection with a quantitative analysis of the dipole mode. Thus, the above description is based on an analysis of the isovector density and potential as a volume effect, but the possibility remains that the interactions may be rather different in the surface region. This problem is connected with the manner in which the field (6-311) cuts off for $r \gtrsim R$; the fact that the main oscillator strength in heavy nuclei is concentrated in a single frequency region approximately given by the above estimate suggests that the average radial matrix element for the one-particle transitions in the dipole mode is not affected in a major way by the cut-off. The additional dipole oscillator strength observed in the energy region above the dipole resonance (see p. 478) appears to be associated with short-range correlations between the nucleons, but the effect of these correlations on the collective dipole mode remains to be explored (see the comments on p. 415). The problem is connected with the velocity dependence of the effective interactions between nucleons in orbits near the Fermi level (see the discussion on pp. 483 ff.).

The strong isovector field implied by the above description of the dipole mode could be directly tested by an analysis of the cross section for inelastic nucleon scattering with excitation of the dipole resonance. (Such experiments, employing high-energy protons, were initiated by Tyrén and Maris, 1958. In the interpretation of the cross sections, one is faced with the problem of distinguishing between the isovector dipole mode and the rather near-lying isoscalar quadrupole mode; Lewis and Bertrand, 1972.)

Another potentially valuable source of evidence on the dipole field in the nucleus is the study of the contribution of the individual nucleonic configurations as revealed, for example, in the direct emission of nucleons in the decay of the dipole mode (Kuchnir *et al.*, 1967). The microscopic structure of the mode, produced by the field F, is characterized by the amplitudes X and Y (see Eq. (6-36)); an example is given in Table 6-4, p. 472. While main features

▲ of the wave function are prescribed by the observed fact that the mode

▼ exhausts the major part of the dipole sum rule, the more detailed configuration structure may provide additional information on the structure of the dipole field (radial variation, spin dependence, etc.).

Effect of velocity-dependent interactions

The presence of velocity-dependent interactions contributes new aspects to the dipole mode (Migdal *et al.*, 1965; Brenig, 1965). Such interactions modify the mass parameter partly through their effect on the energies of the one-particle excitations and partly through the additional coupling of the velocities of the particles to the collective flow.

If the velocity-dependence of the one-particle potential can be expressed in terms of an effective mass M^* for excitations near the Fermi surface (see Vol. I, p. 147), the dipole mass parameter $D^{(0)}$ for independent-particle motion is multiplied by the factor M^*/M (see Eq. (6-312)),

$$\begin{aligned}\frac{1}{D^{(0)}} &= \frac{A}{M^*}\\ &= \frac{A}{M}(1+k_0)\\ k_0 &\equiv \frac{M}{M^*}-1\end{aligned} \tag{6-316}$$

where the dimensionless parameter k_0 measures the magnitude of the velocity-dependent potential in units of the kinetic energy.

The additional interactions associated with the velocity dependence are analogous to those encountered in the analysis of the isoscalar dipole mode (pushing model; see pp. 445ff.). In the latter case, the structure of the coupling is obtained from the requirement of Galilean invariance in the presence of the effective mass term. Using a collective coordinate $\boldsymbol{\alpha}_{\tau=0}$ equal to the isoscalar dipole moment, in analogy to the definition of the isovector dipole coordinate in Eq. (6-311), the coupling (6-262) can be written

$$H' = -\frac{k_0}{M}\boldsymbol{\pi}_{\tau=0}\cdot\mathbf{p} \tag{6-317}$$

where $\boldsymbol{\pi}_{\tau=0} = D\dot{\boldsymbol{\alpha}}_{\tau=0} = A^{-1}M\dot{\boldsymbol{\alpha}}_{\tau=0} = A^{-1}\mathbf{P}$ is the momentum conjugate to $\boldsymbol{\alpha}_{\tau=0}$. (The relation between the effective coupling (6-317) and the effective mass was pointed out by Landau, 1956.)

In the isovector mode, there can occur a velocity-dependent coupling that is the analog of (6-317)

$$H' = -\frac{k_1}{M}\boldsymbol{\pi}_{\tau=1}\cdot\mathbf{p}\tau_z \tag{6-318}$$

▲ involving an isovector coupling constant k_1. In Eq. (6-318), we have assumed

▼ the simple form (6-311) for the collective coordinate of the isovector mode, which corresponds to

$$\pi_{\tau=1} = \frac{1}{A} \sum_k (\mathbf{p}\tau_z)_k \tag{6-319}$$

The interaction (6-318) summed over the particles gives a term proportional to $(\pi_{\tau=1})^2$ and thus a contribution to the mass parameter for the isovector dipole mode. With inclusion of the effective mass term (6-316), we obtain, for the total mass parameter

$$\frac{1}{D} = \frac{A}{M}(1 + k_0 - k_1) \tag{6-320}$$

Whereas for a system with only one type of particle, the interactions do not contribute to the mass parameter of the dipole mode (center-of-mass motion), as a result of Galilean invariance, the velocity dependence of the interactions does contribute to the mass parameter of the isovector mode in the two-fluid system, through the combination $k_0 - k_1$ representing the neutron-proton interaction. (In the Landau theory (see Migdal *et al.*, 1965), the coupling constants k_0 and k_1 are conventionally expressed in terms of the parameters $f_1 = (f_1^{\mathrm{nn}} + f_1^{\mathrm{np}})/2 = -(3k_0/2)(1+k_0)^{-1}$ and $f_1' = (f_1^{\mathrm{nn}} - f_1^{\mathrm{np}})/2 = -(3k_1/2)(1+k_0)^{-1}$.)

The result (6-320) implies that the oscillator strength of the dipole mode is multiplied by the factor $(1 + k_0 - k_1)$, as a consequence of the velocity-dependent interaction. It should be emphasized that this modification is a result of the effective interactions between particles with energies in the neighborhood of the Fermi surface and thus appears in the collective dipole mode. In contrast, the modification of the $E1$-oscillator strength implied by the charge and velocity dependence of the basic nucleonic interaction (see, for example, Eq. (6-200)), refers to the total oscillator sum of the system and includes the strength associated with photoabsorption processes with frequencies large compared to that of the dipole resonance.

The modification in the current implied by the velocity-dependent effective interactions also manifests itself in the magnetic moment associated with the orbital motion of the particles.[20] As in the discussion of the dipole mass parameter, one must consider the contributions from the effective mass term in the one-particle potential as well as from the velocity-dependent interactions. The first contribution implies that the orbital g factor, measured in units of $e\hbar/2Mc$, receives the increment

▲
$$\delta g_l^{(1)} = \tfrac{1}{2}(1 - \tau_z)k_0 \tag{6-321}$$

[20]The present discussion is equivalent to that of Migdal (1966) based on the Landau theory of Fermi liquids (see also the discussion and references in Sec. 3C-6). The contribution of mesonic exchange effects to the orbital g factor was pointed out by Miyazawa (1951); the relationship of these effects to the modification of the dipole oscillator sum has been discussed by Fujita and Hirata (1971).

▼ reflecting the modified velocity of a quasiparticle (proton) of given momentum. The couplings (6-317) and (6-318) imply that, in the presence of a particle i with momentum $\mathbf{p}_i$, each of the other particles k receives the velocity increment

$$\delta\dot{\mathbf{x}}_k = \nabla_p^{(k)} H' = -\frac{k_0}{M}\left(\mathbf{p}_i \cdot \nabla_p^{(k)}\right)\boldsymbol{\pi}_0 - \frac{k_1}{M}\tau_{zi}\left(\mathbf{p}_i \cdot \nabla_p^{(k)}\right)\boldsymbol{\pi}_1 \tag{6-322}$$

and a summation over the particles k yields

$$\begin{aligned} \sum_k \delta\dot{\mathbf{x}}_k &= -\frac{k_0}{M}\mathbf{p}_i \\ \sum_k \tau_{zk}\delta\dot{\mathbf{x}}_k &= -\frac{k_1}{M}\mathbf{p}_i\tau_{zi} \end{aligned} \tag{6-323}$$

If the velocity-dependent couplings can be attributed to effective interactions with a range small compared with the nuclear radius, the extra current (6-323) is localized in the neighborhood of the particle with momentum $\mathbf{p}_i$, and can be included as a renormalization of the local current associated with this particle. The resulting effect on the orbital g factor is

$$\delta g_l^{(2)} = -\tfrac{1}{2}(k_0 - k_1\tau_z) \tag{6-324}$$

and the total modification of g_l becomes

$$\delta g_l = \delta g_l^{(1)} + \delta g_l^{(2)} = -\tfrac{1}{2}(k_0 - k_1)\tau_z \tag{6-325}$$

The result (6-325) involves only the neutron-proton interaction $(k_0 - k_1)$, corresponding to the fact that in a system with only a single type of particle, the interactions, if Galilean invariant, do not modify the total current for given momentum. The same argument shows that in the two-fluid system, with $N = Z$, the interaction effect on g_l must be isovector (proportional to τ_z), since the isoscalar current (associated with a motion of the center of mass) is not affected by the velocity-dependent interactions. Since δg_l vanishes when averaged over the particles, it follows that, in a system with neutron excess, the factor τ_z in Eq. (6-325) is to be replaced by $\tau_z - \langle\tau_z\rangle = \tau_z - (N-Z)/A$. (The contributions to δg_l from the velocity-dependent interactions are analogous to the corrections to the $M1$ operator that result from a spin-orbit force; see the discussion in Sec. 3C-6e, based on a Galilean invariant two-body interaction, the average of which reproduces the spin-orbit coupling in the one-body potential.)

In the presence of a neutron excess, the velocity-dependent interactions also lead to a difference in the effective mass of neutrons and protons that is linear in $(N-Z)/A$; thus, the requirement of Galilean invariance implies ▲ that the interaction (6-318) must be associated with an isovector term in the one-body potential corresponding to $(M/M^*)_{\tau=1} = k_1(N-Z)A^{-1}\tau_z$; however, an additional isovector term linear in $(N-Z)/A$ may be associated with a coupling between π_0 and π_1, corresponding to a difference in the effective nn and pp interactions at the top of the Fermi distributions.

Although the observed energy differences between single-particle states near the Fermi surface appear to be consistent with a velocity-independent potential, this interpretation is uncertain, since a velocity dependence may be masked by self-energy effects (see the discussion on p. 430). The appropriate value of k_0 to be employed in the present context therefore remains an open question.

The analysis of empirical magnetic moments in the region of ^{208}Pb has provided tentative evidence for a renormalization of g_l by about $\delta g_l \sim -0.1\tau_z$ (Yamazaki *et al.*, 1970; Maier *et al.*, 1972; Nakai *et al.*, 1972). This evidence may suggest $k_0 - k_1 \approx 0.2$ (see Eq. (6-325)), and hence an oscillator strength in the dipole resonance about 20% larger than the classical value (see Eq. (6-320)). Such a value of the oscillator strength is compatible with the available empirical evidence on the dipole mode in heavy nuclei (see the caption to Fig. 6-20).

E1-polarization charge

The coupling of the dipole mode to the single-particle motion produces a renormalization of the isovector dipole moment of the particle, described by the polarizability coefficient (see Eqs. (6-216), (6-313), and (6-314)),

$$\chi = -\frac{\kappa}{C}\frac{(\hbar\omega)^2}{(\hbar\omega)^2 - (\Delta E)^2}$$
$$\frac{\kappa}{C} = \frac{\kappa}{C^{(0)}+\kappa} \approx 0.7 \qquad (6\text{-}326)$$

where ΔE is the transition energy.

The electric dipole moment is a combination of isoscalar and isovector moments (see Eq. (6-123)). However, the isoscalar dipole moment represents a center-of-mass displacement and does not contribute to intrinsic excitations. Hence, the effective charge for $E1$ transitions is

$$e_{\text{eff}}(E1) = -\tfrac{1}{2}e\tau_z(1+\chi) \qquad (6\text{-}327)$$

The estimate (6-326) for χ implies a major reduction of the low-frequency $E1$-transition strength.

For transition energies ΔE comparable with or greater than the resonance energy $\hbar\omega$, the coupling gives rise to an increase in the single-particle transition strength, which may account for the observed enhancement of the cross section for direct radiative capture of nucleons (Brown, 1964; Clement

▼ *et al.*, 1965; the direct capture resulting from the polarization charge is sometimes referred to as "semidirect capture"). For $\Delta E \approx \hbar\omega$, the damping of the dipole mode becomes of importance and can be taken into account by adding the imaginary parts $\mp i\Gamma/2$ (where Γ is the resonance width) to the factors $(\hbar\omega \mp \Delta E)$ in the energy denominator in Eq. (6-326); see Eq. (6-286). In the capture process, the dipole transition moment may receive appreciable contributions from the region outside the nucleus, and thus the polarization effect may be sensitive to the radial form factor for the dipole field (see, for example, Zimányi *et al.*, 1970).

Effects of neutron excess

The above analysis of the properties of the dipole mode refers to nuclei with $N = Z$. Because of the symmetric role of neutrons and protons, the frequency and $\tau = 1$ oscillator strength of the dipole mode are even functions of $N - Z$; hence, the leading-order corrections for neutron excess are of order $(N-Z)^2/A^2$ and thus negligible in the present context. Moreover, as a consequence of the translational invariance, the dipole mode, as is the case for any internal excitation, carries no $\tau = 0$ dipole moment; hence, also the $E1$ moment of the collective mode is unaffected to first order in $N - Z$. (In contrast, the charge exchange modes are strongly modified by the neutron excess; see the example on pp. 493 ff.)

However, the neutron excess implies that the field in the dipole mode acts more strongly on the protons than on the neutrons, and this difference leads to terms in the effective charge that are linear in $(N-Z)/A$. In the schematic model, in which the dipole field is taken to be proportional to the x coordinate, the requirement that the dipole field should only depend on coordinates relative to the center of mass X leads to the form

$$\begin{aligned} F &= \sum_k (x_k - X)(\tau_z)_k \\ &= \sum_k x_k \left(\tau_z - \frac{N-Z}{A} \right)_k \end{aligned} \tag{6-328}$$

Since $(N-Z)/A$ represents the average value $\langle \tau_z \rangle$ of the nucleonic isospin, the field (6-328) can be seen to act on the deviation of the isospin from its average value, and thus to vanish at each space point when averaged over the total density.

Since, as argued above, the $E1$ moment and the coupling constant for the dipole mode are not affected to first order in the neutron excess, the coupling of the particle motion to the field (6-328) yields the polarization charge

$$e_{\text{pol}}(E1) = -\tfrac{1}{2} e\chi \left(\tau_z - \frac{N-Z}{A} \right) \tag{6-329}$$

▲ with χ given by Eq. (6-326).

▼ An additional contribution to e_{eff}, linear in the neutron excess, is contained in the recoil effect (correction for center-of-mass motion). This contribution, given by Eq. (3C-35), combined with the polarization charge (6-329), yields the total effective charge for $E1$ transitions

$$e_{\text{eff}}(E1) = -\tfrac{1}{2}e\left(\tau_z - \frac{N-Z}{A}\right)(1+\chi) \tag{6-330}$$

(The recoil contribution can be viewed as a polarization charge resulting from the coupling between the particle motion and the "spurious" center-of-mass vibrational mode. This coupling, given by Eq. (6-256), is proportional to $-\partial V/\partial x = M\ddot{x}$ and hence to the dipole moment of the particle multiplied by $(\Delta E)^2$. The zero-frequency center-of-mass mode thus contributes a frequency-independent polarization charge (see Eq. (6-216)) of magnitude $\delta e_{\text{pol}} = -Ze/A$ (as can be obtained from the diagrams in Fig. 6-8, noting that the center-of-mass mode has the electric dipole moment $Ze\alpha$).)

The analysis of the observed strength of low-energy $E1$ transitions in the region of ^{208}Pb has led to the estimates $|(e_{\text{eff}})_{E1}| \approx 0.15e$ for neutrons and $\approx 0.3e$ for protons (Hamamoto, 1973) to be compared with the values $0.12e$ and $0.18e$ given by Eq. (6-330).

Comparison with liquid-drop description of dipole mode

The properties of the dipole mode can also be described by the liquid-drop model, in terms of oscillations of the neutron fluid with respect to the proton fluid, characterized by the isovector density $\rho_1(\mathbf{r}) = \rho_n(\mathbf{r}) - \rho_p(\mathbf{r})$ (see Sec. 6A-4). The lowest dipole mode has the frequency and oscillator strength (see Eqs. (6A-65) and (6A-70))

$$\hbar\omega = 2.08\left(\frac{\hbar^2}{MR^2}b_{\text{sym}}\right)^{1/2} \approx 78A^{-1/3}\ \text{MeV}$$

$$B(E1; 0+ \rightarrow 1-)\,\hbar\omega = 0.86\,S(E1)_{\text{class}} \tag{6-331}$$

$$(R = 1.2A^{1/3}\,\text{fm}; \qquad b_{\text{sym}} = 50\ \text{MeV})$$

where the coefficient b_{sym} is the nuclear symmetry energy obtained from the empirical mass formula.

The liquid-drop model is based on the assumption of a local energy density and is therefore not in general expected to describe the dynamics of a system with a shell structure (see the comments on p. 351). The dipole oscillation, however, is a special case, since this single mode approximately exhausts the oscillator sum rule; the properties of this mode vary smoothly with N and Z, and can therefore be expressed in terms of macroscopic nuclear parameters. ▲

▼ In the comparison of the parameters of the liquid-drop model with those implied by the above microscopic treatment in terms of particle excitations, we shall ignore the rather minor effects associated with the fact that the radial form factor in the liquid-drop description is proportional to the spherical Bessel function $j_1(2.08r/R)$ (see Eqs. (6A-62) and (6A-64)) and thus deviates somewhat from the linear expression

$$\delta\rho_1 = \frac{\rho_0}{A\langle x^2\rangle}\alpha x \tag{6-332}$$

which corresponds to the field (6-311).

The mass parameter of the dipole mode is directly given by the oscillator strength and is thus model independent, provided the sum rule is exhausted. (See also the discussion of the relation of irrotational flow to the mass parameter of a single mode with the full sum rule oscillator strength, p. 405.)

The restoring force for the dipole mode represents the energy required to produce a static isovector deformation of the form (6-332). In the liquid-drop model, this energy is expressed in terms of an energy density proportional to $(\delta\rho_1)^2$ (see Eq. (6A-61)), and the resulting restoring force parameter is

$$C = \frac{b_{\text{sym}}}{A\langle x^2\rangle} \approx 175A^{-5/3}\ \text{MeV fm}^{-2} \tag{6-333}$$

In the microscopic description, the restoring force is obtained as a sum of two terms, $C = C^{(0)} + \kappa$. The first term results from the one-particle excitation energies, while the second term gives the effect of the interactions. This separation corresponds to the analysis of b_{sym} in terms of a kinetic and a potential energy part (see Vol. I, p. 142). The potential energy part of b_{sym} reflects the contribution to the nuclear energy resulting from the field V_1, and the relation $(b_{\text{sym}})_{\text{pot}} = \frac{1}{4}V_1$ (see Eq. (2-28)) implies that the potential energy part of the restoring force (6-333) is equal to κ, as in the microscopic description (see Eq. (6-314)). The above estimate of the kinetic energy part of the restoring force, $C^{(0)}$, is based on the available empirical evidence on the one-particle excitation spectrum and differs somewhat from the kinetic part of expression (6-333). Thus, the value of $C^{(0)}$ in Eq. (6-313) corresponds to $(b_{\text{sym}})_{\text{kin}} \approx 12$ MeV, and together with the value $(b_{\text{sym}})_{\text{pot}} \approx \frac{1}{4}V_1 \approx 32$ MeV leads to a total $b_{\text{sym}} \approx 44$ MeV, which is somewhat less than the value $b_{\text{sym}} = 50$ MeV obtained from the empirical mass formula. Hence, the restoring force employed in the above microscopic analysis is about 12% smaller than the value (6-333). The difference is relatively small, since the main part of C is associated with the interactions. The close agreement between the resonance frequencies given by Eqs. (6-315) and (6-331) results from the fact that the mass parameter for the liquid-drop model is about 15% larger than the value (6-313), as a consequence of the distribution of the oscillator strength over a number of different normal modes.

The above separation of b_{sym} into its kinetic and potential parts differs somewhat from the analysis in Vol. I, p. 142, which is based on the Fermi gas model. An improved estimate of the average spacing of levels near the Fermi surface may be based on the observed one-particle spectra, which give $g_0 \approx 2(g_0)_n \approx 2(g_0)_p \approx 0.060\,A$ MeV^{-1}, as obtained from the one-particle level densities of neutrons and protons separately (see, for example, Figs. 5-2 to 5-6, pp. 222 ff. as well as the estimate based on the degeneracy of the oscillator shells as employed in Eq. (2-125a)). Since $(b_{\text{sym}})_{\text{kin}} = A(g_0)^{-1}$, we obtain $(b_{\text{sym}})_{\text{kin}} \approx 16$ MeV, which yields $(b_{\text{sym}})_{\text{pot}} = b_{\text{sym}} - (b_{\text{sym}})_{\text{kin}} \approx 34$ MeV, in good agreement with the value $V_1 \approx 130$ employed in the above analysis. ▲

▼ *Splitting of resonance in deformed nuclei (Fig. 6-21; Tables 6-6 and 6-7)*

In a deformed nucleus with axial symmetry, the dipole resonance is split into two components with $\nu=0$ and 1 (Danos, 1958; Okamoto, 1958). The splitting between the two modes is, to leading order, proportional to the deformation (see Eq. (6-277))

$$(\bar{E})^{-1}(E(\nu=1)-E(\nu=0))\approx\delta\approx\frac{\Delta R}{R} \tag{6-334}$$

where $\bar{E}$ is the mean resonance energy. The constant of proportionality between $\Delta E/E$ and δ is somewhat model dependent (see the discussion on p. 454).

While the $\nu=0$ mode represents oscillations in the direction of the nuclear symmetry axis, the $\nu=1$ mode is associated with oscillations in the two perpendicular directions. Thus, if the dipole sum rule is exhausted by single $\nu=0$ and $\nu=1$ resonances, it is expected that two thirds of the oscillator strength will be contained in the $\nu=1$ mode and one third in the mode with $\nu=0$. For a deformed nucleus without axial symmetry, the photoabsorption line would be split into three components with equal oscillator strength.

The photoeffect in deformed nuclei has been found to give rise to cross sections with two major peaks with approximately the expected intensity ratio (Fuller and Hayward, 1962). An example is illustrated in Fig. 6-21, which shows the photoabsorption cross section for a series of Nd isotopes. The onset of a static deformation for ^{150}Nd (as revealed by the rotational spectrum, see Fig. 4-3, p. 27) is seen to give rise to a splitting of the dipole resonance. The resonance parameters for a fit to the Lorentzian line shape (6-309) are listed in Table 6-6; the oscillator strength is proportional to $\int\sigma\, dE$ and hence to $\sigma_0\Gamma$, and the ratio of the oscillator strength in the two peaks is therefore about 2:1. The observed energy separation of the peaks can be used to obtain an estimate of the nuclear deformation by means of Eq. (6-334). The values obtained in this manner are compatible with those obtained from measured $E2$ moments, as illustrated by the data in Table 6-7. The significance of the differences in the two determinations of δ is at present difficult to assess, in view of the uncertainty in the numerical value of the coefficient in the relation (6-277), as well as in the treatment of higher-order terms in the deformation. (Thus, if one assumes the resonance frequencies to be strictly proportional to the inverse major and minor axes, the deformation δ evaluated from Eq. (6-334) equals $\Delta R/R-(\Delta R)^2/3R^2$, while the deformation evaluated from Q_0 by means of Eq. (4-72) equals $\Delta R/R+(\Delta R)^2/6R^2$ (see Eq. (4-73)); the difference in the second-order terms in $(\Delta R/R)$ amounts to about 15% in the determination of δ, which represents a significant part of the discrepancies in Table 6-7.) ▲

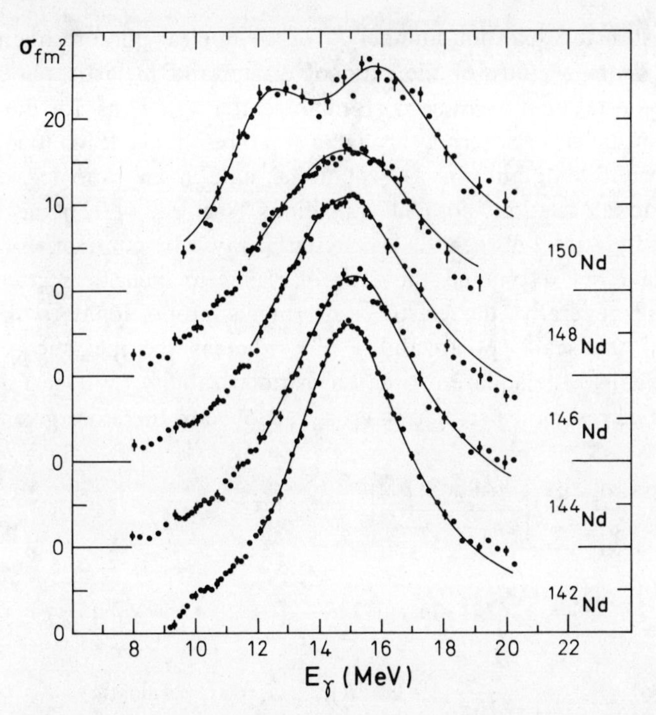

Figure 6-21 Photoabsorption cross section for even isotopes of neodymium. The experimental data are from P. Carlos, H. Beil, R. Bergère, A. Lepretre, and A. Veyssière, *Nuclear Phys.* **A172**, 437 (1971). The solid curves represent Lorentzian fits with the parameters given in Table 6-6.

▼ A direct test of the interpretation of the splitting of the photoresonance line in terms of a deformation effect can be obtained by a measurement of the dependence of the absorption cross section on the orientation of the nucleus with respect to the direction of the incident photon beam. Such a test of the expected photoanisotropy of ^{165}Ho has been performed by Ambler *et al.* (1965) and Kelly *et al.* (1969). ▲

	^{142}Nd	^{144}Nd	^{146}Nd	^{148}Nd	^{150}Nd	
E_0(MeV)	14.9	15.0	14.8	14.7	12.3	16
σ_0(fm^2)	36	32	31	26	17	22
Γ(MeV)	4.4	5.3	6	7.2	3.3	5.2

Table 6-6 Parameters for the dipole resonance in even neodymium isotopes. The table gives the parameters for the Lorentzian resonance curves drawn in Fig. 6-21. The cross section for ^{150}Nd has been fitted to the sum of two resonance functions.

▼ A test of the quantum number K for the nuclear photoresonances can be obtained from a study of the ratio of elastic and inelastic photoscattering leading to rotational excitations (Raman scattering). Thus, for the case of an even-even nucleus, scattering through a $K=0$ resonance leads to a population of the ground state and the $2+$ rotational state in the ratio 1:2, as given by the rotational relations for $E1$ transitions (see Eq. (4-92)), neglecting the energy difference between the scattered γ rays. In contrast, for scattering through a $K=1$ resonance, the ratio of elastic to inelastic scattering equals 2:1. More generally, the elastic scattering is proportional to the absolute square of the scalar polarizability $p^{(0)}$, whereas the inelastic scattering is proportional to the square of the tensor polarizability $p^{(2)}$ (see Eqs. (6-289)). The relative resonance scattering cross sections are therefore given by

$$\frac{\sigma_{\text{el}}}{\sigma_{\text{inel}}}=2\left|\frac{p_1+p_2+p_3}{2p_3-p_1-p_2}\right|^2 \tag{6-335a}$$

$$W(\theta)=\sum_{\mu}\langle I_f\,1-\mu\,1\mu|11\rangle^2\frac{3}{8\pi}\left(|\mathscr{D}^1_{\mu 1}|^2+|\mathscr{D}^1_{\mu -1}|^2\right)$$

$$=\begin{cases}\dfrac{3}{16\pi}(1+\cos^2\theta) & I_f=0 \quad \text{(elastic)}\\[2mm] \dfrac{3}{160\pi}(13+\cos^2\theta) & I_f=2 \quad \text{(inelastic)}\end{cases} \tag{6-335b}$$

where the angular distributions (6-335b) follow from the fact that the photon can only excite states with $I=1$, $M=\pm 1$. In Eq. (6-335a), the polarizabilities p_κ with respect to the principal axes of the nucleus are given by Eq. (6-287). In the estimate of the elastic scattering, one must include the contribution from the Thomson scattering, which arises from the center-of-mass motion of the nucleus and is equivalent to resonance scattering from an oscillator with zero frequency and oscillator strength Z/N, relative to the classical sum-rule value (6-176); see the discussion following this equation.

The Raman scattering of mono-energetic γ rays with energies of the order of 10 MeV has been observed for ^{232}Th and ^{238}U, and the intensity relative to the elastic scattering is found to be in approximate agreement with the resonance parameters obtained from the analysis of the photoabsorption cross section (Hass *et al.*, 1971; Jackson and Wetzel, 1972). Thus, for the 90° scattering of 10.8 MeV photons on ^{238}U, one estimates, on the basis of Eq. (6-335), a ratio of 1.0 between inelastic and elastic scattering, to be compared with the measured value of 0.8 (Jackson and Wetzel, *loc. cit.*). In the theoretical estimate, we have employed the resonance parameters for ^{238}U given by Veyssières *et al.*, 1973 ($E_0=11.0$ MeV, $\Gamma=2.9$ MeV, $\sigma_0=30$ fm^2, for $\nu=0$; $E_0=14.0$ MeV, $\Gamma=4.5$ MeV, $\sigma_0=37$ fm^2, for $\nu=1$) and have included the Thomson scattering, which reduces the elastic scattering by about 20%.

▲ (For the energy considered, the Thomson amplitude is comparable, and of

Nucleus	$E(\nu=0)$	$E(\nu=1)$	$\delta(E1)$	$\delta(Q_0)$
^{153}Eu	12.3	15.8	0.24	0.34
^{159}Tb	12.2	15.8	0.25	0.34
^{160}Gd	12.2	16.0	0.26	0.35
^{165}Ho	12.2	15.7	0.24	0.33
^{181}Ta	12.5	15.2	0.19	0.26
^{186}W	12.6	14.9	0.16	0.22
^{232}Th	11.1	14.1	0.23	0.24
^{235}U	10.8	14.1	0.25	0.26
^{237}Np	11.1	14.2	0.24	0.25
^{238}U	11.0	14.0	0.23	0.26

Table 6-7 Comparison of deformation parameters derived from splitting of $E1$ resonance and from $E2$-matrix elements. The energies $E(\nu=0)$ and $E(\nu=1)$ of the two resonance peaks in the photoabsorption cross sections are taken from: ^{153}Eu, ^{159}Tb, ^{160}Gd, ^{165}Ho, ^{181}Ta, ^{186}W (B. L. Berman, M. A. Kelly, R. L. Bramblett, J. T. Caldwell, H. S. Davis, and S. C. Fultz, *Phys. Rev.* **185**, 1576, 1969); ^{235}U (C. D. Bowman, G. F. Auchampaugh, and S. C. Fultz, *Phys. Rev.* **133**, B676, 1964); ^{159}Tb, ^{165}Ho, ^{181}Ta (R. Bergère, H. Beil, and A. Veyssière, *Nuclear Phys.* **A121**, 463, 1968); ^{232}Th, ^{237}Np, ^{238}U (A. Veyssière, H. Beil, R. Bergère, P. Carlos, A. Lepretre, and K. Kernbath, *Nuclear Phys.* **A199**, 45, 1973). The deformation parameters $\delta(Q_0)$ are from Fig. 4-25, p. 133, and the additional references: ^{235}U (S. A. De Wit, G. Backenstoss, C. Daum, J. C. Sens, and H. L. Acker, *Nuclear Phys.* **87**, 657, 1967); ^{237}Np (J. O. Newton, *Nuclear Phys.* **5**, 218, 1958).

▼ opposite sign, to the real part of the resonance scattering amplitude, but the main part of the elastic cross section is contributed by the imaginary part of the resonance scattering amplitude.)

Isobaric quantum number. Charge exchange modes (Figs. 6-22 to 6-24)

The nuclear photoresonance can be seen in a wider perspective when it is recognized as a single component (with $\mu_\tau=0$) of the isovector dipole modes. The additional components (with $\mu_\tau=\pm1$) represent charge exchange modes, and the study of the multiplet of excitations leads to a variety of phenomena associated with the coupling of the isospin in the dipole motion to that of the neutron excess. (The spectrum of the dipole modes in nuclei with neutron excess has been treated by Fallieros *et al.*, 1965; Novikov and Urin, 1966; Petersen and Veje, 1967; Goulard and Fallieros, 1967; Ejiri et al., 1968.)

Effect of neutron excess on dipole frequencies and transition strengths

In a nucleus with $N=Z$ and $T_0=0$, the charge exchange modes of dipole excitation are related to the $\mu_\tau=0$ mode by isobaric symmetry (see Fig. 6-4, p. 376). In the presence of a neutron excess ($T_0\neq0$), the excitation of an ▲

▼ isovector quantum leads to nuclear states with $T=T_0+1, T_0$, and T_0-1 (see Fig. 6-5, p. 382).

The effect of the neutron excess on the dipole modes with the different values of $\Delta T = T - T_0$ can be seen most easily by considering the dominant transitions with $\mu_\tau = \Delta T$, leading to the fully aligned states $M_T = T$ (see Fig. 6-5, where these transitions are indicated by arrows). The properties of the excitations with $M_T < T$ can be determined from those of the fully aligned states by exploiting rotational invariance in isospace (see Eq. (6-132)). The transitions with $\mu_\tau = \Delta T$ are associated with the particle-hole excitations illustrated in Fig. 6-22. As can be seen from Fig. 6-22, the neutron excess implies a reduction in the number of particle-hole dipole excitations with $\mu_\tau = +1$ and a corresponding increase in those with $\mu_\tau = -1$. The particle-hole excitations with $\mu_\tau = 0$ and -1 are not in general eigenstates of the total isospin (see, for example, the analogous situation for one-particle states considered in Fig. 3-1, Vol. I, p. 315). However, for $T_0 \gg 1$, the components with $T \neq T_0 + \mu_\tau$ are small (amplitudes of order $T_0^{-1/2}$ or smaller) and can be neglected in the analysis of the collective modes. (For small values of T_0, the effect of the neutron excess is only a perturbation, and the analysis on pp. 500ff. yields correctly the terms of leading order in T_0.) ▲

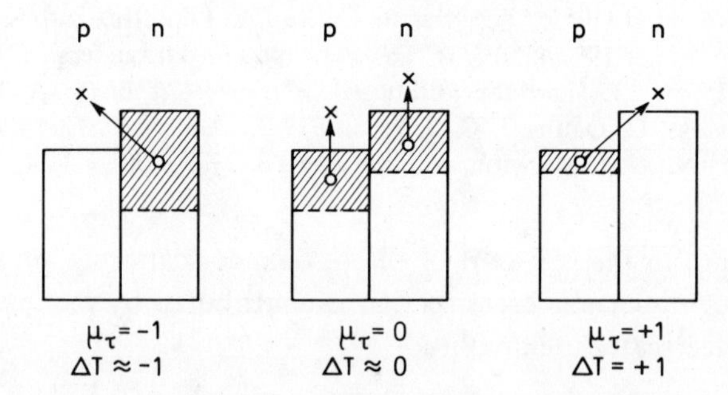

Figure 6-22 Particle-hole excitations associated with dipole modes in a nucleus with neutron excess. The hatched domains represent the particle orbits that can be excited by the dipole field.

▼ The generalization of the dipole field considered for $\mu_\tau = 0$ (see Eq. (6-311)) leads to the isotriplet

$$F_{\mu_\tau} = \sum_k (x\tau_{\mu_\tau})_k \qquad (6\text{-}336)$$

with $\tau_{\mu_\tau} = 2t_{\mu_\tau}$ and t_{μ_τ} given by Eq. (6-120). The single-particle excitations produced by the fields (6-336) are approximately degenerate (see Fig. 6-16), ▲

▼ with energies (see Eq. (2-26) for the symmetry potential)

$$\hbar\omega_{\mu_\tau}^{(0)} = \hbar\omega_0 + \mu_\tau\left(V_1 \frac{N-Z}{2A} - E_{\text{Coul}}\right) \tag{6-337}$$

where E_{Coul} is the average Coulomb energy for a single proton. (The charge exchange frequencies in Eq. (6-337) are defined in terms of binding energy differences rather than mass differences; see comment on p. 376.)

The strength of the one-particle dipole transitions that contribute to the collective modes is characterized by the matrix elements of F_{μ_τ} leading from the unperturbed ground state $|\mathrm{v}=0\rangle$ to the coherent single-particle excitation $|n_{\mu_\tau}^{(0)}=1\rangle$; compare Eq. (6-21). In the following, we shall use the notation

$$if_{\mu_\tau} \equiv \langle n_{\mu_\tau}^{(0)} = 1 | F_{\mu_\tau} | \mathrm{v}=0\rangle \tag{6-338}$$

(The magnitude of these zero-point amplitudes might also have been denoted by $(\alpha_{\mu_\tau}^{(0)})_0$.) In the standard representation, the matrix elements of F are imaginary and the quantities f therefore real. For $N=Z$, the matrix elements f are independent of μ_τ, but the neutron excess implies an increase in the strength of the n→p transitions relative to that of the p→n transitions (see Fig. 6-22). The decrease in f_{+1}^2 is compensated by the increase in f_{-1}^2, and, to first order in $(N-Z)/A$, we have (see Eq. (6-312))

$$\tfrac{1}{2}(f_{+1}^2 + f_{-1}^2) = f_0^2 = \frac{\hbar A}{2M\omega_0} \tag{6-339}$$

The difference between f_{-1}^2 and f_{+1}^2 can be evaluated in terms of the commutator of F_{-1} and $F_{+1} = -F_{-1}^\dagger$,

$$\begin{aligned} f_{-1}^2 - f_{+1}^2 &= \langle \mathrm{v}=0 | [F_{-1}, F_{+1}] | \mathrm{v}=0\rangle \\ &= \langle \mathrm{v}=0 | 2\sum_k (\tau_z x^2)_k | \mathrm{v}=0\rangle = 2(N-Z)\langle x^2\rangle_{\mathrm{n\,exc}} \end{aligned} \tag{6-340}$$

where $\langle x^2\rangle_{\mathrm{n\,exc}}$ represents an average for the excess neutrons.

The interaction is the charge-independent extension of that assumed for the $\mu_\tau=0$ mode (compare Eq. (6-24))

$$\begin{aligned} H' &= \tfrac{1}{2}\kappa \sum_{\mu_\tau} F_{\mu_\tau}^\dagger F_{\mu_\tau} \\ &= -\tfrac{1}{2}\kappa(F_{+1}F_{-1} + F_{-1}F_{+1}) + \tfrac{1}{2}\kappa F_0^2 \end{aligned} \tag{6-341}$$

with the coupling constant κ given by Eq. (6-314). The field F_{+1} creates particle-hole quanta with $\mu_\tau = +1$ and annihilates quanta with $\mu_\tau = -1$; hence, the charge exchange interaction involves partly terms that are diagonal in the number of quanta $n_{\pm 1}^{(0)}$ and partly terms that create and annihilate pairs of ▲

▼ quanta with $\mu_\tau = +1$ and $\mu_\tau = -1$. The latter terms imply a coupling between the two charge exchange modes, but the fact that the interaction creates and annihilates the quanta in pairs leads to a conservation of the difference in the number of quanta of the two modes

$$n_{+1} - n_{-1} = n_{+1}^{(0)} - n_{-1}^{(0)} \tag{6-342}$$

The normal modes arising from the interaction (6-341) are derived in the small print below; the frequencies $\omega_{\pm 1}$ and the transition probabilities $|\langle n_{\pm 1}=1|F_{\pm 1}|0\rangle|^2$ are given by Eqs. (6-355) and (6-358); see also Eq. (6-353).

The pattern of the dipole modes with $\mu_\tau = 0, \pm 1$ is shown in Fig. 6-23 as a function of the neutron excess. The properties of the $\mu_\tau = 0$ mode are not affected to first order in $(N-Z)/A$ (see p. 487); for this mode, the values in Fig. 6-23 are those given by Eq. (6-315). In Fig. 6-23, the difference between the single-particle strengths f_{+1}^2 and f_{-1}^2 is characterized by the parameter ν,

$$f_{\pm 1}^2 = f_0^2(1 \mp \nu) \tag{6-343}$$

The difference in the single-particle charge exchange frequencies is taken to be proportional to the difference in the transition strength

$$\omega_{\pm 1}^{(0)} = \omega_0(1 \mp \nu) \tag{6-344}$$

as would apply to nuclei close to the β-stability line, assuming the difference $f_{-1}^2 - f_{+1}^2$ to be proportional to the difference between the Fermi energies of neutrons and protons. (For a given nucleus, the appropriate values of $\omega_{\pm 1}$ can be obtained from those in Fig. 6-23 by noting that a change in $\omega_{+1}^{(0)} - \omega_{-1}^{(0)}$ leads to an equal change in $\omega_{+1} - \omega_{-1}$, as a consequence of the conservation law (6-342).)

The quantity ν expresses the neutron excess in units of the number of particles in a major shell of the harmonic oscillator. Thus, from Eqs. (2-151) and (2-152),[21] we obtain

$$\begin{aligned} \nu &= (3N)^{1/3} - (3Z)^{1/3} \\ &\approx 0.76 A^{-2/3}(N-Z) \end{aligned} \tag{6-345}$$

For nuclei along the line of β stability, ν approaches unity for $A \approx 200$.

In terms of the parameter ν, the charge exchange frequencies and transition probabilities (6-355) and (6-358) take the form

$$\hbar\omega_{\pm 1} = \left(\mp \nu(1+\zeta) + (1 + 2\zeta + \zeta^2\nu^2)^{1/2}\right)\hbar\omega_0 \tag{6-346a}$$

$$|\langle n_{\pm 1}=1|F_{\pm 1}|0\rangle|^2 = \left(\mp \nu + \frac{1+\zeta\nu^2}{(1+2\zeta+\zeta^2\nu^2)^{1/2}}\right) f_0^2 \tag{6-346b}$$

▲

[21] In the last step of Eq. (2-152), a factor $\frac{1}{3}$ has lamentably been omitted.

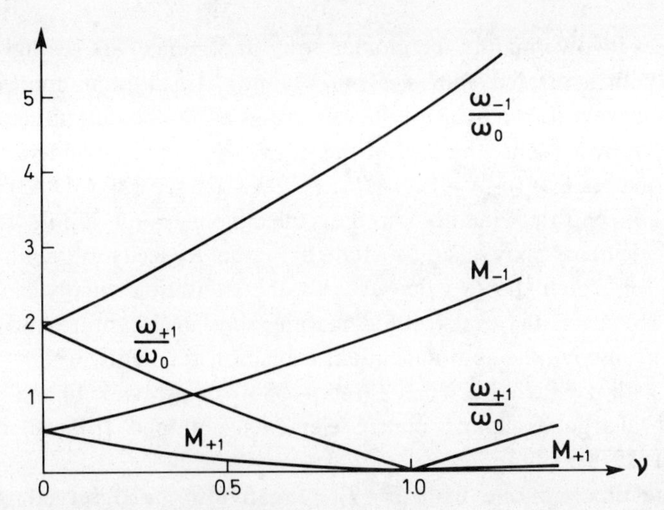

Figure 6-23 Properties of charge-exchange dipole modes as a function of neutron excess. The frequencies and dipole matrix elements for the charge exchange modes are plotted as a function of the parameter ν that measures the magnitude of the neutron excess. For $\nu > 1$, the mode labeled $+1$ represents an excitation in the $\mu_\tau = -1$ channel. The matrix elements $M_{\pm 1}$ are defined by $if_0 M_{\pm 1} = \langle n_{\pm 1} = 1 | F_{\pm 1} | 0 \rangle$ and equal the zero-point amplitudes $(\alpha_{\pm 1})_0$, in units of f_0. (The frequency unit ω_0 is the unperturbed oscillator frequency $(=\omega^{(0)}_{\mu_\tau = 0})$).

▼ with

$$\zeta \equiv \frac{\kappa f_0^2}{\hbar \omega_0} \tag{6-347}$$

The quantities in Fig. 6-23 are plotted for $\zeta = 1.35$, which is approximately the value obtained from estimate (6-314) for κ.

It is seen from Fig. 6-23 that the charge exchange modes are quite sensitive to the symmetry breaking implied by the neutron excess. Thus, for example, the nucleus considered in Fig. 6-16, with $Z = 46$ and $N = 60$, has $\nu \approx 0.48$, for which $\omega_{-1} : \omega_{+1} \approx 3$ and $M_{-1} : M_{+1} \approx 7$. For $\nu > 1$, the $\mu_\tau = +1$ mode is replaced by a second $\mu_\tau = -1$ mode associated with transitions leading to a decrease by 1 unit in the harmonic oscillator quantum number. The frequency and transition strength of the second $\mu_\tau = -1$ mode are obtained from those given for the $\mu_\tau = +1$ mode in Eq. (6-346) by an inversion of the sign, and are also illustrated in Fig. 6-23.

The charge exchange modes could be studied by reaction processes such as (pn), (^{3}He ^{3}H), ($\pi^+ \pi^0$), etc., and the charge conjugate processes leading to $\mu_\tau = +1$ excitations. At present, there appear to be no experimental data on the dipole excitations produced in such processes.

The collective dipole modes with $\mu_\tau = \pm 1$ play a significant role in weak interaction processes, such as in the capture of μ^- mesons (Balashov *et al.*, 1964; Barlow *et al.*, 1964; Foldy and Walecka, 1964). Allowed μ^-capture ▲

processes involve mainly the moment $\boldsymbol{\sigma}\tau_+$ and, hence, are strongly reduced in nuclei with saturated spins, as in ^{16}O and ^{40}Ca, and in nuclei with large neutron excess. In such cases, the capture rate may be determined by the first forbidden transitions; the part of these involving the spin-independent transition moments ($\mathscr{M}(\rho_V,\lambda=1)$, $\mathscr{M}(j_V,\kappa=0,\lambda=1)$; see Eq. (3D-43)) is expected to be concentrated mainly on the collective $\mu_\tau=+1$ dipole mode. These matrix elements may also be studied in the β decay of neutron-deficient nuclei, for which the Q value exceeds the excitation energy of the $\mu_\tau=+1$ mode. However, the first forbidden transitions in μ capture and β decay, in addition, involve the spin-dependent transition moments, which populate the modes with $\sigma=1,\kappa\approx 1,\lambda=0,1,2$ (see p. 383; for a review of the evidence on nuclear charge exchange matrix elements obtained from μ^-capture, see Überall, 1974).

The dipole mode with $T=T_0+1$ can also be observed in the $\mu_\tau=0$ transitions leading to the isobaric analog state with $M_T=T-1=T_0$. This transition is weak in comparison with the $T=T_0$ transitions, partly as a result of the factor $(T_0)^{-1}$ arising from the vector addition coefficient in Eq. (6-128) (see also Eq. (6-133)), and partly from the reduction in the number of excitations that can contribute in the presence of the neutron excess (see Fig. 6-22). For not too large values of the neutron excess, the energy separation between the $T=T_0$ and $T=T_0+1$ dipole modes can be expressed in terms of a coupling of the general form (6-130), leading to

$$E(T=T_0+1)-E(T=T_0)=a(T_0+1) \tag{6-348}$$

and the coefficient a can be obtained from Eqs. (6-355), (6-337), and (6-340),

$$a=A^{-1}V_1-2\kappa\langle x^2\rangle_{\text{nexc}} \tag{6-349}$$

The first term in a represents the effect of the symmetry potential, which favors states of lower isospin (see Eq. (6-129)). The second term results from the reduced dipole interaction energy in the mode with $T=T_0+1$; this effect can be directly obtained from the conservation law (6-342) by noting that the first-order terms in $(N-Z)$ in the interaction (6-341) (see also Eq. (6-351)) are contained in the energy that is diagonal in $n^{(0)}_{\pm 1}$. (The second term in a in Eq. (6-349) can also be viewed as a second-order effect of the particle-vibration coupling, which gives rise to an interaction between the excess neutrons and the dipole quantum; see Fig. 6-24 and Eq. (6-360).) If the value of $\langle x^2\rangle$ for the excess neutrons is taken to be the same as for the total density distribution ($\langle x^2\rangle\approx(1.2\ A^{-1/3}\ \text{fm})^2/5$), the estimate (6-314) for κ leads to $a\approx V_1/2A\approx 65A^{-1}$ MeV.

The $T=T_0+1$ dipole mode has been identified in a wide range of nuclei by photoprocesses and inelastic electron scattering (Axel *et al.*, 1967; Shoda *et al.*, 1969; Hasinoff *et al.*, 1969; see also the review by Paul, 1973). The energy difference between the $T=T_0$ and $T=T_0+1$ modes is found to be in

▼ accordance with the relation (6-348), with the parameter $a=(55\pm 15)A^{-1}$ MeV (Paul, 1973).

An important further test of the effects of the dipole charge exchange fields would be provided by a determination of the $E1$ strength for the excitation of the $T=T_0+1$ mode. The expected reduction of the $\mu_\tau=+1$ strength with increasing neutron excess, illustrated in Fig. 6-23, results partly from the effect of the exclusion principle in reducing the unperturbed strength, and partly from the ground-state correlations associated with the coupling between the $\mu_\tau=+1$ and $\mu_\tau=-1$ particle-hole excitations.

The interaction (6-341) may be treated in terms of the operators $c_{\mu_\tau}^{(0)}$ (and Hermitian conjugates) that annihilate (and create) the coherent single-particle excitations ($n_{\mu_\tau}^{(0)}=1$). The fields have the form

$$\begin{aligned} F_{+1}&=i\left(f_{+1}(c_{+1}^{(0)})^\dagger+f_{-1}c_{-1}^{(0)}\right)\\ F_{-1}&=i\left(f_{+1}c_{+1}^{(0)}+f_{-1}(c_{-1}^{(0)})^\dagger\right)=-F_{+1}^\dagger \\ F_0&=if_0\left((c_0^{(0)})^\dagger-c_0^{(0)}\right)=F_0^\dagger \end{aligned} \tag{6-350}$$

and the Hamiltonian becomes

$$\begin{aligned} H&=H_0+V\\ H_0&=\sum_{\mu_\tau}\hbar\omega_{\mu_\tau}^{(0)}(c_{\mu_\tau}^{(0)})^\dagger c_{\mu_\tau}^{(0)} \\ V&=\tfrac{1}{2}\kappa\left\{f_{+1}(c_{+1}^{(0)})^\dagger+f_{-1}c_{-1}^{(0)},\, f_{+1}c_{+1}^{(0)}+f_{-1}(c_{-1}^{(0)})^\dagger\right\}-\tfrac{1}{2}\kappa f_0^2\left((c_0^{(0)})^\dagger-c_0^{(0)}\right)^2 \end{aligned} \tag{6-351}$$

This Hamiltonian is a quadratic form in the $(c^{(0)})^\dagger$, $c^{(0)}$ operators, and can be diagonalized by a linear transformation to new boson variables $c^\dagger, c$,

$$\begin{aligned} (c_{+1}^{(0)})^\dagger&=Xc_{+1}^\dagger+Yc_{-1} && \\ & && (X^2-Y^2=1)\\ (c_{-1}^{(0)})^\dagger&=Xc_{-1}^\dagger+Yc_{+1} && \\ & && (X_0^2-Y_0^2=1)\\ (c_0^{(0)})^\dagger&=X_0c_0^\dagger+Y_0c_0 && \end{aligned} \tag{6-352}$$

with (see Eqs. (6-337) and (6-339))

$$X^2+Y^2=K^{-1}(\hbar\omega_0+\kappa f_0^2) \qquad 2XY=-K^{-1}\kappa f_{+1}f_{-1} \tag{6-353a}$$

$$K\equiv\left((\hbar\omega_0+\kappa f_0^2)^2-\kappa^2f_{+1}^2f_{-1}^2\right)^{1/2}$$

$$\begin{aligned} X_0^2+Y_0^2&=K_0^{-1}(\hbar\omega_0+\kappa f_0^2) \qquad 2X_0Y_0=K_0^{-1}\kappa f_0^2\\ K_0&\equiv\left((\hbar\omega_0+\kappa f_0^2)^2-\kappa^2f_0^4\right)^{1/2}=\left(\hbar\omega_0(\hbar\omega_0+2\kappa f_0^2)\right)^{1/2} \end{aligned} \tag{6-353b}$$

▲

▼ The transformed Hamiltonian is

$$H = E_0 + \sum_{\mu_\tau} \hbar\omega_{\mu_\tau} c^\dagger_{\mu_\tau} c_{\mu_\tau} \tag{6-354}$$

with the eigenfrequencies

$$\begin{aligned} \hbar\omega_{\pm 1} &= K \pm \tfrac{1}{2}((\hbar\omega^{(0)}_{+1} - \hbar\omega^{(0)}_{-1}) + \kappa(f^2_{+1} - f^2_{-1})) \\ \hbar\omega_{\mu_\tau = 0} &= K_0 \end{aligned} \tag{6-355}$$

The simple result (6-355) for the difference between the eigenfrequencies $\omega_{\pm 1}$ is a consequence of relation (6-342). The zero-point energy of the correlated ground state $|0>$ is given by

$$E_0 = (K - \hbar\omega_0) + \tfrac{1}{2}(K_0 - \hbar\omega_0) \tag{6-356}$$

The matrix elements of the fields F for exciting quanta of the normal modes may be obtained by transforming the expression (6-350) to the new variables $c^\dagger, c$,

$$\begin{aligned} F_{+1} &= -F^\dagger_{-1} = i((Xf_{+1} + Yf_{-1})c^\dagger_{+1} + (Yf_{+1} + Xf_{-1})c_{-1}) \\ F_0 &= i(X_0 - Y_0)f_0(c^\dagger_0 - c_0) \end{aligned} \tag{6-357}$$

Hence, from the relations (6-353), we find

$$\begin{aligned} |\langle n_{\pm 1} = 1|F_{\pm 1}|0\rangle|^2 &= \pm\tfrac{1}{2}(f^2_{+1} - f^2_{-1}) + K^{-1}(\hbar\omega_0 f_0^2 + \kappa(f_0^4 - f^2_{+1}f^2_{-1})) \\ |\langle n_0 = 1|F_0|0\rangle|^2 &= K_0^{-1}\hbar\omega_0 f_0^2 \end{aligned} \tag{6-358}$$

For the $\mu_\tau = 0$ mode, the result (6-358) expresses the conservation of oscillator strength. For the $\mu_\tau = \pm 1$ modes, the interaction does not commute with the field operator, and the oscillator strength is therefore not conserved for these modes. However, as seen from Eq. (6-358), the difference between the transition intensities with $\mu_\tau = \pm 1$ is not affected by the interaction; this result can be obtained directly by evaluating the commutator $[F_{-1}, F_{+1}]$ in the uncorrelated and correlated ground states. (See the discussion of this class of sum rules in Sec. 6-4c, p. 412.)

As noted above (p. 494), the present treatment ignores effects of relative order T_0^{-1}, since it is based on particle-hole excitations that are not exact eigenstates of T $(= T_0 + \mu_\tau)$. For small values of T_0, the components with $T > T_0 + \mu_\tau$ may become appreciable, but the corrections can be derived by exploiting the rotational invariance in isospace. In fact, to first order in T_0, the effect of the neutron excess on frequencies and matrix elements is given by relations of the general form (6-130) and (6-131), and the parameters in these expressions can be determined by considering values of T_0 large compared with unity, for which the results of the present section apply. Thus, the coefficient a in Eq. (6-130) has the value (6-349), while the coefficients m_0 and m_1 in Eq. (6-131) for $\mathscr{M}(\tau = 1, \mu_\tau) = F_{\mu_\tau}$ have the values (see Eqs. (6-358), (6-355), and (6-340))

$$\begin{aligned} m_0 &= if_0\left(\frac{\omega_0}{\omega_{\mu_\tau = 0}}\right)^{1/2} \\ \frac{m_1}{m_0} &= -\frac{\langle x^2\rangle_{\text{n exc}}}{2|m_0|^2} \end{aligned} \tag{6-359}$$

▲

▼ The effects of the neutron excess can also be described in terms of the particle-vibration coupling involving the dipole quanta and the excess neutrons ($H' = \kappa\alpha_{+1}F^{\dagger}_{+1} + \text{H. conj.}$). Thus, for example, the energy shift resulting from the interaction between a dipole quantum and an excess neutron is represented by the diagrams in Fig. 6-24, which yield

$$\delta\hbar\omega_{-1} - \delta\hbar\omega_{+1}$$
$$= \kappa^2\alpha_0^2\left(\frac{1}{\hbar\omega - \hbar\omega_0} + \frac{1}{\hbar\omega + \hbar\omega_0}\right)\left(\sum_{\nu_p\,\text{unocc}} |\langle\nu_p|F_{-1}|\nu_n\rangle|^2 + \sum_{\nu_p\,\text{occ}} |\langle\nu_p^{-1}\nu_n|F_{+1}|\text{v}=0\rangle|^2\right) \tag{6-360}$$

for a neutron in the orbit ν_n. Since the estimate (6-360) represents the first-order effect in $N-Z$, the properties of the vibrational mode have been taken for $N=Z$, for which they are independent of μ_τ. The sums over occupied and unoccupied proton orbits in Eq. (6-360) yield the expectation value $\langle\nu_n|2x^2|\nu_n\rangle$; one therefore obtains, by means of Eqs. (6-355) and (6-358), a result corresponding to the second term in Eq. (6-349). ▲

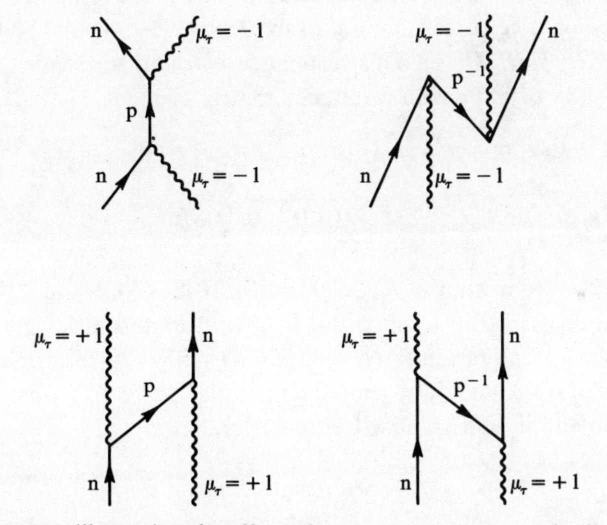

Figure 6-24 Diagrams illustrating the effect of an excess neutron on the frequencies of the charge exchange modes.

▼ *Effective moments for charge exchange dipole transitions*

The coupling of a single particle to the collective dipole modes with $\mu_\tau = \pm 1$ leads to a renormalization of the single-particle charge exchange moments, analogous to the effective charges for the $E1$ moments discussed on pp. 486ff. Thus, for the β-decay moment $\mathscr{M}(\rho_V\lambda=1)$ (see Eq. (3D-30)), ▲

▼ one obtains an effective vector coupling constant

$$g_V(\rho_V, \mu_\tau = \pm 1, \lambda = 1)_{\text{eff}} = g_V\left(1 - \kappa\left(\frac{|\langle n_{+1}=1|F_{+1}|0\rangle|^2}{\hbar\omega_{+1} \mp \Delta E} + \frac{|\langle n_{-1}=1|F_{-1}|0\rangle|^2}{\hbar\omega_{-1} \pm \Delta E}\right)\right) \tag{6-361}$$

where ΔE is the transition energy. For $\Delta E \approx 0$, corresponding to transitions between nuclei close to the β-stability line, the renormalization implied by Eq. (6-361) is the same for $\mu_\tau = \pm 1$ and is independent of the neutron excess, to first order in $(N-Z)$; using the coupling constant (6-314), one obtains $(g_V)_{\text{eff}} \approx 0.3\ g_V$. This estimate is consistent with that obtained from the analysis of the β^- decay of ^{207}Tl (see Vol. I, p. 353, solution 2).

The charge exchange dipole matrix element can also be obtained from the study of the γ decay of isobaric analog states $(T = M_T + 1)$, since the $E1$ moment for this process is related by rotational invariance in isospace to the β-transition moment $\mathscr{M}(\rho_V, \lambda = 1)$; see Eq. (3D-35). An example is provided by the analysis of the proton capture in ^{140}Ce proceeding through the isobaric analog resonance corresponding to the $I\pi = 7/2-$ ground state of ^{141}Ce (Ejiri *et al.*, 1969; Ejiri, 1971). This resonance is found to decay to the $I\pi = 5/2+$ ground state of ^{141}Pr with a reduced matrix element

$$M \equiv \langle I\pi = \tfrac{7}{2}-;\ T = T_0 + \tfrac{1}{2}, M_T = T_0 - \tfrac{1}{2} \| \mathscr{M}(E1) \| I\pi = \tfrac{5}{2}+;\ T = M_T = T_0 - \tfrac{1}{2}\rangle$$
$$= (0.18 \pm 0.04)e\,\text{fm} \tag{6-362}$$

where T_0 is the isospin of $^{140}_{58}Ce$ $(T_0 = 12)$. If the ^{141}Ce and ^{141}Pr ground states could be described in terms of an $f_{7/2}$ neutron and a $d_{5/2}$ proton added to a closed-shell configuration $(N = 82, Z = 58 = 50 + (g_{7/2})^8)$, one would obtain, from Eq. (3D-35) and by employing radial wave functions in the Woods-Saxon potential with standard parameters,

$$M = e(2T_0 + 1)^{-1/2} \langle f_{7/2} \| rY_1 \| d_{5/2} \rangle$$
$$\approx 0.85e\ \text{fm} \tag{6-363}$$

which exceeds the empirical value (6-362) by about a factor of 5. A correction must be applied for the pair correlations, which especially affect the protons; the resulting reduction factor in the matrix element is given by the probability amplitude u for the $d_{5/2}$ proton orbit to be empty in ^{140}Ce. Distributing the eight protons equally over the single-particle states of the approximately degenerate $d_{5/2}$ and $g_{7/2}$ configurations, the amplitude $u(d_{5/2}) \approx (6/14)^{1/2} \approx 0.65$. Hence, the reduction in the transition matrix element that may be attributed to polarization effects is about 0.3, in accordance with the estimate ▲ (6-361).

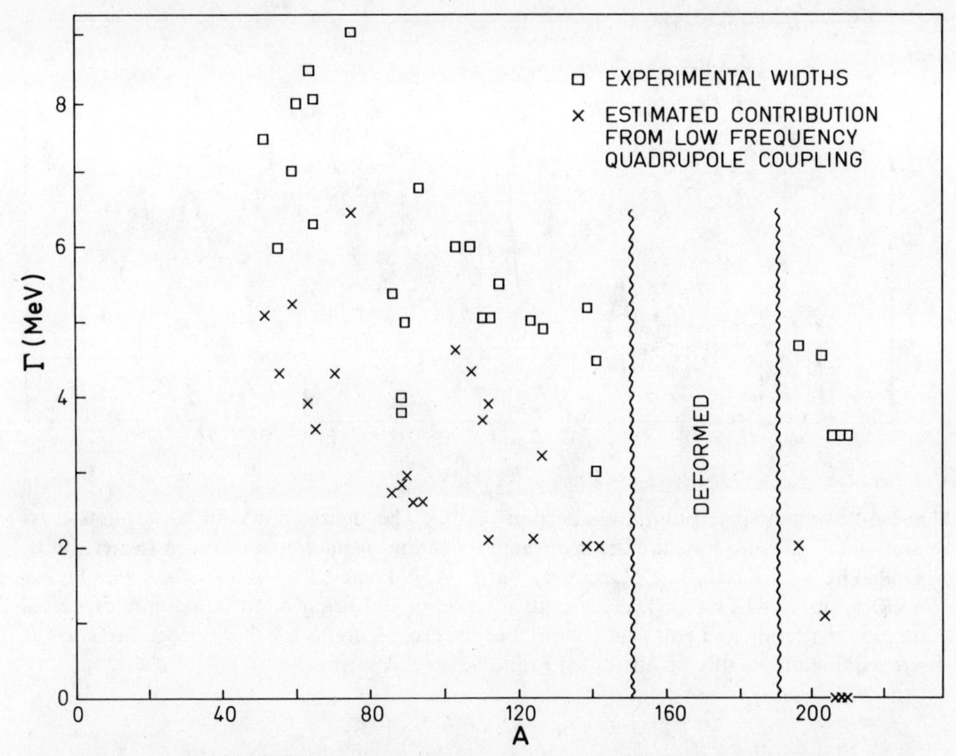

Figure 6-25 Systematics of the width of the dipole resonance. The experimental data are taken from the review article by E. G. Fuller and E. Hayward (*Nuclear Reactions*, Vol. 2, p. 113, eds. Endt and Smith, North Holland Publ. Co., Amsterdam, 1962). The quantity plotted is the width Γ of the resonance line at half maximum. In many cases, the experimental data have considerable uncertainties, and different laboratories often report rather different values for the width. The calculated widths are taken from J. Le Tourneux, *Mat. Fys. Medd. Dan. Vid. Selsk.* **34**, no. 11 (1965), but have been increased by a factor of 1.4 so as to give the full width at half maximum of the function (6-284), rather than 2Δ, where Δ is the second moment of the strength function given by Eq. (6-285).

Width of resonance (Figs. 6-25 and 6-26)

▼ The dipole absorption strength in the photoresonance is found to be distributed over an energy interval, which is typically of the order of 5 MeV (see Fig. 6-25). In some nuclei, this distribution appears to have a fairly simple form, such as is illustrated in Figs. 6-18, p. 475, and 6-21, p. 491. In light nuclei, however, the line shape has considerably greater structure. Partly, there is a high-energy shoulder containing a significant amount of oscillator strength (see p. 482); partly, the lower-energy absorption cross section exhibits fine structure with components having widths of a fraction of an MeV ▲ (see, for example, the spectrum of ^{16}O in Fig. 6-26). The interpretation of the

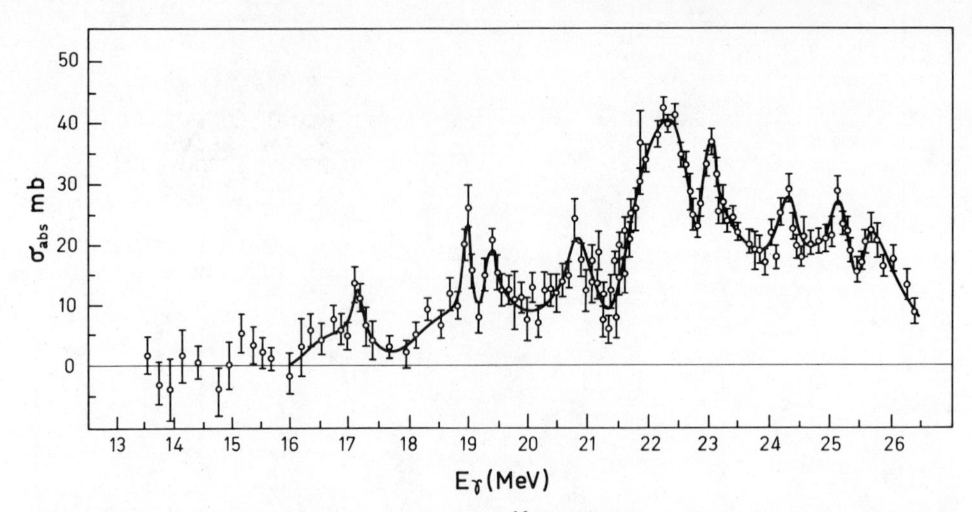

Figure 6-26 Photoabsorption cross section of ^{16}O. The figure shows the absorption cross section obtained from measurements of the attenuation of photons in oxygen (water) by B. S. Dolbilkin, V. I. Korin, L. E. Lazareva, and F. A. Nikolaev, *Zh. Eksper. Teor. Fiz. Pis'ma (USSR)* **1**, no. 5, 47 (1965). The main attenuation of the beam is due to atomic processes (Compton scattering and pair production), but the cross sections for these effects are known; the remaining attenuation is attributed to the nuclear absorption.

▼ width and fine structure in the dipole resonance poses many unsettled questions. We shall here briefly consider some of the coupling mechanisms that appear to play a significant role.

One of the sources of broadening of the dipole resonance is associated with the zero-point oscillations in the quadrupole shape oscillations. In many nuclei, the amplitude of these oscillations is not much smaller than the static eccentricities of the deformed nuclei. Since the latter are found to split the dipole resonance into two components with a separation of 3-4 MeV, it may be expected that the zero-point oscillations lead to an appreciable broadening. This effect can be seen very directly from Fig. 6-21, which shows the dipole resonance for a sequence of Nd isotopes starting from ^{142}Nd, with closed neutron shells ($N=82$) and resulting rather high-frequency, small-amplitude quadrupole mode and ending with ^{150}Nd, which has a static deformation ($\delta \approx 0.3$).

The effect on the dipole mode of the zero-point quadrupole oscillations can be described in terms of a coupling between the dipole and quadrupole oscillations, as discussed in Sec. 6-6b. The strength of the coupling is characterized by the dimensionless coupling parameter η, which can be determined from the observed strengths and frequencies of the dipole and quadrupole vibrations (see Eq. (6-281)). In most nuclei, η is large compared to ▲ unity, and the dipole strength is distributed on a number of components

▼ extending over an energy interval of order $\eta\hbar\omega_2$, as illustrated in Fig. 6-15, p. 457 (see also Eq. (6-285) for the second moment of the strength function).

The resonance line shape resulting from the dipole-quadrupole coupling has a rather complicated structure, and thus there is some ambiguity as to the quantity to be compared with the measured line width. In Fig. 6-25, the quantity chosen for this comparison corresponds to the full width at half maximum of the theoretical strong coupling strength function (6-284)

$$\Gamma \approx 1.67\eta\hbar\omega_2 \tag{6-364}$$

It is seen from Fig. 6-25 that, in most nuclei, the coupling to the low-frequency quadrupole mode makes a major contribution to the width of the dipole resonance, but other couplings must also contribute, as clearly indicated by the closed-shell nuclei, for which the coupling considered only accounts for a very small part of the observed width. (For attempts to interpret fine structure in the photoabsorption in terms of the expected components resulting from dipole-quadrupole coupling, see Fielder *et al.*, 1965, and Huber *et al.*, 1967.) The data in Fig. 6-25 refer to spherical nuclei; in deformed nuclei, similar effects result from the coupling to the β and γ vibrations (Tipler *et al.*, 1963; Arenhövel *et al.*, 1967).

Another source of structure in the dipole strength function may arise from the spread of the unperturbed single-particle dipole frequencies. A single-particle response function such as that given in Fig. 6-16, which has no strength in the region of the collective dipole mode, does not contribute to the width of this mode, and only leads to rather small residual oscillator strength in the region of the unperturbed frequencies. (This strength can be obtained from Eqs. (6-244) and (6-245) and is found to be of relative magnitude $(\Gamma^{(0)}/\Delta E)^4$, where $\Gamma^{(0)}$ is the spread of the unperturbed frequencies and ΔE the shift of the collective mode due to the dipole interaction; see also the example in Table 6-4, p. 472.) However, in a more detailed analysis of the single-particle response function that takes into account the finite binding of the nuclear potential, there will be a continuum extending through the region of the dipole resonance. The presence of such a continuum gives rise to a broadening of the collective mode, since the coupling through the dipole field implies a decay of the dipole oscillation with direct emission of a nucleon (see the diagram in Fig. 6-7, p. 419, and the expression (6-247) for the resulting width). The study of this effect would thus provide significant information on the particle-vibration coupling. However, the contribution to the total width is expected to be small, since the relevant energy region is far from those in which the significant single-particle transitions are in resonance. This conclusion receives some support from the observation that the direct emission of nucleons represents only about 10% of the decay of the photoresonance (see, for example, Kuchnir *et al.*, 1967). ▲

▼ The dipole field considered in the present discussion is assumed to be spin independent ($\kappa=1, \sigma=0, \lambda=1$). However, the spin-dependent nuclear interactions couple fields of this symmetry with the spin-dependent field ($\kappa=1, \sigma=1, \lambda=1$); see the discussion on p. 384. The latter field gives rise to spin-flip transitions of the type $j_1 = l_1 + 1/2 \rightarrow j_2 = l_2 - 1/2$ with energies that, especially in the light nuclei, may be in the region of the dipole resonance. (See, for example, the $p_{3/2}^{-1}\, d_{3/2}$ excitation in ^{16}O which, according to Fig. 3-2b, Vol. I, p. 321, is expected at an excitation energy of about 23 MeV and thus in the region of strong dipole absorption (see Fig. 6-26).) The interactions that couple the $\sigma=0$ and $\sigma=1$ dipole fields may therefore play a significant role in the broadening and line structure of the dipole mode, though at present there is no direct evidence concerning the strength of these couplings.

The coupling between the dipole mode and other degrees of freedom may lead eventually to very complicated states of motion, which finally decay predominantly by neutron emission. A small fraction of emitted neutrons has rather high energies and can be interpreted in terms of a direct emission process (referred to on p. 505), but the large majority of the observed neutrons has an energy distribution corresponding to that expected for evaporation from a compound nucleus (see, for example, Kuchnir *et al.*, 1967). One may attempt to describe the compound nucleus formation as "collision damping" of the single-particle motion represented by the imaginary term W in the optical potential (Dover *et al.*, 1972). For each of the two quasiparticles, the resulting width is $2|W|$(see Eq. (2-139)), and the total width is therefore $4|W|$. The imaginary potential depends on the energy of the particle (see Fig. 2-29, Vol. I, p. 237), and the available energy for each of the two quasiparticles created by the dipole field is equal to the total energy $\hbar\omega$ minus the excitation energy of the other quasiparticle, which on the average equals $\hbar\omega_0/2$. Thus, for example, for ^{208}Pb ($\hbar\omega \approx 14$ MeV, $\hbar\omega_0 \approx 7$ MeV), the damping of each quasiparticle should be similar to that of neutrons with an incident energy of a few MeV (since the separation energy is ≈ 8 MeV). The imaginary potential of ^{208}Pb for neutrons of this energy is not well determined, but the evidence appears to be compatible with the observed value of about 4 MeV for the width of the dipole resonance, corresponding to about 1 MeV for the average matrix element of W. For other nuclei, the empirically determined absorptive potentials are appreciably larger (see, for example, Vol. I, Figs. 2-26, p. 230, and 2-29, p. 237). However, for these nuclei, the effective value of W receives important contributions from the coupling of the one-particle motion to the low-frequency quadrupole shape oscillations. For the dipole mode, this coupling, which is significantly affected by the coherence in the particle motion, has already been included in terms of
▲ the contribution (6-364) to the dipole width.

▼ **Features of Quadrupole Modes in Spherical Nuclei**

The occurrence of strongly collective quadrupole transitions is a dominant feature of the low-energy nuclear spectra. The energies and $E2$-transition probabilities for the first excited $2+$ states of even-even nuclei are shown in Figs. 2-17, Vol. I, pp. 196–197, and 4-5, p. 48. For nuclei with sufficiently many particles outside closed shells, these low-lying $2+$ states represent the first member of the ground-state rotational band (see, for example, Fig. 4-4, p. 28). For configurations with fewer particles outside of closed shells, the quadrupole excitations can be approximately described in terms of vibrations in the neighborhood of the spherical shape. This interpretation is based on the occurrence of multiple excitations with properties that correspond qualitatively to those expected from a superposition of independent quanta.[22]

In order to explore the connection between the quadrupole shape oscillations and the nuclear shell structure, we consider the particle excitations produced by fields resulting from a deformation of the static central potential. For the present qualitative survey, we assume a field proportional to the quadrupole moment

$$F = \sum_k \left(r^2 Y_{20}(\vartheta)\right)_k \tag{6-365}$$

The spectrum of independent-particle excitations produced by this field is illustrated in Fig. 6-17a, p. 466. The transitions can be approximately characterized by the change in the harmonic oscillator total quantum number N, and one obtains a group of low-frequency transitions with $\Delta N = 0$ as well as transitions with $\Delta N = 2$ and energies of order $2\hbar\omega_0$.

The occurrence of two distinct characteristic frequencies in the single-particle quadrupole response function implies the possibility of two different modes of quadrupole shape oscillation. This feature of the nuclear shape oscillations, which results from the shell structure, is a quantal effect that lies outside the scope of the liquid-drop model. (In terms of the motion of the individual particles, the two frequencies can be identified with those characterizing the oscillation of the quadrupole moment of a particle moving in a spherical harmonic oscillator potential; the classical orbit described by the particle is an ellipse, and the frequency $2\omega_0$ derives from the orbital period, while the zero frequency reflects the stationarity of the closed orbit.) ▲

---

[22]On the basis of the liquid-drop model, it was anticipated that the quadrupole surface oscillations would represent the fundamental mode of nuclear collective motion (Bohr and Kalckar, 1937). The experimental evidence for the systematic occurrence of low-frequency quadrupole vibrational excitations in spherical nuclei was collected and interpreted by Scharff-Goldhaber and Weneser (1955).

▼ The strength of the low-frequency quadrupole excitations depends sensitively on the number of particles outside closed shells, and vanishes for a closed-shell configuration in a harmonic oscillator potential. (The special properties of quadrupole oscillations of closed-shell nuclei were emphasized by Gallone and Salvetti, 1953, and Inglis, 1955.) The collective mode generated by the transitions within a shell is therefore expected to show strong variations associated with the shell fillings, as is observed for the lowest 2+ excitations in nuclei.

The strength of the $\Delta N=2$ transitions is not sensitive to the degree of shell filling, and the properties of the high-frequency modes are expected to vary smoothly with A, as for the dipole mode considered in the previous examples. There is at present only tentative evidence concerning the high-frequency quadrupole modes.

In the present set of examples, we first consider some of the expected properties of the high-frequency quadrupole modes, as well as their effect on the low-frequency excitations through the particle-vibration coupling. The following examples deal with a variety of properties of the low-frequency quadrupole mode.

High-frequency modes and effective charges (Tables 6-8 and 6-9)

High-frequency isoscalar mode

The field (6-365) acts symmetrically on neutrons and protons and therefore generates a collective mode with $\tau=0$. Since the estimated interaction effects are rather large, we shall neglect the spread in the unperturbed excitation energies and consider a response function corresponding to that of independent-particle motion in a harmonic oscillator potential. The total one-particle transition strength to the degenerate excitations with $\omega^{(0)}=2\omega_0$ can then be obtained from the oscillator sum rule. The transitions with $\Delta N=0$ and $\omega^{(0)}=0$ have no oscillator strength, and hence, for the field component with $\mu=0$, which contributes the fraction $(2\lambda+1)^{-1}$ of the oscillator sum (6-172),

$$S(F)=2\hbar\omega_0\left(\alpha_0^{(0)}\right)^2=\frac{5}{4\pi}\frac{\hbar^2}{M}A\langle r^2\rangle \tag{6-366}$$

corresponding to restoring force and mass parameters for the unperturbed excitation energies (see Eqs. (6-21) and (6-23)),

$$\begin{aligned} C^{(0)}&=\frac{4\pi}{5}\frac{2M\omega_0^2}{A\langle r^2\rangle} \\ D^{(0)}&=\frac{4\pi}{5}\frac{M}{2A\langle r^2\rangle} \end{aligned} \tag{6-367}$$

▲

▼ The field coupling constant κ associated with shape oscillations may be estimated from the assumption that the eccentricity of the potential is the same as that of the density, which yields for the harmonic oscillator potential (see Eq. (6-78))

$$\kappa = -\frac{4\pi}{5}\frac{M\omega_0^2}{A\langle r^2\rangle} = -\tfrac{1}{2}C^{(0)} \tag{6-368}$$

and we therefore obtain the collective excitation energy (see Eq. (6-27))

$$\begin{aligned}\hbar\omega &= 2\hbar\omega_0\left(\frac{C^{(0)}+\kappa}{C^{(0)}}\right)^{1/2} = \sqrt{2}\,\hbar\omega_0 \\ &\approx 58A^{-1/3} \quad \text{MeV} \qquad (\tau=0, \lambda=2)\end{aligned} \tag{6-369}$$

Recently, evidence for the systematic occurrence of a high-frequency quadrupole mode has become available from the study of inelastic electron scattering (Pitthan and Walcher, 1971 and 1972, Nagao and Torizuka, 1973) as well as by scattering of nuclear projectiles (Lewis and Bertrand, 1972, Moalem *et al.*, 1973). The observed energy corresponds to the estimate (6-369), and the $E2$ strength is found to represent the major part of the $E2$-oscillator sum associated with $\tau \approx 0$ excitations.

The above description of the high-frequency quadrupole mode implies a zero-point amplitude of order $\alpha_0 \sim A^{1/3}R^2$ corresponding to a deformation parameter $\delta \sim A^{-2/3}$, since $\alpha \sim AR^2\delta$. Such deformations are small compared with the observed static nuclear deformations ($\delta \sim A^{-1/3}$). The potential energy is of order $\varepsilon_F A\delta^2$ and hence large compared with the deformation energy associated with the increase in the surface area, which is of order $b_{\text{surf}}A^{2/3}\delta^2$ (see Eqs. (6A-18) and (6A-19)). Correspondingly, the frequency (6-369) is proportional to $A^{-1/3}$, while the frequency of the liquid-drop oscillation is proportional to $A^{-1/2}$. Thus, the detailed structure of the surface region is not expected to play a major role for the high-frequency mode; the same applies to the effect of the Coulomb repulsion, which is of great importance in the treatment of shape oscillations in heavy nuclei in terms of the liquid-drop model (see, for example, Eq. (6A-24)).

The high frequency of the shape oscillations that carry the main part of the oscillator strength is a consequence of the nuclear shell structure, which implies that the nucleonic excitations involved in building these shape oscillations have energies of the same order of magnitude as those involved in compressive oscillations. In a more detailed analysis, it may thus be necessary to consider the coupling between surface and volume oscillations. (See, in this connection, the discussion in Sec. 6A-3c.)

The coupling of the quadrupole mode to single-particle excitations ▲ implies a renormalization of the one-particle quadrupole operator described by the polarizability coefficient (see Eq. (6-216))

$$\chi(\tau=0, \lambda=2) = -\frac{\kappa}{C^{(0)}+\kappa} \frac{(\hbar\omega)^2}{(\hbar\omega)^2-(\Delta E)^2}$$

$$\approx \frac{(\hbar\omega)^2}{(\hbar\omega)^2-(\Delta E)^2} \tag{6-370}$$

Thus, the present estimate implies that in low-frequency quadrupole transitions ($\Delta E \ll \hbar\omega$), the $\tau=0$ quadrupole moment is doubled as a result of the coupling to the high-frequency mode. (This result can also be obtained by considering the equilibrium shape for a configuration involving a single particle outside of closed shells. The equilibrium shape is obtained from the condition of self-consistency of potential and density; for a harmonic oscillator potential, it is found that the total quadrupole moment is contributed half by the single-particle and half by the deformation of the closed shells, as can be seen from Eq. (4-186) by comparing the value of δ with that obtained for the same configuration Σ_κ, in a spherical harmonic oscillator ($\omega_\kappa = \omega_0$).)

The collective coordinate α considered as a function of the particle variables and the collective flow associated with the vibrational motion have an especially simple character for modes such as the high-frequency quadrupole vibration that exhaust the strength of the field F. Thus, when acting on a closed-shell configuration, the quadrupole operator F given by Eq. (6-365) excites only the collective motion of the mode considered, and in the space of the vibrational states, the operators α and F have the same matrix elements, in the harmonic approximation (see p. 336). To this approximation, therefore, the collective coordinate α is identical with the quadrupole moment F, and the separation of the collective mode may be achieved by a point transformation from the particle coordinates x_k to a new set of coordinates $\alpha(x_k)$, $q_i(x_k)$, with q_i representing the degrees of freedom orthogonal to α (including the quadrupole modes with $\mu \neq 0$).

The corresponding transformation of the momenta is given by (Bohr, 1954)

$$\mathbf{p}_k = \boldsymbol{\nabla}_k \cdot F \frac{\hbar}{i}\frac{\partial}{\partial\alpha} + \sum_i \boldsymbol{\nabla}_k q_i \frac{\hbar}{i}\frac{\partial}{\partial q_i} \tag{6-371a}$$

$$\sum_k \boldsymbol{\nabla}_k F \cdot \boldsymbol{\nabla}_k q_i = 0 \tag{6-371b}$$

where the condition (6-371b) on the functions $q_i(x_k)$ is imposed in order that the transformed kinetic energy contain no terms bilinear in the collective momentum π and the momenta conjugate to q_i (condition of separability). The relations (6-371) yield the collective momentum

$$\pi = \frac{\hbar}{i}\frac{\partial}{\partial\alpha} = \frac{D}{M}\sum_k \mathbf{p}_k \cdot \boldsymbol{\nabla}_k F \tag{6-372a}$$

$$D^{-1}M = \sum_k \boldsymbol{\nabla}_k F \cdot \boldsymbol{\nabla}_k F = \frac{5}{2\pi} A\langle r^2\rangle \tag{6-372b}$$

▼ where the last relation may be obtained from Eq. (6-171). The quantity D is the mass parameter for the vibrational mode (as follows from a transformation of the total kinetic energy by means of Eq. (6-371a)) and is seen to have the value (6-367) derived above from other considerations.

The collective flow $\mathbf{u}(\mathbf{r})$ associated with the vibrational motion can be obtained from the current density $(2M)^{-1}\Sigma_k\{\mathbf{p}_k,\delta(\mathbf{r}-\mathbf{r}_k)\}$ the collective part of which yields

$$M\rho(\mathbf{r})\mathbf{u}(\mathbf{r}) = \sum_k \nabla_k F\delta(\mathbf{r}-\mathbf{r}_k)\pi = \nabla F(\mathbf{r})\pi \sum_k \delta(\mathbf{r}-\mathbf{r}_k) \tag{6-373a}$$

$$\mathbf{u}(\mathbf{r}) = M^{-1}\nabla F(\mathbf{r})\pi \tag{6-373b}$$

The velocity field $\mathbf{u}(\mathbf{r})$ is irrotational, as expected for a mode that exhausts the multipole oscillator strength (see p. 405; with this condition, the flow (6-373b) could have been obtained from the continuity equation $\dot{\rho}+\nabla\cdot(\rho_0\mathbf{u})=0$ and the expression for the density variations discussed on p. 356, following Eq. (6-77)).

A velocity dependence of the one-particle potential modifies the analysis of the collective mode. For the high-frequency quadrupole mode, the resulting effects can be exhibited in a rather simple form on the basis of the explicit representation of the collective flow discussed above.

If the velocity dependence in the one-particle potential is expressed in terms of an effective mass M^*, the vibrational mass parameter $D^{(0)}$ is multiplied by the factor M^*/M (see Eq. (6-367)). Such a modification of $D^{(0)}$ affects the oscillator strength of the collective mode by the factor M/M^*, in violation of the classical sum rule. As discussed on p. 402, this sum rule is expected to be approximately valid as a consequence of Galilean invariance. The inconsistency is removed by recognizing that the nucleon velocities entering into the velocity-dependent single-particle potential $V(\mathbf{p})$ are to be measured relative to the velocity $\mathbf{u}$ of the collective flow (local Galilean invariance). By a transformation to a coordinate system moving with velocity $\mathbf{u}$, the particle momenta transform as $\mathbf{p}\rightarrow\mathbf{p}-M\mathbf{u}$ (see Eq. (1-16)), and the one-particle potential becomes, in the effective mass approximation,

$$\begin{aligned} V(\mathbf{p}-M\mathbf{u}) &= \frac{1}{2}\left(\frac{1}{M^*}-\frac{1}{M}\right)(\mathbf{p}-M\mathbf{u})^2 \\ &\approx k_0\frac{p^2}{2M} - k_0(\mathbf{p}\cdot\mathbf{u}) \\ k_0 &\equiv \frac{M}{M^*}-1 \end{aligned} \tag{6-374}$$

including only the term linear in $\mathbf{u}$, which represents a coupling between the single-particle and collective motion. With a collective flow given by Eq. (6-373), this coupling takes the form

$$H' = -\frac{k_0}{M}(\mathbf{p}\cdot\nabla F)\pi \tag{6-375}$$

Summing the energy (6-375) over all the particles (and including a factor $1/2$ to take into account that the coupling (6-375) is a result of two-particle interactions), one ▲ obtains an energy proportional to π^2 (see Eq. (6-372a)) and a total vibrational mass parameter given by

$$D^{-1} = \frac{5}{2\pi}\frac{A\langle r^2\rangle}{M}\left(\frac{M}{M^*} - k_0\right) = \frac{5}{2\pi}\frac{A\langle r^2\rangle}{M} \tag{6-376}$$

involving the bare mass M rather than the effective mass M^*.

While the mass parameter has the classical value (6-376) as a consequence of local Galilean invariance, the restoring force parameter is modified by an effective mass term in the one-particle potential. If the oscillator frequency is adjusted so as to retain the spatial dimensions, as given by $\langle r^2\rangle$, this frequency is multiplied by M/M^*; hence, the restoring force $C^{(0)}$ as well as the coupling constant κ and the total restoring force C are multiplied by the same factor (see Eqs. (6-367) and (6-368)). Consequently, the frequency of the collective mode is multiplied by the factor $(M/M^*)^{1/2}$.

High-frequency isovector mode

Besides the symmetric $\tau=0$ mode, the high-frequency quadrupole excitations can also generate a $\tau=1$ mode, in which the neutrons and protons move in opposite phase. This mode is similar to the dipole mode considered above and, as in the treatment of the dipole mode (see Eq. (6-311)), we assume a field proportional to the isovector multipole moment

$$F = \sum_k (r^2 Y_{20}(\vartheta)\tau_z)_k \tag{6-377}$$

The restoring force and mass parameter for the unperturbed $\tau=1$ quadrupole oscillation are the same as for the $\tau=0$ mode given by Eq. (6-367), and the coupling constant can be obtained from the estimate (6-127)

$$\kappa = \frac{\pi V_1}{A\langle r^4\rangle} \tag{6-378}$$

We thus obtain

$$\frac{\kappa}{C^{(0)}} = \frac{5}{8}\frac{V_1\langle r^2\rangle}{M\omega_0^2\langle r^4\rangle} \approx 1.8 \tag{6-379}$$

The numerical value is based on the approximate estimates

$$\langle r^2\rangle \approx 0.87A^{2/3}\ \text{fm}^2 \qquad \langle r^4\rangle \approx 0.95A^{4/3}\ \text{fm}^4 \qquad V_1 \approx 130\ \text{MeV} \tag{6-380}$$

where the radial moments are based on the expression (2-65), with the parameters (2-69) and (2-70) for $A \sim 100$.

▼ From Eqs. (6-27) and (6-379), the collective excitation energy is found to be

$$\hbar\omega = 2\hbar\omega_0\left(1+\frac{\kappa}{C^{(0)}}\right)^{1/2}$$

$$\approx 135 A^{-1/3} \text{ MeV} \qquad (\tau=1, \lambda=2) \qquad \text{(6-381)}$$

As for the dipole mode, the estimate based on coherent shell-model excitations is rather close to the estimate of the corresponding eigenfrequency obtained from the hydrodynamical treatment of polarization modes (see Eq. (6A-65) and the discussion on pp. 488 ff.).

The polarizability coefficient describing the renormalization of the isovector quadrupole moment of a single particle is obtained from Eqs. (6-216) and (6-379)

$$\chi(\tau=1, \lambda=2) = -0.64\frac{(\hbar\omega)^2}{(\hbar\omega)^2-(\Delta E)^2} \qquad \text{(6-382)}$$

where ΔE is the single-particle transition energy.

The schematic model considered above treats the isoscalar and isovector modes in terms of fields with the same radial form factor, although the modes are assumed to be of a rather different character. The isoscalar mode is envisaged as a shape oscillation with a potential concentrated in the region of the nuclear surface, while the isovector mode is treated as a volume oscillation, with a density and potential confined to the interior of the nucleus. The crucial point in the treatment is the construction of the modes primarily out of the single-particle excitations with $\Delta N=2$. The different structure of the modes is reflected in the fact that the ratio of the effective coupling constants ($\kappa_1/\kappa_0 \approx -3.6$) is much larger than the ratio between the coefficients in the average potentials in the nucleus ($V_1/4V_0 \approx -0.6$).

Effects of neutron excess on high-frequency modes

In the above discussion of the high-frequency quadrupole modes, we have assumed $N=Z$. The frequencies as well as the $\tau=0$ and $\tau=1$ transition probabilities do not depend on the sign of $(N-Z)$, and the leading-order corrections to these quantities are, therefore, of order $(N-Z)^2/A^2$. However, terms linear in $(N-Z)$ are encountered in the interference effects between $\tau=1$ and $\tau=0$ matrix elements.

The effect of the neutron excess on the collective modes involves features of the average fields that are not determined by considerations of symmetry. In the following, we shall consider a simple description of the normal modes based on the assumption that the isospin structure of these modes is dominated by the strong neutron-proton force in the nucleus.

In the shape oscillations ($\tau \approx 0$), the neutron-proton forces will tend to ▲ preserve the local ratio of neutrons to protons, in which case the mode gives

▼ rise to an isovector moment proportional to the neutron excess,

$$\mathscr{M}(\tau=1,\lambda=2) \approx \frac{N-Z}{A}\mathscr{M}(\tau=0,\lambda=2)$$
$$= \frac{N-Z}{A}\alpha_{\tau=0} \tag{6-383}$$

Moreover, since the static nuclear field acts differently on neutrons and protons (symmetry potential; see Eq. (2-26)), there will be a corresponding isovector term in the field coupling associated with the shape oscillation

$$\delta V \approx \kappa_{\tau=0}\alpha_{\tau=0}\left(1+\frac{V_1}{4V_0}\frac{N-Z}{A}\tau_z\right)r^2 Y_{20} \tag{6-384}$$

where $V_1/V_0 \approx -2.6$ is the ratio of isovector to isoscalar components in the static nuclear potential (assuming $V_0 \approx -50$ MeV, $V_1 \approx 130$ MeV).

For the mode corresponding to oscillations of neutrons with respect to protons ($\tau \approx 1$), we shall consider, as for the dipole mode, a field that acts on the difference of the isospin from its mean value

$$F_{\tau=1} = \sum_k \left(r^2 Y_{20}(\vartheta)\left(\tau_z - \frac{N-Z}{A}\right)\right)_k \tag{6-385}$$

For the dipole field, the corresponding form (6-328) was derived on the basis of translational invariance, but in the present case, an additional assumption is involved, which is equivalent to the requirement that the field does not act on the total density at any point. Thus, the form (6-385) ignores possible couplings to compressive modes of quadrupole symmetry. The field (6-385) can also be characterized by the fact that such a collective mode is not excited by a projectile with the same mass to charge ratio as the target. (The mode can, however, be excited by $T=0$ projectiles, due to the neutron excess of the target.)

Effective charge for E2 transitions

Evidence on the coupling of the particle motion to the high-frequency quadrupole modes is provided by the $E2$ effective charge for low-energy transitions (see Sec. 3-3a). From the isoscalar and isovector polarizabilities (6-370) and (6-382), and with inclusion of the linear terms in the neutron

▲ excess discussed above, we obtain the total polarization charge for static

▼ quadrupole moments

$$e_{\text{pol}}^{\text{std}}(E2,\Delta E=0)=\frac{Ze}{A}\chi(\tau=0,\Delta E=0)\left(1+\frac{V_1}{4V_0}\frac{N-Z}{A}\tau_z\right)$$

$$-\frac{e}{2}\chi(\tau=1,\Delta E=0)\left(\tau_z-\frac{N-Z}{A}\right) \qquad \text{(6-386a)}$$

$$=e\left(\frac{Z}{A}-0.32\frac{N-Z}{A}+\left(0.32-0.3\frac{N-Z}{A}\right)\tau_z\right) \qquad \text{(6-386b)}$$

The first term in Eq. (6-386a) gives the contribution of the $\tau \approx 0$ mode with the coupling (6-384) and with a ratio of electric to isoscalar quadrupole moment of Ze/A, as implied by Eq. (6-383). The second term of Eq. (6-386a) gives the contribution of the $\tau \approx 1$ mode with the coupling proportional to the field (6-385) and with a ratio of electric to isovector quadrupole moments of $-e/2$, as follows from the assumption that this mode involves no changes in the total density and thus does not generate a $\tau=0$ moment (see Eq. (6-123)).

As indicated by the superscript std, the estimate (6-386) is a standard value, which neglects the possibility of differences in the radial distributions of the transition moment and the field coupling. In particular, these differences may be significant for loosely bound orbits with large intensity of the wave function outside the nuclear surface resulting in very large values of $\langle r^2 \rangle$, and for certain transitions involving a change in the number of radial nodes, resulting in a partial cancellation in $\langle r^2 \rangle$. (Examples of these effects can be seen in Table 3-2, Vol. I, pp. 341–342.) Such large variations are not expected for the radial matrix elements of the field coupling (6-69), which is mainly effective in the surface region, where the wave functions of the relevant single-particle orbits have similar magnitude (see Fig. 3-4, Vol. I, p. 327). An improved estimate of e_{pol} may therefore be obtained by including the variation of the radial matrix elements of the multipole moment

$$e_{\text{pol}}(E2,\Delta E=0)\approx e\left(\frac{Z}{A}-0.32\frac{N-Z}{A}+\left(0.32-0.3\frac{N-Z}{A}\right)\tau_z\right)\frac{\frac{3}{5}R^2}{\langle j_2|r^2|j_1\rangle} \qquad \text{(6-387)}$$

In a comparison with the empirical data, one must take into account that the closed-shell nuclei are found to possess $\tau=0$ quadrupole modes with energies appreciably below the estimate (6-369). Even though the oscillator strength of these transitions constitutes only a small fraction ($\lesssim 10\%$) of the total, they contribute significantly to the polarizability, which is inversely proportional to the energy E of the excited mode. (The configurations responsible for these low-frequency modes in closed-shell nuclei are considered in the fine print below.) The evidence on these relatively low-energy ▲

▼ quadrupole excitations in the closed-shell nuclei is collected in Table 6-8, and the contributions δe_{pol} to the polarization charge are estimated by assuming δe_{pol} to be proportional to the quantity $E^{-1}B(E2;\, 0\rightarrow 2)$, as in Eq. (6-218), and by normalizing with respect to the value $(Z/A-0.32(N-Z)A^{-1}\tau_z)e$ for
▲ the high-frequency shape oscillation.

Nucleus	E_i MeV	$B(E2;0\rightarrow 2)$ $e^2\text{fm}^4$	$\dfrac{E_i B(E2)(A/Z)^2}{S(\lambda=2,\tau=0)_{\text{class}}}$	$\dfrac{(\delta e_{\text{pol}}^{\text{std}})_i}{e}$	$\dfrac{\Sigma(\delta e_{\text{pol}}^{\text{std}})_i}{e}$
^{16}O	6.92	37	0.11	0.62	
	9.85	0.7	0.003	0.01	
	11.52	19	0.09	0.19	
	(23)	(100)	(1)	(0.5)	1.32
^{40}Ca	3.90	90	0.035	0.32	
	5.63	8	0.005	0.02	
	6.91	70	0.05	0.14	
	(17)	(600)	(1)	(0.5)	0.98
^{208}Pb	4.07	3.0×10^3	0.15	0.31	
				0.40	
	(10)	(8×10^3)	(1)	(0.34)	0.65 (n)
				(0.44)	0.84 (p)

Table 6-8 Contributions of $\tau=0$ excitations of closed shells to $E2$ effective charge. The experimental data in columns two and three are from: ^{16}O (compilation by Ajzenberg-Selove *Nuclear Phys.* **A166**, 1, 1971); ^{40}Ca (J. R. MacDonald, D. H. Wilkinson, and D. E. Alburger, *Phys. Rev.* **C3**, 219, 1971 and references quoted there); ^{208}Pb (J. F. Ziegler and G. A. Peterson, *Phys. Rev.* **165**, 1337, 1968). The value of the classical oscillator sum $S(\lambda=2,\tau=0)$ employed in column four is obtained from Eq. (6-179), using the values $\langle r^2\rangle = 7.0\ \text{fm}^2$ (^{16}O), 12.4 fm^2 (^{40}Ca), and 31 fm^2 (^{208}Pb); see the review by Collard *et al.*, 1967. The last line for each nucleus gives the theoretically estimated properties of the high-frequency $\tau=0$ quadrupole mode, in parentheses. For ^{208}Pb, the two values of δe_{pol} refer to neutrons and protons, respectively.

▼ The empirical evidence on $E2$-polarization charges for single particles outside closed shells is shown in Table 6-9 and compared with the estimate obtained from the expression

$$e_{\text{pol}}(E2) = \frac{\frac{3}{5}R^2}{\langle j_2|r^2|j_1\rangle}\sum_i (\delta e_{\text{pol}}^{\text{std}})_i \frac{E_i^2}{E_i^2-(\Delta E)^2} \tag{6-388}$$

where the values of $\delta e_{\text{pol}}^{\text{std}}$ for the $\tau=0$ excitations are taken from Table 6-8,
▲ while $\delta e_{\text{pol}}^{\text{std}}$ for the $\tau\approx 1$ mode is taken to have the value $0.32\tau_z e$ (see Eq.

Nucleus	j_1	j_2	$\langle j_2 \| \mathscr{M}(E2) \| j_1 \rangle^2$ e^2 fm^4	ΔE MeV	$\langle j_2 \| r^2 \| j_1 \rangle$ fm^2	e_{pol}/e Obs.	e_{pol}/e Calc.
^{17}O	$d_{5/2}$	$d_{5/2}$	11	0	11.5	0.4	1.0
	$s_{1/2}$	$d_{5/2}$	13	0.9	12.0	0.4	1.0
^{17}F	$s_{1/2}$	$d_{5/2}$	128	0.5	13.4	0.2	0.5
^{39}K	$d_{3/2}$	$d_{3/2}$	160	0	13.0	0.7	0.6
^{41}Ca	$p_{3/2}$	$f_{7/2}$	264	1.9	13.9	1.3	1.3
^{41}Sc	$p_{3/2}$	$f_{7/2}$	440	1.7	13.9	0.7	0.7
^{207}Tl	$d_{3/2}$	$s_{1/2}$	~ 800	0.35	26	0.9	0.7
^{207}Pb	$f_{5/2}$	$p_{1/2}$	420 ± 20	0.57	32	0.9	0.9
	$p_{3/2}$	$p_{1/2}$	244 ± 10	0.89	37	0.8	0.8
^{209}Pb	$d_{5/2}$	$g_{9/2}$	1100 ± 270	1.57	40	0.8	0.7
	$s_{1/2}$	$d_{5/2}$	310 ± 15	0.47	57	0.4	0.5
^{209}Bi	$h_{9/2}$	$h_{9/2}$	2600 ± 200	0	35	0.5	0.5
	$f_{7/2}$	$h_{9/2}$	200 ± 70	0.89	17	2.3	1.1
	$f_{5/2}$	$h_{9/2}$	2900 ± 1000	2.81	20	1.6	1.5

Table 6-9 Polarization charges for $E2$-matrix elements in single-particle configurations. The data are taken from the references quoted in Table 3-2, Vol. I, p. 341 and the additional more recent measurements: ^{207}Tl (S. Gorodetzky, F. Beck, and A. Knipper, *Nuclear Phys.* **82**, 275, 1966; A. P. Komar, A. A. Vorobiev, Yu. K. Zalite, and G. A. Korolev, *Dokl. Akad. Nauk SSSR* **191**, 61, 1970); ^{207}Pb (E. Grosse, M. Dost, K. Haberkant, J. W. Hertel, H. V. Klapdor, H. J. Körner, D. Proetel, and P. von Brentano, *Nuclear Phys.* **A174**, 525, 1971; O. Häusser, F. C. Khanna, and D. Ward, *Nuclear Phys.* **A194**, 113, 1972); ^{209}Pb (Häusser *et al.*, *loc.cit.*); ^{209}Bi, quadrupole moment (G. Eisele, I. Koniordos, G. Müller, and R. Winkler, *Phys. Letters* **28B**, 256, 1968); ^{209}Bi, transitions (R. A. Broglia, J. S. Lilley, R. Perazzo, and W. R. Phillips, *Phys. Rev.* **C1**, 1508, 1970; J. W. Hertel, D. G. Fleming, J. P. Schiffer, and H. E. Gove, *Phys. Rev. Letters* **23**, 488, 1969; Häusser *et al.*, *loc.cit.;* W. Kratschmer, H. V. Klapdor, and E. Grosse, *Nuclear Phys.* **A201**, 179, 1973).

▼ (6-387)). The assumed values of the radial matrix elements $\langle j_2|r^2|j_1\rangle$ are given in column six of Table 6-9 and are the same as those employed in Table 3-2. The assumed values of the mean-square radii $\langle r^2\rangle = 3R^2/5$ are given in the caption to Table 6-8. In Table 6-9, we have omitted the observed $E2$-matrix element for the $p_{3/2}\rightarrow p_{1/2}$ transition in ^{15}N (see Table 3-2); for this transition, the frequency is 6.3 MeV and therefore so close to the 6.9 MeV 2+ excitation in ^{16}O (see Table 6-8) that a more detailed analysis of the coupling between the two excitation modes is required.

The data in Table 6-9 agree, in order of magnitude, with the estimated polarization charges. The main contribution to the polarization charge is isoscalar on account of the relatively low frequencies of the $\tau\approx 0$ modes, but the data also provide rather definite evidence for effects resulting from isovector forces. In the quantitative analysis of the isovector polarization charge, contributions arising from the relatively small differences between the coupling of neutrons and protons to the isoscalar modes may be of significance. The expression (6-387) includes the effect of the symmetry potential, but assumes the same radial form factor for neutrons and protons, as may be suggested by the observed similarity in the radial distribution of protons and neutrons in heavy nuclei. (The approximate equality of the mean-square radii for the neutron and proton is inferred from the analysis of the electron scattering data (which determines the proton distribution), combined with the Coulomb energy difference between isobaric analog states (which measures the density distribution of the excess neutrons); this equality contrasts with the densities obtained for protons and neutrons moving in potentials of equal radii, since the symmetry potential implies a significant reduction in the radius of the proton density distribution (Nolen and Schiffer, 1969; see also the discussion of isotope shifts in Vol. I, p. 163). The analysis of the isovector interactions that are responsible for restoring the approximate equality of neutron and proton density radii is an issue that may have important bearings on the isovector collective modes and their coupling to single-particle motion.)

As regards the detailed comparison in Table 6-9, it may be remarked that

(a) the largest disagreement in the region of ^{208}Pb is associated with the $f_{7/2}\rightarrow h_{9/2}$ transition, which is very weak on account of the spin flip and therefore sensitive to small spin-dependent components in the potential:

(b) the observed rather small isoscalar polarization charges in the region of ^{16}O may indicate that the coupling has been overestimated for these light nuclei.

The lowest-energy quadrupole excitations observed in ^{16}O and ^{40}Ca (see Table 6-8) appear to involve large static deformations, as suggested by the evidence for rotational band structure ($K=0$; $I=0,2,4,\ldots$) with strongly enhanced $E2$ transitions ▲ between the members of the band (shape isomerism; see the references on p. 26).

▼ These quadrupole excitations involve primarily two-particle, two-hole and four-particle, four-hole configurations.

In the case of ^{208}Pb, the 2+ mode at 4.07 MeV appears to be primarily associated with the $\Delta N=0$ particle-hole excitations $1i_{13/2}\to 2g_{9/2}$ (neutrons) and $1h_{11/2}\to 2f_{7/2}$ (protons), with energies 5.0 MeV and 6.2 MeV, respectively (see Fig. 3-3, Vol. I, p. 325, and the estimate (6-454) for the Coulomb interaction in the (proton, proton-hole) configuration). The quadrupole couplings will produce linear combinations of these two excitations, of which the lower will be predominantly isoscalar and the higher mainly isovector. One may thus attempt to interpret the 4.07 MeV state as the isoscalar combination, which implies a $\tau=0$ oscillator strength associated with these transitions of $1.6\,(1+\chi(\Delta N=2,\,\tau=0))^2\times 10^4\ \text{fm}^4\ \text{MeV}$ (see Eqs. (3C-34) and (6-304); the radial matrix elements of r^2 have been taken to be the $3R^2/5$ with $R=1.2A^{1/3}$ fm). Using Eq. (6-370) for the $\tau=0$ polarizability, we obtain a resulting $\tau=0$ oscillator strength of about $8\times 10^4\ \text{fm}^4$ MeV. From the empirical $B(E2)$ value for excitation of the 4.07 MeV level in ^{208}Pb (see Table 6-8), one obtains the estimate $\hbar\omega B(\tau=0,\lambda=2)\approx(A/Ze)^2\hbar\omega B(E2)\approx 7.8\times 10^4\ \text{fm}^4$ MeV, in agreement with the above interpretation in terms of the $i_{13/2}$ and $h_{11/2}$ excitations.

A further test of the interpretation of the 2+ excitation in ^{208}Pb is obtained by comparing the observed energy with that implied by the quadrupole field couplings. Since the energy shifts are comparable with the separation between the unperturbed configurations, an approximate analysis may be obtained in terms of a purely isoscalar coupling acting in degenerate particle-hole configurations with $\hbar\omega^{(0)}\approx 5.6$ MeV, corresponding to the mean value of the proton and neutron excitation energies. If we express the coupling in the form (6-68), the amplitudes X_i and Y_i associated with the particle-hole configurations $i=j_1^{-1}j_2$ are given by (see Eq. (6-253))

$$X_i=\frac{\alpha_0\langle i|k(r)Y_{20}|\text{v}=0\rangle}{\hbar(\omega^{(0)}-\omega)}$$
$$Y_i=\frac{\alpha_0\langle i|k(r)Y_{20}|\text{v}=0\rangle}{\hbar(\omega^{(0)}+\omega)} \tag{6-389}$$

where the zero-point amplitude α_0 can be determined from the measured $B(E2)$ value

$$\alpha_0=\left(\frac{3}{4\pi}ZeR^2\right)^{-1}(B(E2;\,2\to 0))^{1/2}\approx 0.025 \tag{6-390}$$

The summation of the matrix elements of $k(r)Y_{20}$ over the two particle-hole configurations referred to above yields

$$\sum_i\langle i|k(r)Y_{20}|\text{v}=0\rangle^2=\sum\frac{2j_1+1}{4\pi}\langle j_1\tfrac{1}{2}20|j_2\tfrac{1}{2}\rangle^2\langle j_2|k(r)|j_1\rangle$$
$$\approx 0.64\langle k\rangle^2 \tag{6-391}$$

where $\langle k\rangle$ is the mean value of the radial matrix elements for the two configurations $i_{13/2}^{-1}g_{9/2}$ and $h_{11/2}^{-1}f_{7/2}$. From the normalization condition $\Sigma_i(X_i^2-Y_i^2)=1$, one finally obtains $\langle k\rangle\approx 75$ MeV for the radial coupling matrix element needed to account for the observed frequency $\hbar\omega=4.1$ MeV. This value of $\langle k\rangle$ is about 50% larger than the radial matrix elements of $R_0\partial V/\partial r$ for the configurations considered (assuming a Woods-Saxon potential with standard parameters). Thus, a significant part of the observed interaction effect in this state is not accounted for in terms of the couplings
▲ considered.

▼ A more detailed analysis of the $\Delta N=0$ quadrupole excitations in ^{208}Pb would involve a treatment of the many different effects that are associated with the neutron excess. Thus, the energies of the neutron and proton excitations differ by an amount comparable with the interaction energies, and hence the normal modes involve a combination of isovector and isoscalar fields. However, a treatment based on the couplings given in the preceding does not alter the main conclusions obtained from the simplified analysis involving a pure $\tau=0$ coupling.

Qualitative analysis of low-frequency mode. Instability effect (Fig. 6-27)

The low-frequency quadrupole mode is characterized by a frequency that decreases as particles are added in unfilled shells. For sufficiently many particles, the spherical shape becomes unstable, and the mode develops into the rotational excitation. In the transition region, one must expect large anharmonicities in the vibrational motion.

The instability phenomenon and the associated anharmonicity may be seen as an effect of the quadrupole field coupling acting among the particles in the partially filled shells. A quantitative treatment of the low-frequency quadrupole mode is sensitive to the detailed features of the nuclear shell structure and will not be attempted in the present context. It is possible, however, to illustrate many of the qualitative relations on the basis of simple considerations concerning the effect of the field coupling in the presence of pair correlations.[23]

Vibrational frequency. Approach to instability

In the pair-correlated system, the particle excitations produced by the quadrupole field correspond to the creation of two quasiparticles. Since the pairing energy is comparable with, or somewhat larger than, the one-particle energies for transitions within a major shell (compare Fig. 6-17a with Eq. (2-94)), the excitation energies are approximately equal to 2Δ,

$$\hbar\omega^{(0)} \approx 2\Delta \approx 25A^{-1/2} \quad \text{MeV} \tag{6-392}$$

If one assumes complete degeneracy of the orbits within a single major oscillator shell with total quantum number N, the unperturbed zero-point amplitude of the quadrupole field is given by

$$\begin{aligned}\left(\alpha_0^{(0)}\right)^2 &= \sum_i \langle \mathrm{v}=2, i|F|\mathrm{v}=0\rangle^2 \\ &= \left(\frac{5}{16\pi}\left(\frac{\hbar}{M\omega_0}\right)^2 N(N+1)(N+2)(N+3)f(1-f)\right)_{\mathrm{n}} + \left(\cdots\right)_{\mathrm{p}}\end{aligned} \tag{6-393}$$

▲

[23]Vibrational motion associated with the competition between the pair correlations and the quadrupole field coupling acting among the particles in an unfilled shell was considered by Belyaev (1959) and Kisslinger and Sorenson (1960).

▼ involving a sum of contributions from neutrons and protons. The derivation of this expression will be given in the small print below (see Eq. (6-411)). The shell-filling parameter f represents the number of particles outside closed shells measured in units of the total degeneracy 2Ω of a major oscillator shell (see Eq. (2-151)),

$$\begin{aligned} f &= \frac{n}{2\Omega} \\ \Omega &= \tfrac{1}{2}(N+1)(N+2) \end{aligned} \tag{6-394}$$

The factors in Eq. (6-393) involving the filling parameter f are seen to represent the product of the probability $v^2=f$ that a given particle orbit is occupied in the initial state times the probability $u^2=(1-f)$ that the final orbit is empty. The subshell structure is expected to somewhat reduce the value of $(\alpha_0^{(0)})^2$, as compared with estimate (6-393). Thus, in the example shown in Fig. 6-17a, the filled $1g_{9/2}$ orbits contribute very little to the $\Delta N=0$ strength, and the neutron contribution to $(\alpha_0^{(0)})^2$ is about a factor of 2 smaller than the value (6-393).

While the $\Delta N=0$ excitations are the basic degrees of freedom for generating the low-frequency quadrupole mode, the properties of this mode are modified to a significant extent by the coupling to the $\Delta N=2$ excitations that are responsible for the high-frequency quadrupole mode. The coupling to the $\Delta N=2$ excitations can be taken into account by a renormalization of the $\Delta N=0$ degrees of freedom. Thus, the quadrupole moment of the low-frequency transitions is increased by the factor $(1+\chi(\Delta N=2))\approx 2$, where $\chi(\Delta N=2)$ is the contribution of the $\Delta N=2$ excitations to the isoscalar polarizability in the static limit (see Eq. (6-370) with $\Delta E\approx 0$). The same renormalization applies to the quadrupole field acting in the low-frequency mode, and the effective coupling constant that describes the interactions within the $\Delta N=0$ configurations is therefore $\kappa_{\text{eff}}=\kappa(1+\chi(\Delta N=2))$, where κ is the unrenormalized coupling constant, an estimate of which is given by Eq. (6-368).

The effect of the field coupling on the collective frequency depends on the ratio between κ_{eff} and the restoring force parameter $C^{(0)}$ (see Eq. (6-27)). From the relation (6-23) for $C^{(0)}$, together with Eqs. (6-392) and (6-393), we therefore obtain, assuming comparable contribution to $\alpha_0^{(0)}$ from neutrons and protons and employing the value (2-157) for $\langle r^2\rangle$ (with $N_{\max}\approx N-1/2$),

$$\begin{aligned} \frac{\kappa_{\text{eff}}}{C^{(0)}} &\approx -\frac{\hbar\omega_0}{\Delta}(1+\chi(\Delta N=2))f(1-f) \\ &= -\frac{f(1-f)}{f_{\text{crit}}(1-f_{\text{crit}})} \end{aligned} \tag{6-395}$$

with

$$f_{\text{crit}}(1-f_{\text{crit}}) = \frac{\Delta}{\hbar\omega_0}(1+\chi(\Delta N=2))^{-1} \approx 0.15A^{-1/6} \tag{6-396}$$ ▲

▼ The expression (6-395), together with the relation (6-27) for the collective frequency, yields the expression

$$\hbar\omega \approx 2\Delta\left(1-\frac{f(1-f)}{f_{\text{crit}}(1-f_{\text{crit}})}\right)^{1/2} \tag{6-397}$$

which describes the decrease of ω as particles are added in the unfilled shell and implies instability of the spherical shape when f reaches the value f_{crit}. The estimate (6-396), which assumes complete degeneracy in the major shell, gives a value of about 0.05–0.1 for f_{crit}, but the subshell effects somewhat reduce the value of $\alpha_0^{(0)}$ and hence increase the value of f_{crit}. A decrease of $(\alpha_0^{(0)})^2$ by a factor of 2 (see p. 521) leads to a value of $f_{\text{crit}} \approx 0.2$, which approximately corresponds to the observed number of particles needed to produce a stable deformation in the nuclear shape (see Fig. 4-3, p. 27). In the above estimate, we have assumed f to have the same value for neutrons and protons; more generally, in an (N,Z) diagram such as Fig. 4-3, the loci of instability become ellipses centered on the points representing half-filled shells of neutrons as well as protons. For $f_{\text{crit}}=0.2$, the ellipses do not intersect the lines representing closed shells of neutrons and protons, and instability does not occur for nuclei that have closed shells of neutrons or protons.

One should emphasize the qualitative nature of the above estimates. Apart from the subshell effects already referred to, it would appear more realistic, for the low-frequency shape oscillations, to consider a coupling more strongly concentrated in the surface region than the simple multipole field (6-365). (Such a field of the type (6-69) will be employed in the more quantitative analysis in the following examples.) Moreover, if the numbers of neutrons and protons outside closed shells are very different, the low-frequency mode may involve a significant coupling to the $\tau=1$ quadrupole field; thus, in the extreme case of configurations with only one type of particles outside closed shells, the effective coupling constant for the low-frequency mode is $\kappa_{\tau=0}(1+\chi(\Delta N=2,\ \tau=0))+\kappa_{\tau=1}(1+\chi(\Delta N=2,\ \tau=1))$, for which the above estimates yield a value of about a third of the $\tau=0$ coupling (see Eqs. (6-368), (6-370), (6-379), and (6-382)).

It can be seen from Eq. (6-396) that the instability phenomenon is a direct consequence of the fact that the binding energy of the pairs is small compared with the energy separation between the shells. Thus, the instability of the spherical shape appears as a general phenomenon for systems in which the residual interactions are not sufficiently strong to destroy the shell structure.

Oscillator strength

The oscillator strength in the low-frequency mode is not affected by the ▲ field coupling acting within the degenerate $\Delta N=0$ excitations and is therefore given by

$$\hbar\omega B(\tau=0,\lambda=2)\approx 2\Delta(2\lambda+1)\left(\alpha_0^{(0)}\right)^2\left(1+\chi(\Delta N=2,\tau=0;\Delta E\approx 0)\right)^2 \quad (6\text{-}398)$$

where we have included the renormalization of the quadrupole moment due to the coupling to the $\Delta N=2$ transitions. Expressing $(\alpha_0^{(0)})^2$ in terms of the quantity $C^{(0)}$ and using Eqs. (6-368) and (6-395) with $\kappa_{\text{eff}}=\kappa(1+\chi(\Delta N=2))$, the oscillator strength in units of the classical oscillator sum (6-179) can be written in the form

$$\frac{\hbar\omega B(\tau=0,\lambda=2)}{S(\tau=0,\lambda=2)_{\text{class}}}=2\left(\frac{\Delta}{\hbar\omega_0}\right)^2(1+\chi(\Delta N=2))\frac{f(1-f)}{f_{\text{crit}}(1-f_{\text{crit}})}$$

$$\approx 0.4A^{-1/3}\frac{f(1-f)}{f_{\text{crit}}(1-f_{\text{crit}})} \quad (6\text{-}399)$$

This estimate implies that, for $A\approx 100$, the oscillator strength of the low-frequency quadrupole mode becomes of the order of 10% of the sum rule unit for configurations approaching instability, $f\approx f_{\text{crit}}$.

Static quadrupole moment

With the approach to instability in the spherical shape, one expects large anharmonicities in the vibrational motion. A measure of the anharmonicity is provided by the static quadrupole moment of the vibrational excitation relative to the transition moment. In the estimate of the static moment, we first consider the contribution Q_1 from the two-quasiparticle states i, out of which the phonon state is built

$$Q_1=-\sum_{ij}X(i)X(j)\langle j|Q|i\rangle$$
$$\langle j|Q|i\rangle\equiv\langle \mathrm{v}=2,j|3z^2-r^2|\mathrm{v}=2,i\rangle \quad (6\text{-}400)$$

where the amplitude $X(i)$ is given by (see Eq. (6-36))

$$X(i)=-\frac{\kappa_{\text{eff}}\langle n=1|\alpha|n=0\rangle\langle \mathrm{v}=2,i|F|\mathrm{v}=0\rangle}{2\Delta-\hbar\omega} \quad (6\text{-}401)$$

In Eq. (6-400), the minus sign results from the fact that Q_1 corresponds to the standard definition of a quadrupole moment representing the matrix element in the state with $I=M$, while the states i and j have been generated with the $M=0$ field; for $I=2$, the expectation values of the quadrupole operator in the $M=0$ and $M=2$ states have the ratio -1.

The transition matrix element of α occurring in Eq. (6-401) can be related to the zero-point amplitude $\alpha_0^{(0)}$ of the quasiparticle excitations (see

▼ Eq. (6-28))

$$\langle n=1|\alpha|n=0\rangle^2=\left(\alpha_0^{(0)}\right)^2\frac{2\Delta}{\hbar\omega} \tag{6-402}$$

and for κ_{eff} we may employ the relation (see Eqs. (6-23), (6-26), and (6-27))

$$\begin{aligned}\kappa_{\text{eff}}\left(\alpha_0^{(0)}\right)^2&=\left(C-C^{(0)}\right)\left(\alpha_0^{(0)}\right)^2\\&=-\Delta\left(1-\left(\frac{\hbar\omega}{2\Delta}\right)^2\right)\end{aligned} \tag{6-403}$$

Using the relations (6-393) and (6-394), and the sum over the quadrupole moments in the two-quasiparticle states evaluated below (see Eq. (6-413)), we obtain from Eqs. (6-400) to (6-403)

$$Q_1=-\frac{1}{10}\Omega^{-1/2}\left(1+\frac{\hbar\omega}{2\Delta}\right)^2\left(\frac{2\Delta}{\hbar\omega}\right)^{1/2}f^{-1/2}(1-f)^{-1/2}(1-2f)|\langle n=1|Q|n=0\rangle| \tag{6-404}$$

where the transition quadrupole moment is given by (see Eq. (3-30))

$$\langle n=1|Q|n=0\rangle=\left(\frac{16\pi}{5}\right)^{1/2}\langle n=1|\alpha|n=0\rangle \tag{6-405}$$

We have assumed equal values of f for neutrons and protons.

The contribution Q_1 to the static moment is illustrated by diagram (i) in Fig. 6-27a. To the same order in the particle-vibration coupling, one must include five other terms, which are illustrated by diagrams (ii)–(vi) in Fig. 6-27a. All six contributions involve the same vertices and differ only in the energy denominators. We therefore obtain

$$\begin{aligned}\sum_{k=1}^{6}Q_k&=Q_1(2\Delta-\hbar\omega)^2\\&\quad\times\left(\frac{1}{(2\Delta-\hbar\omega)^2}+\frac{1}{(2\Delta+\hbar\omega)^2}+\frac{2}{2\Delta(2\Delta-\hbar\omega)}+\frac{2}{2\Delta(2\Delta+\hbar\omega)}\right)\\&=6Q_1\frac{1-\frac{1}{3}(\hbar\omega/2\Delta)^2}{(1+\hbar\omega/2\Delta)^2}\end{aligned} \tag{6-406}$$

The evaluation of the static moment is similar to the analysis of the coupling between states of one and two phonons considered in Sec. 6-5g; see Fig. 6-12, p. 434.
▲

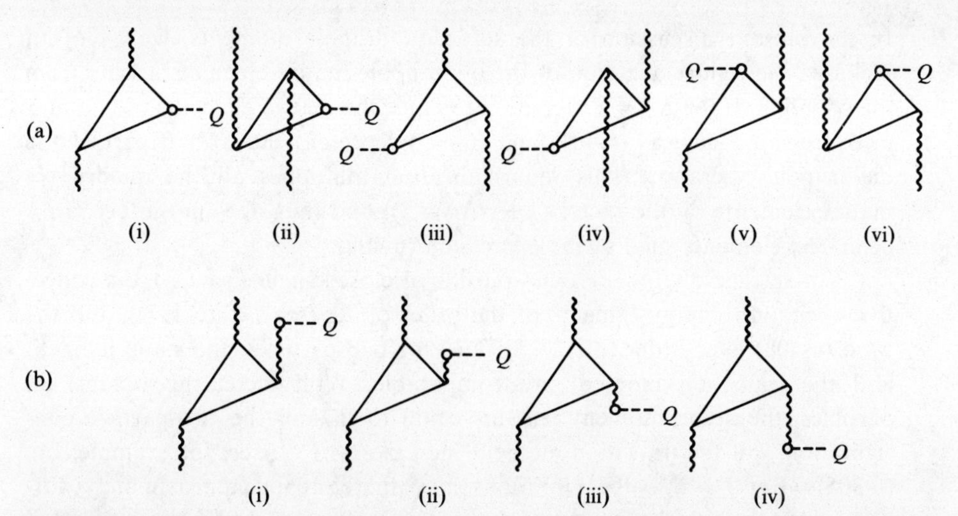

Figure 6-27 Diagrams contributing to the static quadrupole moment of a phonon. The lines represent quasiparticles. The six diagrams in (a) give the leading-order contributions produced by the bare moments of the quasiparticles, while the four diagrams in (b) represent the renormalization of diagram (i) in (a) that results from the coupling to the quadrupole mode itself.

▼ In the above estimate, we have included only the quadrupole moment of the bare particles, but this moment is amplified by the coupling to the quadrupole mode itself, which leads to an increase of static moments by the factor (see Eqs. (6-216) and (6-26))

$$(1+\chi(\Delta E=0))=1-\frac{\kappa_{\text{eff}}}{C}=\frac{C^{(0)}}{C}=\left(\frac{2\Delta}{\hbar\omega}\right)^2 \tag{6-407}$$

The polarizability χ appearing in Eq. (6-407) represents the effect of the coupling to the low-frequency mode ($\Delta N=0$). The polarization effect is illustrated by diagrams (i)–(iv) in Fig. 6-27b, which are all associated with diagram (i) in Fig. 6-27a. The diagrams in Fig. 6-27b can be seen to yield a contribution equal to $\chi(\Delta E=0)$ multiplied by the contribution of diagram (i) in Fig. 6-27a.

The final result for the static moment, obtained from Eqs. (6-404), (6-406), and (6-407), can be expressed in the form

$$\frac{\langle n=1|Q|n=1\rangle}{|\langle n=1|Q|n=0\rangle|}=(1+\chi(\Delta E=0))\sum_{k=1}^{6}Q_k|\langle n=1|Q|n=0\rangle|^{-1}$$

$$=-\frac{6}{10}\Omega^{-1/2}f^{-1/2}(1-f)^{-1/2}(1-2f)\left(1-\frac{1}{3}\left(\frac{\hbar\omega}{2\Delta}\right)^2\right)\left(\frac{2\Delta}{\hbar\omega}\right)^{5/2} \tag{6-408}$$

▲

▼ In the present discussion of the static quadrupole moments, we have not included the renormalization of the quadrupole matrix elements arising from the coupling to the $\Delta N=2$ excitations. (Thus, the matrix elements of Q and α in expressions such as (6-401) and (6-405) represent the $\Delta N=0$ part of the quadrupole operator.) This renormalization multiplies all the quadrupole matrix elements by the factor $(1+\chi(\Delta N=2))$ and thus does not affect ratios of matrix elements such as the expression (6-408).

For sufficiently few particles outside of closed shells ($f \ll f_{\text{crit}}$), the collective excitation energy remains of the order of 2Δ (see Eq. (6-397)), and the ratio (6-408) is of order $(f\Omega)^{-1/2}$. Thus, for two particles, the static moment and the transition moment are comparable. With increasing number of particles, the static moment remains equal to that of the two-particle configuration, but the transition moment increases as $f^{1/2}$ (see, for example, Eq. (6-305)). For $f \gg \Omega^{-1}$, the vibrations may therefore be approximately harmonic, but as f approaches f_{crit}, the collective frequency decreases (see Eq. (6-397)), and the last factor in Eq. (6-408) increases strongly, mainly as a result of the polarizability associated with the self-coupling (see Eq. (6-407)). The condition that the static moment remains small compared with the transition moments restricts the harmonic description to a region in which $\hbar\omega/2\Delta \gg \Omega^{-1/5}$; thus, for shells of the dimensions encountered in nuclei, the static moment is never appreciably smaller than the transition moment. This conclusion ignores the factor $(1-2f)$ in Eq. (6-408), which expresses the antisymmetry of the quadrupole moment with respect to the interchange of particles and holes, and which becomes small in the middle of the shell. However, in situations where the static moment would have been large except for this symmetry, one must expect higher-order anharmonic effects to be important.

Sums over two-quasiparticle excitations within a harmonic-oscillator shell, such as are employed in the above discussion, can be conveniently evaluated by expressing the particle quadrupole moment in terms of the operators $c_m^\dagger$ and c_m that create and annihilate a quantum of oscillation with angular momentum component m. Thus, the $\mu=2$ component of the quadrupole moment takes the form

$$
\begin{aligned}
r^2 Y_{22} &= \frac{1}{4}\left(\frac{15}{2\pi}\right)^{1/2}(x+iy)^2 \\
&= -\left(\frac{15}{8\pi}\right)^{1/2}\frac{\hbar}{2M\omega_0}\left(c_{+1}^\dagger + c_{-1}\right)^2
\end{aligned}
\tag{6-409}
$$

(We have here used the same phase relations as for the cylindrical representation (5-24) employed in the analysis of intrinsic motion in deformed nuclei; in the limit of spherical symmetry, the operators $c_+^\dagger$, $c_3^\dagger$, and $c_-^\dagger$ are equal to the operators $c_{+1}^\dagger$, $c_0^\dagger$, $c_{-1}^\dagger$ employed in the present discussion.) The moment (6-409) induces $\Delta N=0$ transitions of the type $(n_{+1}, n_0, n_{-1}) \rightarrow (n_{+1}+1, n_0, n_{-1}-1)$, where n_m is the number of quanta of type m, and the transition probability has the value

$$
|\langle n_{+1}+1, n_0, n_{-1}-1 | r^2 Y_{22} | n_{+1}, n_0, n_{-1}\rangle|^2 = \frac{15}{8\pi}\left(\frac{\hbar}{M\omega_0}\right)^2 (n_{+1}+1)n_{-1} \tag{6-410}
$$

▲

These individual transitions only contribute if the initial orbit is filled and the final orbit empty. In the paired coupling scheme, each orbit has the same probability of occupancy, $v^2=f$, and for the total transition strength we therefore obtain, considering only one type of particles (neutrons or protons),

$$
\begin{aligned}
(\alpha_0^{(0)})^2 &= 4\frac{15}{8\pi}\left(\frac{\hbar}{M\omega_0}\right)^2 f(1-f)\sum_{n_0=0}^{N}\sum_{n_{+1}=0}^{N-n_0}(n_{+1}+1)n_{-1} \\
&= \frac{5}{16\pi}\left(\frac{\hbar}{M\omega_0}\right)^2 N(N+1)(N+2)(N+3)f(1-f)
\end{aligned}
\tag{6-411}
$$

The factor 4 in the first line of Eq. (6-411) results partly from the spin degeneracy, which gives a factor 2; the additional factor 2 is associated with the fact that each excited state i, specified by two quasiparticles in the orbits 1 and 2, can be produced by two different particle transitions $\bar{1}\rightarrow 2$ and $\bar{2}\rightarrow 1$, where the bar denotes time reversal. (In terms of the pairing coefficients u, v, the matrix element $\langle \text{v}=2,i|F|\text{v}=0\rangle$ involves the factor $2uv$ (see Eq. (6-610b)), while the sum over i is restricted to sets of quasiparticles with $1<2$.)

In the (n_{+1},n_0,n_{-1}) representation, the $\Delta N=0$ matrix elements of the quadrupole operator are diagonal, and the static moments of the two-quasiparticle states are given by

$$
\begin{aligned}
\langle j|Q|i\rangle &= \langle \text{v}=2,j|3z^2-r^2|\text{v}=2,i\rangle \\
&= \frac{2\hbar}{M\omega_0}(3n_0-N)(1-2f)\,\delta(i,j)
\end{aligned}
\tag{6-412}
$$

where the factor $(1-2f)$ results from the fact that the quasiparticle is particle-like with probability $1-f$, and hole-like, with opposite Q, with probability f (see the factor (u^2-v^2) in Eq. (6-610a)). The sum involved in the evaluation of the static moment of the quadrupole excitation is therefore

$$
\begin{aligned}
&\sum_i \langle \text{v}=2,i|r^2Y_{22}|\text{v}=0\rangle^2\langle i|Q|i\rangle\left(=-\sum_i\langle \text{v}=2,i|r^2Y_{20}|\text{v}=0\rangle^2\langle i|Q|i\rangle\right) \\
&= 4\frac{15}{8\pi}\left(\frac{\hbar}{M\omega_0}\right)^2 f(1-f)\frac{2\hbar}{M\omega_0}(1-2f)\sum_{n_0=0}^{N}\sum_{n_{+1}=0}^{N-n_0}(n_{+1}+1)n_{-1}(3n_0-N) \\
&= -\frac{1}{4\pi}\left(\frac{\hbar}{M\omega_0}\right)^3 f(1-f)(1-2f)N(N+1)(N+\tfrac{3}{2})(N+2)(N+3)
\end{aligned}
\tag{6-413}
$$

Systematics of restoring force and mass parameter for low-frequency quadrupole mode (Figs. 6-28 and 6-29)

The low-frequency 2+ mode is strongly excited in electromagnetic processes as well as in the inelastic scattering of nuclear projectiles (see, for example, Fig. 6-2, p. 353). The comparison between the cross sections for these different processes confirms that the mode is predominantly isoscalar ($\tau\approx 0$), as expected for a shape oscillation (see Table 6-2, p. 354). The assumption that the local neutron-proton ratio remains constant leads to an

▼ isovector moment proportional to $(N-Z)/A$, but the available data are not sufficiently accurate to test this rather small effect. (Somewhat larger isovector moments may be expected in the low-frequency quadrupole modes, when the numbers of neutrons and protons outside closed shells are very different.) An especially direct measurement of the $\tau=1$ moment of a vibrational mode could be obtained from a charge exchange excitation of the isobaric analog $I\pi=2+$ state with $M_T=T_0-1$; only the isovector field contributes to this excitation (since $\mu_\tau=-1$), and the ratio of the isovector multipole matrix elements leading to the two different components of the isobaric multiplet is equal to the ratio of the vector addition coefficients (see Eq. (6-132)). Experiments of this type have been reported (see, for example, Rudolph and McGrath, 1973), but the interpretation in terms of the isovector moments of the vibrational excitations remains uncertain due to the difficulty in a quantitative estimate of the contributions from two-step processes.

The most precise information on the transition amplitude is provided by the measured $B(E2)$ values (see Fig. 4-5, p. 48), and we therefore employ the normalization (6-63) for the vibrational amplitude in terms of the $E2$ moment. From the frequencies and $B(E2)$ values, one can determine the restoring force parameters C_2 and the mass parameters D_2 (see Eq. (6-65))

$$\begin{aligned} C_2 &= \frac{5}{2}\hbar\omega_2\left(\frac{3}{4\pi}ZeR^2\right)^2 B(E2;0\rightarrow 2)^{-1} \\ D_2 &= \omega_2^{-2}C_2 \end{aligned} \tag{6-414}$$

The restoring force parameters C_2 obtained from Eq. (6-414) are shown in Fig. 6-28. The figure also gives the values of C_2 derived from the liquid-drop model (see Eq. (6A-24)); the observed values exhibit variations by more than a factor of 10 both above and below the liquid-drop estimate and thus reveal the profound importance of the nuclear shell structure for the low-frequency quadrupole mode. The main trends in C_2 can be understood in terms of the approach to instability as particles are added to the closed shells (see the discussion in the previous example).

The interpretation of the quantity C_2 in Fig. 6-28 as the restoring force assumes harmonic vibrations. More generally, the quantity plotted can be related to the contribution of the excitation mode to the static polarizability for a weak external field acting on the quadrupole moment. (This interpretation also applies to the rotational excitations, which have therefore been included in Fig. 6-28.) The polarizability coefficient χ is given by the ratio of κ and C_2 (see Eq. (6-216), where the amplitude of the vibration is normalized to the quadrupole field ($\alpha=\langle r^2Y_{20}\rangle$)). From the estimate (6-368) for κ, we

▲ obtain, taking into account the changed normalization of the amplitude

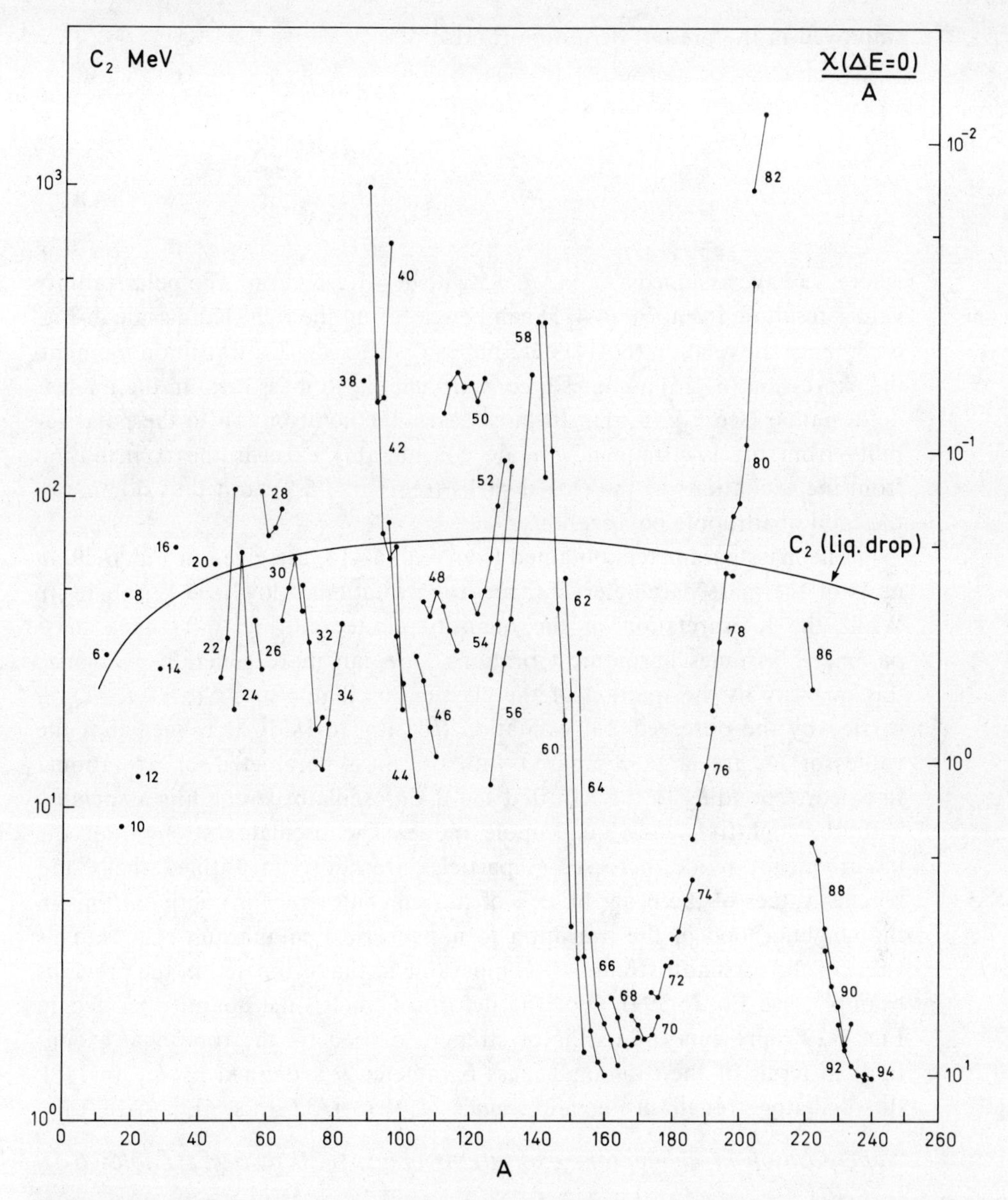

Figure 6-28 Systematics of the restoring force parameter C_2 for the low-frequency quadrupole mode. The quantity C_2 is calculated by means of Eq. (6-414) with $R = 1.2A^{1/3}$ fm. The frequencies $\hbar\omega_2$ and the $B(E2)$ values are taken from the compilation by Stelson and Grodzins (1965); see also Fig. 2-17, Vol. I, pp. 196–197, and Fig. 4-5, p. 48. Additional data on lighter nuclei are taken from Table 4-15, p. 134, and from the compilation by Skorka *et al.*, 1967. The interpretation of the quantity plotted in terms of a restoring force is only appropriate for harmonic vibrations about a spherical equilibrium. More generally, the quantity represents the quadrupole polarizability contributed by the lowest 2+ excitation mode, as given by the right-hand scale.

▼ employed in the present definition (6-414) of C_2,

$$\chi(\tau=0,\lambda=2;\Delta E=0)=\frac{3}{4\pi}\frac{AM\omega_0^2R^2}{C_2}$$

$$\approx\frac{14A}{C_2(\mathrm{MeV})} \tag{6-415}$$

where we have assumed $\langle r^2\rangle=3R^2/5$ with $R=1.2A^{1/3}$ fm. The polarizability values resulting from Eq. (6-415) can be read from the right-hand scale in Fig. 6-28. Since the relation (6-414) determines C_2 from the $E2$-transition moment, the expression (6-415) assumes a constant charge to mass ratio in the nuclear deformation (see Eq. (6-61)). In most cases, the contribution to the polarizability from the low-frequency mode considerably exceeds the contribution from the excitations of the closed shells (see Table 6-8), and thus dominates the total quadrupole polarizability.

The mass parameters obtained from Eq. (6-414) are given in Fig. 6-29, in units of the mass parameter D_2(irrot) for irrotational flow (see Eq. (6-181)). While the interpretation of the quantity plotted in Fig. 6-29 as a mass parameter assumes harmonic vibrations, one can more generally recognize this quantity as the fraction of the classical oscillator sum $S(\tau=0,\lambda=2)_{\mathrm{class}}$ carried by the observed 2+ excitation (see Eq. (6-183)). It is seen that the values of D_2 are large compared with the mass parameter for irrotational flow, corresponding to the fact that the main oscillator strength is associated with the high-frequency quadrupole mode. The oscillator strength in the low-frequency mode increases as particles are added in unfilled shells and reaches values of the order of 10% of the sum rule unit for configurations in the neighborhood of the transition to nonspherical equilibrium shapes. This value of the oscillator strength is comparable to that estimated in the previous example; see Eq. (6-399). (For the deformed nuclei, the quantity plotted in Fig. 6-29 represents the oscillator strength carried by the rotational excitation; in terms of the rotational mass parameter D_{rot} defined by Eq. (6-188), the oscillator strength in Fig. 6-29 equals $2D_2(\mathrm{irrot})/5D_{\mathrm{rot}}$; see Eq. (6-187).)

Superposition of elementary excitations (Figs. 6-30 to 6-37; Tables 6-10 to 6-13)

A rather extensive body of data is available concerning the superposition of quadrupole quanta giving rise to higher excitations in the even-even nuclei, and regarding the superposition of quadrupole quanta and quasiparticles in the spectra of odd-A nuclei. For even-even nuclei, it is found that the $n=1$, $I\pi=2+$ state is connected with large $E2$-matrix elements to higher states that can approximately be interpreted as $n=2$ excitations. In many cases, all the members of the expected triplet ($I\pi=0+,2+,4+$) have been identified; in

▲ other cases, only some of the members of the triplet are known, but the

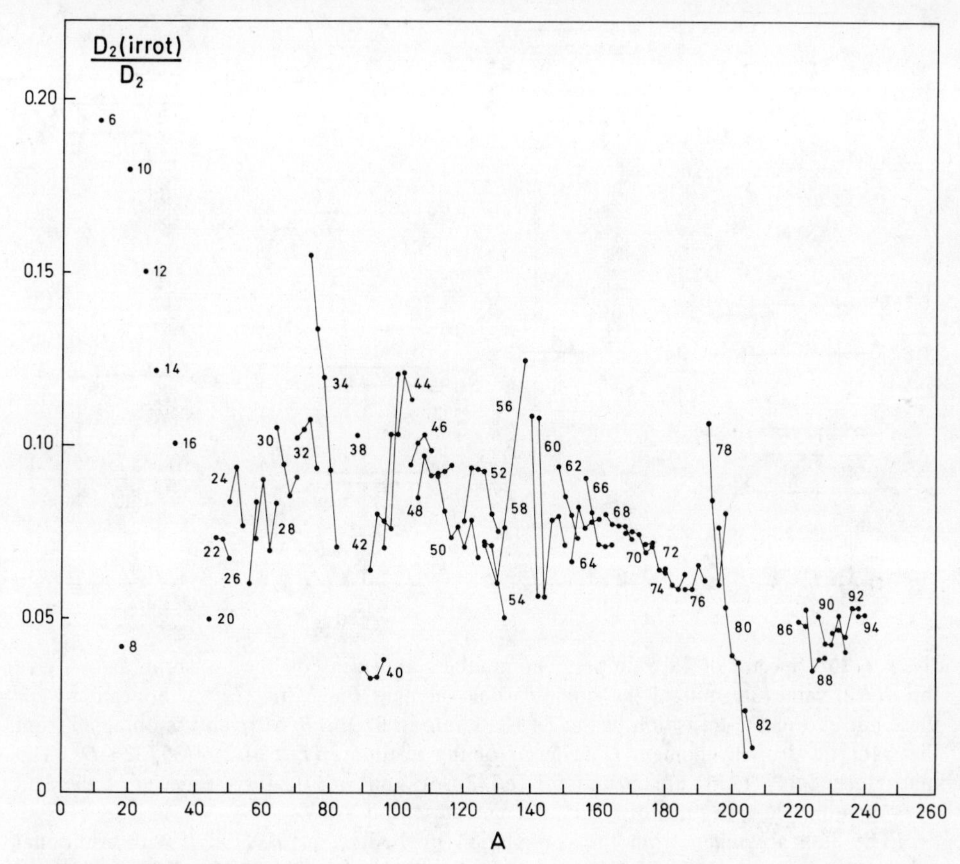

Figure 6-29 Systematics of the mass parameter D_2 for the low-frequency quadrupole mode. The quantity D_2 is calculated from Eq. (6-414) on the basis of the data referred to in the caption to Fig. 6-28.

▼ remaining states may have been missed in the experimental studies so far reported. Examples of $n=2$ states are shown in Figs. 6-30, 6-31, and 6-32. As regards states with $n>2$, the information is very incomplete, but the yrast states of the vibrational band ($I=2n$), with rather large values of n, have been found in many nuclei by means of (heavy iron, $xn\gamma$) reactions; see Fig. 6-33.

In odd-A nuclei, one quite generally observes $E2$ transitions with strengths and frequencies comparable to those in neighboring even-even nuclei (see, for example, Fig. 6-30). In many cases, the pattern of these excitations can be qualitatively interpreted in terms of the superposition of a quasiparticle with spin $I_0=j$ and a quadrupole quantum giving a multiplet of states with $I=|I_0-2|,\ldots,I_0+2$.

In both the odd-A and even-even spectra, rather large deviations from ▲ the approximation of independent elementary excitations are revealed by the

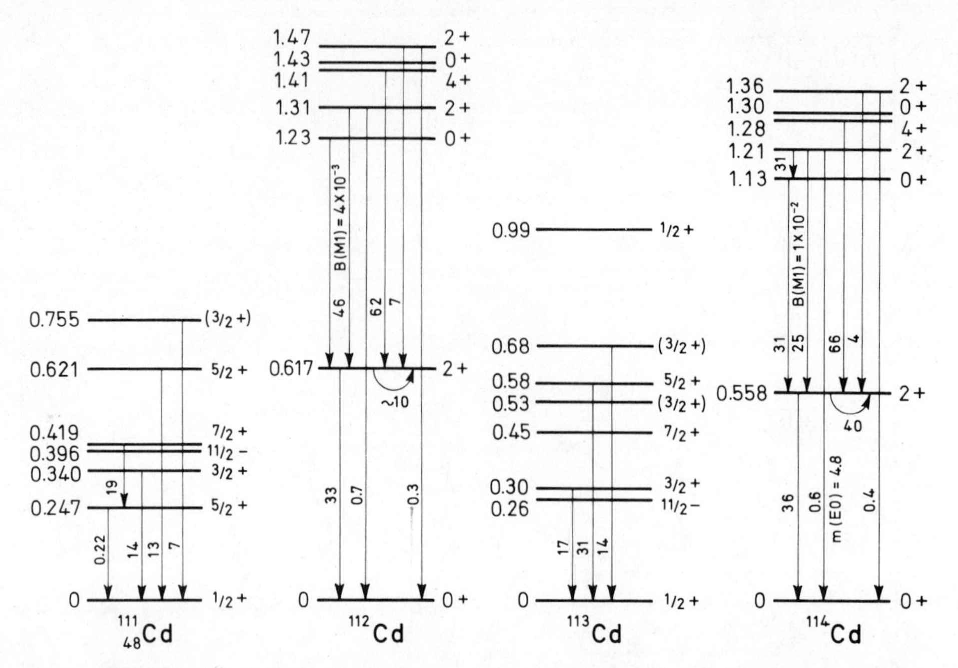

Figure 6-30 Spectra of Cd isotopes. The numbers appearing on the transition arrows give the $B(E2)$ values in units of the single-particle moment $B_W(E2) = 32e^2\text{fm}^4$ appropriate for these nuclei. For the transition of the 2+ state into itself, the $B(E2)$ value is obtained from the static quadrupole moment Q in terms of the relation $B(E2;2\rightarrow 2) = (35/32\pi)Q^2$. The matrix elements $B(M1)$ are in units of $(e\hbar/2Mc)^2$ and $m(E0)$, for the decay of the 1.13 MeV state in ^{114}Cd, in units of e fm^2. The energies are in MeV.

The data are taken from the compilation by Lederer *et al.* (1967), with additional information from: ^{111}Cd (J. McDonald and D. Porter, *Nuclear Phys.* **A109**, 529, 1968); ^{112}Cd and ^{114}Cd (F. K. McGowan, R. L. Robinson, P. H. Stelson, and J. L. C. Ford Jr., *Nuclear Phys.* **66**, 97, 1965; W. T. Milner, F. K. McGowan, P. H. Stelson, R. L. Robinson, and R. O. Sayer, *Nuclear Phys.* **A129**, 687, 1969; see also the compilation by Groshev *et al.*, 1968). The quadrupole moment of the 2+ state in ^{114}Cd ($Q = -60$ fm^2) is taken from the review by de Boer and Eichler (1968), but there remain unresolved discrepancies between different experimental determinations of this quantity (see, for example, Berant *et al.*, 1972). The value of $B(E2;\ 2\rightarrow 2)$ for ^{112}Cd is obtained from the ratio of the ^{114}Cd to ^{112}Cd moments measured by S. G. Steadman, A. M. Kleinfeld, G. G. Seaman, J. de Boer, and D. Ward, *Nuclear Phys.* **A155**, 1 (1970).

▼ energies as well as by the $E2$-matrix elements (see the evidence in Figs. 6-30 to 6-33). The occurrence of such rather large anharmonicity effects is to be expected on the basis of the analysis of the field coupling acting among the particles in the unfilled shells, as discussed above (pp. 523 ff).

In the following sections of the present example, we shall explore the extent to which the observed anharmonicity phenomena can be described in ▲ terms of the leading-order effects of the particle-vibration coupling. In

▼ favorable cases, such an approach should provide a rather quantitative description, but in many cases, the anharmonicities encountered are so large that a quantitative analysis would require the inclusion of higher-order terms.

Effective charge

A rather direct manifestation of the particle-vibration coupling is the effective charge for quadrupole moments and $E2$ transitions involving one-particle configurations. Table 6-10 gives examples of such $E2$-matrix elements for configurations with a single particle or hole with respect to the closed-shell configuration $Z=50$. In these spectra, it appears that the ground state of In ($I\pi=9/2+$) and the low-lying $I\pi=5/2+$ and $7/2+$ states of Sb can be rather well described as single-hole ($g_{9/2}^{-1}$) and single-particle ($d_{5/2}$ and $g_{7/2}$) states, respectively. (See, for example, the evidence on the one-particle transfer processes on Sn targets, Barnes *et al.*, 1966.) The effective charges in Table 6-10 are obtained by dividing the observed moments in column two by the single-particle estimate (see Eqs. (3-27) and (3-32)); the radial matrix element $\langle j_2|r^2|j_1\rangle$ appearing in the single-particle moments is given in column three and has been obtained from wave functions in a Woods-Saxon potential with standard parameters (see Vol. I, pp. 239 and 326). The coupling matrix element $\langle j_2|k(r)|j_1\rangle$ in column four is obtained from the expression (6-213), with the same wave functions and potentials. The polarization charge resulting from the coupling to the low-frequency quadrupole vibrations has been estimated from Eq. (6-218), on the basis of the frequency ($\hbar\omega_2=1.17$ MeV) and $B(E2)$ value ($2.0\times10^3\ e^2\text{fm}^4$) observed for ^{120}Sn (Stelson *et al.*, 1970).
▲ The polarization charge resulting from the coupling to the high-frequency

Nucleus	Matrix element e fm^2	$\langle j_2\|r^2\|j_1\rangle$ fm^2	$\langle j_2\|k\|j_1\rangle$ MeV	e_{eff}/e Obs.	e_{eff}/e Calc.
^{115}In	$Q(g_{9/2}^{-1})=83$	26	68	4.4	5.4
^{121}Sb	$Q(d_{5/2})=-53$	24	58	3.9	5.0
^{123}Sb	$Q(g_{7/2})=-68$	23	58	4.4	5.2
	$\|\langle g_{7/2}\|\| \mathcal{M}(E2)\|\|d_{5/2}\rangle\|=19$	14	47	4.5	6.7

Table 6-10 Effective $E2$ charges in In and Sb. The experimental data in column two are taken from the compilation of nuclear moments by Fuller and Cohen (1969) and from the Coulomb excitation measurements in ^{123}Sb by P. D. Barnes, C. Ellegaard, B. Herskind, and M. C. Joshi, *Phys. Letters* **23B**, 266 (1966). The quadrupole moments of the Sb isotopes are those obtained from the measurements of atomic spectra; atomic beam measurements have given values about a factor of 2 smaller (Fuller and Cohen, *loc.cit.*). For the quadrupole moment of ^{115}In, see also Lee *et al.* (1969).

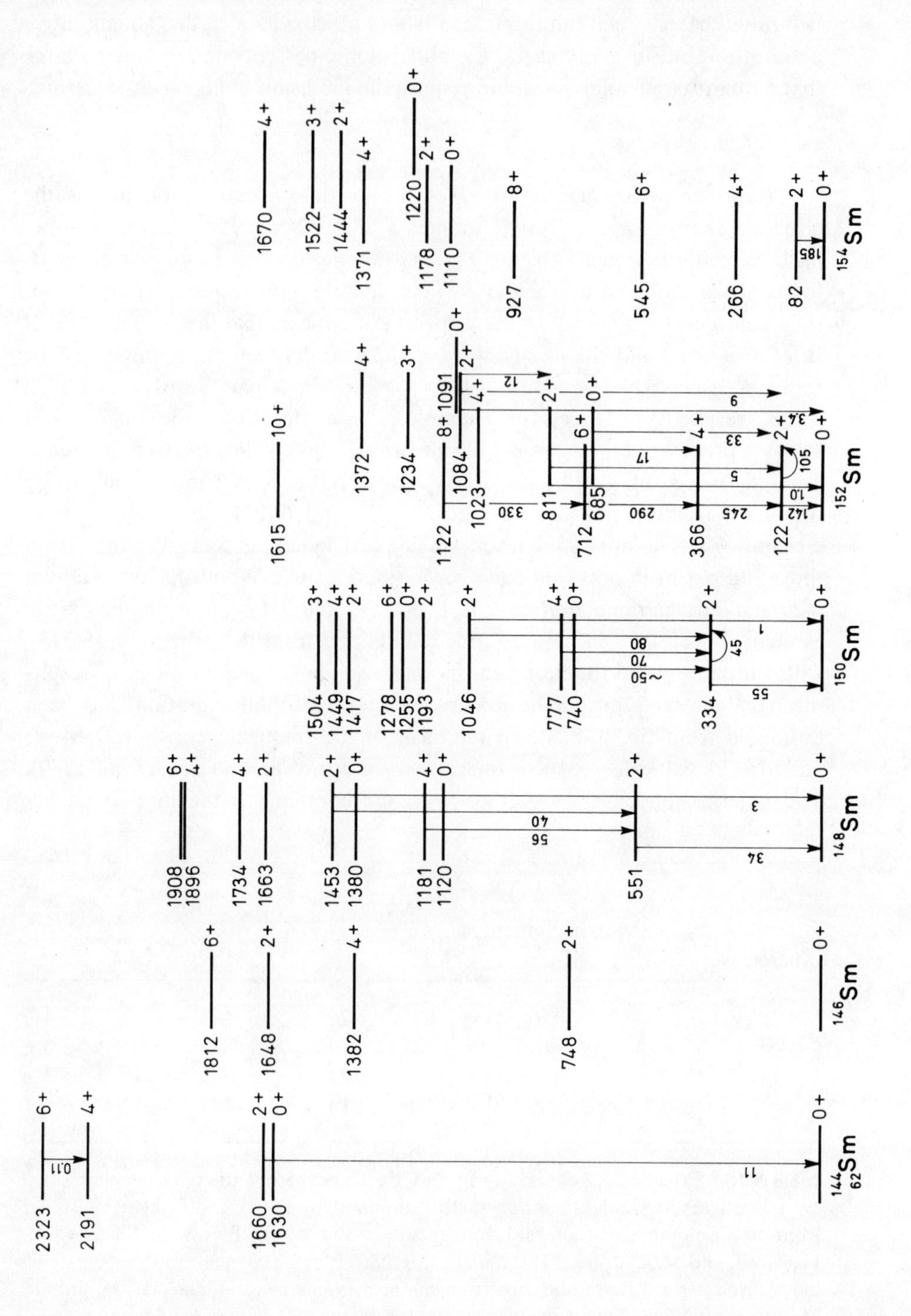

$^{144}_{62}$Sm
^{146}Sm
^{148}Sm
^{150}Sm
^{152}Sm
^{154}Sm

▼ modes is taken from the estimate (6-386). The total estimated effective charge in the last column includes the charge of the bare particle. The comparison between observed and calculated effective charges in Table 6-10 indicates that the assumed coupling is approximately of the right magnitude; however, there appears to be considerable uncertainty in the experimental data. It should also be noted that the coupling is rather strong; the dimensionless coupling constant given by Eq. (6-212) has the value $f_2 \approx 0.8$ for $\langle k \rangle \approx 60$ MeV. Thus, the first-order perturbation treatment may somewhat overestimate the induced quadrupole moment.

The examples considered in Table 6-10 are somewhat special in that the proton configuration involves a single particle in addition to closed shells. When the odd-particle configuration involves several particles outside closed shells, the single-particle degrees of freedom are expressed in terms of the quasiparticle states. The $E2$-matrix elements between states of a single quasiparticle may be strongly reduced as compared with the one-particle value, corresponding to the fact that quasiparticles are combinations of particles and holes, which have opposite sign for the $E2$ moment (see the factor $u_1u_2 - v_1v_2$ in Eq. (6-610a)). In such a situation, the estimate of the $E2$ moment for low-lying states may depend sensitively on the details of the one-particle spectrum. ▲

Figure 6-31 Spectra of even-parity states in even-even Sm isotopes (for odd-parity levels, see Fig. 6-44, p. 577). The numbers appearing on the transition arrows give the $B(E2)$ values in units of the single-particle moment $B_W(E2) = 48e^2\text{fm}^4$ appropriate for these nuclei. The values of $B(E2; 2 \rightarrow 2)$ are obtained from the static quadrupole moments (see the caption to Fig. 6-30); the sign of the moment is found to be negative in ^{152}Sm as well as in ^{150}Sm. All energies are in keV.

The spectra of ^{152}Sm and ^{154}Sm can be described rather well in terms of rotational band sequences corresponding to an axially symmetric deformation, and the levels associated with different bands are displaced horizontally with respect to each other.

The experimental data are taken from the compilation by Sakai (1970) and from: ^{144}Sm (J. Kownacki, H. Ryde, V. O. Sergejev, and Z. Sujkowski, *Nuclear Phys.* **A196**, 498, 1972; ^{148}Sm (B. Harmatz and T. H. Handley, *Nuclear Phys.* **A121**, 481, 1968; D. J. Buss and R. K. Smithers, *Phys. Rev.*, **C2**, 1513, 1970); (α2n) reactions leading to ^{152}Sm and ^{154}Sm (H. Morinaga, *Nuclear Phys.* **75**, 385, 1966; O. Lönsjö and G. B. Hagemann, *Nuclear Phys.* **88**, 624, 1966); (tp) reaction on ^{150}Sm (S. Hinds, J. H. Bjerregaard, O. Hansen, and O. Nathan, *Phys. Letters* **14**, 48, 1965); (tp) reaction on ^{152}Sm (J. H. Bjerregaard, O. Hansen, O. Nathan, and S. Hinds, *Nuclear Phys.* **86**, 145, 1966); (pt) studies (Y. Ishizaki, Y. Yoshida, Y. Saji, T. Ishimatsu, K. Yagi, M. Matoba, C. Y. Huang, and Y. Nakajima, *Contributions to International Conference on Nuclear Structure* (Tokyo), p. 133, 1967); Coulomb excitation studies (J. J. Simpson, D. Eccleshall, M. J. L. Yates, and N. J. Freeman, *Nuclear Phys.* **A94**, 177, 1967; I. A. Fraser, J. S. Greenberg, S. H. Sie, R. G. Stokstad, and D. A. Bromley, *Contributions to International Conference on Properties of Nuclear States* (Montreal), pp. 13 and 14, 1969; G. B. Hagemann, B. Herskind, M. C. Olesen, and B. Elbek *ibid.* p. 29); lifetime measurements in ^{152}Sm (R. M. Diamond, F. S. Stephens, R. Nordhagen, and K. Nakai *ibid.*, p. 7); (nγ) studies of ^{150}Sm (review by Groshev *et al.*, 1968).

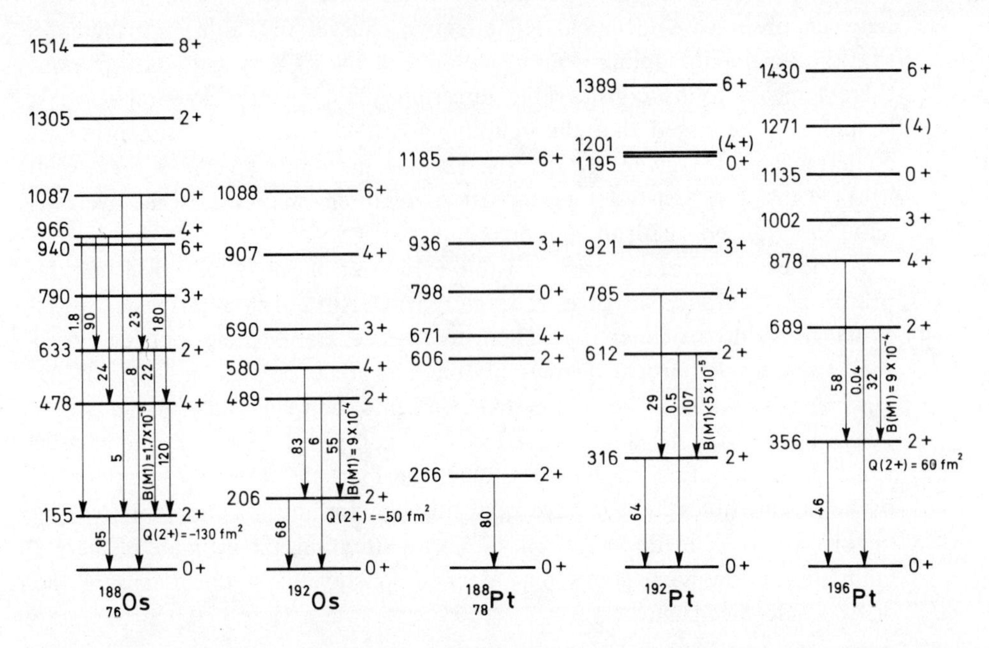

Figure 6-32 Quadrupole vibrational spectra of even-even nuclei in the region $A \approx 190$. The numbers appearing on the transition arrows give the $B(E2)$ values in units of the single-particle value $B_W(E2) = 65e^2\text{fm}^4$ appropriate to these nuclei. The $B(M1)$ values are given in units of $(e\hbar/2Mc)^2$. The data are taken from the compilation by Lederer *et al.* (1967) and from: spectrum of ^{188}Pt (B. R. Erdal, M. Finger, R. Foucher, J. P. Husson, J. Jastrzebski, A. Johnson, N. Perrin, R. Henck, R. Regal, P. Siffert, G. Astner, A. Kjelberg, P. Patzelt, A. Höglund, and S. Malmskog, *International Conference on Properties of Nuclei Far from the Region of Beta-Stability,* p. 1031, CERN 70-30, Geneva 1970); quadrupole moment in ^{196}Pt (J. E. Glenn, R. J. Pryor, and J. X. Saladin, *Phys. Rev.* **188**, 1905, 1969); Coulomb excitation of Pt isotopes (W. T. Milner, F. K. McGowan, R. L. Robinson, P. H. Stelson, and R. O. Sayer, *Nuclear Phys.* **A177**, 1, 1971); Coulomb excitation of Os isotopes (R. F. Casten, J. S. Greenberg, S. H. Sie, G. A. Burginyon, and D. A. Bromley, *Phys. Rev.* **187**, 1532, 1969); quadrupole moments of Os isotopes (S. A. Lane and J. X. Saladin, *Phys. Rev.* **C6**, 613, 1972).

▼ *One-particle transfer leading to vibrational excitations*

Another test of the particle-vibration coupling is provided by the one-particle transfer processes with excitation of a vibrational quantum. For the transfer to an odd-A target leading to the $2+$ vibrational state in the even-even final nucleus, there are two amplitudes, as illustrated in Fig. 6-34; the evaluation of these amplitudes is similar to that of the transfer amplitudes considered in Sec. 6-5c. With the inclusion of the effect of the pair correla- ▲

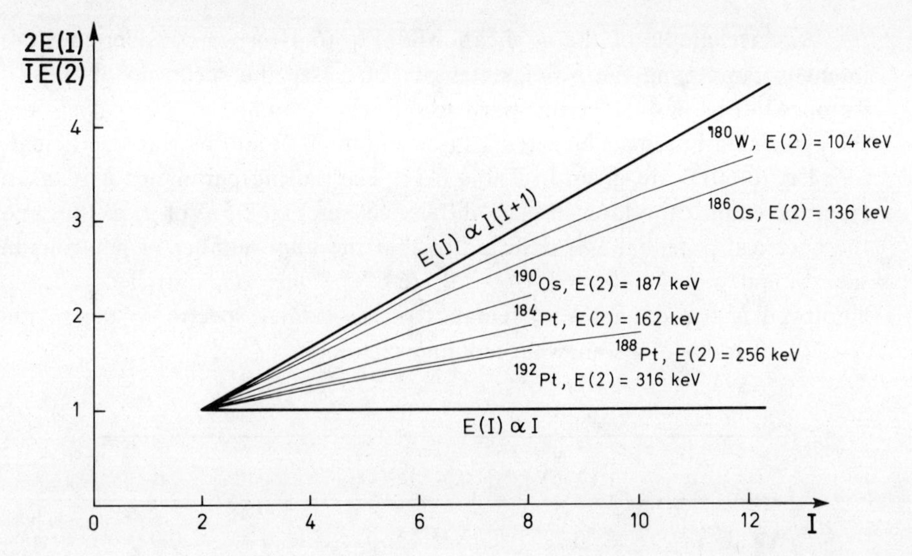

Figure 6-33 Yrast spectra for even-even nuclei in the region $180 \lesssim A \lesssim 190$. The experimental data are taken from the compilations by Sakai (1970) and Saethre *et al.* (1973).

▼ tions (see Eqs. (6-599) and (6-610)), one obtains the parentage factor

$$\langle n_2=1\|a^\dagger(j_2)\|\nu=1,j_1\rangle=(2j_1+1)^{1/2}h(j_1,j_2\lambda=2)$$
$$\times\left(u(j_2)\frac{u(j_1)v(j_2)+v(j_1)u(j_2)}{E(j_1)+E(j_2)-\hbar\omega}+v(j_2)\frac{u(j_1)u(j_2)-v(j_1)v(j_2)}{E(j_2)-E(j_1)+\hbar\omega}\right) \tag{6-416}$$

If the target is in its ground state, the relative sign of the two amplitudes is positive if $u(j_1)u(j_2)-v(j_1)v(j_2)>0$, corresponding to a particle-like transition between the states j_1 and j_2. ▲

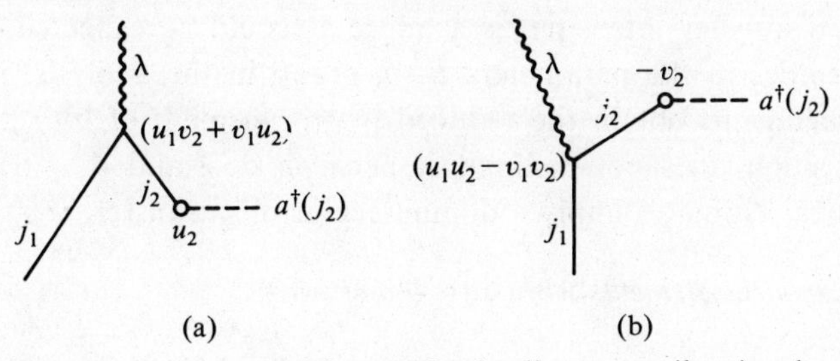

Figure 6-34 Amplitudes for one-particle transfer leading to a vibrational excitation. The factors appearing beside the vertices represent the effect of the pair correlations.

▼ As an example of the application of Eq. (6-416), we consider the (dp) intensity populating the $n_2 = 1$ states of ^{112}Cd (see the spectrum of the Cd isotopes in Fig. 6-30). The one-particle spectrum assumed in the analysis and the associated quasiparticle energies (see Eq. (6-602)) and pairing coefficients (see Eq. (6-601)) are given in Table 6-11. The pairing parameter Δ is taken from the empirical odd-even mass difference (see Fig. 2-5, Vol. I, p. 170) and the chemical potential λ is adjusted so that the total number of neutrons in the ground state is $N = 63$ (see Eq. (6-611)). The one-particle spectrum employed is based on the systematics of the odd-A spectra in the region ▲ $N \approx 60-70$ and is of a somewhat tentative character.

Orbit	ε (MeV)	E (MeV)	u	v
$d_{5/2}$	0.2	1.73	0.39	0.92
$g_{7/2}$	0.8	1.39	0.53	0.85
$s_{1/2}$	2.2	1.48	0.88	0.48
$h_{11/2}$	2.8	1.88	0.93	0.36
$d_{3/2}$	4.6	3.44	0.98	0.19

Table 6-11 Pairing parameters for neutron configurations in ^{111}Cd. The quasiparticle energies E and occupation parameters u and v are calculated for the chemical potential, $\lambda = 1.4$ MeV, and gap parameter, $\Delta = 1.25$ MeV.

▼ The estimate (6-416) of the parentage factor is compared in Table 6-12 with the empirical value deduced from the (dp) cross sections. The radial matrix element $\langle j_2|k(r)|j_1\rangle$ in column two is obtained from the standard Woods-Saxon potential (see Vol. I, pp. 239 and 326) and the vibrational amplitude $((\hbar\omega_2/2C_2)^{1/2} = 8 \times 10^{-2})$ is derived from the data in Fig. 6-30. The amplitudes in columns three and four correspond to the two diagrams in Fig. 6-34. The agreement to within a factor of 2 between the calculated and observed transfer intensities is as close as could be expected, in view of the uncertainties in the parameters E, u, v and in the analysis of the observed cross section to obtain the reduced matrix element of $a^{\dagger}(j_2)$. The observed cross section does not distinguish between $d_{3/2}$ and $d_{5/2}$ transfer, but the theoretical estimates imply a dominance of $d_{3/2}$ transfer.

E2-matrix elements with $\Delta n = 0$ and $\Delta n = 2$

In the harmonic approximation, the $E2$-matrix elements in the vibrational spectrum are governed by the selection rule $\Delta n = 1$ (see p. 349). The ▲ forbiddenness of the $\Delta n = 2$ transition $n = 2, I = 2 \rightarrow n = 0$ is a systematic fea-

j_2	$\langle j_2\|k\|j_1\rangle$ MeV	$\langle n_2=1\|a^\dagger(j_2)\|v=1,j_1\rangle$ Term a	Term b	$\langle n_2=1\|a^\dagger(j_2)\|v=1,j_1\rangle^2$ Calc.	Obs.
$d_{3/2}$	45	-0.31	-0.12	0.18	0.36
$d_{5/2}$	45	-0.40	0.25	0.02	

Table 6-12 One-particle transfer leading to 2+ vibrational state of ^{112}Cd from ground state of ^{111}Cd ($j_1 = s_{1/2}$). The observed parentage factor is obtained from the experimental cross sections for ^{111}Cd (dp)^{112}Cd and their analysis given by P. D. Barnes, J. R. Comfort, and C. K. Bockelman, *Phys. Rev.* **155**, 1319 (1967). These authors have also given an interpretation of this cross section based on the particle-vibration coupling with results essentially the same as those obtained in the present analysis.

▼ ture of the empirical data (see the examples in Figs. 6-30, 6-31, and 6-32).[24] Evidence on the $\Delta n=0$ matrix element that gives the static quadrupole moment of the first excited 2+ state has become available as a result of the development of tools for the study of multiple Coulomb excitation (de Boer *et al.*, 1965). It is found that these matrix elements are usually of a similar magnitude as the $\Delta n=1$ transition moments and thus represent a major violation of the leading-order selection rule (see Figs. 6-30 to 6-32).

The occurrence of static moments comparable in magnitude to the transition moments has already been discussed as a characteristic feature of the low-frequency quadrupole mode associated with the trend toward instability of the spherical shape, when particles are added in unfilled shells (see pp. 523 ff.). The different behavior of the $\Delta n=2$ moments can be qualitatively interpreted in terms of the effect of the leading-order anharmonic term in the vibrational potential energy, proportional to $(\alpha_2\alpha_2\alpha_2)_0$, which gives the ratio $-3\sqrt{2}$ for the reduced matrix elements $\langle n=1, I=2\| \mathscr{M}(E2)\| n=1, I=2\rangle$ and $\langle n=2, I=2\| \mathscr{M}(E2)\| n=0\rangle$ (see Eq. (6-425)), corresponding to a factor of 18, for the ratio of the reduced transition probabilities. A rather large value for the ratio of the matrix elements can be understood in terms of the frequency dependence of the quadrupole polarizability associated with the low-frequency mode itself; for the static moment, the appropriate polarizability is $\chi(\Delta E=0)$ (see the discussion on p. 525), while for the $\Delta n=2$ transition, the polarizability coefficient is $\chi(\Delta E=2\hbar\omega_2) = -\frac{1}{3}\chi(\Delta E=0)$, see Eq. (6-216); ▲

---

[24]This selection rule was recognized at an early stage of nuclear spectroscopy as a striking feature of the spectra of even-even nuclei (Kraushaar and Goldhaber, 1953).

▼ the additional factor $\sqrt{2}$ results from the counting of the diagrams and the inclusion of the boson factor (see Fig. 6-37a).

Particle-phonon interaction. Special favoring of states with $I=j-1$

In the odd-A spectra, the states involving a vibrational quantum are usually found to have energies and transition probabilities differing significantly from those of noninteracting excitations (see examples in Fig. 6-30). The particle-phonon interaction resulting from the particle-vibration coupling acting in second order gives rise to rather large energy shifts, but the anharmonicity in the vibrational motion itself may lead to additional effects of comparable magnitude, which appear as higher-order terms in the particle-vibration coupling. Thus, if the phonon possesses a static quadrupole moment comparable to the transition moment, one must take into account the interaction between this moment and that of the quasiparticle (see the analysis on p. 576 of the corresponding term for an octupole phonon).

Despite the complexity of the quadrupole spectra of the odd-A nuclei, it is possible to identify a rather striking feature, which receives a simple qualitative interpretation in terms of the particle-vibration coupling. One finds quite regularly that, in odd-N (or odd-Z) nuclei with several neutrons (or protons) outside of closed shells, there occur very low-lying states with $I=j-1$, where j ($\geqslant 5/2$) is the angular momentum of a quasiparticle close to the Fermi surface. (See, for example, the $I\pi=7/2+$ states in Fig. 2-24, Vol. I, p. 225, for nuclei with N or Z equal to 43, 45, and 47, for which the $g_{9/2}$ shell is being filled.[25])

The leading order of particle-phonon interaction is illustrated by the diagrams in Fig. 6-10, p. 427. For $j_1=j_2=j$, the diagrams in Figs. 6-10a and b are always repulsive and act only in the $I=j$ state (see Eq. (6-224)). Figures 6-10c and d give rise to the same pattern of splittings within the particle-phonon multiplet, but have opposite sign. From Eq. (6-224) and the quasi- ▲

[25]The systematic occurrence of these "anomalous" states for configurations with several equivalent particles was recognized by Goldhaber and Sunyar (1951) as a part of the elucidation of the "islands of isomerism." The interpretation of the states played a significant role in the early discussion of nuclear coupling schemes, since it was recognized that for $(j)^n$ configurations, short-range attractive forces favor the single state with $I=j$ (Mayer, 1950a), characterized by the seniority $v=1$ (Racah, 1950). It was found that for $(j)^3$ configurations with $j\geqslant 5/2$, the $I=j-1$ state can occur very close to or even below the $I=j$ state for forces of sufficiently long range (Kurath, 1950). An alternative interpretation could be given in terms of the coupling to the quadrupole deformations; for spherical nuclei, these couplings give rise to an enhancement of the quadrupole component of the effective nuclear forces, which favors the $I=j-1$ state; for deformed nuclei, the lowest state of $(j)^3$ also has $I(=K)=j-1$ (Bohr and Mottelson, 1953, pp. 34ff.; see also the erratum quoted in footnote 11 of Bohr and Mottelson, 1955a). The possibility of basing the argument on the quasiparticle-phonon coupling was recognized by Sherwood and Goswami (1966).

▼ particle factors given by Eq. (6-207), one obtains

$$\Delta E^{(2)}(j,n=1;I)=h^2(j,j2)\left(\delta(j,I)\left(\frac{\left(u^2(j)-v^2(j)\right)^2}{\hbar\omega}+\frac{4u^2(j)v^2(j)}{2E(j)+\hbar\omega}\right)\right.$$

$$\left.+(2j+1)\begin{Bmatrix}2 & j & j\\ 2 & j & I\end{Bmatrix}\left(\frac{\left(u^2(j)-v^2(j)\right)^2}{\hbar\omega}-\frac{4u^2(j)v^2(j)}{2E(j)-\hbar\omega}\right)\right) \quad (6\text{-}417)$$

where $E(j)$ is the quasiparticle energy.

The first and third terms in Eq. (6-417), corresponding to the diagrams (a) and (c) in Fig. 6-10, have the form of a coupling between a single particle and the nuclear deformation and favor the state $I=j+1$, which develops into the first rotational excitation for sufficiently strong coupling. (In the special case of $j=5/2$, the second-order interaction energy (c) is equally large for $I=j-2=1/2$ and $I=j+1=7/2$.) The fact that the vibrational excitation is built out of the same degrees of freedom as the particle motion is reflected in the second and fourth terms in Eq. (6-417) (corresponding to the diagrams in Figs. 6-10b and d), which favor the state with $I=j-1$. For $j\geqslant 5/2$, the $6j$ symbol has a positive value only for $I=j-1$.) The diagrams (b) and (d) in Fig. 6-10 are specific to configurations with several equivalent particles. For sufficiently many particles (and holes) in the j shell, these terms dominate over the interactions (a) and (c), and the particle-vibration coupling then favors the state $I=j-1$.

Interaction of phonons and correction to E2-intensity relations

The leading-order anharmonic terms in the vibrational energy spectrum can be described in terms of an interaction between pairs of phonons (see Eqs. (6-269)). In this approximation, the energies of the states with $n\geqslant 3$ can be calculated from the observed energies of the $n=1$ and $n=2$ excitations. The relationship is especially simple for the states with $I=2n$ (see Eqs. (6-272))

$$E(n,I=2n)=nE(n=1)+\tfrac{1}{2}n(n-1)\left(E(n=2,I=4)-2E(n=1)\right)$$

$$=I\left(E(n=1)-\tfrac{1}{4}E(n=2,I=4)\right)+\tfrac{1}{8}I^2\left(E(n=2,I=4)-2E(n=1)\right) \quad (6\text{-}418)$$

The available data on such sequences of states agree rather well with the expression (6-418); see the examples in Fig. 6-33, where the relation (6-418) corresponds to a straight line. (The possibility of representing this sequence of states by an expression involving linear and quadratic terms in I was ▲ discussed by Nathan and Nilsson, 1965.)

▼ For other states with $n \geqslant 3$, there is only little evidence to test the corresponding relations. A few examples are provided by observed $I\pi = 3+$ states. If interpreted as $n=3$ excitations, the energy can be obtained from Eq. (6-271) and the parentage coefficients in Table 6B-1, p. 691,

$$E(n=3, I=3) = 3E(n=1) + \left(\tfrac{15}{7}E(n=2, I=2) + \tfrac{6}{7}E(n=2, I=4) - 6E(n=1)\right) \tag{6-419}$$

The energies of $I\pi = 3+$ states in isotopes of Sm, Os, and Pt are collected in Table 6-13, and it is seen that the relation (6-419) is not well satisfied. If the $n=3$ assignment for these states is confirmed by the observation of large $E2$-matrix elements connecting the states with the $n=2$ excitations, it would appear that the three-phonon interaction is of rather large magnitude. ▲

Nucleus	$n=1$	$n=2$		$n=3, I=3$	
	$I=2$	$I=2$	$I=4$	Calc.	Obs.
^{150}Sm	0.334	1.047	0.777	1.91	1.50
^{152}Sm	0.122	0.811	0.366	1.69	1.23
^{188}Os	0.155	0.633	0.478	1.30	0.79
^{192}Os	0.206	0.489	0.580	0.93	0.69
^{192}Pt	0.316	0.612	0.785	1.04	0.92
^{196}Pt	0.356	0.689	0.878	1.16	1.00

Table 6-13 Analysis of energies of $I\pi = 3+$, $n=3$ states. The data are taken from the compilation by Sakai (1972); see also Figs. 6-31 and 6-32. The energies are in MeV.

▼ A similar analysis of the $E2$-matrix elements leads to an expansion of the effective $E2$ operator in powers of $c_2^\dagger$ and c_2 (see Eq. (6-273)). Thus, to leading order, the $\Delta n = 0$ and $\Delta n = 2$ matrix elements each involve a single parameter (m_{11} and m_{20}), while the corrections to the $\Delta n = 1$ matrix elements involve three parameters ($m_{21,R}$). The relation between the $\Delta n = 0$ matrix elements can be tested by the ratio between the $E2$-matrix element for the $\Delta n = 0$ transition $n=2, I=2 \rightarrow n=2, I=0$ and the static quadrupole moment of the $n=1$ state in ^{114}Cd. From the reduced matrix elements in Table 6-3, p. 450, one obtains

$$\begin{aligned} B(E2; n=2, I=2 \rightarrow n=2, I=0) &= \frac{4}{5} B(E2; n=1, I=2 \rightarrow n=1, I=2) \\ &= \frac{7}{8\pi} (Q(n=1))^2 \end{aligned} \tag{6-420}$$

▲

▼ The experimental data for ^{114}Cd in Fig. 6-30 are seen to be consistent with this relation. (A search for the $n=2, I=4$ to $n=2, I=2$ transitions in ^{108}Pd and ^{134}Ba has yielded upper limits for the $E2$-transition probability, which are an order of magnitude smaller than expected on the basis of the static quadrupole moments of these nuclei (Stelson *et al.*, 1973)).

Another way of viewing the anharmonicity effects is in terms of an expansion of the vibrational Hamiltonian in powers of the vibrational amplitudes and their time derivatives. In the adiabatic situation ($\hbar\omega \ll 2\Delta$), the anharmonic terms in the potential energy are expected to be especially large (see p. 451). With inclusion of cubic and quartic terms in the potential energy, the anharmonic part of the Hamiltonian is given by

$$H' = k_3(\alpha_2\alpha_2\alpha_2)_0 + k_4(\alpha_2\alpha_2)_0(\alpha_2\alpha_2)_0 \tag{6-421}$$

(The occurrence of only a single term in third and fourth order is equivalent to the fact that the spectrum of quadrupole phonons contains a single $I=0$ state for $n=3$, as well as for $n=4$; see Table 6-1, p. 347. The invariants that occur in Eq. (6-421) are expressed in terms of the intrinsic deformation parameters β and γ in Eq. (6B-3).) The leading-order effects of the anharmonicity (6-421) on the energy and the $E2$ moments are evaluated in the small print below.

Since the anharmonicity (6-421) involves only two parameters, there are restrictions on the pattern of energy shifts within the $n=2$ triplet (see Eqs. (6-426) and (6-428)),

$$\begin{gathered} E(n=2, I=0) - 2E(n=1) = \tfrac{7}{2}\big(E(n=2, I=4) - 2E(n=1)\big) \\ E(n=2, I=2) > E(n=2, I=4) \end{gathered} \tag{6-422}$$

As can be seen from Figs. 6-30, 6-31, and 6-32, the relations (6-422) are often violated.

The contribution of the cubic and quartic potential energy terms to the $E2$-matrix elements is given by Eq. (6-429). A number of these matrix elements have been determined for ^{114}Cd (see Fig. 6-30) and are analyzed, together with the energies of the $n=2$ states, in Fig. 6-35. Each of the quantities considered is a linear function of k_3^2 and k_4, and the empirical value therefore defines a straight line in the k_3^2, k_4 plane. In a consistent interpretation, all the lines would intersect at a single point. In addition to the discrepancies among the energies already noted (see Eq. (6-422)), the $E2$-matrix elements among themselves and in comparison with the energies provide further evidence for the shortcomings of the analysis based on the leading-order effects of the potential energy terms. (More detailed phenomenological studies of anharmonicity effects in ^{114}Cd, including higher-
▲ order effects of the cubic terms in the Hamiltonian, confirm that it is not

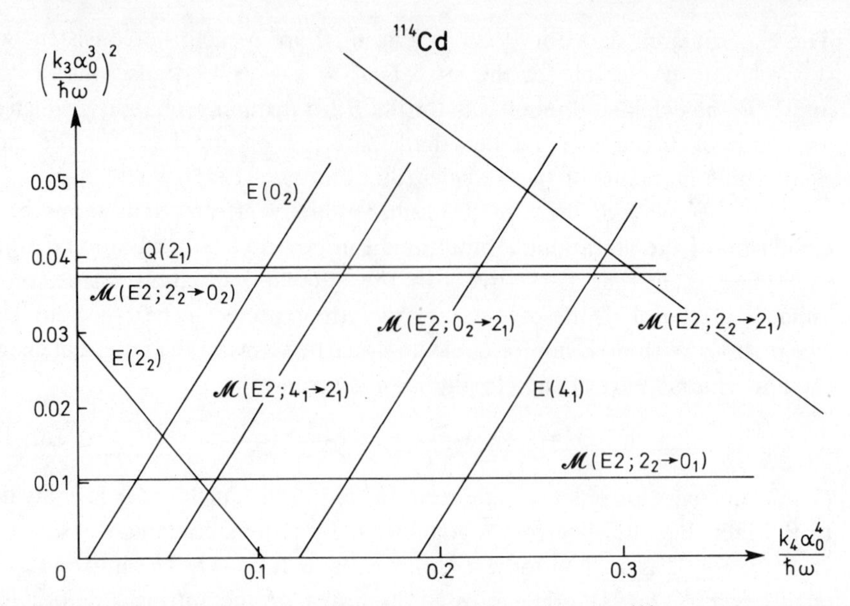

Figure 6-35 Anharmonic effects in the quadrupole spectrum of ^{114}Cd. The empirical data are taken from Fig. 6-30. The $E2$-matrix elements describing the transitions $n=2 \to n=1$ receive contributions from the leading-order (harmonic) as well as from the anharmonic terms, and it has been assumed that the sign of the total matrix element is the same as that of the leading-order term; the assumption of the opposite phase would have led to values of (k_3^2, k_4) lying considerably outside the region illustrated in the figure.

▼ possible on this basis to account for the observed energies and $E2$-matrix elements (Sørensen, 1966; Bès and Dussel, 1969).) At present, it is not clear to what extent these discrepancies result from further anharmonic terms in the quadrupole mode or are to be attributed to the neglect of additional degrees of freedom and interactions that might play a role in the low-energy excitation spectra.

As an alternative approach to the analysis of the higher states in the vibrational spectra, one may attempt to identify rotational trajectories. The separation between rotational and intrinsic vibrational motion is expected as a general feature of the yrast region (see the discussion in Sec. 6B-3). Thus, the leading trajectory with $I=2n$ can be viewed as a rotational sequence associated with a triaxial rotor with an equilibrium deformation changing smoothly with I. This circumstance may account for the especially regular behavior of this sequence of states, as illustrated in Fig. 6-33. Further evidence testing the expected simple pattern of the spectrum in the yrast region would be very valuable.

It is also possible that, for a large class of nuclei, the potential energy surfaces represent situations intermediate between those of harmonic vibrations ($V \approx \frac{1}{2} C\beta^2$) and those of nuclei with stable equilibrium deformations. In ▲

such a situation, one may obtain qualitative insight from an interpolation between these two limiting coupling schemes (Sheline, 1960; Sakai, 1967; Davydov, 1967). The classification of the transition spectra may depend sensitively on the shape and stability of the equilibrium form toward which the system is developing. Thus, for example, for a transition toward an axially symmetric deformation, the second $I=0$ and $I=2$ states will be members of the same $n_\beta=1$, $K=0$ rotational band, if the deformed shape is softer with respect to variations in β than in γ, while these states belong to different bands ($n_\beta=1$, $K=0$ and $n_\gamma=1$, $K=2$) with the $I=2$ state lower in energy , if the deformed shape is especially soft with respect to variations in γ. It appears that these two possibilities are illustrated by the sequences of observed spectra in Figs. 6-31 and 6-32, respectively. (For investigations of the different types of spectra that can occur in connection with such transitions between vibrational and rotational coupling schemes, see, for example, Kumar, 1967; Dussel and Bès, 1970; Gneuss and Greiner, 1971.)

It should be borne in mind that the mass parameters for the β- and γ-vibrational motion in the deformed nuclei are found to be appreciably larger than those associated with the rotational motion (see the discussion on pp. 552 and 550); thus, variations of the mass parameters with β and γ may have a significant influence on the structure of the transition spectra.

The leading-order anharmonic term in the potential energy is given by the cubic coupling

$$\begin{aligned} H' &= k_3(\alpha_2\alpha_2\alpha_2)_0 \\ &= -k_3\alpha_0^3\left((c_2^\dagger c_2^\dagger c_2^\dagger)_0 + 3(c_2^\dagger c_2^\dagger \bar{c}_2)_0 + \text{H. conj.}\right) \end{aligned} \tag{6-423}$$

where α_0 is the zero-point amplitude (see Eqs. (6-51) and (6-52)). The basic interactions produced by the coupling (6-423) are illustrated in Fig. 6-36a. The identity of the three phonons involved in the coupling implies a factor $(n!)^{1/2}$ for an initial or final state with n phonons; each diagram represents a sum of terms corresponding to the permutation of phonons that have distinguishable roles, such as the distinction between phonons entering or leaving the vertex. On this account, the second diagram in Fig. 6-36a gives a factor of 3, as for the coefficient of the $(c_2^\dagger c_2^\dagger \bar{c}_2)_0$ term in Eq. (6-423). The last factor $(5)^{-1/2}$ for this diagram is associated with the angular momentum coupling in the expression (6-423).

In first order, the coupling (6-423) gives rise to $\Delta n=0$ and $\Delta n=2$ terms in the $E2$ moment illustrated by the diagrams in Fig. 6-37a, and the corresponding coefficients in the effective operator (6-273) have the values

$$\begin{aligned} m_{11} &= \frac{12}{\sqrt{5}} \frac{k_3\alpha_0^3}{\hbar\omega} m_{10} \\ m_{20} &= -\frac{2\sqrt{2}}{\sqrt{5}} \frac{k_3\alpha_0^3}{\hbar\omega} m_{10} \end{aligned} \tag{6-424}$$

where m_{10} is the $E2$-matrix element for the creation of a phonon. In the diagrams in

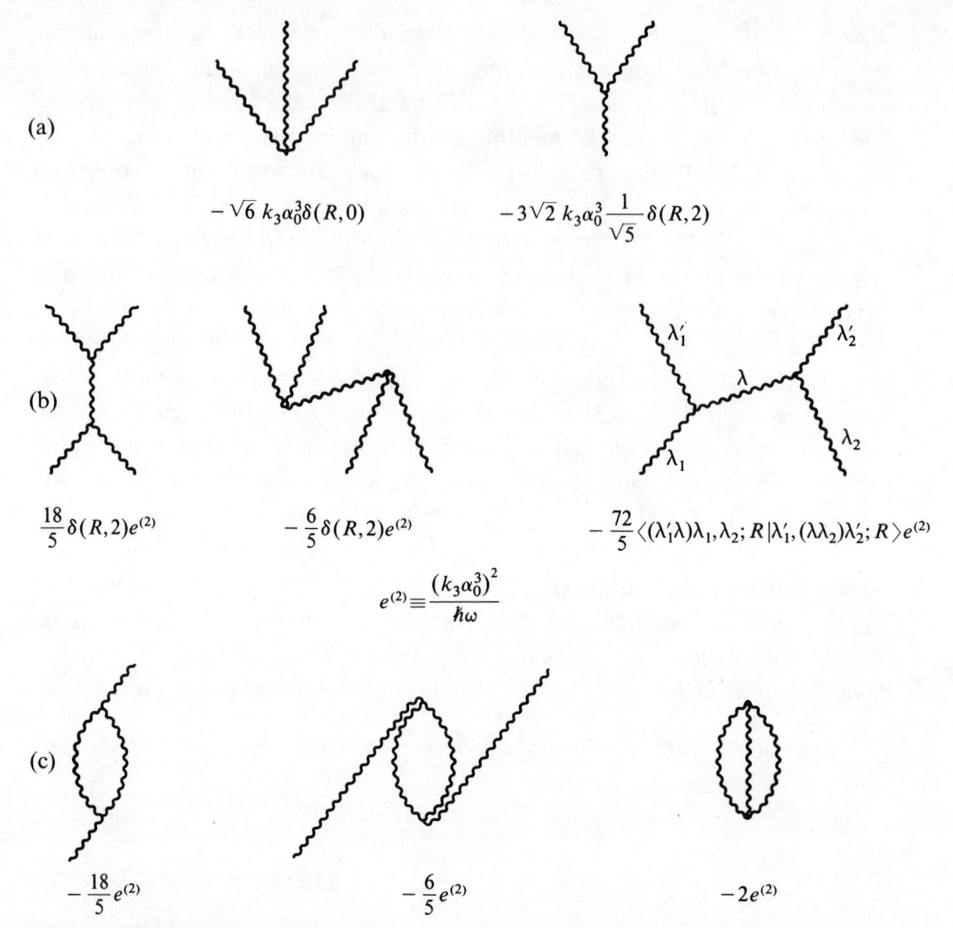

Figure 6-36 Diagrams representing interaction effects associated with cubic anharmonic term in the quadrupole mode.

▼ Fig. 6-37a, the phonon that is created or annihilated by the $E2$ moment is distinguishable from the other phonons; hence, the matrix element involves the permutation factors 6 for the $\Delta n = 0$ diagrams and 3 for the $\Delta n = 2$ diagrams.

The $\Delta n = 0$ term produced by the cubic coupling has a considerably larger coefficient than the $\Delta n = 2$ term; thus, for example (see Table 6-3, p. 450, and Eq. (6-274))

$$\frac{\langle n=1, I=2 \| \mathscr{M}(E2) \| n=1, I=2\rangle}{\langle n=2, I=2 \| \mathscr{M}(E2) \| n=0, I=0\rangle} = -3\sqrt{2} \tag{6-425}$$

A large static quadrupole moment for the $n=1$ state combined with a weak $n=2 \rightarrow n=0$ crossover transition is in fact a characteristic feature of the observed spectra (see p. 539). ▲

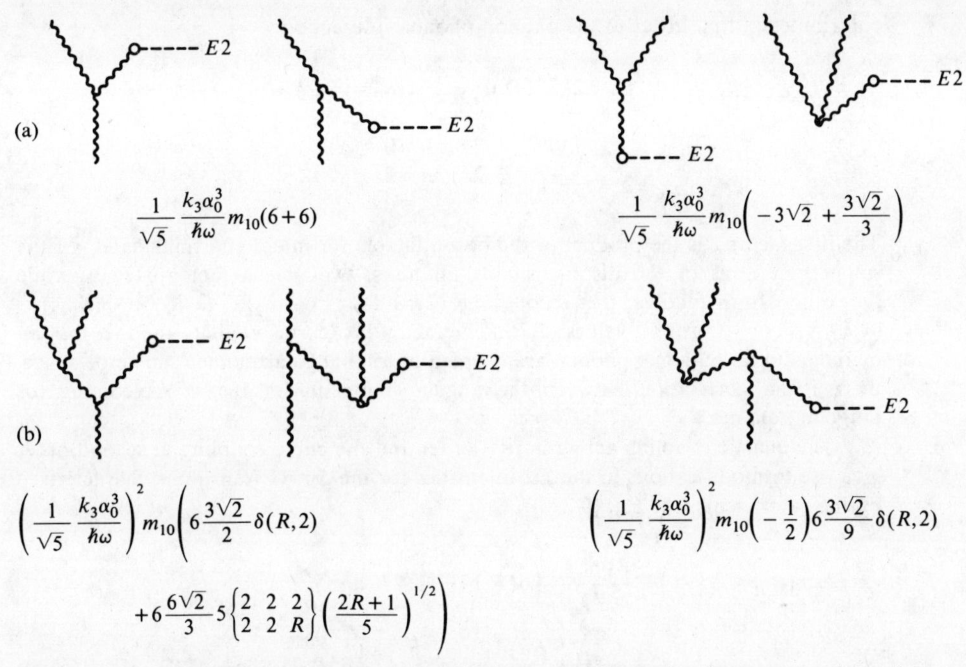

Figure 6-37 Diagrams representing contributions to $E2$-matrix elements resulting from cubic anharmonicity in the quadrupole mode.

▼ In second order, the coupling (6-423) gives rise to phonon-phonon interactions (see Fig. 6-36b), which can be expressed in the form (6-269) with

$$V_R=\left(\frac{12}{5}\delta(R,2)-72\begin{Bmatrix}2&2&2\\2&2&R\end{Bmatrix}\right)\frac{(k_3\alpha_0^3)^2}{\hbar\omega}=c_R\frac{24}{35}\frac{(k_3\alpha_0^3)^2}{\hbar\omega} \tag{6-426}$$

$$c_R=\begin{cases}-21 & R=0\\ +8 & R=2\\ -6 & R=4\end{cases}$$

The cubic coupling acting in second order also gives rise to a renormalization of the phonon energy and of the vacuum state, as illustrated by the diagrams in Fig. 6-36c. The indistinguishable roles of the phonons in the closed loops imply the boson factors 2, 2, and 6 for the three diagrams of Fig. 6-36c, and these factors have been included in the values given in the figure. (The self-energy and interaction effects are separated by evaluating only linked diagrams, as in the treatment of the particle-vibration coupling; see pp. 427 ff.)

The quartic coupling

▲
$$H'=k_4(\alpha_2\alpha_2)_0(\alpha_2\alpha_2)_0 \tag{6-427}$$

contributes in first order to the phonon-phonon interaction

$$V_R = 2\left(2\delta(R,0) + \frac{4}{5}\right)k_4\alpha_0^4 = c_R\frac{4}{5}k_4\alpha_0^4$$

$$c_R = \begin{cases} 7 & R=0 \\ 2 & R=2 \\ 2 & R=4 \end{cases} \tag{6-428}$$

The first factor 2 is the product of the boson factors for initial and final states. Of the six permutations of the distinguishable phonons, two give a factor $\delta(R,0)$, while the other four involve the recoupling $(2R+1)^{-1/2}\langle(\lambda_1\lambda_2)0,\ (\lambda_3\lambda_4)0;\ 0|(\lambda_1\lambda_3)R,\ (\lambda_2\lambda_4)R;\ 0\rangle = 1/5$ for all values of R. (The factor $1/5$ corresponds to the fact that an ingoing and an outgoing phonon are coupled to total angular momentum zero; hence, there is no correlation between the angular momenta of the two incoming (or outgoing) quanta.)

The quartic coupling acting in first order and the cubic coupling in second order give rise to modifications in the $E2$ intensities for the $\Delta n = 1$ transitions characterized by the coefficients (see Eq. (6-273))

$$m_{21,R} = -\left(\frac{8}{5}\right)^{1/2}\frac{k_4\alpha_0^4}{\hbar\omega}m_{10}\left(\delta(R,0) + \frac{2}{5}(2R+1)^{1/2}\right)$$
$$+\left(\frac{2}{5}(2R+1)\right)^{1/2}\left(\frac{k_3\alpha_0^3}{\hbar\omega}\right)^2 m_{10}\left(-\frac{14}{5}\delta(R,2) + 36\begin{Bmatrix} 2 & 2 & 2 \\ 2 & 2 & R \end{Bmatrix}\right) \tag{6-429}$$

The term proportional to k_4 can be obtained by the same procedure as in the derivation of the energy contribution (6-428). The terms of second order in k_3 in Eq. (6-429) are illustrated by the diagrams in Fig. 6-37b. For each of the first two diagrams there are three others, differing only in the energy denominators. The third diagram is associated with the renormalization of the wave function of the final state (with two phonons) that arises as a second-order effect of the coupling to the intermediate state (with five phonons); the amplitude of the final state is renormalized by the factor $(1-(c_i)^2)^{1/2} \approx 1 - \frac{1}{2}(c_i)^2$, where c_i is the first-order amplitude of the admixed five-phonon state. The contribution of the diagram is therefore obtained by including a factor $-1/2$ and setting both of the energy denominators equal to the virtual excitation energy $(-3\hbar\omega)$ of the intermediate state. There are three additional diagrams associated with wave-function renormalization of the final state; one of these is proportional to $\delta(R,2)$, while the two others involve the same recoupling coefficient as in the second diagram of Fig. 6-37.

Features of Quadrupole Oscillations in Deformed Nuclei

In a spheroidal nucleus, the low-frequency quadrupole shape oscillations give rise to two modes with $\nu = 0$ and $\nu = 2$ (see Fig. 6-3, p. 363). Low-lying bands with $K\pi = 0+$ and $K\pi = 2+$ have been identified in many even-even nuclei, and the properties of these bands are compared, in the following examples, with those expected for small-amplitude shape oscillations. The high-frequency $(\Delta N = 2)$ quadrupole modes are expected to separate into modes with $\nu = 0$, 1, and 2, in analogy to the separation between the $\nu = 0$ and $\nu = 1$ components of the dipole mode discussed on pp. 490ff.

▼ *Gamma vibrations (Fig. 6-38)*

The occurrence of a rather low-lying $K\pi = 2+$ band (γ vibration) is a systematic feature in the spectra of the even-even deformed nuclei. A detailed discussion of this excitation is given for the case of ^{166}Er on pp. 164ff.; other examples are illustrated in Fig. 4-7, p. 63 (^{168}Er), Fig. 6-31, p. 534 (^{152}Sm, ^{154}Sm), and Fig. 6-32, p. 536 (^{188}Os). The evidence on the excitation energy of
▲ this mode is collected in Fig. 6-38.

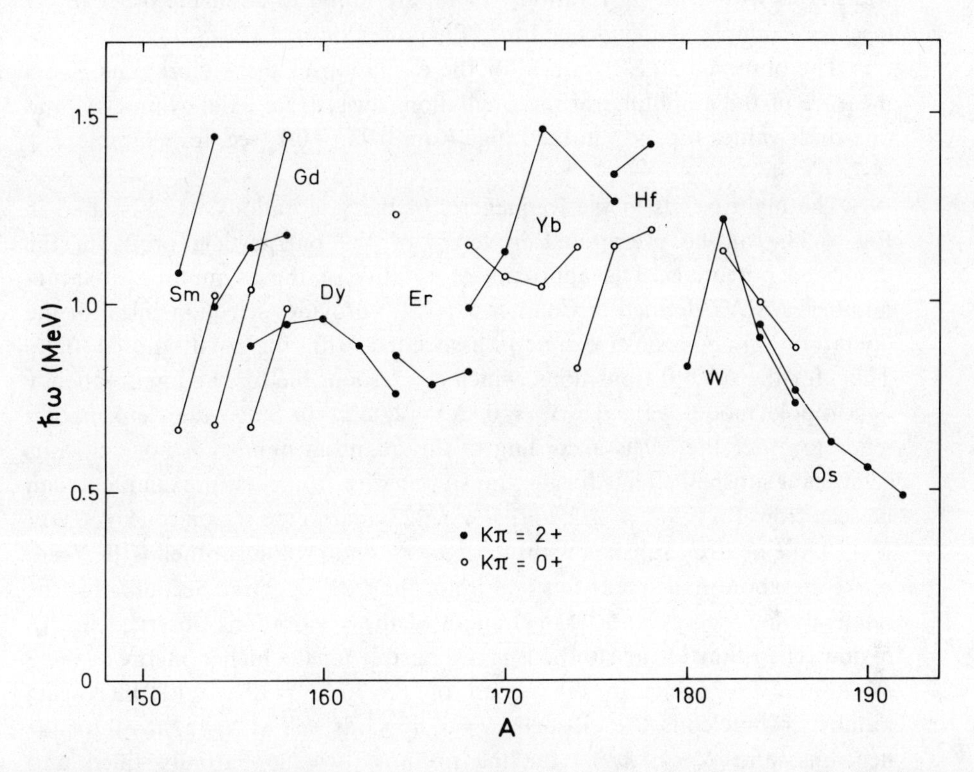

Figure 6-38 Systematics of β- and γ- vibrational frequencies for $150 \leqslant A \leqslant 192$. The figure shows the energies of the lowest $K\pi = 0+$, $I = 0$, and $K\pi = 2+$, $I = 2$ intrinsic excitations. The data are taken from the compilation by Sakai (1970) and from: ^{176}Hf (F. M. Bernthal, J. O. Rasmussen, and J. M. Hollander in *Radioactivity in Nuclear Spectroscopy*, p. 337, eds. J. H. Hamilton and J. C. Manthuruthil, Gordon and Breach, New York, 1972); ^{182,184,186}W (C. Günther, P. Kleinheinz, R. F. Casten, and B. Elbek, *Nuclear Phys.* **A172**, 273, 1971).

▼ The γ-vibrational bands are excited in inelastic processes with enhanced matrix elements. For an $E2$ excitation $K = 0, I = 0 \rightarrow K = 2, I = 2$, the appropriate single-particle unit for the transition probability is $2B_W(E2)$. (See Eq. (3C-38) for the definition of $B_W(E2)$ and Eq. (4-92) for matrix elements in the rotational coupling scheme.) Values of the observed $E2$-transition
▲ probabilities are given in the examples referred to above, and are seen to be

▼ an order of magnitude larger than this single-particle unit. Similar enhancements have been observed for the excitation of this mode in inelastic deuteron scattering (Elbek *et al.*, 1968). This evidence supports the interpretation in terms of a $\tau \approx 0$, $\sigma \approx 0$ collective mode, as for a shape oscillation. Further evidence for these symmetry quantum numbers is provided by the measured g factors for the excitation, which are rather close to those of the collective rotational motion. The values of $g_K - g_R$, as determined from $M1$ transitions within the γ-vibrational band, are found to be of the order of 0.1 (see, for example, the evidence for ^{166}Er quoted on p. 163).

The observed $B(E2)$ values for the excitation of the γ vibrations give a measure of the amplitude of the oscillations away from axial symmetry, and one finds values for $\langle \gamma^2 \rangle$ in the range from 0.02 to 0.1 (see, for example, Eq. (4-248)).

The main trends in the frequencies of the γ vibrations, as illustrated in Fig. 6-38, can be interpreted in terms of the one-particle orbits in the deformed potentials. The approximate validity of the asymptotic quantum numbers $Nn_3\Lambda\Sigma$ defined in Chapter 5, p. 217, provides selection rules for the operators $r^2 Y_{2\pm 2} = \text{const}(x_1 \pm i x_2)^2$ associated with the γ-vibrational field. Thus, for the $\Delta N = 0$ transitions, which are responsible for the low-frequency quadrupole mode, we have $\Delta n_3 = 0$, $\Delta\Lambda = 2$, $\Delta\Sigma = 0$. Since the deformation tends to order the levels according to the quantum number n_3, one obtains significant subshell effects for the transitions with $\Delta n_3 = 0$. For example, it can be seen from Figs. 5-2, p. 222, and 5-3, p. 223, that in the region of $Z = 66$ and $N = 98$, the neutron subshell with $N = 5$, $n_3 = 2$ and proton subshell with $N = 4$, $n_3 = 1$ are about half filled; this feature of the shell structure accounts for the relatively low energy and large strength of the γ vibrations observed in this region. The filling of the subshells is responsible for the higher energy of the γ excitations for nuclei in the region of $Z = 70$, $N = 102$. With increasing numbers of nucleons, the subshells $N = 5$, $n_3 = 1$ as well as $N = 5$, $n_3 = 0$ for the neutrons, and $N = 4$, $n_3 = 0$ for the protons, become partially filled and account for the low-energy γ vibrations in the region of $Z = 76$, $N = 112$.

The oscillator strength carried by the γ vibrations is 2 to 4 times smaller than for the rotations (see the examples in Figs. 4-30, p. 160 (^{166}Er), 6-31, p. 534 (^{152}Sm), and 6-32, p. 536 (^{188}Os)). In units of the oscillator sum $S(\tau = 0, \lambda = 2)_{\text{class}}$, the γ-vibrational strength, $\hbar\omega_\gamma B(E2;\, 0 \to n_\gamma = 1)(A/eZ)^2$, is about 2% (see, for comparison, the oscillator strengths for rotations and for vibrations in spherical nuclei given in Fig. 6-29, p. 531).

The oscillator strength is a measure of the collective mass parameters, and the above relationships imply $D_\gamma \approx 3D_{\text{rot}} \approx 1.5 D_2$, where D_2 is the mass parameter for vibrations in spherical nuclei with many particles outside closed shells (see Eqs. (6-180) and (6-187)). If the effect of the deformation on the intrinsic motion could be treated as a small perturbation, the collective
▲ kinetic energy would be of the form $\frac{1}{2} D_2 \Sigma_\mu \dot{\alpha}^\dagger_{2\mu} \dot{\alpha}_{2\mu}$ (see Eqs. (6-44) and (6-45)),

▼ corresponding to $D_\gamma = D_{\text{rot}}$ ($= D_2$) (see Eq. (6B-17)). The large difference in the observed values of these parameters reflects a major change in the nucleonic coupling scheme associated with the nuclear deformation. The value of D_γ can be qualitatively estimated in a manner similar to the analysis of the oscillator strength for quadrupole vibrations in spherical nuclei (see Eq. (6-399)). Such an estimate yields a mass parameter inversely proportional to the number of particles effectively participating in the collective motion. For a deformed nucleus with half-filled subshells (neutrons with $N=5$, $n_3=2$; protons with $N=4$, $n_3=1$), the number of particles participating in a γ vibration is about 8, which is about half the corresponding number for quadrupole vibrations in spherical nuclei that are close to instability with respect to the onset of static deformations. (See Fig. 4-3, p. 27, as well as the theoretical estimate on p. 522 of the critical particle number for instability. For a discussion of the mass parameter for rotation, see p. 408.)

The measured energies and $B(E2)$ values for the vibrations make possible an estimate of the contribution of this mode to the $\nu = \Delta K = 2$ electric quadrupole polarizability. Employing the same approximations as in Eq. (6-415), we obtain

$$\begin{aligned} \delta e_{\text{pol}}(E2, \Delta K=2, \Delta E=0) &\approx \frac{eZ}{A}\chi(\tau=0,\lambda=2,\Delta K=2,\Delta E=0) \\ &\approx e\frac{14Z}{C_\gamma(\text{MeV})} \\ &= e\frac{14Z}{\hbar\omega_\gamma(\text{MeV})}\frac{B(E2; n_\gamma=0\to n_\gamma=1)}{\left(\frac{3}{4\pi}ZeR^2\right)^2} \end{aligned} \tag{6-430}$$

for the static polarizability. The observed frequencies and $B(E2)$ values for γ vibrations in the rare-earth region imply contributions to the effective charge of the order of $3e$.

Beta vibrations (Fig. 6-38)

In many nuclei, low-lying $K\pi = 0+$ intrinsic excitations have been found with properties approximately corresponding to those of the expected quadrupole shape oscillations with $\nu = 0$ (β vibrations). The systematics of the lowest observed $K\pi = 0+$ excitations is illustrated in Fig. 6-38; see also the examples in Fig. 6-31. The $K\pi = 0+$ band in ^{174}Hf is discussed in more detail in connection with Fig. 4-31, p. 168. In these examples, the $E2$-transition probabilities $B(E2;\, n_\beta=0, I=0 \to n_\beta=1, I=2)$ are about ten times the single-particle unit (which is taken to be $B_W(E2)$, as is appropriate for a $K=0$,
▲ $I=0 \to K=0$, $I=2$ transition in a deformed nucleus). A transition moment of

▼ this magnitude implies a vibrational amplitude $\langle n_\beta = 1 | \beta - \beta_0 | n_\beta = 0 \rangle \sim 0.12\beta_0 \sim 0.04$ (see Eq. (4-263)). Since the energy of the β vibration is about ten times larger than the rotational excitation energy, the oscillator strength carried by the β vibration is about one fifth of that associated with the rotation, corresponding to the mass parameter $D_\beta \approx 3D_{\text{rot}} \approx D_\gamma$ (see Eqs. (6-187)).

One may attempt to obtain a qualitative understanding of the effect of the shell structure on the β-vibrational mode by considering the one-particle excitations induced by the field $r^2 Y_{20} = \text{const.}(2x_3^2 - x_1^2 - x_2^2)$. In terms of the asymptotic quantum numbers (see Chapter 5, p. 233), all the nondiagonal matrix elements of this field have $\Delta N = 2$ and are thus associated with the high-frequency quadrupole mode (see p. 507). However, when the pair correlations are included, the quadrupole field can give rise to excitation of states $(\text{v} = 2, \nu\bar{\nu})$. (The single-particle label ν represents a set of quantum numbers, such as $Nn_3\Lambda\Sigma$, and $\bar{\nu}$ is the time reverse of ν.)

This effect of the pair correlations can be illustrated by considering two-particle configurations involving two different pairs of single-particle orbits $(\nu_1\bar{\nu}_1)$ and $(\nu_2\bar{\nu}_2)$. From these pairs of orbits, one can form two orthogonal linear combinations,

$$
\begin{aligned}
|0\rangle &= (a^2 + b^2)^{-1/2}(a|\nu_1\bar{\nu}_1\rangle + b|\nu_2\bar{\nu}_2\rangle) \\
|0'\rangle &= (a^2 + b^2)^{-1/2}(-b|\nu_1\bar{\nu}_1\rangle + a|\nu_2\bar{\nu}_2\rangle)
\end{aligned}
\tag{6-431}
$$

The quadrupole matrix element connecting these two states is

$$
\langle 0' | \sum_k (r^2 Y_{20})_k | 0 \rangle = \frac{2ab}{a^2 + b^2}(\langle \nu_2 | r^2 Y_{20} | \nu_2 \rangle - \langle \nu_1 | r^2 Y_{20} | \nu_1 \rangle) \tag{6-432}
$$

Thus, the possibility of low-frequency quadrupole fluctuations depends on the occurrence of near-lying single-particle orbits with different quadrupole moments. (Quite generally, it can be seen that if all the contributing single-particle states have the same quadrupole moment, the quadrupole operator becomes proportional to the number operator and thus does not contribute to transition matrix elements.)

One may expect especially large fluctuations in the $r^2 Y_{20}$ field, when orbits from different major shells cross each other near the Fermi level. In the region illustrated in Fig. 6-38, this situation occurs for $N \approx 90$ and again in the latter part of this deformed region (see Figs. 5-2 and 5-3, pp. 222 and 223).

A collective excitation mode with $\nu\pi = 0+$ may also be generated by the monopole pair addition and removal fields; in the superfluid system, the nucleon transfer quantum number α is not in general a constant of the motion for the internal excitations, and thus a given mode may carry enhanced $\alpha = \pm 2$ moments as well as $\alpha = 0$ moments (see the discussion on p. 394 of the breaking of gauge symmetry in the "intrinsic" coordinate system). ▲

▼ Indeed, the observed low-lying $K\pi = 0+$ excitations that are characterized by enhanced $E2$ moments ($\alpha = 0$) are also found to be selectively populated in α decay (Bjørnholm *et al.*, 1963) and in (pt) reactions (Maher *et al.*, 1972). In the example illustrated in Fig. 6-39 ($^{238}\text{Pu} \rightarrow {}^{234}\text{U} + \alpha$), the hindrance factor F_α for the α branch to the β-vibrational excitation is $F_\alpha \approx 4$; this intensity represents a considerable enhancement in comparison with transitions to two-quasiparticle states, which are expected to be of order $10^2 - 10^3$, as for the unfavored transitions in odd-A nuclei. (The hindrance factors F_α express the reduced α-transition lifetimes relative to those for the ground-state transitions in even-even nuclei (see pp. 115ff.); for a discussion of the effect of the pair correlations on the α-decay transition probabilities, see pp. 270ff.) The possibility of studying the β-vibrational excitations in the variety of two-particle transfer processes involving addition and removal of pairs of neutrons and of protons provides the opportunity for a detailed analysis of the interplay between the $\alpha = \pm 2$ and $\alpha = 0$ degrees of freedom, as well as the combination of the different multipole orders λ that may contribute to a $\nu = 0$ mode.

Rather strong $E0$ transitions have been observed as a characteristic feature of the decay of the β-vibrational excitations (see, for example, Table 4-21, p. 171 (^{174}Hf) and Fig. 6-39, p. 555 (^{234}U)). In discussing monopole matrix elements for low-energy transitions, there is no generally valid one-particle transition strength with which to compare, since there are no transitions within a single major shell. Low-energy $E0$ transitions occur as a result of the pair correlations, and thus one may compare with the matrix elements for transitions between two proton states of the type (6-431),

$$\langle 0'|m(E0)|0\rangle = e\frac{2ab}{a^2+b^2}\left(\langle \nu_2|r^2|\nu_2\rangle - \langle \nu_1|r^2|\nu_1\rangle\right) \qquad (6\text{-}433)$$

Assuming $a = b$ in Eq. (6-433) and taking for the difference in r^2 the value corresponding to two successive shells in the harmonic oscillator potential (see Eq. (2-153)), we obtain an effective single-particle unit for the $E0$-matrix element,

$$\langle m(E0)\rangle_{\text{sp}} = \frac{e\hbar}{M\omega_0}$$

$$\approx 1.0A^{1/3}\, e\,\text{fm}^2 \qquad (6\text{-}434)$$

using the estimate $\hbar\omega_0 = 41A^{-1/3}$ MeV. The observed $E0$-matrix elements for the β-vibrational excitations referred to above are seen to be somewhat greater than the unit (6-434).

It is also instructive to compare the observed monopole matrix elements with those evaluated for a shape deformation with volume conservation (see ▲ Eq. (4-264)). The values of $m(E0)$ obtained from this model agree rather well

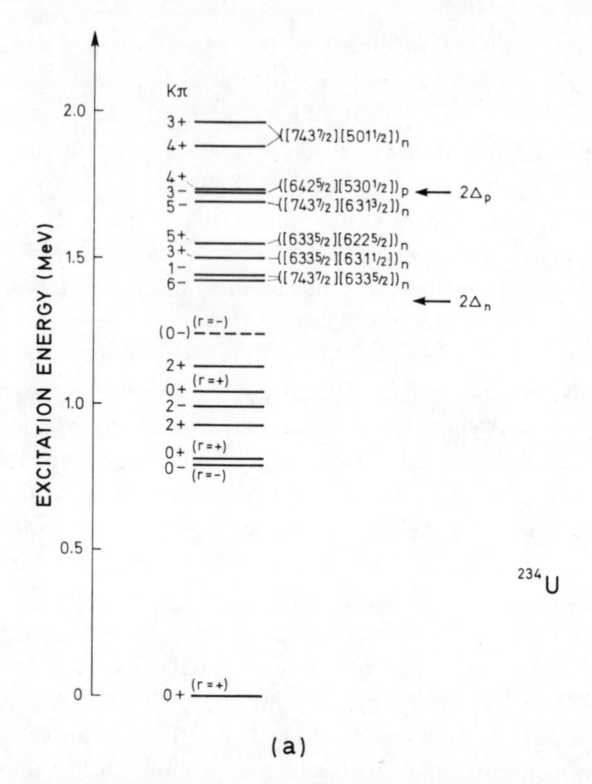

Figure 6-39 Low-energy excitation spectrum of ^{234}U. The identified intrinsic excitations are shown in (a) together with tentative assignments for the quasiparticle quantum numbers (only a single configuration with two proton quasiparticles has been identified so far); the intrinsic excitations below 1.3 MeV are interpreted as collective modes. More detailed information on the rotational band structure of the low-lying even-parity bands is given in (b). The indicated $E2$ and $E0$ transitions represent the main decay modes of the corresponding bands, and the values of F_α give the observed hindrance factors for population by α decay of ^{238}Pu (for definition of the hindrance factors, see p. 115). The data are taken from: one-particle transfer (S. Bjørnholm, J. Dubois, and B. Elbek, *Nuclear Phys.* **A118**, 241, 1968); β decay of ^{234}Pa (S. Bjørnholm, J. Borggreen, D. Davies, N. J. S. Hansen, J. Pedersen, and H. L. Nielsen, *Nuclear Phys.* **A118**, 261, 1968); α decay of ^{238}Pu (C. M. Lederer, F. Asaro, and I. Perlman, quoted in *Nuclear Data* **B4**, 652, 1970).

▼ with those observed, but deviations in the quantitative comparison would not be surprising in view of the assumption of incompressibility underlying the estimate (4-264).

Evidence for additional collective modes with $K\pi = 0+$ and $2+$ (Fig. 6-39)

The properties of the β and γ vibrations discussed above appear to correspond approximately with the pattern expected for small-amplitude quadrupole shape oscillations. In some of the spectra, however, features have ▲ been encountered that would not have been expected in such a description.

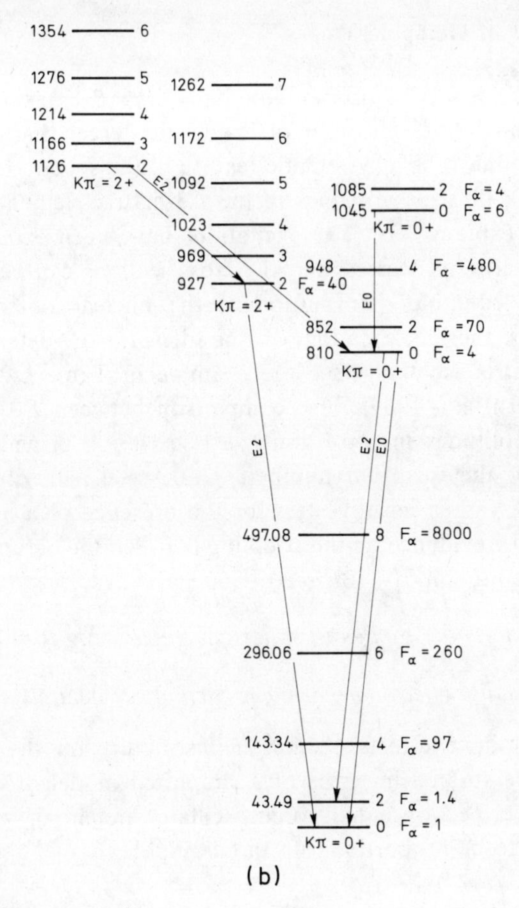

Figure 6-39 continued

▼ Especially significant in this connection is the observation in a number of deformed nuclei of additional intrinsic excitations with $K\pi = 0+$ and $2+$ and with energies well below the threshold for two-quasiparticle excitations. An example is illustrated in Fig. 6-39. In some cases, as in Fig. 6-39, the additional excitations are found to decay predominantly to the lower-lying intrinsic excitations rather than to the ground-state band, suggesting appreciable couplings to the β and γ modes. Thus, the interpretation of the additional $K\pi = 0+$ and $2+$ excitations is crucial to the understanding of the β and γ vibrations. At present, there is no definite evidence as to the degrees of freedom involved, though one may speculate on the possibility of additional vibrational modes involving, for example, the different pair fields, or the occurrence of secondary equilibrium deformations, possibly with triaxial symmetry. The additional degrees of freedom may also be of significance in connection with the interpretation of the rather large second-order rotational couplings observed for the β and γ vibrations (see p. 463). ▲

▼ **Features of Octupole Modes**

The occurrence of rather low-lying $I\pi = 3-$ excitations with $E3$-transition probabilities an order of magnitude larger than the single-particle estimate is found to be a systematic feature of the spectra of even-even nuclei (see Fig. 6-40; illustrative values of the $E3$-matrix elements for these modes are given in Table 6-14).[26] The excitations have been extensively studied by inelastic scattering processes, and the cross sections can be accounted for by assuming an octupole deformation of the nuclear potential of the same magnitude as the deformation of the density, as determined from $E3$-transition matrix elements (see the example in Table 6-2, p. 354, and the review by Bernstein, 1969). This comparison between $E3$ moments and the transition amplitudes in (pp′) and ($\alpha\alpha'$) scattering provides support to the assignment of the quantum numbers $\tau \approx 0$, $\sigma \approx 0$, as expected for a shape oscillation. (A rather sensitive test for the presence of a $\sigma \approx 1$ component is provided by the evidence on the coupling between the octupole excitation and single-particle motion in ^{209}Bi; see p. 566.)

Structure of octupole modes in spherical nuclei (Fig. 6-40 and Table 6-14)

Shell-structure effects for harmonic oscillator potential

The qualitative effects of the shell structure on the nuclear octupole modes may be studied in terms of a simplified model in which the particle motion is described by a harmonic oscillator potential, while the octupole field is taken to be proportional to the multipole moment

$$F = \sum_k (r^3 Y_{30}(\vartheta))_k \tag{6-435}$$

The particle excitations produced by such a field are governed by the selection rule $\Delta N = 1$ or 3 and have the energies $\hbar\omega_0$ and $3\hbar\omega_0$. For a single completed shell with total quantum number N, the transition strength is given by (see the derivation of the corresponding equation (6-411) for $\lambda = 2$)

$$\sum_{\nu_N, \nu_{N+1}} |\langle \nu_{N+1}| r^3 Y_{30} |\nu_N\rangle|^2 = \frac{21}{64\pi}\left(\frac{\hbar}{M\omega_0}\right)^3 N(N+1)(N+2)(N+3)(N+4)$$

$$\sum_{\nu_N, \nu_{N+3}} |\langle \nu_{N+3}| r^3 Y_{30} |\nu_N\rangle|^2 = \frac{7}{64\pi}\left(\frac{\hbar}{M\omega_0}\right)^3 (N+1)(N+2)(N+3)(N+4)(N+5) \tag{6-436}$$ ▲

[26]These states were first observed as "anomalous inelastic peaks" in (pp′) scattering (Cohen, 1957). A number of different interpretations were considered. The evidence in favor of octupole vibrations was systematically presented by Lane and Pendlebury (1960).

where ν_N represents the quantum numbers needed to specify the single-particle states in the shell N. The expression (6-436) includes a factor 4 for the spin-isospin degeneracy. Since the three last shells contribute to the total $\Delta N=3$ strength, one obtains, for large quantum numbers, approximately equal values for the single-particle transition strengths with frequencies ω_0 and $3\omega_0$. Expressing these transition strengths in terms of the classical oscillator sum (6-179), we obtain for the $\tau=0$ transition probabilities

$$B^{(0)}(\tau=0,\lambda=3;\ \Delta N=1) \approx B^{(0)}(\tau=0,\lambda=3;\ \Delta N=3)$$
$$\approx \frac{1}{4}\frac{S(\tau=0,\lambda=3)_{\text{class}}}{\hbar\omega_0} = \frac{147}{32\pi}\frac{\hbar}{M\omega_0}A\langle r^4\rangle \tag{6-437}$$

The octupole density fluctuations give rise to an octupole field with a coupling parameter that can be estimated from the general expression (6-78) for shape oscillations,

$$\kappa(\tau=0,\lambda=3) = -\frac{4\pi}{7}\frac{M\omega_0^2}{A\langle r^4\rangle} \tag{6-438}$$

This interaction partly couples the single-particle transitions with $\Delta N=1$ and $\Delta N=3$ among themselves and partly provides a coupling between the two octupole frequencies. The resulting normal modes can be obtained from the general expression (6-244), which applies to any given single-particle response function. We here give a more elementary derivation, which is an extension of the treatment in Sec. 6-2c to the case of two coupled oscillators.

For the uncoupled octupole oscillators, the restoring force and mass parameters have the values (see Eq. (6-23) and note that the transition strength in this equation refers to a single value of μ and is therefore a seventh of the B values in Eq. (6-437))

$$\begin{aligned} C_1^{(0)} &= \frac{16\pi}{21}\frac{M\omega_0^2}{A\langle r^4\rangle} & C_3^{(0)} &= 3C_1^{(0)} \\ D_1^{(0)} &= \frac{16\pi}{21}\frac{M}{A\langle r^4\rangle} & D_3^{(0)} &= \frac{1}{3}D_1^{(0)} \end{aligned} \tag{6-439}$$

where the subscripts 1 and 3 refer to the value of ΔN. The potential energy (6-24) of the field interaction involves the total deformation, which is the sum of the amplitudes α_1 and α_3 for the separate oscillators,

$$H' = \tfrac{1}{2}\kappa(\alpha_1+\alpha_3)^2 \tag{6-440}$$

The normal modes of octupole oscillation are represented by amplitudes of the form

$$\alpha = c_1\alpha_1 + c_3\alpha_3 \tag{6-441}$$

where c_1 and c_3 are constants, which can be determined from the condition

$$\ddot{\alpha} = -\omega^2\alpha \tag{6-442}$$

The eigenfrequencies ω are the roots of the secular equation

$$-\frac{\kappa}{D_1^{(0)}}\frac{1}{\omega_0^2-\omega^2} - \frac{\kappa}{D_3^{(0)}}\frac{1}{(3\omega_0)^2-\omega^2} = \frac{3}{4}\omega_0^2\left(\frac{1}{\omega_0^2-\omega^2} + \frac{3}{9\omega_0^2-\omega^2}\right) = 1 \tag{6-443}$$

and one obtains

$$\omega(\tau=0,\lambda=3) = \begin{cases} 0 \\ \sqrt{7}\,\omega_0 \end{cases} \tag{6-444}$$

The corresponding oscillator strength is obtained from the mass parameters characterizing the normal modes

$$\hbar\omega B(\tau=0,\lambda=3;\, n=0 \rightarrow n=1) = S(\tau=0,\lambda=3)_{\text{class}} \begin{cases} 3/7 & \omega=0 \\ 4/7 & \omega=\sqrt{7}\,\omega_0 \end{cases} \tag{6-445}$$

in units of the oscillator sum in Eq. (6-437).

The occurrence of an eigenmode with zero frequency (see Eq. (6-444)) implies that the system is on the verge of instability with respect to static octupole deformations. However, this result was obtained in the limit of very large systems ($N \gg 1$). For finite values of A, the terms of relative order N^{-1} in Eq. (6-436) imply that the fraction of the oscillator strength on the $\Delta N=1$ transitions is always less than the limiting value corresponding to Eq. (6-437); in fact, for the $N=0$ shell (α particle), there are no $\Delta N=1$ transitions. Thus, the model remains stable with respect to octupole deformations, but with a collective frequency ω, which decreases faster with increasing A than does the oscillator frequency ($\omega \propto A^{-1/2}$; $\omega_0 \propto A^{-1/3}$). Such a behavior is characteristic of a surface mode; see, for example, the liquid-drop estimate in Sec. 6A-1, which in the absence of Coulomb forces yields a frequency proportional to $A^{-1/2}$. However, the present model cannot be expected to predict correctly the magnitude of the low-frequency mode, since many other effects, including contributions to the surface energy as well as the more detailed features of the shell structure, must be expected to give contributions to the restoring force of order $A^{-1/3}C^{(0)}$.

The $\tau=1$ octupole modes can be analyzed in a manner similar to the $\tau=0$ modes, by considering the field

$$F = \sum_k (r^3 Y_{30}\tau_z)_k \tag{6-446}$$

The C and D parameters of the uncoupled independent-particle excitations are the same as for the $\tau=0$ modes (see Eq. (6-439)), and the coupling

▼ constant (6-127) leads to the frequencies

$$\omega(\tau=1,\lambda=3)=\begin{cases}1.55\,\omega_0\\ 4.76\,\omega_0\end{cases} \tag{6-447}$$

and the oscillator strengths

$$\hbar\omega B(\tau=1,\lambda=3;n=0\rightarrow n=1)=S(\tau=1,\lambda=3)_{\text{class}}\begin{cases}0.032 & \omega=1.55\,\omega_0\\ 0.968 & \omega=4.76\,\omega_0\end{cases} \tag{6-448}$$

with $S(\tau=1,\lambda=3)_{\text{class}}$ given by Eq. (6-179).

Systematics of low-frequency mode

The observed low-energy octupole excitations (see Fig. 6-40) correspond qualitatively to the $\tau=0$ mode associated with the $\Delta N=1$ particle excitations. However, the deviation of the single-particle excitation spectrum from the schematic form assumed above leads to significant variations in the properties of this mode during the filling of the major shells.

For closed-shell nuclei, the subshell structure implies that the $\Delta N=1$ particle excitations have a spread in energy that is comparable to $\hbar\omega_0$. (For example, in ^{208}Pb, the $\Delta N=1$ excitation energies for transitions without spin flip ($\Delta j=\Delta l$) range from about 4 to 9 MeV; see Fig. 3-3, Vol. I, p. 325). Hence, one expects that only a fraction of the low-frequency oscillator strength will be found in the observed collective mode. From the data in Table 6-14 and the assumption $B(\tau=0,\lambda=3)\approx(A/Ze)^2B(E3)$, it follows that the observed $\tau=0$ oscillator strength amounts to 10%, 13%, and 21% of the $\tau=0$ sum rule unit, for ^{16}O, ^{40}Ca, and ^{208}Pb, respectively. The schematic model considered above implies about 40% of $S(\tau=0)_{\text{class}}$ in the low-frequency mode, but this number should be corrected for the finite value of N (see Eq. (6-436)), which leads to the estimates 25%, 33%, and 39% for ^{16}O, ^{40}Ca, and ^{208}Pb, respectively. Thus, in all three cases, about half of the expected low-frequency oscillator strength is concentrated in the observed collective mode, which is consistent with the fact that the observed octupole frequencies are displaced with respect to the $\Delta N=1$ single-particle excitation frequencies by an amount of similar magnitude as the spread in the $\Delta N=1$ frequencies.

For configurations with particles in unfilled shells, the spin-orbit interaction implies that for $A\gtrsim 60$, there may occur strong octupole transitions between orbits within a major shell ($2p_{3/2}\rightarrow 1g_{9/2}$ for the shell 28–50; $2d_{5/2}\rightarrow 1h_{11/2}$ for the shell 50–82; $2f_{7/2}\rightarrow 1i_{13/2}$ for the shell 82–126, and $2g_{9/2}\rightarrow 1j_{15/2}$ for the shell above 126; an example of the octupole response
▲ function, corresponding to $Z=46$, $N=60$, is illustrated in Fig. 6-17b, p. 467).

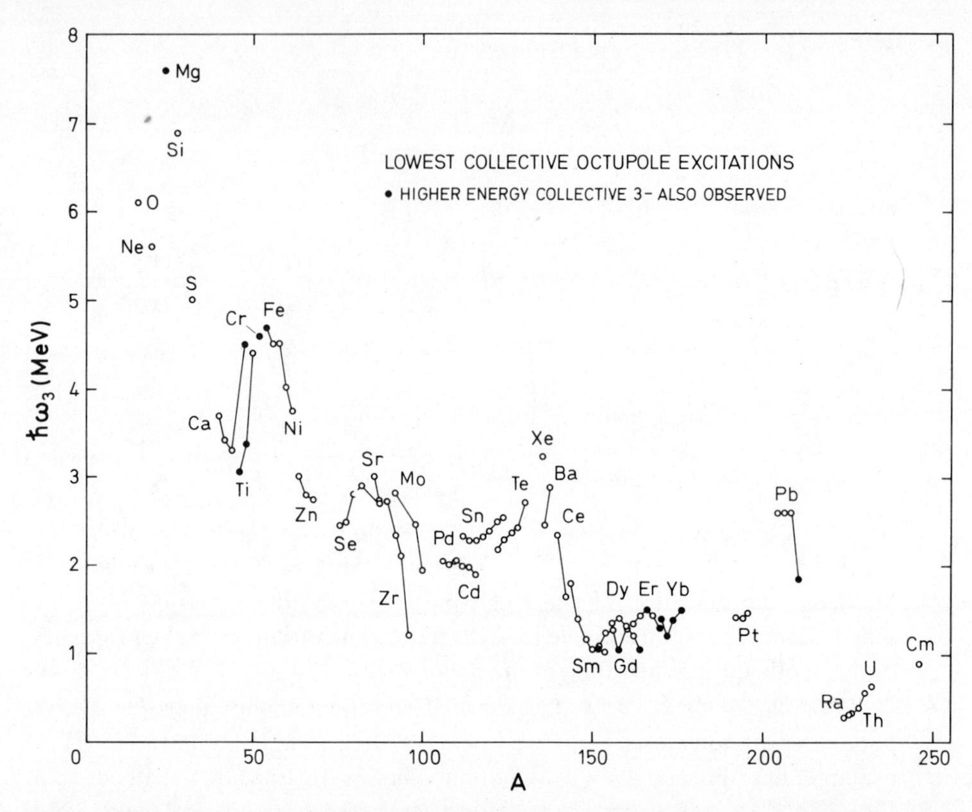

Figure 6-40 Systematics of collective octupole vibrational energies. The figure gives the observed excitation energies of the low-energy collective $I\pi = 3-$ states. Solid circles indicate nuclei in which higher-energy octupole excitations have been identified with a strength not less than half the strength of the excitation that is plotted. The data are from the compilations by Lederer *et al.* (1967) and Bernstein (1970).

▼ The occurrence of these strong low-frequency octupole transitions can be seen as a systematic feature of single-particle motion in a potential with a well-defined surface; these transitions are associated with the occurrence of classical periodic orbits with triangular symmetry (see the discussion on pp. 591 ff.). The observed variation in the octupole frequencies in Fig. 6-40 can be qualitatively understood in terms of this feature of the nuclear shell structure. Thus, for example, in the region from Zr to Ba, the frequency as a function of neutron number decreases as one adds particles in the $d_{5/2}$ orbit at the beginning of the shell, and then increases as the $h_{11/2}$ orbit becomes filled at the end of the shell; the same transition, $d_{5/2} \rightarrow h_{11/2}$ for the protons, accounts for the decrease of the octupole frequency with increasing Z in the region beyond Sn. For neutron numbers $N > 82$, the filling of the $f_{7/2}$ orbit is seen to imply a strong decrease in the octupole frequency for Ce and Sm isotopes. For spherical shape, a minimum in the octupole frequencies would be
▲ expected for $Z \approx 64$ and N between 90 and 100 (see, for example, the level

Nucleus	$\hbar\omega_3$ MeV	$B(E3;0\to 3)$ $e^2\text{fm}^6$	$\frac{B(E3)}{B_{sp}(E3)}$	$(\frac{\hbar\omega_3}{2C_3})^{1/2}$	$\frac{\hbar\omega B(E3;0\to 3)}{S(E3)_{\text{class}}}$
^{16}O	6.13	1.5×10^3	14	0.28	0.05
^{40}Ca	3.73	1.7×10^4	26	0.15	0.06
^{60}Ni	4.05	2.8×10^4	19	0.09	0.06
^{112}Cd	1.97	1.0×10^5	20	0.057	0.03
^{152}Sm	$1.04(\nu=0)$	1.2×10^5	12	0.087	0.011
	$1.58(\nu=1)$	0.7×10^5	7	0.047	0.009
^{208}Pb	2.61	$7\ \times 10^5$	39	0.045	0.08
^{238}U	$0.73(\nu=0)$	$5\ \times 10^5$	21	0.078	0.013

Table 6-14 Transition strength of collective octupole excitations. The listed matrix elements have been obtained from $E3$-transition probabilities except for ^{152}Sm, for which the value has been derived from an analysis of the cross section for inelastic deuteron scattering. The observed $B(E3)$ values are compared in column four with the single-particle unit $B_{sp}(E3)$ $(=7B_W(E3))=(7/16\pi)e^2R^6=0.42A^2e^2\text{fm}^6$. Column five gives the amplitude of the octupole oscillation (with the normalization (6-63)), as deduced from the $B(E3)$ values (see Eq. (6-65) for spherical nuclei and Eq. (6-91) for deformed nuclei). Column six gives the observed oscillator strength in units of the classical $E3$-oscillator sum rule (Eq. (6-177)); in the sum rule, the value of $\langle r^4\rangle$ has been estimated for a density distribution of the form (2-62) with the parameters (2-69) and (2-70). The data are from: ^{16}O (T. K. Alexander and K. W. Allen, *Can. J. Phys.* **43**, 1563, 1965); ^{40}Ca (compilation by Skorka *et al.*, 1967); ^{60}Ni (M. A. Duguay, C. K. Bockelman, T. H. Curtis, and R. A. Eisenstein, *Phys. Rev.* **163**, 1259, 1967); ^{112}Cd (F. K. McGowan, R. L. Robinson, P. H. Stelson, and J. L. C. Ford, *Nuclear Phys.* **66**, 97, 1965); ^{152}Sm (E. Veje, B. Elbek, B. Herskind, and M. C. Olesen, *Nuclear Phys.* **A109**, 489, 1968); ^{208}Pb (J. F. Ziegler and G. A. Peterson, *Phys. Rev.* **165**, 1337, 1968); ^{238}U (Th. W. Elze and J. R. Huizenga, *Nuclear Phys.* **A187**, 545, 1972).

▼ schemes in Figs. 5-2 and 5-3, pp. 222 and 223); in this region, the spherical shape may become unstable with respect to octupole deformations. However, the onset of static quadrupole deformations at $N=90$ prevents the direct observation of this instability. In the region beyond ^{208}Pb, the addition of neutrons in the $g_{9/2}$ orbit may be expected to strongly reduce the octupole frequency, possibly leading to instability of the spherical shape. The relevant configurations around $N\approx 130$ are rather difficult to study due to the short α-decay lifetimes in this region, but evidence for the expected strong octupole effects is provided by the low frequency of the odd-parity excitations at the ▲ beginning of the deformed region ($A\gtrsim 222$; see Fig. 6-40)[27] and by the strong

[27]These $K\pi=0-$ excitations with remarkably low frequency were the first octupole modes to be identified in the nuclear spectra. They were discovered as a systematic feature in the α-decay fine structure (Stephens *et. al.*, 1955), and the interpretation as octupole modes was given by R. F. Christy (private communication, 1954).

coupling of the extra neutron to the octupole vibrations in ^{209}Pb (see the discussion on p. 564).

As indicated in Fig. 6-40, the low-frequency octupole strength is observed in a number of cases to be divided between two or more excitations with comparable intensity. Such a multiplicity of octupole modes may arise as a result of the coupling to quadrupole deformations (see pp. 577ff.), and can also occur in the nuclei that exhibit an especially low-frequency component in the single-particle octupole response. (See, for example, the occurrence of the strong octupole excitations in the spectrum of ^{210}Pb (Ellegaard *et al.*, 1971), reflecting the strong coupling of the octupole mode to the $g_{9/2}\rightarrow j_{15/2}$ single-particle excitation, as discussed in connection with the spectrum of ^{209}Pb on p. 564.)

One-particle transfer to octupole excitation in ^{208}Pb

Evidence on the structure of the octupole vibrations in terms of the particle-hole excitations can be obtained from the study of one-particle transfer reactions or isobaric analog resonances. For example, the ^{209}Bi (d ^{3}He) reaction is found to populate the $I\pi=3-$ state in ^{208}Pb with an intensity of 4.5×10^{-2} of that observed for the ^{208}Pb (d^{3}He) reaction leading to the $d_{3/2}^{-1}$ configuration of ^{207}Tl (McClatchie *et al.*, 1970). Assuming that the process involves pickup of a $d_{3/2}$ proton, we obtain the reduced matrix element for the particle annihilation operator $a(\bar{d}_{3/2})$ (see Eq. (3E-9))

$$\frac{d\sigma(h_{9/2}\rightarrow 3-)}{d\sigma(0\rightarrow d_{3/2}^{-1})}=\frac{1}{10}\frac{\langle 3-\|a(\bar{d}_{3/2})\|h_{9/2}\rangle^2}{\langle d_{3/2}^{-1}\|a(\bar{d}_{3/2})\|0\rangle^2}$$

$$=\frac{1}{40}\langle 3-\|a(\bar{d}_{3/2})\|h_{9/2}\rangle^2\approx 0.045 \qquad (6\text{-}449)$$

A theoretical estimate of this matrix element can be obtained from the particle-vibration coupling, in analogy to the amplitude described by the first diagram in Fig. 6-34, p. 537. (In the absence of pair correlations, the second diagram in Fig. 6-34 does not contribute, if we neglect the small contribution from the $d_{3/2}$ states in higher shells.) Thus, we obtain (see Eqs. (6-210) and (6-211))

$$\langle 3-\|a(\bar{j})\|j_1\rangle=(2j_1+1)^{1/2}\frac{h(j_1 j_2\lambda)}{\hbar\omega_3-E(j_2^{-1}j_1)} \qquad (6\text{-}450)$$

where the matrix element h for the particle-vibration coupling is given by Eq. (6-209). With the radial matrix element

$$\langle h_{9/2}|k(r)|d_{3/2}\rangle\approx 51 \quad \text{MeV} \qquad (6\text{-}451)$$

▼ as obtained from wave functions in a Woods-Saxon potential with the parameters used in Fig. 3-3, Vol. I, p. 325, and the zero-point amplitude

$$\left(\frac{\hbar\omega_3}{2C_3}\right)^{1/2} = 0.045 \tag{6-452}$$

as obtained from the $B(E3;\, 0\rightarrow 3)$ value in Table 6-14 together with Eq. (6-65), one finds

$$h(\mathrm{h}_{9/2}, \mathrm{d}_{3/2}3) \approx -0.75 \quad \mathrm{MeV} \tag{6-453}$$

The energy $E(\mathrm{d}_{3/2}^{-1}\mathrm{h}_{9/2})$ in Eq. (6-450) is that of the particle-hole excitation, which according to the observed binding energies in Fig. 3-3, Vol. I, p. 325 would be estimated as $\varepsilon(\mathrm{h}_{9/2}) - \varepsilon(\mathrm{d}_{3/2}) \approx 4.6$ MeV. This energy, however, is somewhat reduced by the Coulomb interaction, since the hole energies refer to the separation of a proton from the nucleus $Z=82$, while the particle energies refer to separation from the nucleus $Z=83$. Hence, in the proton particle-hole interaction, one must include an attractive Coulomb force; the resulting energy shift is approximately independent of the particle configuration and may be estimated by assuming constant charge distributions (see Eq. (2-19))

$$\delta E_{\mathrm{Coul}}(\mathrm{ph}) = -\frac{6}{5}\frac{e^2}{R} \approx -1.4A^{-1/3} \quad \mathrm{MeV} \tag{6-454}$$

which amounts to -0.2 MeV for $^{208}\mathrm{Pb}$. With $E(\mathrm{d}_{3/2}^{-1}\mathrm{h}_{9/2}) = 4.4$ MeV and $\hbar\omega_3 = 2.6$ MeV (see Table 6-14), one obtains from Eqs. (6-450) and (6-453)

$$\langle 3- \| a(\bar{\mathrm{d}}_{3/2}) \| \mathrm{h}_{9/2} \rangle = -1.3 \tag{6-455}$$

in agreement with the observed value (see Eq. (6-449)).

The reduced matrix element (6-455) can be viewed as expressing the amplitude for the particle-hole component $(\mathrm{d}_{3/2}^{-1}\mathrm{h}_{9/2})$ in the phonon wave function (see the diagram in Fig. 6-13, p. 442 and Eqs. (6-210) and (6-211)),

$$\begin{aligned} X(\mathrm{d}_{3/2}^{-1}\mathrm{h}_{9/2}) &= -\left(\frac{10}{7}\right)^{1/2} \frac{h(\mathrm{h}_{9/2}, \mathrm{d}_{3/2}3)}{\hbar\omega_3 - E(\mathrm{d}_{3/2}^{-1}\mathrm{h}_{9/2})} \\ &\approx -0.50 \end{aligned} \tag{6-456}$$

The $^{209}\mathrm{Bi}(\mathrm{d}\ ^3\mathrm{He})$ reaction can also populate the octupole state in $^{208}\mathrm{Pb}$ by pickup of a $\mathrm{d}_{5/2}$ or $\mathrm{g}_{7/2}$ proton, though these cross sections are expected to be appreciably smaller than for $\mathrm{d}_{3/2}$. With an estimate similar to that given by ▲ Eq. (6-456), one obtains $X(\mathrm{d}_{5/2}^{-1}\mathrm{h}_{9/2}) \approx 0.11$ and $X(\mathrm{g}_{7/2}^{-1}\mathrm{h}_{9/2}) \approx -0.17$.

▼ *Effective charges (Table 6-15)*

The low frequency of the collective octupole mode, as compared with the characteristic frequencies of the constituent particle-hole excitations, implies a rather large value of the isoscalar octupole polarizability. Experimental evidence concerning the effective charge for single-particle octupole transitions is available for nuclei in the region of ^{208}Pb and is listed in Table 6-15. The empirical values of e_{eff} are obtained by dividing the observed matrix element with the single-particle value (3C-33), employing the radial matrix elements listed in column four. The 2.61 MeV octupole state of ^{208}Pb contributes the major part of the effective charge; the value listed in column seven has been obtained from Eq. (6-218) with the value of $B(E3;\, 0\rightarrow 3)$ in Table 6-14 and the coupling matrix element $\langle j_2|k(r)|j_1\rangle$, listed in column six. The radial matrix elements of r^3 and $k(r) = R_0\partial V/\partial r$ are based on wave functions in a Woods-Saxon potential with standard parameters (see Vol. I, pp. 239 and 326). Additional contributions to the polarization charge arise from the coupling to higher-lying isovector as well as isoscalar octupole excitations; these terms, each amounting to a few tenths of a unit, are discussed in the fine print below, and their combined contribution is given in column eight. In the approximation considered, based on the first-order perturbation treatment of the particle-vibration coupling, the total effective charge is the sum of the bare charge and the contributions to the polarization charge in columns seven and eight.

For some of the low-lying single-particle configurations in ^{209}Pb and ^{209}Bi, the coupling to the octupole mode is so strong that significant contributions to the polarization charge may arise from higher-order effects of the particle-vibration coupling. Thus, for ^{209}Pb, the coupling matrix element between the $j_{15/2}$ and $(g_{9/2}3-)$ configurations, as obtained from Eq. (6-209) together with the radial matrix element $\langle j_2|k|j_1\rangle$ in Table 6-15 and the zero-point amplitude (6-452), has the value $h(j_{15/2}, g_{9/2}3-) = -0.88$ MeV, which is comparable with the energy difference between the two configurations. A diagonalization of the coupling in the space of the particle and particle plus phonon states yields the renormalized single-particle states

$$
\begin{aligned}
|\hat{g}_{9/2}\rangle &= 0.97|g_{9/2}\rangle + 0.24|(j_{15/2}3-)9/2\rangle \\
|\hat{j}_{15/2}\rangle &= 0.85|j_{15/2}\rangle + 0.52|(g_{9/2}3-)15/2\rangle
\end{aligned}
\qquad (6\text{-}457)
$$

In obtaining the states (6-457), the energy difference between the unperturbed $j_{15/2}$ and $g_{9/2}$ energies has been taken as $\Delta\varepsilon = 1.7$ MeV, so as to yield the observed difference of 1.4 MeV for the renormalized states. The value of e_{eff} in the last column of Table 6-15 is calculated from the states (6-457), with the single-particle charge δe_{pol} (high freq.) and with the vibrational $E3$-matrix element in Table 6-14. This value of e_{eff} is seen to be reduced by about 30%, in comparison with the first-order perturbation result. ▲

Nucleus	Transition	$B(E3)$ $10^4 e^2\,\text{fm}^6$	$\langle j_2 \lvert r^3 \rvert j_1 \rangle$ fm^3	$(e_{\text{eff}})_{\text{exp}}$ e	$\langle j_2 \lvert k \rvert j_1 \rangle$ MeV	$\delta e_{\text{pol}}/e$		$(e_{\text{eff}})_{\text{th}}$ e
						2.6 MeV	High freq.	
^{209}Pb	$j_{15/2} \rightarrow g_{9/2}$ ($\Delta E = 1.42$ MeV)	7 ± 2	254	2.8 ± 0.5	53	3.3	0.6	2.8
^{209}Bi	$i_{13/2} \rightarrow h_{9/2}$ ($\Delta E = 1.61$ MeV)	1.5 ± 0.5	238	6 ± 1	68	5.0	0.2	5.6

Table 6-15 Effective octupole charges in ^{209}Pb and ^{209}Bi. The experimental $B(E3)$ values in column three are from: ^{209}Pb (C. Ellegaard, J. Kantele, and P. Vedelsby, *Phys. Letters* **25B**, 512, 1967); ^{209}Bi (J. W. Hertel, D. G. Fleming, J. P. Schiffer and H. E. Gove, *Phys. Rev. Letters* **23**, 488, 1969, and R. A. Broglia, J. S. Lilley, R. Perazzo, and W. R. Phillips, *Phys. Rev.* **C1**, 1508, 1970). In the latter case, the listed value represents a weighted average of the two experiments, each corrected by a factor adjusted so that the observed total $E3$-excitation probability for the $(h_{9/2}3-)$ septuplet agrees with the assumed value $7 \times 10^5 e^2\text{fm}^6$ for the $3-$ excitation of ^{208}Pb.

▼ In ^{209}Bi, the coupling between the two configurations in Table 6-15 is much weaker ($h(\mathrm{i}_{13/2}, \mathrm{h}_{9/2}3-) = -0.25$ MeV) on account of the spin flip in the single-particle transition, and the higher-order effects of this coupling may be neglected. Somewhat larger corrections to the $E3$-matrix element in ^{209}Bi arise from the coupling of the $\mathrm{i}_{13/2}$ state to the $(\mathrm{f}_{7/2}3-)$ configuration, which has the large matrix element $h(\mathrm{i}_{13/2}, \mathrm{f}_{7/2}3-) = -1.1$ MeV. A diagonalization of this coupling in analogy to the treatment for ^{209}Pb gives the renormalized $\mathrm{i}_{13/2}$ state

$$|\hat{\mathrm{i}}_{13/2}\rangle = 0.91|\mathrm{i}_{13/2}\rangle + 0.41|(\mathrm{f}_{7/2}3-)13/2\rangle \tag{6-458}$$

The value of $(e_{\mathrm{eff}})_{\mathrm{th}}$ for ^{209}Bi in Table 6-15 has been obtained by multiplying the perturbation value of 6.2 by the factor 0.91, which represents the amplitude of the $\mathrm{i}_{13/2}$ component in the renormalized state (6-458).

The rather good agreement between the observed and estimated $E3$-matrix elements in Table 6-15 lends support to the assumed coupling of the single-particle motion to the 2.61 MeV octupole excitation in ^{208}Pb. In the case of the $\mathrm{i}_{13/2} \rightarrow \mathrm{h}_{9/2}$ transition in ^{209}Bi, the coupling matrix element is rather small and therefore sensitive to the possible occurrence of spin-dependent fields associated with the octupole excitation; the empirical value of e_{eff} may therefore suggest that the spin-dependent fields are rather weak.

The high-frequency contributions to the effective octupole charge may be expressed in terms of the isoscalar and isovector polarizability coefficients by a relation analogous to Eq. (6-386), which refers to the $E2$ effective charge. The high-frequency excitations that are expected to contribute involve:

(a) the $\Delta N = 3$ mode with $\tau = 0$. Using the expression (6-216) for χ together with the estimate (6-438) for the coupling strength and the value of the restoring force derived from Eqs. (6-444) and (6-445), we obtain

$$\delta e_{\mathrm{pol}}^{\mathrm{std}} \approx \frac{12}{49} \frac{Ze}{A} \left(1 - 0.65 \frac{N-Z}{A} \tau_z\right) \tag{6-459}$$

(b) the additional $\Delta N = 1$, $\tau = 0$ modes implied by the fact that the 2.6 MeV state in ^{208}Pb accounts for only about half of the expected $\Delta N = 1$, $\tau = 0$ oscillator strength (see p. 559). Since the remainder of this oscillator strength is expected in the region of $\hbar\omega_0 \approx 7$ MeV, the resulting polarizability, which is inversely proportional to the square of the frequency for constant oscillator strength, will be about an order of magnitude smaller than that contributed by the 2.6 MeV mode. A more detailed estimate may be obtained from the expression (6-243) for the response function. Using the unperturbed single-particle energies in Fig. 3-2f and a coupling constant κ that gives a lowest collective mode with $\hbar\omega = 2.6$ MeV, one finds

$$\delta e_{\mathrm{pol}}^{\mathrm{std}} \approx (0.5 - 0.1\tau_z)e \tag{6-460}$$

(c) the $\tau = 1$ modes with $\Delta N = 1$ and 3. From the estimates (6-127), (6-447), and (6-448), we obtain

$$\delta e_{\mathrm{pol}}^{\mathrm{std}} \approx 0.4\left(\tau_z - \frac{N-Z}{A}\right)e \tag{6-461}$$ ▲

▼ For ^{208}Pb, the sum of these three contributions gives $\delta e_{\mathrm{pol}}^{\mathrm{std}} \approx (0.5+0.3\tau_z)e$. This result is a standard value based on average radial matrix elements for the one-particle states. The values of δe_{pol} in column eight of Table 6-15 include the additional factor $\frac{1}{2}R^3\langle j_2|r^3|j_1\rangle^{-1}$; see the analogous equation (6-387).

Anharmonic effects (Fig. 6-41)

Although, at present, there is only little evidence concerning anharmonic effects in the octupole mode, these effects are potentially of considerable interest. In contrast to the situation for the low-frequency quadrupole mode, the anharmonic effects in the low-frequency octupole mode are in many cases expected to be rather small and thus amenable to a perturbation expansion starting from the harmonic approximation.

Phonon interaction arising from octupole coupling

The lowest-order anharmonic terms in the octupole energy spectrum are the phonon-phonon interactions, represented by fourth-order diagrams like those in Fig. 6-41. As a first step, we consider diagram (a) in Fig. 6-41 for particle motion in a harmonic oscillator potential. For the case in which all four octupole phonons are in the state $M=3$, each of the four vertices must involve the same particle-hole pair (associated with a particle excitation with ▲ $\Delta n_{+1}=2$, $\Delta n_{-1}=-1$, and $\Delta n_0=0$), if we consider only the $\Delta N=1$ excitations.

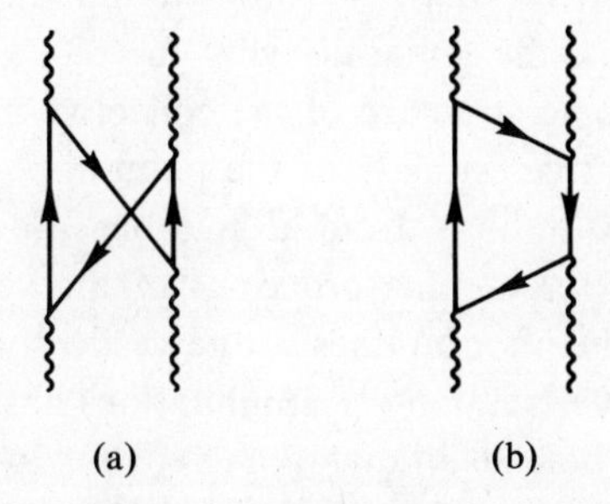

Figure 6-41 Diagrams illustrating phonon-phonon interaction in the octupole mode.

▼ The diagram, therefore, gives an energy shift $X_i^4(\hbar\omega_0-\hbar\omega)$, where X_i is the amplitude of the configuration i in the octupole mode, whose magnitude can be estimated from the normalization condition (see Eqs. (6-249a) and (6-253))

$$\sum_i (X_i^2-Y_i^2)=\frac{4\omega_0\omega}{(\omega_0+\omega)^2}\sum_i X_i^2=1 \qquad (6\text{-}462)$$

▲ Summing over the different configurations, we obtain a contribution to the

▼ fourth-order energy shift of order

$$\delta E^{(4)} \sim \Omega^{-1}(\hbar\omega_0 - \hbar\omega)\frac{(\omega_0+\omega)^4}{(4\omega_0\omega)^2} \tag{6-463}$$

where Ω is the number of particles in the last filled shell, corresponding to the number of configurations i in the sum (6-462).

The interaction (6-463) can be seen to reflect the role of the exclusion principle in preventing the second phonon from exploiting the configurations that appear simultaneously in the microscopic structure of the first phonon. The interaction energy $(\hbar\omega - \hbar\omega_0)$ in the phonon is therefore reduced by a term of relative order Ω^{-1}. The last factor in Eq. (6-463) is connected with the ground-state correlations, which imply that the zero-point amplitude $(\alpha_3)_0$ increases in proportion to $\omega^{-1/2}$; in the adiabatic limit, $\omega \ll \omega_0$, the ω dependence of the expression (6-463) is contained in the factor $(\alpha_3)_0^4$, as is characteristic of a fourth-order interaction.

The quantity Ω^{-1}, which governs the magnitude of the anharmonicity, may also be recognized as characterizing the corrections to the boson commutation relations for the operators $c^{(0)}$ and $(c^{(0)})^\dagger$ associated with the unperturbed motion (see in this connection the discussion on p. 335). The resulting anharmonicity, which is of relative order $A^{-2/3}$, is large compared with the value for the liquid-drop model; in a macroscopic description of a shape oscillation, the anharmonic terms in the Hamiltonian are of relative order $(\alpha_3)_0^2$, assuming an amplitude with the normalization (6A-1) (or (6-63)), and hence of relative order $A^{-4/3}$ for the mode considered (see, for example, Eq. (6-437)). Thus, the anharmonicity (6-463) is a quantal effect associated with the microscopic structure of the collective motion.

In a quantitative estimate of the phonon-phonon interactions, one must include the contributions from a number of additional diagrams, partly resulting from different time orderings of the vertices, partly from different ways of joining the phonon lines to the vertices, and partly from diagrams in which particle-hole creation (or annihilation) has been replaced by scattering of a particle or a hole, as in diagram (b) of Fig. 6-41. The latter diagram gives a contribution of the same order of magnitude as (6-463), but of opposite sign.

As regards order of magnitude, the total phonon-phonon interaction energy is expected to remain as given by Eq. (6-463). For ^{208}Pb, with $\Omega \approx 75$, $\hbar\omega = 2.6$ MeV, and $\hbar\omega_0 \approx 7$ MeV, this estimate implies $\delta E^{(4)} \sim 0.1$ MeV. (A corresponding estimate for the low-frequency quadrupole mode in spherical nuclei yields appreciably larger values, partly as a result of the smaller number of particles involved in the quadrupole mode, and partly as a result of the smallness of the quadrupole frequency, as compared with that of the unperturbed motion $(\omega \ll \omega^{(0)})$.) ▲

▼ An evaluation of the fourth-order interaction for particle motion in a harmonic oscillator potential leads to a cancellation between the different diagrams enumerated above, so that the resulting anharmonicity becomes of order Ω^{-1} relative to the term (6-463) and thus comparable to the liquid-drop value. The very small anharmonicity for the harmonic oscillator model is connected with the fact that the single-particle response function does not depend on the degree of filling of the shells, but varies smoothly with A. However, the octupole mode in the nucleus exhibits a considerable dependence on the shell structure (which can be connected with the triangular single-particle orbits, see p. 591), and therefore the magnitude of the anharmonicity is expected to remain of the order of the estimate (6-463).

Quadrupole moment of octupole phonon

The static moment of the $n_3=1$ state is illustrated by third-order diagrams similar to those considered in the evaluation of the quadrupole moment for the $n_2=1$ state (see Fig. 6-27, p. 525). For particle motion in a harmonic oscillator potential, the term corresponding to diagram (i) in Fig. 6-27a can be simply evaluated, since each of the particle-hole states i, with $M=3$ (and $\Delta n_{+1}=2$, $\Delta n_{-1}=-1$), has the value $-\hbar/(M\omega_0)$ for the $\tau=0$ quadrupole moment. Thus, from Eqs. (6-400) and (6-462), we obtain

$$Q_1=-\frac{(\omega_0+\omega)^2}{4\omega_0\omega}\,\frac{\hbar}{M\omega_0} \tag{6-464}$$

The terms corresponding to the additional diagrams in Fig. 6-27a imply contributions that increase the quadrupole moment by less than a factor of 2 (the diagrams corresponding to those in Fig. 6-27a, (iii) to (vi), involve the $\Delta N=2$ matrix elements of the quadrupole operator and are reduced by the resulting large energy denominators). The bare isoscalar quadrupole moment of the particles in the octupole excitation is enhanced by the polarization effect (see the diagrams in Fig. 6-27b) by a factor $(1+\chi(\tau=0,\lambda=2,\Delta E=0))$, which for ^{208}Pb is about 3 (see Table 6-8, p. 516); multiplying by a factor Z/A to obtain the electric quadrupole moment, one thus finds $Q_{el}\sim-10\ \text{fm}^2$ for the $3-$ state of ^{208}Pb.

The estimated magnitude of the quadrupole moment in the $n_3=1$ state is only of order $A^{-1/3}Q_{sp}$, due to the near cancellation of the contributions from particles and holes in the collective state. The estimated value is comparable to the quadrupole moment that arises in the liquid-drop model as a second-order effect in the amplitude of the octupole deformation yielding $Q\sim-(\alpha_3)_0^2 A R_0^2\sim-A^{-1/3}Q_{sp}$, for a state with one octupole phonon.

Experimental evidence for a quadrupole moment of order $-100\ \text{fm}^2$ for the 2.6 MeV octupole excitation in ^{208}Pb has been reported (Barnett and Phillips, 1969). Such a large value would appear difficult to understand in terms of the microscopic structure of the octupole mode underlying the present discussion; see also the discussion on pp. 570 and 576 of the interaction effects that would be implied by the large reported moment. ▲

Quadrupole interaction of octupole phonons

A large static quadrupole moment of the 3− state would imply major contributions to the effective interaction between octupole phonons (Blomqvist, 1970). If we normalize the amplitude $\alpha_{2\mu}$ of the quadrupole deformation to the total quadrupole moment of all the particles in the nucleus ($\alpha_{2\mu} = \langle \Sigma r^2 Y_{2\mu} \rangle$), the quadrupole interaction between two octupole phonons can be expressed in the form $5^{1/2}\kappa(\alpha_2 F_2)_0$, where κ is given by Eq. (6-368), while $\alpha_{2\mu}$ is the quadrupole deformation associated with the first phonon, and $F_{2\mu} = r^2 Y_{2\mu}$ the bare moment of the particles involved in the second phonon. This bare moment is smaller than $\alpha_{2\mu}$ by the factor $(1+\chi)$, where χ is the $\lambda = 2$ static isoscalar polarizability. The expectation value of $\alpha_{2\mu}$ in the phonon state is related to the quadrupole moment by

$$\begin{aligned} Q_{\tau=0}(n_3=1) &= \left(\frac{16\pi}{35}\right)^{1/2} \langle 3320|33\rangle \langle n_3=1\|\alpha_2\|n_3=1\rangle \\ &\approx \frac{A}{Z} Q_{\mathrm{el}}(n_3=1) \end{aligned} \tag{6-465}$$

and the contribution of the quadrupole coupling to the phonon-phonon interaction is therefore

$$\begin{aligned} \delta E(n_3=2,I) &= \frac{21}{4\pi}\frac{\kappa}{1+\chi}\left(Q_{\tau=0}(n_3=1)\right)^2 \begin{Bmatrix} 3 & 3 & 2 \\ 3 & 3 & I \end{Bmatrix} \\ &\approx -1.6\left(Q_{\mathrm{el}}(n_3=1)\right)^2 \begin{Bmatrix} 3 & 3 & 2 \\ 3 & 3 & I \end{Bmatrix} \mathrm{keV\ fm}^{-4} \end{aligned} \tag{6-466}$$

for $A = 208$, $Z = 82$, $\chi \approx 2$. An electric quadrupole moment of 10^2 fm^2 for the 3− state of ^{208}Pb would imply energy splittings of several MeV, within the $n_3 = 2$ multiplet.

The estimate of $Q(n_3=1)$ given on p. 569 refers to the closed-shell nucleus ^{208}Pb, but considerably larger values may be expected for nuclei with a number of particles outside closed shells, for which the quadrupole polarizability may become an order of magnitude larger (see Fig. 6-28, p. 529). For deformed nuclei, the low-frequency octupole mode is modified in an essential manner as a result of the quadrupole-octupole interaction (see pp. 577 ff.).

The ($h_{9/2}$ 3−) septuplet in ^{209}Bi (Figs. 6-42 and 6-43; Table 6-16)

The addition of an octupole quantum to the $h_{9/2}$ ground state of ^{209}Bi is expected to give rise to a septuplet of states $(h_{9/2}3-)I$ with $I = 3/2, 5/2, \ldots, 15/2$. High-resolution studies of inelastic scattering and the resulting γ-decay spectra have resolved seven close-lying states that can be identified

▼ with the expected multiplet on the basis of their large cross sections for octupole excitation (see, for example, Fig. 6-42, which compares the spectrum of inelastically scattered deuterons from ^{208}Pb and ^{209}Bi).[28]

The observed small splitting of the multiplet components implies a weak coupling between the odd proton and the octupole quantum; thus, one expects that the transition probabilities for exciting the individual components are approximately proportional to the statistical weights (see Eq. (6-86))

$$B(E3;\, n_\lambda=0,j\rightarrow(n_\lambda=1,j)I)=\frac{2I+1}{(2\lambda+1)(2j+1)}B(E3;\, n_\lambda=0\rightarrow n_\lambda=1) \tag{6-467}$$

and a similar relation for the cross sections for inelastic scattering reactions. A comparison of experimental cross sections with the weak coupling relation is shown in Table 6-16. It is seen that the agreement is rather good, except for the weak $I=3/2$ level, which receives only about two thirds of the estimated intensity. (The 2.958 MeV level in Table 6-16 arises from the proton-hole, proton-pair configuration $(d_{3/2}^{-1}0+)3/2+$; see the discussion below.)

A number of different coupling effects give rise to small deviations from the description in terms of independent modes of excitation; major effects may arise from couplings to near-lying configurations. Of near-lying one-particle configurations, only the $i_{13/2}$ state has an appropriate spin and parity for coupling to a state of the $(h_{9/2}3-)$ septuplet. However, the matrix element for this coupling is strongly reduced by the spin flip ($h(i_{13/2}, h_{9/2}3-)=-0.25$ MeV; see p. 566; thus, the admixture of the $i_{13/2}$ and $(h_{9/2}3-)13/2$ states is only about 6%, and the associated energy shift of the $(h_{9/2}3-)13/2$ state is $\approx +60$ keV. (Experimental evidence for this admixture is provided by the one-particle transfer process ^{208}Pb(^{3}He d), which leads to an excitation at 2.60 MeV with an intensity that is about 10% of that leading to the $i_{13/2}$ level at 1.6 MeV (Ellegaard and Vedelsby, 1968); however, the 2.60 MeV excitation represents the unresolved $I=11/2$ and $I=13/2$ components of the multiplet, and the distribution of the one-particle transfer intensity between these two components has not been established.)

Additional low-lying configurations arise from the superposition of a ▲ proton hole and a pair quantum ($\alpha=2$) involving two protons corresponding

---

[28]The occurrence of a weak coupling $(h_{9/2}3-)$ multiplet in ^{209}Bi could be inferred from the observation of an octupole excitation of strength and energy similar to that of ^{208}Pb (see Alster, 1966). High-resolution (pp′) studies resolved the multiplet and led to spin assignments on the basis of the $(2I+1)$ rule (Hafele and Woods, 1966). The discovery of the weak-coupling multiplet stimulated theoretical developments concerning the coupling between particle and octupole motion as well as couplings involving other elementary modes of excitation of the ^{208}Pb core (Hamamoto, 1969 and 1970; Bès and Broglia, 1971; Broglia et al., 1971; see also Mottelson, 1968, for a discussion of the subject as treated in earlier versions of the present volume).

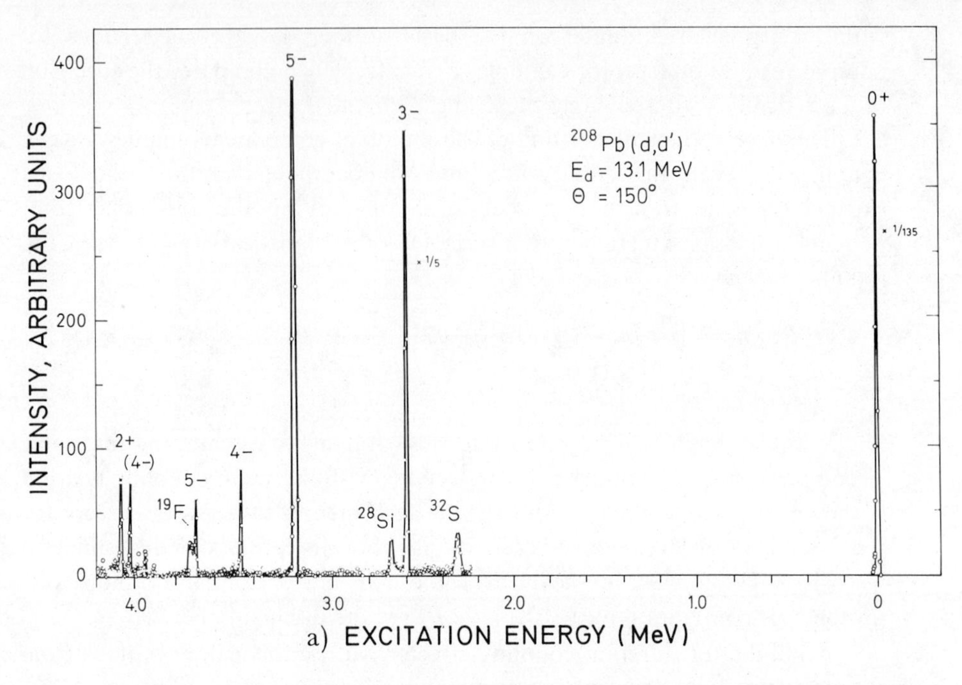

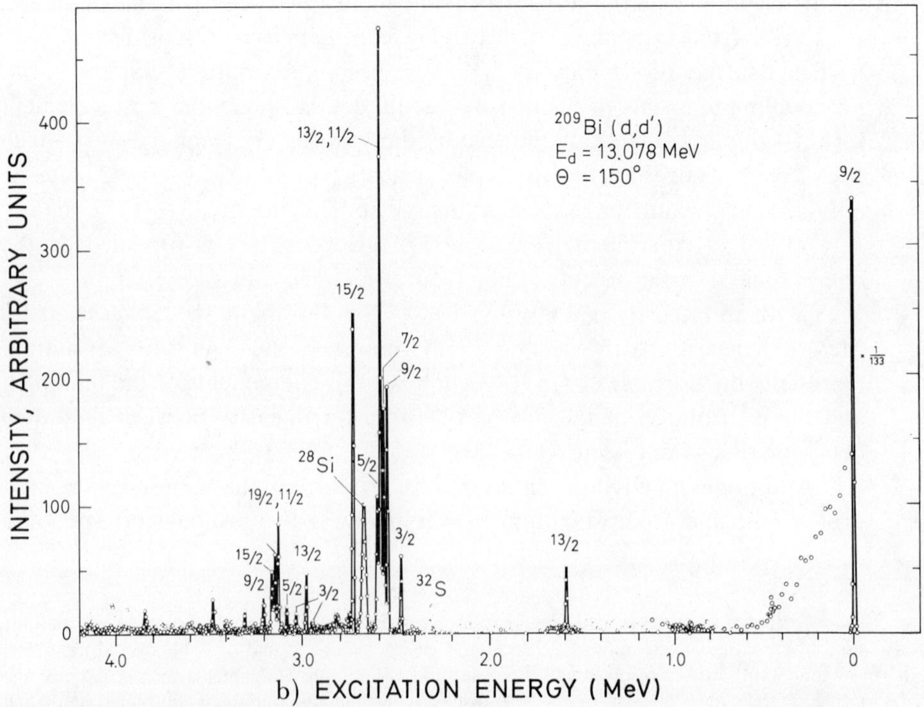

Figure 6-42 Excitation of octupole mode in ^{208}Pb and ^{209}Bi by inelastic scattering of deuterons. The figure is based on experimental data by J. Ungrin, R. M. Diamond, P. O. Tjøm, and B. Elbek, *Mat. Fys. Medd. Dan. Vid. Selsk.* **38**, no. 8, 1971.

Energy MeV	$I\pi$	$\frac{\sigma(h_{9/2}\to I\pi)}{\sigma(0\to 3-)}$	$\frac{\sigma}{\sigma_{wc}}$
2.494	3/2+	0.036	0.63
2.566	9/2+	0.120	0.84
2.585	7/2+	0.107	0.94
2.598	11/2+ }	0.325	0.87
2.600	13/2+ }		
2.618	5/2+	0.079	0.92
2.744	15/2+	0.206	0.90
2.958	3/2+	0.010	0.18

Table 6-16 Cross sections for inelastic deuteron scattering populating states in the $(h_{9/2}3-)$ multiplet in ^{209}Bi. The table gives the ratio between ^{209}Bi(dd′) cross sections for the processes $h_{9/2}\to(h_{9/2}3-)I\pi$ in ^{209}Bi and the ^{208}Pb(dd′) cross section for exciting the 3− octupole excitation at 2.614 MeV with deuterons of 13 MeV. The last column gives the observed cross sections relative to those expected in the weak-coupling approximation ($\sigma_{wc}=\sigma(0\to 3-)(2I+1)/70$; see Eq. (6-467)); no corrections have been made for differences in Q values arising from energy shifts in the multiplet; the Q-value effect on the cross sections is estimated to be about 7% per 100 keV of excitation. The experimental data in the table are from J. Ungrin, R. M. Diamond, P. O. Tjøm, and B. Elbek, *Mat. Fys. Medd. Dan. Vid. Selsk.* **38**, no. 8 (1971), except for the energies of the resolved 11/2+ and 13/2+ states, which are taken from R. A. Broglia, J. S. Lilley, R. Perazzo, and W. R. Phillips, *Phys. Rev.* **C1**, 1508 (1970).

▼ to the low-lying states in ^{210}Po. The lowest of these configurations that can couple to the $(h_{9/2}3-)$ multiplet is $(d_{3/2}^{-1}0+)3/2$. This configuration is characterized by a strong intensity in the proton pickup reaction ^{210}Po$(t\alpha)$, and it is found that the strength is distributed between a 3/2+ level at 2.958 MeV, which receives 55% of the intensity, and the 2.494 MeV 3/2+ member of the $(h_{9/2}3-)$ septuplet, which receives 45% of the intensity (Barnes *et al.*, 1972). This evidence for strong mixing of the $(h_{9/2}3-)3/2$ and $(d_{3/2}^{-1}0+)3/2$ configurations fits in well with the reduction in the strength of the inelastic excitation observed for the $(h_{9/2}3-)3/2$ state at 2.494 MeV (see Table 6-16); the $(d_{3/2}^{-1}0+)3/2$ level at 2.958 MeV is also excited with significant strength in the inelastic scattering (see Table 6-16), though the intensity is somewhat less than might have been inferred from the relative one-particle transfer intensities to the two $I\pi=3/2+$ states.

The coupling between the configurations $(h_{9/2}3-)3/2$ and $(d_{3/2}^{-1}0+)3/2$
▲ is illustrated by the perturbation diagram in Fig. 6-43. However, perturbation

▼ theory is inadequate to describe the strong mixing of states encountered in the present example, and we shall treat the problem in terms of a diagonalization of the effective coupling between the two states. Such a treatment ignores terms of relative magnitude corresponding to the energy separation of the two states divided by the excitation energy of the three-quasiparticle intermediate states shown in Fig. 6-43. (Compare the similar condition for the treatment of second-order rotational perturbations in terms of an effective $\Delta K=2$ coupling, as discussed on p. 149.) ▲

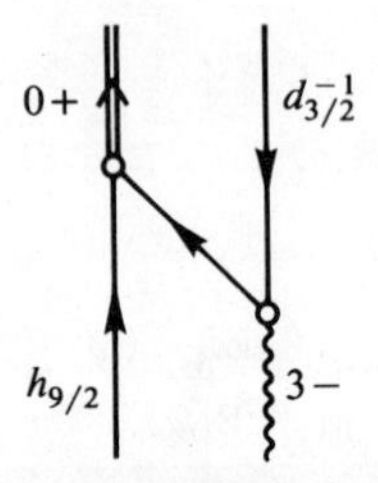

Figure 6-43 Diagram representing coupling between configurations involving a particle with a shape-vibrational quantum and a hole with a quantum of the pair mode.

▼ The effective second-order coupling matrix element obtained from Fig. 6-43 is

$$\langle (\mathrm{h}_{9/2}3-)3/2|H^{(2)}|(\mathrm{d}_{3/2}^{-1}0+)3/2\rangle$$

$$= -\left(\frac{10}{7}\right)^{1/2} \frac{h(\mathrm{h}_{9/2},\mathrm{d}_{3/2}3)\,GM(10)^{1/2}}{E_0 - E\left(\mathrm{d}_{3/2}^{-1}\mathrm{h}_{9/2}^2\right)} \langle (\tfrac{3}{2}\tfrac{9'}{2})3,\tfrac{9}{2};\tfrac{3}{2}|\tfrac{3}{2},(\tfrac{9}{2}\tfrac{9'}{2})0;\tfrac{3}{2}\rangle \quad (6\text{-}468)$$

where E_0 is the energy of the initial (or final) states. In Eq. (6-468), the factor $-(10/7)^{1/2}h(\mathrm{h}_{9/2},\mathrm{d}_{3/2}3)$ represents the coupling matrix element for the transition of the octupole phonon into the $(\mathrm{d}_{3/2}^{-1}\mathrm{h}_{9/2})$ particle-hole configuration (see Eq. (6-211)), and the estimated value of h is given by Eq. (6-453). The quantity $(10)^{1/2}$ GM in Eq. (6-468) represents the matrix element of the pair field coupling (6-148) for the transition of the 0+ pair boson into the configuration with two $\mathrm{h}_{9/2}$ particles, which play distinguishable roles (as also indicated by the necessity of distinguishing the two $j=9/2$ angular momenta in the recoupling coefficient). The amplitude $M=(\alpha_2)_0$ for the creation of the pair boson and the pair coupling constant G can be obtained from an analysis similar to that discussed on pp. 641 ff. for the ^{206}Pb ground state. The observed energy of the ^{210}Po ground state, relative to the $\mathrm{h}_{9/2}^2$ configuration, is $E(0+)-2\varepsilon(\mathrm{h}_{9/2})\approx -1.2$ MeV; the pair binding energy is larger than this ▲

▼ value by the Coulomb repulsion between the two protons, which we take to be $\delta E_{\text{Coul}}(\text{pp}) \approx +0.2$ MeV (compare Eq. (6-454)). Using the one-particle spectrum from Fig. 3-2f, Vol. I, p. 324, one obtains from Eqs. (6-590), (6-592), and (6-593) the values $G \approx 0.14$ MeV and $M \approx -3.8$. It may be noted that the corresponding amplitude of the $(\text{h}_{9/2}^2)0+$ configuration in the ^{210}Po ground state is (see Eq. (6-592))

$$c(\text{h}_{9/2}) = \frac{(5)^{1/2} GM}{E(0+) - E(\text{h}_{9/2}^2)} \approx 0.85 \qquad (6\text{-}469)$$

The matrix element for creation of the $(\text{h}_{9/2}^2)_0$ configuration $((5)^{1/2}GM = 2^{-1/2}(2j+1)^{1/2}GM$ includes the factor $2^{-1/2}$, expressing the effect of the indistinguishability of the two $\text{h}_{9/2}$ particles in the structure of the boson.

The energy denominator in Eq. (6-468) represents the excitation energy of the intermediate two-particle one-hole state, which in the approximation considered may be measured from either the $(\text{h}_{9/2}3-)$ or the $(\text{d}_{3/2}^{-1}0+)$ level; using the mean value for these two levels gives $E_0 - E(\text{d}_{3/2}^{-1}\text{h}_{9/2}^2) \approx -1.7$ MeV (including a Coulomb interaction of -0.2 MeV for the two-particle one-hole state relative to the $(\text{h}_{9/2}3-)$ state). With the above estimates of the parameters in Eq. (6-468) and with the value $-(7/40)^{1/2}$ for the recoupling coefficient, we obtain a coupling matrix element $H^{(2)} \approx 370$ keV. A coupling of this magnitude together with the observed energy separation between the two $I\pi = 3/2+$ states implies that the eigenstates are approximately equal mixtures of the $(\text{d}_{3/2}^{-1}0+)$ and $(\text{h}_{9/2}3-)$ configurations, as indicated by the proton pickup intensities. However, one should not attach too much significance to the quantitative comparison in view of the approximations involved in the treatment in terms of an effective coupling.

The energy separations within the $(\text{h}_{9/2}3-)$ septuplet are shown in Table 6-16; while most of the levels are shifted by less than 50 keV, the $I = 3/2$ level is shifted by -120 keV and the $I = 15/2$ level is shifted by $+130$ keV. The downward shift of the $I = 3/2$ level appears to reflect the coupling to the $(\text{d}_{3/2}^{-1}0+)$ configuration discussed above. The estimate of the effective coupling $H^{(2)}$ is about a factor of 2 larger than that indicated by the observed position of the $I = 3/2$ levels. It is not clear at the present time whether this discrepancy is to be attributed to a somewhat weaker particle-vibration coupling than that employed or whether it reflects couplings to other degrees of freedom.

A relatively large upward shift of the $I = 15/2$ level can be understood in terms of the particle-phonon coupling through the octupole field. The largest energy shifts are associated with the intermediate states involving the $\text{d}_{3/2}^{-1}$ configuration, which is the closest particle configuration that can couple to the $(\text{h}_{9/2}3-)$ multiplet (aside from the $\text{i}_{13/2}$ one-particle state discussed ▲ above). This interaction is described by diagram (d) in Fig. 6-10, p. 427, and

▼ is given by Eq. (6-224),

$$\delta E\big((h_{9/2}3)I\big) = \frac{h^2(h_{9/2}, d_{3/2}3)}{\hbar\omega_3 - E\big(d_{3/2}^{-1}h_{9/2}\big)} 10 \begin{Bmatrix} 3 & 9/2 & 3/2 \\ 3 & 9/2 & I \end{Bmatrix}$$

$$\approx (-1)^{I-15/2} \frac{(I-1/2)(I+1/2)(I+3/2)}{7\cdot 8\cdot 9} 300 \text{ keV} \quad (6\text{-}470)$$

where, in the last line, we have employed the value of $h(h_{9/2}, d_{3/2}3)$ given by Eq. (6-453) and the energy denominator -1.8 MeV, as discussed on p. 563. It is seen that the energy shift (6-470) has its largest value for $I=15/2$ and is positive for this state; the magnitude, however, exceeds the observed shift of the $I=15/2$ state by about a factor of 2.

In a more quantitative analysis of the energy separations within the $(h_{9/2}3-)$ septuplet, one must take into account the contributions from the many different more distant intermediate states in the second-order particle-vibration coupling, as illustrated by the four diagrams in Fig. 6-10. Thus, each member of the multiplet receives a large number of contributions, some of which are as large as 50 to 100 keV (Hamamoto, 1969). Additional terms may arise from interactions involving other moments of the octupole phonon. Thus, for example, the static quadrupole moment of the phonon gives rise to an interaction with the quadrupole moment of the $h_{9/2}$ particle that can be estimated in analogy to the phonon-phonon interaction effect given by Eq. (6-466).

$$\delta E\big((h_{9/2}3)I\big)$$

$$= \kappa Q_{el}(n_3=1)\frac{A}{Z}\frac{Q_{el}(h_{9/2})}{1+(e_{pol}/e)}\frac{5}{16\pi}(-1)^{I-1/2}2(77)^{1/2}\begin{Bmatrix} 3 & 9/2 & I \\ 9/2 & 3 & 2 \end{Bmatrix}$$

$$= 1.4 Q_{el}(n_3=1)Q_{el}(h_{9/2})(-1)^{I+1/2}\begin{Bmatrix} 3 & 9/2 & I \\ 9/2 & 3 & 2 \end{Bmatrix} \text{keV fm}^{-4} \quad (6\text{-}471)$$

where we have used the estimate (6-368) for κ and the value $e_{pol} \approx 0.5e$ (see Table 6-8, p. 516, and Eq. (6-386) for the isovector contributions). Thus, a phonon electric quadrupole moment of -10 fm^2, corresponding to the order of magnitude estimate on p. 569, together with the observed $Q_{el}(h_{9/2}) \approx -40$ fm^2 (see Table 3-2, Vol. I, p. 341), would imply a contribution to the energy splitting of the order of 100 keV. (The estimate (6-471) together with the observed energy separations in the $(h_{9/2}3-)$ septuplet makes it unlikely that the phonon quadrupole moment could be as large as the tentative experimental value, referred to on p. 569.) ▲

▼ *Octupole modes in deformed nuclei (Fig. 6-44)*

In the deformed nuclei, the octupole shape oscillations are expected to give rise to modes with $\nu = 0$, 1, 2, and 3. A considerable body of evidence exhibits the occurrence of enhanced octupole excitations in deformed nuclei with energies of the order of 1 MeV. (See Table 6-14, p. 561, and the systematic evidence obtained from (dd′) studies (Elbek *et al.*, 1968).) The oscillator strength associated with these excitations is only a small fraction of the total expected octupole strength for $\Delta N = 1$ transitions; thus, the proper- ▲

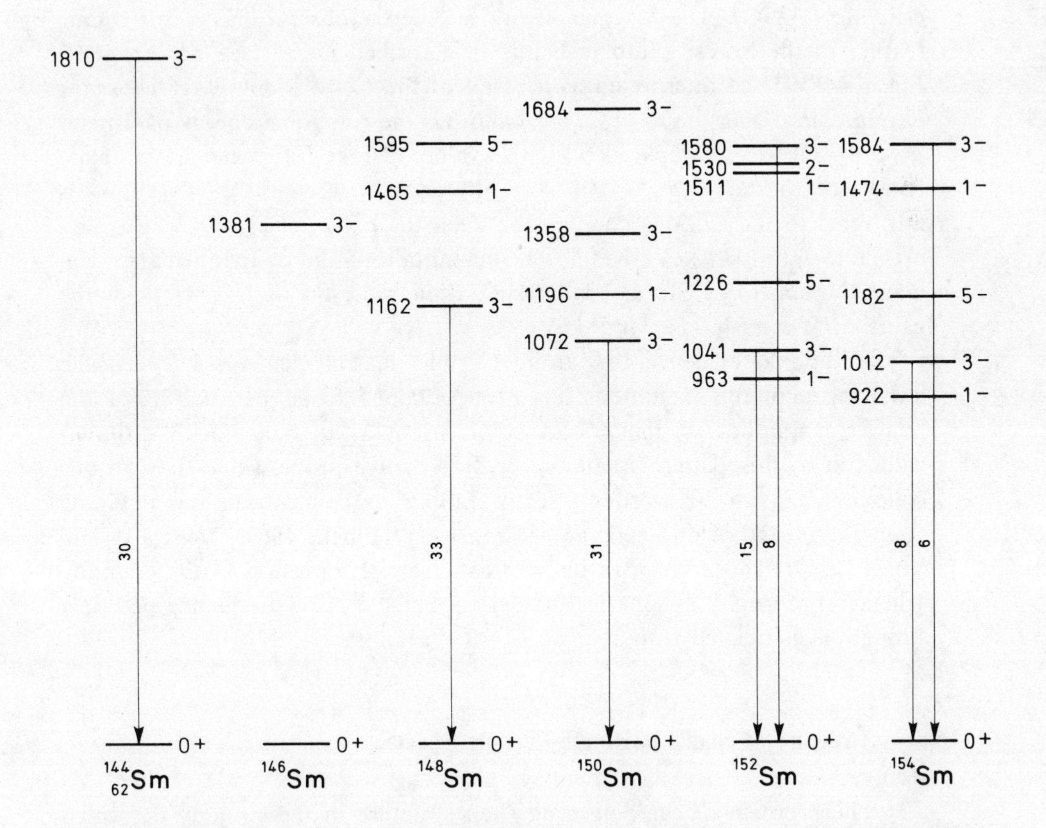

Figure 6-44 Odd-parity states in even-even Sm isotopes. The numbers appearing on the transition arrows give the octupole strength as determined from the analysis of inelastic deuteron scattering by E. Veje, B. Elbek, B. Herskind, and M. C. Olesen, *Nuclear Phys.* **A109**, 489 (1968); these cross sections are found to be approximately proportional to the $E3$-transition probability, and the quantity given in the figure is the estimated $B(E3)$ value in units of $B_W(E3) = 1.3 \times 10^3\ e^2\text{fm}^6$. Coulomb excitation studies of these nuclei have been made by Seaman *et al.* (1966). The level schemes are taken from the quoted references and from the compilation by Sakai (1970).

▼ ties of the states may depend rather sensitively on the occurrence of low-lying two-quasiparticle excitations with the appropriate quantum numbers.

The transition from spherical to deformed nuclei is illustrated by the sequence of Sm isotopes shown in Fig. 6-44. For these nuclei, one sees the gradual change of the pattern of the odd-parity excitations, starting from spherical nuclei with $n_{\lambda=3}=1; I\pi=3-$ lowest, followed by $n_3=1, n_2=1$; $I\pi=1-,2-,\ldots,5-$, to the deformed nuclei, in which the sequence is $n_{\lambda=3,\nu=0}=1; I\pi=1-,3-,5-,\cdots$ followed by $n_{31}=1; I\pi=1-,2-,3-,\cdots$.

The octupole excitations with different values of ν are expected to be rather strongly coupled by the Coriolis interaction, due to the large intrinsic angular momentum involved in the octupole vibrations (see the qualitative estimate (6-294)). The large moments of inertia observed as a systematic feature of the lowest octupole bands in deformed nuclei (see, for example, Fig. 6-44) can be interpreted as a result of the Coriolis coupling. Thus, for example, in ^{154}Sm, the $n_{\lambda=3,\nu=0}=1$ band has the rotational energy parameter $\hbar^2/2\mathscr{I}=8.9$ keV, compared with the value 13.7 keV for the ground-state band (see Fig. 6-31, p. 534). If the difference between these two values is attributed to the Coriolis coupling between the $n_{30}=1$ ($\hbar\omega_{30}=922$ keV) and $n_{31}=1$ ($\hbar\omega_{31}=1474$ keV) bands, the magnitude of the coupling matrix element is found to be $51(I(I+1))^{1/2}$ keV, while the estimate (6-294) yields the value $\langle H' \rangle = -67(I(I+1))^{1/2}$ keV.

Additional evidence for strong Coriolis matrix elements between the low-lying octupole excitations has been obtained from the analysis of the relative magnitude of the $E3$-transition strengths to bands with different values of ν; this pattern is an especially sensitive probe, since the Coriolis coupling leads to first-order effects tending to concentrate the octupole strength onto the lowest octupole transition (Elbek, 1969; Neergård and Vogel, 1970). This effect can be seen as a partial transition to the coupling scheme appropriate to spherical nuclei, in which the low-frequency octupole strength is associated with a single $I\pi=3-$ excitation.

Structure of Shells in Single-Particle Spectra

The problem of characterizing shell structure in the strongly deformed shapes associated with the fission process requires a more systematic analysis of the conditions for occurrence of shells in the single-particle spectrum than that presented in Chapter 2. Concepts involved in such an analysis are introduced in the following example, as a basis for order of magnitude estimates of shell-structure effects in different potentials. Applications to spherical and strongly deformed nuclei are considered in the two subsequent
▲ examples.

▼ *Characterization of shell structure*[29] *(Figs. 6-45 and 6-46; Table 6-17)*

Spherical potentials

The central concepts involved in the analysis of shell-structure effects in one-particle spectra can be illustrated by a treatment of the familiar case of potentials with spherical symmetry. For such a potential, the motion separates into radial and angular components, and the one-particle energies depend on the two quantum numbers l and n, where l is the orbital angular momentum, while the radial quantum number n orders the levels with given l. (With the conventional notation, n is greater by one unit than the number of radial nodes, $n=1,2,3,\ldots$.) Each level (n,l) has a degeneracy of $2l+1$, as a consequence of the rotational invariance, which implies that the energy is independent of the azimuthal quantum number m. (In the present section, the spin degree of freedom and the spin-orbit coupling are omitted, since these features do not affect the qualitative results of the discussion.)

Shell structure occurs if the one-particle energy $\varepsilon(n,l)$ is approximately stationary with respect to certain variations of the quantum numbers. The conditions for stationarity can be simply exhibited if the function $\varepsilon(n,l)$ can be expanded around a given point (n_0,l_0) in the (n,l) plane,

$$\varepsilon(n,l)=\varepsilon(n_0,l_0)+(n-n_0)\left(\frac{\partial\varepsilon}{\partial n}\right)_0+(l-l_0)\left(\frac{\partial\varepsilon}{\partial l}\right)_0$$
$$+\tfrac{1}{2}(n-n_0)^2\left(\frac{\partial^2\varepsilon}{\partial n^2}\right)_0+(n-n_0)(l-l_0)\left(\frac{\partial^2\varepsilon}{\partial n\,\partial l}\right)_0+\tfrac{1}{2}(l-l_0)^2\left(\frac{\partial^2\varepsilon}{\partial l^2}\right)_0+\cdots \tag{6-472}$$

where the subscript 0 indicates that the derivatives are evaluated at the point (n_0,l_0). (Such an analysis of $\varepsilon(n,l)$ as a function of continuous variables (n,l) is a generalization of the definition of Regge trajectories, which is obtained from a study of $\varepsilon(n,l)$ considered as an analytic function of l for fixed n; see, for example, De Alfaro and Regge, 1965.) While it appears appropriate to employ the expansion (6-472) for potentials of the type encountered in the nucleus, the mathematical characterization of the conditions for such a representation remains to be explored. An example of a situation, where the ▲ expansion (6-472) may be of limited usefulness, is provided by a potential

[29]The evaluation of shell-structure effects in the single-particle level density, on the basis of an asymptotic expansion of the single-particle Green function, has been considered by Balian and Bloch (1971). This analysis brought into focus the intimate relationship between shell structure and the occurrence of closed classical orbits. The significance of the commensurability of the radial and angular frequencies in the classical motion has also been discussed by Wheeler (1971) and Swiatecki (private communication, 1971).

▼ with two radial minima separated by a finite barrier; in such a potential, the radial motion changes its character when the energy is in the neighborhood of the barrier (see, for example, the comments concerning the atomic potential on p. 584).

Series of approximately degenerate levels occur in the spectrum (6-472) if the first derivatives of ε are in the ratio of two (small) integers a and b,

$$b\left(\frac{\partial \varepsilon}{\partial n}\right)_0 = a\left(\frac{\partial \varepsilon}{\partial l}\right)_0 \tag{6-473}$$

When this relation is fulfilled, the levels with constant value of $an+bl$ differ ▲ in energy only by terms involving the second and higher derivatives of ε.

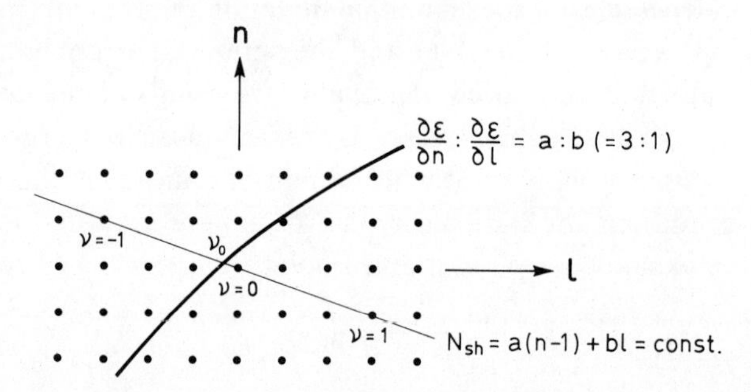

Figure 6-45 Characterization of shells in spherical potentials. The figure illustrates the definition of the shells and the associated trajectories in the (n,l) plane. The shells considered correspond to a ratio of 3:1 for the first derivatives of the energy with respect to n and l. The lines of constant N_{sh} that connect approximately degenerate levels are tangent to the curves of constant energy at the point of intersection with the shell trajectory. The point labeled $\nu=0$ has the quantum numbers $[n_0]$, $[l_0]$, while the intersection of the shell trajectory (thick curve) with the line of constant N_{sh} is labeled ν_0 and has the coordinates $n_0=[n_0]-\nu_0 b$, $l_0=[l_0]+\nu_0 a$.

▼ The locus of points (n_0, l_0) that satisfy the relation (6-473) define a "shell trajectory" in the (n,l) plane (see Fig. 6-45). The successive shells associated with a trajectory with a given value of $a:b$ can be labeled by the quantum number

$$N_{sh} = a(n-1) + bl \tag{6-474}$$

(With this choice of origin, the shell quantum number N_{sh} equals the oscillator quantum number N for the harmonic oscillator potential ($a:b = 2:1$; see p. 583).) The shells with different values of N_{sh} occur with a periodicity in energy of

$$\hbar\omega_{sh} = \frac{1}{a}\left(\frac{\partial \varepsilon}{\partial n}\right)_0 = \frac{1}{b}\left(\frac{\partial \varepsilon}{\partial l}\right)_0 \tag{6-475}$$

▲

The states within a shell, characterized by a constant value of N_{sh}, can be labeled by a quantum number ν,

$$\left.\begin{aligned} l &= [l_0] + \nu a \\ n &= [n_0] - \nu b \end{aligned}\right\} \nu = 0, \pm 1, \pm 2, \ldots \tag{6-476}$$

where $([n_0], [l_0])$ denotes the state in the shell nearest to the point of intersection with the shell trajectory (see Fig. 6-45). For the states (6-476), the energies are approximately of the form

$$\varepsilon(\nu) \approx \varepsilon(n_0, l_0) + \tfrac{1}{2}\beta(\nu - \nu_0)^2 \tag{6-477}$$

with

$$\beta = a^2\left(\frac{\partial^2\varepsilon}{\partial l^2}\right)_0 - 2ab\left(\frac{\partial^2\varepsilon}{\partial n\,\partial l}\right)_0 + b^2\left(\frac{\partial^2\varepsilon}{\partial n^2}\right)_0 \tag{6-478}$$

Examples of shell trajectories are shown in Fig. 6-46, which describes the energy surface $\varepsilon(n,l)$ for the infinite square-well potential. The detailed characterization of the analytic function $\varepsilon(n,l)$ for this model is discussed in the fine print below (pp. 587 ff.). The trajectory with $a:b = 2:1$ coincides with the vertical line $l = -1/2$ in Fig. 6-46. The shell structure associated with this trajectory becomes relatively unimportant for large quantum numbers, due to the low $(2l+1)$ degeneracy of the levels in the corresponding shells.

The order of magnitude of the shell-structure effects associated with a given trajectory can be estimated by noting that, for a system with a large number of particles A, the quantum numbers n and l are typically of order $A^{1/3}$. The first and second derivatives of ε occurring in Eqs. (6-475) and (6-478) are therefore expected to be of order $\varepsilon_F A^{-1/3}$ and $\varepsilon_F A^{-2/3}$, respectively, where ε_F is the Fermi energy. Hence, the spacing of the shells is $\hbar\omega_{sh} \sim \varepsilon_F A^{-1/3}$, and the number of levels (n,l) in a shell occurring within an energy interval of about $\hbar\omega_{sh}$ is of order $(\beta^{-1}\hbar\omega_{sh})^{1/2} \sim A^{1/6}$. The total number Ω of single-particle states in the shell involves the additional $(2l+1)$ degeneracy and is therefore $\Omega \sim A^{1/2}$ (since $l \sim A^{1/3}$), except in the special case of trajectories with l_0 of order unity, for which the shells contain a total number of levels $\Omega \sim A^{1/3}$. It should be noted that the individual shells with specified $a:b$ and N_{sh} contain only a small fraction of the total number of single-particle levels in the energy interval $\hbar\omega_{sh}$; this number is of order $A^{2/3}$, since the total single-particle level density is $g_0 \sim A/\varepsilon_F$. Asymptotically, the shells represent a modulation of the single-particle level density, with a relative amplitude $A^{-1/6}$ (or $A^{-1/3}$).

A special situation occurs if the function $\varepsilon(n,l)$ depends only on a single linear combination of the two quantum numbers, such as for the harmonic oscillator potential ($\varepsilon = \hbar\omega_0(2n + l - 1/2)$) and the Coulomb potential ($\varepsilon = -Ry(n+l)^{-2}$). In these cases, Eq. (6-473) is fulfilled for all values of n

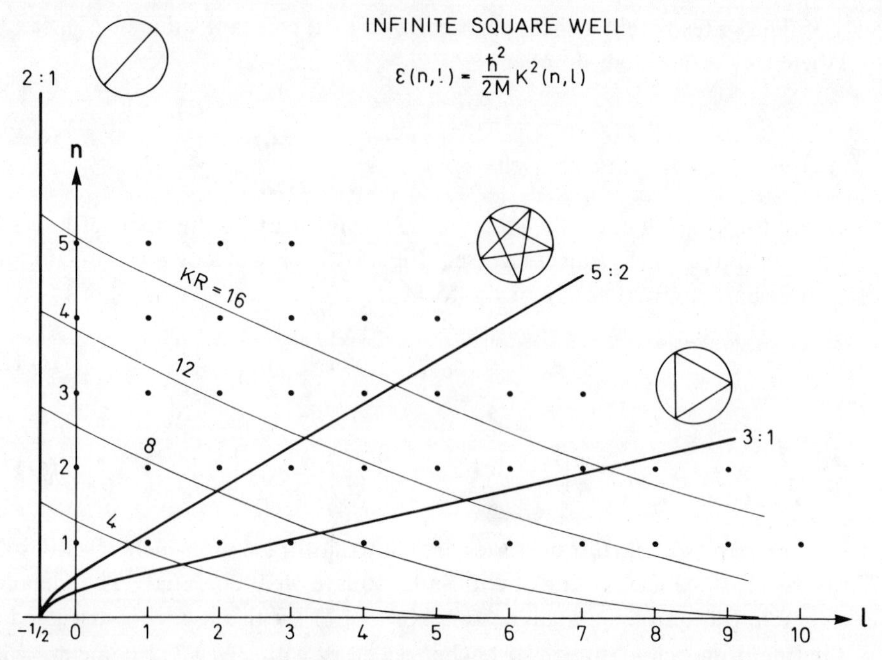

Figure 6-46 Energy surface in the (n,l) diagram for the infinite square well. The physical states correspond to a square lattice in the region $n \geqslant 1$, $l \geqslant 0$. The lightly drawn curves correspond to contour lines of equal energy and are labeled by the value of KR. For large values of l, they approach asymptotically the line $n=0$. The heavy lines represent shell trajectories labeled by the values of $a:b$ and by the geometrical figure that corresponds to the classical orbit. The trajectory with $a:b=2:1$, corresponding to the pendulating orbits, remains slightly outside the physical region ($l=-1/2$).

▼ and l, and the shells contain fully degenerate sets of levels with $|\nu|$ ranging up to values of order $A^{1/3}$, corresponding to a total number of orbits in each shell of order $\Omega \sim A^{2/3}$. In such a situation, all the single-particle states belong to a single sequence of shells.

Relation to periodic classical orbits

The occurrence of shell structure is intimately connected with features of the classical orbits. The derivatives $\partial\varepsilon/\partial n$ and $\partial\varepsilon/\partial l$ are recognized as giving the radial and angular frequencies of the orbits, and the stationarity criterion (6-473) therefore corresponds to the condition that the classical orbits close upon themselves after a radial and b angular oscillations. The frequency of the motion in the closed orbit is ω_{sh}, given by Eq. (6-475). (The quantum numbers (n,l,m) represent the action variables for the orbits, in units of Planck's constant. The energy considered as a function of the quantum numbers, therefore, corresponds to the Hamiltonian expressed in terms of the ▲ action variables. Such an analysis of the classical orbits provided the basis for

▼ the treatment of multiply periodic systems in the early quantal theory (Sommerfeld, 1915; see also Bohr, 1918, and Sommerfeld, 1922)).

Some of the characteristic features of the shell trajectories follow directly from simple properties of the classical orbits. Thus, for any potential that is not singular at the origin, the orbits with vanishing angular momentum correspond to a pendulating motion that passes through the center of the potential, executing two radial oscillations for each angular period ($\omega_r:\omega_\varphi=2:1$). In such potentials, therefore, the $a:b=2:1$ trajectory coincides approximately with the $l=0$ axis. (The arguments for the use of $(l+1/2)^2$ as the coefficient of the centrifugal term in the WKB approximation (see, for example, Fröman and Fröman, 1965) may suggest that the 2:1 trajectory passes along the $l=-1/2$ axis, as in Fig. 6-46.)

Another simple property of the energy surface is obtained by considering the circular orbits ($n \ll l$). For such an orbit, with radius r_1, the angular frequency ω_φ is given by

$$\omega_\varphi^2=\frac{\hbar^2 l^2}{M^2 r_1^4}=\frac{1}{Mr_1}\left(\frac{\partial V}{\partial r}\right)_{r_1} \tag{6-479}$$

For small radial oscillations with respect to the circular orbit, the period ω_r can be determined by expanding the total radial potential, including the centrifugal term, in powers of $(r-r_1)$, and one finds

$$\begin{aligned}\omega_r^2&=\frac{1}{M}\left(\left(\frac{\partial^2 V}{\partial r^2}\right)_{r_1}+\frac{3\hbar^2 l^2}{Mr_1^4}\right)=\omega_\varphi^2\left(3+\frac{r_1(\partial^2 V/\partial r^2)_{r_1}}{(\partial V/\partial r)_{r_1}}\right)\\&=\omega_\varphi^2(2+p)\qquad \text{for}\quad V=kr^p\end{aligned} \tag{6-480}$$

For the harmonic oscillator potential ($p=2$), we obtain the familiar result $\omega_r=2\omega_\varphi$ that holds for all orbits in this potential, corresponding to motion in ellipses centered on the origin. A steeper increase of the potential ($p>2$) leads to values of $\omega_r:\omega_\varphi$ that increase with l and thus to lines of constant energy that are concave upward in the (n,l) plane, as in Fig. 6-46.

In the limit of a potential with a sharp boundary, the classical expression (6-480) implies that the ratio $\omega_r:\omega_\varphi$ increases indefinitely as one approaches the circular orbit, corresponding to the fact that the slope of lines of constant energy in Fig. 6-46 vanishes in the limit of large l and $n \approx 1$. For finite l, the value of the slope at $n=1$ is of order $l^{-1/3}$ (as can be seen from Eq. (6-490) or Eq. (6-495)).

For the Newtonian (Coulomb) potential with its r^{-1} singularity, all orbits have equal angular and radial frequencies ($a=b=1$); the center of force is at the focus of the Kepler ellipse, and the radial motion therefore executes only a single oscillation for each angular period. In atoms, the
▲ screening of the nuclear field by the electrons implies that the average

▼ potential varies more rapidly than r^{-1}, except in the region of the origin; in such a situation, the angular frequency becomes larger than the radial frequency (see Eq. (6-480)).

In the characterization of the atomic shell structure by the present methods, one must take into account that, except for the s and p orbits, the radial potential, with inclusion of the centrifugal term, has two minima separated by a rather large potential barrier (Mayer, 1941). For large distances, the Coulomb field of the ion, together with the centrifugal potential, gives rise to a minimum outside the screening cloud; as the electron penetrates into the atom, the electrostatic attraction may, for a while, increase more rapidly than the centrifugal potential, with the result that a second minimum occurs inside the screening cloud. In such a situation, the analytic structure of $\varepsilon(n,l)$ acquires new features associated with the occurrence of two distinct types of classical orbits for given ε and l.

Separable potentials of more general type

The characterization of shell structure considered above for spherical potentials can be extended to any potential that permits a separation of the motion in the three different dimensions. Familiar examples of potentials that lead to separation of the motion in Cartesian coordinates are provided by the anisotropic harmonic oscillator and the rectangular box with infinite walls. The two-center Newtonian potential and the ellipsoidal box with infinite walls permit separation in ellipsoidal coordinates. (For the systematic enumeration of the coordinate systems that may be employed in order to separate variables and a discussion of the wave equation in these coordinates, see, for example, Morse and Feshbach, 1953 (pp. 494ff.). For a discussion of classical orbits in separable potentials, see, for example, Born, 1925, and Pars, 1965.)

The separability implies that the eigenstates for the one-particle motion can be characterized by three quantum numbers $(n_1n_2n_3)$. An expansion of the energy around a particular point $(n_1n_2n_3)_0$, in analogy to Eq. (6-472), exhibits the occurrence of shells of maximal degeneracy, when the first derivatives of the energy with respect to all three quantum numbers are in rational ratios

$$\left(\frac{\partial\varepsilon}{\partial n_1}\right)_0 : \left(\frac{\partial\varepsilon}{\partial n_2}\right)_0 : \left(\frac{\partial\varepsilon}{\partial n_3}\right)_0 = a:b:c \tag{6-481}$$

where a, b, and c are (small) integers, or zero, and where the subscript 0 implies that the derivatives are evaluated at the point $(n_1n_2n_3)_0$. When the condition (6-481) is satisfied, the energy is stationary with respect to two linear combinations ν_1 and ν_2 of the three quantum numbers. The levels in a shell, specified by ν_1 and ν_2, have the same value of the quantum number

$$N_{sh} = an_1 + bn_2 + cn_3 \tag{6-482}$$

which labels the number of the shell, and the energy spacing between the

▼ shells is given by

$$\hbar\omega_{\text{sh}} = \frac{1}{a}\left(\frac{\partial\varepsilon}{\partial n_1}\right)_0 = \frac{1}{b}\left(\frac{\partial\varepsilon}{\partial n_2}\right)_0 = \frac{1}{c}\left(\frac{\partial\varepsilon}{\partial n_3}\right)_0 \tag{6-483}$$

corresponding to the period of the classical orbit.

Since each of the quantum numbers is typically of order $A^{1/3}$, the energy $\hbar\omega_{\text{sh}}$ is of order $\varepsilon_F A^{-1/3}$. The energy of the levels in the shell depends quadratically on ν_1 and ν_2 with coefficients that are given by the second derivatives of ε with respect to n_1, n_2, and n_3 and therefore typically of order $\varepsilon_F A^{-2/3}$. Hence, each shell comprises a set of levels, with $|\nu_1|$ and $|\nu_2|$ independently ranging up to values of order $A^{1/6}$, and the total number of orbits in the shell is $\Omega \sim A^{1/3}$. Degeneracies of higher order occur if the energy is independent of ν_1 or ν_2 (or a linear combination of ν_1 and ν_2), such as in the spherical potential, where ε is independent of m.

Among the nonspherical potentials, the anisotropic harmonic oscillator

$$V = \tfrac{1}{2}M\left(\omega_1^2 x_1^2 + \omega_2^2 x_2^2 + \omega_3^2 x_3^2\right) \tag{6-484}$$

occupies a special position, since the energy in this potential is a linear function of all three quantum numbers

$$\varepsilon = \hbar\omega_1(n_1 + \tfrac{1}{2}) + \hbar\omega_2(n_2 + \tfrac{1}{2}) + \hbar\omega_3(n_3 + \tfrac{1}{2}) \tag{6-485}$$

Hence, if the frequencies are in the ratios of integers, the energy depends only on the quantum number N_{sh}, given by Eq. (6-482), and each shell has a degeneracy of the same order of magnitude ($\Omega \sim A^{2/3}$) as in the spherical oscillator potential. If the ratios of the frequencies in the harmonic oscillator potential are irrational, there are no shells in the limit of large quantum numbers.

Survey of degeneracies

The order of magnitude of the shell structure in the single-particle spectrum can be associated with the degeneracy of the closed classical orbits. In the neighborhood of a closed trajectory, the classical orbits can be characterized by six parameters. Two of these may be taken to be the energy and the epoch (origin of the time coordinate), while the remaining four are associated with displacements and momenta transverse to the selected orbit. For a separable potential, a displacement in either of the two transverse directions affects only the relative phase of the motion in the three different coordinates and therefore again results in a closed orbit. Further degeneracies of the closed orbits may occur for potentials with additional symmetry. Thus, in a spherical potential, the classical orbits are confined to a plane perpendicular to the angular momentum, and an increment of momentum perpendicular to the orbital plane only leads to a tipping of this plane. Hence, the ▲

▼ periodic orbits in a spherical potential have a threefold degeneracy, which corresponds to the fact that three angular variables (the Euler angles) are needed to specify the orientation of a planar figure. The pendulating orbits represent a special case, since the angular momentum vanishes; these orbits do not exhibit degeneracy with respect to increments in the transverse momenta. In a harmonic oscillator potential (either spherical or with integer frequency ratios), as well as in a Newtonian potential, all classical trajectories form closed orbits, corresponding to a fourfold degeneracy in the transverse coordinates.

The above results are summarized in Table 6-17, which includes degeneracies for motion in two as well as in three dimensions. The degeneracy of the classical orbits with respect to transverse displacements and momenta is characterized by the degeneracy indices s_q and s_p, respectively. The order of magnitude of the degeneracies Ω can be expressed in the form

$$\Omega \sim (A)^{s/2d} \qquad (s = s_q + s_p) \tag{6-486}$$

where d is the number of spatial dimensions. The result (6-486) reflects the fact that the total phase space has $2d$ dimensions; each degree of degeneracy in the transverse coordinates thus contributes a factor $(A)^{1/2d}$ to the degeneracy of the shells. One can also recognize Ω as the number of linearly independent wave packets that can be associated with the classical closed ▲ trajectories having an energy lying within an interval of order $\hbar\omega_{sh}$.

	Symmetry of potential	Orbital degeneracy s_q	Orbital degeneracy s_p	Shell degeneracy Ω
Two dimensions	elliptical	1	0	$A^{1/4}$
	rectangular	1	0	$A^{1/4}$
	circular	1	0	$A^{1/4}$
	harmonic osc.	1	1	$A^{1/2}$
Three dimensions	ellipsoidal	2	0	$A^{1/3}$
	rectangular	2	0	$A^{1/3}$
	spherical	2	1	$A^{1/2}$
	harmonic osc.	2	2	$A^{2/3}$
	Newtonian	2	2	$A^{2/3}$

Table 6-17 Degeneracy of shells in separable potentials. The table gives the maximum degeneracy occurring in a number of different types of separable potentials. The quantities s_q and s_p give the numbers of dimensions in the transverse motion with respect to which the closed classical orbits are degenerate.

▼ Mathematical tools for analyzing the degree of regularity in the eigenvalue spectrum for arbitrary potentials do not seem to be available; however, it appears likely that the connection between the classical closed orbits and the degeneracies of the quantal shells will apply under rather general conditions. For an arbitrary potential, closed classical orbits can occur for any given energy, since the number of parameters in the initial conditions is equal to the number of conditions to be satisfied for periodicity; however, if the potential has no special symmetry, the closed orbits will not, in general, be members of degenerate families (see, for example, Whittaker, 1937, p. 396). The relation (6-486) therefore suggests a very uniform single-particle level density with local variations that are smaller than any (positive) power of A. (The consequences of a distribution of one-particle levels corresponding to the eigenvalues of a random matrix (see Appendix 2C) have been considered in connection with the electronic properties of small metallic particles; Gor'kov and Éliashberg, 1965.)

In the intermediate case of partial separability, such as for an axially symmetric potential, the closed classical orbits have a onefold degeneracy ($s_q = 1$) corresponding to $\Omega \sim A^{1/6}$. For nonseparable (or partially separable) motion, the question arises whether the closed classical orbits can be associated with definite stationary quantum states as for the fully separable motion or whether the correspondence involves a strength function. It has been suggested that the answer to this question depends on whether the classical orbits are stable or unstable with respect to small perturbations in the transverse coordinates (Gutzwiller, 1971).

The analytic function $\varepsilon(n,l)$ for spherical infinite square-well potential

The energy $\varepsilon(n,l)$ as an analytic function of l may be obtained from a solution of the radial wave equation, in which l is treated as a continuous variable. For the square-well potential, a continuation in n may be obtained by generalizing the boundary condition for the radial wave function $\mathscr{R}$ at the origin ($r=0$),

$$\mathscr{R}_{nl}(r) \underset{r\to 0}{=} c(j_l(Kr)\cos n\pi + y_l(Kr)\sin n\pi) \qquad \text{(6-487)}$$

where c is a constant and $K(\varepsilon)$ the inside wave number, while j_l and y_l are the spherical Bessel and Neumann functions, considered as functions of the argument Kr as well as of the continuously varying order l. The condition (6-487), together with the usual boundary condition at infinity, provides the desired relation between the energy and the variables n and l.

For the infinite square-well potential, the boundary condition at the radius R of the potential implies

$$j_l(KR)\cos n\pi + y_l(KR)\sin n\pi = 0 \qquad \text{(6-488)}$$

It may be noted that the quantum number n defined by Eq. (6-488) is equivalent to ▲ the phase shift $\xi_l(KR)$ for scattering from an infinitely repulsive sphere of radius R

▼ (see Eq. (3F-38))[30],

$$\xi_l(KR) = -\pi n(l, KR) \tag{6-489}$$

For $l \gg 1$, one can employ the asymptotic expressions for Bessel functions of large order (Debye expansions; see, for example, Abramowitz and Stegun, 1964, p. 366), which yield for the relation (6-488)

$$(n - \tfrac{1}{4})\pi = KR(\sin\varphi - \varphi\cos\varphi) \tag{6-490}$$

where the angle φ is defined by

$$l + \tfrac{1}{2} = KR\cos\varphi \tag{6-491}$$

The asymptotic relation (6-490) is found to apply with a rather high accuracy over the entire physical region ($n \geqslant 1$, $l \geqslant 0$).

The shell trajectories correspond to rational ratios $a:b$ between the derivatives of ε (or K) with respect to n and l. In the approximation (6-490), one obtains

$$\varphi = \frac{b}{a}\pi \tag{6-492}$$

corresponding to straight lines in the (n, l) diagram. The periodicity of the shells is given by Eq. (6-475), and from Eq. (6-490), one finds

$$\hbar\omega_{\text{sh}} = KR\frac{\hbar^2}{MR^2}\frac{\pi}{a\sin(\pi(b/a))} \tag{6-493}$$

Within each shell, the leading-order energy separations have the dependence (6-477) on the quantum number ν, with the second-order coefficient (6-478) given by (in the approximation (6-490))

$$\beta = -\frac{\hbar^2}{MR^2}\frac{a^2}{\sin^2(\pi(b/a))} \tag{6-494}$$

The asymptotic relations employed above are equivalent to the WKB approximation for the quantization condition

$$(n - \tfrac{1}{4})\pi = \int_{r_{\min}}^{r_{\max}} dr\left(\frac{2M}{\hbar^2}(\varepsilon - V(r)) - \frac{(l+\frac{1}{2})^2}{r^2}\right)^{1/2}$$

$$= \int_{l+1/2}^{KR}\frac{dx}{x}\left(x^2 - (l+\tfrac{1}{2})^2\right)^{1/2} \tag{6-495}$$

The evaluation of the integral is seen to yield the relation (6-490). In the classical limit, the angle 2φ defined by Eq. (6-491) represents the angular interval between consecutive reflections, and the condition (6-492) follows from the fact that this section of the ▲ orbit represents a full radial period.

---

[30]The spherical Neumann function n_l occurring in Eq. (3F-31) is equal to the negative of the function y_l, which is employed in the present context and which is characterized by the asymptotic behavior $y_l(x) \approx x^{-1}\sin(x - (l+1)\pi/2)$, for large x.

▼ *Shell structure in spherical nuclei (Fig. 6-47)*

In a spherical nucleus, the central part of the one-particle nuclear potential can be approximately represented by the function (Woods-Saxon potential)

$$V(r)=\left(1+\exp\left\{\frac{r-R}{a}\right\}\right)^{-1}V_0 \tag{6-496}$$

where V_0 and a are approximately independent of the particle number A, while R increases as $A^{1/3}$. The spectrum of bound states in such a potential with the parameters

$$\begin{aligned} V_0 &= -50 \quad \text{MeV} \\ a &= 0.67 \quad \text{fm} \end{aligned} \tag{6-497}$$

is illustrated in Fig. 6-47. (The effects of the Coulomb and spin-orbit potentials are considered below.) The quantity plotted in the figure is the value of the radius R for which the given orbit (n,l) has a binding energy of 7 MeV. Small differences in the ordinates in Fig. 6-47 can be converted to energy differences by means of the derivative $\partial\varepsilon/\partial R$. This derivative is of order V_0R^{-1}, and numerical values can be obtained from the coefficients given in the insert to Fig. 6-47. Thus, the present figure can be viewed as a representation of the $\varepsilon(n,l)$ energy surface that focuses attention on the shell structure at the Fermi level as a function of particle number (or radius), rather than as a function of energy for a particular nucleus.

The most important shell-structure effects in the spectrum of Fig. 6-47 can be associated with the ratios 2:1 and 3:1 for the angular and radial frequencies; the figure shows the corresponding trajectories as well as examples of the approximately parabolic functions (6-477) that describe the energy levels within each shell. (The trajectories have been obtained by numerical interpolation between the eigenvalues for integer n and l; for the infinite square-well potential, the results of this procedure agree with those given in Fig. 6-46 obtained on the basis of the analytic function (6-488).)

The gross features of the spectrum of the Woods-Saxon potential resemble rather closely those of the infinite square-well potential, since the penetration into the classically forbidden regions and the effect of the diffuseness of the surface can only produce shifts in the quantum number n, for fixed ε and l, amounting to a fraction of a unit. The less abrupt reflection at the surface of the Woods-Saxon potential implies that the ratio of radial to angular frequencies (for fixed energy) increases less rapidly with l than for the infinite square-well potential (see Eq. (6-480)).

Thus, in Fig. 6-47, the 2:1 shells (harmonic oscillator shell structure) have higher degeneracies than in the infinite square-well potential, while the 3:1 trajectory enters the physical region at a later point than in Fig. 6-46. ▲

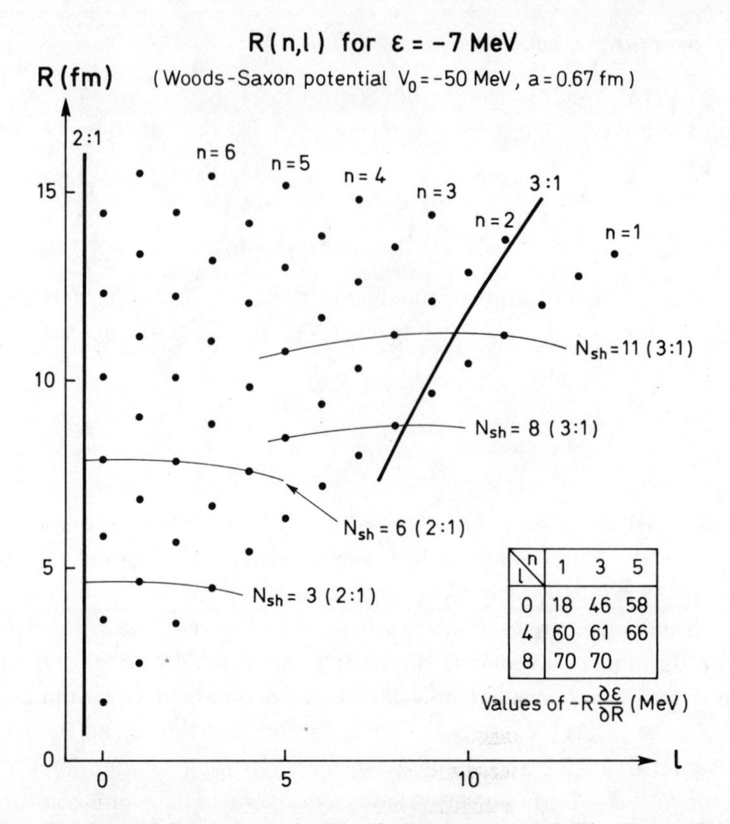

l \ n	1	3	5
0	18	46	58
4	60	61	66
8	70	70	

Figure 6-47 Single-particle spectrum in Woods-Saxon potential. The figure gives the value of the radius R for which the orbit (n,l) has a binding energy of 7 MeV. We wish to thank I. Hamamoto for help in the preparation of the figure.

▼ (The latter figure can be converted into the representation corresponding to Fig. 6-47 by assigning to the Fermi wave number the value $K=1.44\ \text{fm}^{-1}$, which is equivalent to a kinetic energy of 43 MeV.)

For protons, the Coulomb barrier implies a considerable reduction in the penetration into the classically forbidden region, and thus the shell structure more nearly resembles that of the infinite square well than is the case for neutrons. For $50 \lesssim Z \lesssim 100$, the 3:1 trajectory for the proton potential is rather close to that of the infinite square well.

The effect of the spin-orbit coupling on the energy spectrum can be approximately described by adding to the energy $\varepsilon(n,l)$ a term proportional to $\mathbf{l}\cdot\mathbf{s}$ (see, for example, Eq. (6-299)),

$$\varepsilon(nlj)=\varepsilon(nl)+v_{ls}\hbar\omega_0\begin{cases}\frac{1}{2}l & j=l+\frac{1}{2}\\ -\frac{1}{2}(l+1) & j=l-\frac{1}{2}\end{cases} \tag{6-498}$$

▲ where $\hbar\omega_0$ is the harmonic oscillator energy spacing, which provides a

▼ convenient unit of energy, while v_{ls} is a (negative) coefficient whose approximate magnitude is given in Table 5-1. The expression (6-498) implies that the angular frequency $\hbar^{-1}\partial\varepsilon/\partial l$ of particles with $j=l+1/2$ is decreased by the amount $\frac{1}{2}|v_{ls}|\omega_0$, while for $j=l-1/2$, the angular frequency is increased by the same amount. This effect leads to a corresponding splitting of the shell trajectories in such a manner that the $j=l+1/2$ branches of the trajectories in Fig. 6-47 are shifted to the left and the $j=l-1/2$ branches to the right by a few units in l. As a result, the degeneracy of the 2:1 shells is increased for $j=l-1/2$; for $j=l+1/2$, the 2:1 degeneracy is decreased, while the 3:1 trajectory enters the physical region for lower quantum numbers. Indeed, in the spectra around ^{208}Pb, the $j=l+1/2$ orbits $2f_{7/2}$, $1i_{13/2}$ (for protons) and $2g_{9/2}$, $1j_{15/2}$ (for neutrons) are separated by only 0.7 MeV and 1.4 MeV, respectively (see Fig. 3-2f, Vol. I, p. 324). Thus, the large shell-structure effects in the region of ^{208}Pb appear as a constructive interplay of the 2:1 shell trajectory for $j=l-1/2$ and the 3:1 trajectory for $j=l+1/2$.

The characterization of the shell structure in terms of the periodic classical orbits has immediate consequences for the deformations that occur for configurations with many particles outside closed shells. The maximal spatial localization that can be achieved by a superposition of the approximately degenerate one-particle states within a shell corresponds to wave packets that resemble the classical orbits. Attractive forces between nucleons imply that a deformation with this geometry will be especially favored. The shells associated with the 2:1 trajectory involve states with $\Delta l=2, 4, 6, \ldots$, from which one can build density distributions with large quadrupole moments (and higher multipole moments of even order). The importance of the spheroidal nuclear equilibrium deformations can thus be attributed to the significance of the 2:1 shell structure.

The orbits associated with the 3:1 trajectory have the symmetry of an equilateral triangle and involve a superposition of states with $\Delta l=3, 6, 9, \ldots$. The increasing importance of this type of shell structure in heavy nuclei is reflected in the approach to octupole instability in the nuclei in the region beyond ^{208}Pb (see the discussion on p. 561).

Nuclear shell structure for very large deformations (Figs. 6-48 to 6-50)

The occurrence of shape isomers in the fission process provides striking evidence for shell structure in nuclear potentials with much larger deformations than those encountered in the ground states of heavy nuclei. In the present example, we consider the characterization of the shells that appear to be responsible for this effect.

The discussion on p. 584ff. suggests that for large deformations, the shell-structure effects are of smaller magnitude than for spherical shape, except for potentials that resemble the harmonic oscillator. This special type

▲ of potential gives rise to degeneracies that are of the same magnitude

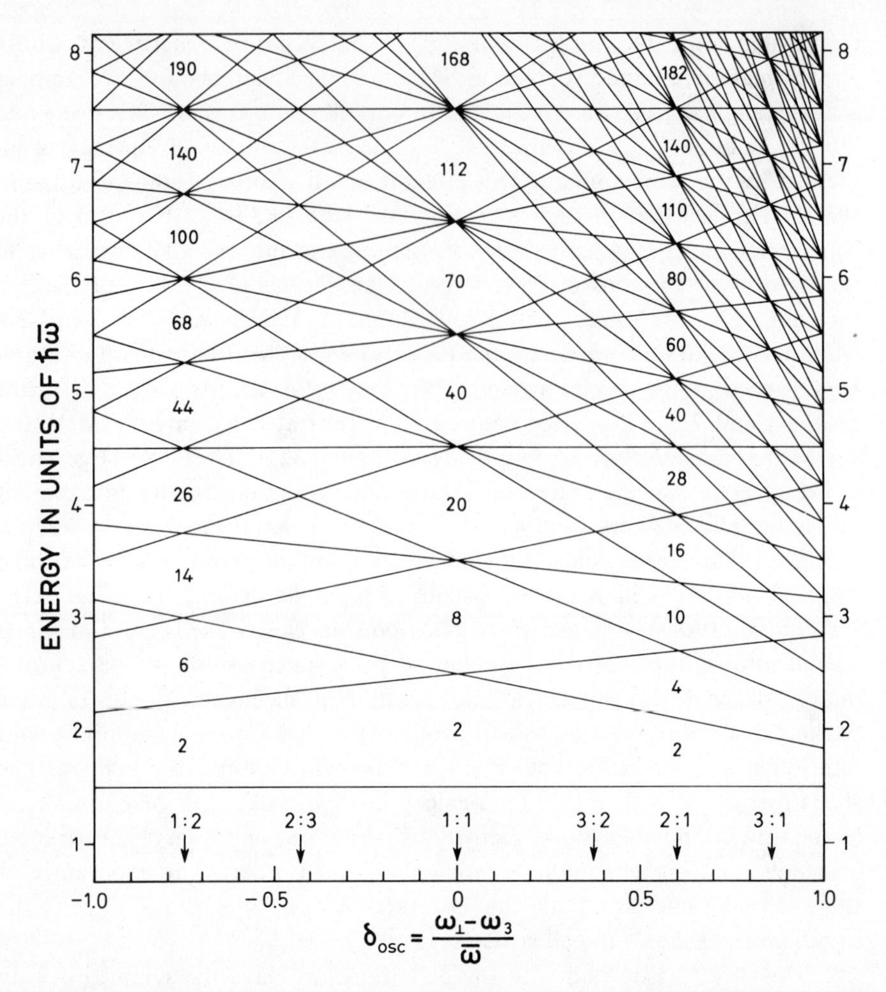

Figure 6-48 Single-particle spectrum for axially symmetric harmonic oscillator potentials. The eigenvalues are measured in units of $\bar{\omega} = (2\omega_\perp + \omega_3)/3$, and the deformation parameter δ_{osc} is that defined by Eq. (5-11). The arrows mark the deformations corresponding to the indicated rational ratios of frequencies $\omega_\perp : \omega_3$.

▼ ($\Omega \sim A^{2/3}$) as for the spherical shape, when the axes are in the ratios of small integers (Geilikman, 1960; see also Wong, 1970).

The single-particle spectrum for axially symmetric oscillator potentials is illustrated in Fig. 6-48. The deformation parameter δ_{osc} employed in the figure is defined by Eq. (5-11), and the eigenvalues, characterized by the quantum numbers n_3 and $n_\perp$, are given in units of the mean frequency $\bar{\omega}$ (see Eq. (5-13)). For prolate deformations, the ordering of the orbits of a spherical oscillator shell corresponds to increasing values of $n_\perp$ $(=0,1,2,\ldots,N)$. The eigenvalues for an axially symmetric oscillator have a degeneracy of $2(n_\perp + 1)$,

▲ due to the nucleon spin and the degeneracy in the perpendicular motion.

▼ The figure gives the total particle number corresponding to completed shells in the spherical potential as well as in the deformed potentials with frequency ratio $\omega_\perp : \omega_3$ equal to 2:1 (prolate) and 1:2 (oblate).

The nuclear potential with its rather well-defined surface implies deviations from the degeneracies in the harmonic oscillator potential. For the spherical shape, the effect of the surface can be approximately represented by a term in the energy that varies as $l(l+1)$ for the levels belonging to the same oscillator shell,

$$\delta\varepsilon(n,l) = v_{ll}\hbar\omega_0\left(l(l+1) - \tfrac{1}{2}(2n+l-\tfrac{1}{2})^2\right) \tag{6-499}$$

This expression corresponds to the term employed in the Hamiltonian (5-10) except for the replacement of $N(N+3)$ by $(N+3/2)^2$. The subtraction of a term depending on the total oscillator quantum number $N=2(n-1)+l$ ensures that the average distance between the oscillator shells is unaffected by the perturbation (6-499). (It may be noted that in a given region of the spectrum, the N-dependent term merely affects the energy scale and is thus equivalent to a renormalization of the oscillator frequency ω_0.)

The coefficient v_{ll} can be determined from the spectrum for the Woods-Saxon potential plotted in Fig. 6-47, and the values in Fig. 6-49 are determined so as to reproduce the separation between the members of the oscillator shell with maximum and minimum values of l. For comparison, Fig. ▲

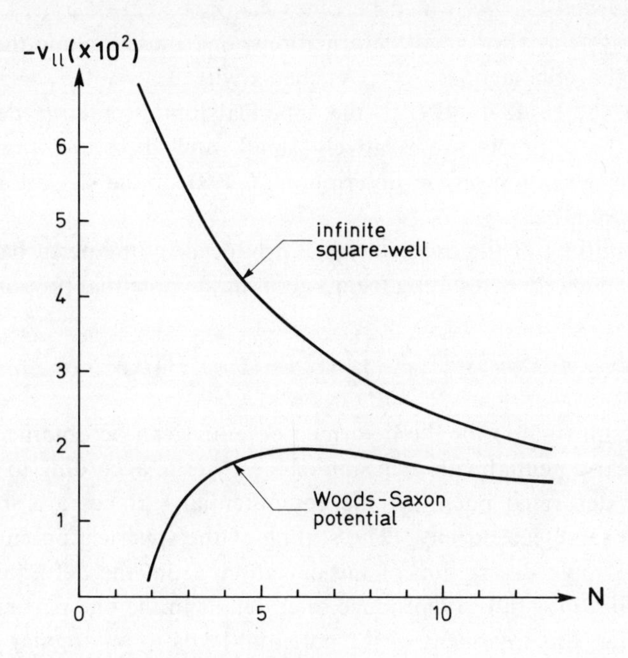

Figure 6-49 Deviations from oscillator degeneracy for a potential of Woods-Saxon shape. The values of v_{ll} for the Woods-Saxon potential are obtained from the spectra shown in Fig. 6-47 and for the infinite square well from Fig. 6-46.

▼ 6-49 also gives the values of v_{ll} appropriate to an infinite square-well potential. For protons, the values of v_{ll} are appreciably larger than for neutrons and, for $4 \leqslant N \leqslant 6$, approach rather closely those for the infinite square-well potential (see the discussion on p. 590). The values of v_{ll} in Table 5-1, which have been obtained from a fitting of the empirical single-particle levels in deformed nuclei, are seen to be in qualitative agreement with the results of the present analysis.

The anisotropic harmonic oscillator can be obtained from the isotropic oscillator by a scaling of the spatial coordinates (see Eq. (5-5))

$$x_1 \rightarrow \frac{\omega_\perp}{\omega_0} x_1 \qquad x_2 \rightarrow \frac{\omega_\perp}{\omega_0} x_2 \qquad x_3 \rightarrow \frac{\omega_3}{\omega_0} x_3 \tag{6-500}$$

while the momentum operators are unaffected. A simple treatment of the deviations from the oscillator potential is obtained by applying the same scaling to the $\mathbf{l}^2$ operator in the perturbation (6-499); the expectation value for the scaled perturbation in the state with quantum numbers $n_3 n_\perp \Lambda$ is given by

$$\delta\varepsilon(n_3 n_\perp \Lambda) = -\frac{\hbar v_{ll}}{\omega_0}\left(\tfrac{1}{2}\left((n_3+\tfrac{1}{2})\omega_3 - (n_\perp+1)\omega_\perp\right)^2 - \omega_\perp^2 \Lambda^2 + \omega_3 \omega_\perp\right) \tag{6-501}$$

The scaling of the term in Eq. (6-499) proportional to $(2n+l-1/2)^2$, which is included in Eq. (6-501), follows directly from the fact that $\hbar\omega_0(2n+l-1/2)$ is the unperturbed energy, which becomes $\hbar\omega_3(n_3+1/2)+\hbar\omega_\perp(n_\perp+1)$ for the deformed system. (For small deformations, one must include the additional effects of the nondiagonal matrix elements with $\Delta n_3 = -\Delta n_\perp = \pm 2$, in order to recover the result (6-499) in the spherical limit; for large deformations, however, these effects are relatively small, and their inclusion would go beyond the accuracy of the description (6-499) of the deviations from the oscillator potential.)

The splitting of the oscillator shell produced by the perturbation (6-499) can also be viewed as resulting from a term in the potential proportional to r^4,

$$\langle nl|r^4|nl\rangle = \left(\frac{\hbar}{M\omega_0}\right)^2 \left(\tfrac{3}{2}(2n+l-\tfrac{1}{2})^2 - \tfrac{1}{2}l(l+1) + \tfrac{3}{8}\right) \tag{6-502}$$

and the result (6-501) for the deformed potential can be obtained by scaling the spherical potential with inclusion of the r^4 term, according to Eq. (6-500). In such a deformed potential, the equipotential surfaces are spheriods, all having the same eccentricity. (The scaling of the spherical potential gives an especially simple description of the deviations from the deformed oscillator potential, but may fail to reproduce finer details in the one-particle spectrum. In particular, this treatment of the potential leads to an angular dependence of the surface steepness and hence to small systematic differences from the results obtained from the leptodermous model (see Eq. (4-188)).) ▲

▼ In a deformed axially symmetric oscillator potential with a rational ratio $a:b$ between the frequencies $\omega_\perp$ and ω_3, the shells correspond to a constant value of the shell quantum number (see Eq. (6-482))

$$N_{\mathrm{sh}} = an_\perp + bn_3 \tag{6-503}$$

while the spacing of the shells is given by (see Eq. (6-483))

$$\hbar\omega_{\mathrm{sh}} = \frac{\hbar\omega_\perp}{a} = \frac{\hbar\omega_3}{b} = \hbar\omega_0(a^2b)^{-1/3} \tag{6-504}$$

assuming a deformation with conservation of volume ($\omega_\perp^2\omega_3 = \omega_0^3$). The total energy with inclusion of the perturbation term (6-501) can thus be written

$$\varepsilon(N_{\mathrm{sh}}n_3\Lambda) = \hbar\omega_{\mathrm{sh}}\Big((N_{\mathrm{sh}} + a + \tfrac{1}{2}b) - v_{ll}(a^2b)^{-1/3}\big(2(b(n_3+\tfrac{1}{2}) - \tfrac{1}{2}(N_{\mathrm{sh}} + a + \tfrac{1}{2}b))^2 - a^2\Lambda^2 + ab\big)\Big) \tag{6-505}$$

The spectrum (6-505) is illustrated in Fig. 6-50, for a prolate deformation with $a:b = 2:1$ and for the value $v_{ll} = -0.019$ that is expected to approximately describe the neutron spectra in a nucleus with $A \approx 240$ (see Fig. 6-49). It is seen that the harmonic oscillator shells are spread over an energy region of about $\hbar\omega_{\mathrm{sh}}$. This situation corresponds approximately to that of spherical potentials where, in the absence of spin-orbit coupling, the deviations from the oscillator potential in heavy nuclei would imply a spreading of the shells over an energy region of about $\hbar\omega_0$ (since $|v_{ll}|N(N+1) \approx 1$).

The removal of the oscillator degeneracy resulting from the more sharply defined surface region implies that the maximum shell-structure effect may no longer correspond to a purely ellipsoidal deformation with ratio of axes 2:1. The perturbation term proportional to Λ^2 in Eq. (6-505) directly reflects the effect of the surface on the perpendicular oscillator motion and cannot be compensated by changes in the nuclear shape (though the term would be changed by a different treatment of the diffuseness), just as the $\mathbf{l}^2$ term in the spherical potential cannot be compensated by a deformation away from spherical symmetry. However, the term in Eq. (6-505) depending on n_3 reflects the fact that the sharpening of the surface acts differently on orbits that are concentrated in different regions of the nucleus. Thus, the orbits with maximum and minimum values of n_3 (and $\Lambda = 0$) correspond to pendulating motion along the major and minor axes, respectively; these orbits have no angular momentum and are equally disfavored by the perturbation of the oscillator potential. The orbits with intermediate values of n_3, which have classical turning points off the axes, carry angular momentum in directions perpendicular to the symmetry axis and are less disfavored. Hence, a shape deformation, which implies a relative extension of the nucleus in the directions of the poles and the equator, will tend to compensate the n_3-dependent ▲

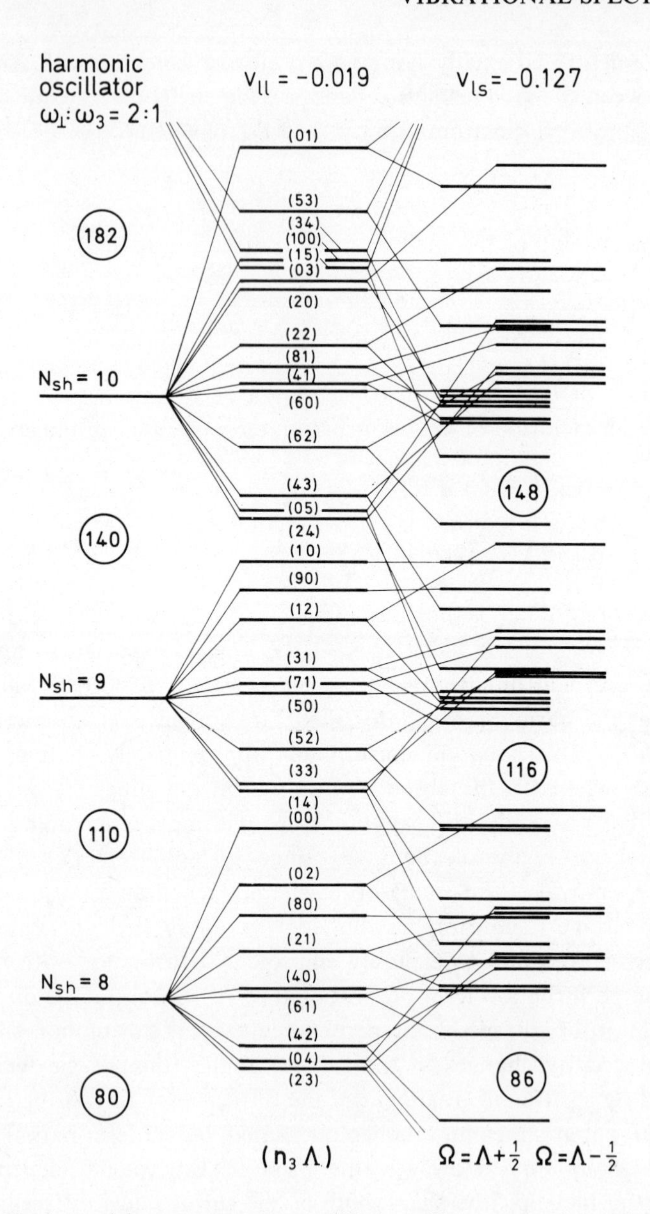

Figure 6-50 Single-particle spectrum for modified oscillator potential with ratios of axes 2:1. The spectrum on the left corresponds to an axially symmetric harmonic oscillator potential with $\omega_\perp:\omega_3=2:1$ with shell quantum number $N_{\mathrm{sh}}=2n_\perp+n_3=2N-n_3$. The spectrum in the center includes the effect of the perturbation (6-505), with $v_{ll}=-0.019$, while the spectrum on the right includes in addition the effect of the spin-orbit coupling (6-506), with $v_{ls}=-0.127$. The states with parallel spin and orbit are displaced slightly to the left, and the $\Lambda=0$ states have been included in this group.

▼ term in Eq. (6-505). Such a deformation is of approximate Y_{40} symmetry; for the parameters in Fig. 6-50, the amplitude required to achieve maximum degeneracy is of order $\beta_4 \sim +0.05$, as may be estimated from the expressions (4-192), (6-77), (6-78), and the relation (4-190) for β_4.

The effect of the deformation on the spin-orbit coupling (6-498) can be obtained by the same scaling as employed in the treatment of the $\mathbf{l}^2$ term; the resulting contribution to the one-particle energies is

$$\delta\varepsilon(N_{\text{sh}} n_3 \Lambda \Sigma) = v_{ls} \hbar\omega_\perp \Lambda\Sigma \qquad (6\text{-}506)$$

where Σ is the projection of the spin on the symmetry axis. The single-particle spectrum for a 2:1 deformation with the inclusion of the term (6-506) with $v_{ls} = -0.127$ (see Table 5-1) is shown in Fig. 6-50. It is seen that the spin-orbit coupling leads to a marked decrease in the spread of the oscillator shells for the levels with $\Omega = \Lambda - 1/2$. For $\Omega = \Lambda + 1/2$, the levels most strongly affected by the spin-orbit force are those with large Λ, and among these the levels with $n_\perp = \Lambda$ are the lowest (see Eq. 6-505) and Fig. 6-50). With the strength of the observed spin-orbit coupling, the levels with large $n_\perp = \Lambda$ and $\Omega = \Lambda + 1/2$ join the shell below. Thus, for $N_{\text{sh}} = 9$, the levels with $n_\perp = \Lambda = 4$, 3, and 2 join the shell with $N_{\text{sh}} = 8$, while the $n_\perp = \Lambda = 5$, 4, 3, 2 levels from $N_{\text{sh}} = 10$ join the $N_{\text{sh}} = 9$ shell. As a result, the closed-shell numbers are changed from 110 and 140 to 116 and 148, respectively, and the magnitude of the shell-structure effect is significantly increased. These effects of the spin-orbit coupling are seen to be similar to those occurring in the spherical potential (see pp. 591 ff.). For the larger deviations from the harmonic oscillator degeneracies that characterize the proton spectra, the shell-structure effects in the 2:1 potential become rather small in the region of the actinides, but the shell closing at $Z \approx 86$ is expected to give a significant effect.

The experimental evidence on the fission isomers can be understood in terms of the formation of a closed shell for neutron number $N = 148$ (as in $^{242}_{94}\text{Pu}$) and with a shape corresponding approximately to a potential with 2:1 symmetry (see the discussion on p. 634).

The shell structure in the deformed nuclear potential can be characterized by the geometry of the closed classical orbits, in the manner discussed on pp. 582 ff. The frequency ω_ρ of the transverse radial motion is associated with the quantum number n_ρ representing the number of radial nodes

$$n_\perp = 2(n_\rho - 1) + \Lambda \qquad (6\text{-}507)$$

For the deformed harmonic oscillator potential with the ratio of axes 2:1, the energy is a linear function of $N_{\text{sh}} = n_3 + 4(n_\rho - 1) + 2\Lambda$, and the frequency ratio is $\omega_3 : \omega_\rho : \omega_\varphi = 1:4:2$. The corresponding orbits are not in general planar; the projection on a plane perpendicular to the symmetry axis is an ellipse with frequency ratio $\omega_\rho : \omega_\varphi = 2:1$, while the projection on a plane containing the symmetry axis has a frequency ratio of $\omega_\rho : \omega_3 = 4:1$ and a geometry varying from that of a figure eight to a banana ▲

▼ shape, depending on the relative phase of the components of the motion parallel and perpendicular to the symmetry axis. The degeneracies that are introduced by the spin-orbit coupling involving levels from adjacent shells correspond to constant values of $n_3+5n_\rho+2\Lambda$, and therefore imply a frequency ratio of 1:5:2. Orbits of this type, when projected on a plane perpendicular to the symmetry axis, describe a rosette with a fivefold symmetry axis.

The geometry of the classical orbits is connected with the shapes of the deformations away from the oscillatory symmetry that will be produced by particles outside closed shells (see the discussion on p. 591). Thus, for example, the banana-shaped orbits in the oscillator potential, with ratio of axes 2:1, may lead to instability in the bending mode; such an instability would imply a tendency for the system on the way toward fission to avoid the 2:1 symmetry by an excursion involving an odd-parity deformation away from axial symmetry.

Shell-Structure Effects in Nuclear Energy

The shell structure in the single-particle spectrum gives rise to collective effects in the total nuclear energy that can be obtained from the sum of the single-particle energies of the occupied orbits, as discussed on pp. 367ff. The following examples illustrate the dependence of the shell-structure energy on particle number, temperature, and deformation.

Shell-structure effect in nuclear masses (Fig. 6-51)

Evidence on the nuclear shell-structure energy is provided by the extensive body of data on the nuclear masses.[31] The main terms in the nuclear binding energy are smooth functions of N and Z that can be described by the semi-empirical mass formula (2-12). The more detailed analysis of the observed masses reveals deviations from the smooth variations with N and Z that are correlated with the nuclear shell structure (see Fig. 2-4, Vol. I, p. 168).

As a first step in a qualitative discussion of the effects of the shell structure on the ground-state masses, we consider particle motion in a spherical harmonic oscillator potential. For this potential, the single-particle eigenvalues have a simple analytic dependence on the quantum numbers, and the separation of the sum of the single-particle energies $\mathscr{E}_{\text{ip}}$ into a smooth part $\tilde{\mathscr{E}}_{\text{ip}}$ and the shell-structure energy $\mathscr{E}_{\text{sh}}$ (see Eq. (6-97)) can be obtained by elementary methods.

Denoting the total number of particles by $\mathscr{N}$ (we consider a single type ▲ of particle, neutrons or protons) and the total oscillator quantum number of

[31] The interpretation of the shell-structure terms in the nuclear masses was considered by Myers and Swiatecki, 1966. Estimates based on single-particle spectra for the nuclear potential were given by Strutinsky, 1967a; Seeger and Perisho, 1967; Nilsson *et al.*, 1969.

▼ the last filled shell by N_0, we have (see Eqs. (2-151) and (2-152))

$$\mathcal{N} = \sum_{N=0}^{N_0} (N+1)(N+2) + x(N_0+2)(N_0+3)$$

$$= \tfrac{1}{3}(N_0+1)(N_0+2)(N_0+3) + x(N_0+2)(N_0+3) \qquad (6\text{-}508a)$$

$$\mathcal{E}_{ip} = \sum_{N=0}^{N_0} (N+1)(N+2)(N+\tfrac{3}{2})\hbar\omega_0 + x(N_0+2)(N_0+3)(N_0+\tfrac{5}{2})\hbar\omega_0$$

$$= \hbar\omega_0\left(\tfrac{1}{4}(N_0+1)(N_0+2)^2(N_0+3) + x(N_0+2)(N_0+\tfrac{5}{2})(N_0+3)\right) \qquad (6\text{-}508b)$$

where x represents the fractional filling of the shell N_0+1, which has a total degeneracy $(N_0+2)(N_0+3)$. The expressions (6-508) include the twofold spin degeneracy. For large quantum numbers, Eq. (6-508a) implies the asymptotic relation

$$N_0+2+x = (3\mathcal{N})^{1/3} + \tfrac{1}{3}(3\mathcal{N})^{-1/3}(1-3x+3x^2) + \cdots \qquad (6\text{-}509)$$

which makes it possible to express $\mathcal{E}_{ip}$ in the form

$$\mathcal{E}_{ip} = \tilde{\mathcal{E}}_{ip} + \mathcal{E}_{sh}$$

$$\tilde{\mathcal{E}}_{ip} = \hbar\omega_0\left(\frac{1}{4}(3\mathcal{N})^{4/3} + \frac{1}{8}(3\mathcal{N})^{2/3}\right) \qquad (6\text{-}510)$$

$$\mathcal{E}_{sh} \approx \frac{1}{24}\hbar\omega_0(3\mathcal{N})^{2/3}(-1+12x(1-x))$$

for large values of $\mathcal{N}$. In Eq. (6-510), we have included a term of order $(3\mathcal{N})^{2/3}$ in $\tilde{\mathcal{E}}_{ip}$, although this term is of the same order of magnitude as $\mathcal{E}_{sh}$ and, therefore, cannot be uniquely separated from $\mathcal{E}_{sh}$ in the present analysis. The x-independent term in $\mathcal{E}_{sh}$ has been chosen in such a manner that $\mathcal{E}_{sh}$ vanishes when the system is deformed away from spherical symmetry (see p. 603) or excited to high temperatures (see p. 609).

The leading term in $\tilde{\mathcal{E}}_{ip}$ in Eq. (6-510) is seen to be of order $\varepsilon_F\mathcal{N}$, corresponding to a volume energy, and to have the value given by the Fermi gas approximation

$$\mathcal{N} \approx \frac{1}{3\pi^2}\left(\frac{2M}{\hbar^2}\right)^{3/2} \int d\tau(\tilde{\varepsilon}_F - V(r))^{3/2}$$

$$= \frac{1}{3}\left(\frac{\tilde{\varepsilon}_F}{\hbar\omega_0}\right)^3 \qquad (6\text{-}511a)$$

▲

$$\tilde{\mathscr{E}}_{\text{ip}} \approx \frac{1}{3\pi^2}\left(\frac{2M}{\hbar^2}\right)^{3/2} \int d\tau \left(\tilde{\varepsilon}_{\text{F}} - V(r)\right)^{3/2} \left(\frac{3}{5}\left(\tilde{\varepsilon}_{\text{F}} - V(r)\right) + V(r)\right)$$

$$= \frac{1}{4}\left(\frac{\tilde{\varepsilon}_{\text{F}}}{\hbar\omega_0}\right)^4 \hbar\omega_0 \approx \frac{1}{4}(3\mathscr{N})^{4/3}\hbar\omega_0 \qquad (6\text{-}511\text{b})$$

where the notation $\tilde{\varepsilon}_{\text{F}}$ refers to the smoothly varying part of ε_{F} defined by the Fermi gas model. The oscillator potential has the special property that the surface term, proportional to $\varepsilon_{\text{F}} \mathscr{N}^{2/3} \sim \hbar\omega_0 \mathscr{N}$, has a vanishing coefficient.

The order of magnitude of the shell-structure energy in Eq. (6-510) is seen to be $\hbar\omega_0\Omega$, where Ω $(\approx N_0^2 \approx (3\mathscr{N})^{2/3})$ is the degeneracy of the shells; the coefficient of the x-dependent term in $\mathscr{E}_{\text{sh}}$ reflects the fact that the average shift of the single-particle levels relative to a uniform energy spectrum is $(\hbar\omega_0)/4$ and that there are $\Omega/2$ particles in a half-filled shell. The total energy (6-508b), with the smooth part $\tilde{\mathscr{E}}_{\text{ip}}$ subtracted, is plotted as a function of particle number (for even values of $\mathscr{N}$), in Fig. 6-51; it is seen that the asymptotic expression (6-510) for $\mathscr{E}_{\text{sh}}$ provides an approximate description, even for moderate values of $\mathscr{N}$.

The harmonic oscillator represents an extreme case with complete degeneracy within each major shell. The spreading of the levels within a shell implies a reduction in the shell-structure energy. Thus, if we assume that the single-particle levels within a single major shell are uniformly distributed over

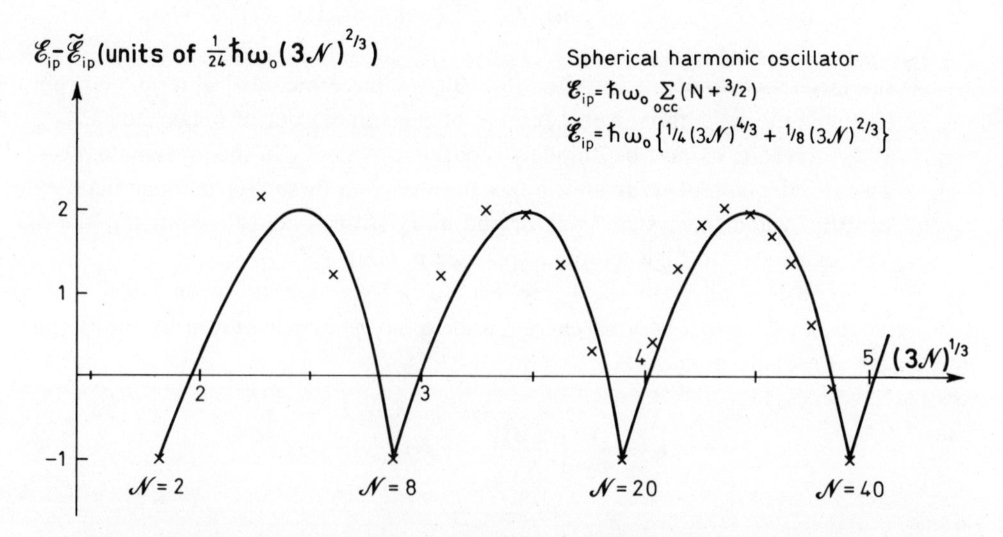

Figure 6-51 Shell-structure energy as a function of particle number for spherical harmonic oscillator. The values of $\mathscr{E}_{\text{sh}} = \mathscr{E}_{\text{ip}} - \tilde{\mathscr{E}}_{\text{ip}}$ are plotted as $\times$, while the asymptotic expression (6-510) for $\mathscr{E}_{\text{sh}}$ is given by the solid curve.

▼ an energy interval W, corresponding to a single-particle level density

$$g(\varepsilon)=\begin{cases}\dfrac{\Omega}{W} & |\varepsilon-(N+\frac{3}{2})\hbar\omega_0|<W/2 \\ 0 & \text{otherwise}\end{cases} \tag{6-512}$$

the relation (6-509) is unaffected, while the shell-structure energy in Eq. (6-510) is multiplied by the factor $(1-W/\hbar\omega_0)$.

The above estimates of the shell-structure effect in the total energy refer to the spherical shape. For configurations with many particles outside closed shells, the nucleus can gain energy by a deformation away from spherical symmetry, and the analysis on pp. 602 ff. (see in particular Fig. 6-52) indicates that (with the x-independent term as in Eq. (6-510)), the resulting shell-structure energy for the equilibrium shape remains negative, even for half-filled shells, but is almost an order of magnitude smaller than for closed shells. Hence, the variation of the shell-structure energy during the filling of the shell is expected to correspond approximately to the value estimated for closed shells. For the harmonic oscillator potential with closed shells of both neutrons and protons, we obtain, from Eq. (6-510),

$$\begin{aligned}\mathscr{E}_{\rm sh}(x_N=x_Z=0)&=-\frac{1}{24}\hbar\omega_0\left((3N)^{2/3}+(3Z)^{2/3}\right)\\ &\approx-4.5A^{1/3}\ \text{MeV}\end{aligned} \tag{6-513}$$

where, in the last expression, we have neglected terms of order $(N-Z)^2/A^2$ and employed the value $\hbar\omega_0=41A^{-1/3}$ MeV.

The empirical evidence on the shell-structure effect in the ground-state masses can be studied under especially favorable conditions in the region of ^{208}Pb, where both neutrons and protons have closed-shell configurations. The observed deviation from the smooth curve in Fig. 2-4 amounts to about 0.06 MeV per particle, corresponding to an increased binding energy of about 13 MeV.

The estimate (6-513) would imply $\mathscr{E}_{\rm sh}\approx-26$ MeV for ^{208}Pb, but a reduction of this value results from the lack of complete degeneracy within the nuclear shells. The observed reduction by about a factor of 2 can be understood from the fact that the single-particle levels within each shell are spread over an energy interval W of about half the separation between shells (see, for example, Vol. I, Figs. 2-30, p. 239, and 3-3, p. 325). The effect of departure from complete degeneracy within a shell can also be treated in terms of a single-particle level density with a harmonic dependence on ε, as in Eq. (6-522); the empirical single-particle spectra in the region of ^{208}Pb correspond to $f\approx g_0$, which implies a reduction of the estimate (6-510) for $\mathscr{E}_{\rm sh}$ by a factor of $6\pi^{-2}$ (see Eq. (6-532b) for $\tau\rightarrow 0$ and $y=x=0$, as given by Eq.
▲ (6-529a)).

▼ *Global features of shell-structure energy as function of deformation; shape isomers (Fig. 6-52)*

The one-particle potential may exhibit major shell-structure effects for a number of different shapes, corresponding to different symmetries for the classical orbits. This feature is especially pronounced for the harmonic oscillator potential, which gives rise to high degeneracies whenever the frequencies are in the ratio of integers (see the discussion on p. 585). The shell-structure energy for axially symmetric oscillator potentials is shown in Fig. 6-52 as a function of particle number $\mathcal{N}$ and the deformation variable δ_{osc} defined in terms of the oscillator frequencies. The value of $\mathcal{E}_{sh}(\mathcal{N}, \delta_{osc})$ has been obtained by a numerical summation of the $\mathcal{N}$ lowest single-particle eigenvalues and subtraction of the smooth function $\tilde{\mathcal{E}}_{ip}$ derived in the small print below. For the spherical potential ($\delta_{osc} = 0$), the variation of $\mathcal{E}_{sh}$ with particle number is the same as that given by Fig. 6-51. The potential energy surface also shows pronounced minima for closed-shell configurations for shapes where the frequencies $\omega_\perp$ and ω_3 are different, but in the ratio of (small) integers. The effects are especially large for shapes with $\omega_\perp : \omega_3 = 1:2$ and $2:1$, and the closed-shell numbers for these potentials are indicated in Fig. 6-52 (see also the single-particle level diagram in Fig. 6-48).

For large quantum numbers, the shell-structure energy in the neighborhood of each symmetry point develops an asymptotic pattern, which is the same for the different symmetry points, apart from a scaling. For a ratio of frequencies $\omega_\perp : \omega_3 = a:b$, the scale factors for the various quantities are

$$\begin{aligned} \hbar\omega_{sh}&: \quad \mathcal{N}^{-1/3}(a^2b)^{-1/3} \\ \Omega&: \quad \mathcal{N}^{2/3}(a^2b)^{-1/3} \\ \mathcal{E}_{sh}&: \quad \mathcal{N}^{1/3}(a^2b)^{-2/3} \\ \delta_{osc}&: \quad \mathcal{N}^{-1/3}(a^2b)^{-1/3} 9ab(2a+b)^{-2} \end{aligned} \tag{6-514}$$

The scaling of the shell spacing is obtained from Eq. (6-504), assuming an oscillator frequency ω_0 proportional to $\mathcal{N}^{-1/3}$. The degeneracy Ω of a shell is equal to $\hbar\omega_{sh}$ divided by the average spacing between single-particle levels, which is proportional to $\mathcal{N}^{-1}$. The scale factor for $\mathcal{E}_{sh}$ is the product of that for $\hbar\omega_{sh}$ and Ω. Finally, the appropriate unit for measuring the deviation of δ_{osc} from the value at the symmetry point is obtained by comparing the separation between successive shells and the energy separation within each shell arising from a deformation away from the symmetry point.

The asymptotic pattern in the neighborhood of each symmetry point involves a system of ridges and valleys that reflect the crossing of orbits from neighboring shells resulting from a deformation away from a symmetry point. Thus, for example, for a closed-shell configuration, the shell-structure energy ▲ increases with deformation until the deformation is of the magnitude of the

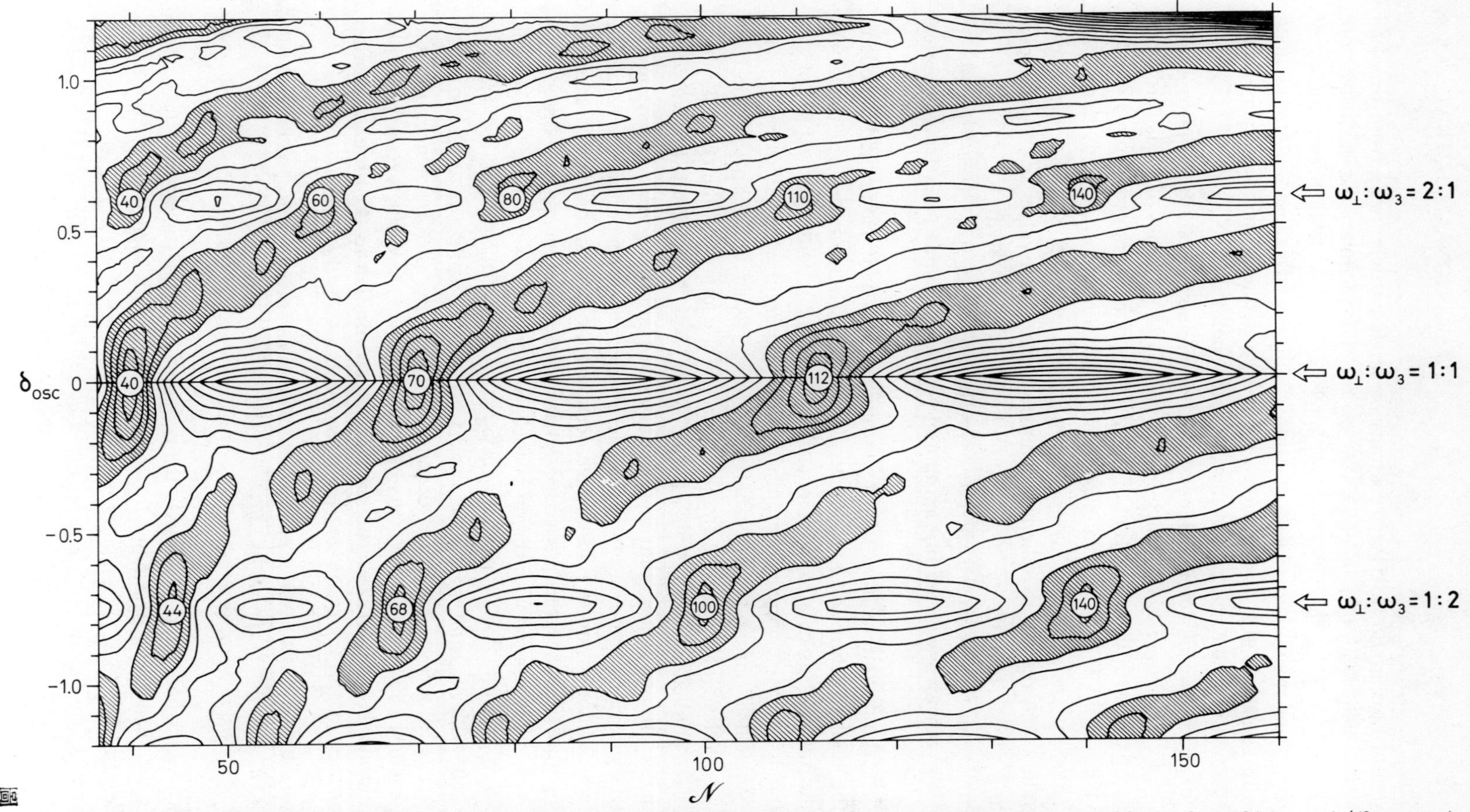

Figure 6-52 Shell-structure energy for axially symmetric oscillator potentials. The deformation variable is $\delta_{osc} = 3(\omega_\perp - \omega_3)/(2\omega_\perp + \omega_3)$. The basic unit of energy in the figure is $\hbar(\omega_\perp^2 \omega_3)^{1/3}(3\mathcal{N})^{2/3}/24$, corresponding to the asymptotic value of the shell-structure energy for a closed-shell configuration in a spherical potential. The shaded area represents configurations with negative values of $\mathscr{E}_{sh}$, and the level lines correspond to values of $\mathscr{E}_{sh}$ differing by one quarter of the basic unit. We thank Jens Damgaard for help in the preparation of the figure.

▼ scale factor in Eq. (6-514); crossings of single-particle levels from the shell above then lead to a decrease in $\mathscr{E}_{sh}$. In this manner, the shell-structure energy for given particle number approaches zero through oscillations of decreasing magnitude. (The formation of ridges and valleys that are inclined with respect to the deformation axis is a consequence of the prolate-oblate asymmetry in the splitting of the oscillator shell.)

The scale of deformations around a symmetry point is proportional to $\mathscr{N}^{-1/3}$ (see Eq. (6-514)), which for sufficiently large values of $\mathscr{N}$ is very small compared with the separation of the different major symmetry points. For the values of $\mathscr{N}$ in Fig. 6-52, however, the separation is only partially developed. Thus, the closed shell at $\mathscr{N}=140$ for the 2:1 potential occurs in the second valley with respect to spherical symmetry. For small values of $\mathscr{N}$, the local and global features of $\mathscr{E}_{sh}$ cannot be at all separated; for example, the minimum for the closed shell at $\mathscr{N}=10$ and $\omega_\perp:\omega_3=2:1$ (see Fig. 6-48, p. 592) coincides with the first minimum for deformations away from spherical symmetry.

The deviations from the asymptotic patterns give rise to modulations of the shell structure for $a:b$ different from 1:1. Thus, for the 2:1 potentials, the maxima associated with partially filled shells with odd values of N_{sh} ($\mathscr{N}\approx 7,22,50,\ldots$) are systematically greater than for even values of N_{sh}. This "supershell" structure is associated with the existence of a special family of closed classical orbits with frequency $\omega_\perp$. While almost all the classical phase space is filled with orbits with frequency $\omega_{sh}=\omega_3$, the restricted set of orbits that correspond to motion in the equatorial plane have twice this frequency and give rise to modulation in the energy spectrum with periodicity $\hbar\omega_\perp$. The reduced dimension of phase space occupied by these special orbits accounts for the fact that the associated shells have a degeneracy of order N_{sh}^{-1} $\sim\mathscr{N}^{-1/3}$compared with the main shell structure in this potential (see p. 586). More generally, for a potential with $\omega_\perp:\omega_3=a:1$ with a integer, the family of shells with $N_{sh}=an+p$, where n is a fixed integer while p takes the values 0, $1,\ldots,a-1$, have the same degeneracy $\Omega=(n+1)(n+2)$; hence, the degeneracy as a function of N_{sh} contains oscillatory terms with a basic period of a shells. This supershell structure gives a contribution to $\mathscr{E}_{sh}$ of relative order $n^{-1}\sim\Omega^{-1/2}\sim(3\mathscr{N})^{-1/3}$, which can be expressed in the form

$$\Delta\mathscr{E}_{sh}=\hbar\omega_0(3\mathscr{N})^{1/3}\frac{a}{3}\left(\frac{p+x}{a}-\frac{1}{2}\right)\left(\left(\frac{p+x}{a}\right)\left(\frac{p+x}{a}-1\right)-3\frac{x(1-x)}{a^2}\right) \tag{6-515}$$

where p labels the partly filled shell occupied by $x\Omega$ particles; thus, $(p+x)/a$ is the shell-filling parameter for the supershell. It is seen that, for $a=2$, the contribution (6-515) vanishes for closed shells ($x=0$) with even ($p=0$) as well as odd ($p=1$) values of N_{sh}; for half-filled shells ($x=1/2$), the value of $\Delta\mathscr{E}_{sh}$ ▲ is equal to $(-1)^p(3/2)(3\mathscr{N})^{-1/3}$, in the units employed in Fig. 6-52.

▼ The potential energy function of Fig. 6-52 brings together many of the qualitative effects of the shell structure that have been identified in the different nuclear equilibrium configurations. The largest effects are those occurring for closed shells in the spherical potential, as discussed in the previous example (pp. 598 ff.). For configurations with particles outside closed shells, the nucleus can gain energy by deformation, and in the middle of shells, the minima occur for deformations $\delta \sim \mathcal{N}^{-1/3}$ (see the discussion in Chapter 4, p. 136). It is seen that the value of $\mathscr{E}_{sh}$ in Fig. 6-52 is always negative at the minimum closest to spherical symmetry; however, the magnitude of $\mathscr{E}_{sh}$ at this minimum decreases rapidly with the addition of particles to closed shells. (Asymptotically, the minimum value of $\mathscr{E}_{sh}$ for half-filled shells is $\mathscr{E}_{sh}(x=1/2) = -0.16$, in the units employed in Fig. 6-52.)

The occurrence of shell structure associated with potentials of different symmetry implies the possibility of shape isomers with widely different equilibrium shapes. Whether or not a minimum in $\mathscr{E}_{sh}$ will give rise to a shape isomer depends on the behavior of the bulk part $\tilde{\mathscr{E}}$ of the nuclear energy (the smoothly varying term in the total nuclear potential energy, see Eq. (6-93)); in general, $\tilde{\mathscr{E}}$ is not stationary at the minimum points of $\mathscr{E}_{sh}$ that lie away from spherical symmetry.

For nuclei in the mass number region $A \approx 240$, the fission saddle point of $\tilde{\mathscr{E}}$ (as obtained from the liquid-drop model) is associated with deformations of $\delta \approx 0.6$. The stationary character of $\tilde{\mathscr{E}}$ at this point is especially favorable for the occurrence of shape isomers associated with closed-shell configurations in the 2:1 potential; see the discussion on pp. 634 ff. of the fission isomers that result from the shell closing at $N = 148$ in the modified 2:1 oscillator potential. (Additional possibilities for shape isomerism may occur as a result of stationary values of $\tilde{\mathscr{E}}$ for finite values of the angular momentum; see the discussion of $\tilde{\mathscr{E}}$ as a function of angular momentum in Sec. 6A-2.)

While the global pattern of the shell-structure energy is specific to the potential considered, some of the qualitative features in the neighborhood of a symmetry point are expected to be of a rather general nature. Thus, the system of ridges and valleys associated with the gradual disappearance of the shell-structure energy for deformations that break the symmetry is a general consequence of the crossings of adjacent shells, as discussed above (Myers and Swiatecki, 1967; Strutinsky, 1968).

The deformations of special significance for configurations with particles outside closed shells are those possessing the symmetry of the classical closed orbits that are responsible for the shells (see the discussion on p. 591). The scale of the deformations is expected to be quite generally of order $\mathcal{N}^{-1/3}$, corresponding to the ratio of $\hbar\omega_{sh}$ and ε_F. However, the magnitude of $\mathscr{E}_{sh}$ is of order $\Omega\hbar\omega_{sh}$ and thus depends on the degeneracy Ω, which involves different powers of the particle number depending on the particular symmetries of the potential considered (see Table 6-17, p. 586). The occurrence of equilibrium deformations of order $\mathcal{N}^{-1/3}$, independently of the magnitude of the degeneracy Ω, may seem surprising, since the particles in unfilled shells can
▲ themselves only produce deformations of order $\Omega\mathcal{N}^{-1}$, which may be much less than

▼ $\mathcal{N}^{-1/3}$ for potentials with lower degeneracy than the oscillator. The large deformations result from the polarization of the particles in closed shells and reflect the relative ease with which the closed shells can be deformed in the absence of the high degeneracies of the harmonic oscillator potential.

Calculation of $\tilde{\mathscr{E}}_{\text{ip}}$ for anisotropic harmonic oscillator potential

A separation of the asymptotic function $\tilde{\mathscr{E}}_{\text{ip}}$ from the sum of single-particle eigenvalues can be based on an analysis of the single-particle level density

$$g(\varepsilon)=\sum_{\nu}\delta(\varepsilon-\varepsilon_{\nu}) \tag{6-516}$$

where ε_ν represents the single-particle eigenvalue labeled by the quantum numbers ν. In a situation in which the parameters of the potential vary smoothly with the particle number $\mathcal{N}$, it is possible to define a function $\tilde{g}(\varepsilon,\mathcal{N})$ which asymptotically, for large $\mathcal{N}$, represents the number of levels per unit energy, in an interval around ε that is large compared to the spacing of the shells, but small compared to the energies over which $\tilde{g}$ varies significantly. For a saturating system, for which the parameters of the potential can be expanded in powers of $\mathcal{N}^{1/3}$, the quantity $\tilde{g}$ is obtained in terms of a corresponding series involving decreasing powers of $\mathcal{N}^{1/3}$, with a leading term of order $\mathcal{N}$.

From the function $\tilde{g}(\varepsilon,\mathcal{N})$, one can determine the asymptotic form of the total independent-particle energy by means of the relations

$$\mathcal{N}=\int_0^{\tilde{\varepsilon}_F}\tilde{g}(\varepsilon,\mathcal{N})\,d\varepsilon \tag{6-517a}$$

$$\tilde{\mathscr{E}}_{\text{ip}}(\mathcal{N})=\int_0^{\tilde{\varepsilon}_F}\tilde{g}(\varepsilon,\mathcal{N})\varepsilon\,d\varepsilon \tag{6-517b}$$

where $\tilde{\varepsilon}_F(\mathcal{N})$, defined by Eq. (6-517a), describes the asymptotic behavior of the Fermi energy.

For the special case of the oscillator potential (or the infinite square-well potential), the spectrum of single-particle energies for different values of $\mathcal{N}$ is obtained by a simple scaling; the function $\tilde{g}(\varepsilon,\mathcal{N})$ can therefore also be expressed as a power series in ε (or $\varepsilon^{1/2}$ for the infinite square-well potential). Such a power series can be conveniently obtained from a Laplace transformation of the single-particle level density (Bhaduri and Ross, 1971). For the anisotropic harmonic oscillator with the energy spectrum (6-485), the transform is given by

$$\begin{aligned}f(\beta)&\equiv\int_0^{\infty}g(\varepsilon)\exp\{-\beta\varepsilon\}\,d\varepsilon\\&=2\sum_{n_1n_2n_3}\exp\left\{-\beta\sum_{\kappa=1}^{3}(n_\kappa+\tfrac{1}{2})\hbar\omega_\kappa\right\}\\&=\frac{1}{4}\prod_{\kappa=1}^{3}\left(\sinh\tfrac{1}{2}\beta\hbar\omega_\kappa\right)^{-1}\end{aligned} \tag{6-518}$$

A term of the form ε^n in $\tilde{g}(\varepsilon)$ gives rise to a term in $f(\beta)$ of the form $n!\beta^{-(n+1)}$; thus, the expression for $\tilde{g}(\varepsilon)$ can be obtained from the expansion of $f(\beta)$ in the neigh-

▼ borhood of the origin,

$$\tilde{g}(\varepsilon)=\hbar^{-3}(\omega_1\omega_2\omega_3)^{-1}\left(\varepsilon^2-\frac{\hbar^2}{12}(\omega_1^2+\omega_2^2+\omega_3^2)\right) \tag{6-519}$$

(The $\mathcal{N}$ dependence is contained in the variation of the frequencies ω_1, ω_2, and ω_3 with particle number.) From Eqs. (6-517) and (6-519), we obtain

$$\tilde{\varepsilon}_F=\hbar(\omega_1\omega_2\omega_3)^{1/3}\left((3\mathcal{N})^{1/3}+\frac{1}{12}\frac{\omega_1^2+\omega_2^2+\omega_3^2}{(\omega_1\omega_2\omega_3)^{2/3}}(3\mathcal{N})^{-1/3}\right) \tag{6-520a}$$

$$\tilde{\mathscr{E}}_{ip}=\hbar(\omega_1\omega_2\omega_3)^{1/3}\left(\frac{1}{4}(3\mathcal{N})^{4/3}+\frac{1}{24}\frac{\omega_1^2+\omega_2^2+\omega_3^2}{(\omega_1\omega_2\omega_3)^{2/3}}(3\mathcal{N})^{2/3}\right) \tag{6-520b}$$

For the isotropic oscillator, the result for $\tilde{\mathscr{E}}_{ip}$ is seen to agree with Eq. (6-510), which was obtained by a direct extraction of the asymptotic terms in $\mathscr{E}_{ip}$.

Effects of shell structure at finite temperature (Figs. 6-53 to 6-55)

Though potentially a topic of considerable scope, the present evidence concerning the effect of shell structure at finite temperatures is rather limited. The most direct evidence is obtained from the total level densities, in particular at excitation energies in the neighborhood of the neutron separation energy; see Fig. 2-12, Vol. I, p. 187.[32]

With increasing temperature, the effects of the shell structure in the one-particle motion are of decreasing importance because of the many configurations that may contribute. For sufficiently high temperatures, the nuclear properties are expected to vary smoothly with particle numbers. In the present example, we shall illustrate the temperature dependence of the shell-structure effects by evaluating the thermodynamic functions for simple schematic forms of the one-particle spectrum.

Shell structure in an extreme form can be represented by a single-particle spectrum consisting of equally spaced, fully degenerate shells, each with the same degeneracy Ω,

$$g(\varepsilon)=\Omega\sum_n \delta(\varepsilon-\varepsilon_0-n\hbar\omega_{sh}) \tag{6-521}$$

More generally, the one-particle spectrum can be expressed in terms of a ▲ Fourier expansion (Ericson, 1958a); we shall illustrate the terms that occur in

---

[32]The effect of shell structure on nuclear level densities was discussed by Rosenzweig (1957) and Ericson (1958a) on the basis of schematic representations of the one-particle spectrum; for evaluations employing empirical one-particle spectra, see, for example, the review by Huizenga and Moretto (1972); for evaluations including rotational excitations in deformed nuclei, see Døssing and Jensen (1974) and Huizenga *et al.*, (1974).

▼ such an analysis by considering a spectrum involving a single harmonic

$$g(\varepsilon)=g_0+f\cos\left\{2\pi\frac{\varepsilon-\varepsilon_0}{\hbar\omega_{\text{sh}}}\right\} \tag{6-522}$$

where g_0 is the average level spacing and f represents the amplitude of the shell-structure modulation.

For a system with independent-particle motion, the thermodynamic functions, for any given single-particle spectrum, can be obtained from the partition function for the grand canonical ensemble as described in Appendix 2B. Such a calculation for the single-particle spectra (6-521) and (6-522) is given in the small print on pp. 612ff. The thermodynamic functions for a given temperature can be written as a sum of a "smooth part" and a shell-structure term that depends on the number of particles outside closed shells. For example, for the energy spectrum (6-522), and a number of particles corresponding to a closed-shell configuration in the ground state ($x=0$), one obtains (see Eqs. (6-532b), (6-534b), and (6-536b))

$$\mathscr{E}_{\text{sh}}(T,x=0)=-\frac{f}{4\pi^2}(\hbar\omega_{\text{sh}})^2\tau^2\frac{\cosh\tau}{\sinh^2\tau} \tag{6-523a}$$

$$S_{\text{sh}}(T,x=0)=-\frac{f}{2}\hbar\omega_{\text{sh}}\frac{\tau\coth\tau-1}{\sinh\tau} \tag{6-523b}$$

$$F_{\text{sh}}(T,x=0)=-\frac{f}{4\pi^2}(\hbar\omega_{\text{sh}})^2\frac{\tau}{\sinh\tau} \tag{6-523c}$$

$$\tau\equiv\frac{2\pi^2T}{\hbar\omega_{\text{sh}}} \tag{6-523d}$$

for the shell-structure terms in the total energy, entropy, and free energy, respectively. The parameter τ measures the temperature T in a scale related to $\hbar\omega_{\text{sh}}$, the energy spacing of the shells. As $T\rightarrow 0$, the canonical ensemble goes over into the ground-state configuration, and the expressions (6-523) acquire the values that would be obtained by summation over the occupied orbits, as in the above discussion of the ground-state masses.

For high temperatures, the contributions $\mathscr{E}_{\text{sh}}$, S_{sh}, F_{sh} approach zero exponentially. Since the parameter τ involves the rather large coefficient $2\pi^2$, the shell-structure effects become quite small, even for temperatures that are only a few tenths of the shell spacing.

The grand canonical ensemble involves an average over systems with different particle numbers, and the validity of the present treatment requires these fluctuations to be small compared with the scale of the shell-structure effects ($\Delta\mathscr{N}\ll g_0\hbar\omega_{\text{sh}}$). The fluctuations increase with increasing excitation energy, but for temperatures at which the shell effects begin to decrease ▲ exponentially ($\tau\sim1$), the fluctuations remain relatively small. (For $T\sim\hbar\omega_{\text{sh}}$,

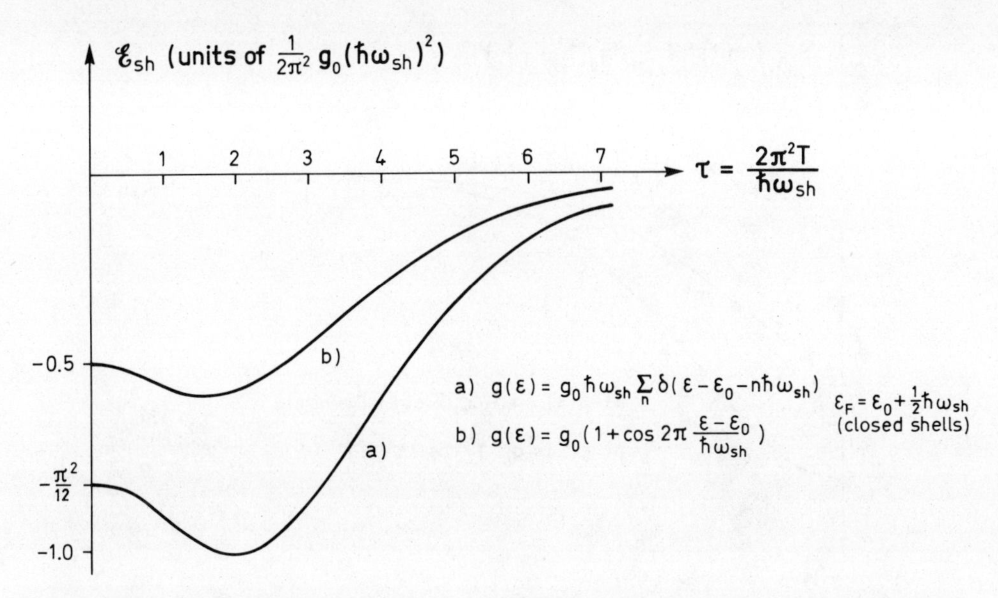

Figure 6-53 Shell-structure energy as a function of temperature. The values in the figure refer to a particle number corresponding to closed shells for $\tau = 0$, and are obtained from Eqs. (6-539a) and (6-523a), for the single-particle specıra a and b, respectively.

▼ the number of excited particles is $n_{ex} \sim g_0 \hbar\omega_{sh}$, and hence $\Delta\mathscr{N} \sim (n_{ex})^{1/2} \ll g_0 \hbar\omega_{sh}$.)

The thermodynamic functions for the two models (6-521) and (6-522) of the single-particle level density are illustrated in Figs. 6-53, 6-54, and 6-55 for systems with a number of particles corresponding to closed shells in the ground state ($T = 0$). The initial decrease of $\mathscr{E}_{sh}$ as a function of temperature in Fig. 6-53 reflects the reduced single-particle level density in the neighborhood of the Fermi energy, for the closed-shell configuration. In fact, for given temperature, the excitation energy is proportional to the number of elementary excitations with energies of order T (see the discussion in Vol. I, pp. 154ff.); thus, the total energy $\mathscr{E} = \tilde{\mathscr{E}} + \mathscr{E}_{sh}$ increases less rapidly with temperature than does the energy $\tilde{\mathscr{E}}$ that corresponds to the average level density g_0. (While $\tilde{\mathscr{E}}$ varies as T^2, the level density (b) in Fig. 6-53 depends on $(\varepsilon - \varepsilon_F)^2$ and thus implies an excitation energy proportional to T^4, as for the photon (or phonon) spectrum. The single-particle spectrum (a) has a gap at the Fermi energy, and the excitation energy therefore involves the exponential factor $\exp\{-\hbar\omega_{sh}/T\}$.)

The free energy represents the generalization of the potential energy to a system with a finite temperature (see the discussion on p. 371). The values of the shell-structure contribution to the free energy, as shown in Fig. 6-54, approach zero uniformly with increasing temperature, corresponding to the fact that the special stability associated with the closed shells has its greatest value for the ground state. ▲

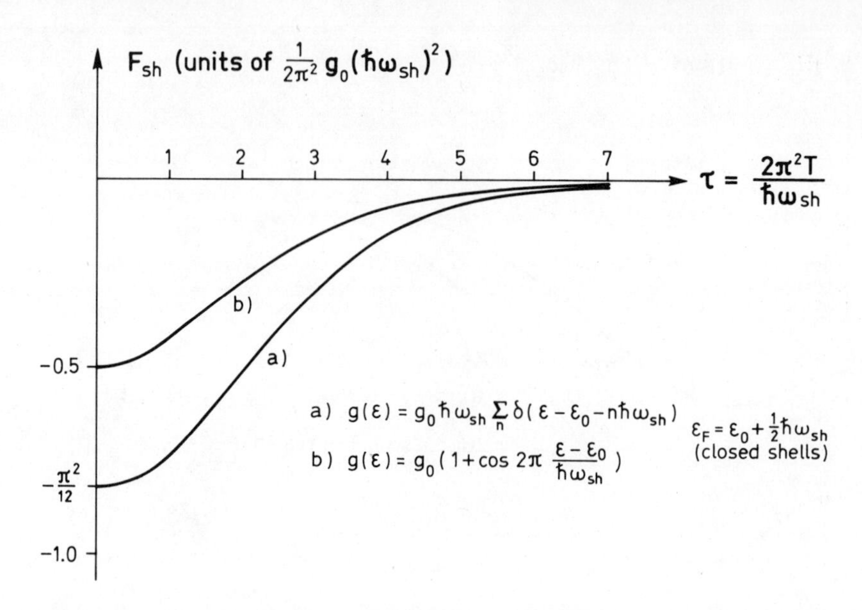

Figure 6-54 Contribution of shell structure to free energy. The values in the figure are obtained from Eqs. (6-539c) and (6-523c), for the single-particle spectra *a* and *b*, respectively.

▼ The total nuclear level density is dominated by the exponential dependence on the entropy (see Eqs. (2B-14) and (2B-37c)). The entropy as a function of energy, for the single-particle spectra (6-521) and (6-522), is compared in Fig. 6-55 with the values corresponding to a constant level density. The exponential decrease of the shell-structure contribution to the entropy at high excitation energy E implies that in this regime, the entropy, and hence the level density, approaches that of a system without shell structure, with an excitation energy equal to $E+\mathscr{E}_{sh}(T=0)$ (Hurwitz and Bethe, 1951; Rosenzweig, 1957).

Experimental evidence concerning the shell-structure effect on the nuclear level densities is especially striking in the region of ^{208}Pb. For these nuclei, the parameter a, as determined from the level density at about 7 MeV of excitation, is about a factor of 3 smaller than the value corresponding to a smooth dependence on A (see Fig. 2-12, Vol. I, p. 187). The shell structure in the single-particle level density can be approximately represented by the form (6-522), with $f \approx g_0$ (see p. 601); the effect on the entropy may thus be obtained from the case (b) in Fig. 6-55. The observed shell-structure energy in the ground state of ^{208}Pb is $\mathscr{E}_{sh}(T=0) \approx -13$ MeV (see p. 601), and for an excitation energy of about $\frac{1}{2}|\mathscr{E}_{sh}(T=0)|$, the entropy of the closed-shell nucleus is expected (see Fig. 6-55) to be about 0.6 of the value obtained for a
▲ system without shell structure at the same excitation energy. Since the

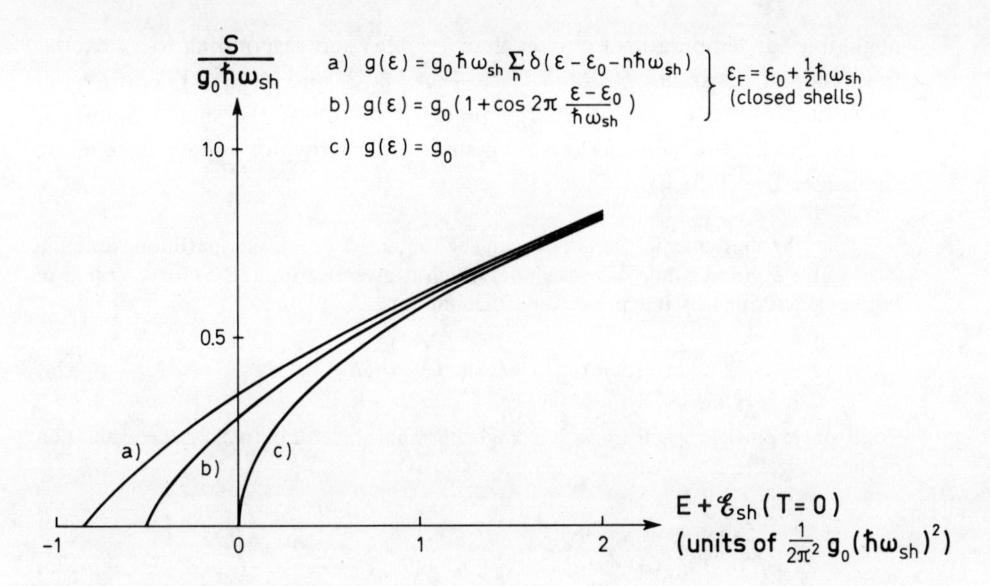

Figure 6-55 Effect of shell structure on entropy, as a function of excitation energy. The entropy of the single-particle motion, as a function of the temperature, is obtained as a sum of a smooth part (6-534a) and a shell-structure term given by Eqs. (6-539b) and (6-523b) for the single-particle spectra (a) and (b), respectively. The relation between temperature T and excitation energy, $E = \mathscr{E} - \mathscr{E}(T=0)$, is given by Eqs. (6-531), (6-532), and (6-539a); see also Fig. 6-53.

▼ parameter a is defined by the relation $S = 2(aE)^{1/2}$, the observed shell-structure effect in the level density is seen to agree well with this estimate. It must be emphasized that the estimate of the average level density in the absence of shell structure, based on the simple interpolation in Fig. 2-12, involves a considerable amount of uncertainty. Thus, in comparing the level densities for closed-shell nuclei with those of nuclei with many particles in unfilled shells, one must take into account that the level density is significantly affected by the contribution from rotational excitations of the deformed nuclei (see pp. 38 ff.) as well as by the pair correlations, which reduce the low-frequency modes of excitation. (A theoretical estimate of the level density in the absence of shell structure (and of collective correlations) can be obtained from the average single-particle level spacing in the region of the Fermi surface; the value given by Eq. (2-125a), which agrees with the empirical average level spacing, would imply that the value of a for ^{208}Pb is decreased by a factor of 2.5 as a result of the shell structure.)

With increasing excitation energy, the shell-structure contributions to the thermodynamic functions decrease. In the region of ^{208}Pb with the parameters

▲ $\hbar\omega_{sh}/2\pi^2 \approx 0.35$ MeV and $g_0\hbar\omega_{sh} \approx 90$, the shell-structure effects become

▼ negligible for temperatures greater than 1.5 MeV, corresponding to an excitation energy of about 50 MeV (see Figs. 6-53 and 6-55). The order of magnitude of the excitation energy required to "melt" the shell structure is fundamentally the same as the shell-structure energy for totally degenerate shells (see Eq. (6-513)).

The thermodynamic functions can be obtained from the partition function $Z(\alpha,\beta)$ for a grand canonical ensemble, which is given by Eq. (2B-5) for a system of fermions described by independent-particle motion,

$$\ln Z(\alpha,\beta)=\int^{\infty} g(\varepsilon)\ln(1+\exp\{\alpha-\beta\varepsilon\})\,d\varepsilon \tag{6-524}$$

For a single-particle spectrum with a single harmonic, as in Eq. (6-522), the evaluation of the integral (6-524) yields

$$\ln Z(\alpha,\beta)=\int^{\alpha/\beta}(\alpha-\beta\varepsilon)g_0\,d\varepsilon+\frac{\pi^2}{6\beta}g_0-\frac{f}{2}\hbar\omega_{\text{sh}}\frac{\cos\left\{\dfrac{2\pi}{\hbar\omega_{\text{sh}}}\left(\dfrac{\alpha}{\beta}-\varepsilon_0\right)\right\}}{\sinh\dfrac{2\pi^2}{\beta\hbar\omega_{\text{sh}}}} \tag{6-525}$$

The terms involving g_0 are given by Eq. (2B-9), while the shell-structure term can be obtained by a contour integration in the complex ε plane. In the evaluation of the thermodynamic properties, we may neglect contributions from the lower limit in the integral (6-524) provided the temperature β^{-1}, as well as the shell spacing $\hbar\omega_{\text{sh}}$, is small compared to the Fermi energy (degenerate Fermi gas approximation; it must also be emphasized that the level density (6-522) is only intended to represent the single-particle spectrum in a region small compared to ε_F).

The values of α and β are determined by the particle number and energy through the stationarity conditions (2B-13). The first of these relations yields the Fermi energy, or chemical potential, $\varepsilon_F=\alpha/\beta$,

$$\begin{aligned}\mathcal{N}&=\frac{\partial}{\partial\alpha}\ln Z\\&=\int^{\varepsilon_F}g_0\,d\varepsilon-\frac{\pi f}{\beta}\frac{\sin 2\pi y}{\sinh\tau}\\\tau&=\frac{2\pi^2}{\beta\hbar\omega_{\text{sh}}}\end{aligned} \tag{6-526}$$

where the parameter y describes the position of the Fermi level with respect to the value corresponding to closed shells

$$\varepsilon_F=\varepsilon_0+\hbar\omega_{\text{sh}}(y-\tfrac{1}{2}) \tag{6-527}$$

Introducing the shell-filling parameter x, defined by

$$\mathcal{N}=\mathcal{N}_{\text{cs}}+g_0\hbar\omega_{\text{sh}}x \tag{6-528}$$

▲ where $\mathcal{N}_{\text{cs}}$ is the particle number for closed shells and $g_0\hbar\omega_{\text{sh}}$ the number of particles

▼ in a shell, the relation (6-526) can be written

$$x=y-\frac{1}{2\pi}\frac{f}{g_0}\frac{\tau}{\sinh\tau}\sin 2\pi y \tag{6-529a}$$

or, alternatively,

$$\begin{aligned}\varepsilon_F&=\tilde{\varepsilon}_F+\frac{\hbar\omega_{sh}}{2\pi}\frac{f}{g_0}\frac{\tau}{\sinh\tau}\sin 2\pi y\\ \tilde{\varepsilon}_F&=(\varepsilon_0-\tfrac{1}{2}\hbar\omega_{sh})+g_0^{-1}(\mathcal{N}-\mathcal{N}_{cs})\end{aligned} \tag{6-529b}$$

where $\tilde{\varepsilon}_F$ is the part of ε_F that varies smoothly with $\mathcal{N}$.

The second of Eqs. (2B-13) yields the energy (representing the sum of single-particle eigenvalues) as a function of temperature $T=\beta^{-1}$,

$$\begin{aligned}\mathcal{E}&=-\frac{\partial}{\partial\beta}\ln Z\\ &=\int^{\varepsilon_F}\varepsilon g_0 d\varepsilon+\frac{\pi^2}{6\beta^2}g_0-\frac{\hbar\omega_{sh}}{2\pi}f\varepsilon_F\tau\frac{\sin 2\pi y}{\sinh\tau}-\left(\frac{\hbar\omega_{sh}}{2\pi}\right)^2 f\tau^2\frac{\cosh\tau}{\sinh^2\tau}\cos 2\pi y\end{aligned} \tag{6-530}$$

As in the analysis of the ground-state energies, the function (6-530) can be expressed as a sum of a smooth term and a shell-structure contribution that contains the dependence on the shell-filling parameter,

$$\mathcal{E}=\tilde{\mathcal{E}}(\mathcal{N},\beta)+\mathcal{E}_{sh} \tag{6-531}$$

with (see Eq. (6-529b))

$$\tilde{\mathcal{E}}(\mathcal{N},\beta)=\tilde{\mathcal{E}}(\mathcal{N},\beta^{-1}=0)+\frac{\pi^2}{6\beta^2}g_0 \tag{6-532a}$$

$$\mathcal{E}_{sh}=-\left(\frac{\hbar\omega_{sh}}{2\pi}\right)^2\left(f\tau^2\frac{\cosh\tau}{\sinh^2\tau}\cos 2\pi y+\frac{f^2}{2g_0}\tau^2\frac{\sin^2 2\pi y}{\sinh^2\tau}\right) \tag{6-532b}$$

The separation between $\tilde{\mathcal{E}}$ and $\mathcal{E}_{sh}$ is made unique by the requirements that $\mathcal{E}_{sh}$ vanishes for $\tau\to\infty$ and that the temperature dependence of $\tilde{\mathcal{E}}$ is that of a Fermi gas with average single-particle level density g_0 (see Eq. (2B-18)).

The entropy may be obtained from the relation (2B-37c)

$$\begin{aligned}S&=-\alpha\mathcal{N}+\beta\mathcal{E}+\ln Z\\ &=\tilde{S}+S_{sh}\end{aligned} \tag{6-533}$$

with

$$\tilde{S}=\frac{\pi^2}{3\beta}g_0 \tag{6-534a}$$

$$S_{sh}=-\hbar\omega_{sh}\frac{f}{2}\frac{\cos 2\pi y}{\sinh\tau}(\tau\coth\tau-1) \tag{6-534b}$$

From $\mathcal{E}$ and S, the free energy follows from the relation (6-102),

$$\begin{aligned}F&=\mathcal{E}-\beta^{-1}S\\ &=\tilde{F}+F_{sh}\end{aligned} \tag{6-535}$$

▲

▼ with

$$\tilde{F}=\tilde{F}(\mathcal{N},\beta^{-1}=0)-\frac{\pi^2}{6\beta^2}g_0 \tag{6-536a}$$

$$F_{\mathrm{sh}}=-\left(\frac{\hbar\omega_{\mathrm{sh}}}{2\pi}\right)^2\left(f\tau\frac{\cos 2\pi y}{\sinh\tau}+\frac{f^2}{2g_0}\tau^2\frac{\sin^2 2\pi y}{\sinh^2\tau}\right) \tag{6-536b}$$

The present treatment of shell-structure effects can be extended to more general single-particle spectra by expanding the level density $g(\varepsilon)$ in a Fourier series. For example, the spectrum (6-521) can be expressed in the form

$$\begin{aligned} g(\varepsilon) &= \Omega\sum_n \delta(\varepsilon-\varepsilon_0-n\hbar\omega_{\mathrm{sh}}) \\ &= \frac{\Omega}{\hbar\omega_{\mathrm{sh}}}\left(1+2\sum_{p=1}^{\infty}\cos\left\{2\pi p\frac{\varepsilon-\varepsilon_0}{\hbar\omega_{\mathrm{sh}}}\right\}\right) \end{aligned} \tag{6-537}$$

and the contribution of the individual Fourier components to $\ln Z$ is given by Eq. (6-525). The particle-number relation, corresponding to Eq. (6-529a), takes the form

$$x=y+\frac{\tau}{\pi}\sum_{p=1}^{\infty}(-1)^p\frac{\sin 2\pi p y}{\sinh p\tau} \tag{6-538}$$

and the shell-structure contributions to the thermodynamic functions are given by

$$\mathscr{E}_{\mathrm{sh}}=\frac{\Omega}{2\pi^2}\hbar\omega_{\mathrm{sh}}\tau^2\left(\sum_{p=1}^{\infty}(-1)^p\frac{\cosh p\tau}{\sinh^2 p\tau}\cos 2\pi p y-\left(\sum_{p=1}^{\infty}(-1)^p\frac{\sin 2\pi p y}{\sinh p\tau}\right)^2\right) \tag{6-539a}$$

$$S_{\mathrm{sh}}=\Omega\sum_{p=1}^{\infty}(-1)^p\frac{\cos 2\pi p y}{\sinh p\tau}\left(\tau\coth p\tau-\frac{1}{p}\right) \tag{6-539b}$$

$$F_{\mathrm{sh}}=\frac{\Omega}{2\pi^2}\hbar\omega_{\mathrm{sh}}\left(\tau\sum_{p=1}^{\infty}(-1)^p\frac{1}{p}\frac{\cos 2\pi p y}{\sinh p\tau}-\tau^2\left(\sum_{p=1}^{\infty}(-1)^p\frac{\sin 2\pi p y}{\sinh p\tau}\right)^2\right) \tag{6-539c}$$

In the limit of zero temperature, the sum over p in Eq. (6-538) exhibits a discontinuity for $y=1/2$,

$$\begin{aligned} \Sigma(y) &\equiv \frac{1}{\pi}\sum_{p=1}^{\infty}(-1)^p\frac{1}{p}\sin 2\pi p y \\ &= \begin{cases} -y & 0\leqslant y<\frac{1}{2} \\ 1-y & \frac{1}{2}<y\leqslant 1 \end{cases} \end{aligned} \tag{6-540}$$

corresponding to the fact that the Fermi energy stays equal to the value for $y=1/2$

▲ during the entire filling of the shell. In the limit $\tau=0$, the evaluation of the shell-

▼ structure energy yields

$$\mathscr{E}_{\mathrm{sh}}(\tau\rightarrow 0)=\Omega\hbar\omega_{\mathrm{sh}}\left(\frac{1}{2\pi^2}\sum_{p=1}^{\infty}\frac{\cos\{2\pi p(y-\frac{1}{2})\}}{p^2}-\frac{1}{2}(\Sigma(y))^2\right)$$

$$=\Omega\hbar\omega_{\mathrm{sh}}\left(\frac{1}{12}-\frac{1}{2}(x-\tfrac{1}{2})^2\right) \qquad (6\text{-}541)$$

The first term has been obtained by setting $y=1/2$ and using the value $\pi^2/6$ for the sum over p^{-2} (see, for example, Eqs. (2B-10b) and (2B-11)); in the second term in Eq. (6-541), the value of $\Sigma(y)$ defined by Eq. (6-540) has been obtained from Eq. (6-538), which yields $\Sigma(y)=x-1/2$, since y stays equal to $1/2$ while x varies from 0 to 1. The result (6-541) for $\mathscr{E}_{\mathrm{sh}}$ as a function of x is seen to be identical to the earlier result (6-510) obtained by a direct summation over the occupied states in the ground-state configuration for the harmonic oscillator potential.

Features of Fission Mode

Fission barrier (Fig. 6-56)

The fission process is characterized by a rather sharply defined threshold energy, referred to as the fission barrier. Below threshold, the fission cross sections depend exponentially on the energy, as a result of the barrier penetration (see, for example, $\sigma(\gamma f)$ for ^{238}U in Fig. 4-23, p. 126), and the barrier may be defined as the energy at which the penetration in the lowest fission channel reaches the value 1/2 (see Eq. (6-109)). Empirical data showing the systematics of the fission barriers for the heavy elements are shown in Fig. 6-56 and are compared with the predictions of the liquid-drop model (see Fig. 6A-1, p. 664). The abscissa in Fig. 6-56 is the fissility parameter x defined by Eq. (6-94), and the barriers are plotted in units of the surface energy $b_{\mathrm{surf}}A^{2/3}$ with the parameter $b_{\mathrm{surf}}=17$ MeV obtained from the analysis of the nuclear masses (see Eq. (2-14)), which implies $(Z^2/A)_{\mathrm{crit}}\approx 49$. It is seen that the observed barriers are similar in magnitude to those given by the model. Since the calculated barriers represent the rather small difference between the much larger surface and Coulomb energies of deformation, the qualitative agreement in Fig. 6-56 constitutes significant support for the estimate of the surface deformation energy obtained from the systematics of the nuclear masses. Thus, as can be seen from Fig. 6-56, a change of the surface energy parameter by 10% would lead to a change by a factor of 2 in the predicted fission barriers for $x\approx 0.75$.

The deviations of the empirical fission barriers in Fig. 6-56 from those calculated on the basis of the liquid-drop model may be associated partly with the assumed bulk properties of the nucleus and partly with shell-structure effects. Thus, there is some uncertainty in the effective surface ▲ tension for heavy nuclei as determined from the nuclear masses, due to the

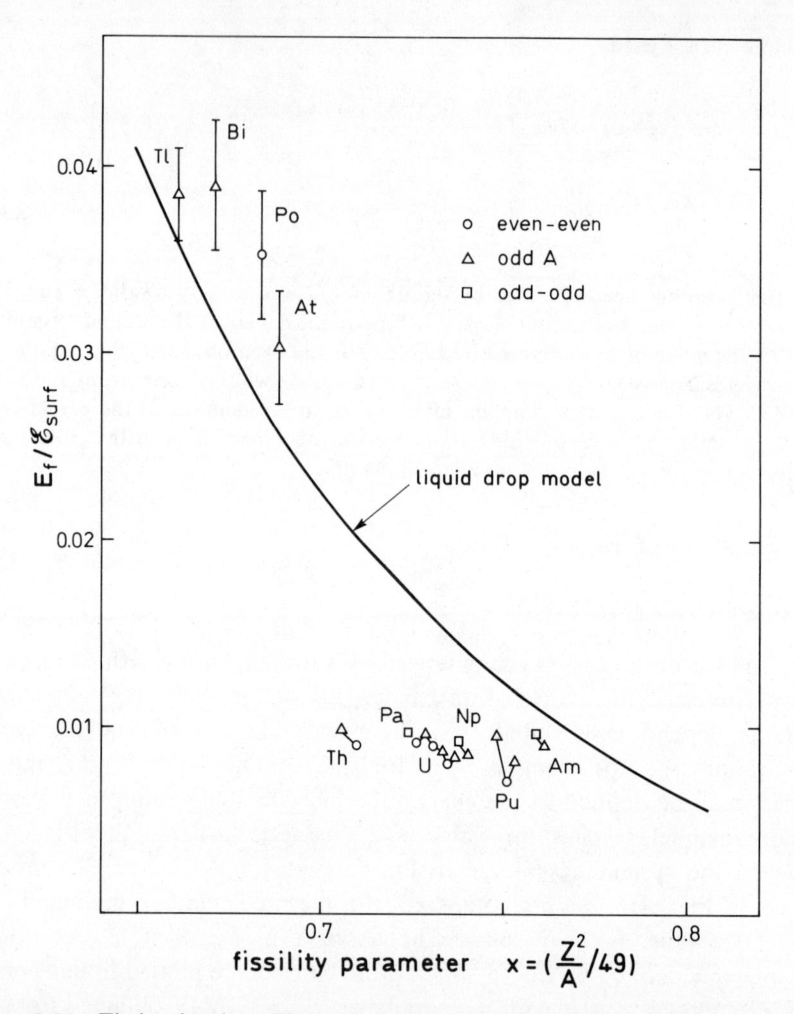

Figure 6-56 Fission barriers. The measured threshold energies E_f for onset of fission are plotted in units of the surface energy $\mathscr{E}_{surf} = 17A^{2/3}$MeV. The solid curve corresponds to the liquid-drop model estimate obtained from Fig. 6A-1. The data are taken from the compilations by Hyde (1964) and Halpern (1959) and from: ^{201}Tl (D. S. Burnett, R. C. Gatti, F. Plasil, P. B. Price, W. J. Swiatecki, and S. G. Thompson, *Phys. Rev.* **134**, B952, 1964); Bi, Po, and At (J. R. Huizenga, R. Chaudhry, and R. Vandenbosch, *Phys. Rev.* **126**, 210, 1962). The values of E_f from the last reference have been increased by 2.6 MeV in order to make them consistent with the more detailed measurements by Burnett *et al.*

▼ influence of the neutron excess and the curvature of the surface; in fact, the evidence on the shape of the nuclei at saddle point, obtained from angular distributions, suggests a value $(Z^2/A)_{crit} \approx 45$ (see pp. 621 ff.). The principal effect of such a modification of the bulk properties would be to shift the empirical points in Fig. 6-56 to the right by about 0.05 units in x. The ▲ prediction of the liquid-drop model would then approximately agree with the

▼ average value of the observed barriers in the region of $Z \gtrsim 90$, but would be about 10 MeV smaller than the observed barriers for nuclei in the neighborhood of Pb. For these nuclei, an increase in the barriers with respect to the liquid-drop estimate is expected as an effect of the shell structure, which implies that the ground-state masses of these approximately closed-shell nuclei are about 10 MeV smaller than the average (see the discussion on p. 601 of the shell-structure energy associated with the closed-shell configuration of ^{208}Pb).

Evidence on saddle-point shape obtained from α-induced fission processes (Fig. 6-57)

Information concerning the shape of the saddle-point configuration can be obtained from the analysis of angular distributions of the fission fragments (Halpern and Strutinsky, 1958). The direction of emission of the fission fragments, relative to that of the total angular momentum vector, is determined by the quantum number K (see Eq. (4-178)), and when many fission channels are open, the statistical distribution of K values is determined by the difference between the moments of inertia parallel and perpendicular to the symmetry axis (see Eq. (4-63c)),

$$\rho(K,I) \approx \rho(K=0,I)\exp\left\{-\frac{K^2}{2K_0^2}\right\} \tag{6-542}$$

with

$$K_0^{-2} = \left(\frac{\hbar^2}{\mathscr{I}_3} - \frac{\hbar^2}{\mathscr{I}_\perp}\right)\frac{1}{T} \tag{6-543a}$$

$$T = \left(\frac{6E}{\pi^2 g_0}\right)^{1/2} \equiv \left(\frac{E}{a}\right)^{1/2} \tag{6-543b}$$

where the expression for the temperature refers to a Fermi gas with a single-particle level density g_0 (see, for example, Eq. (2-57)). Empirical evidence on the parameter a appearing in Eq. (6-543b) is discussed in Vol. I, p. 187. The characterization of the individual channels by a definite value of K and the statistical distribution (6-542) assume axial symmetry of the saddle point; the consequences of a deviation from this symmetry are discussed in the small print on p. 622.

At sufficiently high excitation energies, the moments of inertia are expected to correspond approximately to those of a rigid body (see p. 37), and the difference between the moments of inertia is then directly related to the shape of the nucleus at saddle point. In a qualitative estimate, we may ▲ compare the nuclear shape with a spheroid with eccentricity parameter δ (see

▼ Eq. (4-72)), for which the rigid-body moments are given by

$$\mathscr{I}_3 = \mathscr{I}_{\text{sph}}(1 - \tfrac{2}{3}\delta) \qquad \mathscr{I}_\perp = \mathscr{I}_{\text{sph}}(1 + \tfrac{1}{3}\delta)$$

$$\frac{\hbar^2}{\mathscr{I}_3} - \frac{\hbar^2}{\mathscr{I}_\perp} = \frac{\hbar^2}{\mathscr{I}_{\text{sph}}} \frac{\delta}{(1 - \frac{2}{3}\delta)(1 + \frac{1}{3}\delta)} \tag{6-544}$$

$$\mathscr{I}_{\text{sph}} = \tfrac{2}{3} AM\langle r^2 \rangle = \tfrac{2}{5} AMR_0^2$$

where $\mathscr{I}_{\text{sph}}$ is the rigid-body moment for a spherical shape with the mean radius R_0.

We shall be concerned with the analysis of reactions initiated by fast projectiles, which produce compound nuclei with high angular momenta. In the expression (4-178) for the angular distribution, we may thus use an asymptotic form for the $\mathscr{D}$ functions valid for $I \gg 1$ and $I \gg M$,

$$\begin{aligned} W_{KI}(\theta) &= \frac{2I+1}{2} |\mathscr{D}^I_{M=0,K}|^2 \\ &\approx \frac{I}{\pi} (I^2 \sin^2\theta - K^2)^{-1/2} \end{aligned} \tag{6-545}$$

$$\int_{\theta_0 = \sin^{-1} K/I}^{\pi - \theta_0} W_{KI}(\theta) \sin\theta \, d\theta = 1$$

This result can be obtained from a classical vector model; the probability distribution of the symmetry axis, specified by the components K and $M = 0$ of the angular momentum vector, is given by $(2\pi)^{-1} d\varphi$, where φ is the azimuthal angle of the symmetry axis with respect to the direction of the angular momentum. The angle θ of the symmetry axis with respect to the z axis is related to φ by $\cos\theta = (1 - (K/I)^2)^{1/2} \cos\varphi$ and covers the interval $\theta_0 < \theta < \pi - \theta_0$ twice, while φ varies from 0 to 2π.

From Eqs. (4-178), (6-542), and (6-545), we obtain for the angular distribution of the fission fragments emitted from compound nuclei with given I

$$W_I(\theta) \propto \frac{2}{\pi} \int_0^{I \sin\theta} dK \exp\left\{ -\frac{K^2}{2K_0^2} \right\} (I^2 \sin^2\theta - K^2)^{-1/2}$$

$$= \exp\left\{ -\frac{I^2 \sin^2\theta}{4K_0^2} \right\} J_0\left(i \frac{I^2 \sin^2\theta}{4K_0^2} \right) \tag{6-546a}$$

$$\approx 1 - \frac{I^2 \sin^2\theta}{4K_0^2} \tag{6-546b}$$

where J_0 is the Bessel function of order zero and where Eq. (6-546b) ▲ represents an approximation valid for small anisotropy ($I \ll K_0$).

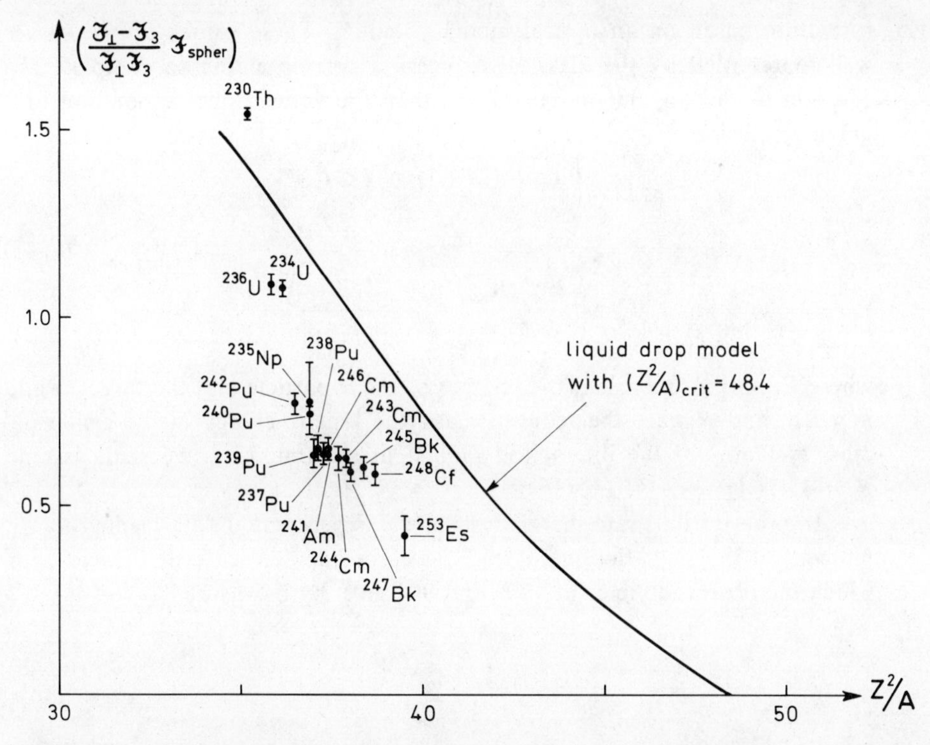

Figure 6-57 Saddle-point shapes as a function of Z^2/A. The figure shows the evidence on the moments of inertia of the fissioning nucleus at saddle point obtained from the analysis of angular anisotropies of the fission fragments following compound nucleus reactions induced by α particles of 42.8 MeV. The data and analysis are taken from R. F. Reising, G. L. Bate, and J. R. Huizenga, *Phys. Rev.* **141**, 1161 (1966). The points are labeled by the compound nucleus involved. The corrections for fission following neutron emission involve uncertainties (in addition to the experimental errors shown in the figure) that are especially large for the compound nuclei of Th and U.

▼ Figure 6-57 shows evidence on the shape of the fissioning nucleus derived from a study of fission induced by α particles of 43 MeV (Reising *et al.*, 1966). The rather high excitation energy of the compound nucleus implies that fission may occur either directly or subsequent to the emission of one or several neutrons. However, in the cases considered, the main yield is associated with first-stage fission ($\Gamma_f \gtrsim \Gamma_n$). Moreover, the temperature and total angular momentum are not much affected by the emission of a neutron. The correction for the effect of fission after neutron emission is thus only of the order of 15% or less; this correction has been included in the analysis given in Fig. 6-57.

The angular distribution depends on the ratio of the angular momentum I to the quantity K_0 (see Eq. (6-546)). The distribution of I values employed ▲ in Fig. 6-57 has been obtained from a calculation of compound nucleus

▼ formation based on an optical model potential. These values can be rather well represented by the classical expression corresponding to complete absorption for all angular momenta less than the value I_m corresponding to a grazing collision

$$\sigma(I) = \begin{cases} \text{const}(2I+1) & I < I_m \\ 0 & I > I_m \end{cases}$$

$$I_m = \left(\frac{2M_{12}R_{12}^2}{\hbar^2} \left(E_{12} - \frac{Z_1 Z_2 e^2}{R_{12}} \right) \right)^{1/2} \tag{6-547}$$

where Z_1 and Z_2 represent the charges of the α particle and the target, while M_{12}, E_{12}, and R_{12} are the reduced mass, the kinetic energy in the center-of-mass system, and the interaction radius; in the following, we shall assume $R_{12} = 1.45(A_1^{1/3} + A_2^{1/3})$ fm.

In order to illustrate the analysis of the experimental data leading to the moment of inertia ratios in Fig. 6-57, we consider the case of $^{238}U(\alpha f)$, for which the observed yields at 170° and 90° give, by means of Eq. (6-546),

$$\frac{\langle I^2 \rangle}{4K_0^2} \approx 0.38 \tag{6-548}$$

for the parameters characterizing first-stage fission. Since the Coulomb barrier is about 24 MeV and $\langle I^2 \rangle \approx 220$ (see Eq. (6-547)), the observed anisotropy implies $K_0^2 \approx 150$. The α particle is unbound by about 5 MeV, and the excitation energy of the compound nucleus is therefore ≈ 37 MeV, which is ≈ 32 MeV above the fission barrier; using $a \approx A/8$ MeV^{-1} (see Fig. 2-12, Vol. I. p. 187), we obtain a temperature of 1.0 MeV (see Eq. (6-543b)). The rotational parameter for ^{242}Pu has the value $\hbar^2/\mathscr{I}_{sph} \approx 7.6$ keV (for $R_0 = 1.2A^{1/3}$ fm), and from Eq. (6-543a), we therefore obtain $\mathscr{I}_{sph}(\mathscr{I}_3^{-1} - \mathscr{I}_\perp^{-1}) \approx 0.9$, corresponding to a deformation parameter $\delta \approx 0.6$ (see Eq. (6-544)). Such a deformation implies a saddle-point shape, in which the ratio of the axes is approximately 2:1, in qualitative agreement with the liquid-drop prediction for a fissility parameter $x = 0.75$, as for ^{242}Pu (see Fig. 6A-1, p. 664).

In the above analysis, we have neglected the effects of the angular momentum on the saddle-point shape and fission barrier. These effects can be characterized by the parameter y, defined by Eq. (6A-35); for $\langle I^2 \rangle \approx 220$ and $A \approx 240$, one obtains $y \approx 1.3 \times 10^{-3}$. In comparison, the critical value of y for which the barrier vanishes is given by Eq. (6A-41), which yields $y_{crit} \approx 0.09$ for $x \approx 0.75$. The effects of the angular momentum on the barrier and the saddle-point deformation are of relative order y/y_{crit} (see Eqs. (6A-40) and (6A-39)), and are thus negligible in the present case.

A more detailed comparison of the observed moment ratios deduced ▲ from the angular distributions with those calculated for the saddle-point

▼ shapes in the liquid-drop model is shown in Fig. 6-57. It is seen that the observed moments imply shapes that are somewhat less deformed than those predicted by the liquid-drop model, assuming $(Z^2/A)_{\text{crit}}=49$ (see Eq. (6-95)). Part of this deviation may result from shell-structure effects, which are found to significantly modify the heights of the fission barriers (see p. 617). However, in the reactions on which the data in Fig. 6-57 are based, the rather large excitation energies imply that only a minor part of the shell-structure effects is expected to remain (Ramamurthy *et al.*, 1970). The significant parameter that characterizes the disappearance of the shell-structure effects with increasing excitation energy is the temperature compared with the energy interval characterizing the periodicity in the single-particle spectrum. The main shell-strucure effects at the saddle-point shape appear to be associated with the degeneracies occurring in an oscillator potential with ratio of frequencies 2:1 (see the discussion on pp. 595 ff.). The distance between such shells, given by Eq. (6-504), is $\hbar\omega_{\text{sh}}\approx 2^{-2/3}\hbar\omega_0\approx 26A^{-1/3}\,\text{MeV}\approx 4$ MeV, for $A\approx 240$. For temperatures of about $T\approx 1$ MeV, the estimate in Fig. 6-54, p. 610, implies that only about 10% of the shell-structure contribution to the free energy remains.

It thus appears that the data in Fig. 6-57 may be interpreted in terms of "macroscopic" nuclear parameters, as in the liquid-drop model, and one may attempt to employ these data to obtain a more precise determination of the fissility parameter x, defined by Eq. (6-94). It is seen that with increasing Z^2/A, the data in Fig. 6-57 extrapolate to a critical value of

$$\left(\frac{Z^2}{A}\right)_{\text{crit}}\approx 45 \tag{6-549}$$

This value is about 10% smaller than that obtained from the parameters of the semi-empirical mass formula as given in Chapter 2. Such a reduction of the effective surface tension could result from the dependence of the surface energy on the neutron excess. This dependence can be expressed in analogy to the volume symmetry energy in the semi-empirical mass formula (2-12),

$$\mathscr{E}_{\text{surf}}=b_{\text{surf}}A^{2/3}+\tfrac{1}{2}b_{\text{surf-sym}}\frac{(N-Z)^2}{A^{4/3}} \tag{6-550}$$

but, as mentioned in Chapter 2, the available data on nuclear masses leave the value of $b_{\text{surf-sym}}$ undetermined. The evidence on the saddle-point shapes could be accounted for by assuming $b_{\text{surf-sym}}\approx -50$ MeV. However, this interpretation of the critical value (6-549) is not unique, since other bulk properties of the nucleus may affect the fissility, to the accuracy considered. (See, for example, the discussion of the effect of a term in the nuclear energy

▲ proportional to the curvature of the nuclear surface (Strutinsky, 1965).)

▼ Additional evidence, consistent with that in Fig. 6-57, has been obtained from the angular distribution in fission processes initiated by heavy ion bombardment leading to compound nuclei with appreciably greater angular momentum (Karamyan *et al.*, 1967). Evidence is also available on the angular distribution of α-induced fission for nuclei with appreciably smaller fissility, corresponding to $Z \approx 82$ (Reising *et al.*, 1966), but this evidence is not included in Fig. 6-57, since the higher fission barriers (≈ 15 MeV) imply that the temperature of the fissioning nucleus is considerably less than in the cases illustrated. Under such conditions, pair correlations and shell structure may significantly influence the nuclear properties involved in the fission process.

In the above discussion, we have assumed that the saddle-point shape possesses axial symmetry, in which case the fission channels can be characterized by the quantum number K. For more general shapes, each channel involves a superposition of different values of K, representing the component of angular momentum in the direction of fission (which is assumed to correspond to a definite direction in the intrinsic coordinate system, such as, for example, the axis of greatest elongation). The density of channels has the form (4-65a), which, for large values of I, can be expressed as an integral over the directions of $\mathbf{I}$ with respect to the intrinsic axes (see the text following Eq. (4-65)); the probability distribution in K, which determines the angular distribution of the fission fragments, is obtained by an integration over the azimuthal angle of the angular momentum vector with respect to the fission direction. For large rotational energies (comparable with T), the resulting distribution in K is not of the Gaussian form (6-542). However, in the limit of small anisotropies, one obtains quite generally an expression of the form (6-546b), where, in the definition of K_0^{-2} (see Eq. (6-543)), the quantity $(\hbar^2/\mathscr{I}_3)$ represents the inertial parameter for rotation about the direction of fission, while $(\hbar^2/\mathscr{I}_\perp)$ represents an average for the perpendicular directions. Thus, in principle, the angular distributions may determine all three moments of inertia separately, but the available empirical data do not appear to be
▲ sufficiently detailed to provide this information.

---

[33]The first evidence was provided by the discovery of an isomer of ^{242}Am that decays by spontaneous fission with a lifetime more than a factor 10^{20} shorter than the spontaneous fission lifetime of the normal state (Polikanov *et al.*, 1962). Around 1968, this type of isomerism had become established as a general feature of the heavy elements (see the review by Michaudon, 1973). The evidence on excitation energy and spin of the isomers provided by studies of the excitation function for a variety of projectiles rendered unlikely an interpretation of the metastability on the basis of a very high spin (see the review by Polikanov, 1968). Attempts to interpret the isomerism in terms of quasistationary states in a second minimum of the potential energy function received significant support from theoretical studies of the influence of the shell structure on the deformation energy (Strutinsky, 1967a; see also footnote 7, p. 367). Experimental evidence for the interpretation in terms of shape isomerism came from the discovery of strength function phenomena in neutron-induced fission. It had been known for many years that there existed rather marked structure in the energy variation of the fission cross section in the threshold region (see the review by Wheeler, 1963). Careful analysis of these cross sections revealed that the resonance-like structure could not be accounted for in terms of the competition between fission and neutron channels, and led to an interpretation in terms of a vibrational state in the second minimum (Lynn, 1968a). Especially striking was the discovery of gross structure in the fission width of slow-neutron resonances below the fission barrier (Paya *et al.*, 1968; Migneco and Theobold, 1968) which could be interpreted in terms of resonant states in the region of the second minimum in the potential energy (Lynn, *loc.cit.*; Weigmann, 1968).

▼ *Shape isomerism in ^{241}Pu (Figs. 6-58 and 6-59; Table 6-18)*

Many different phenomena associated with the fission process have provided evidence for the occurrence of a second minimum in the nuclear potential energy surface (shape isomerism).[33] As an example, we shall consider the evidence obtained from the study of fission of the compound nucleus ^{241}Pu.

The results of slow-neutron resonance studies in n + ^{240}Pu are illustrated in Fig. 6-58. The fission barrier for the process ^{240}Pu(nf) corresponds to $E_n = 710$ keV (see Back *et al.*, 1974), and the slow-neutron region is thus well below the fission barrier. The resonances have $I\pi = 1/2+$ and an average spacing $D = 15$ eV, as determined from the lower-energy part of the total cross-section data, where the levels are all resolved (Kolar and Börkhoff, 1968).

It is seen from Fig. 6-58 that there are very large variations in the fission widths of the individual levels. Most of the levels have an almost vanishing ▲ fission yield, but there occur groups of resonances for which an appreciable

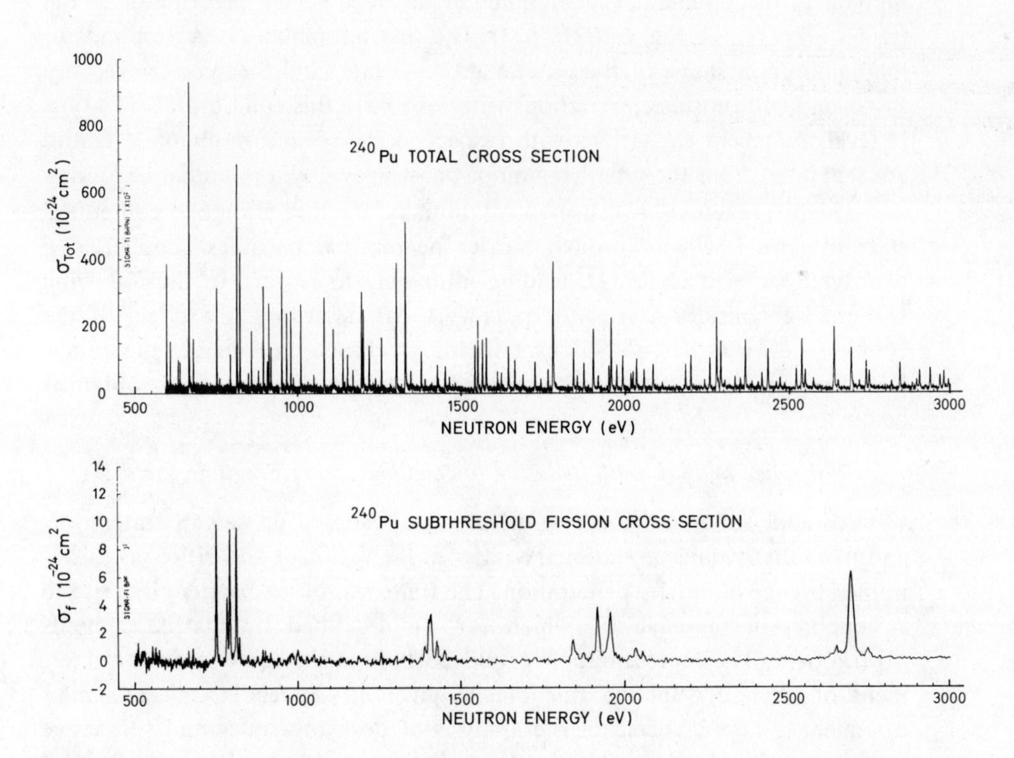

Figure 6-58 Neutron cross sections for ^{240}Pu. The figure gives the total cross section determined by W. Kolar and K. H. Böckhoff, *J. Nuclear Energy* **22**, 299 (1968) and the fission cross section measured by E. Migneco and J. P. Theobald, *Nuclear Phys.* **A112**, 603 (1968).

▼ fraction of the total cross section is associated with the fission mode. The variations are much greater than can be accounted for by random fluctuations (see, for example, the analysis of random fluctuations in the neutron width discussed in Vol. I, p. 182). Such a gross structure in the distribution of fission widths suggests an interpretation in terms of resonances associated with an intermediate stage in the fission process.

The average spacing of the gross-structure resonances, which we shall refer to as type II states, as distinct from the fine-structure resonances, referred to as type I states, is seen from Fig. 6-58 to be

$$D(\mathrm{II}) \approx 700 \quad \mathrm{eV} \tag{6-551}$$

This spacing corresponds to a characteristic period large compared with that of the elementary modes of excitation in the nucleus (vibrations, rotations, single-particle motion) and thus indicates the occurrence of complicated states resembling those of a compound nucleus, but only weakly coupled to the compound states associated with neutron capture. Such a double set of compound states can be interpreted by assuming the occurrence of two minima in the potential energy function involved in the description of the fission process (see Fig. 6-59, p. 633). The first mimimum (I) corresponds to the equilibrium shape of the nuclear ground state, and the type I levels are associated with intrinsic excitations with respect to this equilibrium. The type II levels represent excitations with respect to the second minimum (II) and are separated from the type I region of phase space by a potential barrier.

The (nf) reaction process involves, initially, the formation of a compound state of type I, which through barrier penetration becomes coupled to a near-lying state in region II, leading ultimately to fission. In the following section, we consider a general framework for describing the effect of the coupling between the two regions. (An analysis based on R-matrix theory has been given by Lynn (1968a), with results for $\mathrm{n}+{}^{240}\mathrm{Pu}$ similar to those obtained below.)

Resonance analysis with inclusion of coupling between regions I and II

An analysis of the coupling effects can be based on sets of states $|\mathrm{I}\alpha\rangle$ and $|\mathrm{II}a\rangle$ that would be stationary states in the regions I and II, respectively, in the absence of barrier penetration. The finiteness of the barrier gives rise to a coupling of these states, which can be described by matrix elements $\langle \mathrm{II}a|V|\mathrm{I}\alpha\rangle$. (The representation employed here is the same as in the treatment of strength function phenomena given in Appendix 2D; a similar treatment is also involved in the analysis of decaying states and resonance scattering in terms of an effective perturbation; see Secs. 3F-1a and 3F-1b.)

The effects of the coupling between the type I and II states take a simple form in the limiting situations corresponding to narrow and broad strength

▲ functions.

▼ *a. Narrow strength function.* If the width of the type II level in the absence of the coupling, as well as the average magnitude of the coupling matrix element $\langle V \rangle$, are small compared with the level spacing $D(\mathrm{I})$, the main coupling effect is confined to a few type I levels in the neighborhood of each type II level. Except for rare fluctuations in energy spacings or coupling matrix elements, a perturbation treatment may be employed and leads to the approximate eigenstates

$$\begin{aligned} |\mathrm{I}\hat{\alpha}\rangle &\approx |\mathrm{I}\alpha\rangle + c(\alpha a)|\mathrm{II}a\rangle \\ |\mathrm{II}\hat{a}\rangle &\approx |\mathrm{II}a\rangle - \sum_{\alpha} c(\alpha a)|\mathrm{I}\alpha\rangle \end{aligned} \tag{6-552}$$

with

$$c(\alpha a) = \frac{\langle \mathrm{II}a|V|\mathrm{I}\alpha\rangle}{E(\mathrm{I}\alpha) - E(\mathrm{II}a)} \tag{6-553}$$

The coupling described by the states (6-552) implies that the type I states acquire a small fission width and the type II states a corresponding neutron width

$$\begin{aligned} \Gamma_{\mathrm{f}}(\mathrm{I}\hat{\alpha}) &= c^2(\alpha a)\Gamma_{\mathrm{f}}(\mathrm{II}a) \\ \Gamma_{\mathrm{n}}(\mathrm{II}\hat{a}) &= |\sum_{\alpha} c(\alpha a) g_{\mathrm{n}}(\mathrm{I}\alpha)|^2 \end{aligned} \tag{6-554}$$

where $|g_{\mathrm{n}}(\mathrm{I}\alpha)|^2 = \Gamma_{\mathrm{n}}(\mathrm{I}\alpha)$. The amplitudes $g_{\mathrm{n}}(\mathrm{I}\alpha)$ and $c(\alpha a)$ are associated with two quite different components in the compound states of type I, and the interference between contributions from different levels, α, in $\Gamma_{\mathrm{n}}(\mathrm{II}\hat{a})$ is therefore expected to vanish on the average.

The fission yield of an individual resonance is given by

$$\int \sigma_{\mathrm{nf}}\, dE = 2\pi^2 \lambda\!\!\!^{-2} \frac{\Gamma_{\mathrm{n}}\Gamma_{\mathrm{f}}}{\Gamma} \tag{6-555}$$

where Γ is the total width. If the total width of the type I states is small compared with that of the type II states ($\langle\Gamma(\mathrm{I})\rangle \ll \Gamma(\mathrm{II})$), the main contribution to the fission yield in each gross-structure resonance arises from the type I levels, while the opposite would be the case for $\langle\Gamma(\mathrm{I})\rangle \gg \Gamma(\mathrm{II})$.

b. Broad strength function. If either the width of the type II level or the average coupling matrix element is large compared with $D(\mathrm{I})$, the strength function extends over a large number of individual resonances. In such a situation, one may employ the general analysis of gross-structure resonances based on cross sections averaged over the type I fine structure (see Sec. 3F-1c). In the analysis of such average cross sections, the coupling V can be described in terms of a coupling width $\Gamma_{\mathrm{c}}(\mathrm{II})$ representing the rate of decay of the state II into the channel associated with the type I compound states.
▲ Assuming that the coupling matrix elements do not exhibit any systematic

▼ variation over the gross-structure resonance, we obtain (see, for example, Eq. (2D-11))

$$\Gamma_c(\mathrm{II}) = 2\pi \frac{\langle V^2 \rangle}{D(\mathrm{I})} \tag{6-556}$$

where $\langle V^2 \rangle$ is the average value of the square of the coupling matrix elements.

The total width of the gross-structure resonance is given by

$$\Gamma(\mathrm{II}) = \Gamma_f(\mathrm{II}) + \Gamma_d(\mathrm{II}) + \Gamma_c(\mathrm{II}) \tag{6-557}$$

where $\Gamma_d(\mathrm{II})$ represents the probability for additional decay modes that do not proceed through either the compound states I or the fission channel; it is expected that the main contribution to $\Gamma_d(\mathrm{II})$ comes from γ decay to other states of type II.

The gross-structure resonance cross section, representing the average fission cross section for s-wave neutrons in the neighborhood of a type II resonance, can be expressed in the form

$$\langle \sigma_{\mathrm{nf}} \rangle = \pi \lambda\!\!\!^{-2} \frac{\Gamma_f(\mathrm{II})\Gamma_c(\mathrm{II})}{(E - E(\mathrm{II}))^2 + \frac{1}{4}(\Gamma(\mathrm{II}))^2} F(\mathrm{c} \rightarrow \mathrm{n}) \tag{6-558}$$

where $F(\mathrm{c} \rightarrow \mathrm{n})$ denotes the fraction of the compound nuclei decaying by neutron emission. (The result (6-558) may be derived by considering the inverse process (fn), for which the cross section for compound nucleus formation can be obtained from Eq. (3F-26); the relation between cross sections for inverse reactions is given by Eq. (1-43).) In the example shown in Fig. 6-58, the (nf) cross section between type II resonances is below the sensitivity of detection, and we have therefore ignored the possibility of a nonresonant contribution, such as is included in the more general expression (3F-23).

The fraction $F(\mathrm{c} \rightarrow \mathrm{n})$ in Eq. (6-558) represents an average over the type I resonances, each weighted by the fission width $\Gamma_f(\mathrm{I})$, which is proportional to the probability of formation of the type I state in the decay of the type II state,

$$F(c \rightarrow n) = \left\langle \frac{\Gamma_f(\mathrm{I})\Gamma_n(\mathrm{I})}{\Gamma(\mathrm{I})} \right\rangle \frac{1}{\langle \Gamma_f(\mathrm{I}) \rangle} \tag{6-559}$$

where

$$\Gamma(\mathrm{I}) = \Gamma_f(\mathrm{I}) + \Gamma_\gamma(\mathrm{I}) + \Gamma_n(\mathrm{I}) \tag{6-560}$$

is the total width of the fine-structure resonance. (In Eq. (6-559) and the ▲ following, we omit the index α labeling the different resonances of type I.)

▼ The average cross section (6-558) can also be expressed as a sum of the contributions from the type I resonances

$$\langle \sigma_{nf} \rangle = 2\pi^2 \lambda\!\!\!^{-2} \frac{1}{D(\mathrm{I})} \left\langle \frac{\Gamma_f(\mathrm{I})\Gamma_n(\mathrm{I})}{\Gamma(\mathrm{I})} \right\rangle \tag{6-561}$$

where $D(\mathrm{I})$ is the average level spacing of the fine-structure resonances. Comparison of Eqs. (6-558), (6-559), and (6-561) yields for the average fission width

$$\langle \Gamma_f(\mathrm{I}) \rangle = \frac{D(\mathrm{I})}{2\pi} \frac{\Gamma_f(\mathrm{II})\Gamma_c(\mathrm{II})}{(E - E(\mathrm{II}))^2 + \frac{1}{4}(\Gamma(\mathrm{II}))^2} \tag{6-562}$$

The expression (6-562) could have been obtained directly from the general arguments for a Breit-Wigner form for such a strength function (see Vol. I, p. 435) together with the normalization

$$\int_{\mathrm{II}} \langle \Gamma_f(\mathrm{I}) \rangle \frac{dE}{D(\mathrm{I})} = \frac{\Gamma_f(\mathrm{II})\Gamma_c(\mathrm{II})}{\Gamma(\mathrm{II})} \tag{6-563}$$

where the integration extends over an energy interval including the contributions from a single type II resonance. The relation (6-563) expresses the fact that the sum of the fission widths of the type I resonances is equal to $\Gamma_f(\mathrm{II})$ multiplied by the factor $\Gamma_c(\mathrm{II})/\Gamma(\mathrm{II})$, which represents the fraction of the type II state that is admixed into the fine-structure resonances of type I.

The present description covers a variety of different coupling situations depending on the relative magnitude of $\Gamma_c(\mathrm{II})$ and the total width $\Gamma(\mathrm{II})$. If $\Gamma_c(\mathrm{II}) \ll \Gamma(\mathrm{II})$ (weak coupling), only a small part of the properties of the type II state, such as its fission width, is transferred to the fine-structure resonances (see Eq. (6-563)). The main part of these properties remains associated with a single broad type II state, which, however, contributes only a negligible part to the (nf) yield. (See the discussion following Eq. (6-555); the assumption of a broad strength function implies $\Gamma(\mathrm{II}) \gg \langle \Gamma(\mathrm{I}) \rangle$. The broad type II state would strongly manifest itself in the hypothetical (ff) scattering process.) In the strong coupling situation ($\Gamma_c(\mathrm{II}) \approx \Gamma(\mathrm{II})$), one can no longer distinguish between individual resonances of type I and type II; the type II property is, in this case, distributed over a large number of fine-structure resonances.

Analysis of gross structure in ^{240}Pu (nf)

The ^{240}Pu (nf) cross section shown in Fig. 6-58 appears to represent a coupling situation that is intermediate between the limiting cases of narrow and broad strength functions discussed above. The gross structures around ▲ 780 and 1400 eV contain only a few fine-structure components, while the

▼ structure around 1900 eV extends over a region containing about 15 individual levels. One may consider the possibility that the structure at 1900 eV consists of two closely spaced type II resonances, but the probability of such a near degeneracy is only of the order of 1% (see Eq. (2-59)).

For the rather broad 1900 eV structure, the above analysis of the average cross sections can be used to determine the parameters of the gross-structure resonance without involving the detailed analysis of the fine-structure resonances, which is somewhat uncertain. As a first step, we estimate the quantity $F(\mathrm{c}\to\mathrm{n})$ given by Eq. (6-559). We shall find that $\langle\Gamma_f(\mathrm{I})\rangle \ll \langle\Gamma(\mathrm{I})\rangle$, even in the center of the type II resonance, and thus we have approximately (assuming no correlation between Γ_n and Γ_f)

$$F(\mathrm{c}\to\mathrm{n}) \approx \left\langle \frac{\Gamma_n(\mathrm{I})}{\Gamma(\mathrm{I})} \right\rangle \tag{6-564}$$

The average values of $\Gamma_n(\mathrm{I})$ and $\Gamma(\mathrm{I})$ have been determined from the resonance analysis of the total cross section (Kolar and Böckhoff, 1968; Weigmann and Schmid, 1968)

$$\begin{aligned} \langle\Gamma_n(\mathrm{I})\rangle &\approx 50(E_n(\mathrm{keV}))^{1/2}\ \mathrm{meV} \\ \langle\Gamma_\gamma(\mathrm{I})\rangle &\approx 23\ \mathrm{meV} \\ D(\mathrm{I}) &\approx 15\ \mathrm{eV} \end{aligned} \tag{6-565}$$

The fluctuations in $\Gamma_n(\mathrm{I})$ have a significant effect on the average of the ratio of Γ_n and Γ in Eq. (6-564); for a Porter-Thomas distribution (see Eq. (2-115)) for $\Gamma_n(\mathrm{I})$ and a constant value for $\Gamma_\gamma(\mathrm{I})$, one obtains

$$\begin{aligned} F(\mathrm{c}\to\mathrm{n}) &= 1-(\tfrac{1}{2}\pi y)^{1/2}\exp\{\tfrac{1}{2}y\} + (2y)^{1/2}\exp\{\tfrac{1}{2}y\}\int_0^{(y/2)^{1/2}}\exp\{-t^2\}\,dt \\ &\approx 0.52 \\ y &= \frac{\Gamma_\gamma}{\langle\Gamma_n\rangle} \approx 0.33 \end{aligned} \tag{6-565a}$$

The fluctuations thus reduce the value of F by a factor of about 0.7.

Under the condition $\langle\Gamma_f(\mathrm{I})\rangle \ll \langle\Gamma(\mathrm{I})\rangle$, as assumed in Eq. (6-564), the quantity F is independent of energy over a single type II resonance, and the strength function (6-558) has the simple Breit-Wigner form. However, the data in Fig. 6-58 are not sufficiently detailed to provide a test of the theoretical line shape.

From the total fission yield shown in Fig. 6-58 (see also the yields given ▲ in Table 6-18, p. 630), together with the value of F given above, we obtain

▼ (see Eq. (6-558))

$$\frac{\Gamma_f(\mathrm{II})\Gamma_c(\mathrm{II})}{\Gamma(\mathrm{II})} \approx 100 \text{ meV} \tag{6-566}$$

for the 1900 eV gross-structure resonance. An approximate value for the total width $\Gamma(\mathrm{II})$ can be obtained by considering the energy interval around the center of the gross-structure resonance that contains half the total fission yield of that resonance. From the data in Fig. 6-58, the value of this width may be roughly estimated to be

$$\Gamma(\mathrm{II}) \approx 50 \text{ eV} \tag{6-567}$$

The magnitude of $\Gamma_d(\mathrm{II})$ is expected to be several orders of magnitude smaller than $\Gamma(\mathrm{II})$, since radiative widths associated with excitations of a few MeV are found to be of the order of 10–100 meV (see, for example, Eq. (6-565)). Thus, the contribution of $\Gamma_d(\mathrm{II})$ to the total width $\Gamma(\mathrm{II})$ may be neglected, and the estimates (6-566) and (6-567) yield

$$\begin{aligned} \Gamma_>(\mathrm{II}) &\approx 50 \text{ eV} \\ \Gamma_<(\mathrm{II}) &\approx 0.1 \text{ eV} \end{aligned} \tag{6-568}$$

where $\Gamma_>(\mathrm{II})$ and $\Gamma_<(\mathrm{II})$ are the greater and lesser of $\Gamma_f(\mathrm{II})$ and $\Gamma_c(\mathrm{II})$. The present analysis of the (nf) cross section does not distinguish the roles of Γ_f and Γ_c.

From the parameters (6-568) and the expression (6-562), one obtains

$$\langle \Gamma_f(\mathrm{I}) \rangle_{max} \approx 20 \text{ meV} \tag{6-569}$$

for the average value of the fission width of type I states at the center of the type II resonances. This value is seen to be rather small compared with the total width $\Gamma(\mathrm{I})$ (see Eq. (6-565)), as assumed in obtaining the approximate expression (6-564).

Values of the fission widths of the individual fine-structure levels have been obtained from a resonance analysis of the cross sections in Fig. 6-58 and are given in Table 6-18. The individual fission widths are expected to exhibit rather large fluctuations, characteristic of a single-channel process (see Eq. (2C-28)). In view of these fluctuations, it appears that the detailed fine-structure analysis of the group of levels around 1900 eV is reasonably consistent with the expression (6-562) and the parameters deduced from the gross-structure analysis.

An analysis of the gross-structure fission cross sections in the 780 and 1400 eV groups of levels gives values for the width parameters that are comparable to those obtained for the 1900 eV group, but such values appear ▲ to be inconsistent with the large $\Gamma_f(\mathrm{I})$ values suggested by the fine-structure

E_{res} eV	Γ_n meV	$\Gamma_n\Gamma_f/\Gamma$ meV	Γ_f meV
750	68 ± 3	5.4	8 ± 1
759	6 ± 1		
778	1.2 ± 0.8		
782	3.0 ± 0.9	2.5	$130\ ^{+(a)}_{-80}$
791	24 ± 2	5.1	13 ± 2
810	213 ± 11	8.3	10 ± 1
820	110 ± 6	0.8	1.0 ± 0.1
1401	5.2 ± 1.0 }	10.5	$110\ ^{+(a)}_{-50}$
1409	11 ± 6 }		
1426	37 ± 4	2.3	4.0 ± 0.5
1841	126 ± 10	8.0	10 ± 1
1853	34 ± 6		
1873	77 ± 7		
1902	209 ± 12		
1917	36 ± 6	15.0	42 ± 7
1943	8 ± 5		
1949	82 ± 8		
1956	260 ± 18	20.7	24 ± 3
1973	68 ± 8		
1992	114 ± 10		
1998	6 ± 4		
2017	52 ± 8		
2023	56 ± 8		
2033	102 ± 10	5.5	7 ± 1
2056	68 ± 7	1.4	1.9 ± 0.2

[a)] upper limit undetermined.

Table 6-18 Resonance parameters for $n + {}^{240}Pu$. The table is based on the analysis by E. Migneco and J. P. Theobald, *Nuclear Phys.* **A112**, 603 (1968), of the fission yield shown in Fig. 6-58. The resonance energies and neutron widths employed in this analysis are derived from total cross sections determined by W. Kolar and K. H. Böckhoff, *J. Nuclear Energy* **22**, 299 (1968), and the radiation width is assumed to be constant, equal to the average value 23.2 meV determined by H. Weigmann and H. Schmid, *J. Nuclear Energy* **22**, 317 (1968).

▼ analysis given in Table 6-18. Thus, one is led to consider an analysis of these groups in terms of the narrow strength function approximation. The levels with large values of Γ_f observed in the middle of the groups may correspond to type I levels that are especially close to the type II level (if $\Gamma(\mathrm{II}) \gg \Gamma(\mathrm{I})$) or they may correspond to the type II level itself (if $\Gamma(\mathrm{II}) \lesssim \Gamma(\mathrm{I})$). It is characteristic of the latter of these two interpretations that the level with large Γ_f has a small Γ_n, and indeed the analysis in Table 6-18 suggests such a correlation, though it cannot be excluded that the observed correlation is a fluctuation effect.

If one accepts the interpretation of the levels with large Γ_f as representing type II states, the fine-structure analysis in Table 6-18 is somewhat modified, since the derivation of Γ_f from the observed fission yields has employed the assumption that Γ_γ is the same as for the average of the type I levels measured at lower energies. For a type II level, however, the value of Γ_γ may be somewhat different; from the available data, we can therefore only conclude that $\Gamma_f(\mathrm{II})$ is large compared with Γ_n and smaller than the energy resolution. From the strength of the fission widths of the neighboring type I levels (see Eqs. (6-553) and (6-554)), one obtains $\Gamma_f(\mathrm{II})\langle V^2\rangle \approx 2.5\ \mathrm{eV}^3$ with, however, a large uncertainty due to the fluctuations in the rather small sample that is available. In terms of the quantity $\Gamma_c(\mathrm{II})$ given by Eq. (6-556), the estimate of the coupling implies $\Gamma_f(\mathrm{II})\Gamma_c(\mathrm{II}) \approx 1\ \mathrm{eV}^2$.

The estimated value of $\Gamma_f(\mathrm{II})\Gamma_c(\mathrm{II})$ for the different gross-structure resonances in ^{240}Pu(nf) are seen to vary by about a factor of 5, which, however, appears not unreasonable, since the fission as well as the coupling to the region I are expected to be dominated by the single lowest channel. Thus, it appears possible to obtain a consistent interpretation of the available data by assuming the average parameters

$$\begin{aligned} \langle \Gamma_f(\mathrm{II}) \rangle &\approx 100\ \mathrm{meV} \\ \langle \Gamma_c(\mathrm{II}) \rangle &\approx 25\ \mathrm{eV} \end{aligned} \qquad (6\text{-}570)$$

It must be emphasized, however, that the available data are not sufficiently detailed to exclude other interpretations with different sets of parameters; in particular, the identification of Γ_c as the main contributor to $\Gamma(\mathrm{II})$ hinges on the rather tentative evidence for very large individual fission widths in the fine-structure analysis.

Characterization of potential energy function for ^{241}Pu

The results of the resonance analysis of the ^{240}Pu(nf) reaction provide information on the structure of the potential energy as a function of the deformation variable α associated with the fission mode. An estimate of the

▲ relative energy of the minima in regions I and II can be obtained from the

▼ ratio of the level spacings $D(\mathrm{I})$ and $D(\mathrm{II})$. From the level density expression (2-57) for the Fermi gas model, one finds, assuming that the single-particle level spacing g_0 is the same for the excitations I and II,

$$\ln\frac{D(\mathrm{II})}{D(\mathrm{I})}\approx 2\ln\frac{E-E(\mathrm{II})}{E-E(\mathrm{I})}+\frac{E(\mathrm{II})-E(\mathrm{I})}{T}$$

$$T\approx a^{-1/2}\left(E-\tfrac{1}{2}(E(\mathrm{I})+E(\mathrm{II}))\right)^{1/2} \tag{6-571}$$

$$a=\frac{\pi^2}{6}g_0\approx\frac{A}{8}\ \mathrm{MeV}^{-1}$$

where $E-E(\mathrm{I})$ is the neutron binding energy ($S_n=5.2$ MeV), and where the estimate of the quantity a is taken from Fig. 2-12, Vol. I, p. 187. From the values $D(\mathrm{I})\approx 15$ eV and $D(\mathrm{II})\approx 700$ eV (see Eqs. (6-565) and (6-551)), one finds a temperature of about 0.4 MeV and $E(\mathrm{II})-E(\mathrm{I})\approx 2$ MeV. It should be emphasized that the use of the Fermi gas level density formula neglects the effects of shell structure and pair correlations that may significantly affect the estimated excitation energy of the isomer. Thus, the appreciable shell-structure energy in the region II (see the estimate on p. 635) implies a lower level density for a given excitation energy (see Fig. 6-55); this effect, however, may be largely compensated by the shell-structure energy and the pair correlations in the ground state. For the deformed nuclei, the level density is increased by a large factor due to the rotational degrees of freedom (see p. 38ff.); this factor, however, is approximately the same for the regions I and II, provided the deformations have the same symmetry. The available evidence on rotational spectra for states in the region II (Specht *et al.*, 1972) is compatible with a shape with axial and inversion symmetry, as implied by the interpretation of the shell-structure effect (see p. 635).

Estimates of the barrier heights may be obtained from the parameters $\Gamma_c(\mathrm{II})$ and $\Gamma_f(\mathrm{II})$ derived from the resonance analysis. For a qualitative orientation, we shall ignore the possibility of gross structure in the vibrational strength function in region II at the excitation energies considered; the widths are then given by the relations (see Eq. (6-111)),

$$\begin{aligned}\Gamma_f(\mathrm{II})&=\frac{D(\mathrm{II})}{2\pi}P(\mathrm{II}\rightarrow\mathrm{III})\\ \Gamma_c(\mathrm{II})&=\frac{D(\mathrm{II})}{2\pi}P(\mathrm{II}\rightarrow\mathrm{I})\end{aligned} \tag{6-572}$$

where the barrier penetration factor P represents the probability that a motion in the deformation variable α is transmitted through the barrier. In ▲ Eq. (6-572), only the contribution from the lowest channel is taken into

▼ account, since the barrier penetration depends exponentially on the threshold energy.

From the level spacing (6-551) and the tentative values (6-570) for the widths, we obtain

$$\begin{aligned} P(\text{II}\rightarrow\text{I}) &\approx 0.2 \\ P(\text{II}\rightarrow\text{III}) &\approx 10^{-3} \end{aligned} \tag{6-573}$$

The expression (6-109), which describes the penetration through a parabolic barrier, thus implies that the barrier between I and II is only slightly above the neutron separation energy for ^{241}Pu, as indicated in Fig. 6-59. The barrier II→III corresponds to the fission threshold as observed in fast neutron fission; this threshold occurs at an energy $S_n + 700$ keV, and from the expression (6-109), one thus obtains the estimate

$$\hbar\omega \approx 600 \text{ keV} \tag{6-574}$$

for the parameter describing the energy dependence of the subbarrier penetrability. The value (6-574) may somewhat underestimate the frequency ω, since it is not known whether the observed threshold at 700 keV is associated with s-wave neutrons; thus, it is possible that the s-wave threshold may occur at an energy that is higher by as much as a few hundred keV. The estimate (6-574) for $\hbar\omega$ is similar to the value implied by the observed energy variation of the threshold function for ^{240}Pu(nf); see *Neutron Cross Sections* (1965). ▲

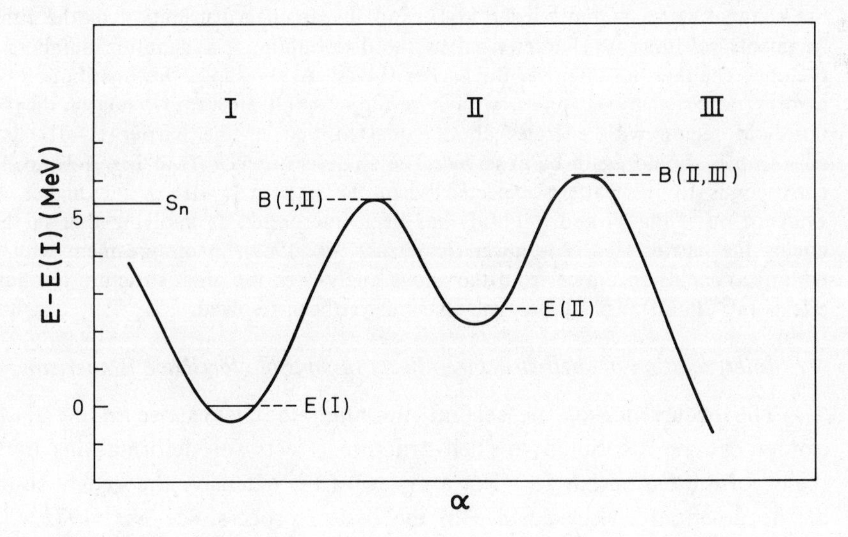

Figure 6-59 Schematic illustration of potential energy function exhibiting a second minimum.

▼ Additional evidence bearing on the consistency of the barrier parameters deduced above is provided by the lifetime of the spontaneous fissioning isomer in ^{241}Pu ($\tau_f = \hbar/\Gamma_f = 4 \times 10^{-5}$ sec; Polikanov and Sletten, 1970). A rough estimate of the lifetime can be obtained by assuming a parabolic barrier, which yields, by means of Eqs. (6-109) and (6-110),

$$\Gamma_f \approx \frac{\hbar\omega_{vib}}{2\pi} \exp\left\{ -2\pi \frac{B(\text{II},\text{III}) - E(\text{II})}{\hbar\omega} \right\} \tag{6-575}$$

where ω_{vib} is the frequency for the collective vibrations around the equilibrium II, which we shall assume to be of the order $\hbar\omega_{vib} \sim 1$ MeV. The height of the barrier between regions II and III is given by the above analysis as $B(\text{II},\text{III}) - E(\text{II}) \approx S_n + 0.7 \text{ MeV} + E(\text{I}) - E(\text{II}) \approx 4$ MeV for $S_n = 5.2$ MeV and $E(\text{I}) - E(\text{II}) \approx -2$ MeV (see p. 632). From Eq. (6-575), one thus obtains the value $\hbar\omega \approx 700$ keV for the parameter describing the penetrability of the barrier separating regions II and III, which is consistent with the value (6-574) obtained from the observed $\Gamma_f(\text{II})$.

A different source of evidence concerning the relative heights of the different barriers is provided by the study of the angular distribution of the fission fragments in the (nf) and (γf) processes (Strutinsky and Bjørnholm, 1968). The analysis of the (γf) distributions given in Chapter 4, pp. 123 ff., assumes that the K quantum number is conserved after passage through the saddle point representing the fission threshold. Such a situation, which leads to large anisotropies in the threshold region, may be expected if the outer barrier II→III is higher than the inner barrier I→II. However, if the threshold is determined by the barrier I→II, the motion of the system through the region II may be damped through coupling to the dense spectrum of compound levels of type II. Such a damping may imply loss of memory of the K quantum number for the channel states at the barrier I→II, and the angular distribution of the fission fragments will then be characterized by the distribution of K quantum numbers for the open channels leading over the barrier II→III. At threshold, this distribution may involve rather many channels, with a resulting small anisotropy, but in the subthreshold region with energies about equal to that of the barrier II→III, large anisotropies should again be observed. The analysis of ^{238}U(γf) in Fig. 4-22, p. 125, corresponds to the pattern expected when the barrier II→III is the higher, but evidence on ^{240}Pu(γf) and ^{242}Pu(γf) has been interpreted as implying that in these nuclei, the barrier II→III is lower (Kapitza *et al.*, 1969), in disagreement with the tentative conclusion drawn from the above analysis of the gross-structure resonance effects in ^{240}Pu(nf). This discrepancy has not yet been resolved.

Interpretation of shell-structure effects responsible for shape isomerism

The occurrence of the second minimum that gives rise to the fission isomer can be attributed to shell-structure effects for deformations in the region of the fission barrier. (For a review of the extensive theoretical studies of the potential energy surface in the fission process, see Nix, 1972.) The effect responsible for the isomer can be understood in terms of the major ▲ shell structure that occurs for a prolate nuclear deformation with a ratio 2:1

▼ between the principal axes (see pp. 595 ff.). This deformation corresponds approximately to the saddle point of the fission barrier as calculated from the liquid-drop model with $x \approx 0.8$ (see Fig. 6A-1). The shell structure is associated with the degeneracy of the eigenvalues for an anisotropic harmonic oscillator potential with a ratio of 2:1 between the frequencies, which leads to the formation of closed shells for particle numbers...80, 110, 140,.... The deviations of the nuclear potential from that of an oscillator are larger for protons than for neutrons and, in the heavy element region, only minor shell-structure effects are expected for the protons at the 2:1 deformation (see p. 597). For neutrons, the effect of the spin-orbit coupling implies that four (doubly degenerate) orbits with $N_{sh}=10$ join the $N_{sh}=9$ shell, which would have been completed at neutron number $N=140$, to produce a major shell closing at $N=148$ (see Fig. 6-50, p. 596). Indeed, this neutron number corresponds to the region of nuclei where the fission isomer has been found as a prominent feature.

This interpretation also provides a qualitative estimate of the magnitude of the shell-structure effect to be expected in the energy of the fission isomer. For the case of complete degeneracy, expression (6-510), together with the scale factors (6-514), yields a value

$$\begin{aligned} \mathscr{E}_{sh} &= -\frac{1}{24}(3N)^{2/3}(a^2b)^{-2/3}\hbar\omega_0 \\ &\approx -6 \text{ MeV} \end{aligned} \tag{6-576}$$

for the closed neutron configuration ($a:b=2:1$, $N=148$, $\hbar\omega_0=41A^{-1/3}$ MeV). As seen from Fig. 6-50, the shells spread over an energy interval of a little more than $\hbar\omega_{sh}/2$; the expected shell-structure energy is therefore of the order of -3 MeV (see the discussion of the effect of incomplete degeneracy on p. 601).

The estimated value of $\mathscr{E}_{sh}$ (II) appears to be consistent with the energy of the second minimum derived from the above analysis, which gives a value for E(II) that is about 3.5 MeV lower than the mean value of the two barriers. ($E(\text{II})-E(\text{I})\approx 2$ MeV, see p. 632; $B(\text{I,II})-E(\text{I})\approx 5.2$ MeV, $B(\text{II,III})-E(\text{I})$ ≈ 5.9 MeV, see p. 634.) In the interpretation of the difference between E(II) and E(I), one must consider contributions from $\tilde{\mathscr{E}}$ as well as from $\mathscr{E}_{sh}$. For $(Z^2/A)_{crit}=45$ (see p. 621), corresponding to $x=0.8$ for ^{241}Pu, the liquid-drop barrier function (6A-33) yields $\tilde{\mathscr{E}}(\text{II})\approx 4$ MeV ($\beta \approx \beta_f = 0.7$) and $\tilde{\mathscr{E}}(\text{I})\approx$ 1 MeV ($\beta \approx 0.23$). For a spherical shape, the shell-structure energy would be large and positive, but the deformation relieves this "stress" in the nuclear structure and is expected to lead to a negative, though relatively small value of $\mathscr{E}_{sh}$. The estimate of this effect based on the oscillator model (see p. 605) would imply $\mathscr{E}_{sh}(\text{I})\approx -4$ MeV, and deviations from complete degeneracy of
▲ the shells are expected to reduce this value by about a factor of 2. Thus, we

▼ obtain the estimate $E(\text{II})-E(\text{I})\approx\tilde{\mathscr{E}}(\text{II})-\tilde{\mathscr{E}}(\text{I})+\mathscr{E}_{\text{sh}}(\text{II})-\mathscr{E}_{\text{sh}}(\text{I})\approx 2$ MeV. While it appears possible on this simple basis to obtain a coherent interpretation of the main features of the potential energy function for ^{241}Pu, the qualitative nature of the various numerical estimates should of course be emphasized.

Features of Spin Excitations

Collective modes with $\lambda\pi = 1+$ (Fig. 6-60)

The exploration of nuclear collective motion involving the intrinsic spin degrees of freedom is still in a very preliminary stage, in spite of the far-reaching significance of this aspect of nuclear dynamics. In the present example, we consider the tentative data on the modes excited by the simple spatially independent fields

$$\mathbf{F}_\tau = \begin{cases} \boldsymbol{\sigma} & \tau=0 \\ \boldsymbol{\sigma}\tau_z & \tau=1 \end{cases} \tag{6-577}$$

The spin-dependent nuclear interactions can couple the fields (6-577), with orbital and spin symmetry $\kappa=0$, $\sigma=1$, to fields with $\kappa=2$, $\sigma=1$, $\lambda\pi=1+$ (see p. 384), but the available data are not sufficiently detailed to establish the role of these couplings. The present discussion is confined to the diagonal interaction involving the fields (6-577), but it must be emphasized that a proper understanding of the spin excitations will require an analysis of the couplings that are nondiagonal in κ.

Analysis in terms of single-particle excitations

For a closed-shell nucleus, the only single-particle transitions that can be generated by the fields (6-577) are of the type $j=l+1/2\rightarrow j=l-1/2$,with a frequency corresponding to the spin-orbit splitting; such transitions can occur if the $j=l+1/2$ shell is filled, while the $j=l-1/2$ shell is empty. Thus, the closed-shell configurations in ^{4}He, ^{16}O, and ^{40}Ca do not give rise to excitation modes associated with the fields (6-577), corresponding to the fact that these configurations have a total spin quantum number $S=0$, for neutrons as well as protons. The closed-shell configurations with particle numbers 28, 50, 82, and 126, however, involve unsaturated spins as a result of the spin-orbit interactions, and a collective spin mode may be expected.

For a nucleus with $N=Z$ ($T_0=0$), the normal modes correspond to isoscalar or isovector excitations. In nuclei with a neutron excess, the unsaturated spins of the neutrons and protons are associated with different orbits; however, when the neutrons and protons have similar spin-orbit ▲ frequencies and transition strengths (as, for example, in ^{208}Pb), it is expected

▼ that the spin-dependent interactions will combine the neutron and proton excitations into collective modes with $\tau \approx 0$ and $\tau \approx 1$.

For a single-particle response function with only a single frequency, the effects of the interactions can be treated as in Sec. 6-2c. The single-particle excitations associated with the fields (6-577) have the total strength

$$\begin{aligned}(\alpha_0^{(0)})^2 &= \sum_i \langle i|\sigma_z|\mathrm{v}=0\rangle^2 = \sum_i \langle i|\sigma_z\tau_z|\mathrm{v}=0\rangle^2 \\ &= \frac{8}{3}\left(\frac{l_\mathrm{n}(l_\mathrm{n}+1)}{2l_\mathrm{n}+1} + \frac{l_\mathrm{p}(l_\mathrm{p}+1)}{2l_\mathrm{p}+1}\right) \\ &\approx \frac{4}{3}(2\bar{l}+1)\end{aligned} \qquad (6\text{-}578)$$

where we have assumed that the neutrons and protons each contribute with a single transition leading from the closed-shell configuration $\mathrm{v}=0$ to the particle-hole configuration $(j=l+1/2)^{-1}(j=l-1/2)^1$; the matrix element of σ_z can be obtained from Eqs. (3A-22) and (3B-25). In the last line of Eq. (6-578), we have introduced the mean value $\bar{l}=(l_\mathrm{n}+l_\mathrm{p})/2$ and dropped terms of relative magnitude $(l)^{-2}$. The restoring force for the unperturbed motion is given by (see Eq. (6-23))

$$C^{(0)} = \frac{\hbar\omega^{(0)}}{2(\alpha_0^{(0)})^2} \approx \frac{3}{8}\frac{\hbar\omega^{(0)}}{2\bar{l}+1} \qquad (6\text{-}579)$$

where the unperturbed energy $\hbar\omega^{(0)}$ is equal to the spin-orbit energy splitting $(\hbar\omega^{(0)} = \varepsilon(j=l-1/2) - \varepsilon(j=l+1/2))$, taken as a mean value for neutrons and protons. The parameters $C^{(0)}$, $\hbar\omega^{(0)}$, and $\alpha_0^{(0)}$ for the unperturbed motion are the same for the $\tau=0$ and $\tau=1$ modes.

With inclusion of the interactions associated with the spin-dependent fields, characterized by the parameters κ_τ, the normal modes have the frequencies and transition strengths (see Eqs. (6-27) and (6-28))

$$\hbar\omega_\tau = \hbar\omega^{(0)}\left(1 + \frac{\kappa_\tau}{C^{(0)}}\right)^{1/2} \qquad (6\text{-}580a)$$

$$\langle n_\tau=1, \lambda\pi=1+, \mu=0|(F_\tau)_z|0\rangle^2 = (\alpha_0^{(0)})^2 \frac{\hbar\omega^{(0)}}{\hbar\omega_\tau} \qquad (6\text{-}580b)$$

The collective excitation $n_\tau=1$ is associated with a strong spin-dependent ▲ nuclear field that could be studied in inelastic scattering processes. Moreover,

▼ the isovector spin excitation gives rise to a collective $M1$ transition with

$$B(M1;0\rightarrow n_1=1,\lambda\pi=1+)=\frac{9}{16\pi}(g_s-g_l)^2_{\tau=1}\left(\frac{e\hbar}{2Mc}\right)^2\langle n_{\tau=1}=1|(F_{\tau=1})_z|0\rangle^2$$

$$\approx\frac{3}{4\pi}(4.21)^2(2\bar{l}+1)\frac{\hbar\omega^{(0)}}{\hbar\omega_{\tau=1}}\left(\frac{e\hbar}{2Mc}\right)^2 \qquad (6\text{-}581)$$

where g_s and g_l are the spin and orbital g factors with values given by Eq. (3-37).

Estimate of coupling strength

Rather little information is available concerning the coupling associated with the isoscalar spin fields, but evidence on the isovector fields can be derived from the observed renormalization of the effective spin g factor in single-particle $M1$-matrix elements (see Table 3-3, Vol. I, p. 343). From Eqs. (6-216) and (6-26), we obtain

$$\left(\frac{\delta g_s}{g_s}\right)_{\tau=1}=-\left(\frac{\kappa}{C}\right)_{\tau=1}=-\frac{\kappa_{\tau=1}}{C^{(0)}+\kappa_{\tau=1}} \qquad (6\text{-}582)$$

for the static polarization contribution δg_s to the effective isovector spin g factor. The observed magnetic moments in the neighborhood of ^{208}Pb suggest a value of $\delta g_s/g_s\approx-0.5$ for the isovector spin g factor (see Vol. I, p. 344), which implies $\kappa_{\tau=1}\approx C^{(0)}$ and hence, from Eqs. (6-578) and (6-579),

$$\kappa_{\tau=1}\approx C^{(0)}\approx\frac{3}{8}\frac{\hbar\omega^{(0)}}{2\bar{l}+1}\approx 0.18\text{ MeV} \qquad (6\text{-}583)$$

where the unperturbed frequency $\hbar\omega^{(0)}=5.7$ MeV is taken as the mean of the $h_{11/2}\rightarrow h_{9/2}$ (proton) and $i_{13/2}\rightarrow i_{11/2}$ (neutron) excitation frequencies (see Fig. 3-3, Vol. I, p. 325). Assuming the coupling to be proportional to A^{-1}, as for a volume effect, we obtain

$$\kappa_{\tau=1}\approx 40A^{-1}\text{ MeV} \qquad (6\text{-}584)$$

Evidence for collective M1 excitations

The coupling strength (6-583) leads to a collective isovector spin mode for ^{208}Pb, with frequency and transition strength

$$\hbar\omega\approx\sqrt{2}\,\hbar\omega^{(0)}\approx 8.1\text{ MeV}$$
$$B(M1;0+\rightarrow n_{\tau=1}=1,\lambda\pi=1+)\approx 36\left(\frac{e\hbar}{2Mc}\right)^2 \qquad (6\text{-}585)$$

▲ Tentative evidence for the expected collective $M1$ excitation in ^{208}Pb has

▼ been obtained from the study of the ^{208}Pb(γn) reaction (see Fig. 6-60). In the region up to about 1 MeV above the neutron threshold ($S_n = 7.38$ MeV), the contributions of the individual resonances can be resolved by means of neutron time of flight spectroscopy. The seven resonances shown in Fig. 6-60 have been assigned $I\pi = 1+$ on the basis of angular distributions; the observed $M1$ strength in the excitation window 7.4–8.3 MeV corresponds to ▲ about 70% of the estimate (6-585).

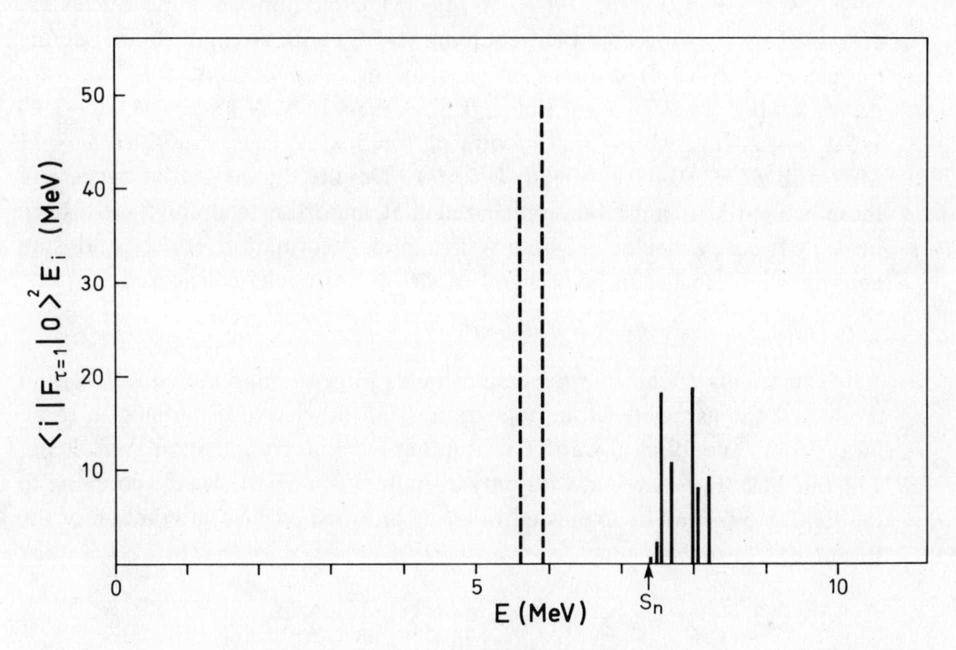

Figure 6-60 Magnetic dipole transitions in ^{208}Pb. The dotted lines give the energies and oscillator strength for the ($\kappa = 0$, $\sigma = 1$, $\tau = 1$) transitions calculated without inclusion of the interaction effects. The observed transition strengths (solid lines) are obtained from measured widths for $M1$ γ-transitions. The experimental data are taken from C. D. Bowman, R. J. Baglan, B. L. Berman, and T. W. Phillips, *Phys. Rev. Letters* **25**, 1302 (1970).

▼ Evidence for the isovector spin mode in Ce has been obtained from electron scattering, which has identified a collective $M1$ transition with a strength $B(M1; 0 \rightarrow 1+) \approx (36 \pm 18)(e\hbar/2Mc)^2$ at an energy $\hbar\omega = 8.7$ MeV and with a total width $\Gamma_{tot} \approx 2$ MeV (Pitthan and Walcher, 1972). The dominant Ce isotope is $^{140}_{58}$Ce with 82 neutrons, which contribute with the $h_{11/2} \rightarrow h_{9/2}$ transition; the proton contribution is expected to be similar to that of the $g_{9/2} \rightarrow g_{7/2}$ excitation of the closed shell at $Z = 50$. The additional protons partly occupy the $g_{7/2}$ orbit, reducing the $g_{9/2}$ contribution, and partly the ▲ $d_{5/2}$ orbit, adding new unsaturated spins. Taking $\hbar\omega^{(0)}$ to be the same as in

▼ ^{208}Pb, the coupling (6-584), and $\bar{l}=4.5$, we obtain $\hbar\omega = 8.6$ MeV and $B(M1; 0\rightarrow 1+) = 29\ (e\hbar/2Mc)^2$.

There is also evidence for major interaction effects in the $M1$-excitation modes of lighter nuclei that have unsaturated spins in their ground states. For example, strong $M1$ excitations are found in ^{12}C ($\hbar\omega = 15.1$ MeV, $B(M1; 0\rightarrow 1+) \approx 3(e\hbar/2Mc)^2$; Ajzenberg-Selove and Lauritsen, 1968) and in ^{28}Si (several lines with $\hbar\omega \approx 11.5$ MeV and a total $B(M1; 0\rightarrow 1+) \approx 7(e\hbar/2Mc)^2$; Fagg *et al.*, 1969). If the $M1$ excitations in these nuclei are described by the isovector field coupling (6-577) with strength (6-584) acting on closed-shell configurations in *jj* coupling, one obtains, for ^{12}C, with $\hbar\omega^{(0)} = 6$ MeV (see Fig. 3-2b, Vol. I, p. 321), $\hbar\omega = 14$ MeV and $B(M1; 0\rightarrow 1+) = 5.4\ (e\hbar/2Mc)^2$, while for ^{28}Si, with $\hbar\omega^{(0)} = 5$ MeV (see Fig. 3-2b), $\hbar\omega = 11$ MeV and $B(M1; 0\rightarrow 1+) = 9.6\ (e\hbar/2Mc)^2$. Despite the qualitative success of these estimates, it must be emphasized that important couplings have been neglected; in particular the interweaving of deformation effects and spin couplings is expected to be of significance for the nuclei considered.

Transition from jj to LS coupling

The effects of the spin-dependent fields in generating the collective spin mode and the associated spin polarization can be viewed in terms of a trend away from *jj* coupling toward *LS* coupling (see the comments in Vol. I, pp. 339 and 348). For a system with only a single type of particle, the coupling to the field (6-577) can be expressed in terms of a two-particle interaction of the form (see Eq. (6-24)),

$$\begin{aligned} H' &= \kappa \sum_{j<k} (\boldsymbol{\sigma}_j \cdot \boldsymbol{\sigma}_k) = \frac{\kappa}{2}\left(\left(\sum_k \boldsymbol{\sigma}_k\right)^2 - \sum_k (\boldsymbol{\sigma}_k)^2\right) \\ &= 2\kappa\left(\mathbf{S}^2 - \tfrac{3}{4}n\right) \end{aligned} \tag{6-586}$$

where $\mathbf{S}$ is the total spin operator and n the number of particles.

For the closed-shell configuration $(j = l + 1/2)^{2l+2}$, it follows from Eq. (6-578) that $\langle \mathbf{S}^2 \rangle \approx l$, for $l \gg 1$; the average of the coupling energy is therefore $\langle H' \rangle \approx -\kappa l$. The spin-spin interaction, with positive κ, favors states of low S and implies a reduction of $\langle \mathbf{S}^2 \rangle$ in the ground state, as exhibited by the last factor in Eq. (6-580b); for sufficiently large values of κ, the ground state approaches a state with $S = 0$ and an interaction energy $\langle H' \rangle \approx -3\kappa l$.

The transition from *jj* to *LS* coupling thus leads to a gain in the spin interaction energy of order κ per particle; at the same time, the change of coupling scheme involves a loss of the binding energy associated with the spin-orbit coupling, which amounts to about $\hbar\omega^{(0)}/2$ per particle. Thus, the transition occurs when $\kappa \sim \hbar\omega^{(0)} \sim lC^{(0)}$, corresponding to $\hbar\omega \sim l^{1/2}\hbar\omega^{(0)}$ and $\langle \mathbf{S}^2 \rangle \sim l^{1/2}$; see Eqs. (6-579) and (6-580). (For negative κ, the spin-spin
▲ coupling would favor states of high spin, and the transition to the

▼ "ferromagnetic" phase would occur already for a coupling strength of $|\kappa| \sim C^{(0)}$, which leads to a collective spin oscillation of zero frequency.)

While κ is expected to vary approximately as A^{-1} (see Eq. (6-584)), the spin-orbit splitting $\hbar\omega^{(0)}$, for the states of largest l in the nucleus, is approximately the same throughout the periodic system. Thus, the validity of jj coupling improves as one goes toward the heavy nuclei. In the very light nuclei ($A \sim 10$), the estimated value of κ becomes comparable with $\hbar\omega^{(0)}$, corresponding to an intermediate coupling situation.

Features of Pair Correlations[34]

In the following examples, we consider some of the simple features associated with the coupling between the pair field and the motion of the individual particles. (The more systematic discussion of the pair correlations and of the quasiparticle coupling scheme is the topic of Chapter 8.)

Ground state of ^{206}Pb (Fig. 6-61 and Table 6-19)

The extensive evidence on the properties of the ground state of ^{206}Pb provides the opportunity for a rather detailed analysis of the pair correlations between the two neutron holes in the closed-shell configuration of ^{208}Pb. The separation energies of ^{208}Pb for one and two neutrons are 7.375 MeV and 14.110 MeV, respectively (see Mattauch *et al.*, 1965); the ground state of ^{206}Pb, with $I\pi = 0+$, is therefore bound by 640 keV with respect to the lowest unperturbed two-hole configuration, $(p_{1/2})^{-2}$. This interaction energy is comparable to the distance between the different configurations $(j^{-2})_0$ (see Table ▲ 6-19) and thus suggests the possible importance of configuration mixing.

[34]The occurrence of a systematic difference between even and odd nuclei, associated with a pairing effect, was recognized at an early stage of nuclear physics (see Heisenberg, 1932a) and provided the basis for understanding the striking difference in the fission of the odd and even isotopes of U (Bohr and Wheeler, 1939). A contribution to the odd-even mass difference arises from the twofold degeneracy of the orbits in the one-particle potential (see Vol. I, p. 143) as well as from the effects of orbital permutation symmetry on the nucleonic interaction energies (Wigner, 1937). These contributions imply a pairing energy proportional to A^{-1}, but the accumulating empirical evidence revealed a less rapid decrease with increasing mass number (see, for example, Green, 1958). A first step in the analysis of the correlation effects involved in the odd-even mass differences resulted from the study of shell-model configurations involving a pairwise coupling of equivalent nucleons to a state of zero angular momentum (seniority coupling scheme; Mayer, 1950a; Racah, 1952; Racah and Talmi, 1953). The study of nuclear collective properties focused attention on the pair correlation as a systematic effect tending to oppose the alignment of the individual orbits in the deformed potential (Bohr and Mottelson, 1955). After the discovery of the appropriate concepts for treating the electronic correlations in superconductors (Bardeen, Cooper, and Schrieffer, 1957), it was recognized that these concepts could provide a basis for analyzing the pair correlations in nuclei (Bohr, Mottelson, and Pines, 1958; Belyaev, 1959; Migdal, 1959). The quantitative interpretation of nuclear phenomena on this basis was initiated by Kisslinger and Sorensen (1960), Griffin and Rich (1960), Nilsson and Prior (1961), Soloviev (1961) and (1962), Yoshida (1961) and (1962), Mang and Rasmussen (1962). The quasiparticle coupling scheme may be viewed as a generalized seniority coupling, and the notation ν for the quasiparticle is that employed by Racah (1943), which derives from the Hebrew word for seniority ("vethek").

Orbit	$\Omega = j+\frac{1}{2}$	$\varepsilon(p_{1/2})-\varepsilon(j)$	$\lvert c(j)\rvert_{exp}$	$c(j)_{calc}$
$p_{1/2}$	1	0	0.66	0.74
$f_{5/2}$	3	0.57	0.51	0.46
$p_{3/2}$	2	0.89	0.32	0.28
$i_{13/2}$	7	1.63	0.28	0.32
$f_{7/2}$	4	2.34	0.22	0.18
$h_{9/2}$	5	3.47	0.3	0.14

Table 6-19 Configuration analysis of the ground state of ^{206}Pb. The table shows the amplitudes $c(j)$ for the configurations j^{-2} involving two neutron holes with respect to the ground state of ^{208}Pb. The experimental amplitudes have been derived from an analysis of the ^{206}Pb(dp) cross sections for populating the j^{-1} configurations in ^{207}Pb (R. A. Moyer, B. L. Cohen, and R. C. Diehl, *Phys. Rev.* **C2**, 1898, 1970); the relative amplitudes obtained from this analysis have been normalized to $\Sigma c^2(j)=1$. The experimental single-particle energies in column three are taken from Fig. 3-2f, Vol. I, p. 324.

▼ The squares of the amplitude $c(j)$ that describe the structure of the ^{206}Pb ground state

$$|^{206}\text{Pb}; 0+\rangle = \sum_j c(j) |(j^{-2})_0\rangle \tag{6-587}$$

are directly measured by the one-particle transfer reactions ^{206}Pb (dp) to the various j^{-1} configurations of ^{207}Pb; the results of such an experiment are given in Table 6-19. The amplitudes $|c(j)|$ can also be determined from a resonance analysis of the elastic scattering of protons on ^{206}Pb proceeding through the isobaric analog states of the j^{-1} configurations of ^{207}Pb, and the results obtained in this manner are consistent with those in Table 6-19 (Richard *et al.*, 1968).

The experiments referred to do not determine the relative phases of the amplitudes $c(j)$. However, the fact that the two-particle transfer reaction ^{208}Pb (pt) ^{206}Pb populates the ground state with an intensity an order of magnitude larger than the excited 0+ states (Holland *et al.*, 1968) implies that the relative phases are such as to produce an enhanced spatial overlap of the two neutrons. With the standard phase convention, the amplitude for spatial coincidence in the state $(j^2)_0$ is (see Eqs. (3A-5) and (1A-43))

$$\langle \mathbf{r}_1 = \mathbf{r}_2 = \mathbf{r}, S=0 | j^2; I=0\rangle = (4\pi)^{-1}\left(j+\tfrac{1}{2}\right)^{1/2} \mathscr{R}^2_{nlj}(r) \tag{6-588}$$

where $\mathscr{R}_{nlj}$ is the radial wave function (and where the angular wave functions are normalized to a solid angle of 4π). Constructive interference in the spatial overlap therefore requires the different coefficients $c(j)$ to have the same sign. ▲

▼ The configuration mixing in the ^{206}Pb ground state can be characterized by the magnitude of the pair field. For the 0+ state, the pair density is spatially isotropic, and the matrix element of the monopole moment (6-145) is

$$M \equiv \langle {}^{206}\text{Pb}; 0+ | M_{\frac{1}{2}}^{\dagger} | {}^{208}\text{Pb}; 0+ \rangle = -\sum_j c(j)(j+\tfrac{1}{2})^{1/2} \tag{6-589}$$

The factor $j+\frac{1}{2}$ represents the number of different components $(\nu\bar{\nu})$ in the coupled state $(j^{-2})_0$. The value of $|M|$ implied by the data in Table 6-19, assuming positive sign for all the $c(j)$, is about 3.8; this may be compared with $|M|=1$ for a pure $p_{1/2}^{-2}$ configuration and with the maximum value $|M|=(22)^{1/2}\approx 4.7$ that can be generated by the orbits in the last filled neutron shell $(82 < N \leqslant 126)$.

The strong enhancement of the pair field suggests an interpretation of the ^{206}Pb ground state in terms of a collective phonon with quantum numbers $\alpha=-2$ and $\lambda\pi = 0+$; see pp. 387 ff. Such a mode may be generated by the coupling between the individual configurations $(\nu\bar{\nu})$ and the collective pair field. Assuming this particle-vibration coupling to have the form (6-148) (with a constant radial form factor), the self-consistency condition for the pair moment illustrated in Fig. 6-61 yields the eigenvalue equation

$$\sum_j \frac{G(j+\frac{1}{2})}{2|\varepsilon(j)| - \hbar\omega} = 1 \tag{6-590}$$

where $|\varepsilon(j)|$ are the energies of the single-hole configurations j^{-1}, while $\hbar\omega$ is the energy of the collective pair mode. The expression (6-590) relates the coupling constant G to the energy of the collective mode, and the observed binding energy of ^{206}Pb and empirical single-particle energies in Table 6-19 yield

$$G = 0.14 \text{ MeV} \tag{6-591}$$

for the pair-field coupling constant.

The amplitudes $c(j)$ that characterize the microscopic structure of the 0+ excitation (see Eq. (6-587)) can be obtained from the matrix element of the operator $A^{\dagger}((j^{-2})_0)$ that creates the normalized state $(j^{-2})_0$. The matrix element between the vacuum state and the one-phonon state, which is represented by a diagram equivalent to the first diagram in Fig. 6-13, yields

$$c(j) = \frac{GM(j+\frac{1}{2})^{1/2}}{\hbar\omega - 2|\varepsilon(j)|} \tag{6-592}$$

Alternatively, one can characterize the amplitude $c(j)$ by the one-particle transfer matrix element illustrated in Fig. 6-61b. The reduced matrix element ▲ for this process has the value $2^{1/2}c(j)$, where the factor $2^{1/2}$ reflects the

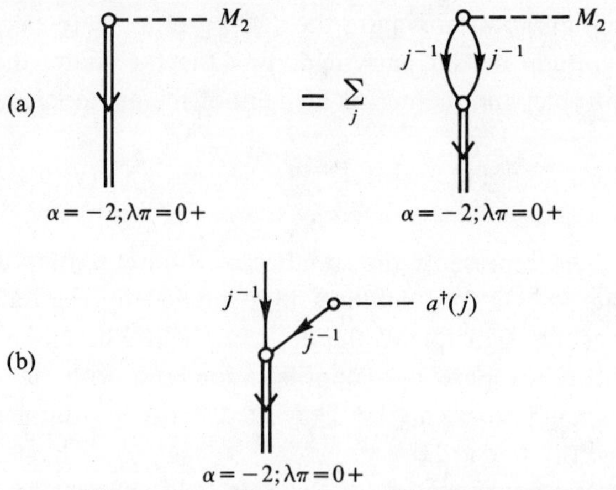

Figure 6-61 Diagrams illustrating properties of pair quanta. Figure 6-61a illustrates the self-consistency relation for the pair moment

$$M_2 = \sum_{\nu>0} a^{\dagger}(\bar{\nu})a^{\dagger}(\nu) = -\sum_j (j+\tfrac{1}{2})^{1/2} A^{\dagger}((j^2)_0)$$

$$A^{\dagger}((j^2)_0) \equiv 2^{-1/2}(a^{\dagger}(j')a^{\dagger}(j))_{(jj')0}$$

The diagram in Fig. 6-61b represents the amplitude for one-particle transfer leading from the pair removal quantum to a single-hole state.

occurrence of two equivalent holes, either of which may be filled by the added neutron; the difference in the weight of the diagram (6-61b) and the corresponding one for $A^{\dagger}((j^{-2})_0)$ can be related to the fact that the two identical neutrons play distinguishable roles in the former but not in the latter. The normalization condition

$$\sum_j c^2(j) = 1 \qquad (6\text{-}593)$$

applied to the expression (6-592) determines the magnitude of the pair moment, which is found to be $M = -3.5$, in rather good agreement with the empirical value quoted above. (We have chosen a phase for the $0+$ excitation that yields positive values of $c(j)$ and, correspondingly, a negative value of M; see Eq. (6-589).) The amplitudes $c(j)$ obtained from Eqs. (6-592) and (6-593) are given in the last column of Table 6-19 and are seen to reproduce the main features of the observed configuration mixing.

The above analysis of the pair correlation has restricted the configurations of the neutron holes to the orbits in the shell $82 < N \leqslant 126$. The inclusion of more distant orbits may significantly renormalize the moment of the pair field and the effective coupling constant. Thus, for example, the contribution of matrix elements by which the pair field creates and annihilates pairs of particles into the next higher shell ($126 < N \leqslant 184$) can be treated by including the amplitude illustrated by the second diagram in Fig. 6-13. The self-consistency condition now includes also terms associated with intermediate states involving a phonon together with two particles in the upper shell (ground-state correlations); see Fig. 6-14. Assuming that the pair field acts with equal strength in both shells, the observed pair binding energy in ^{206}Pb is accounted for by $G = 0.094$ MeV, which is significantly smaller than the value (6-591). However, the value of GM is much less affected, and the estimated single-particle amplitudes remain close to those given in Table 6-19.

The coupling to the collective pair field (6-148) is equivalent to a two-body interaction of the form (pairing force)

$$H_{\mathrm{pair}} = -G \sum_{\nu'>0} a^{\dagger}(\bar{\nu}')a^{\dagger}(\nu') \sum_{\nu>0} a(\nu)a(\bar{\nu}) \tag{6-594}$$

In a spherical potential, the interaction (6-594) only affects the two-body configurations with $J=0$,

$$\langle (j')^2; JM | H_{\mathrm{pair}} | (j)^2; JM \rangle = -G(j+\tfrac{1}{2})^{1/2}(j'+\tfrac{1}{2})^{1/2}\delta(J,0) \tag{6-595}$$

A diagonalization of this interaction in the space of the two-hole configurations is seen to yield the secular equation (6-590) for the eigenvalues and Eq. (6-592) for the wave function.

Neutron pair excitations in the region of ^{208}Pb (Fig. 6-62)

The pair addition and removal modes represent elementary quanta of excitation that are found to approximately retain their identity when superposed on other excitations of the same or different type. This feature is illustrated in Fig. 6-62, which shows the observed states in the Pb isotopes that arise from superposition of quanta of the neutron pair addition and removal modes based on ^{208}Pb. In the harmonic approximation, the energy is linear in the number of quanta (see Eq. (6-137)), and the corresponding values are indicated in Fig. 6-62 by dashed lines. The leading-order anharmonicity can be described in terms of the interactions between pairs of quanta (compare the treatment of anharmonicities for quadrupole phonons on pp. 448 ff.) and give rise to energy shifts.

$$\delta E(n_-, n_+) = \tfrac{1}{2}V_{--}\, n_-(n_- - 1) + V_{-+}\, n_- n_+ + \tfrac{1}{2}V_{++}\, n_+(n_+ - 1) \tag{6-596}$$

(We here employ the abbreviated notations n_- and n_+ to denote the quantities $n_{\alpha=-2}$ and $n_{\alpha=+2}$.) With the values of the interaction parameters ($V_{--} = 710$ keV, $V_{-+} = -130$ keV, and $V_{++} = 170$ keV) determined from the empirical energies of the states $(n_-, n_+) = (2,0)$, $(1,1)$, and $(0,2)$, the estimated energies for $n_- + n_+ \geqslant 3$ are represented in Fig. 6-62 by dotted lines. It is seen that these estimated energies describe the main features of the available data,

▼ despite the fact that the interaction for the $\alpha = -2$ mode is comparable with the binding energy of the neutron-hole pair. The occurrence of an especially large anharmonicity for this mode can be understood in terms of the large amplitude for the $p_{1/2}$ configuration, which can only accommodate a single pair (see Table 6-19).

The states in Fig. 6-62 are linked by collectively enhanced two-particle transfer reactions. The observed transitions are indicated by arrows and the intensities are given in units of the observed cross sections for exciting the one-phonon states. The intensities approximately obey the harmonic relation (6-138), but the deviations appear to be outside the experimental uncertainties ▲ and thus to provide evidence on the anharmonicity.

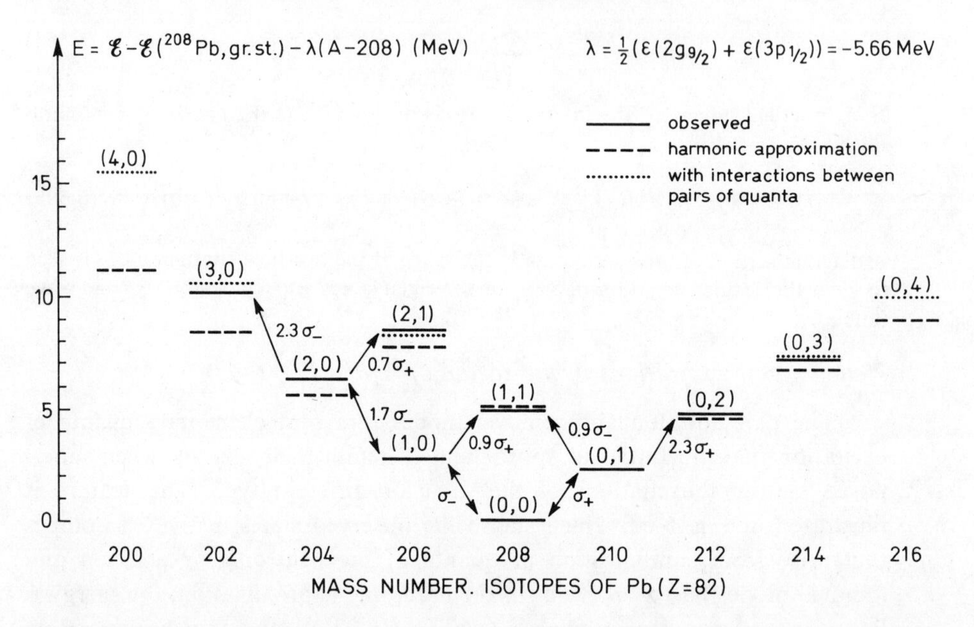

Figure 6-62 Neutron pair vibrational spectrum based on ^{208}Pb. The levels are labeled by the number of quanta (n_-, n_+) of the pair removal and pair addition modes; the energy scale is similar to that of Fig. 6-6, p. 388. The arrows show the transitions that have been observed as collective two-particle transfer processes ((tp) or (pt)), and the number on the arrows gives the cross section, in units of the cross sections σ_- and σ_+ for the processes $(0,0) \leftrightarrow (1,0)$ and $(0,0) \leftrightarrow (0,1)$, respectively, for the same projectile energy. (No corrections have been made for the small differences in Q values.)

The empirical ground-state binding energies are from the compilation by Mattauch *et al.*, 1965. The energies of the pair vibrational excited states and the transfer cross sections are from: ^{204}Pb(pt) and ^{206}Pb(pt) (W. A. Lanford and J. B. McGrory, *Phys. Letters* **45B**, 238, 1973); ^{204}Pb(tp) (E. R. Flynn, G. J. Igo, and R. A. Broglia, *Phys. Letters* **41B**, 397, 1972, and E. R. Flynn, R. A. Broglia, R. Liotta, and B. S. Nilsson, *Nuclear Phys.* **A221**, 509, 1974); ^{206}Pb(tp) and ^{210}Pb(tp) (G. J. Igo, P. D. Barnes , and E. R. Flynn, *Ann. Phys.* **66**, 60, 1971); ^{208}Pb(tp) (E. R. Flynn, G. J. Igo, R. A. Broglia, S. Landowne, V. Paar, and B. Nilsson, *Nuclear Phys.* **A195**, 97, 1972); ^{210}Pb(tp) (C. Ellegaard, P. D. Barnes, and E.R. Flynn, *Nuclear Phys.* **A170**, 209, 1971).

▼ The data in Fig. 6-62 refer to the superposition of pair quanta; these quanta are also found to behave as elementary modes in the presence of particle or hole excitations. Examples are provided by the $I\pi = 9/2+$ state at $E = 2.71$ MeV in ^{207}Pb $((n_- = 1, g_{9/2})9/2+)$ and the $I\pi = 1/2-$ state at $E = 2.15$ MeV in ^{209}Pb $((n_+ = 1, p_{1/2}^{-1})1/2-)$; see Fig. 3-2f, Vol. I, p. 324. These states are populated with approximately the full single-particle strength, in the one-particle transfer reactions ^{206}Pb(dp) (Mukherjee and Cohen, 1962) and ^{210}Pb(dt) (Igo *et al.*, 1971). The $(n_+ = 1, p_{1/2}^{-1})$ state has also been observed in the reaction ^{207}Pb(tp) with a strength similar to that between the ground states of ^{208}Pb and ^{210}Pb (Flynn *et al.*, 1971).

Single-particle motion in potential with pair deformation; quasiparticles (Figs. 6-63 and 6-64)

In nuclei with many particles outside closed shells, an approximately static deformation of the pair field may be established (see pp. 392 ff). The present example brings together some of the basic features of nucleonic motion in the presence of such a pair condensate.

If the pair field is assumed to be approximately constant over the nuclear volume, the pair potential acting on the individual nucleons has the form (see Eq. (6-148))

$$V_{\text{pair}} = -\Delta \sum_{\nu>0} \left(a^\dagger(\bar{\nu})a^\dagger(\nu) + a(\nu)a(\bar{\nu})\right) \tag{6-597}$$

where the constant Δ is given by $\Delta = G\langle M_2\rangle$ in terms of the coupling parameter G and the pair moment M_2.

The creation and annihilation of pairs of nucleons implied by the potential (6-597) are associated with a corresponding removal and addition of particles from the collective pair motion (the condensate), and the energy involved in this transfer of pairs to the condensate must be included in the analysis of the one-particle motion in the presence of the pair field. If the energy increase of the condensate per added particle (chemical potential) is denoted by λ, the nucleonic motion can be described by the Hamiltonian

$$\begin{aligned} H' &= H - \lambda\mathcal{N} = H_0 + V_{\text{pair}} - \lambda\mathcal{N} \\ &= \sum_\nu (\varepsilon(\nu) - \lambda)a^\dagger(\nu)a(\nu) + V_{\text{pair}} \\ &= \sum_{\nu>0} \left((\varepsilon(\nu)-\lambda)(a^\dagger(\nu)a(\nu) + a^\dagger(\bar{\nu})a(\bar{\nu})) - \Delta(a^\dagger(\bar{\nu})a^\dagger(\nu) + a(\nu)a(\bar{\nu}))\right) \end{aligned} \tag{6-598}$$

where $\mathcal{N}$ is the particle-number operator and where H is the energy of the nucleons. The one-particle eigenvalue $\varepsilon(\nu)$ in the absence of the pair potential refers to nucleonic motion in the average potential (spherical or deformed).

The Hamiltonian (6-598) does not conserve particle number, and the ▲ eigenstates of H' are characterized by a distribution of particle number,

▼ corresponding to the possibility of transfer of particles to and from the condensate. The mean value $\langle \mathcal{N} \rangle$ for an eigenstate of H' depends on the chemical potential λ relative to the single-particle spectrum $\varepsilon(\nu)$. Inversely, one can view λ as a parameter to be so adjusted as to give an eigenstate with a prescribed value of $\langle \mathcal{N} \rangle$; hence, λ appears as a Lagrange multiplier associated with the constraint on the average number of particles. (The term $-\lambda\mathcal{N}$ in the Hamiltonian (6-598) can also be viewed as the Coriolis interaction associated with the rotational motion in gauge space; see Eq. (6-151), which exhibits the particle number as the angular momentum in gauge space, and Eq. (6-155), which relates λ to the rotational frequency in this dimension.)

The Hamiltonian (6-598) is a quadratic form in the fermion variables $a^\dagger$ and a and can therefore be diagonalized by a linear transformation among these variables (Bogoliubov, 1958; Valatin, 1958),

$$\begin{aligned} \alpha^\dagger(\nu) &= u(\nu)a^\dagger(\nu) + v(\nu)a(\bar{\nu}) \\ \alpha^\dagger(\bar{\nu}) &= u(\nu)a^\dagger(\bar{\nu}) - v(\nu)a(\nu) \end{aligned} \tag{6-599a}$$

$$\begin{aligned} a^\dagger(\nu) &= u(\nu)\alpha^\dagger(\nu) - v(\nu)\alpha(\bar{\nu}) \\ a^\dagger(\bar{\nu}) &= u(\nu)\alpha^\dagger(\bar{\nu}) + v(\nu)\alpha(\nu) \end{aligned} \tag{6-599b}$$

(For fermion operators, $a^\dagger(\bar{\bar{\nu}}) = -a^\dagger(\nu)$; see Eq. (3B-1).) The canonical character of the transformation (6-599) requires

$$u^2(\nu) + v^2(\nu) = 1 \tag{6-600}$$

and the condition that the transformed Hamiltonian is in diagonal form yields

$$\begin{aligned} u(\nu) &= 2^{-1/2}\left(1 + \frac{\varepsilon(\nu) - \lambda}{E(\nu)}\right)^{1/2} \\ v(\nu) &= 2^{-1/2}\left(1 - \frac{\varepsilon(\nu) - \lambda}{E(\nu)}\right)^{1/2} \end{aligned} \tag{6-601}$$

with

$$E(\nu) = \left((\varepsilon(\nu) - \lambda)^2 + \Delta^2\right)^{1/2} \tag{6-602}$$

With the values (6-601) for the coefficients $u(\nu)$ and $v(\nu)$, the transformed Hamiltonian takes the form

$$H' = \mathcal{E}_0' + \sum_\nu E(\nu)\alpha^\dagger(\nu)\alpha(\nu) \tag{6-603}$$

▲

▼ with

$$\mathscr{E}_0' = \sum_{\nu>0} (2v^2(\nu)(\varepsilon(\nu)-\lambda) - 2u(\nu)v(\nu)\Delta)$$

$$= \sum_{\nu>0} ((\varepsilon(\nu)-\lambda) - E(\nu)) \tag{6-604}$$

The Hamiltonian (6-603) describes a system of independent "quasiparticles" represented by the new fermion operators $\alpha^\dagger, \alpha$. The number of quasiparticles is denoted by v, and the ground state is the quasiparticle vacuum state, $\mathrm{v}=0$, with the property

$$\alpha(\nu)|\mathrm{v}=0\rangle = 0 \tag{6-605}$$

for all ν. When expressed in terms of the particle operators, the quasiparticle vacuum state takes the form (see Eqs. (6-599) and (6-605))

$$|\mathrm{v}=0\rangle = \prod_{\nu>0} (u(\nu) + v(\nu)a^\dagger(\bar{\nu})a^\dagger(\nu))|\mathscr{N}=0\rangle \tag{6-606}$$

where $|\mathscr{N}=0\rangle$ is the state with no particles. Thus, $v(\nu)$ represents the probability amplitude for pairwise occupancy of the orbits $(\nu\bar{\nu})$ in the paired state $\mathrm{v}=0$. For a normal Fermi system, $v(\nu)$ drops abruptly from 1 to 0 at the Fermi level λ, but in the superfluid system, the pair correlations give rise to a diffuseness of the Fermi surface extending over an energy interval of order Δ (see Fig. 6-63a). A wave function with the correlation structure (6-606) was first employed in the theory of superconductivity (Bardeen, Cooper, and Schrieffer, 1957).

The one-quasiparticle states are obtained by operating on the $\mathrm{v}=0$ state with the quasiparticle creation operators,

$$|\mathrm{v}=1,\nu\rangle = \alpha^\dagger(\nu)|\mathrm{v}=0\rangle \tag{6-607}$$

and have the excitation energies $E(\nu)$ given by Eq. (6-602). The quasiparticle energies $E(\nu)$ are compared in Fig. 6-63b with the energies $|\varepsilon(\nu)-\lambda|$ for particles and holes in the uncorrelated system.

In nuclei with even $\mathscr{N}$, the ground state is the $\mathrm{v}=0$ state, and the lowest intrinsic excitations have $\mathrm{v}=2$ and energies $\gtrsim 2\Delta$ (see Eq. (6-602)). Thus, 2Δ is the energy gap associated with the pairing. In odd-$\mathscr{N}$ nuclei, the lowest states have $\mathrm{v}=1$, and Δ is seen to represent the odd-even mass difference.

The modification in the single-particle degrees of freedom implied by the pair correlation can be directly tested in the one-particle transfer reactions, which measure the matrix elements of $a^\dagger(\nu)$ and $a(\nu)$. In the absence of pair correlations, these matrix elements are governed by the selection rules that restrict the states produced by $a^\dagger$ to be particle states and those produced by a to be holes, assuming the target to be an even-even nucleus in its ground ▲ state. In the pair-correlated system, the quasiparticles are hybrids of particles

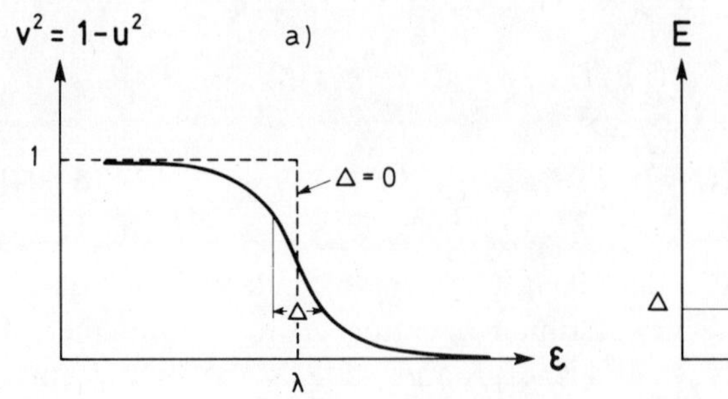

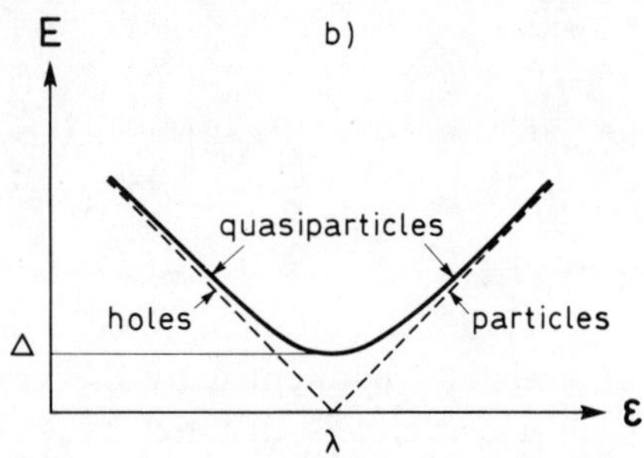

Figure 6-63 Occupation probabilities and quasiparticle excitation energies for the pair-correlated state.

▼ and holes (see Eq. (6-599)); the comparison of the cross sections for one-particle addition and one-particle removal leading to the same quasiparticle state thus provides a direct measure of the coefficients u and v appearing in the quasiparticle transformation. An example of such an analysis involving a sequence of Yb isotopes is given in Fig. 5-11, p. 264.

Quite generally, matrix elements involving quasiparticles can be obtained by a transformation of the operators from the $a^\dagger, a$ variables to the $\alpha^\dagger, \alpha$ variables. Thus, a one-particle operator F can be expressed in the form (see Eq. (2A-24))

$$\begin{aligned} F &= \sum_{\nu_1\nu_2} \langle \nu_2|F|\nu_1\rangle a^\dagger(\nu_2)a(\nu_1) \\ &= \sum_{\nu>0} (1-c)v^2(\nu)\langle \nu|F|\nu\rangle \\ &+ \sum_{\nu_1\nu_2} (u_1u_2+cv_1v_2)\langle \nu_2|F|\nu_1\rangle \alpha^\dagger(\nu_2)\alpha(\nu_1) \\ &+ \sum_{\nu_1<\nu_2} (v_1u_2-cu_1v_2)(\langle \nu_2|F|\bar{\nu}_1\rangle \alpha^\dagger(\nu_2)\alpha^\dagger(\nu_1)+\langle \bar{\nu}_1|F|\nu_2\rangle \alpha(\nu_1)\alpha(\nu_2)) \end{aligned} \tag{6-608}$$

with the notation $u_1 = u(\nu_1)$, etc. The phase factor c characterizes the transformation of F under particle-hole conjugation, which involves a combination of time reversal and Hermitian conjugation (see Eqs. (3-11) to (3-14)),

$$\begin{aligned} \langle \bar{\nu}_1|F|\bar{\nu}_2\rangle &= -c\langle \nu_2|F|\nu_1\rangle \\ \langle \nu_1|F|\bar{\nu}_2\rangle &= c\langle \nu_2|F|\bar{\nu}_1\rangle \end{aligned} \tag{6-609}$$

where the second equation follows from the first, since $|\bar{\bar{\nu}}\rangle = -|\nu\rangle$. Under particle-hole conjugation, electric moments change sign ($c=-1$), while magnetic moments are invariant ($c=+1$). ▲

▼ From Eq. (6-608), one obtains the matrix elements

$$\langle \mathrm{v}=1, \nu_2 | F | \mathrm{v}=1, \nu_1 \rangle = (u_1 u_2 + c v_1 v_2) \langle \nu_2 | F | \nu_1 \rangle \tag{6-610a}$$

$$\langle \mathrm{v}=2, \nu_1 \nu_2 | F | \mathrm{v}=0 \rangle = (v_1 u_2 - c u_1 v_2) \langle \nu_2 | F | \bar{\nu}_1 \rangle \tag{6-610b}$$

These relations are illustrated in Fig. 6-64 and can be seen to reflect the definition of the quasiparticle as a hybrid of a particle with amplitude u, and a hole with amplitude v (see Eq. (6-599a) and the particle-hole transformation (3B-3)). For $\nu_1 = \nu_2$, the diagonal matrix elements (6-610a) receive an additional contribution from the first term in Eq. (6-608), which expresses the expectation value of F in the condensate.

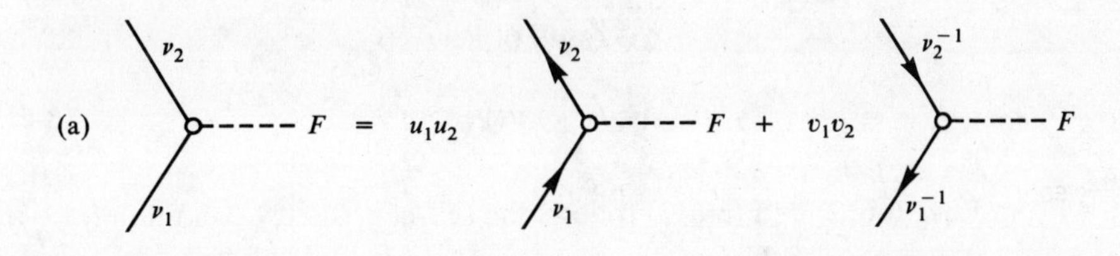

Figure 6-64 Matrix elements of one-particle operator between quasiparticle states. The quasiparticles are hybrids of particles and holes and are therefore drawn without arrows. For the evaluation of the graphs involving hole states, see Fig. 3B-1, Vol. I, p. 371.

▼ In the application of the quasiparticle coupling scheme, one may take the parameter Δ as empirically determined from the spectra or masses (see, for example, the systematics of the odd-even mass differences in Fig. 2-5, Vol. I, p. 170). The value of λ can also be estimated empirically from the two-particle separation energies. However, many effects are rather sensitive to the position of λ relative to the one-particle energies $\varepsilon(\nu)$, and one can ensure consistency in these relative energies by determining λ in such a manner that the average number of particles has the appropriate value. Thus, for the $\mathrm{v}=0$ state, we have

$$\begin{aligned}\langle \mathrm{v}=0 | \mathcal{N} | \mathrm{v}=0 \rangle &= \sum_{\nu>0} 2v^2(\nu) \\ &= \sum_{\nu>0} \left(1 - \frac{\varepsilon(\nu) - \lambda}{E(\nu)}\right)\end{aligned} \tag{6-611}$$

▲ from which λ can be obtained if Δ and $\varepsilon(\nu)$ are known.

▼ One may also attempt to estimate the value of Δ on the basis of the coupling constant G that characterizes the relation between the pair density and potential (see Eq. (6-148)). Such an estimate of Δ can be obtained from a self-consistency condition for the monopole pair moment M_2 given by Eq. (6-145). In the $\mathrm{v}=0$ state, M_2 has the expectation value (see Eqs. (6-599) and (6-601))

$$\begin{aligned}\langle \mathrm{v}=0|M_2|\mathrm{v}=0\rangle &= \sum_{\nu>0} u(\nu)v(\nu) \\ &= \sum_{\nu>0} \frac{\Delta}{2E(\nu)}\end{aligned} \tag{6-612}$$

where the strength Δ of the pair potential (see Eq. (6-598)) is itself proportional to $\langle M_2^\dagger\rangle$ (see Eq. (6-148)),

$$\begin{aligned}\Delta &= G\langle \mathrm{v}=0|M_2|\mathrm{v}=0\rangle \\ &= G\sum_{\nu>0} u(\nu)v(\nu)\end{aligned} \tag{6-613}$$

From Eqs. (6-612) and (6-613) follows the self-consistency condition for $\langle M_2\rangle$

$$G\sum_{\nu>0}\frac{1}{2E(\nu)}=1 \tag{6-614}$$

which determines Δ for specified single-particle spectrum $\varepsilon(\nu)$ and coupling constant G.

In a spherical nucleus, the relation (6-614) between Δ and G depends strongly on the degeneracy of the single-particle levels in the region of the Fermi level. For a deformed nucleus, with a more uniform single-particle spectrum, an approximate relationship between Δ and G may be obtained by considering a spectrum $\varepsilon(\nu)$ of equally spaced levels, each with twofold degeneracy $(\nu\bar{\nu})$. If the spacing d between these levels is small compared with Δ, the sum in Eq. (6-614) can be replaced by an integral

$$G\int_{-S/2}^{+S/2}\frac{d\varepsilon}{2d}(\varepsilon^2+\Delta^2)^{-1/2}=\frac{G}{d}\sinh^{-1}\left(\frac{S}{2\Delta}\right)=1 \tag{6-615}$$

where it has been assumed that the constant pair coupling extends over the single-particle states lying within an energy interval $S/2$ on either side of the Fermi level λ. From Eq. (6-615), we obtain

$$\begin{aligned}\Delta &= \frac{S}{2}\left(\sinh\frac{d}{G}\right)^{-1} \\ &\approx S\exp\left\{-\frac{d}{G}\right\}\end{aligned} \tag{6-616}$$

where the last expression is valid for $G \ll d$, which is equivalent to $\Delta \ll S$. For ▲ example, for $A\approx160$, the neutron correlations are characterized by the

▼ parameters $d \approx 0.4$ MeV (see Fig. 5-3, p. 223) and $\Delta \approx 0.8$ (see Fig. 2-5, Vol. I, p. 170); thus, including single-particle orbits within an interval $S = 2\hbar\omega_0 \approx 15$ MeV, we obtain from Eq. (6-616) the estimate $G \approx 140$ keV. Since the matrix element G represents a volume effect and is therefore expected to be approximately proportional to A^{-1}, the present estimate of G agrees rather well with the value obtained on p. 645 for ^{206}Pb, with the assumption of a similar energy interval S.

The sum occurring in the self-consistency condition (6-614) must be restricted to a finite energy interval, since the contribution of distant levels would diverge logarithmically (see, for example, Eq. (6-615)). This divergence results from the assumption of a sharply local pair field and would be removed by taking into account the finite range of the effective interactions. The details of the contributions from the distant levels are not important for the low-energy quasiparticle spectra, since these contributions can be simulated by a renormalization of the effective coupling constant G (see, for example, the discussion on p. 645 of the effect of the distant orbits on the pair correlations in ^{206}Pb).

The correlation energy in the $\mathrm{v}=0$ state can be obtained from the constant $\mathscr{E}_0'$ in the diagonalized Hamiltonian (6-603),

$$\mathscr{E}_{\mathrm{corr}}(\mathrm{v}=0) = \mathscr{E}_0' + \lambda\langle\mathscr{N}\rangle + \frac{\Delta^2}{G} - \sum_{\substack{\nu>0 \\ \varepsilon(\nu)\leqslant\varepsilon_F}} 2\varepsilon(\nu) \tag{6-617}$$

where the second and third terms reflect the fact that the energy differs from the Hamiltonian H' partly by the term $\lambda\mathscr{N}$ (see Eq. (6-598)) and partly because the pair potential (6-597), when summed over all the particles, counts the two-body interaction energy twice. The last term in Eq. (6-617) is the energy in the uncorrelated system in which the single-particle orbits are filled up to the Fermi level with energy ε_F. In the uniform model considered above, we have $\varepsilon_F = \lambda$, and the expression (6-604) for $\mathscr{E}_0'$ yields

$$\begin{aligned}\mathscr{E}_{\mathrm{corr}}(\mathrm{v}=0) &= 2\int_0^{S/2} \frac{d\varepsilon}{d}\left(\varepsilon - (\varepsilon^2+\Delta^2)^{1/2}\right) + \frac{\Delta^2}{G} \\ &\approx -\frac{\Delta^2}{2d} \qquad (\Delta \ll S)\end{aligned} \tag{6-618}$$

where we have used the relation (6-616) for Δ. In the above example ($A \approx 160$), the neutron pair correlation energy (6-618) is $\mathscr{E}_{\mathrm{corr}} = -0.8$ MeV, which is seen to be only about half the binding energy 2Δ of a single pair. The rather small value of the correlation energy reflects the fact that only a few single-particle levels lie within the interval of strong pair correlations. Thus, the condition for the treatment in terms of a static pair field is only marginally satisfied. ▲

APPENDIX

6A

Liquid-Drop Model of Vibrations and Rotations

In the present appendix, we consider the normal modes of a drop of non-viscous fluid described in terms of macroscopic properties such as surface tension, electrostatic energy, compressibility, and polarizability. The fission degree of freedom (Sec. 6A-2) appears as an extension of the discussion of small-amplitude surface oscillations in Sec. 6A-1. The compression and polarization modes are the topics of Secs. 6A-3 and 6A-4, and the rotational motion that can be built out of the surface oscillations is considered in Sec. 6A-5.[35]

6A-1 Surface Vibrations about Spherically Symmetric Equilibrium

6A-1a Normal coordinates

The surface vibrations of a liquid drop can be described by a set of normal coordinates $\alpha_{\lambda\mu}$ obtained by expanding the surface in spherical harmonics,

$$R(\vartheta,\varphi)=R_0\left(1+\sum_{\lambda\mu}\alpha_{\lambda\mu}Y^*_{\lambda\mu}(\vartheta,\varphi)\right)$$
$$\alpha_{\lambda\mu}=(-1)^{\mu}\alpha^*_{\lambda-\mu}=R_0^{-1}\int d\Omega R(\vartheta,\varphi)Y_{\lambda\mu}(\vartheta,\varphi) \tag{6A-1}$$

where R_0 is the equilibrium radius and $R(\vartheta,\varphi)$ the distance of the surface from the

[35]The liquid-drop model of the nuclear dynamics was introduced (Bohr and Kalckar, 1937) in order to describe the consequences of the strong coupling in the motion of the individual nucleons, as revealed by the slow-neutron resonances. This picture of the nucleus provided the framework for interpreting basic features of the fission process (Meitner and Frisch, 1939; Bohr and Wheeler, 1939). The extension of the model to include the degrees of freedom associated with the motion of neutrons with respect to protons, as observed in the nuclear photoeffect, was developed by Steinwedel and Jensen (1950). Many features of the analysis of normal modes of a liquid drop go back to the classical work of Rayleigh (1877).

origin. The coordinates $\alpha_{\lambda\mu}$ are spherical tensor components (transform as spherical harmonics under rotation of the coordinate system; see Vol. I, p. 81). The term in Eq. (6A-1) with $\lambda=0$ represents a compression (or dilatation) without change of shape, and the terms with $\lambda=1$ are associated with a displacement of the drop as a whole. The surface oscillations of lowest order are therefore the quadrupole modes, with $\lambda=2$.

A deformation of order $\lambda\mu$ gives rise to a multipole moment of the same symmetry

$$\mathscr{M}(\lambda\mu)=\int\rho(\mathbf{r})r^{\lambda}Y_{\lambda\mu}\,d\tau\approx\rho_0 R_0^{\lambda+3}\alpha_{\lambda\mu} \tag{6A-2}$$

where $\rho(\mathbf{r})$ is the particle density and ρ_0 its value at equilibrium. In Eq. (6A-2), we have neglected higher-order terms in $\delta R=R-R_0$.

6A-1b Hamiltonian

For small amplitudes of oscillation, the energy to leading order is a quadratic expression in the coordinates $\alpha_{\lambda\mu}$ and their time derivatives $\dot{\alpha}_{\lambda\mu}$,

$$E(\alpha_{\lambda\mu},\dot{\alpha}_{\lambda\mu})=T+V$$

$$V=\tfrac{1}{2}\sum_{\lambda\mu}C_\lambda|\alpha_{\lambda\mu}|^2=\tfrac{1}{2}\sum_{\lambda}C_\lambda(-1)^{\lambda}(2\lambda+1)^{1/2}(\alpha_\lambda\alpha_\lambda)_0 \tag{6A-3}$$

$$T=\tfrac{1}{2}\sum_{\lambda\mu}D_\lambda|\dot{\alpha}_{\lambda\mu}|^2=\tfrac{1}{2}\sum_{\lambda}D_\lambda(-1)^{\lambda}(2\lambda+1)^{1/2}(\dot{\alpha}_\lambda\dot{\alpha}_\lambda)_0$$

where the restoring force parameters C_λ and the mass parameters D_λ depend on the properties of the liquid (see Secs. 6A-1e to 6A-1g). The quantities $(\alpha_\lambda\alpha_\lambda)_0$ and $(\dot{\alpha}_\lambda\dot{\alpha}_\lambda)_0$ are scalar products of the tensors α_λ and $\dot{\alpha}_\lambda$, respectively (see Eq. (1A-71)), and the form (6A-3) follows from rotational invariance. (Time-reversal invariance may be invoked to eliminate mixed terms of the type $(\alpha_\lambda\dot{\alpha}_\lambda)_0$.)

The momenta $\pi_{\lambda\mu}$ conjugate to the coordinates $\alpha_{\lambda\mu}$ are given by the canonical relation

$$\pi_{\lambda\mu}=\frac{\partial}{\partial\dot{\alpha}_{\lambda\mu}}(T-V)$$

$$=D_\lambda\dot{\alpha}^*_{\lambda\mu}=(-1)^{\mu}D_\lambda\dot{\alpha}_{\lambda-\mu} \tag{6A-4}$$

$$\pi^*_{\lambda\mu}=(-1)^{\mu}\pi_{\lambda-\mu}$$

The quantities $\pi^*_{\lambda\mu}$ (but not the momenta themselves) form spherical tensors. The Hamiltonian obtained from Eqs. (6A-3) and (6A-4) takes the form

$$H=\sum_{\lambda\mu}\left(\tfrac{1}{2}D_\lambda^{-1}|\pi_{\lambda\mu}|^2+\tfrac{1}{2}C_\lambda|\alpha_{\lambda\mu}|^2\right) \tag{6A-5}$$

and yields the equations of motion

$$\ddot{\alpha}_{\lambda\mu} + \omega_\lambda^2 \alpha_{\lambda\mu} = 0 \qquad \omega_\lambda = \left(\frac{C_\lambda}{D_\lambda}\right)^{1/2} \tag{6A-6}$$

To the approximation considered, the amplitudes $\alpha_{\lambda\mu}$ perform harmonic oscillations with the frequency ω_λ.

6A-1c Traveling and standing waves

The general solution to the equations of motion (6A-6) has the form

$$\begin{aligned} \alpha_{\lambda\mu}(t) &= a_{\lambda\mu+} \exp\{i\omega_\lambda t + i\delta_{\lambda\mu+}\} + a_{\lambda\mu-} \exp\{-i\omega_\lambda t + i\delta_{\lambda\mu-}\} \qquad \mu > 0 \\ \alpha_{\lambda-\mu}(t) &= (-1)^\mu \alpha_{\lambda\mu}^* \\ \alpha_{\lambda 0}(t) &= a_{\lambda 0} \cos(\omega_\lambda t - \delta_{\lambda 0}) \end{aligned} \tag{6A-7}$$

where the amplitudes $a_{\lambda\mu\pm}$ and phases $\delta_{\lambda\mu\pm}$ are real parameters. For the motion of the surface, we obtain, from Eq. (6A-1),

$$\begin{aligned} R(t) - R_0 = R_0 \sum_\lambda \Bigg(& a_{\lambda 0} \cos(\omega_\lambda t - \delta_{\lambda 0}) Y_{\lambda 0}(\vartheta) \\ & + 2 \sum_{\mu=1}^{\lambda} \big(a_{\lambda\mu+} \cos(\mu\varphi - \omega_\lambda t - \delta_{\lambda\mu+}) + a_{\lambda\mu-} \cos(\mu\varphi + \omega_\lambda t - \delta_{\lambda\mu-})\big) Y_{\lambda\mu}(\vartheta, \varphi = 0) \Bigg) \end{aligned} \tag{6A-8}$$

It is seen that the amplitudes $a_{\lambda\mu\pm}$ (with $\mu > 0$) represent traveling waves that rotate around the polar axis.

One can also describe the oscillations of the surface in terms of standing waves, the amplitudes of which are obtained by decomposing $\alpha_{\lambda\mu}$ for $\mu \neq 0$ into real and imaginary parts,

$$\begin{aligned} \alpha_{\lambda\mu} &= 2^{-1/2}(\alpha_{\lambda\mu,1} + i\alpha_{\lambda\mu,2}) \\ \alpha_{\lambda-\mu} &= 2^{-1/2}(\alpha_{\lambda\mu,1} - i\alpha_{\lambda\mu,2})(-1)^\mu \end{aligned} \tag{6A-9}$$

where μ is taken to be positive. The real coordinates $\alpha_{\lambda\mu,1}$ and $\alpha_{\lambda\mu,2}$ correspond to surface deformations that are proportional to $\cos\mu\varphi \cos(\omega_\lambda t + \delta_{\lambda\mu,1})$ and $\sin\mu\varphi \cos(\omega_\lambda t + \delta_{\lambda\mu,2})$, respectively (as follows from Eq. (6A-8)).

The possibility of combining the amplitudes $\alpha_{\lambda\mu}$ and $\alpha_{\lambda-\mu}$ into standing or traveling waves is a consequence of the degeneracy of the two conjugate modes, which in turn is a consequence of the spherical symmetry of the equilibrium shape. (Axial symmetry combined with invariance with respect to a rotation of 180° about an axis perpendicular to the symmetry axis ($\mathscr{R}$ symmetry), or combined with time reversal invariance, would also give the $\pm\mu$ degeneracy.) Symmetry-violating perturbations lead to eigenmodes that are definite combinations of the conjugate components. For example, the eigenmodes of oscillations of the rotating earth are the traveling waves with definite μ (since axial symmetry is not violated by the rotation), and the modes with opposite μ have slightly different frequencies. The fundamental quadrupole oscillation of the earth, with a period of about one hour, is observed to be split into five components with frequency shifts varying approximately linearly in μ; the separations amount to a few percent, reflecting the ratio of rotational to vibrational frequency (Backus and Gilbert, 1961; see also Slichter, 1967.)

6A-1d Angular momentum

The traveling waves carry angular momentum. To leading order in the amplitudes, the angular momentum vector $\mathbf{M}$ $(=\hbar\mathbf{I})$ is a bilinear expression in α and π, the form of which can be obtained from arguments of rotational invariance. Thus, the Poisson bracket of the spherical component M_μ and the amplitude $\alpha_{\lambda\mu'}$ is the classical analog of the commutator relation between the angular momentum and a spherical tensor (see Eq. (1A-64)),

$$\{M_\mu, \alpha_{\lambda\mu'}\} = -\frac{\partial M_\mu}{\partial \pi_{\lambda\mu'}} = -i(\lambda(\lambda+1))^{1/2}\langle\lambda\mu' 1\mu|\lambda\mu+\mu'\rangle\alpha_{\lambda,\mu+\mu'}$$

$$\{A,B\} \equiv \sum_{\lambda\mu}\left(\frac{\partial A}{\partial \alpha_{\lambda\mu}}\frac{\partial B}{\partial \pi_{\lambda\mu}} - \frac{\partial A}{\partial \pi_{\lambda\mu}}\frac{\partial B}{\partial \alpha_{\lambda\mu}}\right) \tag{6A-10}$$

By integration, we obtain (see the symmetry relations (1A-10) and (1A-11) for the vector addition coefficients)

$$\begin{aligned} M_\mu &= i\sum_{\lambda\mu'\mu''}(\lambda(\lambda+1))^{1/2}\langle\lambda\mu' 1\mu|\lambda\mu''\rangle\pi_{\lambda\mu'}\alpha_{\lambda\mu''} \\ &= -i\sum_\lambda(-1)^\lambda\left(\tfrac{1}{3}\lambda(\lambda+1)(2\lambda+1)\right)^{1/2}(\pi_\lambda^*\alpha_\lambda)_{1\mu} \end{aligned} \tag{6A-11}$$

where the last factor represents the coupling of the tensors π_λ^* and α_λ to form a vector.

The component of $\mathbf{M}$ along the polar axis can be written in the forms

$$\begin{aligned} M_z = M_{\mu=0} &= i\sum_{\lambda\mu}\mu\pi_{\lambda\mu}\alpha_{\lambda\mu} \\ &= \sum_{\lambda,\mu>0}2\mu D_\lambda\omega_\lambda\left(a_{\lambda\mu+}^2 - a_{\lambda\mu-}^2\right) \\ &= \sum_{\lambda,\mu>0}\mu D_\lambda\left(\dot{\alpha}_{\lambda\mu,2}\alpha_{\lambda\mu,1} - \dot{\alpha}_{\lambda\mu,1}\alpha_{\lambda\mu,2}\right) \end{aligned} \tag{6A-12}$$

While the traveling waves carry angular momentum, the eigenmodes corresponding to standing waves have vanishing expectation value for the angular momentum.

6A-1e Surface energy for incompressible liquid

The increase in surface energy associated with the deformation is given by the increment in the surface area multiplied with the surface tension parameter $\mathscr{S}$. In the direction (ϑ,φ), the plane tangent to the surface $R(\vartheta,\varphi)$ forms an angle $|\nabla R|$ with the plane tangent to the spherical surface, and the increase in surface energy is therefore, to second order in the deformation parameters,

$$\begin{aligned} V_{\text{surf}} &= \mathscr{S}\int d\Omega\left(R^2 - R_0^2 + \tfrac{1}{2}R_0^2(\nabla R)^2\right) \\ &= \mathscr{S}\int d\Omega\left(2R_0(R-R_0) + (R-R_0)^2 - \tfrac{1}{2}R_0^2 R\nabla^2 R\right) \\ &= \mathscr{S}R_0^2\left(2(4\pi)^{1/2}\alpha_0 + \sum_{\lambda\mu}|\alpha_{\lambda\mu}|^2 + \tfrac{1}{2}\sum_{\lambda\mu}\lambda(\lambda+1)|\alpha_{\lambda\mu}|^2\right) \end{aligned} \tag{6A-13}$$

For an incompressible fluid, the conservation of volume implies a relation between α_0 and the coefficients $\alpha_{\lambda\mu}$ with $\lambda \neq 0$, which may be obtained by writing the density in the form

$$\rho(r,\vartheta,\varphi) = \rho_0 S(r - R(\vartheta,\varphi)) \tag{6A-14a}$$

$$S(x) \equiv \begin{cases} 1 & x<0 \\ 0 & x>0 \end{cases} \tag{6A-14b}$$

where S is the step function. For small deformations, $|R-R_0| \ll R_0$, we can employ the expansion

$$\rho(r,\vartheta,\varphi) = \rho_0\left(S(r-R_0) + (R-R_0)\delta(r-R_0) - \tfrac{1}{2}(R-R_0)^2\delta'(r-R_0) + \cdots\right) \tag{6A-15}$$

The volume of the fluid is given by the integral of the density (6A-15), from which we obtain the relation

$$\alpha_0 = -(4\pi)^{-1/2}\sum_{\lambda\mu}|\alpha_{\lambda\mu}|^2 \tag{6A-16}$$

valid to leading order. In a similar manner, the condition that the center of mass remain at rest leads to the constraint

$$\alpha_{1\mu} = \sum_{\lambda>0}(-1)^{\lambda+1}3\left(\frac{\lambda+1}{4\pi}\right)^{1/2}(\alpha_\lambda\alpha_{\lambda+1})_{1\mu} \tag{6A-17}$$

The surface energy (6A-13), together with the relation (6A-16) for α_0, gives the contribution

$$(C_\lambda)_{\text{surf}} = (\lambda-1)(\lambda+2)R_0^2\mathscr{S} \tag{6A-18}$$

to the restoring force parameter. In the comparison of the nucleus with a liquid drop, the surface tension is related to the parameter b_{surf} in the mass formula (2-12)

$$4\pi r_0^2 \mathscr{S} = b_{\text{surf}} \approx 17 \quad \text{MeV} \tag{6A-19}$$

where r_0 is the radius parameter ($R_0 = r_0 A^{1/3}$).

6A-1f Coulomb energy

For a charged fluid, the change in the electrostatic energy associated with the deformation is given by

$$\begin{aligned} V_{\text{Coul}} &= \tfrac{1}{2}\int \rho^{\text{el}}(\mathbf{r})\varphi^{\text{el}}(\mathbf{r})\,d\tau - \tfrac{1}{2}\int \rho_0^{\text{el}}(\mathbf{r})\varphi_0^{\text{el}}(\mathbf{r})\,d\tau \\ &= \int \delta\rho^{\text{el}}(\mathbf{r})\varphi_0^{\text{el}}(\mathbf{r})\,d\tau + \tfrac{1}{2}\int \delta\rho^{\text{el}}(\mathbf{r})\delta\varphi^{\text{el}}(\mathbf{r})\,d\tau \end{aligned} \tag{6A-20}$$

where $\rho_0^{\text{el}}(\mathbf{r})$ and $\varphi_0^{\text{el}}(\mathbf{r})$ are the charge density and Coulomb potential for the spherical equilibrium shape, while $\rho^{\text{el}} = \rho_0^{\text{el}} + \delta\rho^{\text{el}}$ and $\varphi^{\text{el}} = \varphi_0^{\text{el}} + \delta\varphi^{\text{el}}$ refer to the deformed shape. The evaluation of the energy (6A-20) to second order in $\alpha_{\lambda\mu}$ involves $\delta\rho$ to second order, but $\delta\varphi$ only to first order. For a total charge Ze uniformly distributed over the volume, we have (see Eq. (6A-15))

$$\delta\rho^{\text{el}} = \frac{3}{4\pi}\frac{Ze}{R_0^3}\left((R-R_0)\delta(r-R_0) - \tfrac{1}{2}(R-R_0)^2\delta'(r-R_0)\right) \tag{6A-21a}$$

$$\delta\varphi^{\text{el}} = \frac{Ze}{R_0}\sum_{\lambda\mu}\frac{3}{2\lambda+1}\alpha_{\lambda\mu}Y_{\lambda\mu}^*(\vartheta,\varphi)\begin{cases}(r/R_0)^{\lambda} & r<R_0\\ (R_0/r)^{\lambda+1} & r>R_0\end{cases} \tag{6A-21b}$$

and hence (see Eq. (6A-16))

$$\begin{aligned} \int \delta\rho^{\text{el}}(\mathbf{r})\varphi_0^{\text{el}}(\mathbf{r})\,d\tau &= \frac{3}{4\pi}\frac{Ze}{R_0^2}(4\pi)^{1/2}\alpha_0\int\left(\delta(r-R_0) + \tfrac{1}{2}R_0\delta'(r-R_0)\right)\varphi_0^{\text{el}}(r)r^2\,dr \\ &= -\frac{3}{8\pi}ZeR_0(4\pi)^{1/2}\alpha_0\left(\frac{\partial\varphi_0^{\text{el}}}{\partial r}\right)_{r=R_0} \\ &= \frac{3}{8\pi}\frac{Z^2e^2}{R_0}(4\pi)^{1/2}\alpha_0 \end{aligned} \tag{6A-22a}$$

$$\tfrac{1}{2}\int \delta\rho^{\text{el}}(\mathbf{r})\delta\varphi^{\text{el}}(\mathbf{r})\,d\tau = \frac{9}{8\pi}\frac{Z^2e^2}{R_0}\sum_{\lambda\mu}(2\lambda+1)^{-1}|\alpha_{\lambda\mu}|^2 \tag{6A-22b}$$

The contribution of the Coulomb energy to the restoring force parameter is therefore

$$(C_\lambda)_{\text{Coul}} = -\frac{3}{2\pi}\frac{\lambda-1}{2\lambda+1}\frac{Z^2e^2}{R_0} \tag{6A-23}$$

(A diffuseness a of the surface implies a decrease of the total Coulomb energy of order $(a/R_0)^2$ (see Eq. (2-66)), but to this order, the correction is independent of the shape and therefore does not affect the parameter C_λ; Myers and Swiatecki, 1966.)

From Eqs. (6A-18) and (6A-23), we obtain the total restoring force parameter

$$\begin{aligned} C_\lambda &= (C_\lambda)_{\text{surf}} + (C_\lambda)_{\text{Coul}} \\ &= \frac{1}{4\pi}(\lambda-1)(\lambda+2)b_{\text{surf}}A^{2/3} - \frac{3}{2\pi}\frac{\lambda-1}{2\lambda+1}\frac{e^2}{r_0}Z^2A^{-1/3} \end{aligned} \tag{6A-24}$$

The Coulomb repulsion counteracts the effect of the surface tension and leads to negative values of C_λ for sufficiently large values of Z^2/A. Instability with respect to the lowest mode ($\lambda=2$) occurs for a critical value of Z^2/A given by (Bohr and Wheeler, 1939)

$$\begin{aligned} \left(\frac{Z^2}{A}\right)_{\text{crit}} &= \frac{10}{3}\frac{b_{\text{surf}}r_0}{e^2} \\ &\approx 49 \end{aligned} \tag{6A-25}$$

where the numerical estimate is based on the value (6A-19) for the surface tension and the radius parameter $r_0 = 1.25$ fm for the Coulomb energy (see Vol. I, p. 145).

6A-1g Mass parameter for irrotational flow

The mass parameters D_λ in Eq. (6A-3) depend on the flow associated with the surface oscillations. For irrotational flow, the velocity field $\mathbf{v}(\mathbf{r})$ can be derived from a potential $\chi(\mathbf{r})$,

$$\mathbf{v}(\mathbf{r}) = -\nabla\chi(\mathbf{r}) \tag{6A-26}$$

and the continuity equation yields, for an incompressible liquid,

$$\nabla\cdot\mathbf{v} = 0 \tag{6A-27a}$$

$$\nabla^2\chi = 0 \tag{6A-27b}$$

The general solution to the Laplace equation (6A-27b) is a linear combination of multipole potentials

$$\chi(r,\vartheta,\varphi) = \sum c_{\lambda\mu}r^\lambda Y^*_{\lambda\mu}(\vartheta,\varphi) \tag{6A-28}$$

and the coefficients $c_{\lambda\mu}$ are related to the surface coordinates $\alpha_{\lambda\mu}$ by the boundary condition at the surface, which equates the components of $\mathbf{v}$ and $\dot{\mathbf{R}}$ in the direction of the normal. For small values of α, the boundary condition implies $v_r = \dot{R}$, for $r = R_0$, and gives

$$c_{\lambda\mu} = -\lambda^{-1}R_0^{2-\lambda}\dot{\alpha}_{\lambda\mu} \tag{6A-29}$$

The kinetic energy of the flow, for small amplitude oscillations, is

$$T=\tfrac{1}{2}M\rho_0\int \mathrm{v}^2 d\tau=\tfrac{1}{2}M\rho_0 R_0^2\int\left(\chi\frac{\partial\chi}{\partial r}\right)_{r=R_0}d\Omega \tag{6A-30}$$

where ρ_0 is the equilibrium density of particles, each having a mass M. The expressions (6A-28) and (6A-29) for χ lead to a kinetic energy of the form (6A-3) with the mass parameters

$$D_\lambda=\lambda^{-1}M\rho_0 R_0^5=\frac{3}{4\pi}\frac{1}{\lambda}AMR_0^2 \tag{6A-31}$$

with A representing the total number of particles in the drop. (A similar derivation of the angular momentum of the flow leads to the expression (6A-11), with the mass parameter (6A-31) for the ratio between $\pi^*_{\lambda\mu}$ and $\dot{\alpha}_{\lambda\mu}$.)

6A-2 Large-Amplitude Deformations. Fission Mode

The treatment of surface oscillations in Sec. 6A-1 applies to deformations of small amplitude ($|\alpha_{\lambda\mu}|\ll 1$). For larger deformations of the surface, it is necessary to include terms in the Hamiltonian involving higher powers of $\alpha_{\lambda\mu}$; these give rise to anharmonicity in the vibrations and to couplings between modes of different multipole order.

6A-2a Potential energy for nonrotating drop

The potential energy for large deformations has been studied in connection with the analysis of the fission process. An electrically charged drop, even if stable against small-amplitude oscillations, becomes unstable if sufficiently strongly deformed. In fact, on account of the long-range character of the Coulomb repulsion, the energy as a function of deformation must eventually decrease as one approaches the point of division into two drops.

The potential energy function $V(\alpha_{\lambda\mu})$ can be expressed in terms of a dimensionless function depending on the ratio between Coulomb and surface energies. This ratio is proportional to the quantity Z^2/A and is usually characterized by the fissility parameter

$$\begin{aligned} x &= \frac{Z^2}{A}\left(\frac{Z^2}{A}\right)^{-1}_{\text{crit}} \\ &\approx 0.0205\frac{Z^2}{A} \end{aligned} \tag{6A-32}$$

where $(Z^2/A)_{\text{crit}}$ is the value of Z^2/A for which the spherical shape becomes instable with respect to quadrupole deformations (see Eq. (6A-25)).

For values of x close to unity, the saddle-point shape and the fission barrier can be obtained by carrying the expansion of the potential energy to third order in the deformation variables. To leading order in $(1-x)$, the saddle-point shape has quadrupole symmetry, and the deformation can be expressed in terms of the shape variables β and γ, defined by Eqs. (6B-1) and (6B-2). The second-order term in the potential energy is proportional to β^2 (see Eq. (6B-3a)), and the third-order term involves the unique rotational invariant $\beta^3 \cos 3\gamma$ that can be formed from a cubic polynomial in $\alpha_{2\mu}$ (see Eq. (6B-3b)). The evaluation of surface and Coulomb energies leads to the potential energy function (Bohr and Wheeler, 1939; see also Swiatecki, 1956)

$$V_{\text{surf}} + V_{\text{Coul}} = \frac{1}{2\pi} b_{\text{surf}} A^{2/3} \left((1-x)\beta^2 - \frac{2}{21}\left(\frac{5}{4\pi}\right)^{1/2} (1+2x)\beta^3 \cos 3\gamma \right) \quad \text{(6A-33)}$$

The saddle-point shape is seen to be prolate ($\gamma=0$); the deformation β_{f} of the saddle point and the fission barrier $E_{\text{f}} = E(\beta_{\text{f}})$ are, to leading order in $(1-x)$,

$$\begin{aligned} \beta_{\text{f}} &= \frac{7}{3}\left(\frac{4\pi}{5}\right)^{1/2}(1-x) \\ E_{\text{f}} &= \frac{98}{135}(1-x)^3 b_{\text{surf}} A^{2/3} \end{aligned} \quad \text{(6A-34)}$$

In the vicinity of the saddle point, the potential energy (6A-33) has a parabolic shape; in the β direction, the coefficient of $(\beta - \beta_{\text{f}})^2$ has the same magnitude (but opposite sign) as for small quadrupole deformations with respect to the spherical shape.

While the saddle-point shape for $(1-x) \ll 1$ is predominantly of quadrupole symmetry, a small axially symmetric hexadecapole deformation, with α_{40} of the order of β^2, arises from the cubic term $((\alpha_2\alpha_2)_4\alpha_4)_0$ in the potential energy. However, the saddle-point shape is stable with respect to odd-parity deformations, such as α_{30}, since the reflection symmetry of the potential energy prevents the occurrence of terms that are linear in odd-parity deformation amplitudes. (For $x \to 1$, the third-order term $((\alpha_3\alpha_3)_2\alpha_2)_0$ becomes vanishingly small compared with the $(\alpha_3\alpha_3)_0$ term, the coefficient of which remains finite and positive in this limit.)

For arbitrary values of x, the potential energy functions have been explored in considerable detail by numerical computation.[36] The fission barriers and saddle-point shapes obtained from such an analysis are illustrated in Fig. 6A-1a. (The quantity $\xi(x)$ is the fission barrier in units of the total surface energy for the spherical shape.) In the range $0.75 < x < 1$, which is especially relevant to the fission of heavy nuclei, the simple approximation (6A-34) is found to reproduce the more exact result, to an accuracy of a few percent. The normal modes for small

---

[36] As soon as high-speed computers became available, an analysis of this type was carried out by Frankel and Metropolis (1947). For more detailed later results, see especially the series of papers by Swiatecki and co-workers (Cohen and Swiatecki, 1963, and references contained therein) and by Strutinsky and co-workers (Strutinsky *et al.*, 1962, and references contained therein).

oscillations with respect to the fission saddle point have been considered by Wheeler, 1963, and Nix, 1967.

The saddle-point shape is found to remain stable with respect to odd-parity deformations for $x>0.4$. For $x\approx 0.4$, instability develops, and the family of paths leading toward fission into comparable fragments merges with those associated with spallation processes; for $x<0.4$, there is no longer a well-defined fission barrier (Cohen and Swiatecki, 1963); the effects of angular momentum tend to reestablish the stability with respect to odd-parity deformations (see Fig. 6A-1b).

6A-2b Influence of angular momentum

For large values of the angular momentum, centrifugal forces may strongly modify the dependence of the energy on the shape and may thus affect the fission barrier.[37] The centrifugal effects depend on the collective rotational flow, which determines the moments of inertia. In the present discussion, we shall assume that the moments of inertia have the values corresponding to rigid rotation, as is expected for a system of fermions with independent-particle motion (see the discussion on p. 78). In the superfluid phase, the moments of inertia are smaller than the rigid-body values and depend more sensitively on the deformation parameters (see p. 82); however, the pair correlations are expected to disappear for sufficiently high excitation energy (see p. 37) or angular momentum (see p. 72).

The effect of the rotational angular momentum $\hbar I$ is conveniently expressed in terms of a parameter y, which measures the ratio of rotational and surface energy for the spherical shape,

$$\begin{aligned} y &= \frac{\hbar^2 I^2}{2\mathscr{I}(\alpha=0)}\left(b_{\text{surf}}A^{2/3}\right)^{-1} \\ &= \frac{5}{4}\frac{\hbar^2}{Mr_0^2}b_{\text{surf}}^{-1}A^{-7/3}I^2 \\ &\approx 2.1A^{-7/3}I^2 \qquad \text{(6A-35)} \\ & (r_0 = 1.2 \text{ fm} \quad b_{\text{surf}} = 17 \text{ MeV}) \end{aligned}$$

With the two parameters x and y, a variety of different structural features and instability phenomena are encountered in the characterization of the energy function for the liquid-drop model. For systems with a large fissility parameter ($x\approx 1$), the equilibrium shapes (both saddle point and ground state) have small deformations of approximate quadrupole symmetry, and the potential energy of deforma-

---

[37]Early studies of the effect of the rotational motion on the fission barrier were made by Pik-Pichak (1958). A number of authors have contributed to the subject; see especially Cohen, Plasil, and Swiatecki (1974), which also contains a survey of the development. The phenomena are related to the classical problem of the shapes of rotating bodies in gravitational equilibrium (see Chandrasekhar, 1969).

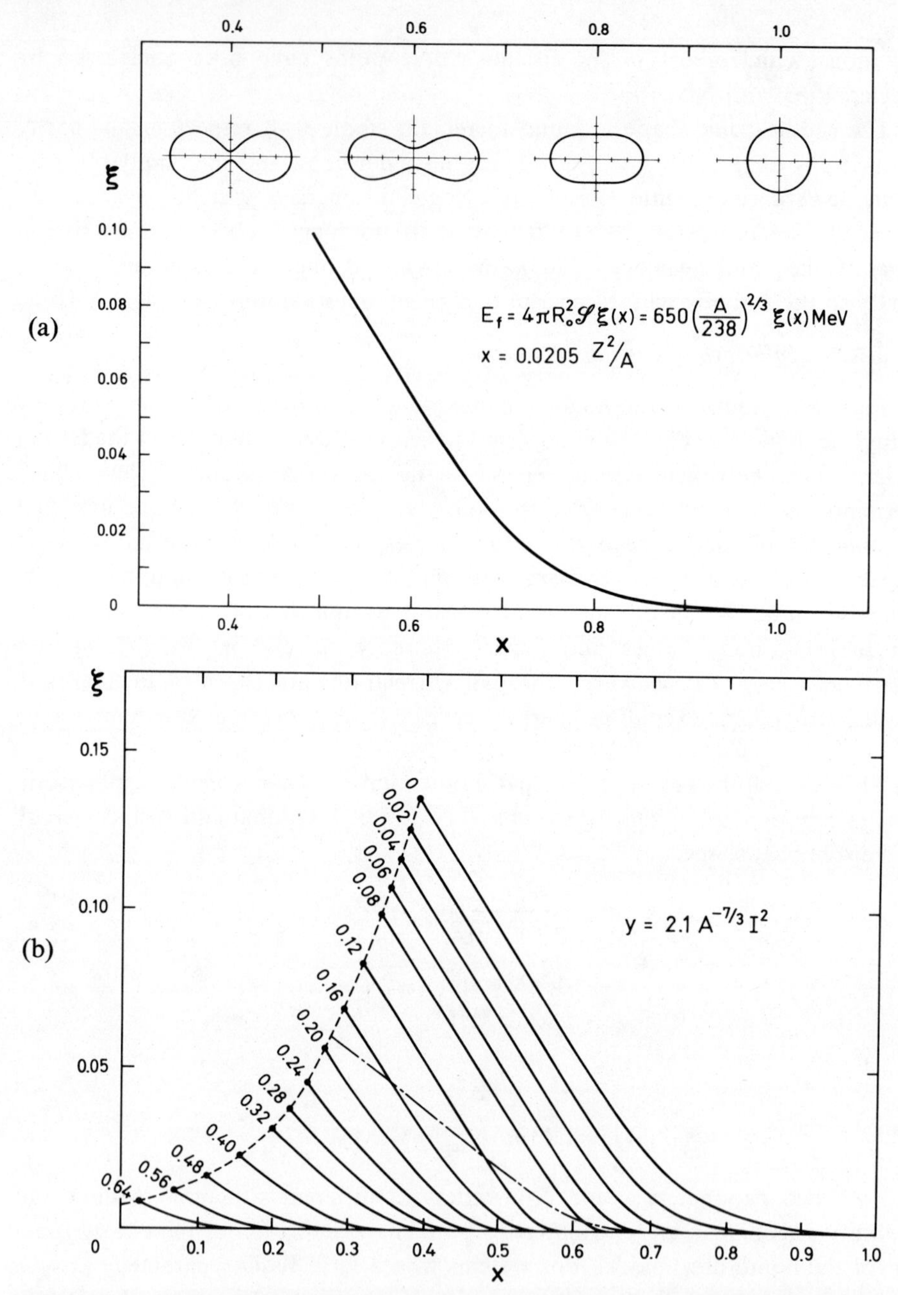

Figure 6A-1 Saddle-point shapes and energies for fission of a charged liquid drop. Fig. 6A-1a refers to a system without angular momentum and shows the shape and the deformation energy E_f at the saddle point for fission for various values of the fissility parameter x. The results are taken from S. Cohen and W. J. Swiatecki, *Ann. Phys.* **22**, 406, (1963). Fig. 6A-1b gives the fission barrier for rotating liquid drops; the numbers labeling the different curves give the value of the parameter y, which is a measure of the angular momentum. The curves terminate along the locus of instability of the saddle-point shape with respect to deformations of odd parity (mass asymmetry mode). The dot-dashed curve represents the limit of instability of the axially symmetric oblate equilibrium shape; below this curve, the equilibrium has triaxial shape. The results are from S. Cohen, F. Plasil, and W. J. Swiatecki, *Ann. Phys.* **82**, 557 (1974).

tion can be represented by the expression (6A-33). The moments of inertia can be obtained from the values of the semi-axes given by Eq. (6B-4); in the segment $0 \leqslant \gamma \leqslant 60°$ (see Fig. 6B-1), the labeling of the axes is such that $\delta R_2 \leqslant \delta R_1 \leqslant \delta R_3$ and, hence, the moment of inertia is largest for rotation about the 2 axis, for which, to leading order in β,

$$\frac{\mathscr{I}_2(\beta,\gamma)}{\mathscr{I}(\beta=0)} - 1 = \frac{1}{R_0}(\delta R_1 + \delta R_3) = -\frac{1}{R_0}\delta R_2$$
$$= -\left(\frac{5}{4\pi}\right)^{1/2} \beta \cos\left(\gamma + \frac{2\pi}{3}\right) \tag{6A-36}$$

The total energy for fixed values of the shape and angular momentum is obtained by adding the surface and Coulomb energies (6A-33) to that of the rotational motion assumed to take place about the axis with the largest moment of inertia,

$$\frac{E(\beta,\gamma) - E(\beta=0)}{b_{\text{surf}} A^{2/3}}$$
$$= \frac{1}{2\pi}\left((1-x)\beta^2 - \frac{2}{7}\left(\frac{5}{4\pi}\right)^{1/2}\beta^3 \cos 3\gamma\right) + y\left(\frac{5}{4\pi}\right)^{1/2} \beta \cos\left(\gamma + \frac{2\pi}{3}\right) \tag{6A-37}$$

for $1 - x \ll 1$. The equilibrium shapes are determined by the derivatives with respect to β and γ

$$(1-x)\beta - \frac{3}{7}\left(\frac{5}{4\pi}\right)^{1/2}\beta^2 \cos 3\gamma + \pi\left(\frac{5}{4\pi}\right)^{1/2} y \cos\left(\gamma + \frac{2\pi}{3}\right) = 0 \tag{6A-38a}$$

$$\sin\left(\gamma + \frac{2\pi}{3}\right)\left(\beta^2\left(\cos^2\left(\gamma + \frac{2\pi}{3}\right) - \frac{1}{4}\right) - \frac{7\pi}{12} y\right) = 0 \tag{6A-38b}$$

The equations (6A-38) have the two solutions

(A)

$$\gamma = \frac{\pi}{3}$$
$$\beta = \frac{7}{6}\left(\frac{4\pi}{5}\right)^{1/2}(1-x)\left(-1 + \left(1 + \frac{15}{7}\frac{y}{(1-x)^2}\right)^{1/2}\right) \approx \left(\frac{5\pi}{4}\right)^{1/2} \frac{y}{1-x}$$

$$\tag{6A-39}$$

(B)

$$\gamma = \frac{\pi}{3} - \arccos\left(4 - \frac{15}{7}\frac{y}{(1-x)^2}\right)^{-1/2} \approx \frac{5\sqrt{3}}{56}\frac{y}{(1-x)^2}$$
$$\beta = \frac{7}{6}\left(\frac{4\pi}{5}\right)^{1/2}(1-x)\left(4 - \frac{15}{7}\frac{y}{(1-x)^2}\right)^{1/2} \approx \frac{7}{3}\left(\frac{4\pi}{5}\right)^{1/2}(1-x)\left(1 - \frac{15}{56}\frac{y}{(1-x)^2}\right)$$

where the approximate expressions are appropriate in the limit of small y. The solution (A) is the equilibrium shape, which is oblate and approaches the spherical shape as $y \to 0$. The solution (B) is the saddle point, which, for $y \to 0$, approaches the prolate shape given by Eq. (6A-34). For both the stationary points, the rotational energy leads to a lowering of the symmetry, since the moment of inertia has a linear variation with respect to the deviations from symmetry.

The inclusion of the rotational energy implies a lowering of the fission barrier (6A-34), which, to leading order in y, can be obtained by evaluating the rotational energy at the two equilibrium shapes for $y=0$,

$$E_{\mathrm{f}}=\frac{98}{135}(1-x)^3 b_{\mathrm{surf}} A^{2/3}\left(1-\frac{45}{28}\frac{y}{(1-x)^2}\right) \tag{6A-40}$$

With increasing y, solution (B) moves toward solution (A), and the two coalesce for a critical value of y, given by

$$y_{\mathrm{crit}}=\frac{7}{5}(1-x)^2 \tag{6A-41}$$

for which the shape is

$$\gamma=\frac{\pi}{3} \qquad \beta=\frac{7}{6}\left(\frac{4\pi}{5}\right)^{1/2}(1-x) \tag{6A-42}$$

For values of $y \geqslant y_{\mathrm{crit}}$, the system has no stable equilibrium.

The results of numerical calculations of the fission barrier, as a function of x and y, are shown in Fig. 6A-1b. For arbitrary values of x, and small values of y, the equilibrium shape has an oblate deformation corresponding to the familiar flattening of the figure of the earth resulting from centrifugal distortion (see also solution A given in Eq. (6A-39)). With increasing angular momentum, the oblate shape becomes unstable with respect to triaxial deformations, as in the transition from Maclaurin spheroids to Jacobi ellipsoids for rotating fluid bodies held together by gravitational forces. The dot-dashed curve in Fig. 6A-1b indicates the values of y for which the equilibrium shape acquires triaxial symmetry; for values of x larger than about 0.8, this instability would occur only for values of y greater than those leading to fission instability.

6A-3 Compression Modes

6A-3a Wave equation

In addition to the surface oscillations, a liquid drop possesses normal modes of vibration involving compression of the density (sound waves). For small deviations of the density $\rho(\mathbf{r})$ from its equilibrium value ρ_0,

$$\rho(\mathbf{r})=\rho_0+\delta\rho(\mathbf{r}) \tag{6A-43}$$

the excess pressure δp is proportional to $\delta\rho$, and it is convenient to express the ratio in the form

$$\frac{\delta p}{\delta \rho} = M u_c^2 \tag{6A-44}$$

where the parameter u_c will appear as the velocity of sound. (The mass of each particle is denoted by M, and ρ is the particle density.) The derivative of the pressure with respect to density is equal to the compressibility coefficient b_{comp} defined by Eq. (2-207), and the relation (6A-44) can therefore also be written

$$u_c = \left(\frac{b_{\mathrm{comp}}}{M} \right)^{1/2} \tag{6A-45}$$

The hydrodynamical equations, to first order in $\delta\rho$, δp, and the flow velocity $\mathbf{v}$, may be written

$$\frac{\partial}{\partial t}\delta\rho(\mathbf{r},t) + \rho_0 \boldsymbol{\nabla}\cdot\mathbf{v}(\mathbf{r},t) = 0 \tag{6A-46a}$$

$$\frac{d\mathbf{v}}{dt} = \frac{\partial}{\partial t}\mathbf{v}(\mathbf{r},t) = -(M\rho_0)^{-1}\boldsymbol{\nabla}\delta p(\mathbf{r},t) \tag{6A-46b}$$

In Eq. (6A-46b), the Coulomb force has been neglected; it plays a lesser role for the compression waves than for the surface oscillations, because of the large restoring forces associated with the compression.

The linearized hydrodynamical equations (6A-46) lead to the wave equation

$$\boldsymbol{\nabla}^2\delta\rho(\mathbf{r},t) - u_c^{-2}\frac{\partial^2}{\partial t^2}\delta\rho(\mathbf{r},t) = 0 \tag{6A-47}$$

where we have used the relation (6A-44). The flow associated with the density variations $\delta\rho$ is irrotational with a velocity potential χ satisfying the relation $\dot{\chi} = \rho_0^{-1} u_c^2 \delta\rho$.

6A-3b *Eigenmodes*

For a system with spherical equilibrium, the eigenmodes of the wave equation (6A-47) correspond to density variations and velocity fields of the form

$$\delta\rho = \rho_0 j_\lambda(k_{n\lambda} r) Y^*_{\lambda\mu}(\vartheta\ \varphi)\alpha_{n\lambda\mu}(t) \tag{6A-48a}$$

$$\mathbf{v} = k_{n\lambda}^{-2}\boldsymbol{\nabla}\big(j_\lambda(k_{n\lambda} r) Y^*_{\lambda\mu}(\vartheta,\varphi)\big)\dot{\alpha}_{n\lambda\mu}(t) \tag{6A-48b}$$

where the radial function j_λ is a spherical Bessel function. The eigenvalues $k_{n\lambda}$ are determined by the boundary condition that $\delta\rho$ vanish at the surface, since no excess pressure can be maintained at a free surface,

$$j_\lambda(k_{n\lambda}R_0)=0 \tag{6A-49}$$

The amplitude $\alpha_{n\lambda\mu}$ in Eq. (6A-48) is a harmonic function of time with frequency

$$\omega_{n\lambda}=k_{n\lambda}u_c=\frac{u_c}{R_0}\begin{cases}3.14 & \lambda=0,\, n=1\\ 4.49 & \lambda=1,\, n=1\\ 5.76 & \lambda=2,\, n=1\end{cases} \tag{6A-50}$$

There is at present only little empirical evidence concerning the nuclear compressibility, but if one assumes $b_{\text{comp}}=15$ MeV (see Vol. I, p. 257), together with the radius $R_0=1.2A^{1/3}$ fm, the excitation energies become

$$\hbar\omega_{n\lambda}=\begin{cases}65A^{-1/3}\ \text{MeV} & \lambda=0,\, n=1\\ 93A^{-1/3}\ \text{MeV} & \lambda=1,\, n=1\\ 120A^{-1/3}\ \text{MeV} & \lambda=2,\, n=1\end{cases} \tag{6A-51}$$

The compression modes do not generate a net multipole moment of the form (6A-2), since the contribution from $r<R$ is compensated by that arising from the small displacement of the nuclear surface. The amplitude $\alpha_{\lambda\mu}$ characterizing this displacement (see Eq. (6A-1)) can be obtained from the radial velocity at the surface

$$(v_r)_{r=R_0}=R_0\dot{\alpha}_{\lambda\mu}Y^*_{\lambda\mu} \tag{6A-52}$$

which gives (see Eq. (6A-48))

$$\alpha_{\lambda\mu}=\alpha_{n\lambda\mu}(k^2_{n\lambda}R_0)^{-1}\left(\frac{\partial}{\partial r}j_\lambda(k_{n\lambda}r)\right)_{r=R_0} \tag{6A-53}$$

The total multipole moment is

$$\mathscr{M}(\lambda\mu)=\int r^\lambda Y_{\lambda\mu}\delta\rho(\mathbf{r})\,d\tau+\rho_0R_0^{\lambda+3}\alpha_{\lambda\mu} \tag{6A-54}$$

and the cancellation of the two terms is seen to follow from the identity

$$\int r^\lambda j_\lambda(kr)r^2\,dr=k^{-2}\left(\lambda r^{\lambda+1}j_\lambda(kr)-r^{\lambda+2}\frac{\partial}{\partial r}(j_\lambda(kr))\right) \tag{6A-55}$$

together with the boundary condition (6A-49).

6A-3c Relation to surface oscillations

The ratio between the frequencies of surface oscillations and compression modes (with $n \sim 1$) is of the order of (see Eqs. (6A-6), (6A-18), (6A-19), (6A-31), (6A-45), and (6A-50))

$$\frac{(\omega_{\text{surf}})_\lambda}{(\omega_{\text{comp}})_\lambda} \sim \left(\frac{b_{\text{surf}}}{b_{\text{comp}}}\right)^{1/2} \lambda^{1/2} A^{-1/6} \tag{6A-56}$$

and thus decreases, although slowly, as the number of particles in the system increases. (For a drop of water ($\mathscr{S} = 75$ dyne cm^{-1}, $u_c \approx 1500$ m sec^{-1}, corresponding to $b_{\text{surf}} \approx 3.5 \times 10^{-13}$ erg and $b_{\text{comp}} = 6.8 \times 10^{-13}$ erg), the eigenfrequencies $(\omega_{\text{surf}})_{\lambda=2}$ and $(\omega_{\text{comp}})_{n=1,\lambda=2}$ for $R_0 = 1$ cm are 24 sec^{-1} and 8.6×10^5 sec^{-1}, respectively; the corresponding value of $A^{1/6}$ equals 7.2×10^3.)

The increase of the ratio (6A-56) with the multipole order λ corresponds to the fact that the velocity of surface waves (capillary waves) is proportional to the square root of the wave number. If the parameters b_{surf} and b_{comp} are of the same order of magnitude, the ratio (6A-56) remains small compared to unity for $\lambda \ll A^{1/3}$. For $\lambda \gtrsim A^{1/3}$, the liquid can no longer be described as a continuous medium.

The sharp distinction between surface and compression modes is only a first approximation, valid when the condition $\omega_{\text{surf}} \ll \omega_{\text{comp}}$ is fulfilled or, equivalently, when the velocity of the surface waves is small compared with the sound velocity. Both types of modes can be considered together as solutions to the wave equation (6A-47) with a modified boundary condition, which takes into account the pressure at $r = R_0$ generated by the surface deformation (6A-53). This pressure can be obtained as the derivative of the deformation energy per unit surface area with respect to a displacement of the surface

$$\begin{aligned}(\delta p)_{r=R_0} &= R_0^{-3} C_\lambda \alpha_{\lambda\mu} Y^*_{\lambda\mu} \\ &= k_{n\lambda}^{-2} R_0^{-4} C_\lambda \alpha_{n\lambda\mu} Y^*_{\lambda\mu} \left(\frac{\partial}{\partial r} j_\lambda(k_{n\lambda} r)\right)_{r=R_0}\end{aligned} \tag{6A-57}$$

where the last expression is obtained by using the relation (6A-53). The pressure (6A-57) must equal that needed to sustain the compression at $r = R_0$, as given by Eqs. (6A-44) and (6A-48); we therefore obtain the eigenvalue equation for $k_{n\lambda}$

$$\left(j_\lambda(k_{n\lambda} R_0)\right)^{-1} \left(\frac{\partial}{\partial r} j_\lambda(k_{n\lambda} r)\right)_{r=R_0} = \frac{\lambda}{R_0} \frac{\omega_{n\lambda}^2}{\omega_\lambda^2} \tag{6A-58}$$

where $\omega_{n\lambda} = u_c k_{n\lambda}$, while ω_λ is the frequency of the surface oscillation for an incompressible fluid, given by Eq. (6A-6), with the mass parameter (6A-31).

The lowest eigenmode for given λ (and $\lambda \geqslant 2$) has no radial node ($n=0$) and can be associated with a surface oscillation. If the ratio (6A-56) is small, the wave number $k_{n=0,\lambda}$ is small compared with R_0^{-1}, and $j_\lambda(kr)$ is approximately proportional to r^λ. The boundary condition (6A-58) then reduces to the usual frequency relation for surface modes. For the higher modes ($n=1,2,\ldots$), the right-hand side of Eq. (6A-58) is large compared with R_0^{-1}, and the boundary condition is approximately given by Eq. (6A-49). The small coupling between the surface and volume vibrations implies that the multipole moment is no longer exclusively associated with the surface vibrations ($n=0$).

6A-4 Polarization Modes in Two-Fluid System

6A-4a Energy of polarization

If the fluid consists of two components, such as neutrons and protons, there are additional modes of vibration involving a relative motion of the two components of the fluid. The polarization modes may be treated in analogy to the compression modes, with the difference that the oscillating density is now the isovector

$$\rho_1(\mathbf{r}) = \delta\rho_{\mathrm{n}}(\mathbf{r}) - \delta\rho_{\mathrm{p}}(\mathbf{r}) \tag{6A-59}$$

instead of the isoscalar $\delta\rho_{\mathrm{n}} + \delta\rho_{\mathrm{p}}$. The equilibrium densities of neutrons and protons are assumed to be the same ($(\rho_{\mathrm{n}})_0 = (\rho_{\mathrm{p}})_0; N=Z$), in which case the normal modes are the symmetric (compression) oscillations and the antisymmetric (polarization) oscillations.

For the polarization waves, the velocity of propagation is

$$u_{\mathrm{s}} = \left(\frac{b_{\mathrm{sym}}}{M}\right)^{1/2} \tag{6A-60}$$

The energy b_{sym} is the analog of the compressibility coefficient and gives the energy density associated with the polarization

$$\mathscr{E}_{\mathrm{sym}} = \frac{b_{\mathrm{sym}}}{2\rho_0} \int (\rho_1(\mathbf{r}))^2 \, d\tau \tag{6A-61}$$

For the nucleus, $\mathscr{E}_{\mathrm{sym}}$ is referred to as the symmetry energy, and the coefficient b_{sym} can be estimated from the empirical nuclear binding energies ($b_{\mathrm{sym}} \approx 50$ MeV; see Eqs. (2-12) and (2-14)).

6A-4b Eigenfrequencies

The polarization waves obey a wave equation analogous to Eq. (6A-47), but the boundary condition is to be replaced by $v_r = 0$ at $r = R_0$, since an excessive energy would be involved in displacing protons beyond the volume occupied by the neutrons. For these modes, therefore, the surface is not free, but effectively clamped. The general solution to the wave equation has the form

$$\rho_1(\mathbf{r}) = \rho_0 j_\lambda(k_{n\lambda} r) Y^*_{\lambda\mu}(\vartheta, \varphi) \alpha_{n\lambda\mu}(t) \tag{6A-62}$$

where ρ_0 is the total particle density, and the boundary condition takes the form

$$\left(\frac{\partial}{\partial r} j_\lambda(k_{n\lambda} r)\right)_{r=R_0} = 0 \tag{6A-63}$$

For the lowest modes, the eigenfrequencies are

$$\omega_{n\lambda} = u_s k_{n\lambda} = \frac{u_s}{R_0} \begin{cases} 2.08 & \lambda=1, n=0 \\ 5.94 & \lambda=1, n=1 \\ 3.34 & \lambda=2, n=0 \\ 4.49 & \lambda=0, n=1 \end{cases} \tag{6A-64}$$

corresponding to the excitation energies

$$\hbar\omega_{n\lambda} = \begin{cases} 79A^{-1/3}\ \text{MeV} & \lambda=1, n=0 \\ 225A^{-1/3}\ \text{MeV} & \lambda=1, n=1 \\ 127A^{-1/3}\ \text{MeV} & \lambda=2, n=0 \\ 170A^{-1/3}\ \text{MeV} & \lambda=0, n=1 \end{cases} \tag{6A-65}$$

where we have assumed $b_{\text{sym}} = 50$ MeV (see Eq. (2-14)) and $R_0 = 1.2A^{1/3}$ fm.

6A-4c Oscillator strength in sum rule units

The isovector multipole moment associated with the polarization oscillations is

$$\begin{aligned} \mathscr{M}(\tau=1, \lambda\mu) &= \int \rho_1(\mathbf{r}) r^\lambda Y_{\lambda\mu}\, d\tau = m_{n\lambda} \alpha_{n\lambda\mu} \\ m_{n\lambda} &\equiv \rho_0 \int_0^{R_0} j_\lambda(k_{n\lambda} r) r^{\lambda+2}\, dr \\ &= \lambda \rho_0 k_{n\lambda}^{-2} R_0^{\lambda+1} j_\lambda(k_{n\lambda} R_0) \end{aligned} \tag{6A-66}$$

In the evaluation of the integral of the Bessel function, we have employed the boundary condition (6A-63).

The multipole moments for the different eigenmodes can be characterized by the oscillator strength, which obeys a sum rule of the type discussed in Sec. 6-4. (The oscillator strength refers to the quantized system, but a quantity of corresponding significance for the classical system may be obtained by dividing the oscillator strength by $\hbar^2$.)

The oscillator strength involves the matrix element of $\alpha_{n\lambda\mu}$ for the excitation of a vibrational quantum (see Eq. (6-15))

$$|\langle n_{n\lambda\mu}=1|\alpha_{n\lambda\mu}|n_{n\lambda\mu}=0\rangle|^2=\frac{\hbar\omega_{n\lambda}}{2C_{n\lambda}} \tag{6A-67}$$

where the restoring force $C_{n\lambda}$ can be obtained from the potential energy density (6A-61)

$$\begin{aligned} C_{n\lambda} &= \rho_0 b_{\text{sym}} \int_0^{R_0} (j_\lambda(k_{n\lambda} r))^2 r^2\, dr \\ &= \tfrac{1}{2}\rho_0 b_{\text{sym}} R_0 k_{n\lambda}^{-2} (j_\lambda(k_{n\lambda} R_0))^2 (k_{n\lambda}^2 R_0^2 - \lambda(\lambda+1)) \end{aligned} \tag{6A-68}$$

Hence, the oscillator strength for the mode $n\lambda$ is given by (see Eqs. (6A-66) and (6A-67))

$$\begin{aligned} \hbar\omega_{n\lambda} B(\tau=1,\lambda;\, n_{n\lambda}=0\rightarrow n_{n\lambda}=1) &= (2\lambda+1) m_{n\lambda}^2 \frac{(\hbar\omega_{n\lambda})^2}{2C_{n\lambda}} \\ &= \frac{3}{4\pi}\frac{\hbar^2}{M} A R_0^{2\lambda-2} \frac{(2\lambda+1)\lambda^2}{k_{n\lambda}^2 R_0^2 - \lambda(\lambda+1)} \\ &= \zeta_{n\lambda} S(\tau=1,\lambda) \end{aligned} \tag{6A-69}$$

where

$$\zeta_{n\lambda} = \frac{2\lambda}{k_{n\lambda}^2 R_0^2 - \lambda(\lambda+1)} = \begin{cases} 0.86 & \lambda=1, n=0 \\ 0.06 & \lambda=1, n=1 \\ 0.78 & \lambda=2, n=0 \end{cases} \tag{6A-70}$$

expresses the oscillator strength in units of the oscillator sum $S(\tau=1,\lambda)$, which has the value given by Eq. (6-179) with $\langle r^{2\lambda-2}\rangle = 3(2\lambda+1)^{-1} R_0^{2\lambda-2}$

$$\begin{aligned} S(\tau=1,\lambda) &\equiv \sum_n \hbar\omega_{n\lambda} B(\tau=1,\lambda; 0\rightarrow n_{n\lambda}=1) \\ &= \frac{3\lambda(2\lambda+1)}{4\pi}\frac{\hbar^2}{2M} A R_0^{2\lambda-2} \end{aligned} \tag{6A-71}$$

The validity of the classical sum rule (6A-71) for the liquid-drop model follows from the derivation in Sec. 6-4, since the interactions assumed in this model are velocity independent. (Velocity-dependent interactions would have implied additional contributions to the mass parameter for the collective flow.)

6A-4d Effect of deformation

If the shape of the system does not possess spherical symmetry, the vibrational motion no longer separates into angular and radial components. However, for small deformations, the eigenmodes can be obtained by a perturbation treatment, starting from the solutions for spherical symmetry.

We illustrate the procedure by considering the polarization mode for a system that possesses a small deformation of axial symmetry.[38] For such a deformation, the eigenmodes can be classified by the multipole component ν associated with the angular momentum with respect to the symmetry axis. The perturbed mode with polarization density ρ_1' and wave number k' satisfies the equations

$$\nabla^2\rho_1' + k'^2\rho_1' = 0 \tag{6A-72a}$$

$$\left(\frac{\partial \rho_1'}{\partial n}\right)_{r=R(\vartheta)} = 0 \tag{6A-72b}$$

where $R(\vartheta)$ describes the deformed surface and where the derivative of ρ_1' is taken in the direction of the normal to the surface. Multiplying Eq. (6A-72a) by ρ_1^*, where ρ_1 is the density in the unperturbed mode, and subtracting the corresponding equation with ρ_1 and ρ_1' interchanged, we obtain

$$\begin{aligned}(k'^2 - k_{n\lambda}^2)\int_{r<R(\vartheta)} \rho_1^*\rho_1'\,d\tau &= \int_{r<R(\vartheta)} (\rho_1'\nabla^2\rho_1^* - \rho_1^*\nabla^2\rho_1')\,d\tau \\ &= \int_{R(\vartheta)} \left(\rho_1'\frac{\partial\rho_1^*}{\partial n} - \rho_1^*\frac{\partial\rho_1'}{\partial n}\right)d\sigma = \int_{R(\vartheta)} \rho_1'\frac{\partial\rho_1^*}{\partial n}\,d\sigma\end{aligned} \tag{6A-73}$$

where the two last integrals are extended over the deformed surface, for which the boundary condition (6A-72b) applies.

For small deformations, the leading-order perturbation, $\delta k_{n\lambda\nu} = k' - k_{n\lambda}$, is obtained from Eq. (6A-73) in terms of the difference between the normal derivatives of ρ_1 for the spherical surface ($\partial\rho_1/\partial n = 0$) and for the deformed surface ($\hat{\mathbf{n}} \approx \hat{\mathbf{r}} - \nabla R$),

$$2k_{n\lambda}\,\delta k_{n\lambda\nu}\int_{r<R_0} |\rho_1|^2\,d\tau = \int_{r=R_0} \rho_1\left((R-R_0)\frac{\partial^2\rho_1^*}{\partial r^2} - \nabla R\cdot\nabla\rho_1^*\right)R_0^2\,d\Omega \tag{6A-74}$$

Evaluating the integrals in Eq. (6A-74) for ρ_1 given by Eq. (6A-62) and for a

---

[38] The effect of a spheroidal deformation on the frequencies of the dipole mode was given by Danos (1958) and Okamoto (1958).

surface deformation of multipole order λ' and amplitude $\alpha_{\lambda'0}$, one obtains

$$\frac{\delta k_{n\lambda\nu}}{k_{n\lambda}} = \frac{\delta\omega_{n\lambda\nu}}{\omega_{n\lambda}}$$

$$= -\left(\frac{2\lambda'+1}{4\pi}\right)^{1/2} \alpha_{\lambda'0}\langle\lambda\nu\lambda'0|\lambda\nu\rangle\langle\lambda 0\lambda'0|\lambda 0\rangle \frac{k_{n\lambda}^2 R_0^2 - \lambda(\lambda+1) + \frac{1}{2}\lambda'(\lambda'+1)}{k_{n\lambda}^2 R_0^2 - \lambda(\lambda+1)} \tag{6A-75}$$

From the results obtained for an axially symmetric deformation, the effects of arbitrary deformations can be derived by exploiting the rotational invariance of the Hamiltonian. The frequency shift $\delta\omega$ in the mode $n\lambda$ can be expressed as a matrix in the $(2\lambda+1)$-dimensional space of the unperturbed modes $n\lambda\nu$. For a deformation of multipole order λ', the matrix $\delta\omega$, to first order in the amplitudes $\alpha_{\lambda'\nu'}$ can be expressed in the form $\delta\omega = \Sigma_{\nu'}\alpha^*_{\lambda'\nu'}t_{\lambda'\nu'}$, where the matrix t is a spherical tensor of order $\lambda'\nu'$, as a consequence of rotational invariance. The matrix elements $\langle n\lambda\nu_2|\delta\omega|n\lambda\nu_1\rangle$ can therefore be obtained from the diagonal terms (6A-75) by replacing $\alpha_{\lambda'0}\langle\lambda\nu\lambda'0|\lambda\nu\rangle$ by $\Sigma_{\nu'}\alpha^*_{\lambda'\nu'}\langle\lambda\nu_1\lambda'\nu'|\lambda\nu_2\rangle$. The eigenmodes follow by a diagonalization of $\delta\omega$.

6A-5 Rotational Motion of Irrotational Fluid

In the present section, we consider the motion of an irrotational (and incompressible) liquid that is held in a nonspherical shape, which is rotated uniformly. The potential flow associated with the rotation can be obtained by considering the velocity $\mathbf{v}'$ relative to the rotating frame

$$\begin{aligned}\mathbf{v}' &= \mathbf{v} - (\boldsymbol{\omega} \times \mathbf{r}) \\ &= -\boldsymbol{\nabla}\chi - (\boldsymbol{\omega} \times \mathbf{r})\end{aligned} \tag{6A-76}$$

where $\boldsymbol{\omega}$ is the angular velocity of the rotational motion and χ the velocity potential. The condition that the surface is stationary in the rotating coordinate system implies

$$(\mathbf{v}' \cdot \mathbf{n})_{r=R} = 0 \tag{6A-77}$$

where $\mathbf{n}$ is the normal to the surface. For any given surface, Eq. (6A-77) together with the Laplace equation (6A-27b) uniquely determines the velocity potential.

An especially simple flow is obtained for an ellipsoidal shape,

$$F(x_1, x_2, x_3) = \frac{x_1^2}{R_1^2} + \frac{x_2^2}{R_2^2} + \frac{x_3^2}{R_3^2} = 1 \tag{6A-78}$$

expressed in terms of the intrinsic coordinates x_κ. The normal to the surface has the direction of $\boldsymbol{\nabla}F$, and the solution satisfying the boundary condition (6A-77) is the quadrupole potential

$$\chi=\frac{R_2^2-R_1^2}{R_1^2+R_2^2}x_1x_2\omega_3+\frac{R_3^2-R_2^2}{R_2^2+R_3^2}x_2x_3\omega_1+\frac{R_1^2-R_3^2}{R_3^2+R_1^2}x_3x_1\omega_2 \qquad \text{(6A-79)}$$

where ω_κ are the components of $\boldsymbol{\omega}$ in the intrinsic system. The flow pattern derived from Eq. (6A-79) is illustrated in Fig. 6A-2, which shows the velocity field $\mathbf{v}$ as seen from a fixed coordinate system, as well as the field $\mathbf{v}'$ seen from the rotating system. For comparison, Fig. 6A-2 also shows the flow patterns for a rigid rotation ($\mathbf{v}'=0$, $\mathbf{v}=\boldsymbol{\omega}\times\mathbf{r}$).

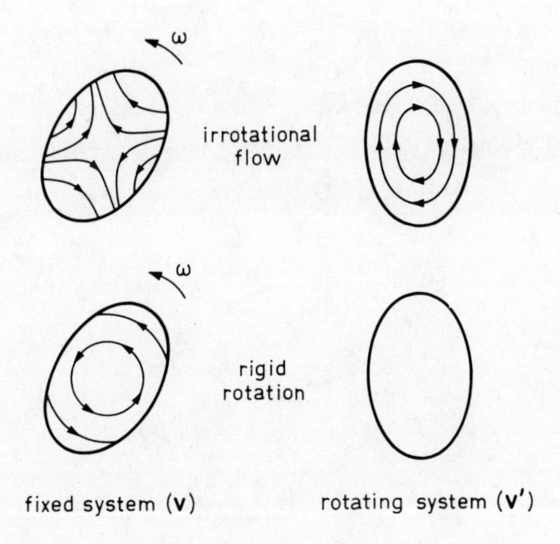

Figure 6A-2 Flow patterns for rotation of ellipsoidal body.

The kinetic energy of the rotational motion generated by the velocity potential (6A-79) is

$$T=\tfrac{1}{2}M\rho_0\int(\boldsymbol{\nabla}\chi)^2\,d\tau$$

$$=\tfrac{1}{2}(\mathcal{I}_1\omega_1^2+\mathcal{I}_2\omega_2^2+\mathcal{I}_3\omega_3^2) \qquad \text{(6A-80)}$$

with the moments of inertia

$$\mathcal{I}_1=\left(\frac{R_2^2-R_3^2}{R_2^2+R_3^2}\right)^2(\mathcal{I}_{\text{rig}})_1 \quad \text{and cyclic permutations} \qquad \text{(6A-81)}$$

expressed in terms of the moments for rigid rotation

$$(\mathcal{I}_{\text{rig}})_1=M\rho_0\int(x_2^2+x_3^2)\,d\tau$$

$$=\tfrac{1}{5}AM(R_2^2+R_3^2) \qquad \text{(6A-82)}$$

where AM is the total mass.

The rotational motion considered in the present section involves the same degrees of freedom as the surface oscillations. Thus, for an ellipsoidal deformation, three of the five quadrupole surface modes combine to form rotational modes and only two remain as "intrinsic" vibrations associated with variations of the shape (β and γ vibrations). The connection between rotational modes and shape oscillations is further discussed in Appendix 6B; the moments of inertia (6B-17) with the value of D $(=D_2)$ given by Eq. (6A-31) equal the moments (6A-81), to leading order in the deformation β. (The relation between δR_κ and the deformation parameters β and γ is given by Eq. (6B-4).)

APPENDIX

6B

The Five-Dimensional Quadrupole Oscillator

The present appendix deals with various features of the spectra associated with shape deformations of quadrupole symmetry. This degree of freedom is capable of exhibiting both rotational and vibrational spectra, and the connection can be exhibited in terms of a transformation to a set of coordinates describing the intrinsic shape and the orientation of the system (Sec. 6B-1). Small-amplitude vibrations about spherical symmetry are considered in Sec. 6B-2, while the classification of the quadrupole spectra in the yrast region is the subject of Sec. 6B-3. Tools for constructing wave functions for many-phonon states are considered in Sec. 6B-4.

6B-1 Shape and Angle Coordinates. Vibrational and Rotational Degrees of Freedom[39]

6B-1a Definition of coordinates

The fundamental mode of deformation of a spherical system is of quadrupole symmetry (see Sec. 6A-1a) and can be characterized by a set of five amplitudes $\alpha_{2\mu}$ that form the components of a spherical tensor (see, for example, Eq. (6-39)). A deformation of this type produces an ellipsoidal shape, for small $\alpha_{2\mu}$. (More generally, the shape possesses D_2 symmetry, with invariance with respect to rotations of 180° about each of three orthogonal axes.) The principal axes of the deformed shape define an intrinsic coordinate system whose orientation angles $\omega=(\phi,\theta,\psi)$ are given by the relations

$$\begin{gathered}\alpha_{2\mu}=\sum_{\nu} a_{2\nu}\mathscr{D}^{2}_{\mu\nu}(\omega)\\ a_{21}=a_{2-1}=0 \qquad a_{22}=a_{2-2}\end{gathered} \tag{6B-1}$$

[39]The analysis of quadrupole vibrations in terms of shape and angle variables was considered by A. Bohr (1952). Such variables were also employed by Hill and Wheeler (1953). The structure of the anharmonic Hamiltonian in these variables was analyzed by Kumar and Baranger (1967).

The two nonvanishing intrinsic deformation variables a_{20} and a_{22} $(=a_{2-2})$ are usually expressed in terms of the parameters β and γ defined by

$$a_{20}=\beta\cos\gamma$$
$$a_{22}=2^{-1/2}\beta\sin\gamma \tag{6B-2}$$

The relations (6B-1) and (6B-2) provide a transformation from the five tensorial variables $\alpha_{2\mu}$ to the three orientation angles ω and the two shape variables β and γ. The intrinsic variables β and γ are rotational invariants and can be related to the basic second- and third-order invariants that can be formed from the tensor $\alpha_{2\mu}$

$$(\alpha_2\alpha_2)_0=5^{-1/2}\beta^2 \tag{6B-3a}$$

$$(\alpha_2\alpha_2\alpha_2)_0=-(2/35)^{1/2}\beta^3\cos 3\gamma \tag{6B-3b}$$

The ellipsoidal shape, for small values of $\alpha_{2\mu}$, can also be characterized by the increments of the three principal axes

$$\delta R_\kappa=\left(\frac{5}{4\pi}\right)^{1/2}\beta R_0\cos\left(\gamma-\kappa\frac{2\pi}{3}\right)\qquad \kappa=1,2,3 \tag{6B-4}$$

where R_0 is the radius of the spherical shape. Since the quadrupole deformation conserves volume, to first order in the amplitudes, we have $\Sigma_\kappa \delta R_\kappa=0$. The relation of the deformations δR_κ and the coordinates β,γ is illustrated in Fig. 6B-1.

The set of angular and shape coordinates (ω,β,γ) is not unique. A given deformation $\alpha_{2\mu}$ only specifies the three symmetry planes of the ellipsoidal shape, but the labeling of the intrinsic axes is arbitrary. Since the new coordinates depend on this labeling, the nuclear Hamiltonian and wave functions, when expressed in terms of the variables (ω,β,γ), must be invariant with respect to the symmetry operations that correspond to a relabeling of the intrinsic axes. (These operations form the octahedral group O, which comprises the 24 rotations that transform the cube into itself.)

The symmetry operations that correspond to a relabeling of the intrinsic axes can be constructed from the two elements $\mathscr{R}_2(\pi/2)$ and $\mathscr{R}_3(\pi/2)$ representing rotations through the angle $\pi/2$ about the intrinsic 2 and 3 axes. The effect of $\mathscr{R}_2(\pi/2)$ and $\mathscr{R}_3(\pi/2)$ on the coordinates β and γ and on the labeling of the three axes is given by (see, for example, Eq. (6B-4))

$$\mathscr{R}_2(\pi/2)\{\beta,\gamma,I_1,I_2,I_3\}\mathscr{R}_2^{-1}(\pi/2)=\{\beta,-\gamma+2\pi/3,I_3,I_2,-I_1\}$$
$$\mathscr{R}_3(\pi/2)\{\beta,\gamma,I_1,I_2,I_3\}\mathscr{R}_3^{-1}(\pi/2)=\{\beta,-\gamma,-I_2,I_1,I_3\} \tag{6B-5}$$

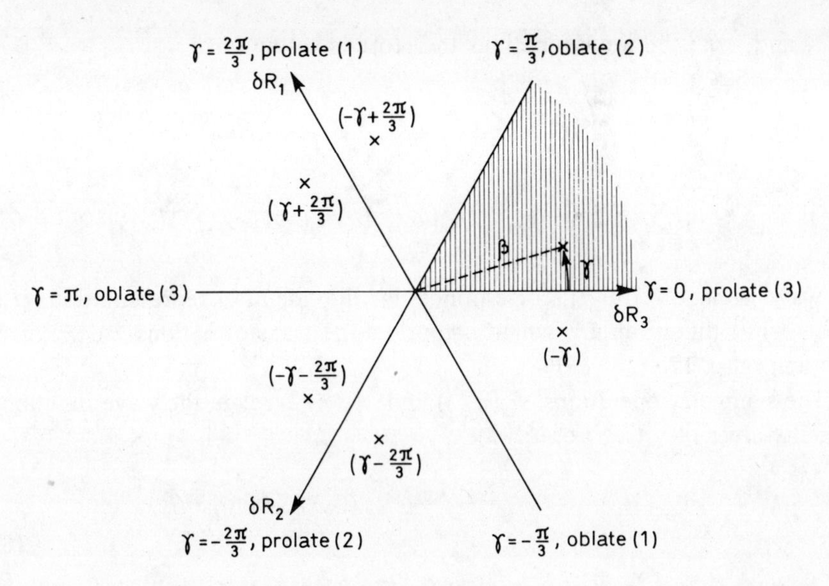

Figure 6B-1 Symmetries in the $\beta\gamma$ plane. The figure shows a polar diagram for the deformation variables β and γ. The projections on the three axes in the figure are proportional to the increments δR_κ in the principal radii of the shape. Points lying on the axes correspond to axially symmetric shapes. The six different points obtained by reflection in the axes represent the same shape of the nucleus.

where I_κ denote the components of the angular momentum with respect to the intrinsic axes.

6B-1b *Symmetry of wave function*

In the coordinates (ω, β, γ), the wave function for a state of total angular momentum IM can be expressed in the general form

$$\Psi_{IM} = \left(\frac{2I+1}{8\pi^2}\right)^{1/2} \sum_{K=-I}^{I} \Phi_{IK}(\beta,\gamma)\mathscr{D}^I_{MK}(\omega) \tag{6B-6}$$

where the $\mathscr{D}$ functions form an orthogonal basis for expanding functions of the orientation angle ω (see Sec. 1A-4). The invariance of Ψ with respect to the operations that relabel the intrinsic axes leads to symmetry conditions on the functions $\Phi_{IK}(\beta,\gamma)$.

We first consider the operations corresponding to a rotation of π about an intrinsic axis. These operations leave β and γ invariant, and their effect on wave functions of the type (6B-6) has been discussed in connection with the eigenstates of the asymmetric rotor (Chapter 4, pp. 177 ff.). The $\mathscr{R}_3(\pi)$ invariance implies that K must be even, while the invariance with respect to $\mathscr{R}_2(\pi)$ relates the components with K and $-K$; the wave functions therefore have the form

$$\Psi_{IM} = \left(\frac{2I+1}{8\pi^2}\right)^{1/2}\left(\tfrac{1}{2}\left(1+(-1)^I\right)\Phi_{I,K=0}(\beta,\gamma)\mathscr{D}^I_{M0}(\omega)\right.$$
$$\left. + \sum_{K=2,4,\ldots} 2^{-1/2}\Phi_{IK}(\beta,\gamma)\left(\mathscr{D}^I_{MK}(\omega)+(-1)^I\mathscr{D}^I_{M-K}(\omega)\right)\right) \tag{6B-7}$$

The wave function (6B-7) corresponds to the identity representation $(r_1r_2r_3) = (+++)$ of the group D_2, which comprises the transformations $\mathscr{R}_\kappa(\pi)$; see the discussion on p. 178.

The symmetry operations $\mathscr{R}_2(\pi/2)$ and $\mathscr{R}_3(\pi/2)$ relate the wave functions for different values of γ (see Eq. (6B-5)),

$$\Phi_{IK}(\beta, -\gamma+2\pi/3) = \sum_{K'} \mathscr{D}^I_{KK'}(0,\pi/2,0)\Phi_{IK'}(\beta,\gamma)$$
$$\Phi_{IK}(\beta,-\gamma) = \exp\{i\pi K/2\}\Phi_{IK}(\beta,\gamma) \tag{6B-8}$$

A knowledge of the wave function in the interval $0 \leqslant \gamma \leqslant \pi/3$ therefore completely specifies the state. The symmetry in the $\beta\gamma$ plane is illustrated in Fig. 6B-1.

6B-1c Potential and kinetic energies

The potential energy of the nuclear deformation is independent of the orientation and depends only on the shape parameters β and γ,

$$V = V(\beta,\gamma) \tag{6B-9}$$

Furthermore, as a consequence of the symmetry relation (6B-5),

$$V(\beta,\gamma) = V(\beta,-\gamma) = V(\beta,\gamma+2\pi/3) \tag{6B-10}$$

which is equivalent to

$$V = V(\beta, \cos 3\gamma) \tag{6B-11}$$

The kinetic energy T involves the time derivatives $\dot\beta$, $\dot\gamma$, and the components $\dot\varphi_\kappa$ of the angular frequency on the three intrinsic axes. The invariance with respect to the transformations $\mathscr{R}_\kappa(\pi)$ implies that T must be even in each of the $\dot\varphi_\kappa$, and considering only terms that are quadratic in the time derivatives, the kinetic energy can therefore be written

$$T = T_{\text{vib}} + T_{\text{rot}} \tag{6B-12}$$

with

$$T_{\text{vib}} = \tfrac{1}{2}D_{\beta\beta}(\beta,\gamma)\dot\beta^2 + D_{\beta\gamma}(\beta,\gamma)\dot\beta\dot\gamma + \tfrac{1}{2}D_{\gamma\gamma}(\beta,\gamma)\dot\gamma^2$$
$$= \left(D_{\beta\beta}D_{\gamma\gamma} - D_{\beta\gamma}^2\right)^{-1}\left(\tfrac{1}{2}D_{\gamma\gamma}p_\beta^2 - D_{\beta\gamma}p_\beta p_\gamma + \tfrac{1}{2}D_{\beta\beta}p_\gamma^2\right) \tag{6B-13}$$

and

$$T_{\text{rot}} = \frac{1}{2}\sum_{\kappa}\mathscr{I}_{\kappa}(\beta,\gamma)\dot{\varphi}_{\kappa}^{2} = \sum_{\kappa}\frac{\hbar^{2}I_{\kappa}^{2}}{2\mathscr{I}_{\kappa}} \tag{6B-14}$$

where p_β and p_γ are the momenta canonically conjugate to β and γ, while I_κ (the conjugates of φ_κ) are the components of the angular momentum with respect to the intrinsic axes. (In Eq. (6B-13), the condition of hermiticity requires a symmetrization of the terms involving noncommutable operators; differences depending on the ordering of factors are of lower power in the momenta and hence equivalent to additional terms in the potential energy. For the construction of a Schrödinger equation in the (ω,β,γ) coordinates, on the basis of the metric defined by the kinetic energy, see Pauli, 1933, p. 120, and Sec. 6B-2.)

The inertial functions $D_{\beta\beta}(\beta,\gamma)$ and $D_{\gamma\gamma}(\beta,\gamma)$ obey the same symmetry as $V(\beta,\gamma)$ and thus depend only on γ through $\cos 3\gamma$ (see Eq. (6B-11)), while $D_{\beta\gamma}(\beta,\gamma)$ is odd in γ and therefore of the form $\sin 3\gamma$ times a function of β and $\cos 3\gamma$. The effective moments of inertia are subject to the relations

$$\begin{aligned}
&\mathscr{I}_3(-\gamma) = \mathscr{I}_3(\gamma)\\
&\mathscr{I}_1(-\gamma) = \mathscr{I}_2(\gamma)\\
&\mathscr{I}_3(\gamma - 2\pi/3) = \mathscr{I}_1(\gamma) = \mathscr{I}_2(\gamma + 2\pi/3)\\
&\mathscr{I}_3(\gamma + 2\pi/3) = \mathscr{I}_2(\gamma) = \mathscr{I}_1(\gamma - 2\pi/3)
\end{aligned} \tag{6B-15}$$

that follow from Eq. (6B-5).

For shapes that possess axial symmetry ($\gamma = 0, \pm\pi/3, \pm 2\pi/3,\ldots$), the azimuthal angle φ_κ with respect to the symmetry axis becomes indeterminate; hence, the moment of inertia with respect to such an axis vanishes. For small deviations from axial symmetry, it follows from the relation (6B-15) that the moment of inertia with respect to the approximate symmetry axis is proportional to the square of the departure from axial symmetry. In the neighborhood of spherical symmetry, all the moments of inertia are proportional to β^2.

The quadrupole Hamiltonian $H = T + V$ describes a variety of different types of collective motion depending on the potential energy function and the inertial parameters. The limit of harmonic vibrations corresponds to a potential energy $V = \frac{1}{2}C\beta^2$ and a kinetic energy involving a single mass parameter D (see Eq. (6B-17)). If the potential energy function has a well-defined minimum (β_0, γ_0) away from the origin, the motion in the $\beta\gamma$ plane may be localized in the neighborhood of this minimum. In such a situation, the quadrupole degrees of freedom separate into vibrational and rotational components. The vibrations correspond to small amplitude oscillations about the equilibrium, and the rotation is approximately described by the kinetic energy (6B-14), with moments of inertia $\mathscr{I}_\kappa(\beta_0, \gamma_0)$. If the equilibrium shape deviates from axial symmetry, the rotational motion involves three degrees of freedom (asymmetric rotor), and the vibrational motion corresponds to two normal modes in the $\beta\gamma$ plane. If the equilibrium has axial symmetry, the rotational motion involves two degrees of freedom, and the vibrational modes comprise oscillations that preserve the axial symmetry (β vibrations) as well as twofold degenerate oscillations away from axial symmetry (γ vibrations).

6B-2 Oscillations about Spherical Equilibrium

For small oscillations about a spherical equilibrium, the potential and kinetic energies can be expressed as a power series in $\alpha_{2\mu}$ and $\dot{\alpha}_{2\mu}$. The leading-order terms have the form

$$\begin{aligned} V &= \tfrac{1}{2}C\sum_{\mu}|\alpha_{2\mu}|^2 = \tfrac{1}{2}C\sum_{\mu}(-1)^{\mu}\alpha_{2\mu}\alpha_{2-\mu} \\ T &= \tfrac{1}{2}D\sum_{\mu}|\dot{\alpha}_{2\mu}|^2 = -\frac{\hbar^2}{2D}\sum_{\mu}(-1)^{\mu}\frac{\partial^2}{\partial\alpha_{2\mu}\,\partial\alpha_{2-\mu}} \end{aligned} \tag{6B-16}$$

To this order, the Hamiltonian represents a set of five degenerate harmonic oscillators; the spectrum and eigenfunctions of such a system are considered in Sec. 6-3a. The construction of many-phonon states is further discussed in Sec. 6B-4.

In the variables (ω,β,γ), the Hamiltonian in the harmonic approximation (6B-16) takes the form (Bohr, 1952)

$$\begin{aligned} H &= T + V = T_{\text{vib}} + T_{\text{rot}} + V \\ V &= \tfrac{1}{2}C\beta^2 \\ T_{\text{vib}} &= \tfrac{1}{2}D(\dot{\beta}^2 + \beta^2\dot{\gamma}^2) \\ &= -\frac{\hbar^2}{2D}\left(\beta^{-4}\frac{\partial}{\partial\beta}\beta^4\frac{\partial}{\partial\beta} + \beta^{-2}(\sin 3\gamma)^{-1}\frac{\partial}{\partial\gamma}\sin 3\gamma\frac{\partial}{\partial\gamma}\right) \\ T_{\text{rot}} &= \tfrac{1}{2}\sum_{\kappa}\mathcal{I}_{\kappa}\dot{\varphi}_{\kappa}^2 = \sum_{\kappa}\frac{\hbar^2 I_{\kappa}^2}{2\mathcal{I}_{\kappa}} \\ \mathcal{I}_{\kappa} &= 4D\beta^2\sin^2(\gamma - \kappa 2\pi/3) \end{aligned} \tag{6B-17}$$

which is seen to be of the general form considered in Sec. 6B-1. The differential operators for the kinetic energy correspond to wave functions normalized with respect to the volume element

$$d\tau = \beta^4 d\beta\,|\sin 3\gamma|\,d\gamma\,\sin\theta\,d\theta\,d\phi\,d\psi \tag{6B-18}$$

where (ϕ,θ,ψ) are the Euler angles.

The motion defined by the Hamiltonian (6B-17) separates in the variables β and (ω,γ),

$$\Psi = \varphi(\beta)\Phi(\omega,\gamma) \tag{6B-19}$$

For example, the ground state ($n=0, I=0$) and the one-quantum state ($n=1, I=2$) have the wave functions

$$\Psi_{n=0,I=0}=(8\pi^2)^{-1/2}(2\pi)^{-1/4}b^{-5/2}\exp\{-\beta^2/4b^2\} \tag{6B-20a}$$

$$\Psi_{n=1,I=2,M}=(8\pi^2)^{-1/2}(2\pi)^{-1/4}b^{-7/2}\beta\exp\{-\beta^2/4b^2\}$$
$$\times\left(\cos\gamma\mathscr{D}^2_{M0}(\omega)+2^{-1/2}\sin\gamma\left(\mathscr{D}^2_{M2}(\omega)+\mathscr{D}^2_{M-2}(\omega)\right)\right) \tag{6B-20b}$$

where

$$b=\left(\frac{\hbar^2}{4CD}\right)^{1/4} \tag{6B-21}$$

is the root mean square zero-point amplitude, and where the normalization corresponds to integration over $0\leqslant\gamma\leqslant\pi/3$. The wave functions are seen to satisfy the symmetry requirements implied by Eqs. (6B-7) and (6B-8).

6B-3 Yrast Region for Harmonic Vibrations

The spectrum of the quadrupole oscillator exhibits a relatively simple pattern when many quanta are present, with a total angular momentum close to the maximum value $I_{max}=2n$ (region of the "yrast" line). In such states, the centrifugal force produces an average deformation that is large compared with the zero-point fluctuations. The system can therefore be described in terms of an approximate separation between rotational motion and vibrations about the equilibrium shape.

In the present section, we consider the spectrum in the yrast region for harmonic quadrupole vibrations. For $I\approx I_{max}=2n\gg1$, the main terms in the Hamiltonian (6B-17) are the potential energy V and the rotational energy T_{rot}. For given I and β, the rotational energy has a minimum for $\gamma=\pi/6$, for which the moment of inertia $\mathscr{I}_1$ takes on the maximum value $4D\beta^2$. To leading order in I, corresponding to the classical value for the rotational energy, the sum of potential and rotational energies thus possesses a minimum for a shape with parameters

$$\begin{aligned}\gamma_0&=\frac{\pi}{6}\\ \beta_0&=\left(\frac{\hbar^2}{4CD}\right)^{1/4}I^{1/2}=bI^{1/2}\end{aligned} \tag{6B-22}$$

where b is the zero-point amplitude (6B-21). The equilibrium deformation (6B-22) corresponds to a triaxial shape with (see Eq. (6B-4))

$$\begin{aligned}\delta R_1&=0\\ \delta R_2&=-\delta R_3\\ \delta R_3&=\left(\frac{15}{16\pi}\right)^{1/2}\beta_0 R_0\end{aligned} \tag{6B-23}$$

The sum of potential and rotational energies possesses a second minimum for $\gamma=0$, corresponding to a prolate shape, but for $I>8$, the triaxial shape with $\gamma=\pi/6$ gives rise to the lowest eigenvalue of $V+T_{\text{rot}}$ for given I.

For the shape (6B-22), the moments of inertia $\mathscr{I}_2$ and $\mathscr{I}_3$ are equal, and the rotational motion is that of a symmetric top, which can be characterized by the quantum number $K=I_1$. An expansion of the Hamiltonian around the equilibrium shape yields, for states with $I_{\max}-I\ll I_{\max}$,

$$H=H_{\text{rot}}+H_K+H_\beta+H_\gamma \tag{6B-24}$$

with

$$H_{\text{rot}}=\tfrac{1}{2}C\beta_0^2+\frac{\hbar^2}{8D\beta_0^2}I^2$$

$$=\tfrac{1}{2}\hbar\omega I \qquad \left(\omega=(C/D)^{1/2}\right) \tag{6B-25a}$$

$$H_K=\frac{\hbar^2}{2D\beta_0^2}(I(I+1)-K^2)-\frac{\hbar^2}{8D\beta_0^2}(I^2-K^2)$$

$$\approx\hbar\omega(\tfrac{3}{2}(I-K)+1) \tag{6B-25b}$$

$$H_\beta\approx-\frac{\hbar^2}{2D}\frac{\partial^2}{\partial\beta^2}+2C(\beta-\beta_0)^2 \tag{6B-25c}$$

$$H_\gamma\approx-\frac{\hbar^2}{2D\beta_0^2}\frac{\partial^2}{\partial\gamma^2}+\tfrac{1}{2}C\beta_0^2(\gamma-\gamma_0)^2 \tag{6B-25d}$$

In the expansion of H, we have neglected terms whose contributions to the energies are of order $\hbar\omega I^{-1}$.

The leading-order term H_{rot} in the Hamiltonian (6B-24) describes the energy along the yrast line, while the terms H_K, H_β, and H_γ describe three normal modes of excitation with respect to the yrast line.

The term H_K is associated with changes in the rotational motion leading to states with $K=I-2, I-4,\ldots$; the excitation energy is linear in $I-K$ and may be viewed as a vibrational excitation corresponding to a wobbling motion of the angular momentum with respect to the intrinsic 1 axis (compare the discussion of the asymmetric rotor in Sec. 4-5e). The eigenvalues of H_K can be written

$$E_K=\hbar\omega(\tfrac{3}{2}n_K+1) \tag{6B-26}$$

with the vibrational quantum number

$$n_K=I-K \tag{6B-27}$$

Since K is an even integer (see p. 679), the excitations are associated with rotational bands with

$$\begin{aligned} I \text{ even} \qquad & n_K = 0, 2, 4, \ldots \\ I \text{ odd} \qquad & n_K = 1, 3, 5, \ldots \end{aligned} \tag{6B-28}$$

The terms H_β and H_γ in the Hamiltonian (6B-24) represent oscillations around the equilibrium shape, with eigenvalues

$$E_\beta = 2\hbar\omega\,(n_\beta + \tfrac{1}{2}) \tag{6B-29a}$$

$$E_\gamma = \hbar\omega\,(n_\gamma + \tfrac{1}{2}) \tag{6B-29b}$$

with vibrational quantum numbers n_β and n_γ. Combining Eqs. (6B-25a), (6B-26), and (6B-29), we obtain the total spectrum in the yrast region

$$E(n_\beta, n_\gamma, n_K, I) = \hbar\omega(2n_\beta + n_\gamma + \tfrac{3}{2}n_K + \tfrac{1}{2}I + \tfrac{5}{2}) \tag{6B-30}$$

The bands in the neighborhood of the yrast line are illustrated in Fig. 6B-2; for $I \geqslant \frac{1}{2}I_{\max} = n$, the enumeration of states is the same as obtained from the exact counting given in Table 6-1, p. 347. (This result may be obtained on the basis of the excitations from the aligned condensate discussed on p. 687.)

The coupling scheme in the yrast region implies simple relations for the $E2$-matrix elements. The intrinsic moments with respect to the 1 axis are given by (see Eqs. (6-63) and (6B-2) and note that the intrinsic moments with respect to the 1 axis are obtained from those with respect to the 3 axis by a cyclic permutation of the axes, corresponding to $\gamma \rightarrow \gamma - 2\pi/3$; see, for example, Eq. (6B-4))

$$\begin{aligned} \mathscr{M}(E2, \nu=0) &\approx \left(\frac{3}{4\pi} ZeR^2\right)\beta_0(\gamma - \gamma_0) \\ \mathscr{M}(E2, \nu=2) &\approx -\left(\frac{3}{4\pi} ZeR^2\right)\frac{1}{\sqrt{2}}\beta \end{aligned} \tag{6B-31}$$

in the neighborhood of the equilibrium shape. Thus, the static intrinsic quadrupole moments are

$$\begin{aligned} Q_0 &= 0 \\ \left(\frac{5}{16\pi}\right)^{1/2} Q_2 &= -\frac{1}{\sqrt{2}}\beta_0\left(\frac{3}{4\pi} ZeR^2\right) \end{aligned} \tag{6B-32}$$

which yield, for the $E2$-matrix elements within a rotational sequence,

$$\begin{aligned} Q &= 0 \\ B(E2; n_\beta n_\gamma n_K I \rightarrow n_\beta n_\gamma n_K, I-2) &\approx \frac{I}{2} b^2\left(\frac{3}{4\pi} ZeR^2\right)^2 \end{aligned} \tag{6B-33}$$

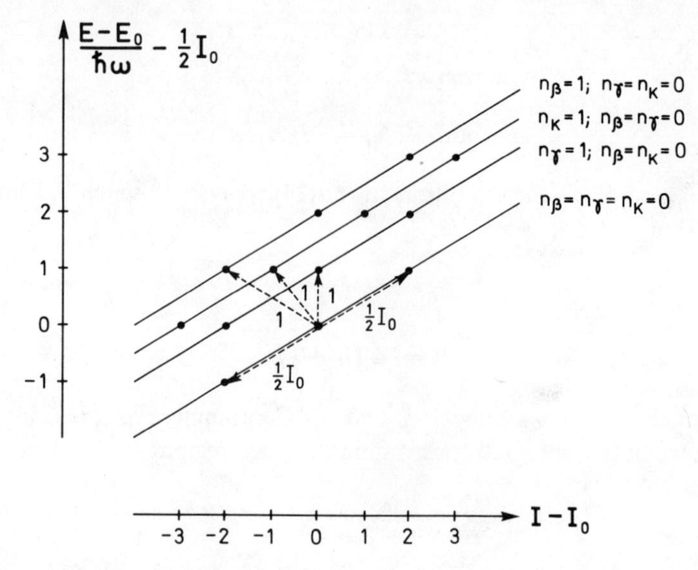

Figure 6B-2 Yrast region for quadrupole vibrations. The figure shows part of the spectrum for harmonic quadrupole vibrations, with angular momentum values $I \approx I_0 \gg 1$ and with energies close to the minimum value for given I (yrast region). The solid lines represent rotational trajectories. The figure includes trajectories associated with the lowest intrinsic state ($n_\beta = n_\gamma = n_K = 0$) as well as trajectories for the intrinsic states involving a single vibrational quantum. (The trajectory with $n_\gamma = 2$, $n_\beta = n_K = 0$ coincides with the $n_\beta = 1$ trajectory, but the states on the $n_\gamma = 2$ trajectory are not shown.) The arrows represent the transitions with nonvanishing $E2$-matrix elements, leading from the state $I = I_0$ on the yrast line; the numbers on the arrows give the $B(E2)$ values for these transitions, in units of $b^2(3ZeR^2/4\pi)^2$.

The vibrational transitions with $\Delta n_K = \pm 1$ have $\Delta K = \mp 2$ and $\Delta I = \mp 1$, and one obtains from the general intensity rule for transitions between rotational bands (see Eq. (4-91))

$$B(E2; n_\beta n_\gamma n_K I \rightarrow n_\beta n_\gamma, n_K - 1, I+1) \approx \tfrac{1}{2}\beta_0^2\left(\frac{3}{4\pi}ZeR^2\right)^2 \langle I\, I - n_K\, 2 2 | I+1\;\; I - n_K + 2\rangle^2$$

$$\approx n_K b^2\left(\frac{3}{4\pi}ZeR^2\right)^2 \qquad \text{(6B-34)}$$

(This result could also have been obtained from Eq. (4-314), which describes the wobbling transitions for the asymmetric rotor.) The vibrational transitions with $\Delta n_\beta = \pm 1$ have $\Delta K = \Delta I = \mp 2$, while the transitions with $\Delta n_\gamma = \pm 1$ have $\Delta K = \Delta I = 0$. From the intrinsic moments (6B-31) and vibrational Hamiltonians (6B-25c) and (6B-25d), we obtain

$$B(E2;n_\beta n_\gamma n_K I\to n_\beta-1,n_\gamma n_K,I+2)\approx n_\beta b^2\left(\frac{3}{4\pi}ZeR^2\right)^2$$
$$B(E2;n_\beta n_\gamma n_K I\to n_\beta,n_\gamma-1,n_K I)\approx n_\gamma b^2\left(\frac{3}{4\pi}ZeR^2\right)^2 \tag{6B-35}$$

The matrix elements for the transitions $\Delta n_\beta=\pm 1$ receive half their value from the contribution resulting from the I dependence of β_0 (see Eq. (6B-22)).

It is seen that the rotational transitions are stronger than those involving a change in the vibrational quantum numbers n_β, n_γ, n_K by a factor of the order of the total angular momentum I. (The pattern of the $E2$ transitions is illustrated in Fig. 6B-2.) The vibrational transitions (6B-34) and (6B-35) constitute only a small fraction of those allowed by the selection rule $\Delta n=1$ for the total number n of vibrational quanta. The additional transitions involve simultaneous changes in several of the quantum numbers n_β, n_γ, n_K and are smaller than those considered by at least a factor of order I^{-1}.

An alternative derivation of the above results can be obtained by noting that in the yrast region, the alignment of the angular momenta of the quadrupole quanta implies the existence of a condensate of identical bosons. In a coordinate system in which the total angular momentum points in the direction of the z axis, the condensate is formed of the quanta with $\mu=\lambda=2$. The strong transitions along the rotational trajectories correspond to the addition of quanta to the condensate, and the enhancement results from the boson factor $(n+1)$ in the transition probability (see Eq. (6-1)). Additional excitations result from adding quanta with $\mu=1$, 0, -1, and -2. The addition of a quantum with $\mu=1$ leads to a state in which the condensate has been rotated with respect to the z axis (since the operator I_x-iI_y acting on the condensate shifts a quantum from $\mu=2$ to $\mu=1$). The addition of a quantum with $\mu=0$, -1, or -2 leads to intrinsic excitations with $\Delta I=\mu$, which can be identified with those labeled above by $n_\gamma=1$, $n_K=1$, and $n_\beta=1$, respectively (see Fig. 6B-2).

The spectrum and transition probabilities derived in the present section refer to harmonic vibrational motion, but the qualitative pattern in the yrast region has a broader range of validity, since it follows directly from the existence of an average deformation that is large compared with the amplitude of the vibrational motion about equilibrium, or equivalently, from the existence of a condensate. The anharmonicity in the vibrational Hamiltonian may affect the value of the equilibrium deformation as well as the parameters associated with the three intrinsic vibrational degrees of freedom, but the enumeration of states and the selection rules for the $E2$ transitions remain valid, provided the vibrational quantum numbers n_β, n_γ, n_K are small compared with the total number of quanta n.

6B-4 Many-Phonon States

In the present section, we consider some of the systematic tools that may be employed in the construction of a complete set of states with a given number of phonons and total angular momentum. The tools considered are not specific to quadrupole quanta, and the formulation is therefore expressed in a form appropriate to phonons of arbitrary multipolarity λ.

6B-4a Parentage coefficients

The construction of states with two quanta of the same mode λ is directly obtained by an angular momentum coupling; the normalized state with total angular momentum I and component M can be written

$$\begin{aligned} |n_\lambda = 2, IM\rangle &= 2^{-1/2}(c^\dagger(\lambda)c^\dagger(\lambda))_{(\lambda\lambda)IM}|0\rangle \\ &= 2^{-1/2}\sum_{\mu'\mu''}\langle\lambda\mu'\lambda\mu''|IM\rangle c^\dagger(\lambda\mu')c^\dagger(\lambda\mu'')|0\rangle \qquad \text{(6B-36)} \\ & I = 0, 2, 4, \ldots, 2\lambda \end{aligned}$$

The factor $2^{-1/2}$ arises from the fact that each substate ($n_{\lambda\mu'} = 1$, $n_{\lambda\mu''} = 1$) occurs twice in the sum in Eq. (6B-36). (If M is even, there is only a single term with $\mu' = \mu'' = M/2$, but the state with $n_{\lambda\mu'} = 2$ involves the normalization factor $2^{-1/2}$ (see Eq. (6-4)).)

For $n_\lambda > 2$, one can proceed by adding further quanta to the states (6B-36). However, one must take into account that states with the same n_λ and I, but constructed in terms of different intermediate angular momenta (such as the angular momentum of the two first quanta), are not in general orthogonal.

A systematic procedure for the construction of many-phonon states and the evaluation of matrix elements may be based on a chain calculation involving the parentage coefficients $\langle \zeta_n I_n \| c^\dagger(\lambda) \| \zeta_{n-1} I_{n-1}\rangle$ connecting states with successive numbers of quanta.[40] The states of n quanta are labeled by the quantum numbers ζ_n, in addition to I_n (and M_n). We have simplified the notation by omitting the subscript λ on the quantum number n, since in the present discussion we consider only states involving a single type of boson.

---

[40] The systematic procedure for calculating parentage coefficients by means of recoupling techniques was developed by Racah (1943) for many-electron configurations of atoms. The coefficient of fractional parentage $\langle \zeta_n I_n M_n \{| \zeta_{n-1} I_{n-1} M_{n-1}, \lambda\mu\rangle$ in the notation of Racah is equal to the quantity $n^{-1/2}\langle \zeta_n I_n M_n | c^\dagger(\lambda\mu) | \zeta_{n-1} I_{n-1} M_{n-1}\rangle$ and, correspondingly, for angular momentum coupled states, $\langle \zeta_n I_n \{| \zeta_{n-1} I_{n-1}, \lambda; I_n\rangle = n^{-1/2}(2I_n+1)^{-1/2}\langle \zeta_n I_n \| c^\dagger(\lambda) \| \zeta_{n-1} I_{n-1}\rangle$. As indicated by the special bracket notation, the state on the right in Racah's fractional parentage coefficient is symmetric with respect to the first $(n-1)$ quanta, but not with respect to the nth, whereas the state on the left is totally symmetric in all n quanta.

We thus assume that an orthogonal set of states $\zeta_{n-1}I_{n-1}$ with $n-1$ quanta has been constructed and proceed by adding a quantum to these states

$$|\zeta_{n-1}I_{n-1},\lambda;I_nM_n\rangle_{\text{unnorm}} \equiv c^\dagger(\lambda)|\zeta_{n-1}I_{n-1}\rangle_{(I_{n-1}\lambda)I_nM_n} \tag{6B-37}$$

The states (6B-37) are in general neither normalized nor orthogonal, but we can evaluate the overlap factors by exploiting the commutation relations (6-47), which can be written

$$\left[c(\bar{\lambda}),c^\dagger(\lambda)\right]_{(\lambda\lambda)\Lambda} = (2\lambda+1)^{1/2}\delta(\Lambda,0) \tag{6B-38}$$

Performing appropriate recouplings (see Sec. 1A-3 and the relation (1A-65) for the scalar product of state vectors), one obtains, for the overlap of two states of the form (6B-37),

$$\begin{aligned}
&\langle\zeta'_{n-1}I'_{n-1},\lambda';I_nM_n|\zeta_{n-1}I_{n-1},\lambda;I_nM_n\rangle_{\text{unnorm}}\\
&\quad=(2I_n+1)^{-1/2}\langle\zeta'_{n-1}\bar{I}'_{n-1}|c(\bar{\lambda}')c^\dagger(\lambda)|\zeta_{n-1}I_{n-1}\rangle_{(I_{n-1}\lambda)I_n,(I'_{n-1}\lambda')I_n;0}\\
&\quad=(2I_n+1)^{-1/2}\langle(I_{n-1}\lambda)I_n,(I'_{n-1}\lambda')I_n;0|(I_{n-1}I'_{n-1})0,(\lambda\lambda')0;0\rangle\\
&\qquad\times\langle\zeta'_{n-1}\bar{I}'_{n-1}|\left[c(\bar{\lambda}'),c^\dagger(\lambda)\right]|\zeta_{n-1}I_{n-1}\rangle_{(I_{n-1}I'_{n-1})0,(\lambda\lambda')0;0}\\
&\qquad+(2I_n+1)^{-1/2}\sum_{I_{n-2}}\langle(I_{n-1}\lambda)I_n,(I'_{n-1}\lambda')I_n;0|(I_{n-1}\lambda')I_{n-2},(I'_{n-1}\lambda)I_{n-2};0\rangle\\
&\qquad\times\langle\zeta'_{n-1}\bar{I}'_{n-1}|c^\dagger(\lambda)c(\bar{\lambda}')|\zeta_{n-1}I_{n-1}\rangle_{(I_{n-1}\lambda')I_{n-2},(I'_{n-1}\lambda)I_{n-2};0}\\
&\quad=\delta(\zeta_{n-1},\zeta'_{n-1})\delta(I_{n-1},I'_{n-1})+\sum_{\zeta_{n-2}I_{n-2}}(2I_{n-2}+1)\begin{Bmatrix}I_{n-1}&\lambda&I_n\\I'_{n-1}&\lambda&I_{n-2}\end{Bmatrix}\\
&\qquad\times\langle\zeta'_{n-1}\bar{I}'_{n-1}|c^\dagger(\lambda)|\zeta_{n-2}I_{n-2}M_{n-2}\rangle_{(I'_{n-1}\lambda)\overline{I_{n-2}M_{n-2}}}\langle\zeta_{n-2}I_{n-2}M_{n-2}|c(\bar{\lambda})|\zeta_{n-1}I_{n-1}\rangle_{(I_{n-1}\lambda)I_{n-2}M_{n-2}}\\
&\quad=\delta(\zeta_{n-1},\zeta'_{n-1})\delta(I_{n-1},I'_{n-1})+\sum_{\zeta_{n-2}I_{n-2}}(-1)^{I_{n-1}-I'_{n-1}}\begin{Bmatrix}I_{n-1}&\lambda&I_n\\I'_{n-1}&\lambda&I_{n-2}\end{Bmatrix}\\
&\qquad\times\langle\zeta'_{n-1}I'_{n-1}\|c^\dagger(\lambda)\|\zeta_{n-2}I_{n-2}\rangle\langle\zeta_{n-1}I_{n-1}\|c^\dagger(\lambda)\|\zeta_{n-2}I_{n-2}\rangle
\end{aligned} \tag{6B-39}$$

The notation λ and λ' distinguishing the two angular momenta of equal magnitude is used when necessary for the specification of the coupling scheme. We have employed the phase relation for the standard representation (compare Eq. (1A-84))

$$\langle \zeta_{n-2} I_{n-2} \| c(\bar{\lambda}) \| \zeta_{n-1} I_{n-1} \rangle = (-1)^{I_{n-2}+\lambda-I_{n-1}} \langle \zeta_{n-1} I_{n-1} \| c^\dagger(\lambda) \| \zeta_{n-2} I_{n-2} \rangle \quad \text{(6B-40)}$$

If the parentage coefficients connecting the states with $n-2$ and $n-1$ phonons are known, one can determine the overlap factors (6B-39) and thus construct an orthogonal basis $|\zeta_n I_n M_n\rangle$ for the states with n phonons. The expansion of the unnormalized states (6B-37) in this basis gives the parentage coefficients for the n-phonon states,

$$\langle \zeta_n I_n \| c^\dagger(\lambda) \| \zeta_{n-1} I_{n-1} \rangle = (2I_n+1)^{1/2} \langle \zeta_n I_n M_n | \zeta_{n-1} I_{n-1}, \lambda; I_n M_n \rangle_{\text{unnorm}} \quad \text{(6B-41)}$$

The parentage factors obey orthonormality relations

$$\sum_{\zeta_{n-1} I_{n-1}} \langle \zeta_n I_n \| c^\dagger(\lambda) \| \zeta_{n-1} I_{n-1} \rangle \langle \zeta'_n I_n \| c^\dagger(\lambda) \| \zeta_{n-1} I_{n-1} \rangle = n(2I_n+1)\delta(\zeta_n, \zeta'_n)$$

$$\sum_{\zeta_n I_n} \langle \zeta_n I_n \| c^\dagger(\lambda) \| \zeta_{n-1} I_{n-1} \rangle \langle \zeta_n I_n \| c^\dagger(\lambda) \| \zeta'_{n-1} I_{n-1} \rangle = (2\lambda+n)(2I_{n-1}+1)\delta(\zeta_{n-1}, \zeta'_{n-1}) \quad \text{(6B-41a)}$$

which follow from the commutation relations (6-47) and the expression (6-48) for the numbers of quanta. (The relations (6B-41a) are the counterpart for bosons to the sum rules (3E-12) and (3E-13) for fermions.)

Matrix elements of the various operators depending on the vibrational variables can be directly evaluated in terms of the parentage coefficients. Thus, the matrix elements of the vibrational coordinates $\alpha_{\lambda\mu}$ and of the associated multipole operators $\mathscr{M}(\lambda\mu)$ are proportional to the parentage coefficient (see, for example, Eq. (6-66)). Operators quadratic in the vibrational coordinates can be expressed in terms of products of two parentage coefficients, as in the example

$$\langle \zeta'_n I'_n \| \left(c^\dagger(\lambda') c(\bar{\lambda}) \right)_{(\lambda\lambda')\Lambda} \| \zeta_n I_n \rangle = \sum_{\zeta_{n-1} I_{n-1}} (-1)^{I_{n-1}+\lambda-I'_n} (2\Lambda+1)^{1/2} \begin{Bmatrix} I_n & \lambda & I_{n-1} \\ \lambda & I'_n & \Lambda \end{Bmatrix}$$

$$\times \langle \zeta'_n I'_n \| c^\dagger(\lambda) \| \zeta_{n-1} I_{n-1} \rangle \langle \zeta_n I_n \| c^\dagger(\lambda) \| \zeta_{n-1} I_{n-1} \rangle \quad \text{(6B-42)}$$

To illustrate the evaluation of parentage coefficients, we consider states with three quadrupole quanta ($n=3$, $\lambda=2$). These states, as well as the states with $n=2$, are completely specified by the total angular momentum I (see Table 6-1, p. 347), and the parentage coefficients are denoted by $\langle I_3 \| c^\dagger(\lambda=2) \| I_2 \rangle$. As a first step, we evaluate the parentage coefficients for $n=2$, which follow from Eq. (6B-36),

$$\langle I_2 \| c^\dagger(\lambda) \| I_1 = \lambda \rangle = (2(2I_2+1))^{1/2} \quad \text{(6B-43)}$$

The coefficients for $n=3$ can now be obtained from Eqs. (6B-39) and (6B-41)

$$(2I_3+1)^{-1}\langle I_3\|c^\dagger(\lambda=2)\|I_2\rangle\langle I_3\|c^\dagger(\lambda=2)\|I_2'\rangle$$

$$=\delta(I_2,I_2')+2(2I_2+1)^{1/2}(2I_2'+1)^{1/2}\begin{Bmatrix} I_2 & 2 & I_3 \\ I_2' & 2 & 2 \end{Bmatrix} \tag{6B-44}$$

Setting $I_2'=I_2$, the absolute magnitude of the parentage coefficients can be determined; the relative phases can be found by considering the relation (6B-44) for $I_2'\neq I_2$. The parentage coefficients $\langle I_3\|c^\dagger(\lambda=2)\|I_2\rangle$ thus obtained are given in Table 6B-1. For each state I_3, there is an arbitrary phase which, in the table, has been chosen so that the parentage coefficients are positive for the smallest contributing value of I_2.

	I_2		
I_3	0	2	4
0		$3^{1/2}$	
2	$7^{1/2}$	$(20/7)^{1/2}$	$(36/7)^{1/2}$
3		$15^{1/2}$	$-6^{1/2}$
4		$(99/7)^{1/2}$	$(90/7)^{1/2}$
6			$39^{1/2}$

Table 6B-1 Parentage coefficients for states with three quadrupole quanta. The table lists the coefficients $\langle I_3\|c^\dagger(\lambda=2)\|I_2\rangle$. These coefficients were first evaluated by D. C. Choudhury, *Mat. Fys. Medd. Dan. Vid. Selsk.* **28**, no. 4 (1954). A tabulation of parentage coefficients extending to states with $n_2\leqslant 7$, $n_3\leqslant 6$, and $n_4\leqslant 5$ is given by Bayman and Landé (1966).

6B-4b *Seniority*[41]

The many-phonon states with specified total angular momentum can be further classified by the number of pairs of quanta with angular momenta coupled to zero. Such a pair is created by the operator (see Eq. (6B-36))

$$\mathscr{C}_\lambda^\dagger=2^{-1/2}(c^\dagger(\lambda)c^\dagger(\lambda))_{(\lambda\lambda)0} \tag{6B-45}$$

---

[41]The seniority classification of many-particle states was introduced by Racah (1943). The application to nuclear vibrations was considered by Rakavy (1957a).

A state that vanishes when acted upon by the pair annihilation operator $\mathscr{C}_\lambda$ is said to consist of only unpaired quanta, and the number of these quanta is referred to as the seniority, v. From a state consisting of v unpaired quanta, one can generate a family of states by acting successively with the pair creation operator $\mathscr{C}_\lambda^\dagger$; these states are labeled by the same seniority v and the same quantum numbers ζ_v of the unpaired particles and differ only in the number of pairs $p=(n-v)/2$. By means of the commutation relation

$$[\mathscr{C}_\lambda,\mathscr{C}_\lambda^\dagger]=(2\lambda+1)^{-1}\sum_\mu\left(c(\lambda\mu)c^\dagger(\lambda\mu)+c^\dagger(\lambda\mu)c(\lambda\mu)\right)$$
$$=1+\frac{2n}{2\lambda+1} \tag{6B-46}$$

it can be verified that the states $|\zeta_v,p\rangle$ form a complete orthonormal set, and the normalized states can be written in the form

$$|\zeta_v,p\rangle=\left(\frac{(2\lambda+1)^p(2\lambda+2v-1)!!}{p!(2\lambda+2v+2p-1)!!}\right)^{1/2}(\mathscr{C}_\lambda^\dagger)^p|\zeta_v,p=0\rangle \tag{6B-47}$$

with the notation

$$(2s+1)!!=(2s+1)(2s-1)\cdots3\cdot1 \tag{6B-48}$$

The number of states $g(nvI)$ of given n,v,and angular momentum I, is equal to the difference $g(n=v,I)-g(n=v-2,I)$, where $g(nI)$ is the number of states with given n and I. From this rule, one can assign seniority quantum numbers to the states of the quadrupole spectrum listed in Table 6-1, p. 347. It is seen that the set of quantum numbers n, v, I, and M is sufficient to specify all states with $n\leqslant5$, but that there occur two states with $n=6$, $v=6$, and $I=6$. (A full classification of the quadrupole spectrum requires five quantum numbers, such as the $n_{2\mu}$.)

From the properties of the states with $n=v$, the corresponding matrix elements involving states with additional pairs can be obtained from the commutation relations (6B-46) and

$$[\mathscr{C}_\lambda,c^\dagger(\lambda\mu)]=\left(\frac{2}{2\lambda+1}\right)^{1/2}c(\overline{\lambda\mu}) \tag{6B-49}$$

In particular, the recursion relations for the parentage coefficients

$$\langle\zeta_{v+1},p\|c^\dagger(\lambda)\|\zeta_v,p\rangle=\left(\frac{2\lambda+2v+2p+1}{2\lambda+2v+1}\right)^{1/2}\langle\zeta_{v+1},p=0\|c^\dagger(\lambda)\|\zeta_v,p=0\rangle$$
$$\langle\zeta_{v-1},p+1\|c^\dagger(\lambda)\|\zeta_v,p\rangle=\left(\frac{2p+2}{2\lambda+2v-1}\right)^{1/2}\langle\zeta_{v-1},p=0\|c(\overline{\lambda})\|\zeta_v,p=0\rangle \tag{6B-50}$$

can be derived on this basis (Le Tourneux, 1965; for a further discussion of recursion relations and the classification of states of quadrupole vibrations, see Weber *et al*., 1966; Kishimoto and Tamura, 1971).

BIBLIOGRAPHY

The bibliography includes the references quoted in this volume as well as in Volume I. The numbers in brackets refer to the pages on which the quotations appear. In the case of textbooks and compilations, some of which are quoted extensively, the *loci citati* have been omitted.

Abramowitz, M., and Stegun I. A. (1964), *Handbook of Mathematical Functions*, National Bureau of Standards, Washington, D.C.

Adler, S. L. (1965), *Phys. Rev.* **140**, B736. [I:401]

Ahrens, J., Borchert, H., Eppler, H. B., Gimm, H., Gundrum, H., Riehn, P., Ram, G. S., Zieger, A., Kröning, M., and Ziegler, B. (1972), in *Nuclear Structure Studies Using Electron Scattering and Photoreactions*, p. 213, eds. K. Shoda and H. Ui, Tohoku Univ., Sendai, Japan. [II:478, 482]

Ajzenberg-Selove, F. (1971), *Nuclear Phys.* **A166**, 1.

Ajzenberg-Selove, F. (1972), *Nuclear Phys.* **A190**. 1.

Ajzenberg-Selove, F., and Lauritsen, T. (1959), *Nuclear Phys.* **11**, 1.

Ajzenberg-Selove, F., and Lauritsen, T. (1966), *Nuclear Phys.* **78**, 1.

Ajzenberg-Selove, F., and Lauritsen, T. (1968), *Nuclear Phys.* **A114**, 1.

Alaga, G. (1955), *Phys. Rev.* **100**, 432. [II:246]

Alaga, G. (1969), *Cargèse Lectures in Physics* **3**, 579. ed. M. Jean, Gordon and Breach, New York, N.Y. [II:433]

Alaga, G., and Ialongo, G. (1967), *Nuclear Phys.* **A97**, 600. [II:433]

Alaga, G., Alder, K., Bohr, A., and Mottelson, B. R. (1955), *Mat. Fys. Medd. Dan. Vid. Selsk.* **29**, no.9. [II:2, 22]

Alburger, D. E., Chasman, C., Jones, K. W., Olness, J. W., and Ristinen, R. A. (1964), *Phys. Rev.* **136**, B916. [II:285]

Alburger, D. E., Gallmann, A., Nelson, J. B., Sample, J. T., and Warburton, E. K. (1966), *Phys. Rev.* **148**, 1050. [I:43]

Alder, K., and Winther, A. (1966), *Coulomb Excitation*, Academic Press, New York, N.Y.

Alder, K., Bohr, A., Huus, T., Mottelson, B., and Winther, A. (1956), *Rev. Mod. Phys.* **28**, 432. [II:2, 67, 359]

Alexander, P., and Boehm, F. (1963), *Nuclear Phys.* **46**, 108. [II:103]

Alexander, T. K., and Allen, K. W. (1965), *Can. J. Phys.* **43**, 1563. [II:561]

Alexander, T. K., Häusser, O., McDonald, A. B., and Ewan, G. T. (1972), *Can. J. Phys.* **50**, 2198. [II:291]

Alexander, T. K., Häusser, O., McDonald, A. B., Ferguson, A. J., Diamond, W. T., and Litherland, A. E. (1972), *Nuclear Phys.* **A179**, 477. [II:98]

Alfvén, H. (1965), *Rev. Mod. Phys.* **37**, 652. [I:202]

Alikhanov, A. I., Galatnikov, Yu. V., Gorodkov, Yu. V., Eliseev, G. P., and Lyubimov, V. A. (1960), *Zhur. Eksp. i Teoret. Fiz.* **38**, 1918; transl. *Soviet Phys. JETP* **11**, 1380. [I:27]

Alkazov, D. G., Gangrskii, Y. P., Lemberg, I. K., and Undralov, Y. I. (1964), *Izv. Akad. Nauk, Ser. Fiz.* **28**, 232. [II:354]

Aller, L. H. (1961), *The Abundance of Elements*, Wiley (Interscience), New York, N.Y. [I:206]

Alster, J. (1966), *Phys. Rev.* **141**, 1138. [II:571]

Alvensleben, H., Becker, U., Bertram, W. K., Chen, M., Cohen, K. J., Knasel, T. M., Marshall, R., Quinn, D. J., Rohde, M., Sanders, G. H., Schubel, H., and Ting, S. C. C. (1970), *Nuclear Phys.* **B18**, 333. [II:480]

Amati, D. (1964), in *Compt. Rend. du Congrès Intern. de Physique Nucléaire*, vol. I, p. 57, ed. P. Gugenberger, C.N.R.S., Paris. [I:241, 268]

Ambler, E., Fuller, E. G., and Marshak, H. (1965), *Phys. Rev.* **138**, B117. [II:491]

Amusia, M. Ya., Cherepkov, N. A., and Cheruysheva, L. V. (1971), *J. Exptl. Theoret. Phys. (USSR)*, **60**, 160; transl. *Soviet Phys. JETP* **33**; 90 [II:352]

Andersen, B. L., (1968) *Nuclear Phys.* **A112**, 443. [II:230]

Andersen, B. L. (1972), *Nuclear Phys.* **A196**, 547. [II:230]

Andersen, B. L., Bondorf, J. P., and Madsen, B. S. (1966), *Phys. Letters* **22**, 651. [I:355]

Anderson, J. D., and Wong, C. (1961), *Phys. Rev. Letters* **7**, 250. [I:36]

Anderson, J. D., Wong, C., and McClure, J. W. (1965), *Phys. Rev.* **138**, B615. [I:145]

Anderson, P. W. (1964), in *Lectures on the Many-Body Problem* **2**, 113, ed. E. R. Caianiello, Academic Press, New York, N.Y. [II:396]

Anderson, P. W. (1969), in *Superconductivity*, vol. **2**, p. 1343, ed. R. D. Parks, Marcel Dekker, New York, N.Y. [II:396]

Andreev, D. S., Gangrskij, Yu. P., Lemberg, I. Ch., and Nabičvižvili, V. A. (1965), *Izv. Akad. Nauk* **29**, 2231. [I:341]

Anyas-Weiss, N., and Litherland, A. E. (1969), *Can. J. Phys.* **47**, 2609. [II:291]

Araújo, J. M. (1956), *Nuclear Phys.* **1**, 259. [II:436]

Araújo, J. M. (1959), *Nuclear Phys.* **13**, 360. [II:436]

Arenhövel, H., Danos, M., and Greiner, W. (1967), *Phys. Rev.* **157**, 1109. [II:505]

Arima, A., and Horie, H. (1954), *Progr. Theoret. Phys. (Kyoto)* **12**, 623. [I:338]

Arndt, R. A., and MacGregor, M. H. (1966), *Phys. Rev.* **141**, 873. [I:265]

Arvieu, R., and Vénéroni, M. (1960), *Compt. rend.* **250**, 992 and 2155; also *ibid.* (1961), **252**, 670. [II:436]

Asaro, F., and Perlman, I. (1952), *Phys. Rev.* **87**, 393. [I:197]

Asaro, F., and Perlman, I. (1953), *Phys. Rev.* **91**, 763. [II:2]

Ascuitto, R. J., King, C. H., McVay, L. J., and Sørensen, B. (1974), *Nuclear Phys.* **A226**, 454. [II:258]

Auerbach, E. H., Dover, C. B., Kerman, A. K., Lemmer, R. H., and Schwarcz, E. H. (1966), *Phys. Rev. Letters* **17**, 1184. [I:51]

Auerbach, E. H., Kahana, S., and Weneser, J. (1969), *Phys. Rev. Letters* **23**, 1253. [II:295]

Austern, N. (1963), in *Selected Topics in Nuclear Theory*, p. 17, ed. F. Janouch, Intern. Atomic Energy Agency, Vienna. [I:425]

Austern, N., and Blair, J. S. (1965), *Ann. Phys.* **33**, 15. [II:322]

Axel, P., Drake, D. M., Whetstone, S., and Hanna, S. S. (1967), *Phys. Rev. Letters* **19**, 1343. [II:498]

Azhgirey, L. S., Klepikov, N. P., Kumekin, Yu. P., Mescheryakov, M. G., Nurushev, S. B., and Stoletov, G. D. (1963), *Phys. Letters* **6**, 196. [I: 263]

Baader, H. A. (1970), *dissertation*, Techn. Univ. München. [II: 463]

Back, B. B., Bang, J., Bjørnholm, S., Hattula, J., Keinheinz, P., and Lien, J. R. (1974), *Nuclear Phys.* **A222**, 377. [II: 302]

Back, B. B., Britt, H. C., Hansen, O., Leroux, B., and Garrett, J. D. (1974), *Phys. Rev.* **C10**, 1948. [II: 623]

Backus, G., and Gilbert, F. (1961), *Proc. Nat. Acad. Sci. USA* **47**, 362. [II: 657]

Bahcall, J. N. (1966), *Nuclear Phys.* **75**, 10. [I: 349]

Bakke, F. H. (1958), *Nuclear Phys.* **9**, 670. [II: 265]

Balashov, V. V., and Eramzhyan, R. A. (1967), *Atomic Energy Review* **5**, 3. [I: 405]

Balashov, V. V., Belyaev, V. B., Eramjian, R. A., and Kabachnik, N. M. (1964), *Phys. Letters* **9**, 168. [II: 497]

Baldin, A. M. (1959), *J. Exptl. Theoret. Phys. (USSR)*, **37**, 202; transl. *Soviet Phys. JETP* (1960), **10**, 142. [II: 458]

Baldwin, G. C., and Klaiber, G. S. (1947), *Phys. Rev.* **71**, 3. [II: 474]

Baldwin, G. C., and Klaiber, G. S. (1948), *Phys. Rev.* **73**, 1156 [II: 474]

Balian, R., and Werthamer, N. R. (1963), *Phys. Rev.* **131**, 1553. [II: 21]

Balian, R., and Bloch, C. (1971), *Ann. Phys.* **69**, 76. [II: 579]

Ballhausen, C. J. (1962), *Introduction to Ligand Field Theory*, McGraw Hill, New York, N.Y.

Baranger, M. (1960), *Phys. Rev.* **120**, 957. [II: 436]

Baranger, M. (1967), in *Proc. Intern. Nuclear Physics Conf.*, Gatlinburg, p. 659, ed.-in-chief R. L. Becker, Academic Press, New York, N.Y. [I: 318]

Baranov, S. A., Kulakov, V. M., and Shatinsky, V. M. (1964), *Nuclear Phys.* **56**, 252. [II: 266]

Bardeen, J., Cooper, L. N., and Schrieffer, J. R. (1957), *Phys. Rev.* **106**, 162 and **108**, 1175. [II: 396, 641, 649]

Bardeen, J., Baym, G., and Pines, D. (1967), *Phys. Rev.* **156**, 207. [II: 416]

Bardin, R. K., Gollon, P. J., Ullman, J. D., and Wu, C. S. (1967), *Phys. Letters* **26B**, 112. [I: 398]

Barlow, J., Sens, J. C., Duke, P. J., and Kemp, M. A. R. (1964), *Phys. Letters* **9**, 84. [II: 497]

Barnard, E., Ferguson, A. T. G., McMurray, W. R., and van Heerden, I. J. (1966), *Nuclear Phys.* **80**, 46. [I: 186]

Barnes, P. D., Ellegaard, C., Herskind, B., and Joshi, M. C. (1966), *Phys. Letters* **23B**, 266. [II: 533]

Barnes, P. D., Comfort, J. R., and Bockelman, C. K. (1967), *Phys. Rev.* **155**, 1319. [II: 539]

Barnes, P. D., Romberg, E., Ellegaard, C., Casten, R. F., Hansen, O., Mulligan, T. J., Broglia, R. A., and Liotta, R. (1972), *Nuclear Phys.* **A195**, 146. [II: 573]

Barnett, A. R., and Phillips, W. R. (1969), *Phys. Rev.* **186**, 1205. [II: 569]

Bartholomew, G. A. (1960), *Nuclear Spectroscopy*, part A, p. 304, ed. F. Ajzenberg-Selove, Academic Press, New York, N.Y. [I: 179]

Batty, C. J., Gillmore, R. S., and Stafford, G. H. (1966), *Nuclear Phys.* **75**, 599. [I: 145]

Baumgartner, E., Conzett, H. E., Shield, E., and Slobodrian, R. J. (1966), *Phys. Rev. Letters* **16**, 105. [I: 241]

Bayman, B. F. (1957), *Groups and their Application to Spectroscopy*, NORDITA Lecture Notes, NORDITA, Copenhagen. See also 2nd edition (1960).

Bayman, B. F. (1966), *Am. J. Phys.* **34**, 216. [I:33]
Bayman, B. F., and Lande, A. (1966), *Nuclear Phys.* **77**, 1. [II:691]
Baz', A. I., Gol'danskii, V. I., and Zel'dovich, Ya. B. (1960), *Usp. Fiz., Nauk* **72**, 211; transl. *Soviet Phys. Uspekhi* **3**, 729. [I:204]
Bearse, R. C., Youngblood, D. H., and Segel, R. E. (1968), *Nuclear Phys.* **A111**, 678. [I:342]
Becker, J. A., and Wilkinson, D. H. (1964), *Phys. Rev.* **134**, B1200. [I:341]
Beer, G. A., Brix, P., Clerc, H.-G., and Laube, B. (1968), *Phys. Letters* **26**, B506. [I:341, 343]
Bég, M. A. B., Lee, B. W., and Pais, A. (1964), *Phys. Rev. Letters* **13**, 514. [I:386]
Bell, J. S. (1959), *Nuclear Phys.* **12**, 117 [I:369]
Bell, R. E., Bjørnholm, S., and Severiens, J. C. (1960), *Mat. Fys. Medd. Dan. Vid. Selsk.* **32**, no. 12. [II:133]
Belote, T. A., Sperduto, A., and Buechner, W. W. (1965), *Phys. Rev.* **139**, B80. [I:356]
Belyaev, S. T. (1959), *Mat. Fys. Medd. Dan. Vid. Selsk.* **31**, no. 11. [II:82, 436, 520, 641]
Belyaev, S. T. (1963), in *Selected Topics in Nuclear Theory*, p. 291, ed. F. Janouch, Intern. Atomic Energy Agency, Vienna. [II:355]
Belyaev, S. T., and Zelevinsky, V. G. (1962), *Nuclear Phys.* **39**, 582. [II:433, 451]
Belyaev, S. T., and Zelevinsky, V. G. (1970), *Yadernaya Fiz.* **11**, 741; transl. *Soviet J. Nuclear Phys.* **11**, 416. [II:412]
Bemis, C. E., Jr., McGowan, F. K., Ford, J. L. C., Jr., Milner, W. T., Stelson, P. H., and Robinson, R. L. (1973), *Phys. Rev.* **C8**, 1466. [II:141]
Benn, J., Dally, E. B., Muller, H. H., Pixley, R. E., and Winkler, H. (1966), *Phys. Letters* **20**, 43. [II:101]
Berant, Z., Eisenstein, R. A., Horowitz, Y., Smilansky, U., Tandon, P. N., Greenberg, J. S., Kleinfeld, A. M., and Maggi, H. G. (1972), *Nuclear Phys.* **A196**, 312. [II:532]
Bergère, R., Beil, H., and Veyssière, A. (1968), *Nuclear Phys.* **A121**, 463. [II:493]
Berggren, T., and Jacob G. (1963), *Nuclear Phys.* **47**, 481. [I:233]
Berman, B. L., Kelly, M. A., Bramblett, R. L., Caldwell, J. T., Davis, H. S., and Fultz, S. C. (1969), *Phys. Rev.* **185**, 1576. [II:493]
Bernardini, G. (1966), in *Proc. Intern. School of Physics "Enrico Fermi"*, Course 32, Academic Press, New York, N.Y. [I:398]
Bernstein, A. M. (1969), *Advances in Nuclear Physics*, **3**, 325, eds. M. Baranger and E. Vogt, Plenum Press, New York, N. Y. [II:383, 556, 560]
Bernstein, J., Feinberg, G., and Lee, T. D. (1965), *Phys. Rev.* **139**, B1650. [I:390]
Bernthal, F. M. (1969), UCRL-18651, University of California, Berkeley, Cal. [II:109]
Bernthal, F. M., and Rasmussen, J. O. (1967), *Nuclear Phys.* **A101**, 513. [II:110, 111]
Bernthal, F. M., Rasmussen, J. O., and Hollander, J. M. (1972), in *Radioactivity in Nuclear Spectroscopy*, p. 337, eds. J. H. Hamilton and J. C. Manthuruthil, Gordon and Breach, New York, N.Y. [II:549]
Bertozzi, W., Cooper, T., Ensslin, N., Heisenberg, J., Kowalski, S., Mills, M., Turchinetz, W., Williamson, C., Fivozinsky, S. P., Lightbody, J. W., Jr., and Penner, S. (1972), *Phys. Rev. Letters* **28**, 1711. [II:135, 140]
Bès, D. R. (1961), *Mat. Fys. Medd. Dan. Vid. Selsk.* **33**, no. 2. [II:451]
Bès, D. R., and Broglia, R. A. (1966), *Nuclear Phys.* **80**, 289. [II:386]
Bès, D. R., and Cho, Yi-chung (1966), *Nuclear Phys.* **86**, 581. [II:255]

Bès, D. R., and Dussel, G. G. (1969), *Nuclear Phys.* **A135**, 1. [II:544]
Bès, D. R., and Broglia, R. A. (1971), *Phys. Rev.* **3C**, 2349 and 2389. [II:571]
Bès, D. R., Dussel, G. G., Broglia, R. A., Liotta, R., and Mottelson, B. R. (1974), *Phys. Letters* **52B**, 253. [II:429]
Bethe, H. A. (1937), *Rev. Mod. Phys.* **9**, 69. [I:154, 156]
Bethe, H. A. (1967), in *Proc. Intern. Nuclear Physics Conf.*, Gatlinburg, p. 625, ed.-in-chief R. L. Becker, Academic Press, New York, N.Y. [I:255, 257]
Bethe, H. A. (1968), in *Proc. Intern. Conf. on Nuclear Structure, J. Phys. Soc. Japan* **24**, Suppl., p. 56. [I:262]
Bethe, H. A., and Bacher, R. F. (1936), *Rev. Mod. Phys.* **8**, 82. [I:141, 152]
Bethe, H. A., and Rose, M. E. (1937), *Phys. Rev.* **51**, 283. [II:475]
Bhaduri, R. K., and Ross, C. K. (1971), *Phys. Rev. Letters* **27**, 606. [II:606]
Biedenharn, L. C., and Brussaard, P. J. (1965), *Coulomb Excitation*, Clarendon Press, Oxford.
Biedenharn, L. C., and van Dam, H. (1965), *Quantum Theory of Angular Momentum*, Academic Press, New York, N.Y.
Bjerregaard, J. H., Hansen, O., Nathan, O., and Hinds, S. (1966), *Nuclear Phys.* **86**, 145. [II:535]
Bjerregaard, J. H., Hansen, O., Nathan, O., and Hinds, S. (1966a), *Nuclear Phys.* **89**, 337. [II:386]
Bjerregaard, J. H., Hansen, O., Nathan, O., Stock, R., Chapman, R., and Hinds, S. (1967), *Phys. Letters* **24B**, 568. [I:323]
Bjerrum, N. (1912), in *Nernst Festschrift*, p. 90, Knapp, Halle. [II:2]
Bjørnholm, S., Boehm, F., Knutsen, A. B., and Nielsen, O. B. (1963), *Nuclear Phys.* **42**, 469. [II:553]
Bjørnholm, S., Dubois, J., and Elbek, B. (1968), *Nuclear Phys.* **A118**, 241. [II:554]
Bjørnholm, S., Borggreen, J., Davies, D., Hansen, N. J. S., Pedersen, J., and Nielsen, H. L. (1968), *Nuclear Phys.* **A118**, 261. [II:554]
Bjørnholm, S., Bohr, A., and Mottelson, B. R. (1974), *Physics and Chemistry of Fission*, vol. I, p. 367, Intern. Atomic Energy Agency, Vienna. [II:38]
Blair, J. S. (1959), *Phys. Rev.* **115**, 928. [II:322]
Blair, J. S. (1960), in *Proc. Intern. Conf. on Nuclear Structure*, p. 824, eds. D. A. Bromley and E. W. Vogt, Univ. of Toronto Press, Toronto, Canada. [II:354]
Blair, J. S., Sharp, D., and Wilets, L. (1962), *Phys. Rev.* **125**, 1625. [II:322]
Blatt, J. M., and Weisskopf, V. F. (1952), *Theoretical Nuclear Physics*, Wiley, New York, N. Y.
Bleuler, K. (1966), *Proc. Intern. School of Physics "Enrico Fermi", Course* 36, p. 464. [II:14]
Blin-Stoyle, R. J. (1960), *Phys. Rev.* **118**, 1605. [I:25, 67]
Blin-Stoyle, R. J. (1964), in *Selected Topics in Nuclear Spectroscopy*, p. 213, ed. B. J. Verhaar, North-Holland, Amsterdam. [I:54]
Blin-Stoyle, R. J., and Perks, M. A. (1954), *Proc. Phys. Soc.* (*London*) **67A**, 885. [I:338]
Blin-Stoyle, R. J., and Rosina, M. (1965), *Nuclear Phys.* **70**, 321. [I:418]
Bloch, C. (1954), *Phys. Rev.* **93**, 1094. [I:281]
Blomqvist, J. (1970), *Phys. Letters* **33B**, 541. [II:570]
Blomqvist, J., and Wahlborn, S. (1960), *Arkiv Fysik* **16**, 545. [I:240, 325, 326, 327, 344, 352]
Bochnacki, Z., and Ogaza, S. (1966), *Nuclear Phys.* **83**, 619. [II:304]
Bodansky, D., Eccles, S. F., Farwell, G. W., Rickey, M. E., and Robinson, P. C. (1959), *Phys. Rev. Letters* **2**, 101. [I:28]
Bodansky, D., Braithwaite, W. J., Shreve, D. C., Storm, D. W., and Weitkamp, W. G. (1966), *Phys. Rev. Letters* **17**, 589. [I:29]

Bodenstedt, E., and Rogers, J. D. (1964), in *Perturbed Angular Correlations*, eds. E. Karlsson, E. Matthias, and K. Siegbahn, North-Holland, Amsterdam. [II : 305]

Boehm, F., Goldring, G., Hagemann, G. B., Symons, G. D., and Tveter, A. (1966), *Phys. Letters* **22B**, 627. [II : 57, 255, 303]

Boerner, H. (1963), *Representations of Groups*, North-Holland, Amsterdam; Wiley, New York, N.Y.

Bogoliubov, N. (1947), *J. Phys. (USSR)* **11**, 23. [II : 396]

Bogoliubov, N. N. (1958), *J. Exptl. Theoret. Phys. (USSR)*, **34**, 58; transl. *Soviet Phys. JETP* **7**, 41 and *Nuovo cimento* **7**, 794. [II : 648]

Bohm, D., and Pines, D. (1953), *Phys. Rev.* **92**, 609. [II : 439]

Bohr, A. (1951), *Phys. Rev.* **81**, 134. [II : 2]

Bohr, A. (1952), *Mat. Fys. Medd. Dan. Vid. Selsk.* **26**, no. 14. [II : vi, 418, 677, 682]

Bohr, A. (1954), *Rotational States of Atomic Nuclei*, Munksgaard, Copenhagen. [II : 510]

Bohr, A. (1956), in *Proc. Intern. Conf. on the Peaceful Uses of Atomic Energy*, vol. 2, p. 151. United Nations, New York, N.Y. [II : 123, 124]

Bohr, A. (1961), in *Lectures in Theoretical Physics*, vol. 3, p. 1, eds. W. E. Britten, B. W. Downs, and J. Downs, Wiley (Interscience), New York, N.Y. [I : 5]

Bohr, A. (1964), *Compt. Rend. du Congrès Intern. de Physique Nucléaire*, vol. I, p. 487, ed. P. Gugenberger, C.N.R.S., Paris. [II : 386]

Bohr, A. (1968), *Nuclear Structure*, Dubna Symposium, p. 179, Intern. Atomic Energy Agency, Vienna. [II : 386]

Bohr, A., and Mottelson, B. R. (1953), *Mat. Fys. Medd. Dan. Vid. Selsk.* **27**, no. 16. [I : 302, 334; II : vi, 33, 359, 540]

Bohr, A., and Mottelson, B. R. (1953a), *Phys. Rev.* **89**, 316. [II : 2]

Bohr, A., and Mottelson, B. R. (1953b), *Phys. Rev.* **90**, 717. [II : 2]

Bohr, A., and Mottelson, B. R. (1955), *Mat. Fys. Medd. Dan. Vid. Selsk.* **30**, no. 1. [II : 78, 641]

Bohr, A., and Mottelson, B. R. (1955a), in *Beta-, and Gamma-Ray Spectroscopy*, p. 468, ed. K. Siegbahn, North-Holland, Amsterdam. [II : 540]

Bohr, A., and Mottelson, B. R. (1958), *Kongel. Norske Vid. Selsk. Forhandl.*, **31**, 1. See also discussion in *Rehovoth Conference on Nuclear Structure*, Proceedings, p. 149, ed. H. J. Lipkin, North-Holland, Amsterdam. [II : 93, 210]

Bohr, A., and Mottelson, B. R. (1962), *Phys. Rev.* **125**, 495. [II : 397]

Bohr, A., and Mottelson, B. (1963), *Atomnaya Energiya* **14**, 41. [II : 23]

Bohr, A., Fröman, P. O., and Mottelson, B. R. (1955), *Mat. Fys. Medd. Dan. Vid. Selsk.* **29**, no. 10. [II : 22, 116]

Bohr, A., Mottelson, B. R., and Pines, D. (1958), *Phys. Rev.* **110**, 936. [II : 395, 641]

Bohr, A., Damgaard, J., and Mottelson, B. R. (1967), in *Nuclear Structure*, p. 1, eds. A. Hossain, Harun-ar-Rashid, and M. Islam, North-Holland, Amsterdam. [I : 174]

Bohr, N. (1911), *Studier over Metallernes Elektrontheori*, dissertation, Thaning og Appel, Copenhagen; transl. in *Niels Bohr, Collected Works*, vol. 1, p. 291, ed. J. Rud Nielsen, North-Holland 1972. [II : 79]

Bohr, N. (1918), *Mat. Fys. Skr. Dan. Vid. Selsk.*, Afd. 8, Raekke IV, 1. [II : 583]

Bohr, N. (1936), *Nature* **137**, 344. [I : 156, 185]

Bohr, N., and Kalckar, F. (1937), *Mat. Fys. Medd. Dan. Vid. Selsk.* **14**, no. 10. [I : 154; II : v, 507, 654]

Bohr, N., and Wheeler, J. A. (1939), *Phys. Rev.* **56**, 426. [II : 365, 372, 374, 641, 654, 660, 662]

Bohr, N., Peierls, R., and Placzek, G. (1939), *Nature* **144**, 200. [I : 186]
Bollinger, L. M., and Thomas, G. E. (1964), *Phys. Letters* **8**, 45. [I : 177]
Bollinger, L. M., and Thomas, G. E. (1970), *Phys. Letters* **32B**, 457. [II : 120]
Bondorf, J. P., Lütken, H., and Jägare, S. (1966), *Phys. Letters* **21**, 185. [I : 51]
Borggreen, J., Hansen, N. J. S., Pedersen, J., Westgård, L., Żylicz, J., and Bjørnholm, S. (1967), *Nuclear Phys.* **A96**, 561. [II : 73]
Born, M. (1925), *Vorlesungen über Atommechanik*, Springer, Berlin. [II : 584]
Born, M., and Oppenheimer, R. (1927), *Ann. Physik* **84**, 457. [II : 3]
Bothe, W., and Gentner, W. (1939), *Z. Physik* **112**, 45. [II : 474]
Bowen, P. H., Scanlon, J. P., Stafford, G. H., Thresher, J. J., and Hodgson, P. E. (1961), *Nuclear Phys.* **22**, 640. [I : 237]
Bowman, C. D., Auchampaugh, G. F., and Fultz, S. C. (1964), *Phys. Rev.* **133**, B676. [II : 493]
Bowman, C. D., Baglan, R. J., Berman, B. L., and Phillips, T. W. (1970), *Phys. Rev. Letters* **25**, 1302. [II : 639]
Bowman, J. D., de Boer, J., and Boehm, F. (1965), *Nuclear Phys.* **61**, 682. [II : 104]
Bowman, J. D., Zawislak, F. C., and Kaufmann, E. N. (1969), *Phys. Letters* **29B**, 226. [II : 357]
Braid, T. H., Chasman, R. R., Erskine, J. R., and Friedman, A. M. (1970), *Phys. Rev.* **C1**, 275. [II : 274]
Bransden, B. H. (1965), in *Advances in Atomic and Molecular Physics*, vol. 1, p. 85, eds. D. R. Bates and I. Estermann, Academic Press, New York, N.Y. [I : 425]
Braunschweig, D., Tamura, T., and Udagawa, T. (1971), *Phys. Letters* **35B**, 273. [II : 293]
Breit, G. (1958), *Rev. Mod. Phys.* **30**, 507. [I : 162, 384]
Breit, G., and Wigner, E. (1936), *Phys. Rev.* **49**, 519. [I : 431]
Breit, G., Condon, E. U., and Present, R. D. (1936), *Phys. Rev.* **50**, 825. [I : 31, 241]
Breit, G., Hull, M. H., Jr., Lassila, K. E., Pyatt, K. D., Jr., and Ruppel, H. M. (1962), *Phys. Rev.* **128**, 826. [I : 265, 271]
Brene, N., Veje, L., Roos, M., and Cronström, C. (1966), *Phys. Rev.* **149**, 1288. [I : 402]
Brenig, W. (1965), *Advances in Theoretical Physics*, vol. 1, ed. K. A. Brueckner, Academic Press, New York, N.Y. [II : 483]
Brentano, P. von, Dawson, W. K., Moore, C. F., Richard, P., Wharton, W., and Wieman, H. (1968), *Phys. Letters* **26B**, 666. [I : 324]
Brink, D. M. (1957), *Nuclear Phys.* **4**, 215. [II : 475]
Brink, D. M., De Toledo Piza, A. F. R., and Kerman A. K. (1965), *Phys. Letters* **19**, 413. [II : 448]
Britt, H. C., and Plasil, F. (1966), *Phys. Rev.* **144**, 1046. [II : 127]
Britt, H. C., Gibbs. W. R., Griffin, J. J., and Stokes, R. H. (1965), *Phys. Rev.* **139**, B354. [II : 127]
Brix, P., and Kopfermann, H. (1949), *Z. Physik* **126**, 344. [I : 163]
Brix, P., and Kopfermann, H. (1958), *Rev. Mod. Phys.* **30**, 517. [I : 162]
Broglia, R. A., Lilley, J. S., Perazzo, R., and Phillips, W. R., (1970), *Phys. Rev.* **C1**, 1508. [II : 517, 565, 573]
Broglia, R. A., Paar, V., and Bès, D. R. (1971), *Phys. Letters* **37B**, 159. [II : 571]
Broglia, R. A., Hansen, O., and Riedel, C. (1973), *Advances in Nuclear Physics* **6**, 287, Plenum Press, New York, N.Y. [II : 249, 389]
Brown, G. E. (1959), *Rev. Mod. Phys.* **31**, 893. [I : 433]
Brown, G. E. (1964), *Nuclear Phys.* **57**, 339. [II : 486]

Brown, G. E. (1967), *Unified Theory of Nuclear Models*, 2nd edition, North-Holland, Amsterdam.

Brown, G. E., and Bolsterli, M. (1959), *Phys. Rev. Letters* **3**, 472. [II:475]

Brueckner, K. A. (1959), *The Many-Body Problem*, Ecole d'Eté de Physique Théorique, Les Houches, 1958, Wiley, New York, N.Y. [I:262]

Brueckner, K. A., Eden, R. J., and Francis, N. C. (1955), *Phys. Rev.* **99**, 76. [I:334]

Bryan, R. A. (1967), in *Proc. Intern. Nuclear Physics Conf.*, Gatlinburg, p. 603, ed.-in-chief R. L. Becker, Academic Press, New York, N.Y. [I:250]

Buck, B. (1963), *Phys. Rev.* **130**, 712. [II:353]

Bühring, W. (1965), *Nuclear Phys.* **61**, 110. [I:411]

Büttgenbach, S., Herschel, M., Meisel, G., Schrödl, E., and Witte, W. (1973), *Phys. Letters* **43B** 479. [II:109, 303]

Bund, G. W., and Wajntal, W. (1963), *Nuovo cimento* **27**, 1019. [I:257]

Bunker, M. E., and Reich, C. W. (1971), *Rev. Mod. Phys.* **43**, 348.

Burbidge, G. (1962), *Ann. Rev. Nuclear Sci.* **12**, 507. [I:199, 206]

Burbidge, E. M., Burbidge, G. R., Fowler, W. A., and Hoyle, F. (1957), *Rev. Mod. Phys.* **29**, 547. [I:199, 207, 208]

Burhop, E. H. S. (1967), *Nuclear Phys.* **B1**, 438. [I:138]

Burke, D. G., Zeidman, B., Elbek, B., Herskind, B., and Olesen, M. (1966), *Mat. Fys. Medd. Dan. Vid. Selsk.* **35**, no. 2. [II:259, 261, 264]

Burnett, D. S., Gatti, R. C., Plasil, F., Price, P. B., Swiatecki, W. J., and Thompson, S. G. (1964), *Phys. Rev.* **134**, B952. [I:142; II:616]

Burr, W., Schütte, D., and Bleuler, K. (1969), *Nuclear Phys.* **A133**, 581. [II:14]

Buss, D. J., and Smither, R. K. (1970), *Phys. Rev.* **C2**, 1513. [II:535]

Butler, S. T. (1951), *Proc. Roy. Soc. (London)* **A208**, 559. [I:421]

Byers, N., and Yang, C. N. (1961), *Phys. Rev. Letters* **7**, 46. [II:398]

Cabibbo, N. (1963), *Phys. Rev. Letters* **10**, 531. [I:402]

Calaprice, F. P., Commins, E. D., Gibbs, H. M., Wick, G. L., and Dobson, D. A. (1967), *Phys. Rev. Letters* **18**, 918. [I:27]

Caldwell, D. O., Elings, V. B., Hesse, W. P., Jahn, G. E., Morrison, R. J., Murphy, F. V., and Yount, D. E. (1969), *Phys. Rev. Letters* **23**, 1256. [II:480]

Cameron, A. G. W. (1968), in *Origin and Distribution of the Elements*, p. 125, ed. L. H. Ahrens, Pergamon Press, Oxford. [I:206]

Camp, D. C., and Langer, L. M. (1963), *Phys. Rev.* **129**, 1782. [I:53]

Campbell, E. J., Feshbach, H., Porter, C. E., and Weisskopf, V. F. (1960), *M. I. T. Techn. Rep.* **73**, Cambridge, Mass. [I:230]

Carlos, P., Beil, H., Bergère, R., Lepretre, A., and Veyssière, A. (1971), *Nuclear Phys.* **A172**, 437. [II:491]

Carlson, B. C., and Talmi, I. (1954), *Phys. Rev.* **96**, 436. [I:146]

Carter, E. B., Mitchell, G. E., and Davis, R. H. (1964), *Phys. Rev.* **133**, B1421. [II:26]

Casimir, H. B. G. (1931), *Rotation of a Rigid Body in Quantum Mechanics*, Wolters, Groningen. [II:3, 149]

Casimir, H. B. G. (1936), *On the Interaction Between Atomic Nuclei and Electrons*, Prize Essay, Teyler's, Tweede, Haarlem. [II:2]

Casten, R. F., Greenberg, S., Sie, S. H., Burginyon, G. A., and Bromley, D. A. (1969), *Phys. Rev.* **187**, 1532. [II:536]

Casten, R. F., Kleinheinz, P., Daly, P. J., and Elbek, B. (1972), *Mat. Fys. Medd. Dan. Vid. Selsk.* **38**, no. 13. [II:258, 301, 318]

Cerny, J., and Pehl, R. H. (1964), *Phys. Rev. Letters* **12**, 619. [I:46]

Cerny, J., Pehl, R. H., Rivet, E., and Harvey, B. G. (1963), *Phys. Letters* **7**, 67. [I:45]

Chakrabarti, A. (1964), *Ann. Institut Henri Poincaré* **1**, 301. [I:73]

Chamberlain, O., Segré, E., Tripp, R. D., Wiegand, C., and Ypsilantis, T. (1957), *Phys. Rev.* **105**, 288. [I:245]

Chan, L. H., Chen, K. W., Dunning, J. R., Ramsey, N. F., Walker, J. K., and Wilson, R. (1966), *Phys. Rev.* **141**, 1298. [I:386]

Chandrasekhar, S. (1969), *Ellipsoidal Figures of Equilibrium*, Yale Univ. Press, New Haven, Conn. [II:663]

Chang, T. H. (1964), *Acta Phys. Sinica* **20**, 159. [II:451]

Chase, D. M., Wilets, L., and Edmonds, A. R. (1958), *Phys. Rev.* **110**, 1080. [II:238]

Chasman, R. R., and Rasmussen, J. O. (1959), *Phys. Rev.* **115**, 1257. [II:117]

Chen, C. T., and Hurley, F. W. (1966), *Nuclear Data*, Sec. B1-13, p. 1.

Chen, M-y (1969), *Contributions to Intern. Conf. on Properties of Nuclear States*, p. 72, Les Presses de l'Université de Montreal, Montreal. [II:135]

Chesler, R. B., and Boehm, F. (1968), *Phys. Rev.* **166**, 1206. [I:165]

Chi, B. E., and Davidson, J. P. (1963), *Phys. Rev.* **131**, 366. [II:288]

Chilosi, G., Ricci, R. A., Touchard, J., and Wapstra, A. H. (1964), *Nuclear Phys.* **53**, 235. [I:344]

Chilosi, G., O'Kelley, G. D., and Eichler, E. (1965), *Bull. Am. Phys. Soc.* **10**, 92. [I:350]

Choudhury, D. C. (1954), *Mat. Fys. Medd. Dan. Vid. Selsk.* **28**, no. 4. [II:691]

Christensen, C. J., Nielsen, A., Bahnsen, B., Brown, W. K., and Rustad, B. M. (1967), *Phys. Letters* **26B**, 11. [I:5, 349, 405]

Christensen, J. H., Cronin, J. W., Fitch, V. L., and Turlay, R. (1964), *Phys. Rev. Letters* **13**, 138. [I:16]

Christensen, P. R., Nielsen, O. B., and Nordby, H. (1963), *Phys. Letters* **4**, 318. [I:351]

Church, E. L., and Weneser, J. (1956), *Phys. Rev.* **103**, 1035. [I:383]

Church, E. L., and Weneser, J. (1960), *Ann. Rev. Nuclear Sci.* **10**, 193. [I:384]

Clement, C. F., Lane, A. M., and Rook, J. R. (1965), *Nuclear Phys.* **66**, 273. [II:486]

Clementel, E., and Villi, C. (1955), *Nuovo cimento* **2**, 176. [I:261]

Clemmow, P. C., and Dougherty, J. P. (1969), *Electrodynamics of Particles and Plasmas*, Addison-Wesley, Reading, Mass.

Cline, J. E. (1968), *Nuclear Phys.* **A106**, 481. [II:274]

Cohen, B. L. (1957), *Phys. Rev.* **105**, 1549. [II:353, 556]

Cohen, B. L. (1963), *Phys. Rev.* **130**, 227. [I:329]

Cohen, B. L. (1968), *Phys. Letters* **27B**, 271. [I:329]

Cohen, B. L., and Rubin, A. G. (1968), *Phys. Rev.* **111**, 1568. [II:353]

Cohen, S., and Swiatecki, W. J. (1963), *Ann. Phys.* **22**, 406. [II:662, 663, 664]

Cohen, S., Plasil, F., and Swiatecki, W. J. (1974), *Ann. Phys.* **82**, 557. [II:663, 664]

Coleman, S., and Glashow, S. L. (1961), *Phys. Rev. Letters* **6**, 423. [I:40]

Collard, H. R., Elton, L. R. B., and Hofstadter, R. (1967), *Nuclear Radii*, in Landolt-Börnstein, New Series, I, vol. 2, Springer, Berlin.

Condon, E. U., and Shortley, G. H. (1935), *The Theory of Atomic Spectra*, Cambridge University Press, London.

Cooper, L. N. (1956), *Phys. Rev.* **104**, 1189. [II:396]

Coor, T., Hill, D. A., Hornyak, W. F., Smith, L. W., and Snow, G. (1955), *Phys. Rev.* **98**, 1369. [I:165]

Coulon, T. W., Bayman, B. F., and Kashy, E. (1966), *Phys. Rev.* **144**, 941. [I:323]
Courant, E. D. (1951), *Phys. Rev.* **82**, 703. [II:474]
Craig, R. M., Dore, J. C., Greenlees, G. W., Lilley, J. S., Lowe J., and Rowe, P. C. (1964), *Nuclear Phys.* **58**, 515. [I:234]
Čujec, B. (1964), *Phys. Rev.* **136**, B1305. [II:294]
Cziffra, P., MacGregor, M. H., Moravcsik, M. J., and Stapp, H. P. (1959), *Phys. Rev.* **114**, 880. [I:5, 245, 250]

Dąbrowski, J., and Sobiczewski, A. (1963), *Phys. Letters* **5**, 87. [I:271]
Dalitz, R. H. (1963), in *Proc. Intern. Conf. on Hyperfragments*, St. Cergue, March 1963, CERN Report 64-1, p. 147, CERN, Geneva. [I:56]
Dalitz, R. H. (1967), in *Proc. 13th Intern. Conf. on High-Energy Physics*, University of California Press, Berkeley, Cal. [I:57, 62]
Damgaard, J. (1966), *Nuclear Phys.* **79**, 374. [I:54]
Damgaard, J., and Winther, A. (1964), *Nuclear Phys.* **54**, 615. [I:350, 352]
Damgaard, J., and Winther, A. (1966), *Phys. Letters* **23**, 345. [I:408]
Damgaard, J., Scott, C. K., and Osnes, E. (1970), *Nuclear Phys.* **A154**, 12. [II:295]
Daniel, H., and Schmitt, H. (1965), *Nuclear Phys.* **65**, 481. [I:54]
Danos, M. (1958), *Nuclear Phys.* **5**, 23. [II:490, 673]
Danysz, M., Garbowska, K., Pniewski, J., Pniewski, T., Zakrzewski, J., Fletcher, E. R., Lemonne, J., Renard, P., Sacton, J., Toner, W. T., O'Sullivan, D., Shah, T. P., Thompson, A., Allen, P., Heeran, Sr. M., Montwill, A., Allen, J. E., Beniston, M. J., Davis, D. H., Garbutt, D. A., Bull, V. A., Kumar, R. C., and March, P. V. (1963). *Nuclear Phys.* **49**, 121. [I:56]
Darriulat, P., Igo, G., Pugh, H. G., and Holmgren, H. D. (1965), *Phys. Rev.* **137**, B315. [II:100]
Davidson, J. P. (1965), *Rev. Mod. Phys.* **37**, 105. [II:176]
Davidson, J. P., and Feenberg, E. (1953), *Phys. Rev.* **89**, 856. [II:33]
Davies, D. W., and Hollander, J. M. (1965), *Nuclear Phys.* **68**, 161. [II:114]
Davis, D. H., Lovell, S. P., Csejthey-Barth, M., Sacton, J., Schorochoff, G., and O'Reilly, M. (1967), *Nuclear Phys.* **B1**, 434. [I:138]
Davydov, A. S. (1967), *Excited States of Atomic Nuclei* (in Russian), Atomizdat, Moscow. See also *Atomic Energy Review*, vol. VI, no. 2, p. 3, Intern. Atomic Energy Agency, Vienna (1968). [II:545]
Davydov, A. S., and Filippov, G. F. (1958), *Nuclear Phys.* **8**, 237. [II:176]
De Alfaro, V., and Regge, T. (1965), *Potential Scattering*, North-Holland, Amsterdam.
Deaver, B. S., Jr., and Fairbank, W. M. (1961), *Phys. Rev. Letters* **7**, 43. [II:398]
de Boer, J. (1957), in *Progress in Low Temperature Physics* **1**, 381, ed. C. J. Gorter, North-Holland, Amsterdam. [I:268]
de Boer, J., and Eichler, J. (1968), *Advances in Nuclear Physics*, **1**, 1, eds. M. Baranger and E. Vogt, Plenum Press, New York, N.Y. [II:458, 532]
de Boer, J., Stokstad, R. G., Symons, G. D., and Winther, A. (1965), *Phys. Rev. Letters* **14**, 564. [II:433, 539]
De Dominicis, C., and Martin, P. C. (1957), *Phys. Rev.* **105**, 1417. [I:256]
De Forest, T., Jr., and Walecka, J. D. (1966), *Advances in Physics* **15**, 1. [I:384]
de Groot, S. R., Tolhoek, H. A., and Huiskamp, W. J. (1965), in *Alpha-, Beta-, and Gamma-Ray Spectroscopy*, vol. 2, p. 1199, ed. K. Siegbahn, North-Holland, Amsterdam. [I:383]
Dehnhard, D., and Mayer-Böricke, C. (1967), *Nuclear Phys.* **A97**, 164. [II:61]
Depommier, P., Duclos, J., Heitze, J., Kleinknecht, K., Rieseberg, H., and Soergel,

V. (1968), *Nuclear Phys.* **B4**, 189. [I:52]
de-Shalit, A. (1961), *Phys. Rev.* **122**, 1530. [II:359]
de-Shalit, A., and Goldhaber, M. (1952), *Phys. Rev.* **92**, 1211. [I:329]
de-Shalit, A., and Talmi, I. (1963), *Nuclear Shell Theory*, Academic Press, New York, N.Y.
Desjardins, J. S., Rosen, J. L., Havens, W. W., Jr., and Rainwater, J. (1960), *Phys. Rev.* **120**, 2214. [I:181; II:38]
Deutch, B. I. (1962), *Nuclear Phys.* **30**, 191. [II:155]
Devons, S., and Duerdoth, I. (1969), *Advances in Nuclear Physics*, eds. M. Baranger and E. Vogt, Plenum Press, New York, N.Y. [II:134]
De Wit, S. A., Backenstoss, G., Daum, C., Sens, J. C., and Acker, H. L. (1967), *Nuclear Phys.* **87**, 657. [II:493]
Dey, W., Ebersold, P., Leisi, H. J., Scheck, F., Boehm, F., Engfer, R., Link, R., Michaelsen, R., Robert-Tissot, B., Schellenberg, L., Schneuwly, H., Schröder, W. U., Vuilleumier, J. L., Walter, H. K., and Zehnder, A. (1973), *Suppl. to J. Phys. Soc. Japan* **34**, 582. [II:276]
Diamond, R. M., and Stephens, F. S. (1967), *Arkiv Fysik* **36**, 221. [II:67]
Diamond, R. M., Elbek, B., and Stephens, F. S. (1963), *Nuclear Phys.* **43**, 560. [II:103, 254]
Diamond, R. M., Stephens, F. S., Kelly, W. H., and Ward, D. (1969a), *Phys. Rev. Letters* **22**, 546. [II:73, 129]
Diamond, R. M., Stephens, F. S., Nordhagen, R., and Nakai, K. (1969b), *Contributions to Intern. Conf. on Properties of Nuclear States*, p. 7, Les Presses de l'Université de Montreal, Montreal. [II:129, 535]
Di Capua, E., Garland, R., Pondrom, L., and Strelzoff, A. (1964), *Phys. Rev.* **133**, B1333. [I:398]
Dirac, P. A. M. (1935), *The Principles of Quantum Mechanics*, 2nd edition, Clarendon Press, Oxford.
Dixon, W. R., Storey, R. S., Aitken, J. H., Litherland, A. E., and Rogers, D. W. O. (1971), *Phys. Rev. Letters* **27**, 1460. [II:289]
Døssing, T., and Jensen, A. S. (1974), *Nuclear Phys.* **A222**, 493. [II:607]
Dolbilkin, B. S., Korin, V. I., Lazareva, L. E., and Nikolaev, F. A. (1965), *Zh. Eksper. Teor. Fiz. Pis'ma (USSR)* **1**, no. 5, 47; transl. *JETP Letters* **1**, 148. [II:479, 504]
Dolginov, A. Z. (1961), in *Gamma Luchi*, Chap. 6, p. 524, ed. L. Sliv, Akademia Nauk SSSR, Moscow. [I:383]
Doll, R., and Näbauer, M. (1961), *Phys. Rev. Letters* **7**, 51. [II:398]
Domingos, J. M., Symons, G. D., and Douglas, A. C. (1972), *Nuclear Phys.* **A180**, 600. [II:160]
Dothan, Y., Gell-Mann, M., and Ne'eman, Y. (1965), unpublished manuscript; see also Weaver *et al.* (1973). [II:410, 411]
Dover, C. B., Lemmer, R. H., and Hahne, F. J. W. (1972), *Ann. Phys.* **70**, 458. [II:506]
Dragt, A. (1965), *J. Math. Phys.* **6**, 533. [I:73]
Dreizler, R. M., Klein, A., Wu, C.-S., and Do Dang, G. (1967), *Phys. Rev.* **156**, 1167. [II:433]
Drell, S. D., and Walecka, J. D. (1960), *Phys. Rev.* **120**, 1069. [I:391]
Drozdov, S. (1955), *J. Exptl. Theoret. Phys. (USSR)* **28**, 734 and 736; transl. *Soviet Phys. JETP* **1**, 588 and 591. [II:322]
Drozdov, S. (1958), *J. Exptl. Theoret. Phys. (USSR)* **34**, 1288; transl. *Soviet Phys. JETP* **7**, 889. [II:322]

Dubois, J. (1967), *Nuclear Phys.* **A104**, 657. [II:286]
Duguay, M. A., Bockelman, C. K., Curtis, T. H., and Eisenstein, R. A. (1967), *Phys. Rev.* **163**, 1259. [II:561]
Dunaitsev, A. F., Petrukhin, V. I., Prokoshkin, Yu. D., and Rykalin, V. I. (1963), *Intern. Conf. on Fundamental Aspects of Weak Interactions*, BNL 837, (C-39), Brookhaven, Upton, N.Y. [I:52]
Durand, L. (1964), *Phys. Rev.* **135**, B310. [I:411]
Dussel, G. G., and Bès, D. R. (1970), *Nuclear Phys.* **A143**, 623. [II:545]
Dussel, G. G., Perazzo, R. P. J., Bès, D. R., and Broglia, R. A. (1971), *Nuclear Phys.* **A175**, 513. [II:395]
Dyson, F. J. (1962), *J. Math. Phys.* **3**, 140, 157, 166. [I:180, 298, 299]
Dyson, F. J. (1962a), *J. Math. Phys.* **3**, 1191. [I:298]
Dyson, F. J. (1962b), *J. Math. Phys.* **3**, 1199. [I:100, 298]
Dyson, F. J. (1966), *Symmetry Groups in Nuclear and Particle Physics*, Benjamin, New York, N.Y. [I:41]
Dyson, F. J., and Mehta, M. L. (1963), *J. Math. Phys.* **4**, 701. [I:181, 299, 300]

Eccleshall, D., and Yates, M. J. L. (1965), in *Physics and Chemistry of Fission*, vol. I, p. 77, Intern. Atomic Energy Agency, Vienna. [II:127]
Edmonds, A. R. (1957), *Angular Momentum in Quantum Mechanics*, Princeton University Press, Princeton, N. J.
Ehrenberg, H. F., Hofstadter, R., Meyer-Berkhout, U., Ravenhall, D. G., and Sobottka, S. E. (1959), *Phys. Rev.* **113**, 666. [II:134]
Ehrman, J. B. (1951), *Phys. Rev.* **81**, 412. [I:43]
Eichler, J. (1963), *Z. Physik* **171**, 463. [I:408]
Eichler, J. (1964), *Phys. Rev.* **133**, B1162. [II:458]
Eichler, J., Tombrello, T. A., and Bahcall, J. N. (1964), *Phys. Letters* **13**, 146. [I:418]
Eisele, G., Koniordos, I., Müller, G., and Winkler, R. (1968), *Phys. Letters* **28B**, 256. [II:517]
Eisenbud, L., and Wigner, E. P. (1941), *Proc. Nat. Acad. Sci. U.S.A.* **27**, 281. [I:65]
Ejiri, H. (1971), *Nuclear Phys.* **A166**, 594. [II:502]
Ejiri, H., and Hagemann, G. B. (1971), *Nuclear Phys.* **A161**, 449. [II:168, 170, 171, 172]
Ejiri, H., Ikeda, K., and Fujita, J.-I. (1968), *Phys. Rev.* **176**, 1277. [II:493]
Ejiri, H., Richard, P., Ferguson, S. M., Heffner, R., and Perry, D. (1969), *Nuclear Phys.* **A128**, 388. [II:502]
Ekström, C., Olsmats, M., and Wannberg, B. (1971), *Nuclear Phys.* **A170**, 649. [II:298]
Ekström, C., Ingelman, S., Wannberg, B., and Lamm, I.-L. (1972), *Phys. Letters* **39B**, 199. [II:298]
Elbek, B. (1963), *Determination of Nuclear Transition Probabilities by Coulomb Excitation*, thesis, Munksgaard, Copenhagen. [II:133]
Elbek, B. (1969), in *Proc. Intern. Conf. on Properties of Nuclear States*, p. 63, eds. Harvey, Cusson, Geiger, and Pearson, Les Presses de l'Université de Montreal, Montreal. [II:298, 578]
Elbek, B., and Tjøm, P. O. (1969), *Advances in Nuclear Physics*, **3**, 259, eds. M. Baranger and E. Vogt, Plenum Press, New York, N.Y. [II:244]
Elbek, B., Nielsen, K. O., and Olesen, M. C. (1957), *Phys. Rev.* **108**, 406. [II:133]
Elbek, B., Olesen, M. C., and Skilbreid, O. (1959), *Nuclear Phys.* **10**, 294. [II:133]
Elbek, B., Olesen, M. C., and Skilbreid, O. (1960), *Nuclear Phys.* **19**, 523. [II:133, 161]

Elbek, B., Grotdal, T., Nybø, K., Tjøm, P. O., and Veje, E. (1968), in *Proc. Intern. Conference on Nuclear Structure*, p. 180, ed. J. Sanada, Phys. Society of Japan. [II:550, 577]
Ellegaard, C., and Vedelsby, P. (1968), *Phys. Letters* **26B**, 155. [I:324; II:571]
Ellegaard, C., Kantele, J., and Vedelsby, P. (1967), *Phys. Letters* **25B**, 512. [I:340; II:565]
Ellegaard, C., Barnes, P. D., Flynn, E. R., and Igo, G. J. (1971), *Nuclear Phys.* **A162**, 1. [II:562]
Ellegaard, C., Barnes, P. D., and Flynn, E. R. (1971a), *Nuclear Phys.* **A170**, 209. [II:646]
Elliott, J. P. (1958), *Proc. Roy. Soc.* (*London*), **A245**, 128 and 562. [II:4, 84, 91, 97, 283, 289]
Elliott, J. P., and Skyrme, T. H. R. (1955), *Proc. Roy. Soc.* (*London*), **A232**, 561. [II:475]
Elliott, J. P., and Flowers, B. H. (1957), *Proc. Roy. Soc.* (*London*), **A242**, 57. [I:22; II:475]
Elliott, J. P., Mavromatis, H. A., and Sanderson, E. A. (1967), *Phys. Letters* **24B**, 358. [I:260]
Elton, L. R. B. (1961), *Nuclear Sizes*, Oxford University Press, Oxford.
Elze, Th. W., and Huizenga, J. R. (1969), *Nuclear Phys.* **A133**, 10. [II:274]
Elze, Th. W., and Huizenga, J. R. (1970), *Phys. Rev.* **C1**, 328. [II:266]
Elze, Th. W., and Huizenga, J. R. (1972), *Nuclear Phys.* **A187**, 545. [II:561]
Endt, P. M., and van der Leun, C. (1967), *Nuclear Phys.* **A105**, 1.
Erb, K. A., Holden, J. E., Lee, I. Y., Saladin, J. X., and Saylor, T. K. (1972), *Phys. Rev. Letters* **29**, 1010. [II:140]
Erba, E., Facchini, U., and Saetta-Menichella, E. (1961), *Nuovo cimento* **22**, 1237. [I:187]
Erdal, B. R., Finger, M. Foucher, R., Husson, J. P., Jastrzebski, J., Johnson, A., Perrin, N., Henck, R., Regal, R., Siffert, P., Astner, G., Kjelberg, A., Patzelt, P., Höglund, A., and Malmskog, S. (1970), in *Intern. Conf. on Properties of Nuclei Far from the Region of Beta-Stability*, p. 1031 (CERN Report 70-30) Geneva. [II:536]
Erdélyi, A. (1953), *Higher Transcendental Functions*, vol. II, McGraw Hill, New York, N.Y.
Ericson, M., and Ericson, T. E. O. (1966), *Ann. Phys.* **36**, 323 [I:219]
Ericson, T. (1958), *Nuclear Phys.* **6**, 62. [II:38]
Ericson, T. (1958a), *Nuclear Phys.* **8**, 265, and **9**, 697. [II:607]
Ericson, T. (1959), *Nuclear Phys.* **11**, 481. [I:187]
Ericson, T. (1960), *Advances in Physics* **9**, 425. [I:281]
Ericson, T., and Mayer-Kuckuk, T. (1966), *Ann Rev. Nuclear Sci.* **16**, 183. [I:435]
Erlandsson, G. (1956), *Arkiv Fysik* **10**, 65. [II: 182, 185]
Erskine, J. R. (1965), *Phys. Rev.* **138**, B66. [II:258]
Erskine, J. R. (1966), *Phys. Rev.* **149**, 854. [I:322]
Erskine, J. R., Marinov, A., and Schiffer, J. P. (1966), *Phys. Rev.* **142**, 633. [I:350]
Euler, H. (1937), *Z. Physik* **105**, 553. [I:262]
Ewan, G. T., Geiger, J. S., Graham, R. L., and MacKenzie, D. R. (1959), *Phys. Rev.* **116**, 950. [II:113]
Evans, H. D. (1950), *Proc. Phys. Soc. (London)* **63A**, 575. [I:351]

Faessler, A. (1966), *Nuclear Phys.* **85**, 679. [II:275]
Fagg, L. W., Bendel, W. L., Jones, E. C., Jr., and Numrich, S. (1969), *Phys. Rev.* **187**, 1378. [II:640]

Fallieros, S., Goulard, B., and Venter, R. H. (1965), *Phys. Letters* **19**, 398. [II:380, 493]
Fano, U. (1961), *Phys. Rev.* **124**, 1866. [I:302]
Fano, U., and Cooper, J. W. (1968), *Rev. Mod. Phys.* **40**, 441. [II:352]
Favro, L. D., and MacDonald, J. F. (1967), *Phys. Rev. Letters* **19**, 1254. [I:298]
Federman, P., Rubinstein, H. R., and Talmi, I. (1966), *Phys. Letters* **22**, 208 [I:59]
Fermi, E. (1934), *Z. Physik* **88**, 161. [I:396]
Ferrell, R. A. (1957), *Phys. Rev.* **107**, 1631. [II:436]
Ferrell, R. A., and Fallieros, S. (1959), *Phys. Rev.* **116**, 660. [II:436]
Feshbach, H. (1967), in *Proc. Intern. Nuclear Physics Conf.*, Gatlinburg, p. 181, ed.-in-chief R. L. Becker, Academic Press, New York, N.Y. [I:231]
Feshbach, H., Porter, C. E., and Weisskopf, V. F. (1954), *Phys. Rev.* **96**, 448. [I:434]
Feshbach, H., Kerman, A. K., and Lemmer, R. H. (1967), *Ann. Phys.* **41**, 230. [I:436]
Feynman, R. P. (1955), *Progress in Low Temperature Physics*, **1**, 17, ed. J. C. Gorter, North Holland, Amsterdam. [II:397]
Feynman, R. P., and Gell-Mann, M. (1958), *Phys. Rev.* **109**, 193. [I:400]
Fielder, D. S., Le Tourneux, J., Min, K., and Whitehead, W. D. (1965), *Phys. Rev. Letters* **15**, 33. [II:505]
Fierz, M. (1943), *Helv. Phys. Acta* **16**, 365. [II:341]
Fisher, T. R., Tabor, S. L., and Watson, B. A. (1971), *Phys. Rev. Letters* **27**, 1078. [II:137]
Flerov, G. N., Oganeayan, Yu. Ts., Lobanov, Yu. V., Kuznetsov, V. I. Druin, V. A., Perelygin, V. P., Gavrilov, K. A., Tretiakova, S. P., and Plotko, V. M. (1964), *Phys. Letters* **13**, 73. [I:205]
Flügge, S. (1941), *Ann. Physik* **39**, 373. [II:341]
Flynn, E. R., Igo, G., Barnes, P. D., Kovar, D., Bès, D., and Broglia, R. (1971), *Phys. Rev.* **C3**, 2371. [II:647]
Flynn, E. R., Igo, G. J., Broglia, R. A., Landowne, S., Paar, V., and Nilsson, B. (1972), *Nuclear Phys.* **A195**, 97. [II:646]
Flynn, E. R., Igo, G. J., and Broglia, R. A. (1972), *Phys. Letters* **41B**, 397. [II:646]
Flynn, E. R., Broglia, R. A., Liotta, R., and Nilsson, B. S. (1974), *Nuclear Phys.* **A221**, 509. [II:646]
Foissel, P., Cassagnou, Y., Laméhi-Rachti, M., Lévi, C., Mittig, W., and Papineau, L. (1972), *Nuclear Phys.* **A178**, 640. [II:261]
Foldy, L. L. (1953), *Phys. Rev.* **92**, 178. [I:384]
Foldy, L. L. (1958), *Rev. Mod. Phys.* **30**, 471. [I:5]
Foldy, L. L., and Milford, F. J. (1950), *Phys. Rev.* **80**, 751. [II:359, 418]
Foldy, L. L., and Walecka, J. D. (1964), *Nuovo cimento* **34**, 1026. [II:497]
Foldy, L. L., and Walecka, J. (1965), *Phys. Rev.* **140**, B1339. [I:405]
Ford, J. L. C., Jr., Stelson, P. H., Bemis, C. E., Jr., McGowan, F. K., Robinson, R. L., and Milner, W. T. (1971), *Phys. Rev. Letters* **27**, 1232. [II:133]
Forker, M., and Wagner, H. F. (1969), *Nuclear Phys.* **A138**, 13. [II:47]
Fowler, W. A., and Hoyle, F. (1964), *Astrophys. J., Suppl.* **91**. [I:200, 202, 397]
Fox, J. D., Moore, C. F., and Robson, D. (1964), *Phys. Rev. Letters* **12**, 198. [I:46]
Franco, V. (1965), *Phys. Rev.* **140**, B1501. [I:167]
Frankel, S., and Metropolis, N. (1947), *Phys. Rev.* **72**, 914. [II:662]
Fraser, I. A., Greenberg, J. S., Sie, S. H., Stokstad, R. G., and Bromley, D. A. (1969), in *Contributions to Intern. Conf. on Properties of Nuclear States* pp. 13, 14, Les Presses de l'Université de Montreal, Montreal. [II:535]

Frauenfelder, H., and Steffen, R. M. (1965), in *Alpha-, Beta-, and Gamma-Ray Spectroscopy*, vol. 2, p. 997, ed. K. Siegbahn, North-Holland, Amsterdam. [I:383]

Freeman, J. M., Murray, G., and Burcham, W. E. (1965), *Phys. Letters* **17**, 317. [I:401]

Freeman, J. M., Jenkin, J. G., Murray, G., and Burcham, W. E. (1966), *Phys. Rev. Letters* **16**, 939. [I:52]

French, J. B. (1964), *Phys. Letters* **13**, 249. [I:424].

Frenkel, J. (1936), *Phys. Z. Sowietunion* **9**, 533. [I:154]

Frenkel, J. (1939), *J. Phys. (USSR)* **1**, 125; see also *Phys. Rev.* **55**, 987. [II:365]

Friedman, F. L., and Weisskopf, V. F. (1955), in *Niels Bohr and the Development of Physics*, p. 134, ed. W. Pauli, Pergamon Press, New York, N.Y. [I:434]

Frisch, R., and Stern, O. (1933), *Z. Physik* **85**, 4. [I:385]

Fröman, P. O. (1957), *Mat. Fys. Skr. Dan. Vid. Selsk.* **1**, no. 3. [II:115, 141, 269]

Fröman, N., and Fröman, P. O. (1965), *JWKB Approximation*, North-Holland, Amsterdam.

Fujita, J. I. (1962), *Phys. Rev.* **126**, 202. [I:408]

Fujita, J. I., and Ikeda, K. (1965), *Nuclear Phys.* **67**, 145. [I:347, 348]

Fujita, J. I., and Hirata, M. (1971), *Phys. Letters* **37B**, 237. [II:484]

Fulbright, H. W., Lassen, N. O., and Poulsen, N. O. R. (1959), *Mat. Fys. Medd. Dan. Vid. Selsk.* **31**, no. 10. [II:353]

Fulbright, H. W., Alford, W. P., Bilaniuk, O. M., Deshpande, V. K., and Verba, J. W. (1965), *Nuclear Phys.* **70**, 553. [I:46]

Fuller, E. G., and Hayward, E. (1962), *Nuclear Phys.* **30**, 613. [II:490]

Fuller, E. G., and Hayward, E. (1962a), in *Nuclear Reactions*, vol. II, p. 713, eds. P. M. Endt and P. B. Smith, North-Holland, Amsterdam. [II:458, 503]

Fuller, G. H., and Cohen, V. W. (1965), *Nuclear Moments,* Appendix 1 to *Nuclear Data Sheets*, Oak Ridge Nat. Lab., Oak Ridge, Tenn.

Fuller, G. H., and Cohen, V. W. (1969), *Nuclear Data* **A5**, 433.

Fulmer, R. H., McCarthy, A. L., Cohen, B. L., and Middleton, R. (1964), *Phys. Rev.* **133**, B955. [I:228]

Fultz, S. C., Bramblett, R. L., Caldwell, J. T., and Kerr, N. A. (1962), *Phys. Rev.* **127**, 1273. [II:475]

Fultz, S. C., Caldwell, J. T., Berman, B. L., Bramblett, R. L., and Harvey, R. R. (1966), *Phys. Rev.* **143**, 790. [II:479]

Fultz, S. C., Bramblett, R. L., Berman, B. L., Caldwell, J. T., and Kelly, M. A. (1967), in *Proc. Intern. Nuclear Physics Conf.*, Gatlinburg, p. 397, ed-in-chief R. L. Becker, Academic Press, New York, N.Y. [II:479]

Gallagher, C. J., Jr., Nielsen, O. B., and Sunyar, A. W. (1965), *Phys. Letters* **16**, 298. [II:160]

Gallone, S., and Salvetti, C. (1953), *Nuovo cimento, Ser. 9*, **10**, 145. [II:508]

Gamba, A., Malvano, R., and Radicati, L. A. (1952), *Phys. Rev.* **87**, 440. [I:44]

Gamow, G. (1934), in *Rapports et Discussions du Septième Conseil de Physique de l'Institut Intern. Solvay*, p. 231, Gauthier-Villars, Paris. [II:v]

Garg, J. B., Rainwater, J., Petersen, J. S., and Havens, W. W., Jr. (1964), *Phys. Rev.* **134**, B985. [I:180, 182]

Gatto, R. (1967), in *High Energy Physics*, vol. II, p. 1, ed. E. H. S. Burhop, Academic Press, New York, N.Y. [I:380]

Gaudin, M. (1961), *Nuclear Phys.* **25**, 447. [I:299]

Geilikman, B. T. (1960), in *Proc. Intern. Conf. on Nuclear Structure*, Kingston,

Canada, p. 874, eds. D. A. Bromley and E. W. Vogt, Univ. of Toronto Press, Toronto. [II:592]

Gell-Mann, M. (1953), *Phys. Rev.* **92**, 833. [I:38]

Gell-Mann, M. (1958), *Phys. Rev.* **111**, 362. [I:414]

Gell-Mann, M. (1961), *Cal. Inst. Tech. Rep.* CTSL-20, Pasadena, Cal., reproduced in Gell-Mann and Ne'eman (1964). [I:40]

Gell-Mann, M. (1962), *Phys. Rev.* **125**, 1067. [I:59]

Gell-Mann, M. (1964), *Phys. Letters* **8**, 214. [I:41]

Gell-Mann, M., and Berman, S. M. (1959), *Phys. Rev. Letters* **3**, 99. [I:415]

Gell-Mann, M., and Ne'eman, Y. (1964), *The Eightfold Way*, Benjamin, New York, N.Y. [I:41, 58, 59]

Gell-Mann, M., Goldberger, M. L., and Thirring, W. E. (1954), *Phys. Rev.* **95**, 1612. [II:416]

Gerholm, T. R., and Pettersson, B. G. (1965), in *Alpha-, Beta-, and Gamma-Ray Spectroscopy*, vol. 2, p. 981, ed. K. Siegbahn, North-Holland, Amsterdam. [I:384]

Giltinan, D. A., and Thaler, R. M. (1963), *Phys. Rev.* **131**, 805. [I:244]

Ginocchio, J. N., and Weneser, J. (1968), *Phys. Rev.* **170**, 859. [II:395]

Ginzburg, V. L. (1964), *The Propagation of Electromagnetic Waves in Plasmas*, Pergamon Press, Oxford.

Ginzburg, V. L., and Landau, L. D. (1950), *J. Exptl. Theoret. Phys. (USSR)* **20**, 1064; transl. in D. ter Haar, *Men of Physics: L. D. Landau*, vol. **I**, p. 138, Pergamon Press, Oxford. [II:396]

Glassgold, A. E., Heckrotte, W., and Watson, K. M. (1959), *Ann. Phys.* **6**, 1. [II:383, 436]

Glauber, R. J. (1959), in *Lectures in Theoretical Physics*, vol. 1, eds. W. E. Brittin and L. G. Dunham, Wiley (Interscience), New York, N.Y. [I:271]

Glendenning, N. K. (1965), *Phys. Rev.* **137**, B102. [I:425]

Glenn, J. E., Pryor, R. J., and Saladin, J. X. (1969), *Phys. Rev.* **188**, 1905. [II:536]

Gneuss, G., and Greiner, W. (1971), *Nuclear Phys.*, **A171**, 449. [II:545]

Goldberger, M. L., and Watson, K. M. (1964), *Collision Theory*, Wiley, New York, N.Y.

Golden, S., and Bragg, J. K. (1949), *J. Chem. Phys.* **17**, 439. [II:190]

Goldhaber, G. (1967), in *Proc. 13th Intern. Conf. on High-Energy Physics*, University of California Press, Berkeley, Cal. [I:64]

Goldhaber, M., and Teller, E. (1948), *Phys. Rev.* **74**, 1046. [II:474]

Goldhaber, M., and Sunyar, A. W. (1951), *Phys. Rev.* **83**, 906. [I:226; II:2, 540]

Goldhaber, M., and Hill, R. D. (1952), *Rev. Mod. Phys.* **24**, 179. [I:329]

Goldstein, H. (1963), in *Fast Neutron Physics*, Part 2, p. 1525, eds. J. B. Marion and J. L. Fowler, Wiley (Interscience), New York, N.Y. [I:185]

Goldstone, J., and Gottfried, K. (1959), *Nuovo cimento* **13**, 849. [II:436]

Gomez, L. C., Walecka, J. D., and Weisskopf, V. F. (1958), *Ann. Phys.* **3**, 241. [I:251]

Gor'kov, L. P., and Éliashberg, G. M. (1965), *J. Exptl. Theoret. Phys. (USSR)* **48**, 1407; transl. *Soviet Phys. JETP* **21**, 940. [II:587]

Gorodetzky, S., Mennrath, P., Benenson, W., Chevallier, P., and Scheibling, F. (1963), *J. phys. radium* **24**, 887. [II:26]

Gorodetzky, S., Beck, F., and Knipper, A. (1966), *Nuclear Phys.* **82**, 275. [II:517]

Gorodetzky, S., Freeman, R. M., Gallmann, A., and Haas, F. (1966), *Phys. Rev.* **149**, 801. [I:44]

Goshal, S. N. (1950), *Phys. Rev.* **80**, 939. [I:186]

Gottfried, K. (1956), *Phys. Rev.* **103**, 1017. [II:212]
Gottfried, K. (1963), *Ann. Phys.* **21**, 29. [I:141]
Gottfried, K., and Yennie, D. R. (1969), *Phys. Rev.* **182**, 1595. [II:480]
Goulard, B., and Fallieros, S. (1967), *Can. J. Phys.* **45**, 3221. [II:493]
Green, A. E. S. (1958), *Rev. Mod. Phys.* **30**, 569. [II:641]
Green, A. E. S., and Engler, N. A. (1953), *Phys. Rev.* **91**, 40. [I:168]
Green, A. E. S., and Sharma, R. D. (1965), *Phys. Rev. Letters* **14**, 380. [I:245]
Greenberg, J. S., Bromley, D. A., Seaman, G. C., and Bishop, E. V. (1963), in *Proc. Third Conf. on the Reactions between Complex Nuclei*, eds. A. Ghiorso et al., p. 295, University of California Press, Berkeley, Cal. [II:254]
Greenlees, G. W., and Pyle, G. J. (1966), *Phys. Rev.* **149**, 836. [I:235, 236, 237, 238]
Greenlees, G. W., Pyle, G. J., and Tang, Y. C. (1966), *Phys. Rev. Letters* **17**, 33. [I:238]
Griffin, J. J. (1957), *Phys. Rev.* **108**, 328. [II:436]
Griffin, J. J., and Wheeler, J. A. (1957), *Phys. Rev.* **108**, 311. [II:436]
Griffin, J. J., and Rich, M. (1960), *Phys. Rev.* **118**, 850. [II:83, 641]
Grin, Yu., T. (1967), *Yadernaya Fizika* **6**, 1181; transl. *Soviet J. Nuclear Phys.* **6**, 858. [II:110]
Grin, Yu. T., and Pavlichenkow, I. M. (1964), *Phys. Letters* **9**, 249. [II:61, 110]
Grodzins, L. (1968), *Ann. Rev. Nuclear Sci.* **18**, 291.
Groshev, L. V., Demidov, A. M., Pelekhov, V. I., Sokolovskii, L. L., Bartholomew, G. A., Doveika, A., Eastwood, K. M., and Monaro, S. (1968), *Nuclear Data* **A5**, 1.
Groshev, L. V., Demidov, A. M., Pelekhov, V. I., Sokolovskii, L. L., Bartholomew, G. A., Doveika, A., Eastwood, K. M., and Monaro, S. (1969), *Nuclear Data* **A5**, 243.
Grosse, E., Dost, M., Haberkant, K., Hertel, J. W., Klapdor, H. V., Körner, H. J., Proetel, D., and von Brentano, P. (1971), *Nuclear Phys.* **A174**, 525. [II:517]
Grotdal, T., Nybø, K., and Elbek, B. (1970), *Mat. Fys. Medd. Dan. Vid. Selsk.* **37**, no. 12. [II:230, 301]
Grotdal, T., Løset, L., Nybø, K., and Thorsteinsen, T. F. (1973), *Nuclear Phys.* **A211**, 541. [II:117]
Grover, J. R. (1967), *Phys. Rev.* **157**, 832. [II:41]
Günther, C., and Parsignault, D. R. (1967), *Phys. Rev.* **153**, 1297. [II:160, 161]
Günther, C., Hübel, H., Kluge, A., Krien, K., and Toschinski, H. (1969), *Nuclear Phys.* **A123**, 386. [II:104, 105]
Günther, C., Kleinheinz, P., Casten, R. F., and Elbek, B. (1971), *Nuclear Phys.* **A172**, 273. [II:463, 549]
Gustafson, C., Lamm, I. L., Nilsson, B., and Nilsson, S. G. (1967), *Arkiv Fysik* **36**, 613. [I:205; II:219, 222, 367, 469]
Gustafson, C., Möller, P., and Nilsson, S. G. (1971), *Phys. Letters* **34B**, 349. [II:213]
Gutzwiller, M. C. (1971), *J. Math. Phys.* **12**, 343. [II:587]

Häusser, O., Hooton, B. W., Pelte, D., Alexander, T. K., and Evans, H. C. (1969), *Phys. Rev. Letters* **23**, 320. [II:287]
Häusser, O., Khanna, F. C., and Ward, D. (1972), *Nuclear Phys.* **A194**, 113. [II:517]
Hafele, J. C., and Woods, R. (1966), *Phys. Letters* **23B**, 579. [II:571]
Hagemann, G. B., Herskind, B., Olesen, M. C., and Elbek, B. (1969), in *Contributions to Intern. Conf. on Properties of Nuclear States*, p. 29, Les Presses de

l'Université de Montreal, Montreal. [II:535]
Hahn, B., Ravenhall, D. G., and Hofstadter, R. (1956), *Phys. Rev.* **101**, 1131. [I:159]
Halbert, M. L., and Zucker, A. (1961), *Phys. Rev.* **121**, 236. [I:45]
Halbleib, J. A., Sr., and Sorensen, R. A. (1967), *Nuclear Phys.* **A98**, 542. [I:347]
Halpern, I. (1959), *Ann. Rev. Nuclear Sci.* **9**, 245. [II:616]
Halpern, I., and Strutinsky, V. M. (1958), *Proc. Second Intern. Conf. on the Peaceful Uses of Atomic Energy* **15**, 408 P/1513, United Nations, Geneva. [II:617]
Hama, Y., and Hoshizaki, N. (1964), *Compt. Rend. Congrès Intern. de Physique Nucléaire*, Paris 1963, vol. II, p. 195, ed. P. Gugenberger, C.N.R.S., Paris. [I:263]
Hamada, T., and Johnston, I. D. (1962), *Nuclear Phys.* **34**, 382. [I:266]
Hamamoto, I. (1969), *Nuclear Phys.* **A126**, 545. [II:571, 576]
Hamamoto, I. (1970), *Nuclear Phys.* **A141**, 1 and *ibid.* **A155**, 362. [II:571]
Hamamoto, I. (1971), *Nuclear Phys.* **A177**, 484. [II:282]
Hamamoto, I. (1972), *Nuclear Phys.* **A196**, 101. [II:423]
Hamamoto, I. (1973), *Nuclear Phys.* **A205**, 225. [II:488]
Hamamoto, I., and Siemens, P. (1974), see I. Hamamoto, in *Proc. of Topical Conf. on Problems of Vibrational Nuclei*, Zagreb. [II:430]
Hamermesh, M. (1962), *Group Theory and its Application to Physical Problems*, Addison-Wesley, Reading, Mass.
Hamilton, J., and Woolcock, W. S. (1963), *Rev. Mod. Phys.* **35**, 737. [I:5]
Hand, L. N., Miller, D. G., and Wilson, R. (1963), *Rev. Mod. Phys.* **35**, 335. [I:5, 386]
Hansen, O., and Nathan, O. (1971), *Phys. Rev. Letters* **27**, 1810. [II:390]
Hansen, O., Olesen, M. C., Skilbreid, O., and Elbek, B. (1961), *Nuclear Phys.* **25**, 634. [II:109, 133]
Hansen, P. G. (1964), *Experimental Investigation of Decay Schemes of Deformed Nuclei*, Risø Report 92. [II:36, 142]
Hansen, P. G., Nielsen, O. B., and Sheline, R. K. (1959), *Nuclear Phys.* **12**, 389. [II:159]
Hansen, P. G., Wilsky, K., Baba, C. V. K., and Vandenbosch, S. E. (1963), *Nuclear Phys.* **45**, 410. [II:144]
Hansen, P. G., Nielsen, H. L., Wilsky, K., Agarwal, Y. K., Baba, C. V. K., and Bhattacherjee, S. K. (1966), *Nuclear Phys.* **76**, 257. [II:143]
Hansen, P. G., Nielsen, H. L., Wilsky, K., and Cuninghame, J. G. (1967), *Phys. Letters* **24B**, 95. [I:53]
Hansen, P. G., Hornshøj, P., and Johansen, K. H. (1969), *Nuclear Phys.* **A126**, 464. [II:154, 155]
Harari, H., and Rashid, M. A. (1966), *Phys. Rev.* **143**, 1354. [I:62]
Harchol, M., Jaffe, A. A., Miron, J., Unna, I., and Zioni, Z. (1967), *Nuclear Phys.* **A90**, 459. [I:146]
Harmatz, B., and Handley, T. H. (1968), *Nuclear Phys.* **A121**, 481. [II:535]
Harris, S. M. (1965), *Phys. Rev.* **138**, B509. [II:25, 31]
Harrison, B. K., Thorne, K. S., Wakano, M., and Wheeler, J. A. (1965), *Gravitation Theory and Gravitational Collapse*, University of Chicago Press, Chicago, Ill. [I:205]
Hasinoff, M., Fisher, G. A., Kuan, H. M., and Hanna, S. S. (1969), *Phys. Letters* **30B**, 337. [II:498]
Hass, M., Moreh, R., and Salzmann, D. (1971), *Phys. Letters* **36B**, 68. [II:492]

Haverfield, A. J., Bernthal, F. M., and Hollander, J. M. (1967), *Nuclear Phys.* **A94**, 337. [II:57, 106, 108, 109, 110]
Haxel, O., Jensen, J. H. D., and Suess, H. E. (1949), *Phys. Rev.* **75**, 1766. [I:189, 209; II:v]
Haxel, O., Jensen, J. H. D., and Suess, H. E. (1950), *Z. Physik* **128**, 295 [I:226]
Hayward, E. (1965), in *Nuclear Structure and Electromagnetic Interactions*, p. 141, ed. N. MacDonald, Oliver and Boyd, Edinburgh. [I:383; II:476, 478]
Hecht, K. T., and Satchler, G. R. (1962), *Nuclear Phys.* **32**, 286. [II:187]
Heestand, G. M., Borchers, R. R., Herskind, B., Grodzins, L., Kalish, R., and Murnick, D. E. (1969), *Nuclear Phys.* **A133**, 310. [II:55]
Heisenberg, J. H., and Sick, I. (1970), *Phys. Letters* **32B**, 249. [II:350]
Heisenberg, W. (1925), *Z. Physik* **33**, 879. [II:400]
Heisenberg, W. (1932), *Z. Physik* **77**, 1. [I:31, 243]
Heisenberg, W. (1932a), *Z. Physik* **78**, 156. [II:641]
Heisenberg, W. (1934), in *Rapports et Discussions du Septième Conseil de Physique de l'Institut Intern. Solvay*, p. 284, Gauthier-Villars, Paris. [II:v]
Heitler, W. (1954), *The Quantum Theory of Radiation*, 3rd edition, Clarendon Press, Oxford.
Helm, R. H. (1956), *Phys. Rev.* **104**, 1466. [II:134]
Helmer, R. G., and Reich, C. W. (1968), *Nuclear Phys.* **A114**, 649. [II:73]
Helmers, K. (1960), *Nuclear Phys.* **20**, 585. [II:22]
Hemmer, P. C. (1962), *Nuclear Phys.* **32**, 128. [II:204]
Hendrie, D. L., Glendenning, N. K., Harvey, B. G., Jarvis, O. N., Duhm, H. H., Mahoney, J., and Saudinos, J. (1968), *Japan. J. Phys., Suppl.* **24**, 306. [II:140]
Hendrie, D. L., Glendenning, N. K., Harvey, B. G., Jarvis, O. N., Duhm, H. H., Saudinos, J., and Mahoney, J. (1968a), *Phys. Letters* **26B**, 127. [II:140]
Henley, E. M. (1966), in *Isobaric Spin in Nuclear Physics*, p. 1, eds. J. D. Fox and D. Robson, Academic Press, New York, N.Y. [I:242]
Henley, E. M., and Jacobsohn, B. A. (1959), *Phys. Rev.* **113**, 225. [I:21]
Henley, E. M., and Thirring, W. (1962), *Elementary Quantum Field Theory*, McGraw Hill, New York, N.Y.
Herczeg, P. (1963), *Nuclear Phys.* **48**, 263. [I:67]
Herman, R., and Hofstadter, R. (1960), *High Energy Electron Scattering Tables*, Stanford University Press, Stanford, Cal. [I:159]
Herring, C. (1966), in *Magnetism*, vol. IV, eds. G. T. Rado and H. Suhl, Academic Press, New York, N.Y. [II:213]
Hertel, J. W., Fleming, D. G., Schiffer, J. P., and Gove, H. E. (1969), *Phys. Rev. Letters* **23**, 488. [II:517, 565]
Herzberg, G. (1945), *Molecular Spectra and Molecular Structure*, vol. II: *Infrared and Raman Spectra of Polyatomic Molecules*, Van Nostrand, New York, N.Y.
Herzberg, G. (1950), *Molecular Spectra and Molecular Structure*, vol. I: *Spectra of the Diatomic Molecules*, 2nd edition, Van Nostrand, New York, N.Y.
Herzberg, G. (1966), *Molecular Spectra and Molecular Structure*, vol. III: *Electronic Spectra and Electronic Structure of Polyatomic Molecules*, Van Nostrand, Princeton, N.J.
Hiebert, J. C., Newman, E., and Bassel, R. H. (1967), *Phys. Rev.* **154**, 898. [I:233]
Hill, D. L., and Wheeler, J. A. (1953), *Phys. Rev.* **89**, 1102. [II:vi, 91, 212, 367, 373, 677]
Hillman, P., Johansson, A., and Tibell, G. (1958), *Phys. Rev.* **110**, 1218. [I:30, 31]
Hinds, S., and Middleton, R. (1966), *Nuclear Phys.* **84**, 651. [I:322]

Hinds, S., Middleton, R., and Litherland, A. E. (1961), in *Proceedings of the Rutherford Jubilee Intern. Conf.*, ed. J. B. Birks, Heywood and Co., London. [II:284]

Hinds, S., Middleton, R., Bjerregaard, J. H., Hansen, O., and Nathan, O. (1965), *Phys. Letters* **17**, 302. [I:324]

Hinds, S., Bjerregaard, J. H., Hansen, O., and Nathan, O. (1965), *Phys. Letters* **14**, 48. [II:535]

Hirzel, O., and Wäffler, H. (1947), *Helv. Phys. Acta* **20**, 373. [II:474]

Hjorth, S. A., and Ryde, H. (1970), *Phys. Letters* **31B**, 201. [II:310]

Hjorth, S. A., Ryde, H., Hagemann, K. A., Løvhøiden, G., and Waddington, J. C (1970), *Nuclear Phys.* **A144**, 513. [II:298]

Hodgson, P. E. (1964), in *Compt. Rend. du Congrès Intern. de Physique Nucléaire*, vol. I, p. 257, ed. P. Gugenberger, C.N.R.S., Paris. [I:215, 238]

Hodgson, P. E. (1971), *Nuclear Reactions and Nuclear Structure*, Clarendon Press, Oxford.

Högaasen-Feldman, J. (1961), *Nuclear Phys.* **28**, 258. [II:386]

Hönl, H., and London, F. (1925), *Z. Physik* **33**, 803. [II:22]

Hoffberg, M., Glassgold, A. E., Richardson, R. W., and Ruderman, M. (1970), *Phys. Rev. Letters* **24**, 775. [II:395]

Hofstadter, R. (1957), *Ann. Rev. Nuclear Sci.* **7**, 231. [I:233]

Hofstadter, R., ed. (1963), *Nuclear and Nucleon Structure*, Benjamin, New York, N.Y.

Holland, G. E., Stein, N., Whitten, C. A., Jr., and Bromley, D. A. (1968), in *Proc. Intern. Conf. on Nuclear Structure*, p. 703, ed. J. Sanada, Phys. Soc. of Japan, see also, Bromley, D. A., *ibid.* p. 251. [II:642]

Horikawa, Y., Torizuka, Y., Nakada, A., Mitsunobu, S., Kojima, Y., and Kimura, M. (1971), *Phys. Letters* **36B**, 9. [II: 141]

Hosono, K. (1968), *J. Phys. Soc. (Japan)* **25**, 36. [II:293]

Howard, A. J., Pronko, J. G., and Whitten, C. A., Jr., *Phys. Rev.* **184**, 1094. [II: 286]

Hoyle, F., and Fowler, W. A. (1963), *Nature* **197**, 533. [I:202]

Huang, K., and Yang, C. N. (1957), *Phys. Rev.* **105**, 767. [I:256]

Huber, M. G., Danos, M., Weber, H. J., and Greiner, W. (1967), *Phys. Rev.* **155**, 1073. [II:453, 505]

Hübel, H., Günther, C., Krien, K., Toschinski, H., Speidel, K.-H., Klemme, B., Kumbartzki, G., Gidefeldt, L., and Bodenstedt, E. (1969), *Nuclear Phys.* **A127**, 609. [II: 108]

Huffaker, J. N., and Laird, C. E. (1967), *Nuclear Phys.* **A92**, 584. [I:415, 416]

Hughes, V. W. (1964), in *Gravitation and Relativity*, Chap. 13, eds. H.-Y. Chiu and W. F. Hoffmann, Benjamin, New York, N.Y. [I:5]

Huizenga, J. R., and Moretto, L. G. (1972), *Ann. Rev. Nuclear Sci.* **22**, 427. [II: 607]

Huizenga, J. R., Chaudhry, R., and Vandenbosch, R. (1962), *Phys. Rev.* **126**, 210. [II:616]

Huizenga, J. R., Behkami, A. N., Atcher, R. W., Sventek, J. S., Britt, H. C., and Freiesleben, H. (1974), *Nuclear Phys.* **A223**, 589. [II:607]

Hull, M. H., Jr., Lassila, K. E., Ruppel, H. M., MacDonald, F. A., and Breit, G. (1962), *Phys. Rev.* **128**, 830. [I:265, 271]

Hulthén, L. M., and Sugawara, M. (1957), *Encyclopedia of Physics*, vol. 39, Springer, Berlin. [I:269]

Humblet, J. (1967), in *Fundamentals of Nuclear Theory*, eds. A. de-Shalit and C. Villi, Intern. Atomic Energy Agency, Vienna. [I:432]

Hund, F. (1927), *Z. Physik* **42**, 93. [II: 12]

Hund, F. (1937), *Z. Physik* **105**, 202. [I: 38]
Hunt, W. E., Mehta, M. K., and Davis, R. H. (1967), *Phys. Rev.* **160**, 782. [II: 97]
Hurwitz, H., Jr., and Bethe, H. A. (1951), *Phys. Rev.* **81**, 898. [II: 610]
Huus, T., and Zupančič, Č. (1953), *Mat. Fys. Medd. Dan. Vid. Selsk.* **28**, no. 1 [II: 2]
Hyde, E. K., Perlman, I., and Seaborg, G. T. (1964), *The Nuclear Properties of the Heavy Elements*, vol. I: *Systematics of Nuclear Structure and Radioactivity*, Prentice Hall, Englewood Cliffs, N.J.
Hyde, E. K. (1964), *The Nuclear Properties of the Heavy Elements*, vol. III: *Fission Phenomena*, Prentice Hall, Englewood Cliffs, N.J.

Igo, G., Barnes, P. D., and Flynn, E. R. (1971), *Ann. Phys.* **66**, 60. [II: 646]
Igo, G. J., Flynn, E. R., Dropesky, B. J., and Barnes, P. D. (1971), *Phys. Rev.* **C3**, 349. [II: 647]
Ikeda, K., Kobayasi, M., Marumori, T., Shiozaki, T., and Takagi, S. (1959), *Progr. Theor. Phys. (Kyoto)* **22**, 663. [II: 436]
Ikeda, K., Marumori, T., Tamagaki, R., Tanaka, H., Hiura, J., Horiuchi, H., Suzuki, Y., Nemoto, F., Bando, H., Abe, Y., Takigawa, N., Kamimura, M., Takada, K., Akaishi, Y., and Nagata, S. (1972), *Progr. Theoret. Phys. (Kyoto)*, Suppl. **52**. [II: 99]
Inglis, D. R. (1954), *Phys. Rev.* **96**, 1059. [II: 76]
Inglis, D. (1955), *Phys. Rev.* **97**, 701. [II: 436, 508]
Inopin, E. V. (1956), *J. Exptl. Theoret. Phys. (USSR)* **30**, 210 and **31**, 901; transl. *Soviet Phys. JETP* **3**, 134 and **4**, 764. [II: 322]
Ishizaki, Y., Yoshida, Y., Saji, Y., Ishimatsu, T., Yagi, K., Matoba, M., Huang, C. Y., and Nakajima, Y. (1967), in *Contributions to Intern. Conf. on Nuclear Structure*, Tokyo, p. 133. [II: 535]
Itzykson, C., and Nauenberg, M. (1966), *Rev. Mod. Phys.* **38**, 121. [I: 114, 126, 129]

Jackson, H. E., and Wetzel, K. J. (1972), *Phys. Rev. Letters* **28**, 513. [II: 492]
Jackson, J. D. (1962), *Classical Electrodynamics*, Wiley, New York, N.Y.
Jackson, J. D., and Blatt, J. M. (1950), *Rev. Mod. Phys.* **22**, 77. [I: 241]
Jackson, K. P., Ram, K. B., Lawson, P. G., Chapman, N. G., and Allen, K. W. (1969), *Phys. Letters* **30B**, 162. [II: 98]
Jacob, G., and Maris, Th. A. J. (1966), *Rev. Mod. Phys.* **38**, 121 [I: 232]
Jacob, M., and Wick, G. C. (1959), *Ann. Phys.* **7**, 404. [I: 77, 429]
Jacobsohn, B. A. (1954), *Phys. Rev.* **96**, 1637. [II: 134]
Jarvis, O. N., Harvey, B. G., Hendrie, D. L., and Mahoney, J. (1967), *Nuclear Phys.* **102**, 625. [II: 353, 354]
Jastrow, R. (1950), *Phys. Rev.* **79**, 389. [I: 245]
Jensen, J. H. D., and Jensen, P. (1950), *Z. Naturforsch.* **5a**, 343. [II: 474]
Jensen, J. H. D., and Mayer, M. G. (1952), *Phys. Rev.* **85**, 1040. [I: 340, 394]
Jett, J. H., Lind, D. A., Jones, G. D., and Ristinen, R. A. (1968), *Bull. Am. Phys. Soc.* **13**, 671. [II: 168]
Jørgensen, M., Nielsen, O. B., and Sidenius, G. (1962), *Phys. Letters* **1**, 321. [II: 105]
Johnson, A., Rensfelt, K.-G., and Hjorth, S. A. (1969), *Annual Report AFI* (Research Institute of Physics, Stockholm), p. 23. [II: 297]
Johnson, A., Ryde, H., and Hjorth, S. A. (1972), *Nuclear Phys.* **A179**, 753. [II: 42]
Jolly, R. K. (1965), *Phys. Rev.* **139**, B318. [II: 354]
Jones, K. W., Schiffer, J. P., Lee, L. L., Jr., Marinov, A., and Lerner, J. L. (1966), *Phys. Rev.* **145**, 894. [I: 323]
Josephson, B. D. (1962), *Phys. Letters* **1**, 251. [II: 396]

Jurney, E. T. (1969), *Neutron Capture Gamma-Ray Spectroscopy*, Proc. Intern. Symposium Studsvik, p. 431, Intern. Atomic Energy Agency, Vienna. [II:274]

Källén, G. (1967), *Nuclear Phys.* **B1**, 225. [I:402]
Kapitza, S. P., Rabotnov, N. S., Smirenkin, G. N., Soldatov, A. S., Usachev, L. N., and Tsipenyuk, Yu. M. (1969), *Zh. Eksper. Fiz. Pis'ma (USSR)* **9**, 128; transl. *JETP Letters* **9**, 73. [II:634]
Karamyan, S. A., Kuznetsov, I. V., Muzycka, T. A., Oganesyan, Yu. Ts., Penionzkevich. Yu. E., and Pustyl'nik, B. I. (1967), *Yadernaya Fiz.* **6**, 494; transl. *Soviet J. Nuclear Phys.* **6**, 360. [II:622]
Karnaukhov, V. A., and Ter-Akopyan, G. M. (1964), *Phys. Letters* **12**, 339. [I:204]
Katz, L., Baerg, A. P., and Brown, F. (1958), in *Proc. Second Intern. Conf. on the Peaceful Uses of Atomic Energy* **15**, 188, United Nations, Geneva. [II:126]
Kaufmann, E. N., Bowman, J. D., and Bhattacherjee, S. K. (1968), *Nuclear Phys.* **A119**, 417. [II:104]
Kavanagh, R. W. (1964), *Phys. Rev.* **133**, B1504. [I:417]
Kazarinov, Yu. M., and Simonov, Yu. N. (1962), *Zhur. Eksp. i Teoret. Fiz.* **43**, 35; transl. *Soviet Phys. JETP* **16**, 24. [I:243]
Kelly, M. A., Berman, B. L., Bramblett, R. L., and Fultz, S. C. (1969), *Phys. Rev.* **179**, 1194. [II:491]
Kelson, I., and Levinson, C. A. (1964), *Phys. Rev.* **134**, B269. [I:330; II:283]
Kemmer, N., Polkinghorne, J. C., and Pursey, D. L. (1959), *Reports on Progress in Physics* **22**, 368. [I:21]
Kerman, A. K. (1956), *Mat. Fys. Medd. Dan. Vid. Selsk.* **30**, no. 15. [II:147, 203]
Kerman, A. K., and Shakin, C. M. (1962), *Phys. Letters* **1**, 151. [II:451]
Kerman, A. K., and Klein, A. (1963), *Phys. Rev.* **132**, 1326. [II:412]
Kerman, A. K., and Quang, H. K. (1964), *Phys. Rev.* **135**, B883. [II:455]
Kerman, A. K., McManus, H., and Thaler, R. M. (1959), *Ann. Phys.* **8**, 551. [I:215, 271]
Kerman, A. K., Svenne, J. P., and Villars, F. M. H. (1966), *Phys. Rev.* **147**, 710. [I:330]
Khan, A. M., and Knowles, J. W. (1972), *Nuclear Phys.* **A179**, 333. [II:126]
King, G. W., Hainer, R. M., and Cross, P. C. (1943), *J. Chem. Phys.* **11**, 27. [II:182, 185]
King, G. W., Hainer, R. M., and Cross, P. C. (1949), *J. Chem. Phys.* **17**, 826. [II:182, 185]
Kirsten, T., Schaeffer, O. A., Norton, E., and Stoenner, R. W. (1968), *Phys. Rev. Letters* **20**, 1300. [I:398]
Kishimoto, T., and Tamura, T. (1971), *Nuclear Phys.* **A163**, 100. [II:692]
Kisslinger, L. (1955), *Phys. Rev.* **98**, 761. [I:219]
Kisslinger, L. S., and Sorensen, R. A. (1960), *Mat. Fys. Medd. Dan. Vid. Selsk.* **32**, no. 9. [II:520, 641]
Kistner, O. C. (1967), *Phys. Rev. Letters* **19**, 872. [I:21]
Knowles, J. W. (1965), in *Alpha-, Beta-, and Gamma-Ray Spectroscopy*, vol. **1**, p. 203, ed. K. Siegbahn, North-Holland, Amsterdam. [II:62]
Kobayasi, M., and Marumori, T. (1960), *Progr. Theor. Phys.* (Kyoto) **23**, 387. [II:436]
Koch, H. R. (1965), thesis subm. to Techn. Hochschule München. [II:63, 65]
Koch, H. R. (1966), *Z. Physik* **192**, 142. [II:63, 65]
Körner, H. J., Auerbach, K., Braunsfurth, J., and Gerdau, E. (1966), *Nuclear Phys.* **86**, 395. [I:341, 343]

Kofoed-Hansen, O. (1965), in *Alpha-, Beta-, and Gamma-Ray Spectroscopy*, vol. 2, p. 1517, ed. K. Siegbahn, North-Holland, Amsterdam. [I:405]
Kohn, W., and Luttinger, J. M. (1965), *Phys. Rev. Letters* **15**, 524. [I:176]
Koike, M., Nonaka, I., Kokame, J., Kamitsubo, H., Awaya, Y., Wada, T., and Nakamura, H. (1969), *Nuclear Phys.* **A125**, 161. [II:350]
Kolar, W., and Böckhoff, K. H. (1968), *J. Nuclear Energy* **22**, 299. [II:623, 628, 630]
Kolesov, V. E., Korotkich, V. L., and Malashkina, V. G. (1963), *Izv. Akad. Nauk* **27**, 903. [I:359]
Komar, A. P., Vorobiev, A. A., Zalite, Yu. K., and Korolev, G. A. (1970), *Dokl. Akad. Nauk (SSSR)* **191**, 61. [II:517]
Konopinski, E. J. (1966), *The Theory of Beta Radioactivity*, Oxford University Press, Oxford.
Konopinski, E. J., and Uhlenbeck, G. E. (1941), *Phys. Rev.* **60**, 308. [I:410]
Konopinski, E. J., and Rose, M. E. (1965), in *Alpha-, Beta- and Gamma-Ray Spectroscopy*, vol. 2, p. 1327, ed. K. Siegbahn, North-Holland, Amsterdam. [I:410]
Kopfermann, H. (1958), *Nuclear Moments*, Academic Press, New York, N.Y.
Kownacki, J., Ryde, H., Sergejev, V. O., and Sujkowski, Z. (1972), *Nuclear Phys.* **A196**, 498. [II:535]
Kramers, H. A. (1930), *Proc. Koninkl. Ned. Akad. Wetenschap.* **33**, 959. [I:19]
Krane, K. S. (1974), *Phys. Rev.* **10C**, 1197. [II:357]
Kratschmer, W., Klapdor, H. V., and Grosse, E. (1973), *Nuclear Phys.* **A201**, 179. [II:517]
Kraushaar, J. J., and Goldhaber, M. (1953), *Phys. Rev.* **89**, 1081. [II:357, 539]
Kuchnir, F. T., Axel, P., Criegee, L., Drake, D. M., Hanson, A. O., and Sutton, D. C. (1967), *Phys. Rev.* **161**, 1236. [II:482, 505, 506]
Kuehner, J. A., and Almqvist, E. (1967), *Can. J. Phys.* **45**, 1605. [II:96, 97]
Kugel, H. W., Kalish, R., and Borchers, R. R. (1971), *Nuclear Phys.* **A167**, 193. [II:55]
Kuhn, H. G., and Turner, R. (1962), *Proc. Roy. Soc. (London)* **265A**, 39. [I:164]
Kuhn, W. (1925), *Z. Physik* **33**, 408. [II:400]
Kumabe, I., Ogata, H., Kim, T. H., Inoue, M., Okuma, Y., and Matoba, M. (1968), *J. Phys. Soc. Japan* **25**, 14. [II:354]
Kumar, K. (1967), *Nuclear Phys.* **A92**, 653. [II:545]
Kumar, K., and Baranger, M. (1968), *Nuclear Phys.* **A122**, 273. [II:433, 677]
Kurath, D. (1950), *Phys. Rev.* **80**, 98. [II:540]
Kurath, D. (1963), *Phys. Rev.* **130**, 1525. [II:383]
Kurath, D., and Pičman, L. (1959), *Nuclear Phys.* **10**, 313. [II:212]

Lamb, H. (1916), *Hydrodynamics*, 4th edition, Cambridge University Press, Cambridge.
Landau, L. (1937), *Physik. Z. Sowjetunion* **11**, 556. [I:154]
Landau, L. D. (1941), *J. Phys. (USSR)* **5**, 71. [II:326]
Landau, L. (1946), *J. Phys. (USSR)* **10**, 25. [II:440]
Landau, L. D. (1956), *Zhur. Eksp. i Teoret. Fiz.* **30**, 1058; transl. *Soviet Phys. JETP* **3**, 920. [I:334; II:483]
Landau, L. D. (1958), *Zhur. Eksp. i Teoret. Fiz.* **35**, 97; transl. *Soviet Phys. JETP* **8**, 70. [I:334]
Landau, L. D., and Lifschitz, E. M. (1958), *Quantum Mechanics, Nonrelativistic Theory*, Pergamon Press, London and Paris.

Lane, A. M. (1962), *Nuclear Phys.* **35**, 676. [I : 149]

Lane, A. M., and Thomas, R. G. (1958), *Rev. Mod. Phys.* **30**, 257. [I : 433, 434]

Lane, A. M., and Pendlebury, E. D. (1960), *Nuclear Phys* **15**, 39. [II : 556]

Lane, A. M., Thomas, R. G., and Wigner, E. P. (1955), *Phys. Rev.* **98**, 693. [I : 302]

Lane, S. A., and Saladin, J. X. (1972), *Phys. Rev.* **C6**, 613. [II : 536]

Lanford, W. A., and McGrory, J. B. (1973), *Phys. Letters* **45B**, 238. [II : 646]

Lang, J. M. B., and Le Couteur, K. J. (1954), *Proc. Phys. Soc. (London)* **67A**, 586. [I : 281]

Lassila, K. E., Hull, M. H., Jr., Ruppel, H. M., MacDonald, F. A., and Breit, G. (1962), *Phys. Rev.* **126**, 881. [I : 266]

Lauritsen, T., and Ajzenberg-Selove, F. (1961), in *Landolt-Börnstein*, Neue Serie, vol. 1, Springer, Berlin. [I : 321]

Lauritsen, T., and Ajzenberg-Selove, F. (1962), *Nuclear Data Sheets*, The Nuclear Data Group, Oak Ridge Nat. Lab., Oak Ridge, Tenn.

Lauritsen, T., and Ajzenberg-Selove, F. (1966), *Nuclear Phys.* **78**, 10.

Lederer, C. M., Poggenburg, J. K., Asaro, F., Rasmussen, J. O., and Perlman, I. (1966), *Nuclear Phys.* **84**, 481. [II : 266]

Lederer, C. M., Hollander, J. M., and Perlman, I. (1967), *Table of Isotopes*, 6th edition, Wiley, New York, N.Y.

Lederer, C. M., Asaro, F., and Perlman, I. (1970) quoted in *Nuclear Data* **B4**, 652. [II : 554]

Lee, L. L., Jr., and Schiffer, J. P. (1964), *Phys. Rev.* **136**, B405. [I : 357]

Lee, L. L., Jr., Schiffer, J. P., Zeidman, B., Satchler, G. R., Drisko, R. M., and Bassel, R. H. (1964), *Phys. Rev.* **136**, B971. [I : 357, 425]

Lee, T. D., and Yang, C. N. (1956), *Phys. Rev.* **104**, 254. [I : 15]

Lee, T. D., and Wu, C. S. (1965), *Ann. Rev. Nuclear Sci.* **15**, 381. [I : 21, 397, 398, 402, 405]

Lee, T. D., and Wu, C. S. (1966), *Ann. Rev. Nuclear Sci.* **16**, 471, 511. [I : 21]

Lee, W. Y., Bernow, S., Chen, M. Y., Cheng, S. C., Hitlin, D., Kast, J. W., Macagno, E. R., Rushton, A. M., Wu, C. S., and Budicky, B. (1969), *Phys. Rev. Letters* **23**, 648. [II : 533]

Legett, A. J. (1972), *Phys. Rev. Letters* **29**, 1227. [II : 395]

Leisi, H. J., Dey, W., Ebersold, P., Engfer, R., Scheck, F., and Walter, H. K. (1973), *Suppl. to J. Phys. Soc. Japan* **34**, 355. [II : 132, 137]

Lemonne, J., Mayeur, C., Sacton, J., Vilain, P., Wilquet, G., Stanley, D., Allen, P., Davis, D. H., Fletcher, E. R., Garbutt, D. A., Shaukat, M. A., Allen, J. E., Bull, V. A., Conway, A. P., and March, P. V. (1965), *Phys. Letters* **18**, 354. [I : 56]

Le Tourneux, J. (1965), *Mat. Fys. Medd. Dan. Vid. Selsk.* **34**, no. 11. [II : 454, 455, 457, 503, 692]

Levinger, J. S. (1951), *Phys. Rev.* **84**, 43. [II : 479]

Levinger, J. S. (1960), *Nuclear Photo-Disintegration*, Oxford University Press, Oxford.

Levinger, J. S., and Bethe, H. (1950), *Phys. Rev.* **78**, 115. [II : 412, 414]

Levinger, J. S., and Simmons, L. M. (1961), *Phys. Rev.* **124**, 916. [I : 204]

Levinson, C. A., Lipkin, H. J., and Meshkov, S. (1963), *Phys. Letters* **7**, 81. [I : 40]

Levi-Setti, R. (1964), *Proc. Intern. Conf. on Hyperfragments*, St. Cergue, 1963, CERN Rep. 64-1, ed. W. O. Lock, CERN, Geneva. [I : 55, 56]

Lewis, M. B., and Bertrand, F. E. (1972), *Nuclear Phys.* **A196**, 337. [II : 482, 509]

Li, A. C., and Schwarzschild, A. (1963), *Phys. Rev.* **129**, 2664. [II : 129]

Lilley, J. S., and Stein, N. (1967), *Phys. Rev. Letters* **19**, 709. [I : 324]

Lindgren, I. (1965), in *Alpha-, Beta- and Gamma-Ray Spectroscopy*, vol. 2, p. 1621, ed. K. Siegbahn, North-Holland, Amsterdam.
Lindhard, J. (1954), *Mat. Fys. Medd. Dan. Vid. Selsk.* **28**, no. 8 [II:436]
Lipkin, H. J. (1965), *Lie Groups for Pedestrians*, North-Holland, Amsterdam.
Lipkin, H. J. (1967), in *Proc. Intern. Nuclear Physics Conf.*, Gatlinburg, p. 450, ed.-in-chief R. L. Becker, Academic Press, New York, N.Y. [I:42]
Listengarten, M. A. (1961), in *Gamma-Luchi*, p. 271, ed. L. A. Sliv, Akad. Nauk, SSSR, Moscow. See also Sliv, L. A., and Band, I. M. (1965), in *Alpha-, Beta-, and Gamma-Ray Spectroscopy*, vol. 2, p. 1639, ed. K. Siegbahn, North-Holland, Amsterdam. [I:384]
Litherland, A. E. (1968), *Third Symposium on the Structure of Low-Medium Mass Nuclei*, p. 92, ed. J. P. Davidson, University Press of Kansas, Lawrence, Kansas. [II:284, 291]
Litherland, A. E., McManus, H., Paul, E. B., Bromley, D. A., and Gove, H. E. (1958), *Can. J. Phys.* **36**, 378. [II:212, 284]
Littlewood, D. E. (1950), *The Theory of Group Characters*, Clarendon Press, Oxford.
Lobashov, V. M., Nazarenko, V. A., Saenko, L. F., Smotritsky, L. M., and Kharkevitch, G. I. (1967), *Phys. Letters* **25B**, 104. [I:24]
Löbner, K. E. G. (1965), *Gamma Ray Transition Probabilities in Deformed Odd-A Nuclei*, thesis, University of Amsterdam, Rototype. [II:275]
Lönsjö, O., and Hagemann, G. B. (1966), *Nuclear Phys.* **88**, 624. [II:535]
Löwdin, P. O. (1966), in *Quantum Theory of Atoms, Molecules and the Solid State*, ed. P. O. Löwdin, Academic Press, New York, N.Y. [II:91]
Lomon, E., and Feshbach, H. (1967), *Rev. Mod. Phys.* **39**, 611 [I:245]
London, F. (1950), *Superfluids*, vol. I, p. 152, Wiley, New York, N.Y. [II:398]
Lushnikov, A. A., and Zaretsky, D. F. (1965), *Nuclear Phys.* **66**, 35. [II:480]
Lynn, J. E. (1968), *The Theory of Neutron Resonance Reactions*, Clarendon Press, Oxford. [I:230]
Lynn, J. E. (1968a), in *Nuclear Structure*, Dubna Symposium. p. 463, Intern. Atomic Energy Agency, Vienna. [II:374, 622, 624]

MacDonald, J. R., Start, D. F. H., Anderson, R., Robertson, A. G., and Grace, M. A. (1968), *Nuclear Phys.* **A108**, 6. [II:49]
MacDonald, J. R., Wilkinson, D. H., and Alburger, D. E. (1971), *Phys. Rev.* **C3**, 219. [II:516]
MacDonald, N. (1964), *Phys. Letters* **10**, 334. [II:458]
MacDonald, W. M. (1956), *Phys. Rev.* **101**, 271. [I:175]
Macfarlane, M. H., and French, J. B. (1960), *Rev. Mod. Phys.* **32**, 567. [I:423, 424]
MacGregor, M. H., Moravcsik, M. J., and Stapp, H. P. (1959), *Phys. Rev.* **116**, 1248. [I:5]
MacGregor, M. H., Arndt, R. A., and Wright, R. M. (1968), *Phys. Rev.* **169**, 1128; *ibid.* **173**, 1272, and *ibid.* **182**, 1714. [I:265]
Mackintosh, R. S. (1970), thesis, University of California, Berkeley, Cal., UCRL-19529. [II:293]
Macklin, R. L., and Gibbons, J. H. (1965), *Rev. Mod. Phys.* **37**, 166. [I:201]
Macklin, R. L., and Gibbons, J. H. (1967), *Astrophys. J.* **149**, 577. [I:201]
Mafethe, M. E., and Hodgson, P. E. (1966), *Proc. Phys. Soc.* (*London*) **87**, 429. [I:418]
Maher, J. V., Erskine, J. R., Friedman, A. M., Siemssen, R. H., and Schiffer, J. P. (1972), *Phys. Rev.* **C5**, 1380. [II:553]
Maier, K. H., Nakai, K., Leigh, J. R., Diamond, R. M., and Stephens, F. S. (1972),

Nuclear Phys. **A183**, 289. [II: 486]
Majorana, E. (1933), *Z. Physik* **82**, 137. [I: 243]
Malmfors, K. G. (1965), in *Alpha-, Beta-, and Gamma-Ray Spectroscopy*, vol. 2, p. 1281, ed. K. Siegbahn, North-Holland, Amsterdam. [I: 383]
Mang, H. J. (1957), *Z. Physik* **148**, 582. [II: 267]
Mang, H. J., and Rasmussen, J. O. (1962), *Mat. Fys. Skr. Dan. Vid. Selsk.* **2**, no. 3. [II: 267, 270, 271, 641]
Margolis, B., and Troubetzkoy, E. S. (1957), *Phys. Rev.* **106**, 105. [II: 234, 235]
Marić, Z., and Möbius, P. (1959), *Nuclear Phys.* **10**, 135. [II: 458]
Mariscotti, M. A. J., Scharff-Goldhaber, G., and Buck, B. (1969), *Phys. Rev.* **178**, 1864. [II: 31]
Marshak, H., Langsford, A., Wong, C. Y., and Tamura, T. (1968), *Phys. Rev. Letters* **20**, 554. [II: 137]
Marshalek, E. R. (1973), *Phys. Letters* **44B**, 5. [II: 433]
Marumori, T. (1960), *Progr. Theor. Phys.* (*Kyoto*) **24**, 331. [II: 436]
Marumori, T., Yamamura, M., and Tokunaga, A. (1964), *Progr. Theor. Phys.* (*Kyoto*) **31**, 1009. [II: 433]
Massey, H. S. W., and Burhop, E. H. S. (1952), *Electronic and Ionic Impact Phenomena*, Oxford University Press, Oxford.
Mattauch, J. H. E., Thiele, W., and Wapstra, A. H. (1965), *Nuclear Phys.* **67**, 1.
Mayer, M. G. (1941), *Phys. Rev.* **60**, 184. [II: 584]
Mayer, M. G. (1949), *Phys. Rev.* **75**, 1969. [I: 189, 209; II: v]
Mayer, M. G. (1950), *Phys. Rev.* **78**, 16. [I: 226]
Mayer, M. G. (1950a), *Phys. Rev.* **78**, 22. [II: 540, 641]
Mayer, M. G., and Jensen, J. H. D. (1955), *Elementary Theory of Nuclear Shell Structure*, Wiley, New York, N.Y.
Mayeur, C., Sacton, J., Vilain, P., Wilquet, G., Stanley, D., Allen, P., Davis, D. H., Fletcher, E. R., Garbutt, D. A., Shaukat, M. A., Allen, J. E., Bull, V. A., Conway, A. P., and March, P. V. (1965), *Univ. Libre de Bruxelles*, Bulletin No. 24, Presses Acad. Europ., Bruxelles. [I: 55]
McClatchie, E. A., Glashausser, C., and Hendrie, D. L. (1970), *Phys. Rev.* **C1**, 1828. [II: 562]
McDaniels, D. K., Blair, J. S., Chen, S. W., and Farwell, G. W. (1960), *Nuclear Phys.* **17**, 614. [II: 353]
McDonald, J., and Porter, D. (1968), *Nuclear Phys.* **A109**, 529. [II: 532]
McGowan, F. K. (1959), in *Compt. Rend. du Congrès Intern. de Physique Nucléaire*, p. 225, ed. P.. Gugenberger, Dunod, Paris. [II: 433]
McGowan, F. K., Robinson, R. L., Stelson, P. H., and Ford, J. L. C., Jr., (1965), *Nuclear Phys.* **66**, 97. [II: 532, 561]
McVoy, K. W. (1967), in *Fundamentals in Nuclear Theory*, p. 419, Intern. Atomic Energy Agency, Vienna. [I: 432]
McVoy, K. W. (1967a), *Ann. Phys.* **43**, 91. [I: 167,, 447]
Mehta, M. L. (1960), *Nuclear Phys.* **18**, 395. [I: 299]
Mehta, M. L. (1967) *Random Matrices*, Academic Press, New York, N.Y. [I: 299]
Mehta, M. L., and Gaudin, M. (1960), *Nuclear Phys.* **18**, 420. [I: 299]
Meitner, L., and Frisch, O. R. (1939), *Nature* **143**, 239. [II: 365, 654]
Meldner, H., Süssmann, G., and Ubrici, W. (1965), *Z. Naturforsch.* **20a**, 1217. [I: 240]
Messiah, A. (1962), *Quantum Mechanics*, North-Holland, Amsterdam.
Meyerhof, W. E., and Tombrello, T. A. (1968), *Nuclear Phys.* **A109**, 1.
Miazawa, H. (1951), *see* Miyazawa, H. (1951).

Michaudon, A. (1973), *Advances in Nuclear Physics* **6**, 1, eds. M. Baranger and E. Vogt, Plenum Press, New York, N.Y. [II:622]
Michel, F. C. (1964), *Phys. Rev.* **133**, B329. [I:25]
Michel, H. V. (1966), *UCRL*-17301, Radiation Lab., University of California, Berkeley, Cal. [II:268]
Middleton, R., and Hinds, S. (1962), *Nuclear Phys.* **34**, 404. [II:293]
Middleton, R., and Pullen, D. J. (1964), *Nuclear Phys.* **51**, 77. [II:386]
Migdal, A. (1944), *J. Phys.* (*USSR.*) **8**, 331. [II:474]
Migdal, A. B. (1959), *Nuclear Phys.* **13**, 655. [II:80, 82, 83, 641]
Migdal, A. B. (1966), *Nuclear Phys.* **75**, 441. [II:484]
Migdal, A. B. (1967), *Theory of Finite Fermi Systems and Applications to Atomic Nuclei*, Wiley (Interscience), New York, N.Y.
Migdal, A. B., Lushnikov., A. A., and Zaretsky, D. F. (1965), *Nuclear Phys.* **66**, 193. [II:483, 484]
Migneco, E., and Theobald, J. P. (1968), *Nuclear Phys.* **A112**, 603. [II:622, 623, 630]
Mikhailov, V. M. (1966), *Izv. Akad. Nauk, ser. Fiz.* **30**, 1334. [II:60, 159]
Miller, P. D., Dress, W. B., Baird, J. K., and Ramsey, N. F. (1967), *Phys. Rev. Letters* **19**, 381. [I:15]
Milner, W. T., McGowan, F. K., Stelson, P. H., Robinson, R. L., and Sayer, R. O. (1969), *Nuclear Phys.* **A129**, 687. [II:532]
Milner, W. T., McGowan, F. K., Robinson, R. L., Stelson, P. H., and Sayer, R. O. (1971), *Nuclear Phys.* **A177**, 1. [II:129, 536]
Minor, M. M., Sheline, R. K., Shera, E. B., and Jurney, E. T. (1969) *Phys. Rev.* **187**, 1516. [II:141, 142]
Miyazawa, H. (1951), *Progr. Theor. Phys.* (*Kyoto*) **6**, 801. [I:391; II:484]
Moalem, A., Benenson, W., and Crawley, G. M. (1973), *Phys. Rev. Letters* **31**, 482. [II:509]
Möller, P., and Nilsson, S. G. (1970), *Phys. Letters* **31B**, 283. [II:369]
Mössbauer, R. L. (1965), in *Alpha-, Beta-, and Gamma-Ray Spectroscopy*, vol. 2, p. 1293, ed. K. Siegbahn, North-Holland, Amsterdam. [I:383]
Moore, C. E. (1949), *Atomic Energy Levels*, Circular 467, vol. 1, p. XL, Nat. Bureau of Standards, Washington, D.C. [I:191]
Moravcsik, M. J. (1963), *The Two-Nucleon Interaction*, Clarendon Press, Oxford.
Morinaga, H. (1956), *Phys. Rev.* **101**, 254. [II:26]
Morinaga, H. (1966), *Nuclear Phys.* **75**, 385. [II:535]
Morinaga, H., and Gugelot, P. C. (1963), *Nuclear Phys.* **46**, 210. [II:68]
Morpurgo, G. (1958), *Phys. Rev.* **110**, 721. [I:45]
Morrison, P. (1953), in *Experimental Nuclear Physics*, vol. 2, ed. E. Segré, Wiley, New York, N.Y. [I:444]
Morse, P. M., and Feshbach, H. (1953), *Methods of Theoretical Physics*, McGraw Hill, New York, N.Y.
Moszkowski, S. A. (1955), *Phys. Rev.* **99**, 803. [II:212]
Moszkowski, S. A. (1965), in *Alpha-, Beta-, and Gamma-Ray Spectroscopy*, vol. 2, p. 863, ed. K. Siegbahn, North-Holland, Amsterdam. [I:382]
Mottelson, B. R. (1968), in *Proc. Intern. Conf. on Nuclear Structure*, p. 87, ed. J. Sanada, *Phys. Soc. of Japan*. [II:571]
Mottelson, B. R., and Nilsson, S. G. (1955), *Phys. Rev.* **99**, 1615. [II:212]
Mottelson, B. R., and Nilsson, S. G. (1959), *Mat. Fys. Skr. Dan. Vid. Selsk.* **1**, no. 8. [II:117, 212, 221, 233, 276, 306]
Mottelson, B. R., and Valatin, J. G. (1960), *Phys. Rev. Letters* **5**, 511. [II:42, 83]

Motz, H. T., Jurney, E. T., Schult, O. W. B., Koch, H. R., Gruber, U., Maier, B. P., Baader, H., Struble, G. L., Kern, J., Sheline, R. K., von Egidy, T., Elze, Th., Bieber, B., and Bäcklin, A. (1967), *Phys. Rev.* **155**, 1265. [II:119, 120, 122]
Moyer, R. A., Cohen, B. L., and Diehl, R. C. (1970), *Phys. Rev.* **C2**, 1898. [II:642]
Mukherjee, P., and Cohen, B. L. (1962), *Phys. Rev.* **127**, 1284. [II:647]
Myers, W. D. (1973), *Nuclear Phys.* **A204**, 465. [II:137]
Myers, W. D., and Swiatecki, W. J. (1966), *Nuclear Phys.* **81**, 1. [I:142, 143, 145, 199, 204; II:367, 598, 660]
Myers, W. D., and Swiatecki, W. J. (1967), *Arkiv Fysik* **36**, 343. [II:605]
Myers, W. D., and Swiatecki, W. J. (1969), *Ann. Phys.* **55**, 395. [II:137]

Nagao, M., and Torizuka, Y. (1973), *Phys. Rev. Letters* **30**, 1068. [II:509]
Nagatani, K., Le Vine, M. J., Belote, T. A., and Arima, A. (1971), *Phys. Rev. Letters* **27**, 1071. [II:96]
Nagel, J. G., and Moshinsky, M. (1965), *J. Math. Phys.* **6**, 682. [I:126]
Nakai, K., Herskind, B., Blomqvist, J., Filevich, A. Rensfelt, K.-G., Sztankier, J., Bergström, I., and Nagamiya, S. (1972), *Nuclear Phys.* **A189**, 526. [II:486]
Nathan, O. (1968), in *Nuclear Structure*, Dubna Symposium, p. 191, Intern. Atomic Energy Agency, Vienna. [II:390]
Nathan, O., and Nilsson, S. G. (1965), in *Alpha-, Beta-, and Gamma-Ray Spectroscopy*, vol. I, p. 601, ed. K. Siegbahn, North-Holland, Amsterdam. [II:541]
Neal, W. R., and Kraner, H. W. (1965), *Phys. Rev.* **137**, B1164. [II:129]
Nedzel, V. A. (1954), *Phys. Rev.* **94**, 174. [I:165]
Ne'eman, Y. (1961), *Nuclear Phys.* **26**, 222. [I:40]
Neergård, K., and Vogel, P. (1970), *Nuclear Phys.* **A145**, 33. [II:578]
Nemirovsky, P. E., and Adamachuk, Yu., V. (1962), *Nuclear Phys.* **39**, 551. [I:171]
Nemirovsky, P. E., and Chepurnov. V. A. (1966), *Yadernaya Fizika* **3**, 998; transl. *Soviet J. Nuclear Phys.* **3**, 730. [II:230]
Neudatchin, V. G., and Smirnow, Yu. F. (1969), *Nukleonie Assatsiatsiiv* Liogkich Jadrach, Nauka, Moscow. [II:99]
Neutron Cross Sections (1964), Sigma Center, Brookhaven Nat. Lab., BNL 325, Suppl. 2, Brookhaven, N.Y.
Newman, E., Hiebert, J. C., and Zeidman, B. (1966), *Phys. Rev. Letters* **16**, 28. [I:323]
Newson, H. W. (1966), in *Nuclear Structure Studies with Neutrons*, p. 195, eds. N. de Mevergies *et al.*, North-Holland, Amsterdam. [I:231]
Newton, J. O. (1958), *Nuclear Phys.* **5**, 218. [II:493]
Newton, J. O., Stephens, F. S., Diamond, R. M., Kelly, W. H., and Ward, D. (1970), *Nuclear Phys.* **A141**, 631. [II:73]
Newton, T. D. (1960), *Can. J. Phys.* **38**, 700. [II:213]
Nilsson, S. G. (1955), *Mat. Fys. Medd. Dan. Vid. Selsk.* **29**, no. 16. [II:212, 219, 228, 239, 464]
Nilsson, S. G., and Prior, O. (1961), *Mat. Fys. Medd. Dan. Vid. Selsk.* **32**, no. 16. [II:55, 83, 641]
Nilsson, S. G., Tsang, C. F., Sobiczewski, A., Szymánski, Z., Wycech, S., Gustafson, C., Lamm, I.-L., Möller, P., and Nilsson, B. (1969), *Nuclear Phys.* **A131**, 1. [II:598]
Nishijima, K. (1954), *Progr. Theor. Phys. (Kyoto)* **12**, 107. [I:38]
Nix, J. R. (1967), *Ann. Phys.* **41**, 52. [II:663]
Nix, J. R. (1972), *Ann. Rev. Nuclear Sci.* **22**, 65. [II:634]

Nolen, J. A., Jr., and Schiffer, J. P. (1969), *Phys. Letters* **29B**, 396. [II:518]
Nordheim, L. A. (1951), *Rev. Mod. Phys.* **23**, 322. [I:226]
Novikov. V. M., and Urin, M. G. (1966), *Yadernaya Fiz.* **3**, 419; transl. *Soviet J. Nuclear Phys.* **3**, 302. [II:493]
Nozières, P. (1964), *Theory of Interacting Fermi Systems*, Benjamin, New York, N.Y.
Nozières, P. (1966), in *Quantum Fluids*, p. 1, ed. D. F. Brewer, North-Holland, Amsterdam. [II:396]
Nozières, P., and Pines, D. (1958), *Phys. Rev.* **109**, 741, 762, and 1062. [II:436]
Nuclear Data Sheets, Nuclear Data Group, Oak Ridge Nat. Lab., Oak Ridge, Tenn.

Oehme, R. (1963), in *Strong Interactions and High Energy Physics*, ed. R. G. Moorhouse, Oliver and Boyd, Edinburgh and London. [I:13]
Ogle, W., Wahlborn, S., Piepenbring, R., and Fredriksson, S. (1971), *Rev. Mod. Phys.* **43**, 424. [II:226]
Okamoto, K. (1958), *Phys. Rev.* **110**, 143; see also *Progr. Theor. Phys. (Kyoto)* **15**, 75. [II:490, 673]
Okubo, S. (1962), *Progr. Theor. Phys. (Kyoto)* **27**, 949. [I:59]
Olesen, M. C., and Elbek, B. (1960), *Nuclear Phys.* **15**, 134. [II:133]
Onsager, L. (1954), in *Proc. Intern. Conf. on Theoretical Physics, Kyoto and Tokyo*, p. 935, Science Council of Japan, Tokyo. [II:397]
Onsager, L. (1961), *Phys. Rev. Letters* **7**, 50. [II:398]
Oothoudt, M. A., and Hintz, N. M. (1973), *Nuclear Phys.* **A213**, 221. [II:272]
Oppenheimer, J. R., and Schwinger, J. (1941), *Phys. Rev.* **60**, 150. [II:20]
Osborne, R. K., and Foldy, L. L. (1950), *Phys. Rev.* **79**, 795. [I:389]
Ottaviani, P. L., Savoia, M., Sawicki, J., and Tomasini, A. (1967), *Phys. Rev.* **153**, 1138. [II:433]

Pars, L. A. (1965), *A Treatise on Analytical Dynamics*, Heinemann, London.
Paul, E. B. (1957), *Phil. Mag.* **2**, 311. [II:97, 212]
Paul, P. (1973), in *Intern. Conf. on Photonuclear Reactions and Applications*, vol. I, p. 407, ed. B. L. Berman, U. S. Atomic Energy Commission, Office of Information Services, Oak Ridge, Tenn. [II:498, 499]
Pauli, H. C., Ledergerber, T., and Brack, M. (1971), *Phys. Letters* **34B**, 264. [II:369]
Pauli, W. (1933), in *Handbuch der Physik* XXIV/1, p. 3, ed. A. Smekal, Springer, Berlin.
Pauli, W., and Dancoff, S. M. (1942), *Phys. Rev.* **62**, 85. [II:20]
Pauly, H., and Toennies, J. P. (1965), in *Advances in Atomic and Molecular Physics*, vol. 1, p. 195, eds. D. R. Bates and I. Estermann, Academic Press, New York, N.Y. [I:166]
Paya, D., Blons, J., Derrien, H., Fubini, A., Michaudon, A., and Ribon, P. (1968), *J. phys. radium*, Suppl. **C1**, 159. [II:622]
Peierls, R. E., and Yoccoz, J. (1957), *Proc. Phys. Soc. (London)* **A70**, 381. [II:91]
Peierls, R. E., and Thouless, D. J. (1962), *Nuclear Phys.* **38**, 154. [II:93]
Peker, L. K. (1960), *Izv. Akad. Nauk*, Ser. Fiz. **24**, 365. [II:121]
Perey, F., and Buck, B. (1962), *Nuclear Phys.* **32**, 353. [I:217]
Perey, F. G., and Schiffer, J. P. (1966), *Phys. Rev. Letters* **17**, 324. [I:163]
Perlman, I., Ghiorso, A., and Seaborg, G. T. (1950), *Phys. Rev.* **77**, 26. [II:116]
Perring, J. K., and Skyrme, T. H. R. (1956), *Proc. Phys. Soc. (London)* **A69**, 600. [II:101]

Persson, B., Blumberg, H., and Agresti, D. (1968), *Phys. Rev.* **170**, 1066. [II: 132]
Petersen, D. F., and Veje, C. J. (1967), *Phys. Letters* **24B**, 449. [II: 493]
Peterson, J. M. (1962), *Phys. Rev.* **125**, 955. [I: 165]
Pik-Pichak, G. A. (1958), *J. Exptl. Theoret. Phys. (USSR)* **34**, 341; transl. *Soviet Phys. JETP* **34**, 238. [II: 663]
Pilt, A. A., Spear, R. H., Elliott, R. V., Kelly, D. T., Kuehner, J. A., Ewan, G. T., and Rolfs, C. (1972), *Can. J. Phys.* **50**, 1286. [II: 286]
Pines, D. (1963), *Elementary Excitations in Solids*, Benjamin, New York, N.Y.
Pines, D., and Nozières, P. (1966), *The Theory of Quantum Liquids*, Benjamin, New York, N.Y.
Pitthan, R., and Walcher, Th. (1971), *Phys. Letters* **36B**, 563. [II: 509]
Pitthan, R., and Walcher, Th. (1972), *Z. Naturforsch.* **27a**, 1683. [II: 509, 639]
Pniewsky, J., and Danysz, M. (1962), *Phys. Letters* **1**, 142. [I: 56]
Poggenburg, J. K., Jr. (1965), UCRL-16187, *thesis* subm. to University of California, Berkeley, Cal. [II: 117, 268]
Poggenburg, J. K., Mang, H. J., and Rasmussen, J. O. (1969), *Phys. Rev.* **181**, 1697. [II: 249, 268]
Polikanov, S. M. (1968), *Uspeki Fiz. Nauk* **94**, 46; transl. *Soviet Phys. Uspekhi* **11**, 22. [II: 622]
Polikanov, S. M., and Sletten, G. (1970), *Nuclear Phys.* **A151**, 656. [II: 634]
Polikanov, S. M., Druin, V. A., Karnaukhov, V. A., Mikheev, V. L., Pleve, A. A., Skobelev, N. K., Subbotin, V. G., Ter-Akop'yan, G. M., and Fomichev, V. A. (1962), *J. Exptl. Theoret. Phys. (USSR)* **42**, 1464; transl. *Soviet Phys. JETP* **15**, 1016. [II: 622]
Pontecorvo, B., and Smorodinski, Y. (1961), *Zhur. Eksp. i Teoret. Fiz.* **41**, 239; transl. *Soviet Phys. JETP* **14**, 173. [I: 398]
Porter, C. E., and Thomas, R. G. (1956), *Phys. Rev.* **104**, 483. [I: 182]
Porter, C. E., and Rosenzweig, N. (1960), *Ann. Acad. Sci. Finland*, **A6**, no. 44. [I: 294]
Prowse, D. J. (1966), *Phys. Rev. Letters* **17**, 782 [I: 56]

Rabotnov, N. S., Smirenkin, G. N., Soldatov, A. S., Usachev, L. N., Kapitza, S. P., and Tsipenyuk, Yu. M. (1970), *Yadernaya Fizika* **11**, 508. [II: 126]
Racah, G. (1942), *Phys. Rev.* **62**, 438. [I: 312]
Racah, G. (1943), *Phys. Rev.* **63**, 367. [II: 641, 688, 691]
Racah, G. (1950), *Phys. Rev.* **78**, 622. [II: 540]
Racah, G. (1951), *Group Theory and Spectroscopy*, Lecture Notes, Inst. for Advanced Study, Princeton, N. J.
Racah, G. (1952), in *Farkas Memorial Volume*, p. 294, eds. A. Farkas and E. P. Wigner, Research Council of Israel, Jerusalem. [II: 641]
Racah, G., and Talmi, I. (1953), *Phys. Rev.* **89**, 913. [II: 641]
Rainwater, J. (1950), *Phys. Rev.* **79**, 432. [I: 334; II: vi, 26]
Rakavy, G. (1957), *Nuclear Phys.* **4**, 375. [II: 212]
Rakavy, G. (1957a), *Nuclear Phys.* **4**, 289. [II: 691]
Ramamurthy, V. S., Kapoor, S. S., and Kataria, S. K. (1970), *Phys. Rev. Letters* **25**, 386. [II: 621]
Ramšak, V., Olesen, M. C., and Elbek, B. (1958), *Nuclear Phys.* **6**, 451. [II: 133]
Ramsey, N. F. (1950), see Ramsey, N. F. (1953), *Nuclear Moments*, Wiley, New York, N.Y.
Ramsey, N. F. (1956), *Molecular Beams*, Clarendon Press, Oxford.
Rand, R. E., Frosch, R., and Yearin, M. R. (1966), *Phys. Rev.* **144**, 859. [II: 134]

Randolph, W. L., Ayres de Campos, N., Beene, J. R., Burde, J., Grace, M.A., Start, D.F.H., and Warner, R. E. (1973), *Phys. Letters* **44B**, 36. [II : 357]
Rayleigh, J. W. S. (1877), *Theory of Sound*, Mac Millan, London. [II : 654]
Raynal, J. (1967), *Nuclear Phys.* **A97**, 572. [I : 362]
Raz, B. J. (1959), *Phys. Rev.* **114**, 1116. [II : 433]
Redlich, M. G. (1958), *Phys. Rev.* **110**, 468. [II : 283]
Reich, C. W., and Cline, J. E. (1965), *Phys. Rev.* **137**, 1424. [II : 158]
Reich, C. W., and Cline, J. E. (1970), *Nuclear Phys.* **A159**, 181. [II : 159, 163]
Reid, R. V. Jr. (1968), *Ann. Phys.* **50**, 411. [I : 266]
Reiner, A. S. (1961), *Nuclear Phys.* **27**, 115. [II : 174]
Reines, F., Cowan, C. L., and Goldhaber, M. (1954), *Phys. Rev.* **96**, 1157. [I : 5]
Reines, F., Cowan, C. L., and Kruse, H. W. (1957), *Phys. Rev.* **109**, 609. [I : 5]
Reising, R. F., Bate, G. L., and Huizenga, J. R. (1966), *Phys. Rev.* **141**, 1161 [II : 619, 622]
Ricci, R. A., Girgis, R. K., and van Lieshout, R. (1960), *Nuclear Phys.* **21**, 177. [I : 53]
Richard, P., Moore, C. F., Becker, J. A., and Fox, J. D. (1966), *Phys. Rev.* **145**, 971. [I : 47, 48]
Richard, P., Stein, N., Kavaloski, C. D., and Lilley, J. S. (1968), *Phys. Rev.* **171**, 1308 [II : 642]
Ridley, B. W., and Turner, J. F. (1964), *Nuclear Phys.* **58**, 497. [I : 234]
Ripka, G. (1968), *Advances in Nuclear Physics* **1**, 183, eds. M. Baranger and E. Vogt, Plenum Press, New York, N.Y. [II : 283]
Robson, D. (1965), *Phys. Rev.* **137**, B535. [I : 50, 437]
Rogers, J. D. (1965), *Ann. Rev. Nuclear Sci.* **15**, 241. [II : 303]
Rojo, O., and Simmons, L. M. (1962), *Phys. Rev.* **125**, 273. [I : 245]
Rood, H. P. C. (1966), *Nuovo cimento*, Suppl. Ser. 1, **4**, 185. [I : 405]
Roos, C. E., and Peterson, V. Z. (1961), *Phys. Rev.* **124**, 1610. [II : 479]
Roos, P. G., and Wall, N. S. (1965), *Phys. Rev.* **140**, B1237. [I : 237]
Rose, M. E. (1955), *Multipole Fields*, Wiley, New York, N. Y.
Rose, M. E. (1957), *Elementary Theory of Angular Momentum*, Wiley, New York, N. Y.
Rose, M. E. (1965), in *Alpha-, Beta-, and Gamma-Ray Spectroscopy*, vol. **2**, p. 887, ed. K. Siegbahn, North-Holland, Amsterdam. [I : 384]
Rosen, L., Beery, J. G., Goldhaber, A. S., and Auerbach, E. H. (1965), *Ann. Phys.* **34**, 96. [I : 237, 238]
Rosenblum, S., and Valadares, M. (1952), *Compt. rend.* **235**, 711. [I : 197]
Rosenfeld, A. H., Barbaro-Galtieri, A., Podolsky, W. J., Price, L. R., Soding, P., Wohl, C. G., Roos, M., and Willis, W. J. (1967), *Rev. Mod. Phys.* **39**, 1.
Rosenfeld, L. (1948), *Nuclear Forces*, North-Holland, Amsterdam.
Rosenzweig, N. (1957), *Phys. Rev.* **108**, 817. [II : 607, 610]
Rosenzweig, N. (1963), in *Statistical Physics*, Brandeis Summer Institute, vol. 3, p. 91, ed. K. W. Ford, Benjamin, New York, N. Y. [I : 294]
Rosenzweig, N., and Porter, C. E. (1960), *Phys. Rev.* **120**, 1968. [I : 299]
Rosenzweig, N., Monahan, J. E., and Mehta, M. L. (1968), *Nuclear Phys.* **A109**, 437. [I : 298]
Ross, A. A., Lawson, R. D., and Mark, H. (1956), *Phys. Rev.* **104**, 401. [I : 240]
Rud, N., and Nielsen, K. B. (1970), *Nuclear Phys.* **A158**, 546. [II : 164]
Rud, N., Nielsen, H. L., and Wilsky, K. (1971), *Nuclear Phys.* **A167**, 401. [II : 463]
Ruderman, M. (1972), *Ann. Rev. Astron. and Astrophys.* **10**, 427. [II : 398]
Rudolph, H., and McGrath, R. L. (1973), *Phys. Rev.* **C8**, 247. [II : 528]

Rutherford, E., Chadwick, J., and Ellis, C. D. (1930), *Radiations from Radioactive Substances*, Cambridge University Press, Cambridge.

Sachs, R. G. (1948), *Phys. Rev.* **74**, 433. [I:392]
Sachs, R. G. (1953), *Nuclear Theory*, Addison-Wesley, Reading, Mass.
Saethre, Ø., Hjorth, S. A., Johnson, A., Jägare, S., Ryde, H., and Szymański, Z. (1973), *Nuclear Phys.* **A207**, 486.
Sakai, M. (1967), *Nuclear Phys.* **A 104**, 301. [II:545]
Sakai, M. (1970), *Nuclear Data Tables* **A8**, 323.
Sakai, M. (1972), *Nuclear Data Tables* **A10**, 511.
Salisbury, S. R., and Richards, H. T. (1962), *Phys. Rev.* **126**, 2147. [I:358]
Salling, P. (1965), *Phys. Letters* **17**, 139. [I:341]
Satchler, G. R. (1955), *Phys. Rev.* **97**, 1416. [II:22, 243]
Satchler, R. (1966), in *Lectures in Theoretical Physics* **VIIIC**, p. 73, eds. Kunz, Lind, and Britten, University of Colorado Press, Boulder, Colorado. [I:425]
Satchler, G. R., Bassel, R. H., and Drisko, R. M. (1963), *Phys. Letters* **5**, 256. [II:353]
Sayer, R. O., Stelson, P. H., McGowan, F. K., Milner, W. T., and Robinson, R. L. (1970), *Phys. Rev.* **1C**, 1525. [II:163]
Scharff-Goldhaber, G. (1952), *Physica* **18**, 1105. [I:197]
Scharff-Goldhaber, G. (1953), *Phys. Rev.* **90**, 587. [I:197]
Scharff-Goldhaber, G., and Weneser, J. (1955), *Phys. Rev.* **98**, 212. [II:433, 507]
Scharff-Goldhaber, G., and Takahashi, K. (1967), *Izv. Akad. Nauk, ser. Fiz.* **31**, 38. [II:121]
Schmorak, M., Bemis, C. E. Jr., Zender, M. J., Gove, N. B., and Dittner, P. F. (1972), *Nuclear Phys.* **A178**, 410. [II:65]
Schneid, E. J., Prakash, A., and Cohen, B. L. (1967), *Phys. Rev.* **156**, 1316. [I:48]
Schopper, H. F. (1966), *Weak Interactions and Nuclear Beta Decay*, North-Holland, Amsterdam.
Schulz, H., Wiebicke, H. J., Fülle, R., Netzband, D., and Schlott, K., (1970), *Nuclear Phys.* **A159**, 324. [II:293]
Schwalm, D., Bamberger, A., Bizzeti, P. G., Povh, B., Engelbertink, G. A. P., Olness, J. W., and Warburton, E. K. (1972), *Nuclear Phys.* **A192**, 449. [II:98]
Schwinger, J. (1950), *Phys. Rev.* **78**, 135. [I:241]
Seaman, G. G., Greenberg, J. S., Bromley, D. A., and McGowan, F. K. (1966), *Phys. Rev.* **149**, 925. [II:577]
Seaman, G. G., Bernstein, E. M., and Palms, J. M. (1967), *Phys. Rev.* **161**, 1223. [II:57]
Seeger, P. A., and Perisho, R. C. (1967), *Los Alamos Scientific Lab. Report* (LA 3751), Los Alamos, N. M. [II:598]
Seeger, P. A., Fowler, W. A., and Clayton, D. D. (1965), *Astrophys. J. Suppl.* **97**, 121. [I:201, 202]
Segel, R. E., Olness, J. W., and Sprenkel, E. L. (1961), *Phys. Rev.* **123**, 1382. [I:22]
Segel, R. E., Olness, J. W., and Sprenkel, E. L. (1961a), *Phil. Mag.* **6**, 163. [I:22]
Segel, R. E., Singh, P. P., Allas, R. G., and Hanna, S. S. (1963), *Phys. Rev. Letters* **10**, 345. [I:321]
Semenko, S. F. (1964), *Phys. Letters* **10**, 182 and **13**, 157. [II:455]
Shapiro, I. S., and Estulin, I. V. (1956), *Zhur. Eksp. i Teoret. Fiz.* **30**, 579; transl. *Soviet Phys. JETP* **3**, 626. [I:5]
Sharpey-Schafer, J. F., Ollerhead, R. W., Ferguson, A. J., and Litherland, A. E. (1968), Can. J. Phys. **46**, 2039. [II:291]
Shaw, G. L. (1959), *Ann. Phys.* **8**, 509. [I:261]

Sheline, R. K. (1960), *Rev. Mod. Phys.* **32**, 1. [II:545]
Sheline, R. K., Watson, C. E., Maier, B. P., Gruber, U., Koch, R. H., Schult, O. W. B., Motz, H. T., Jurney, E. T., Struble, G. L., Egidy, T. v., Elze, Th., and Bieber, E. (1966), *Phys. Rev.* **143**, 857. [II:36]
Sheline, R. K., Bennett, M. J., Dawson, J. W., and Sluda, Y. (1967), *Phys. Letters* **26B**, 14. [II:230]
Shenoy, G. K., and Kalvius, G. M. (1971), in *Hyperfine Interactions in Excited Nuclei*, vol. 4, p. 1201, eds. G. Goldring and R. Kalish, Gordon and Breach, New York, N. Y.
Sherif, H. (1969), *Nuclear Phys.* **A131**, 532. [II:215, 355]
Sherwood, A. I., and Goswami, A. (1966), *Nuclear Phys.* **89**, 465. [II:540]
Shirley, D. A. (1964), *Rev. Mod. Phys.* **36**, 339. [I:384]
Shoda, K., Sugawara, M., Saito, T., and Miyase, H. (1969), *Phys. Letters* **28B**, 30. [II:498]
Shull, C. G., and Nathans, R. (1967), *Phys. Rev. Letters* **19**, 384. [I:15]
Siegbahn, K., ed. (1965), *Alpha-, Beta-, and Gamma-Ray Spectroscopy*, North-Holland, Amsterdam.
Siegert, A. J. F., (1937), *Phys. Rev.* **52**, 787. [I:390]
Siemens, P. J., and Sobiczewski, A. (1972), *Phys. Letters* **41B**, 16. [II:368]
Silverberg, L. (1962), *Arkiv Fysik* **20**, 341. [I:329]
Silverberg, L. (1964), *Nuclear Phys.* **60**, 483. [I:329]
Simpson, J. J., Eccleshall, D., Yates, M. J. L., and Freeman, N. J. (1967), *Nuclear Phys.* **A94**, 177. [II:535]
Singh, P. P., Segel, R. E., Siemssen, R. H., Baker, S., and Blaugrund, A. E. (1967), *Phys. Rev.* **158**, 1063. [I:341]
Skorka, S. J., Hartel, J., and Retz-Schmidt, T. W. (1966), *Nuclear Data* **A2**, 347.
Slater, J. C. (1929), *Phys. Rev.* **34**, 1293. [I:149]
Slichter, L. B. (1967), in *Intern. Dictionary of Geophysics*, vol. 1, p. 331, ed. S. K. Runcorn, Pergamon Press, Oxford. [II:657]
Sliv, L. A., and Kharitonov, Yu. I. (1965), *Phys. Letters* **16**, 176. [I:175]
Smulders, P. J. M., Broude, C., and Litherland, A. E. (1967), *Can. J. Phys.* **45**, 2133. [II:96]
Sørensen, B. (1966), *Phys. Letters* **21**, 683. [II:544]
Sørensen, B. (1967), *Nuclear Phys.* **A97**, 1. [II:433]
Soldatov, A. S., Smirenkin, G. N., Kapitza, S. P., and Tsipeniuk, Y. M., (1965), *Phys. Letters*, **14**, 217. [II:124, 125]
Soloviev, V. G. (1961), *Mat. Fys. Skr. Dan. Vid. Selsk.* **1**, no. 11. [II:307, 641]
Soloviev, V. G. (1962), *Phys. Letters* **1**, 202. [II:270, 641]
Soloviev, V. G., and Vogel, P. (1967), *Nuclear Phys.* **A92**, 449. [II:255, 641]
Sommerfeld, A. (1915), *Sitzungsber. Bayer. Akad. Wiss., München*, pp. 425 ff. [II:583]
Sommerfeld, A. (1922), *Atombau und Spektrallinien*, Vieweg, Braunschweig.
Sood, P. C., and Green, A. E. S. (1957), *Nuclear Phys.* **5**, 274. [I:146]
Sorensen, R. (1966), *Phys. Letters* **21**, 333. [I:164]
Sosnovsky, A. N., Spivak, P. E., Prokofiev, Yu. A., Kutikov, I. E., and Dobrinin, Yu. P. (1959), *Nuclear Phys.* **10**, 395. [I:5]
Spalding, I. J., and Smith, K. F. (1962), *Proc. Phys. Soc. (London)* **79**, 787. [II:132]
Specht, H. J., Weber, J., Konecny, E., and Heunemann, D. (1972), *Phys. Letters* **41B**, 43. [II:26, 83, 632]
Spruch, L. (1950), *Phys. Rev.* **80**, 372. [I:392]
Stähelin, P., and Preiswerk, P. (1951), *Helv. Phys. Acta* **24**, 623. [I:197]
Stahl, R. H., and Ramsey, N. F. (1954), *Phys. Rev.* **96**, 1310. [I:243]

Stanford, C. P., Stephenson, T. E., and Bernstein, S. (1954), *Phys. Rev.* **96**, 983. [I:5]

Stanford Conference on Nuclear Sizes and Density Distributions (1958), *Rev. Mod. Phys.* **30**, 412. [I:138]

Stapp, H. P., Ypsilantis, T. J., and Metropolis, N. (1957), *Phys. Rev.* **105**, 302. [I:245, 265]

Steadman, S. G., Kleinfeld, A. M., Seaman, G. G., de Boer, J., and Ward, D. (1970), *Nuclear Phys.* **A155**, 1. [II:532]

Stech, B., and Schülke, L. (1964), *Z. Physik* **179**, 314. [I:403]

Stein, P. C., Odian, A. C., Wattenberg, A., and Weinstein, R. (1960), *Phys. Rev.* **119**, 348. [II:479]

Steinwedel, H., and Jensen, J.H.D. (1950), *Z. Naturforsch.* **5a**, 413. [II:474, 654]

Stelson, P. H., and Grodzins, L. (1965), *Nuclear Data* **A1**, 21.

Stelson, P. H., Robinson, R. L., Kim, H. J., Rapaport, J., and Satchler, G. R. (1965), *Nuclear Phys.* **68**, 97. [II:354]

Stelson, P. H., McGowan, F. K., Robinson, R. L., and Milner, W. T. (1970), *Phys. Rev.* **C2**, 2015. [II:354, 533]

Stelson, P. H., Raman, S., McNabb, J. A., Lide, R. W., and Bingham, C. R. (1973), *Phys. Rev.* **8C**, 368. [II:543]

Stephens, F. (1960), quoted by E. Hyde, I. Perlman, and G. T. Seaborg (1964) in *The Nuclear Properties of the Heavy Elements,* vol. II, p. 732, Prentice Hall, Englewood Cliffs, N. J. [II:252]

Stephens, F. S., Jr., Asaro, F., and Perlman, I. (1955), *Phys. Rev.* **100**, 1543. [II:61, 561]

Stephens, F. S., Asaro, F., and Perlman, I. (1959), *Phys. Rev.* **113**, 212. [II:212]

Stephens, F. S., Elbek, B., and Diamond, R. M. (1963), in *Proc. Third Conf. on the Reactions between Complex Nuclei*, p. 303. eds. A. Ghiorso, R. M. Diamond, and H. E. Conzett, Univ. of California Press, Berkeley, Calif. [II:463]

Stephens, F. S., Lark, N. L., and Diamond, R. M. (1965), *Nuclear Phys.* **63**, 82. [II:68, 72]

Stephens, F. S., Holtz, M. D., Diamond, R. M., and Newton, J. O. (1968), *Nuclear Phys.* **A115**, 129 [II:274, 275, 277]

Stephens, F. S., Ward, D., and Newton, J. O. (1968), *Japan J. Phys., Suppl.* **24**, 160. [II:72]

Stephens, F. S., and Simon, R. S. (1972), *Nuclear Phys.* **A183**, 257. [II:43]

Stevens, R.R., Jr., Eck, J. S., Ritter, E. T., Lee, Y. K., and Walker, J. C. (1967), *Phys. Rev.* **158**, 1118. [II:132]

Stodolsky, L. (1967), *Phys. Rev. Letters* **18**, 135. [II:480]

Streater, R. F., and Wightman, A. S. (1964), *PCT, Spin and Statistics, and All That*, Benjamin, New York, N.Y. [I:21]

Strömgren, B. (1968). *NORDITA Lectures*, NORDITA, Copenhagen. [I:199]

Stroke, H. H., Blin-Stoyle, R. J., and Jaccarino, V. (1961), *Phys. Rev.* **123**, 1326. [I:384]

Strominger, D., Hollander, J. M., and Seaborg, G. T. (1958), *Rev. Mod. Phys.* **30**, 585.

Struble, G. L., Kern, J., and Sheline, R. K. (1965), *Phys. Rev.* **137**, B772. [II:121]

Strutinsky, V. M. (1956), *Zhur. Eksp. i Teoret. Fiz.* **30**, 606, transl. *Soviet Phys.* JETP **3**, 638. [II:123]

Strutinsky, V. M. (1958), *unpublished lectures*, The Niels Bohr Institute, Copenhagen. [I:281]

Strutinsky. V. M. (1965), *Yadernaya Fiz.* **1**, 821; transl. *Soviet J. Nuclear Phys.* **1**, 588. [II:621]

Strutinsky, V. M. (1966), *Yadernaya Fiz.* **3**, 614; transl. *Soviet J. Nuclear Phys.* **3**, 449. [II:367, 368]
Strutinsky, V. M. (1967), *Arkiv Fysik* **36**, 629. [I:205; II:598]
Strutinsky, V. M. (1967a), *Nuclear Phys.* **A95**, 420. [II:598, 622]
Strutinsky, V. M. (1968), *Nuclear Phys.* **A122**, 1. [II:370, 605]
Strutinsky, V. M., and Bjørnholm, S. (1968), *Nuclear Structure*, Dubna Symposium, p. 431, Intern. Atomic Energy Agency, Vienna. [II:634]
Strutinsky, V. M., Lyaschenko, N. Ya., and Popov, N. A. (1962), *J. Exptl. Theoret. Phys. (USSR)* **43**, 584; transl. *Soviet Phys. JETP* **16**, 418. [II:662]
Suess, H. E., and Urey, H. C. (1956), *Rev. Mod. Phys.* **28**, 53. [I:206]
Sugimoto, K., Mizobuchi, A., Nakai, K., and Matuda, K. (1965), *Phys. Letters* **18**, 38. [I:343]
Swiatecki, W. J. (1951), *Proc. Phys. Soc. (London)* **A64**, 226. [II:368]
Swiatecki, W. J. (1956), *Phys. Rev.* **104**, 993. [II:662]
Swift, A., and Elton, L. R. B. (1966), *Phys. Rev. Letters* **17**, 484. [I:163]

Tabakin, F. (1964), *Ann. Phys.* **30**, 51. [I:266]
Takagi, S. (1959), *Progr. Theoret. Phys. (Kyoto)* **21**, 174. [II:436]
Taketani, M., and articles by Iwadare, J., Otsuki, S., Tamagaki, R., Machida, S., Toyoda, T., Watari, W., Nishijima, K., Nakamura, S., and Sasaki, S. (1956), *Progr. Theoret. Phys.* (Kyoto), Suppl. 3. [I:250]
Talman, J. D. (1970), *Nuclear Phys.* **A141**, 273. [II:227]
Talman, J. D. (1971), *Nuclear Phys.* **A161**, 481. [II:196]
Tamagaki, R. (1967), *Rev. Mod. Phys.* **39**, 629. [I:245]
Tamura, T. (1966), *Phys. Letters* **22**, 644. [I:51]
Tamura, T., and Udagawa, T. (1964), *Nuclear Phys.* **53**, 33. [II:433]
Tanaka, S. (1960), *J. Phys. Soc. Japan* **15**, 2159. [I:186]
Taylor, B. N., Parker, W. H., and Langenberg, D. N. (1969), *Rev. Mod. Phys.* **41**, 375. [II:396]
Teller, E., and Wheeler, J.A. (1938), *Phys. Rev.* **53**, 778 [II:2]
Terasawa, T. (1960), *Prog. Theoret. Phys.* (Kyoto) **23**, 87. [I:260]
Tewari, S. N., and Banerjee, M. K. (1966), *Nuclear Phys.* **82**, 337. [II:213]
Thomas, R. G. (1952), *Phys. Rev.* **88**, 1109. [I:43]
Thomas, W. (1925), *Naturwissenschaften* **13**, 627. [II:400]
Thouless, D. J. (1961), *The Quantum Mechanics of Many-Body Systems*, Academic Press, New York, N.Y.
Thouless, D. J. (1961), *Nuclear Phys.* **22**, 78. [II:436]
Tipler, P.A., Axel, P., Stein, N., and Sutton, D. C. (1963), *Phys. Rev.* **129**, 2096. [II:505]
Tjøm, P.O., and Elbek, B. (1967), *Mat. Fys. Medd. Dan. Vid. Selsk.* **36**, no. 8. [II:230, 301]
Tjøm, P.O., and Elbek, B. (1969), *Mat. Fys. Medd. Dan. Vid. Selsk.* **37**, no. 7. [II:301]
Townes, C. H., and Schawlow, A. L. (1955), *Microwave Spectroscopy*, McGraw Hill, New York, N.Y.
Trainer, L. E. H. (1952), *Phys. Rev.* **85**, 962. [I:44]
Tsukada, K., Tanaka, S., Maruyama, M., and Tomita, Y. (1966), *Nuclear Phys.* **78**, 369. [I:183, 184]
Turner, J. F., Ridley, B. W., Cavanagh, P. E., Gard, G. A., and Hardacre, A. G. (1964), *Nuclear Phys.* **58**, 509. [I:235]
Tveter, A., and Herskind, B. (1969), *Nuclear Phys.* **A134**, 599. [II:132]
Tyrén, H., and Maris, Th. A. J. (1958), *Nuclear Phys.* **7**, 24. [II:482]

Tyrén, H., Kullander, S., Sundberg, O., Ramachandran, R., Isacsson, P., and Berggren, T. (1966), *Nuclear Phys.* **79**, 321. [I:232]

Überall, H. (1974), *Springer Tracts in Modern Physics*, **71**, 1. [II:498]
Uher, R. A., and Sorensen, R. A. (1966), *Nuclear Phys.* **86**, 1. [I:163]
Ungrin, J., Diamond, R. M., Tjøm, P. O., and Elbek, B. (1971), *Mat. Fys. Medd. Dan. Vid. Selsk.* **38**, no. 8. [II:572, 573]
Urey, H. C. (1964), *Rev. Geophysics* **2**, 1. [I:206]

Valatin, J. G. (1956), *Proc. Roy. Soc. (London)*, **238**, 132. [II:85, 204]
Valatin, J. G. (1958), *Nuovo cimento* **7**, 843, [II:648]
Van de Hulst, H. C. (1957), *Light Scattering by Small Particles*, Wiley, New York, N. Y.
Vandenbosch, R., and Huizenga, J. R. (1958), *Proc. Second Intern. Conf. on the Peaceful Uses of Atomic Energy*, vol. **15**, p. 284, United Nations, Geneva. [II:374]
Vandenbosch, S. E., and Day, P. (1962), *Nuclear Phys.* **30**, 177. [II:144]
Van Leeuwen, H. J. (1921), *J. phys. radium* **2**, 361. [II:79]
Van Oostrum, K. J., Hofstadter, R., Nöldeke, G. K., Yearian, M. R., Clark, B. C., Herman, R., and Ravenhall, D. G. (1966), *Phys. Rev. Letters* **16**, 528. [I:165]
Van Vleck, J. H. (1929), *Phys. Rev.* **33**, 467: [II:33]
Van Vleck, J. H. (1932), *The Theory of Electric and Magnetic Susceptibilities*, Oxford University Press, Oxford.
Veje, E., Elbek, B., Herskind, B., and Olesen, M. C. (1968), *Nuclear Phys.* **A109**, 489. [II:561, 577]
Veltman, M. (1966), *Phys. Rev. Letters* **17**, 553. [I:408]
Vergnes, M. N., and Sheline, R. K. (1963), *Phys. Rev.* **132**, 1736. [II:244]
Vergnes, M. N., and Rasmussen, J. O. (1965), *Nuclear Phys.* **62**, 233. [II:110]
Veyssière, A., Beil, H., Bergère, R., Carlos, P., and Lepretre, A. (1970), *Nuclear Phys.* **A159**, 561. [II:479]
Veyssière, A., Beil, H., Bergère, R., Carlos, P., Lepretre, A., and Kernbath, K. (1973), *Nuclear Phys.* **A199**, 45. [II:492, 493]
Viggars, D. A., Butler, P. A., Carr, P. E., Gadeken, L. L., James, A. N., Nolan, P. J., and Sharpey-Schafer, J. F. (1973), *J. Phys. A*, **6**, L67. [II:287]
Villars, F. (1947), *Helv. Phys. Acta* **20**, 476. [I:337]
Villars, F. (1957), *Ann. Rev. Nuclear Sci.* **7**, 185. [II:91]
Vogt, E. (1967), in *Proc. Intern. Nuclear Physics Conf.*, Gatlinburg, p. 748, ed.-in-chief R. L. Becker, Academic Press, New York, N.Y. [I:445]

Wahlborn, S. (1965), *Phys. Rev.* **138**, B530. [I:25]
Wapstra, A. H. (1953), *Arkiv Fysik* **6**, 263. [I:352]
Warburton, E. K., Parker, P. D., and Donovan, P. F. (1965), *Phys. Letters* **19**, 397. [I:320]
Watson, R. E., and Freeman, A. J. (1967), in *Hyperfine Interactions*, p. 53, eds. A. J. Freeman and R. B. Frankel, Academic Press, New York, N.Y. [II:213]
Way, K., and Hurley, F. W. (1966), *Nuclear Data* **A1**, 473.
Weaver, L., Biedenharn, L. C., and Cusson, R. Y. (1973), *Ann. Phys.* **77**, 250. [II:411]
Weber, H. J., Huber, M. G., and Greiner, W. (1966), *Z. Physik* **190**, 25, and *ibid.* **192**, 182. [II:692]
Weidenmüller, H. A. (1960), *Nuclear Phys.* **21**, 397. [I:418]
Weigmann, H. (1968), *Z. Physik* **214**, 7. [II:622]

Weigmann, H., and Schmid, H. (1968), *J. Nuclear Energy* **22**, 317. [II:628, 630]
Weinberg, S. (1958), *Phys. Rev.* **112**, 1375. [I:400]
Weinberg, S. (1962), *Phys. Rev.* **128**, 1457. [I:398]
Weinberg, S. (1962a), *Nuovo cimento* **25**, 15. [I:398]
Weisberger, W. I. (1965), *Phys. Rev. Letters* **14**, 1047. [I:401]
Weise, W. (1974), *Phys. Letters C* **13**, 55. [II:480]
Weisskopf, V. F. (1937), *Phys. Rev.* **52**, 295. [I:154, 185]
Weisskopf, V. F. (1955), in *Proc. Glasgow Conf. on Nuclear and Meson Physics*, p. 167, eds. E. H. Bellamy and R. G. Moorhouse, Pergamon, Press, London. [II:474]
Weisskopf, V. F. (1957), *Nuclear Phys.* **3**, 423. [I:147]
Weizsäcker, von, C. F. (1935), *Z. Physik* **96**, 431. [I:141]
Wendin, G. (1973), *J. Phys. B: Atom. Molec. Phys.* **6**, 42. [II:352]
Wentzel, G. (1940), *Helv. Phys. Acta* **13**, 269. [II:20]
Wheatley, J. C. (1966), in *Quantum Fluids*, p. 183, ed. D. F. Brewer, North-Holland, Amsterdam. [I:255]
Wheatley, J. C. (1970), in *Progress in Low Temperature Physics*, **6**, 77, ed. C. J. Gorter, North-Holland, Amsterdam. [II:385]
Wheeler, J. A. (1955), in *Niels Bohr and the Development of Physics*, ed. W. Pauli, Pergamon Press, London. [I:199, 204]
Wheeler, J. A. (1963), in *Fast Neutron Physics*, Part II, p. 2051, eds. J. B. Marion and J. L. Fowler, Interscience, New York, N.Y. [II:123, 373, 622, 663]
Wheeler, J. A. (1964), in *Gravitation and Relativity*, Chap. 10, eds. H.-Y. Chiu and W. F. Hoffmann, Benjamin, New York, N.Y. [I:205]
Wheeler, J. A. (1971), in *Atti del Convegno Mendeleeviano*, p. 189, ed. M. Verde, Accademia delle Science di Torino, Torino. [II:579]
Whineray, S., Dietrich, F. S., and Stokstad, R. G. (1970), *Nuclear Phys.* **A157**, 529. [II:259, 261, 262, 301]
Whittaker, E. T. (1937), *Analytical Dynamics*, 4th. edition, Cambridge University Press, Cambridge.
Wick, C. G. (1955), *Rev. Mod. Phys.* **27**, 339. [II:429]
Wigner, E. P. (1937), *Phys. Rev.* **51**, 106. [I:38; II:641]
Wigner, E. P. (1939), *Phys. Rev.* **56**, 519. [I:348]
Wigner, E. P. (1958), *Ann. Math.* **67**, 325; see also *ibid.* **62**, 548 (1955) and **65**, 203 (1957). [I:294]
Wigner, E. P. (1959), *Group Theory and its Applications to the Quantum Mechanics of Atomic Spectra*, Academic Press, New York, N.Y.
Wild, W. (1955), *Sitzber. Bayer. Akad. Wiss., Math.-Naturw. Klasse*, München, **18**, 371. [II:383]
Wildermuth, K., and Kanellopoulos, Th. (1958), *Nuclear Phys.* **7**, 150. [II:99]
Wilets, L. (1954), *Mat. Fys. Medd. Dan. Vid. Selsk.*, **29**, no. 3. [II:134]
Wilets, L., Hill, D. L., and Ford, K. W. (1953), *Phys. Rev.* **91**, 1488. [I:384]
Wilkins, B. D., Unik, J. P., and Huizenga, J. R. (1964), *Phys. Letters* **12**, 243. [II:127]
Wilkinson, D. H. (1955), in *Proc. Glasgow Conf. on Nuclear and Meson Physics*, p. 161, eds. E. H. Bellamy and R. G. Moorhouse, Pergamon Press, London. [II:474]
Wilkinson, D. H. (1956), *Physica* **22**, 1039. [II:474]
Wilkinson, D. H., and Mafethe, M. E. (1966), *Nuclear Phys.* **85**, 97. [I:42]
Williamson, C. F., Ferguson, S. M., Shepherd, B. J., and Halpern, I. (1968), *Phys. Rev.* **174**, 1544. [II:73]
Wilmore, D., and Hodgson, P. E. (1964), *Nuclear Phys.* **55**, 673. [I:217]

Wilson, R. (1963), *The Nucleon-Nucleon Interaction*, Wiley (Interscience), New York, N.Y.

Winhold, E. J., Demos, P. T., and Halpern, I. (1952), *Phys. Rev.* **87**, 1139. [II: 123]

Winner, D. R., and Drisko, R. M. (1965), *Techn. Rep., Univ. of Pittsburgh*, Sarah Mellon Scaife Rad. Lab., Pittsburgh, Pa. [I: 237]

Winther, A. (1962), *On the Theory of Nuclear Beta-Decay*, Munksgaard, Copenhagen. [I: 403]

Witsch, von W., Richter, A., and von Brentano, P. (1967), *Phys. Rev. Letters* **19**, 524. [I: 29]

Wong, C. Y. (1970), *Phys. Letters* **32B**, 668. [II: 592]

Wood, R. W., Borchers, R. R., and Barschall, H. H. (1965), *Nuclear Phys.* **71**, 529. [I: 186]

Wu, C. S. (1964), *Rev. Mod. Phys.* **36**, 618. [I: 419]

Wu, C. S. (1967), in *Proc. Intern. Nuclear Physics Conf.*, Gatlinburg, p. 409, ed.-in-chief R. L. Becker, Academic Press, New York, N.Y. [I: 165, 384]

Wu, C. S. (1968), in *Physics of the One and Two Electron Atoms*, and *Arnold Sommerfeld Centennial Meeting*, p. 429, eds. F. Bopp and H. Kleinpoppen, North-Holland, Amsterdam. [I: 65]

Wu, C. S., and Moszkowski, S. A. (1966), *Beta Decay*, Wiley (Interscience), New York, N.Y.

Wu, C. S., and Wilets, L. (1969), *Ann. Rev. Nuclear Sci.* **19**, 527. [II: 134]

Wu, C. S., Ambler, E., Hayward, R. W., Hoppes, D. D., and Hudson, R. F. (1957), *Phys. Rev.* **105**, 1413. [I: 16, 26]

Wyatt, P. J., Wills, J. G., and Green, A. E. S. (1960), *Phys. Rev.* **119**, 1031. [I: 217, 240]

Yamanouchi, T. (1937), *Proc. Phys.-Math. Soc. Japan* **19**, 436. [I: 112]

Yamazaki, T., Nomura, T., Nagamiya, S., and Katou, T. (1970), *Phys. Rev. Letters* **25**, 547. [II: 486]

Yennie, D. R., Ravenhall, D. G., and Wilson, R. N. (1954), *Phys. Rev.* **95**, 500. [I: 159]

Yntema, J. L., and Zeidman, B. (1959), *Phys. Rev.* **114**, 815. [II: 353]

Yoccoz, J. (1957), *Proc. Phys. Soc. (London)*, **A70**, 388. [II: 91]

Yoshida, S. (1961), *Phys. Rev.* **123**, 2122. [II: 641]

Yoshida, S. (1962), *Nuclear Phys.* **33**, 685. [II: 641]

Youngblood, D. H., Aldridge, J. P., and Class, C. M. (1965), *Phys. Letters* **18**, 291. [I: 341]

Yule, H. P. (1967), *Nuclear Phys.* **A94**, 442. [I: 344]

Zeldes, N., Grill, A., and Simievic, A. (1967), *Mat. Fys. Skr. Dan. Vid. Selsk*, **3**, no. 5. [I: 143, 170, 171]

Ziegler, J. F., and Peterson, G. A. (1968), *Phys. Rev.* **165**, 1337. [II: 49, 350, 516, 561]

Zimányi, J., Halpern, I., and Madsen, V. A. (1970), *Phys. Letters* **33B**, 205. [II: 487]

Zweig, A. (1964), *CERN Reports TH*-401 *and TH*-412, CERN, Geneva; see also *Proc. Intern. School of Physics "Ettore Majorana"*, ed. A. Zichichi, Academic Press, New York, N.Y. (1965). [I: 41]

Żylicz, J., Hansen, P. G., Nielsen, H. L., and Wilsky, K. (1966), *Nuclear Phys.* **84**, 13. [II: 318]

Żylicz, J., Hansen, P. G., Nielsen, H. L., and Wilsky, K. (1967), *Arkiv Fysik* **36**, 643. [II: 307]

Zyryanova, L. N. (1963), *Once-Forbidden Beta-Transitions*, Pergamon Press, New York, N.Y.

INDEX

The Subject Index is cumulative, covering topics discussed in Volume I as well as in Volume II. The references to subject matter treated in Volume I have been somewhat expanded as compared with the Index to Volume I. References to individual nuclei are collected at the end of the Index.

NUCLEI

The following index contains references to the more extensive discussions of properties of individual nuclei.